# Essential Environment

## The Science Behind the Stories

5TH EDITION

**Jay Withgott**

**Matthew Laposata**

PEARSON

Boston Columbus Indianapolis New York San Francisco Upper Saddle River
Amsterdam Cape Town Dubai London Madrid Milan Munich Paris Montréal Toronto
Delhi Mexico City São Paulo Sydney Hong Kong Seoul Singapore Taipei Tokyo

Acquisitions Editor: Alison Rodal
Project Manager: Margaret Young
Program Manager: Anna Amato
Editorial Assistants: Libby Reiser, Alison Cagle
Text Permissions Project Manager: William Opaluch
Development Manager: Ginnie Simione-Jutson
Program Management Team Lead: Michael Early
Project Management Team Lead: David Zielonka
Production Management: Thistle Hill Publishing Services
Compositor: Cenveo® Publisher Services
Design Manager and Cover Designer: Derek Bacchus
Interior Designer: Yvo Riezebos
Illustrator: Imagineering
Photo Permissions Management: Lumina Datamatics
Photo Research: Steve Merland, Lumina Datamatics
Photo Lead: Donna Kalal
Manufacturing Buyer: Maura Zaldivar-Garcia
Executive Marketing Manager: Lauren Harp

Cover Photo Credit: Kathryn Hansen/Earth Observatory, NASA

Credits and acknowledgments for materials borrowed from other sources and reproduced, with permission, in this textbook appear on the appropriate page within the text or on page CR-1.



**Library of Congress Cataloging-in-Publication Data**
Withgott, Jay.
Essential environment : the science behind the stories / Jay Withgott, Matthew Laposata. — Fifth edition.
pages cm
Includes bibliographical references and index.
ISBN 978-0-321-98457-9 — ISBN 0-321-98457-9
1. Environmental sciences. I. Laposata, Matthew. II. Title.
GE105.B74 2016
363.7—dc23
2014026297

ISBN 10: 0-321-98457-9; ISBN 13: 978-0-321-98457-9 (Student edition)
ISBN 10: 0-133-90074-6; ISBN 13: 978-0-133-90074-3 (Instructor's Review Copy)
ISBN 10: 0-321-97650-9; ISBN 13: 978-0-321-97650-5 (a la Carte)

PEARSON www.pearsonhighered.com 1 2 3 4 5 6 7 8 9 10—V303—18 17 16 15 14

# About the Authors

**Jay Withgott** has authored *Essential Environment* as well as its parent volume, *Environment: The Science behind the Stories*, since their inception. In dedicating himself to these books, he works to keep abreast of a diverse and rapidly changing field and continually seeks to develop new and better ways to help today's students learn environmental science.

As a researcher, Jay has published scientific papers in ecology, evolution, animal behavior, and conservation biology in journals ranging from *Evolution* to *Proceedings of the National Academy of Sciences*. As an instructor, he has taught university lab courses in ecology and other disciplines. As a science writer, he has authored articles for numerous journals and magazines including *Science, New Scientist, BioScience, Smithsonian,* and *Natural History*. By combining his scientific training with prior experience as a newspaper reporter and editor, he strives to make science accessible and engaging for general audiences. Jay holds degrees from Yale University, the University of Arkansas, and the University of Arizona.

Jay lives with his wife, biologist Susan Masta, in Portland, Oregon.

**Matthew Laposata** is a professor of environmental science at Kennesaw State University (KSU). He holds a bachelor's degree in biology education from Indiana University of Pennsylvania, a master's degree in biology from Bowling Green State University, and a doctorate in ecology from The Pennsylvania State University.

Matt is the coordinator of KSU's two-semester general education science sequence titled Science, Society, and the Environment, which enrolls over 5000 students per year. He focuses exclusively on introductory environmental science courses and has enjoyed teaching and interacting with thousands of nonscience majors during his career. He is an active scholar in environmental science education and has received grants from state, federal, and private sources to develop and evaluate innovative curricular materials. His scholarly work has received numerous awards, including the Georgia Board of Regents' highest award for the Scholarship of Teaching and Learning.

Matt resides in suburban Atlanta with his wife, Lisa, and children, Lauren, Cameron, and Saffron.


## ABOUT OUR SUSTAINABILITY INITIATIVES

This book is carefully crafted to minimize environmental impact. The materials used to manufacture this book originated from sources committed to responsible forestry practices. The paper is Forest Stewardship Council® (FSC®) certified. The printing, binding, cover, and paper come from facilities that minimize waste, energy consumption, and the use of harmful chemicals.

Pearson closes the loop by recycling every out-of-date text returned to our warehouse. We pulp the books, and the pulp is used to produce items such as paper coffee cups and shopping bags. In addition, Pearson has become the first climate-neutral educational publishing company.

The future holds great promise for reducing our impact on Earth's environment, and Pearson is proud to be leading the way. We strive to publish the best books with the most up-to-date and accurate content—and to do so in ways that minimize our environmental impact.

FSC
www.fsc.org
MIX
Paper from responsible sources
FSC® C100141

PEARSON

# Contents

# Preface

## Dear Student,

You are coming of age at a unique and momentous time in history. Within your lifetime, our global society must chart a promising course for a sustainable future. The stakes could not be higher.

Today we live long lives enriched with astonishing technologies, in societies more free, just, and equal than ever before. We enjoy wealth on a scale our ancestors could hardly have dreamed of. Yet we have purchased these wonderful things at a price. By exploiting Earth's resources and ecological services, we are depleting our planet's bank account and running up its credit card. More than ever before, the future of our society rests with how we treat the world around us.

Your future is being shaped by the phenomena you will learn about in your environmental science course. Environmental science gives us a big-picture understanding of the world and our place within it. Environmental science also offers hope and solutions, revealing ways to address the problems we face. Environmental science is not simply some subject you study in college. Rather, it provides you basic literacy in the foremost issues of the 21st century, and it relates to everything around you throughout your lifetime.

We have written this book because today's students will shape tomorrow's world. At this unique moment in history, the decisions and actions of your generation are key to achieving a sustainable future for our civilization. The many environmental challenges that face us can seem overwhelming, but you should feel encouraged and motivated. Remember that each dilemma is also an opportunity. For every problem that human carelessness has created, human ingenuity can devise a solution. Now is the time for innovation, creativity, and the fresh perspectives that a new generation can offer. Your own ideas and energy *will* make a difference.

**—Jay Withgott and Matthew Laposata**

## Dear Instructor,

You perform one of our society's most vital jobs by educating today's students—the citizens and leaders of tomorrow—on the fundamentals of the world around them, the nature of science, and the most central issues of our time. We have written this book to assist you in this endeavor because we feel that the crucial role of environmental science in today's world makes it imperative to engage, educate, and inspire a broad audience of students.

In *Essential Environment: The Science behind the Stories*, we strive to implement a diversity of modern teaching approaches and to show how science can inform efforts to bring about a sustainable society. We aim to encourage critical thinking and to maintain a balanced approach as we flesh out the vibrant social debate that accompanies environmental issues. As we assess the challenges facing our civilization and our planet, we focus on providing forward-looking solutions, for we truly feel there are many reasons for optimism.

As environmental science has grown, so has the length of textbooks that cover it. With this volume, we aim to meet the needs of introductory courses that favor a more succinct and affordable book. We have distilled the most essential content from our full-length book, *Environment: The Science behind the Stories*, now in its fifth edition. We have streamlined our material, updated our coverage, and carefully crafted our writing to make *Essential Environment* every bit as readable, informative, and engaging as its parent volume.

## New to This Edition

In this fifth edition of *Essential Environment*, we have incorporated the most current information from a fast-moving field. A number of major changes new to this edition enhance our presentation while strengthening our commitment to teach science in an engaging and accessible way.

- **CENTRAL CASE STUDY**

  **Seven of our 17 *Central Case Studies* are new to this edition,** providing a wealth of fresh stories and new ways to frame issues in environmental science. Students will travel from Pennsylvania to Hawaii and from Africa to Mexico to Costa Rica as they learn how debates over hydraulic fracturing, air pollution, deforestation, and biodiversity conservation are affecting people's lives. In addition, two new case studies focus on campus sustainability initiatives, showcasing some of the many innovative steps students are taking to advance sustainable solutions on campus.

- **Chapter 3:** Saving Hawaii's Native Forest Birds
- **Chapter 5:** Costa Rica Values Its Ecosystem Services
- **Chapter 7:** Farm to Table—And Back Again—The Commons at Kennesaw State University
- **Chapter 8:** Will We Slice through the Serengeti?
- **Chapter 13:** Clearing the Air in L.A. and Mexico City
- **Chapter 15:** Hydrofracking the Marcellus Shale
- **Chapter 17:** A Mania for Recycling on Campus

■ THE SCIENCE BEHIND THE STORY

**Eight of our 18 *Science Behind the Story* features are new to this edition,** giving you a current and exciting selection of scientific studies to highlight. Students will follow researchers as they evaluate energy sources, design climate models, monitor animal populations, assess impacts of smog and forest fragmentation, and seek solutions to pollution from hydraulic fracturing and plastic in the ocean.

- **Chapter 3:** Monitoring Bird Populations at Hakalau Forest
- **Chapter 8:** Wildlife Declines in African Reserves
- **Chapter 9:** Forest Fragmentation in the Amazon
- **Chapter 11:** Can Acid Mine Drainage Reduce Fracking's Environmental Impact?
- **Chapter 12:** Predicting the Oceans' "Garbage Patches"
- **Chapter 13:** Measuring the Health Impacts of Mexico City's Air Pollution
- **Chapter 14:** How Do Climate Models Work?
- **Chapter 16:** Comparing Energy Sources

Selected *Science behind the Story* features are now supported by new "Process of Science" exercises online in *MasteringEnvironmentalScience* that use these examples to help students explore how scientists conduct their work.

■ DATA Q **Each chapter now contains *Data Q* data-analysis questions** that help students to actively engage with graphs and other data-driven figures. This new feature accompanies several figures in each chapter, challenging students to practice quantitative skills of interpretation and analysis. To encourage students to test their understanding as they read, answers are provided in Appendix A, and students can practice data analysis skills further with new *Interpreting Graphs and Data: DataQs* in *MasteringEnvironmentalScience*.

■ **Currency and coverage of topical issues** To live up to our book's hard-won reputation for currency, we have incorporated the most recent data possible and have enhanced coverage of issues now gaining prominence. As climate change and energy concerns play ever-larger roles in today's world, our coverage continues to evolve. This edition highlights how renewable energy is growing, yet also how we are reaching further for fossil fuels with deep offshore drilling, Arctic exploration, oil sands extraction, and hydraulic fracturing for tight oil and shale gas. Climate change connections continue to proliferate throughout our text, and our climate change chapter includes new coverage of climate modeling, Hurricane Sandy, research into jet stream effects on extreme weather, the latest predictions for America and the world, and political responses at all levels.

This edition also expands coverage of a diversity of topics including the valuation of ecosystem services, introduced species and their ecological impacts on islands, prospects for nuclear power and safety after Fukushima, ocean acidification, advanced biofuels, sustainable agriculture, hormone-disrupting substances, impacts on coastal wetlands, marine plastic pollution, and green-collar jobs. We continue to use sustainability as an organizing theme, and we aid this by moving primary coverage of sustainable development to Chapter 5 and highlighting campus sustainability initiatives in several chapters.

■ **Enhanced style elements** We have updated and improved the look and clarity of our visual presentation throughout the text. A more open layout, more engaging photo treatments, and new designs for tables and for each feature all make the book more inviting and accessible for learning. This edition includes more than 30% entirely new photos, graphs, and illustrations, while most existing figures have been revised to reflect current data or for better clarity or pedagogy.

# Existing Features

We have also retained the major features that made the first four editions of our book unique and that are proving so successful in classrooms across North America:

■ **A focus on science and data analysis** We have maintained and strengthened our commitment to a rigorous presentation of modern scientific research while at the same time making science clear, accessible, and engaging to students. Explaining and illustrating the *process* of science remains a foundational goal of this endeavor. We also continue to provide an abundance of clearly cited data-rich graphs, with accompanying tools for data analysis. In our text, our figures, and numerous print and online features, we aim to challenge students and to assist them with the vital skills of data analysis and interpretation.

- **An emphasis on solutions** For many students, today's deluge of environmental dilemmas can lead them to believe that there is no hope or that they cannot personally make a difference. We have aimed to counter this impression by highlighting innovative solutions being developed on campuses and around the world. While being careful not to paint too rosy a picture of the challenges that lie ahead, we demonstrate that there is ample reason for optimism, and we encourage action.

- ***Central Case Studies* integrated throughout the text** We integrate each chapter's *Central Case Study* into the main text, weaving information and elaboration throughout the chapter. In this way, compelling stories about real people and real places help to teach foundational concepts by giving students a tangible framework with which to incorporate novel ideas.

- ***The Science Behind the Story*** Because we strive to engage students in the scientific process of testing and discovery, we feature *The Science Behind the Story* in each chapter. By guiding students through key research efforts, this feature shows not merely *what* scientists discovered, but *how* they discovered it.

- **FAQ** The *FAQ* feature highlights questions frequently posed by students in introductory environmental science courses, thereby helping to address widely held misconceptions and to fill in common conceptual gaps in knowledge. By also including questions students sometimes hesitate to ask, the *FAQs* show students that they are not alone in having these questions, thereby fostering a spirit of open inquiry in the classroom. A number of new *FAQs* have been added in this edition.

- ***Weighing the Issues*** These questions aim to help develop the critical-thinking skills students need to navigate multifaceted issues at the juncture of science, policy, and ethics. They serve as stopping points for students to reflect on what they have read, wrestle with complex dilemmas, and engage in spirited classroom discussion.

- **Diverse end-of-chapter features** *Testing Your Comprehension* provides concise study questions on main topics, while *Seeking Solutions* encourages broader creative thinking aimed at finding solutions. "Think It Through" questions place students in a scenario and empower them to make decisions to resolve problems. *Calculating Ecological Footprints* enables students to quantify the impacts of their choices and measure how individual impacts scale up to the societal level.

# MasteringEnvironmental Science

With this edition we are thrilled to offer expanded opportunities through *MasteringEnvironmentalScience*, our powerful yet easy-to-use online learning and assessment platform. With Pearson's talented staff and extensive resources, we have developed new content and activities specifically to support features in the textbook, thus strengthening the connection between online and print resources. This approach encourages students to practice their science literacy skills in an interactive environment with a diverse set of automatically graded exercises. Students benefit from self-paced activities that feature immediate wrong-answer feedback, while instructors can gauge student performance with informative diagnostics. By enabling assessment of student learning outside the classroom, *MasteringEnvironmentalScience* helps the instructor to maximize the impact of classroom time. As a result, both educators and learners benefit from an integrated text and online solution.

- **New to this edition** Informed by instructor feedback and motivated by the desire to help students to build science literacy skills, we have made the following additions to *MasteringEnvironmentalScience*. The first two were created specifically for the fifth edition by co-author Matthew Laposata:
  - *Process of Science* activities help students navigate the scientific method, guiding them through explorations of experimental design using *Science Behind the Story* features from the fifth edition. These activities encourage students to think like a scientist and to practice basic skills in experimental design.
  - *First Impressions Pre-Quizzes* help instructors determine their students' existing knowledge of core content areas in environmental science at the outset of the academic term, providing class-specific data that can then be employed for powerful teachable moments throughout the term. Assessment items in the Test Bank connect to each quiz item, so instructors can formally assess student understanding.
  - *Interpreting Graphs and Data: Data Q* activities pair with the new in-text *Data Analysis Questions* and coach students to further develop skills related to presenting, interpreting, and thinking critically about environmental science data.
  - Five more *Video Field Trips* have been added to the existing library in *MasteringEnvironmentalScience*. With these new videos, you can now kick off your class period with a field trip to a wind farm or a water desalination plant, a visit with researchers tackling invasive species or bee colony collapse disorder, or a tour of a sustainable college campus.

- **Existing features** *MasteringEnvironmentalScience* also retains its popular existing features. *Interpreting Graphs and Data* exercises and the interactive *GraphIt!* program each guide students in exploring how to present and interpret data and how to create graphs. Interactive *Causes and Consequences* exercises let students probe the causes behind major issues, their consequences, and possible solutions. *Viewpoints* offers paired essays authored by invited experts who present divergent points of view on topical questions.

*Essential Environment* has grown from our experiences in teaching, research, and writing. We have been guided in our efforts by input from hundreds of instructors across North America who have served as reviewers and advisors. The participation of so many learned, thoughtful, and committed experts and educators has improved this volume in countless ways.

We sincerely hope that our efforts are worthy of the immense importance of our subject matter. We invite you to let us know how well we have achieved our goals and where you feel we have fallen short. Please write to us in care of our editor, Alison Rodal (*alison.rodal@pearson.com*) at Pearson Education. We value your feedback and are eager to know how we can serve you better.

**—Jay Withgott and Matthew Laposata**

# Instructor Supplements

## Instructor Resource Center on DVD with TestGen (0-133-89259-X)

This powerful media package is organized chapter-by-chapter and includes all teaching resources in one convenient location. You'll find Video Field Trips, PowerPoint presentations, Active Lecture questions to facilitate class discussions (for use with or without clickers), and an image library that includes all art and tables from the text.

Included on the IRDVD, the Test Bank includes hundreds of multiple-choice questions plus unique graphing, and scenario-based questions to test students' critical-thinking abilities.

## Instructor Guide (0-133-90122-X)

This comprehensive resource provides chapter outlines, key terms, and teaching tips for lecture and classroom activities.

## Blackboard Open Access (0-133-97474-X)

MasteringEnvironmentalScience™ for Essential Environment: The Science Behind the Stories (0-321-97688-6)

The *MasteringEnvironmentalScience* platform is the most effective and widely used online tutorial, homework, and assessment system for the sciences.

# Acknowledgments

A textbook is the collective product of many minds and hearts. The two of us are exceedingly fortunate to be supported and guided by a tremendous publishing team.

Our acquisitions editor, Alison Rodal, coordinated our team's work for this fifth edition of *Essential Environment*. Alison's skills, focus, dedication, sound judgment, and experience in multiple aspects of publishing greatly enhanced our efforts. Project manager Margaret Young led us ably through the complex choreography of the textbook process. As program manager, Anna Amato lent her insight and her steady hand, building on her past contributions. Development editor Julia Osborne added perceptive and valuable feedback on our art program and the layout of our chapters. We appreciate their patience with us, and we admire their commitment to top-quality work.

We thank our editor-in-chief, Beth Wilbur, for her steady support of this book through its five editions, as well as Pearson's upper management for continuing to invest in the resources and top-notch personnel that our books have enjoyed for the past decade.

We thank executive editorial manager Ginnie Simione Jutson. Editorial assistant Libby Reiser was amazingly responsive and managed a smooth review process. Sally Peyrefitte, as always, provided reliable and meticulous copy editing, and photo researcher Steve Merland helped acquire quality photos. Wynne Au Yeung did a terrific job executing the art program, Yvo Riezebos designed our text interior, and Derek Bacchus designed our cover. We send a big thank-you to Angela Urquhart and Andrea Archer for their excellent work managing the production process.

We also remain grateful for lasting contributions to the book's earlier editions by Nora Lally-Graves, Mary Ann Murray, Susan Teahan, Tim Flem, Deborah Gale, Dan Kaveney, Etienne Benson, Russell Chun, Jonathan Frye, April Lynch, Kristy Manning, and of course Scott Brennan. And our books continue to owe a great deal to the vision, guidance, and heartfelt commitment of Chalon Bridges.

As we expand our online offerings with *MasteringEnvironmentalScience*, we thank Tania Mlawer, Juliana Tringali, Sarah Jensen, Lee Ann Doctor, Daniel Ross, Jana Pratt, and Todd Brown for their work on the *Mastering* website and our media supplements. A special thanks to Jellyfish Smack Productions for their fabulous work on our *Video Field Trips*.

As always, a select number of top instructors have teamed with us to produce the supplementary materials, and we remain deeply grateful for their work. Our thanks go to Heidi Marcum for working on EnvSci dynamic study modules, Danielle DuCharme for updating our Instructor's Guide, Daniel Pavuk for his help with the Test Bank, Reggie Cobb for revising the PowerPoint lectures and clicker questions, Steve Fitzpatrick for revising the reading quizzes, and Donna Bivans for correlating the shared media.

We give a big thanks to marketing manager Amee Mosley, as well as to Lauren Harp. And of course, the many sales representatives who help communicate our vision, deliver our product, and work with instructors to ensure their satisfaction are absolutely vital, and we deeply admire and appreciate their tireless work and commitment.

Lastly, Jay thanks his parents and his many teachers and mentors over the years for making his own life and education so enriching. He gives loving thanks to his wife, Susan, who has patiently provided caring support throughout this book's writing and revision over the years. Matt thanks his family, friends, and colleagues, and he is grateful for his children, who give him three reasons to care passionately about the future. Most important, he thanks his wife, Lisa, for enriching his life with her keen insight, passion for life, unconscious grace, and effortless beauty—and for faithfully reminding him to treasure and cherish the truly valuable things in life. The talents, input, and advice of Susan and of Lisa have been vital to this project, and without their support our own contributions would not have been possible.

We dedicate this book to today's students, who will shape tomorrow's world.

**—Jay Withgott and Matthew Laposata**

# Reviewers

We wish to express special thanks to the dedicated reviewers who shared their time and expertise to help make this edition the best it could be. Over 600 instructors and outside experts have reviewed material for the previous four editions of this book and the five editions of this book's parent volume, where they are acknowledged in full. Below we thank those who contributed in particular to this fifth edition of *Essential Environment*—in most cases with multiple in-depth chapter reviews despite busy teaching schedules. Our sincere gratitude goes out to all of them. If the thoughtfulness and thoroughness of these reviewers are any indication, we feel confident that the teaching of environmental science is in excellent hands!

Marc Albrecht, *University of Nebraska at Kearney*
Eric C. Atkinson, *Northwest College*
Martin Baranowski, *Passaic County Community College*
Stefan Becker, *Lehman College, CUNY*
Tom Campbell, *Northeastern Illinois University*
Daniel Capuano, *Hudson Valley Community College*
John B. Dunning, Jr., *Purdue University*
Kathleen A. Enseñat, *South Puget Sound Community College*
Brad C. Fiero, *Pima Community College*
Jeffrey French, *North Greenville University*
Wendy R. Hartman, *Palm Beach State College*
Keith R. Hench, *Kirkwood Community College*
Leslie Hendon, *University of Alabama at Birmingham*
Robert Hickey, *Central Washington University*
Jason Hlebakos, *Mount San Jacinto College*
Kelley Hodges, *Gulf Coast State College*
Jennifer Latimer, *Indiana State University*
Kurt M. Leuschner, *College of the Desert*
Mark Manteuffel, *St. Louis Community College and Washington University*
John B. McGill, *York Technical College*
Kiran Misra, *Edinboro University of Pennsylvania*
Martha Murphy, *Santa Rosa Junior College*
Troy Mutchler, *Kennesaw State University*
Eric R. Myers, *South Suburban College*
Douglas Nesmith, *Baylor University*
Dorna Sakurai, *Santa Monica College*
Amanda Senft, *Bellevue College*
Jamey Thompson, *Hudson Valley Community College*
Melanie Trecek-King, *Massasoit Community College*
Marie Trone, *Valencia College, Osceola Campus*
Candice Weber, *College of the Desert*
Nine (9) anonymous reviewers

1

# Science and Sustainability: An Introduction to Environmental Science

## Upon completing this chapter, you will be able to:

- ☐ Describe the field of environmental science
- ☐ Explain the importance of natural resources and ecosystem services to our lives
- ☐ Discuss the scale and consequences of population growth and resource consumption
- ☐ Describe the steps of the scientific method
- ☐ Comprehend the nature and importance of science, and characterize aspects of the process of science
- ☐ Appreciate the role of ethics in environmental science, and compare and contrast major approaches in environmental ethics
- ☐ Diagnose and illustrate major pressures on the global environment
- ☐ Articulate the concept of sustainability and describe campus sustainability efforts

**Photo: Our Island, Earth**

# Our Island, Earth

Viewed from space, our home planet resembles a small blue marble suspended in a vast inky-black void. Earth may seem enormous to us as we go about our lives on its surface, but the astronaut's view reveals that our planet is finite and limited. With this perspective, it becomes clear that as our population, technological power, and resource consumption all increase, so does our capacity to alter our surroundings and damage the very systems that keep us alive. Finding ways to live peacefully, healthfully, and sustainably on our diverse and complex planet is our society's prime challenge today. The field of environmental science is crucial in this endeavor.

## Our environment surrounds us

A photograph of Earth offers a revealing perspective, but it cannot convey the complexity of our environment. Our **environment** consists of all the living and nonliving things around us. It includes the continents, oceans, clouds, and ice caps you can see in the photo of Earth from space, as well as the animals, plants, forests, and farms of the landscapes surrounding us. In a more inclusive sense, it also encompasses the structures, urban centers, and living spaces that people have created. In its broadest sense, our environment includes the complex webs of social relationships and institutions that shape our daily lives.

People commonly use the term *environment* in the narrowest sense—to mean a nonhuman or "natural" world apart from human society. This is unfortunate, because it masks the vital fact that people exist within the environment and are part of nature. As one of many species on Earth, we share dependence on a healthy, functioning planet. The limitations of language make it all too easy to speak of "people and nature," or "humans and the environment," as though they were separate and did not interact. However, the fundamental insight of environmental science is that we are part of the "natural" world and that our interactions with the rest of it matter a great deal.

## Environmental science explores our interactions with the world

Understanding our relationship with the world around us is vital because we depend on our environment for air, water, food, shelter, and everything else essential for living. Moreover, we modify our environment. Many of our actions have enriched our lives, bringing us better health, longer life spans, and greater material wealth, mobility, and leisure time—yet they have often degraded the natural systems that sustain us. Air and water pollution, soil erosion, species extinction, and other impacts compromise our well-being and jeopardize our ability to build a society that will survive and thrive in the long term.

**Environmental science** is the scientific study of how the natural world works, how our environment affects us, and how we affect our environment. By understanding these interactions, we may be able to devise solutions to our many pressing challenges. It can be daunting to reflect on the sheer magnitude of dilemmas that confront us today, but these problems also bring countless opportunities for creative solutions.

Environmental scientists study the issues most centrally important to our world and its future. Right now, global conditions are changing more quickly than ever. Right now, we are gaining scientific knowledge more rapidly than ever. And right now there is still time to tackle society's biggest challenges. With such bountiful opportunities, this moment in history is an exciting time to be alive—and to be studying environmental science.

## We rely on natural resources

Islands are finite and bounded, and their inhabitants must cope with limitations in the materials they need. On our island—planet Earth—there are limits to many of our **natural resources,** the substances and energy sources we take from our environment and that we rely on to survive (**FIGURE 1.1**).

**(a) Inexhaustible renewable natural resources**

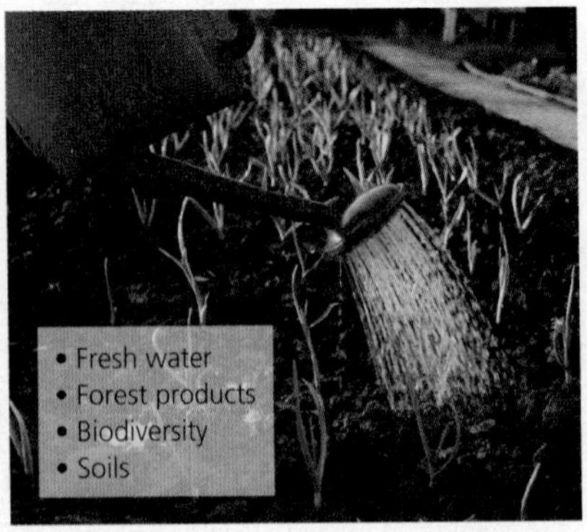


**(b) Exhaustible renewable natural resources**

**(c) Nonrenewable natural resources**

**FIGURE 1.1 Natural resources may be renewable or nonrenewable.** Perpetually renewable, or inexhaustible, resources such as sunlight and wind energy **(a)** will always be there for us. Renewable resources such as timber, soils, and fresh water **(b)** are replenished on intermediate time scales, if we are careful not to deplete them. Nonrenewable resources such as minerals and fossil fuels **(c)** exist in limited amounts that could one day be gone.

Natural resources that are replenished over short periods are known as **renewable natural resources.** Some renewable natural resources, such as sunlight, wind, and wave energy, are perpetually renewed and essentially inexhaustible. Others, such as timber, water, animal populations, and fertile soil, renew themselves over months, years, or decades. These types of renewable resources may be used at sustainable rates, or they may be depleted if we consume them faster than they are replenished. **Nonrenewable natural resources,** such as minerals and crude oil, are in finite supply and are formed far more slowly than we use them. Once we deplete a nonrenewable resource, it is no longer available.

## We rely on ecosystem services

If we think of natural resources as "goods" produced by nature, then it is also true that Earth's natural systems provide "services" on which we depend. Our planet's ecological systems purify air and water, cycle nutrients, regulate climate, pollinate plants, and recycle our waste. Such essential services are commonly called **ecosystem services** (**FIGURE 1.2**). Ecosystem services arise from the normal functioning of natural systems and are not meant for our benefit, yet we could not survive without them. The ways that ecosystem services support our lives and civilization are countless and profound (pp. 35, 97).

Just as we may deplete natural resources, we may degrade ecosystem services when, for example, we destroy habitat or generate pollution. In recent years, our depletion of nature's goods and our disruption of nature's services have intensified, driven by rising affluence and a human population that grows larger every day.

**FIGURE 1.2 We rely on the ecosystem services that natural systems provide.** For example, forested hillsides help people living below by purifying water and air, cycling nutrients, regulating water flow, preventing flooding, and reducing erosion, as well as by providing game, wildlife, timber, recreation, and aesthetic beauty.

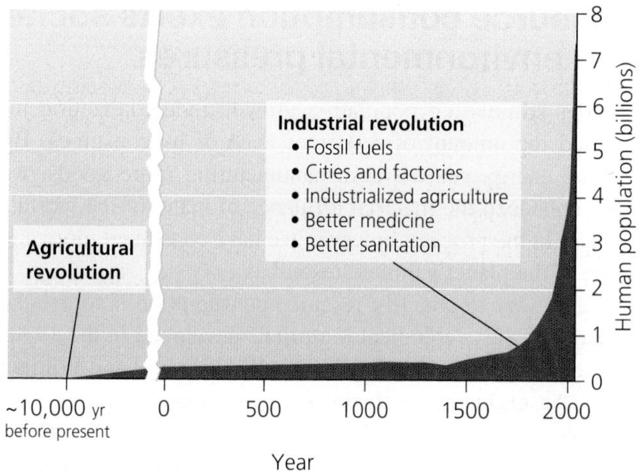


**FIGURE 1.3 The global human population increased after the agricultural revolution and then skyrocketed as a result of the industrial revolution.** *Data from U.S. Census Bureau, U.N. Population Division, and other sources.*

DATA Q For every person alive in the year 1800, about how many are alive today?

GO TO **INTERPRETING GRAPHS & DATA** ON MasteringEnvironmentalScience®

## Population growth amplifies our impact

For nearly all of human history, fewer than a million people populated Earth at any one time. Today our population has grown beyond 7 *billion* people. For every one person who used to exist, several thousand people exist today! **FIGURE 1.3** shows just how recently and suddenly this monumental change has taken place.

Two phenomena triggered our remarkable increase in population size. The first was our transition from a hunter-gatherer lifestyle to an agricultural way of life. This change began around 10,000 years ago and is known as the **agricultural revolution.** As people began to grow crops, domesticate animals, and live sedentary lives on farms and in villages, they produced more food to meet their nutritional needs and began having more children.

The second phenomenon, known as the **industrial revolution,** began in the mid-1700s. It entailed a shift from rural life, animal-powered agriculture, and handcrafted goods toward an urban society provisioned by the mass production of factory-made goods and powered by **fossil fuels** (nonrenewable energy sources such as coal, oil, and natural gas; pp. 337–338). Industrialization brought dramatic advances in technology, sanitation, and medicine. It also enhanced food production through the use of fossil-fuel-powered equipment and synthetic pesticides and fertilizers (pp. 137, 147).

The factors driving population growth have brought us better lives in many ways. Yet as our world fills with people, population growth has begun to threaten our well-being. We must ask how well the planet can accommodate the 9 billion people forecast by 2050. Already our sheer numbers, unparalleled in history, are putting unprecedented stress on natural systems and the availability of resources.

## Resource consumption exerts social and environmental pressures

Besides stimulating population growth, industrialization increased the amount of resources each of us consumes. By mining energy sources and manufacturing more goods, we have enhanced the material affluence of many of the world's people. In the process, however, we have consumed more and more of the planet's limited resources.

One way to quantify resource consumption is to use the concept of the "ecological footprint," developed in the 1990s by environmental scientists Mathis Wackernagel and William Rees. An **ecological footprint** expresses the cumulative area of biologically productive land and water required to provide the resources a person or population consumes and to dispose of or recycle the waste the person or population produces (**FIGURE 1.4**). It measures the total area of Earth's biologically productive surface that a given person or population "uses" once all direct and indirect impacts are totaled up.

For humanity as a whole, Wackernagel and his colleagues at the Global Footprint Network calculate that we are now using 50% more of the planet's resources than are available on a sustainable basis. That is, we are depleting renewable resources by using them 50% faster than they are being replenished. This is like drawing the money out of a bank account rather than living off the interest the money makes. To look at this another way, it would take 1.5 years for the planet to regenerate the renewable resources that people use in 1 year. This excess use is termed **overshoot** because we are overshooting, or surpassing, Earth's capacity to sustainably support us (**FIGURE 1.5**).

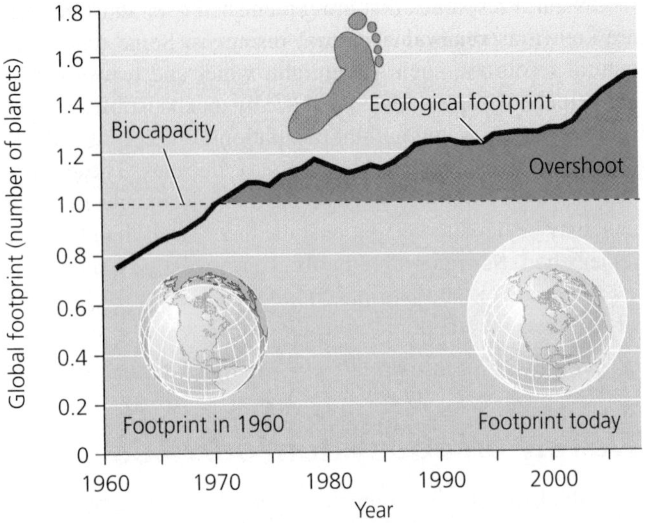


**FIGURE 1.5 Analyses by one research group indicate that we have overshot Earth's biocapacity—its capacity to support us—by 50%.** We are using renewable natural resources 50% faster than they are being replenished. *Data from WWF, 2012.* Living planet report 2012. *WWF International, Gland, Switzerland.*

DATA Q How much larger is the global ecological footprint today than it was half a century ago?

GO TO **INTERPRETING GRAPHS & DATA** ON MasteringEnvironmentalScience®

Some scientists have criticized the methods by which the Global Footprint Network calculates footprints, and many question how well its methods measure overshoot. Indeed, any attempt to boil complicated issues down to a single number is fraught with peril, even if the general concept is sound and useful. Yet some things are clear; for instance, people from wealthy nations such as the United States have much larger ecological footprints than do people from poorer nations. Using the Global Footprint Network's calculations, if all the world's people consumed resources at the rate of Americans, we would need the equivalent of four planet Earths!

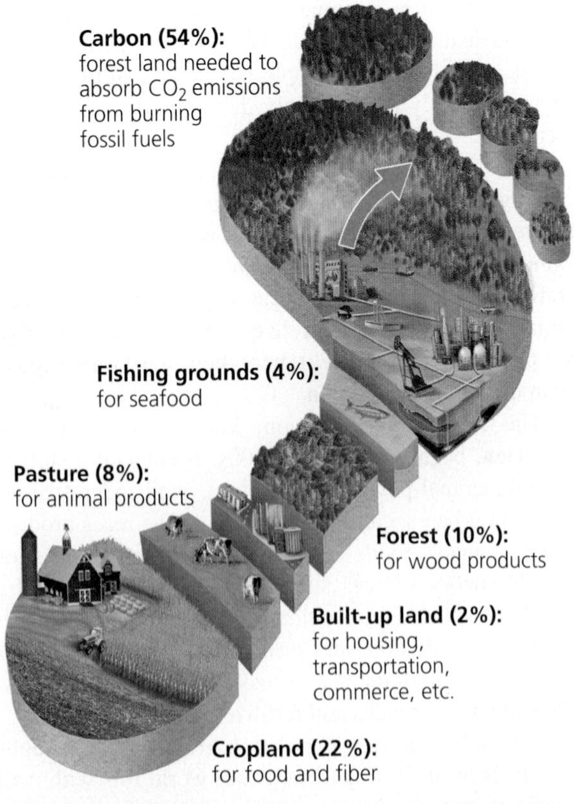


**FIGURE 1.4 An ecological footprint shows the total area of biologically productive land and water used by a given person or population.** Shown is a breakdown of major components of the average person's footprint. *Data from WWF, 2012.* Living planet report 2012. *WWF International, Gland, Switzerland.*

## Environmental science can help us avoid past mistakes

Historical evidence suggests that civilizations can crumble when pressures from population and consumption overwhelm resource availability. Historians have inferred that environmental degradation contributed to the fall of the Greek and Roman empires; the Angkor civilization of Southeast Asia; and the Maya, Anasazi, and other civilizations of the Americas. In Iraq, Syria, and elsewhere in the Middle East, areas that today are barren desert had earlier been lush enough to support the origin of agriculture and thriving ancient societies. Easter Island has long been held up as a society that self-destructed after depleting its resources, although new research paints a more complex picture (see **THE SCIENCE BEHIND THE STORY**, pp. 6–7).

In today's globalized society, the stakes are higher than ever because our environmental impacts are global. If we cannot forge sustainable solutions to our problems, then the resulting societal collapse will be global. Fortunately, environmental science holds keys to building a better world. By studying environmental science, you will learn to evaluate the many changes happening around us and to think critically and creatively about ways to respond.

## The Nature of Environmental Science

Environmental scientists examine how Earth's natural systems function, how these systems affect people, and how we influence these systems. Many environmental scientists are motivated by a desire to develop solutions to environmental problems. These solutions (such as new technologies, policies, or resource management strategies) are *applications* of environmental science. The study of such applications and their consequences is, in turn, also part of environmental science.

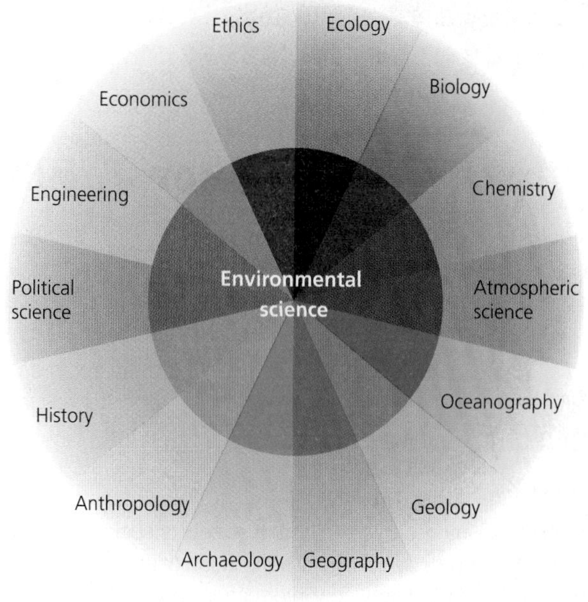


**FIGURE 1.6 Environmental science is an interdisciplinary pursuit.** It draws from many different established fields of study across the natural sciences and social sciences.

### Environmental science is interdisciplinary

Studying our interactions with our environment is a complex endeavor that requires expertise from many disciplines, including ecology, earth science, chemistry, biology, geography, economics, political science, demography, ethics, and others. Environmental science is **interdisciplinary,** bringing techniques, perspectives, and research results from multiple disciplines together into a broad synthesis (**FIGURE 1.6**).

Traditional established disciplines are valuable because their scholars delve deeply into topics, developing expertise in particular areas and uncovering new knowledge. Interdisciplinary fields are valuable because their practitioners consolidate and synthesize the specialized knowledge from many disciplines and make sense of it in a broad context to serve the multifaceted interests of society.

Environmental science is especially broad because it encompasses not only the **natural sciences** (disciplines that examine the natural world), but also the **social sciences** (disciplines that address human interactions and institutions). Most environmental science programs focus more on the natural sciences, whereas programs that emphasize the social sciences often use the term **environmental studies.** Whichever approach one takes, these fields bring together many diverse perspectives and sources of knowledge.

An interdisciplinary approach to addressing environmental problems can produce effective solutions for society. For example, we used to add lead to gasoline to make cars run more smoothly, even though research showed that lead emissions from tailpipes caused health problems, including brain damage and premature death. In 1970 air pollution was severe, and motor vehicles accounted for 78% of U.S. lead emissions. In response, environmental scientists, engineers, medical researchers, and policymakers all merged their knowledge and skills into a process that eventually brought about a ban on leaded gasoline. By 1996 all gasoline sold in the United States was unleaded, and the nation's largest source of atmospheric lead pollution had been completely eliminated.

### Environmental science is not the same as environmentalism

Although many environmental scientists are interested in solving problems, it would be incorrect to confuse environmental science with environmentalism or environmental activism. They are very different. Environmental science involves the scientific study of the environment and our interactions with it. In contrast, **environmentalism** is a social movement dedicated to protecting the natural world—and, by extension, people—from undesirable changes brought about by human actions.

**FAQ** **Aren't environmental scientists also environmentalists?**

Not necessarily. Although environmental scientists search for solutions to environmental problems, they strive to keep their research rigorously objective and free from advocacy. Of course, like all human beings, scientists are motivated by personal values and interests—and like any human endeavor, science can never be entirely free of social influence. Yet, although personal values and social concerns may help shape the questions scientists ask, scientists do their utmost to carry out their work impartially and to interpret their results with wide-open minds. Remaining open to whatever conclusions the data demand is a hallmark of the effective scientist.

## What Are the Lessons of Easter Island?

**Easter Island's immense statues**

A mere speck of land in the vast Pacific Ocean, Easter Island is one of the most remote spots on the globe. Yet this far-flung island—called Rapa Nui by its inhabitants—is the focus of an intense debate among scientists seeking to solve its mysteries and decipher the lessons it can offer us. The debate shows how, in science, new information can challenge existing ideas—and also how interdisciplinary research helps us to tackle complex questions.

Ever since European explorers stumbled upon Rapa Nui on Easter Sunday, 1722, outsiders have been struck by the island's barren landscape. Early European accounts suggested that the 2000–3000 people living on the island seemed impoverished, subsisting on a few meager crops and possessing only stone tools. Yet the forlorn island also featured hundreds of gigantic statues of carved rock. How could people without wheels or ropes, on an island without trees, have moved 90-ton,10 m (33 ft) statues as far as 10 km (6.2 mi) from the quarry where they were chiseled to the coastal sites where they were erected? Apparently some calamity must have befallen a once-mighty civilization on the island.

Researchers who set out to solve Rapa Nui's mysteries soon discovered that the island had once been lushly forested. Scientist John Flenley and his colleagues drilled cores deep into lake sediments and examined ancient pollen grains preserved there, seeking to reconstruct, layer by layer, the history of vegetation in the region. Finding a great deal of palm pollen, they inferred that when Polynesian people colonized the island (between A.D. 300–900, they estimated), it was covered with palm trees similar to the Chilean wine palm—which can live for centuries.

By studying pollen and the remains of wood from charcoal, archaeologist Catherine Orliac found that at least 21 other plant species—now gone—had also been common. Clearly the island had once supported a diverse forest. Forest plants would have provided fuelwood, building material for houses and canoes, fruit to eat, fiber for clothing—and, researchers guessed, logs and fibrous rope to help move statues.

But pollen analysis showed that trees began declining after human arrival, and ferns and grasses became more common. Then between 1400 and 1600, pollen levels plummeted. Charcoal in the soil proved the forest had been burned, likely for slash-and-burn farming. Researchers concluded that the islanders, desperate for forest resources and cropland, had deforested their own island.

With the forest gone, soil eroded away (data from lake bottoms showed a great deal of accumulated sediment). Erosion would have lowered yields of bananas, sugarcane, and sweet potatoes, perhaps leading to starvation and population decline.

Further evidence indicated that wild animals disappeared. Archaeologist David Steadman analyzed 6500 bones and found that at least 31 bird species provided food for the islanders. Today, only one native bird species is left. Remains from charcoal fires show that early islanders feasted on fish, sharks, porpoises, turtles, octopus, and shellfish—but in later years they consumed little seafood.

As resources declined, researchers concluded, people fell into clan warfare, revealed by unearthed weapons and skulls with head wounds. Rapa Nui appeared to be a tragic case of ecological suicide: A once-flourishing civilization depleted its resources and destroyed itself. In this interpretation—popularized by scientist Jared Diamond in his best-selling 2005 book *Collapse*—Rapa Nui seemed to offer a clear lesson: We on our global island, planet Earth, had better learn to use our limited resources sustainably.

When Terry Hunt and Carl Lipo began research on Rapa Nui in 2001, they expected simply to help fill gaps in a well-understood history. But science is a process of discovery, and sometimes evidence leads researchers far from where they anticipated. For Hunt, an anthropologist at University of Hawaii at Manoa, and Lipo, an archaeologist at California State University, Long Beach, their work led them to conclude that the traditional "ecocide" interpretation didn't tell the whole story.

First, their radiocarbon dating (dating of items using radioisotopes of carbon; p. 28) indicated that people had not colonized the island until about A.D. 1200, suggesting that deforestation occurred rapidly after their arrival. How could so few people have destroyed so much forest so fast? Hunt and Lipo's answer: rats. When Polynesians settled new islands, they brought crop plants and chickens and other domestic animals.

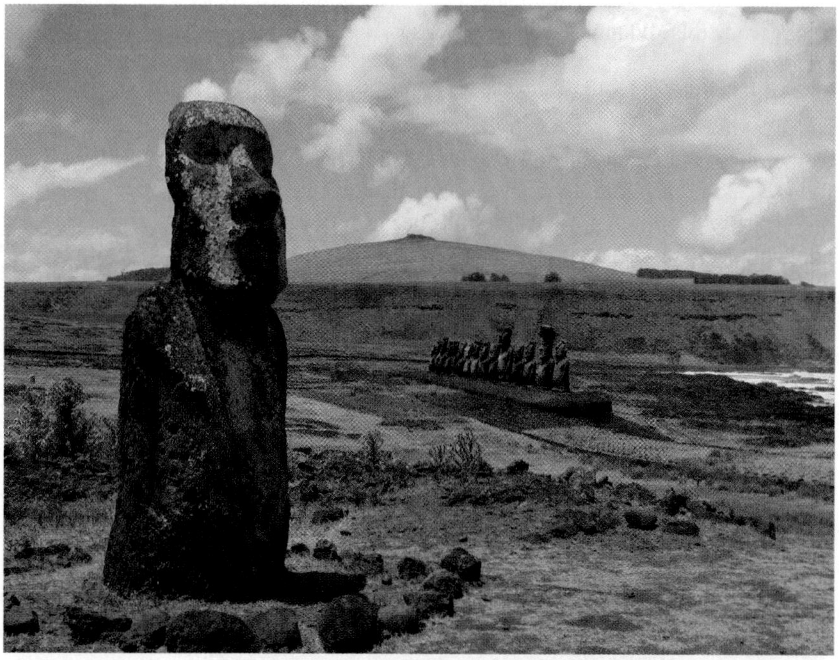

**Were the haunting statues of Rapa Nui built by a civilization that collapsed after devastating its environment, or by a sustainable civilization that fell because of outside influence?**

They also brought rats—intentionally as a food source or unintentionally as stowaways. In either case, rats can multiply quickly, and they soon overran Rapa Nui.

Researchers find rat tooth marks on old nut casings, and Hunt and Lipo suggested that rats ate so many palm nuts and shoots that the trees could not regenerate. With no young trees growing, the palm went extinct once mature trees died.

Diamond and others counter that plenty of palm nuts on Easter Island escaped rat damage, that most plants on other islands survived rats introduced by Polynesians, and that over 20 additional plant species went extinct on Rapa Nui. Moreover, people brought the rats, so even if rats destroyed the forest, human colonization was still to blame.

Despite the forest loss, Hunt and Lipo argue that islanders were able to persist and thrive. Archaeology shows how islanders adapted to Rapa Nui's poor soil and windy weather by developing rock gardens to protect crop plants and nourish the soil. Hunt and Lipo contended that tools that previous researchers viewed as weapons were actually farm implements; lethal injuries were rare; and no evidence of battle or defensive fortresses was uncovered.

Hunt, Lipo, and others also unearthed old roads and inferred how the statues were transported. It had been thought that a powerful central authority forced armies of laborers to roll them over countless palm logs, but Hunt and Lipo concluded that small numbers of people could move them by tilting and rocking them upright —the same way we move refrigerators today. Indeed, the distribution of statues on the island suggested the work of family groups. Islanders had adapted to their resource-poor environment by becoming a peaceful and cooperative society, with the statues providing a harmless outlet for competition over status and prestige.

Altogether, the evidence led Hunt and Lipo to propose that far from destroying their environment, the islanders had acted as responsible stewards. The collapse of this sustainable civilization, they argue, came with the arrival of Europeans, who unwittingly brought contagious diseases to which the islanders had never been exposed. Indeed, historical journals of sequential European voyages depict a society falling into disarray as if reeling from epidemics, its statues tumbling around it.

Peruvian ships then began raiding Rapa Nui and taking islanders away into slavery. Foreigners acquired the land, forced the remaining people into labor, and introduced thousands of sheep, which destroyed the few native plants left on the island. Thus, the collapse of Rapa Nui's civilization resulted from a barrage of disease, violence, and slave raids following foreign contact. Before that, Hunt and Lipo say, Rapa Nui's people boasted 500 years of a peaceful and resilient society.

Hunt and Lipo's interpretation, put forth in a 2011 book, *The Statues That Walked*, would represent a paradigm shift (p. 12) in how we view Easter Island. Debate between the two camps over these complex issues remains heated, however, and interdisciplinary research continues as scientists look for new ways to test the differing hypotheses. This is how science advances, and in the long-term, data from additional studies should lead us closer and closer to the truth.

Like the people of Rapa Nui, we are all stranded together on an island with limited resources. What is the lesson of Easter Island for our global island, Earth? Perhaps there are two: Any island population must learn to live within its means—but with care and ingenuity, there is hope that we can. ☐

# The Nature of Science

**Science** is a systematic process for learning about the world and testing our understanding of it. The term *science* is also used to refer to the accumulated body of knowledge that arises from this dynamic process of observing, questioning, testing, and discovery.

Knowledge gained from science can be applied to address society's needs—for instance, to develop technology or to inform policy and management decisions (**FIGURE 1.7**). From the food we eat to the clothing we wear to the health care we rely on, virtually everything in our lives has been improved by the application of science. Many scientists are motivated by the potential for developing useful applications. Others are motivated simply by a desire to understand how the world works.

## Scientists test ideas by critically examining evidence

Science is all about asking and answering questions. Scientists examine how the world works by making observations, taking measurements, and testing whether their ideas are supported by evidence. The effective scientist thinks critically and does not simply accept conventional wisdom from others. The scientist becomes excited by novel ideas but is skeptical and judges ideas by the strength of evidence that supports them. In these ways, scientists are good role models for the rest of us, because we can all benefit from learning to think critically in our everyday lives.

A great deal of scientific work is **observational science** or **descriptive science,** research in which scientists gather basic information about organisms, materials, systems, or processes that are not yet well known. In this approach, researchers explore new frontiers of knowledge by observing and measuring phenomena to gain a better understanding of them. Such research is common in traditional fields such as astronomy, paleontology, and taxonomy, and also in newer, fast-growing fields such as molecular biology and genomics.

Once enough basic information is known about a subject, scientists can begin posing questions that seek deeper explanations about how and why things are the way they are. At this point they may pursue **hypothesis-driven science,** research that proceeds in a more targeted and structured manner, using experiments to test hypotheses within a framework traditionally known as the scientific method.

(a) Chevy Volt, an electric hybrid car

(b) Prescribed burning

**FIGURE 1.7 Scientific knowledge is applied in engineering and technology and in policy and management decisions.** Energy-efficient electric automobiles **(a)** are technological advances made possible by materials and energy research. Prescribed burning **(b)** is a forest management practice informed by ecological research.

## The scientific method is a traditional approach to research

The **scientific method** is a technique for testing ideas with observations. There is nothing mysterious or intimidating about the scientific method; it is merely a formalized version of the way any of us might use logic to resolve a question. Because science is an active, creative process, innovative researchers may depart from the traditional scientific method when particular situations demand it. Moreover, scientists in different fields approach their work differently because they deal with dissimilar types of information. Nonetheless, scientists of all persuasions broadly agree on fundamental elements of the process of scientific inquiry. As practiced by individual researchers or research teams, the scientific method (**FIGURE 1.8**) typically follows the steps outlined below.

**Make observations** Advances in science usually begin with the observation of some phenomenon that the scientist wishes to explain. Observations set the scientific method in motion and play a role throughout the process.

**Ask questions** Curiosity is in our human nature. Just observe young children exploring a new environment—they want to touch, taste, watch, and listen to everything, and as soon as they can speak, they begin asking questions. Scientists, in this respect, are kids at heart. Why is the ocean salty? Why are storms becoming more severe? What is causing algae to cover local ponds? When pesticides poison fish or frogs, are people also affected? How can we help restore populations of plants and animals? All of these are questions environmental scientists ask.

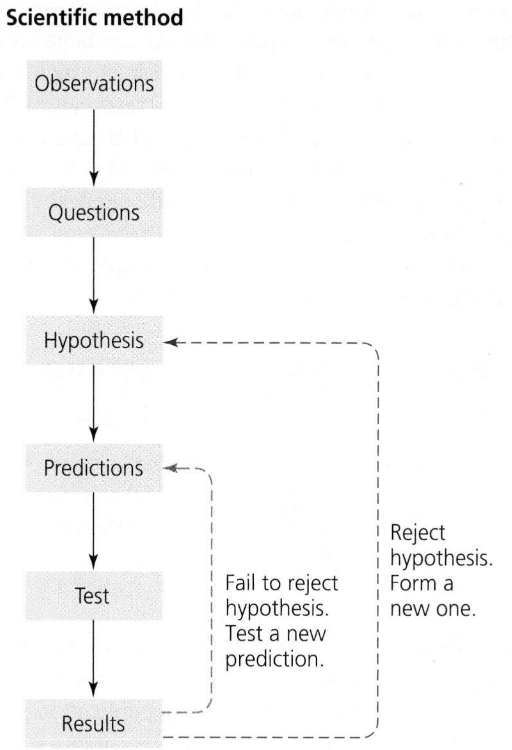


**FIGURE 1.8 The scientific method is the traditional experimental approach that scientists use to learn how the world works.**

**Develop a hypothesis** Scientists address their questions by devising explanations that they can test. A **hypothesis** is a statement that attempts to explain a phenomenon or answer a scientific question. For example, a scientist investigating why algae are growing excessively in local ponds might observe that chemical fertilizers are being applied on farm fields nearby. The scientist might then propose a hypothesis as follows: "Agricultural fertilizers running into ponds cause the amount of algae in the ponds to increase."

**Make predictions** The scientist next uses the hypothesis to generate **predictions,** specific statements that can be directly and unequivocally tested. In our algae example, a researcher might predict: "If agricultural fertilizers are added to a pond, the quantity of algae in the pond will increase."

**Test the predictions** Scientists test predictions by gathering evidence that could potentially refute them and thus disprove the hypothesis. The strongest form of evidence comes from experiments. An **experiment** is an activity designed to test the validity of a prediction or a hypothesis. It involves manipulating **variables,** or conditions that can change.

For example, a scientist could test the prediction linking algal growth to fertilizer by selecting two identical ponds and adding fertilizer to one of them. In this example, fertilizer input is an **independent variable,** a variable the scientist manipulates, whereas the quantity of algae that results is the **dependent variable,** a variable that depends on the fertilizer input. If the two ponds are identical except for a single independent variable (fertilizer input), then any differences that arise between the ponds can be attributed to changes in the independent variable. Such an experiment is known as a **controlled experiment** because the scientist controls for the effects of all variables except the one he or she is testing. In our example, the pond left unfertilized serves as a **control,** an unmanipulated point of comparison for the manipulated **treatment** pond.

Whenever possible, it is best to replicate one's experiment; that is, to stage multiple tests of the same comparison. Our scientist could perform a replicated experiment on, say, 10 pairs of ponds, adding fertilizer to one of each pair.

**Analyze and interpret results** Scientists record **data,** or information, from their studies (**FIGURE 1.9**). They particularly value quantitative data (information expressed using numbers), because numbers provide precision and are easy to compare. The scientist conducting the fertilization experiment, for instance, might quantify the area of water surface covered by algae in each pond or might measure the dry weight of algae in a certain volume of water taken from each. It is vital, however, to collect data that is representative.

**FIGURE 1.9 Researchers gather data in order to test predictions in experiments.** Here, Dr. Jennifer Smith of the Scripps Institution of Oceanography in San Diego photographs coral at a remote reef in the South Pacific. Data from analysis of the photos will help her test hypotheses about how human impacts affect the condition and community structure of coral reefs.

Because it is impractical to measure a pond's total algal growth, our researcher might instead sample from multiple areas of each pond. These areas must be selected in a random manner; choosing areas with the most growth or the least growth, or areas most convenient to sample, would not provide a representative sample.

Even with the precision that numbers provide, experimental results may not be clear-cut. Data from treatments and controls may vary only slightly, or replicates may yield different results. The researcher must therefore analyze the data using statistical tests. With these mathematical methods, scientists can determine objectively and precisely the strength and reliability of patterns they find.

If experiments disprove a hypothesis, the scientist will reject it and may formulate a new hypothesis to replace it. If experiments fail to disprove a hypothesis, this lends support to the hypothesis but does not *prove* it is correct. The scientist may choose to generate new predictions to test the hypothesis in different ways and further assess its likelihood of being true. Thus, the scientific method loops back on itself, giving rise to repeated rounds of hypothesis revision and experimentation (see Figure 1.8).

If repeated tests fail to reject a hypothesis, evidence in favor of it accumulates, and the researcher may eventually conclude that the hypothesis is well supported. Ideally, the scientist would want to test all possible explanations. For instance, our researcher might formulate an additional hypothesis, proposing that algae increase in fertilized ponds because chemical fertilizers diminish the numbers of fish or invertebrate animals that eat algae. It is possible, of course, that both hypotheses could be correct and that each may explain some portion of the initial observation that local ponds were experiencing algal blooms.

## We test hypotheses in different ways

An experiment in which the researcher actively chooses and manipulates the independent variable is known as a *manipulative experiment.* A manipulative experiment provides strong evidence because it can reveal causal relationships, showing that changes in an independent variable cause changes in a dependent variable. In practice, however, we cannot run manipulative experiments for all questions, especially for processes involving large spatial scales or long time scales. For example, in studying global climate change (Chapter 14), we cannot run a manipulative experiment adding carbon dioxide to 10 treatment planets and 10 control planets and then compare the results! Thus, it is common for researchers to run *natural experiments,* which compare how dependent variables are expressed in naturally different contexts, and to search for **correlation,** or statistical association among variables.

For instance, let's suppose our scientist studying algae surveys 50 ponds, 25 of which happen to be fed by fertilizer runoff from nearby farm fields and 25 of which are not. Let's say he or she finds seven times more algae in the fertilized ponds. The scientist may conclude that algal growth is correlated with fertilizer input; that is, that one tends to increase along with the other.

This type of evidence is not as strong as the causal demonstration that manipulative experiments can provide, but often a natural experiment is the only feasible approach for a subject of immense scale, such as an ecosystem. Many questions in environmental science are complex and exist at large scales, so they must be addressed with correlative data. As such, environmental scientists cannot always provide clear-cut, black-and-white answers to questions from policymakers and the public. Nonetheless, good correlative studies can make for very strong science, and they preserve the real-world complexity that manipulative experiments often sacrifice. Whenever possible, scientists try to integrate natural and manipulative experiments to gain the advantages of each.

## The scientific process continues beyond the scientific method

Scientific research takes place within the context of a community of peers. To have impact, a researcher's work must be published and made accessible to this community. Thus, the scientific method is embedded within a larger process involving the scientific community as a whole (**FIGURE 1.10**).

**Peer review** When a researcher's work is complete and the results analyzed, he or she writes up the findings and submits them to a journal (a scholarly publication in which scientists share their work). The journal's editor asks several other scientists who specialize in the subject area to examine the manuscript, provide comments and criticism (generally anonymously), and judge whether the work merits publication in the journal. This procedure, known as **peer review,** is an essential part of the scientific process.

Peer review is a valuable guard against faulty research contaminating the literature (the body of published studies) on which all scientists rely. However, because scientists are human, personal biases and politics can sometimes creep into the review process. Fortunately, just as individual scientists strive to remain objective in conducting their research, the scientific community does its best to ensure fair review of all work. Winston Churchill once called democracy the worst form of government, except for all the others that had been tried. The same might be said about peer review; it is an imperfect system, yet it is the best we have.

It is important to note that scientists are not paid money for peer review; their services are entirely voluntary. Moreover, researchers generally have to pay the journals that publish their papers.

**Conference presentations** Scientists also frequently present their work at professional conferences, where they interact with colleagues and receive comments on their research. Such interactions can help improve a researcher's work and foster collaboration among researchers.

**Grants and funding** To fund their research, most scientists need to spend a great deal of time requesting money from private foundations or from government agencies such as the National Science Foundation. Grant applications undergo peer review just as scientific papers do, and competition for funding is generally intense.

Scientists' reliance on funding sources can occasionally lead to conflicts of interest. A researcher who obtains data

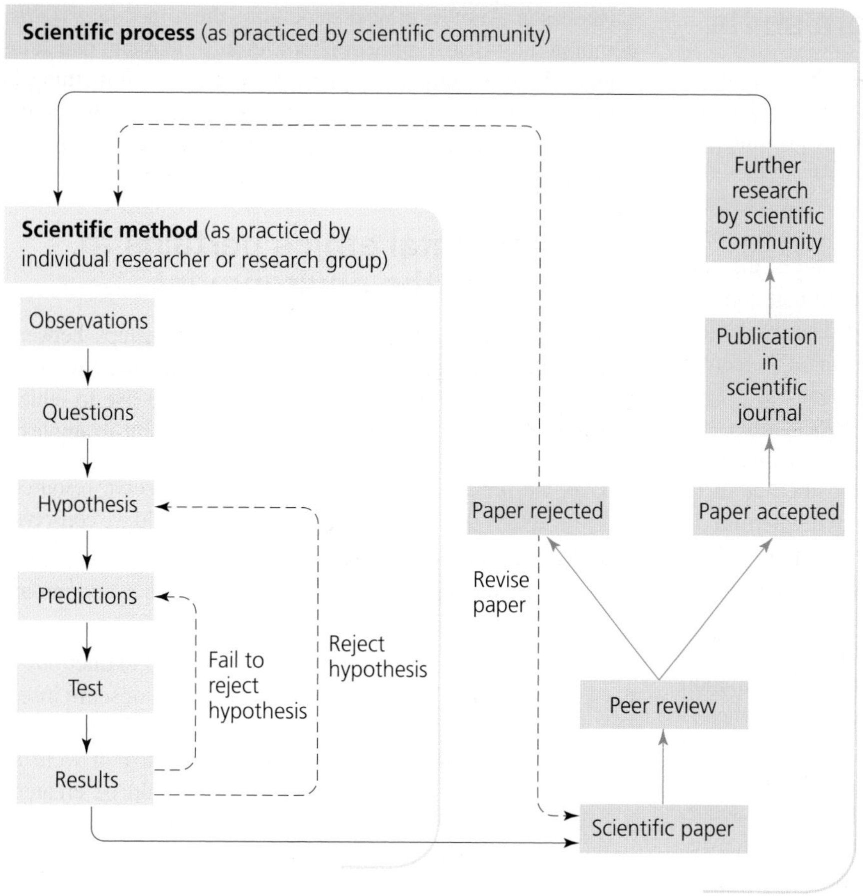


**FIGURE 1.10 The scientific method that research teams follow is part of a larger framework—the overall process of science carried out by the scientific community.** This process includes peer review and publication of research, acquisition of funding, and the elaboration of theory through the cumulative work of many researchers.

showing his or her funding source in an unfavorable light may be reluctant to publish the results for fear of losing funding—or worse yet, could be tempted to doctor the results. This situation can arise, for instance, when an industry funds research to test its products for health or safety. Most scientists resist these pressures, but whenever you are assessing a scientific study, it is always a good idea to note where the researchers obtained their funding.

## WEIGHING THE ISSUES

**Follow the Money** Let us say you are a research scientist wanting to study the impacts of chemicals released into lakes by pulp-and-paper mills. Obtaining research funding has been difficult. Then a large pulp-and-paper company contacts you and offers to fund your research examining how its chemical effluents affect water bodies. What are the benefits and drawbacks of this offer? Would you accept the offer?

**Repeatability** The careful scientist may test a hypothesis repeatedly in various ways. Following publication, other scientists may attempt to reproduce the results in their own experiments. Scientists are inherently cautious about accepting a novel hypothesis, so the more a result can be reproduced by different research teams, the more confidence scientists will have that it provides a correct explanation.

**Theories** If a hypothesis survives repeated testing by numerous research teams and continues to predict experimental outcomes and observations accurately, it may be incorporated into a theory. A **theory** is a widely accepted, well-tested explanation of one or more cause-and-effect relationships that has been extensively validated by a great amount of research. Whereas a hypothesis is a simple explanatory statement that may be disproved by a single experiment, a theory consolidates many related hypotheses that have been supported by a large body of data.

Note that scientific use of the word *theory* differs from popular usage of the word. In everyday language when we say something is "just a theory," we are suggesting it is a speculative idea without much substance. However, scientists mean just the opposite when they use the term. To them, a theory is a conceptual framework that explains a phenomenon and has undergone extensive and rigorous testing, such that confidence in it is extremely strong.

For example, Darwin's theory of evolution by natural selection (pp. 48–50) has been supported and elaborated by many thousands of studies over 150 years of intensive research. Research has shown repeatedly and in great detail how plants and animals change over generations, or evolve, expressing characteristics that best promote survival and reproduction. Because of its strong support and explanatory power, evolutionary theory is the central unifying principle of modern biology. Other prominent scientific theories include atomic theory, cell theory, big bang theory, plate tectonics, and general relativity.

## Science undergoes paradigm shifts

As the scientific community accumulates data in an area of research, interpretations sometimes may change. Thomas Kuhn's influential 1962 book *The Structure of Scientific Revolutions* argued that science goes through periodic upheavals in thought, in which one scientific **paradigm,** or dominant view, is abandoned for another. For example, before the 16th century, European scientists believed that Earth was at the center of the universe. Their data on the movements of planets fit that concept somewhat well—yet the idea eventually was disproved after Nicolaus Copernicus showed that placing the sun at the center of the solar system explained the data much better.

Another paradigm shift occurred in the 1960s, when geologists accepted plate tectonics (pp. 227–228). By this time, evidence for the movement of continents and the action of tectonic plates had accumulated and become overwhelmingly convincing. Paradigm shifts demonstrate the strength and vitality of science, showing science to be a process that refines and improves itself through time.

Understanding how science works is vital to assessing how scientific interpretations progress through time as information accrues. This is especially relevant in environmental science—a young discipline that is changing rapidly as we attain vast amounts of new information. However, to understand and address environmental problems, we need more than science. We also need to consider ethics. People's ethical perspectives, worldviews, and cultural backgrounds influence how we apply scientific knowledge. Thus, our examination of ethics (and of economics and policy in Chapter 5) will help us learn how values shape human behavior and how information from science is interpreted and put to use in our society.

# Environmental Ethics

**Ethics** is a branch of philosophy that involves the study of good and bad, of right and wrong. The term *ethics* can also refer to the set of moral principles or values held by a person or a society. Ethicists help clarify how people judge right from wrong by elucidating the criteria that people use in making these judgments. Such criteria are grounded in values—for instance, promoting human welfare, maximizing individual freedom, or minimizing pain and suffering.

People of different cultures or worldviews may differ in their values and thus in the specific actions they consider to be right or wrong. This is why some ethicists are **relativists,** who believe that ethics do and should vary with social context. However, different societies show a remarkable extent of agreement on what moral standards are appropriate. For this reason, many ethicists are **universalists,** who maintain that there are objective notions of right and wrong that hold across cultures and contexts. For both relativists and universalists, ethics is not just descriptive, but prescriptive; it prescribes how we *ought* to behave.

**Ethical standards** are the criteria that help differentiate right from wrong. One classic ethical standard is the *categorical imperative* proposed by German philosopher Immanuel Kant, which advises us to treat others as we would prefer to be treated ourselves. In Christianity this standard is called the "Golden Rule," and most of the world's religions teach this same lesson. Another ethical standard is the *principle of utility,* elaborated by British philosophers Jeremy Bentham and John Stuart Mill. The utilitarian principle holds that something is right when it produces the greatest practical benefits for the most people. We all employ ethical standards as tools for making countless decisions in our everyday lives.

## Environmental ethics pertains to people and the environment

The application of ethical standards to relationships between people and nonhuman entities is known as **environmental ethics.** Our interactions with our environment can give rise to ethical questions that are difficult to resolve. Consider some examples:

1. Is the present generation obligated to conserve resources for future generations? If so, how much should we conserve?
2. Can we justify exposing some communities to a disproportionate share of pollution? If not, what actions are warranted to prevent this?
3. Are humans justified in driving species to extinction? If destroying a forest would drive extinct an obscure insect species but would create jobs for 10,000 people, would that action be ethically admissible? What if it were an owl species? What if only 100 jobs would be created? What if it were a species harmful to people, such as a mosquito, bacterium, or virus?

Answers to such questions depend partly on what ethical standard(s) a person adopts. They also depend on the breadth of the person's domain of ethical concern. A person who feels responsibility for the welfare of insects would answer the third set of questions very differently from a person whose domain of ethical concern ends with human beings. We can think about how peoples' domains of ethical concern can vary by dividing the continuum of attitudes toward the natural world into three ethical perspectives, or worldviews: anthropocentrism, biocentrism, and ecocentrism (**FIGURE 1.11**).

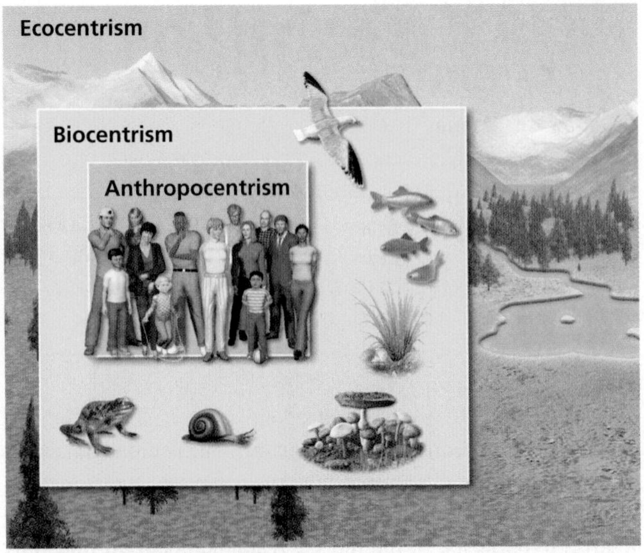


**FIGURE 1.11 We can categorize people's ethical perspectives as anthropocentric, biocentric, or ecocentric.**

**Anthropocentrism** decribes a human-centered view of our relations with the environment. An anthropocentrist denies, overlooks, or devalues the notion that nonhuman entities have rights and inherent value. An anthropocentrist evaluates the costs and benefits of actions solely according to their impact on people. For example, if cutting down a forest for farming or ranching would provide significant economic benefits while doing little harm to aesthetics or human health, the anthropocentrist would conclude this was worthwhile, even if it would destroy many plants and animals. Conversely, if protecting the forest would provide greater economic, spiritual, or other benefits to people, an anthropocentrist would favor its protection. In the anthropocentric perspective, anything not providing benefit to people is considered to be of negligible value.

In contrast, **biocentrism** ascribes inherent value to certain living things or to the biotic realm in general. In this perspective, human life and nonhuman life both have ethical standing. A biocentrist might oppose clearing a forest if this would destroy a great number of plants and animals, even if it would increase food production and generate economic growth for people.

**Ecocentrism** judges actions in terms of their effects on whole ecological systems, which consist of living and nonliving elements and the relationships among them. An ecocentrist values the well-being of entire species, communities, or ecosystems (we study these in Chapters 2–4) over the welfare of a given individual. Implicit in this view is that preserving systems generally protects their components, whereas protecting components may not safeguard the entire system. An ecocentrist would respond to a proposal to clear forest by broadly assessing the potential impacts on water quality, air quality, wildlife populations, soil structure, nutrient cycling, and ecosystem services. Ecocentrism is a more holistic perspective than biocentrism or anthropocentrism. It encompasses a wider variety of entities at a larger scale and seeks to preserve the connections that tie them together into functional systems.

## Conservation and preservation arose with the 20th century

As industrialization proceeded, raising standards of living but amplifying human impacts on the environment, more people began adopting biocentric and ecocentric worldviews. In the 19th and 20th centuries, worldviews of people in the United States evolved as the nation pushed west, urbanized, and exploited the continent's resources, boosting affluence and dramatically altering the landscape in the process.

A key voice for restraint during this period of rapid growth and change was **John Muir** (1838–1914), a Scottish immigrant to the United States who made California's Yosemite Valley his wilderness home. Although Muir chose to live in isolation in his beloved Sierra Nevada for long stretches of time, he also became politically active and won fame as a tireless advocate for the preservation of wilderness (**FIGURE 1.12**).

Muir was motivated by the rapid environmental change he witnessed during his life and by his belief that the natural world should be treated with the same respect we give to cathedrals. Today he is associated with the **preservation ethic,** which holds that we should protect the natural environment in a pristine, unaltered state. Muir argued that nature deserved protection for its own sake (an ecocentrist argument), but he also maintained that nature promoted human happiness (an anthropocentrist argument based on the principle of utility). "Everybody needs beauty as well as bread," he wrote, "Places to play in and pray in, where nature may heal and give strength to body and soul alike."

**FIGURE 1.12 A pioneering advocate of the preservation ethic, John Muir helped establish the Sierra Club, a leading environmental organization.** Here Muir **(right)** is shown with President Theodore Roosevelt in Yosemite National Park in 1903. After this wilderness camping trip with Muir, the president expanded protection of areas in the Sierra Nevada.

Some of the factors that motivated Muir also inspired **Gifford Pinchot** (1865–1946; **FIGURE 1.13**). Pinchot founded what would become the U.S. Forest Service and served as its chief in President Theodore Roosevelt's administration. Like Muir, Pinchot opposed the deforestation and unregulated development of American lands. However, Pinchot took a more anthropocentric view of how and why we should value nature. He espoused the **conservation ethic,** which holds that people should put natural resources to use but that we have a responsibility to manage them wisely. The conservation ethic employs a utilitarian standard, stating that we should allocate

**FIGURE 1.13 Gifford Pinchot was a leading proponent of the conservation ethic.** This ethic holds that we should use natural resources in ways that ensure the greatest good for the greatest number of people for the longest time.

resources so as to provide the greatest good to the greatest number of people for the longest time. Whereas preservation aims to preserve nature for its own sake and for our aesthetic and spiritual benefit, conservation promotes the prudent, efficient, and sustainable extraction and use of natural resources for the good of present and future generations.

Pinchot and Muir came to represent different branches of the American environmental movement, and their contrasting ethical approaches often pitted them against one another on policy issues of the day. Nonetheless, they both represented reactions against a prevailing "development ethic," which holds that people should be masters of nature and which promotes economic development without regard to its negative consequences. Both Pinchot and Muir left legacies that reverberate today in the ethical approaches to environmentalism.

## Aldo Leopold's land ethic inspires many people

As a young forester and wildlife manager, **Aldo Leopold** (1887–1949; **FIGURE 1.14**) began his career in the conservationist camp after graduating from Yale Forestry School, which Pinchot had helped found just as Roosevelt and Pinchot were advancing conservation on the national stage. As a forest manager in Arizona and New Mexico, Leopold embraced the government policy of shooting predators, such as wolves, to increase populations of deer and other game animals.

At the same time, Leopold followed the advance of ecological science. He eventually ceased to view certain species as "good" or "bad" and instead came to see that healthy ecological systems depend on protecting all their interacting parts. Drawing an analogy to mechanical maintenance, he wrote, "to keep every cog and wheel is the first precaution of intelligent tinkering."

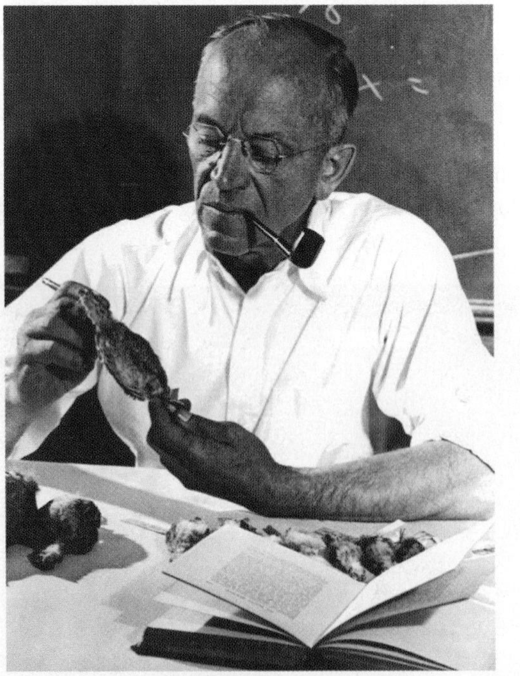

**FIGURE 1.14 Aldo Leopold, a wildlife manager, author, and philosopher, articulated a new relationship between people and the environment.** In his essay "The Land Ethic" he called on people to embrace their environment in their ethical outlook.

It was more than science that pulled Leopold from an anthropocentric perspective toward a more holistic one. One day he shot a wolf, and when he reached the animal, Leopold was transfixed by "a fierce green fire dying in her eyes." The experience remained with him for the rest of his life and helped lead him to an ecocentric ethical outlook. Years later, as a University of Wisconsin professor, Leopold argued that people should view themselves and "the land" as members of the same community and that we are obligated to treat the land in an ethical manner. In his 1949 essay "The Land Ethic," he wrote:

> All ethics so far evolved rest upon a single premise: that the individual is a member of a community of interdependent parts. . . . The land ethic simply enlarges the boundaries of the community to include soils, waters, plants, and animals, or collectively: the land. . . . A land ethic changes the role of *Homo sapiens* from conqueror of the land-community to plain member and citizen of it. . . . It implies respect for his fellow-members, and also respect for the community as such.

Leopold intended the land ethic to help guide decisions. "A thing is right," he wrote, "when it tends to preserve the integrity, stability, and beauty of the biotic community. It is wrong when it tends otherwise." Leopold died before seeing "The Land Ethic" and his best-known book, *A Sand County Almanac*, in print, but today many view him as the most eloquent philosopher of environmental ethics.

## Environmental justice seeks fair treatment for all people

Our society's domain of ethical concern has been expanding from rich to poor, and from majority races and ethnic groups to minority ones. This ethical expansion involves applying a standard of fairness and equality, and it has given rise to the environmental justice movement. **Environmental justice** involves the fair and equitable treatment of all people with respect to environmental policy and practice, regardless of their income, race, or ethnicity.

The struggle for environmental justice has been fueled by the recognition that poor people tend to be exposed to a greater share of pollution, hazards, and environmental degradation than are richer people. Environmental justice advocates also note that racial and ethnic minorities tend to suffer more than their share of exposure to most hazards. Indeed, studies repeatedly document that poor and nonwhite communities each tend to bear heavier burdens of air pollution, lead poisoning, pesticide exposure, toxic waste exposure, and workplace hazards. This is thought to occur because lower-income and minority communities often have less access to information on environmental health risks, less political power with which to protect their interests, and less money to spend on avoiding or alleviating risks.

A protest in the 1980s by residents of Warren County, North Carolina, against a toxic waste dump in their community helped to ignite the environmental justice movement. The state had chosen to site the dump in the county with the highest percentage of African Americans. Warren County residents lost their battle, but won the war; the dump was established but the protest made the nation aware of environmental justice concerns and has inspired countless efforts elsewhere since that time.

Like African Americans, Native Americans have encountered many environmental justice issues. Uranium mining on lands of the Navajo nation employed many Navajo in the 1950s and 1960s. Although uranium mining had been linked to health problems and premature death, neither the mining industry nor the U.S. government provided the miners information or safeguards against radiation and its risks. As cancer began to appear among Navajo miners, a later generation of Americans perceived negligence and discrimination. They sought relief through the Radiation Exposure Compensation Act of 1990, a federal law compensating Navajo miners who suffered health impacts from unprotected work in the mines.

Likewise, low-income white residents of the Appalachian region have long been the focus of environmental justice concerns. Mountaintop coal mining practices (pp. 241, 344) in this economically neglected region provide jobs to local residents but also pollute water, bury streams, destroy forests, and cause flooding. Low-income residents of affected Appalachian communities continue to have little political power to voice complaints over the impacts of these mining practices.

Today our economies have grown, but the gaps between rich and poor have widened. And despite much progress toward racial equality, significant inequities remain. Environmental laws have proliferated, but minorities and the poor still suffer substandard environmental conditions (**FIGURE 1.15**). Still, today more people are fighting environmental hazards in their communities and winning.

One ongoing story involves the Latino farm workers in California's San Joaquin Valley. These workers harvest much of the U.S. food supply of fruits and vegetables yet suffer some of the nation's worst air pollution. Industrial agriculture generates pesticide emissions, dairy feedlot emissions, and windblown dust from eroding farmland, yet this pollution was not being regulated. Valley residents enlisted the help of organizations including the Center on Race, Poverty, and the Environment, a San Francisco–based environmental justice law firm. Together they persuaded California regulators to enforce Clean Air Act provisions and convinced California legislators to pass new laws regulating agricultural emissions. Today the farm workers and their advocates continue working to strengthen and enforce clean air regulations.

As we explore environmental issues from a scientific standpoint throughout this book, we will also encounter the social, political, ethical, and economic aspects of these issues, and the concept of environmental justice will arise again and again. Environmental justice is a key component in pursuing the environmental, economic, and social goals of the modern drive for sustainability and sustainable development.

**FIGURE 1.15 Hurricane Katrina revealed our ongoing need for environmental justice.** Many of the people most affected by the storm were poor and nonwhite. These children are playing in the Lower Ninth Ward of New Orleans, where many homes were destroyed and water remained unsafe to drink long afterward.

## WEIGHING THE ISSUES

**Environmental Justice?** Consider the place where you grew up. Where were the factories, waste dumps, and polluting facilities located, and who lived closest to them? Who lives nearest them in the town or city that hosts your campus? Do you think the concerns of environmental justice advocates are justified? If so, what could be done to ensure that poor communities do not suffer more hazards than wealthy ones?

# Sustainability and Our Future

Recall the ethical question posed earlier (p. 12): "Is the present generation obligated to conserve resources for future generations?" This question cuts to the core of **sustainability,** a guiding principle of modern environmental science and a concept you will encounter throughout this book.

Sustainability means living within our planet's means, such that Earth can sustain us—and all life—for the future. It means leaving our children and grandchildren a world as rich and full as the world we live in now. Sustainability means conserving Earth's resources so that our descendants may enjoy them as we have. It means developing solutions that work in the long term. Sustainability requires maintaining fully functioning ecological systems, because we cannot sustain human civilization without sustaining the natural systems that nourish it.

We can think of our planet's vast store of resources and ecosystem services—Earth's **natural capital**—as a bank account. To keep a bank account full, we need to leave the principal intact and spend just the interest, so that we can continue

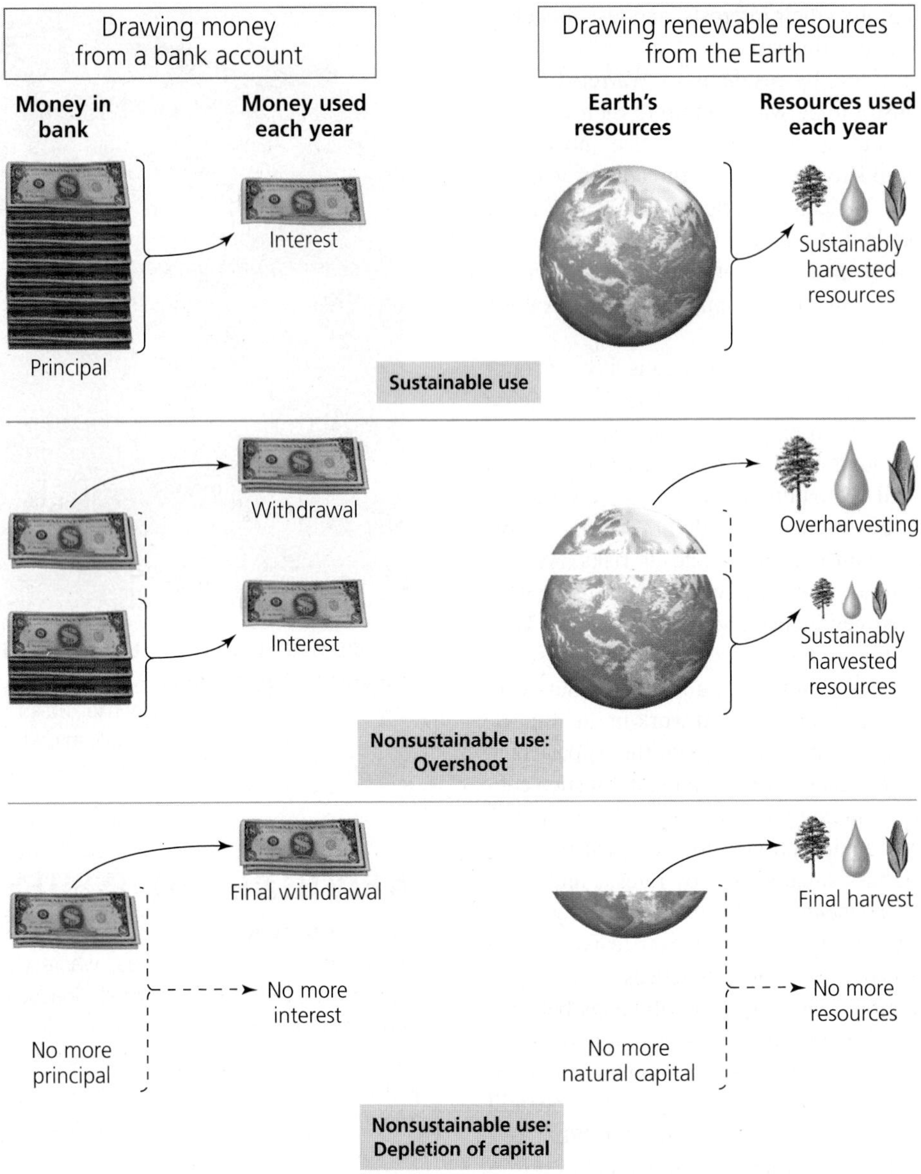


**FIGURE 1.16 Earth's natural capital is like a bank account.** Spending only the interest from a bank account is sustainable; as long as the principal remains, it will forever produce interest. But if we make withdrawals and deplete the principal, then our financial capital disappears. The same is true of Earth's renewable resources. We can harvest a certain amount sustainably, but if we harvest too much, we may deplete those resources and destroy Earth's natural capital.

living off the interest far into the future. If we begin depleting the principal, we draw down the bank account. Currently we are drawing down Earth's natural capital. Recall (p. 4) that one research group tracking this question calculates that we are withdrawing our planet's natural capital 50% faster than it is being replenished. To live off nature's interest—its renewable resources—is sustainable. To draw down resources faster than they are replaced is to eat into nature's capital—the bank account for our planet and our civilization—and we cannot get away with this for long (**FIGURE 1.16**).

## Population and consumption drive environmental impact

Every day, we add over 200,000 people to the planet. This is like adding a city the size of Augusta, Georgia, on Monday; Akron, Ohio, on Tuesday; Richmond, Virginia, on Wednesday; Rochester, New York, on Thursday; Amarillo, Texas, on Friday; and on and on, day after day. This ongoing rise in human population (Chapter 6) amplifies nearly all of our environmental impacts.

Our consumption of resources has risen even faster than our population. The modern rise in affluence has been a positive development for humanity, and our conversion of the planet's natural capital has made life better for most of us so far. However, like rising population, rising per capita consumption magnifies the demands we make on our environment.

The world's citizens have not benefited equally from our overall rise in affluence. Today the 20 wealthiest nations boast over 55 times the per capita income of the 20 poorest nations—three times the gap that existed just two generations ago. In the United States, the richest one-fifth of people now claim over

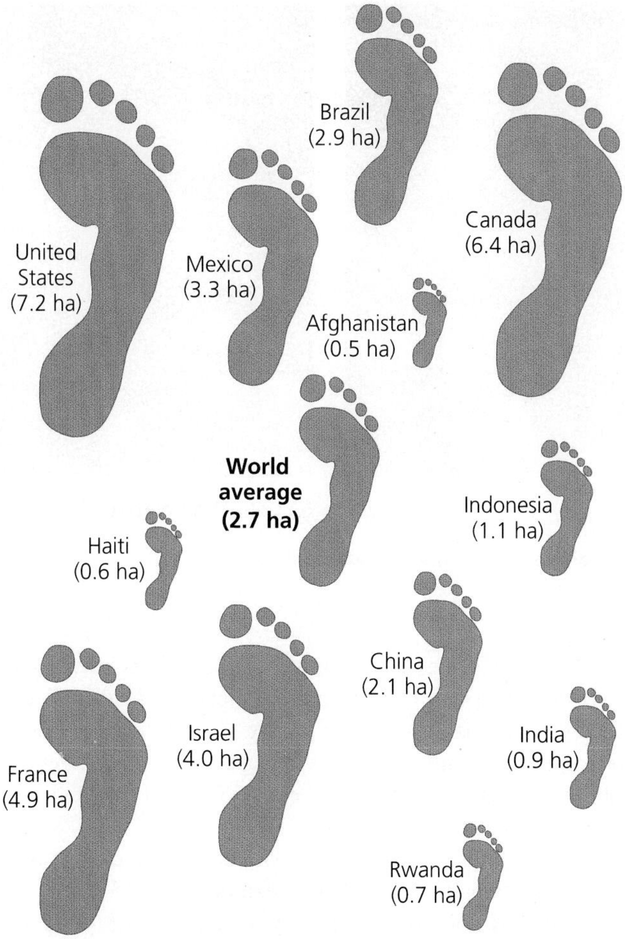


**FIGURE 1.17 People of some nations have much larger ecological footprints than people of others.** Shown are ecological footprints for average citizens of several nations, along with the world's average per capita footprint of 2.7 hectares. One hectare (ha) = 2.47 acres. *Data from Global Footprint Network, in: WWF, 2012.* Living planet report 2012. *WWF International, Gland, Switzerland.*

**DATA Q** Which nation has the largest footprint, and how many times larger is it than that of the nation with the smallest footprint?

GO TO **INTERPRETING GRAPHS & DATA** ON MasteringEnvironmentalScience®

half the income, whereas the poorest one-fifth claim just 3.2%. The ecological footprint of the average citizen of a wealthy developed nation such as the United States is considerably larger than that of the average resident of a developing country (**FIGURE 1.17**).

Our growing population and consumption are intensifying the many environmental impacts we examine in this book, including erosion and other impacts from agriculture (Chapter 7), deforestation (Chapter 9), toxic substances (Chapter 10), mineral extraction and mining impacts (Chapter 11), fresh water depletion (Chapter 12), fisheries declines (Chapter 12), air and water pollution (Chapters 12 and 13), waste generation (Chapter 17), and of course, global climate change (Chapter 14). These impacts degrade our health and quality of life, and they alter the landscapes in which we live. They also are driving the loss of Earth's biodiversity (Chapter 8)—perhaps our greatest problem, because extinction is irreversible. Once a species becomes extinct, it is lost forever.

## WEIGHING THE ISSUES

**Leaving a Large Footprint** What do you think accounts for the variation in per capita ecological footprints among societies? Do you feel that people with larger footprints have an ethical obligation to reduce their environmental impact, so as to leave more resources available for people with smaller footprints? Why or why not?

## Energy choices will shape our future

Our reliance on fossil fuels intensifies virtually every impact we exert on our environment. Yet fossil fuels have also helped to bring us the material affluence we enjoy. By exploiting the richly concentrated energy in coal, oil, and natural gas, we have been able to power the machinery of the industrial revolution, produce chemicals that boost crop yields, run vehicles and transportation networks, and manufacture and distribute countless consumer products (Chapter 15).

However, in extracting coal, oil, and natural gas, we are splurging on a one-time bonanza, for these fuels are nonrenewable and in finite supply. Attempts to reach further for new fossil fuel sources (through hydraulic fracturing, oil sands extraction, deepwater and Arctic drilling, and other avenues) all seem to threaten more impacts for relatively less fuel. The energy choices we make now will greatly influence the nature of our lives in the 21st century.

## Sustainable solutions abound

Humanity's challenge is to develop solutions that enhance our quality of life while protecting and restoring the environment that supports us. Fortunately, many workable solutions are at hand:

- Renewable energy sources (such as solar, wind, geothermal, and ocean power) are being developed to replace fossil fuels (Chapter 16).
- Energy-efficiency efforts continue to gain ground (Chapter 15).
- Scientists and farmers are pursuing soil conservation, high-efficiency irrigation, and organic agriculture (Chapter 7).
- Laws and new technologies have reduced air pollution in wealthier nations (Chapters 5 and 13).
- Conservation biologists are helping to protect habitat and safeguard endangered species (Chapter 8).
- Better waste management is helping us to conserve resources (Chapter 17).
- Governments, businesses, and individuals are taking steps to reduce emissions of the greenhouse gases that drive climate change (Chapter 14).

These are a few of the many efforts we will examine while exploring sustainable solutions in the course of this book. At the global level, today's search for sustainable solutions

(a) Urging divestment from fossil fuels (b) Recycling at Davidson (c) Collecting electronic waste

**FIGURE 1.18** **Students are helping to make their campuses more sustainable in all kinds of ways.**

centers on **sustainable development,** the use of resources for economic advancement in a manner that satisfies our current needs but does not compromise the future availability of resources. Sustainable development (Chapter 5) aims to enhance people's quality of life while preserving environmental quality so that future generations can also enjoy an enhanced quality of life. At the local level, meanwhile, every individual person can help create sustainable solutions in his or her own community—including college and university campuses.

## Students are promoting solutions on campus

As a college student, you can help to design and implement sustainable solutions on your own campus. Proponents of **campus sustainability** seek ways to help colleges and universities reduce their ecological footprints. For although we tend to think of colleges and universities as enlightened and progressive institutions that benefit society, they are also centers of lavish resource consumption. Classrooms, offices, research labs, dormitories, dining halls, sports arenas, vehicle fleets, and road networks all consume resources and generate waste. Together the 4500 campuses in the United States generate about 2% of U.S. carbon emissions. Reducing the ecological footprint of a campus can be challenging, yet students, faculty, staff, and administrators on thousands of campuses are working together to make the operations of educational institutions more sustainable (**FIGURE 1.18**).

Students are running recycling programs, promoting efficient transportation options, planting trees and restoring native plants, growing organic gardens, and fostering sustainable dining halls. They are finding ways to improve energy efficiency and water conservation and are pressing for new buildings to meet certification guidelines for sustainable construction. To help address climate change, students are urging their institutions to reduce greenhouse gas emissions, divest from fossil fuel corporations, and use and invest in renewable energy.

In response, nearly 700 university presidents have signed onto the American College and University Presidents' Climate Commitment—a public pledge to inventory emissions, set target dates and milestones for becoming carbon-neutral, take immediate steps to lower emissions, and integrate sustainability into the curriculum.

Throughout this book you will encounter examples of campus sustainability efforts (for example, pp. 135, 389, and 422). Should you wish to pursue such efforts on your own campus, information and links in the Selected Sources and References online at MasteringEnvironmentalScience point you toward organizations and resources that can help.

## Environmental science prepares you for the future

By taking a course in environmental science, you are preparing yourself for a lifetime in a world increasingly dominated by concerns over sustainability. The course for which you are using this book right now likely did not exist a generation ago. But as society's concerns have evolved, colleges and universities have adapted their curricula. As our society comes to appreciate the challenges of creating a sustainable future, colleges and universities are teaching students to tackle these challenges.

Yet at most schools, fewer than half of students take even a single course on the basic functions of Earth's natural systems, and still fewer take courses on the links between human activity and sustainability. As a result, many educators worry that most students graduate lacking basic "environmentally literacy." This means that you are in a privileged minority, benefiting from a valuable education that most of your peers are missing. Your environmental science course will equip you with a better understanding of how the world works. You will be better qualified for the green-collar job opportunities of today and tomorrow. And you will be better prepared to navigate the many challenges of creating a sustainable future.

## Conclusion

Finding effective ways of living peacefully, healthfully, and sustainably on our diverse and complex planet requires a solid ethical grounding as well as a sound scientific understanding of natural and social systems. Environmental science helps us comprehend our intricate relationship with our environment and informs our attempts to solve and prevent environmental problems. Although many of today's trends may cause concern, a multitude of inspiring success stories give us reason for optimism. Identifying a problem is the first step toward devising a solution, and addressing environmental problems can move us toward health, longevity, peace, and prosperity. Science in general, and environmental science in particular, can help us develop balanced, workable, sustainable solutions and create a better world now and for the future.

# Testing Your Comprehension

1. How and why did the agricultural revolution affect human population size? How and why did the industrial revolution affect human population size? Explain what benefits and what environmental impacts have resulted.
2. What is an *ecological footprint*? Explain what is meant by the term *overshoot*.
3. What is *environmental science*? Name several disciplines that environmental science draws upon.
4. Compare and contrast the two meanings of the term *science*. Name three applications of science.
5. Describe the scientific method. What is its typical sequence of steps? What needs to occur before a researcher's results are published? Why is this process important?
6. Compare and contrast anthropocentrism, biocentrism, and ecocentrism. Explain how individuals with each perspective might evaluate the development of a shopping mall atop a wetland in your town or city.
7. Differentiate the preservation ethic from the conservation ethic. Explain the contributions of John Muir and Gifford Pinchot in the history of environmental ethics.
8. Describe Aldo Leopold's land ethic. How did Leopold define the "community" to which ethical standards should be applied?
9. Explain the concept of environmental justice. Give an example of an inequity relevant to environmental justice that you believe exists in your city, state, or country.
10. Describe in your own words what you think is meant by the term *sustainability*. Name three ways that students, faculty, or administrators are seeking to make their campuses more sustainable.

# Seeking Solutions

1. Resources such as soils, timber, fresh water, and biodiversity are renewable if we use them in moderation, but they can become nonrenewable if we overexploit them (see Figure 1.1). For each of these four resources, describe one way we sometimes overexploit it, and one thing we could do to conserve it. For each, what might constitute sustainable use? (Feel free to look ahead and peruse coverage of these issues throughout this book.)
2. What do you think is the lesson of Easter Island? What similarities do you perceive between Easter Island and our own modern society? What more would you like to learn or understand about this island and its people? What differences do you see between their predicament and ours?
3. Describe your ethical perspective, or worldview, as it pertains to your relationship with your environment. Do you feel that you fit into any particular category discussed in this chapter? How do you think your culture has influenced your worldview? How has your personal experience shaped it? What environmental problem do *you* feel most acutely yourself?
4. Find out what sustainability efforts are being made on your campus. What results have these efforts produced so far? What further efforts would you like to see pursued on your campus? Do you foresee any obstacles to these efforts? How could these obstacles be overcome? How could you become involved?
5. **THINK IT THROUGH** You have become head of a major funding agency that grants money to scientists pursuing research in environmental science. You must give your staff several priorities to determine what types of scientific research to fund. What environmental problems would you most like to see addressed with research? Describe the research you think would need to be completed in order to develop workable solutions. What else, beyond scientific research, might be needed to develop sustainable solutions?

# Calculating Ecological Footprints

Mathis Wackernagel and his colleagues at the Global Footprint Network continue to refine the method of calculating ecological footprints—the amount of biologically productive land and water required to produce the energy and natural resources we consume and to absorb the wastes we generate. According to their most recent data, there are 1.8 hectares (4.4 acres) available for each person in the world, yet we use on average 2.7 ha (6.7 acres) per person, creating a global ecological deficit, or overshoot (p. 4), of 50%.

Compare the ecological footprints of each nation listed in the table. Calculate their proportional relationships to the world population's average ecological footprint and to the area available globally to meet our ecological demands.

| Nation | Ecological footprint (hectares per person) | Proportion relative to world average footprint | Proportion relative to world area available |
|---|---|---|---|
| Bangladesh | 0.7 | 0.3 (0.7 ÷ 2.7) | 0.4 (0.7 ÷ 1.8) |
| Tanzania | 1.2 | | |
| Colombia | 1.8 | | |
| Thailand | 2.4 | | |
| Mexico | 3.3 | | |
| Sweden | 5.7 | | |
| United States | 7.2 | | |
| World average | 2.7 | 1.0 (2.7 ÷ 2.7) | 1.5 (2.7 ÷ 1.8) |
| Your personal footprint (see Question 4) | | | |

*Data from Global Footprint Network, in: WWF, 2012.* Living planet report 2012. *WWF International, Gland, Switzerland.*

1. Why do you think the ecological footprint for people in Bangladesh is so small?
2. Why is it so large for people in the United States?
3. Based on the data in the table, how do you think average per capita income is related to ecological footprints? Name some ways in which you believe a wealthy society can decrease its ecological footprint.
4. Go to an online footprint calculator such as the one at www.myfootprint.org or www.footprintnetwork.org/en/index.php/GFN/page/personal_footprint and take the test to determine your own personal ecological footprint. Enter the value you obtain in the table, and calculate the other values as you did for each nation. How does your footprint compare to that of the average U.S. citizen? How does it compare to that of people from other nations? Name three actions you could take to reduce your footprint.

# 2

# Environmental Systems: Matter, Energy, and Ecosystems

## Upon completing this chapter, you will be able to:

- ☐ Describe the nature of environmental systems
- ☐ Explain the fundamentals of matter and chemistry, and apply them to real-world situations
- ☐ Differentiate among forms of energy and explain the basics of energy flow
- ☐ Distinguish photosynthesis from cellular respiration and summarize the importance of both processes to living things
- ☐ Define ecosystems and evaluate how living and nonliving entities interact in ecosystem-level ecology
- ☐ Outline the fundamentals of landscape ecology and ecological modeling
- ☐ Assess ecosystem services and how they benefit our lives
- ☐ Describe how water, carbon, nitrogen, and phosphorus cycle through the environment, and explain how human activities affect these cycles

**Photo: An oysterman unloads his catch on the shores of the Chesapeake Bay.**

# CENTRAL CASE STUDY

# The Vanishing Oysters of the Chesapeake Bay

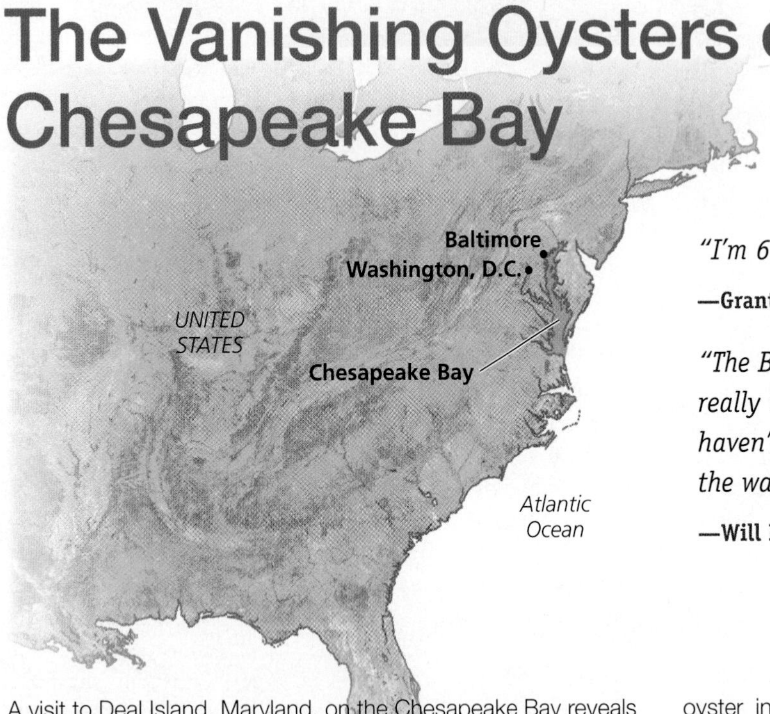


*"I'm 60. Danny's 58. We're the young ones."*

**—Grant Corbin, oysterman in Deal Island, Maryland**

*"The Bay continues to be in serious trouble. And it's really no question why this is occurring. We simply haven't managed the Chesapeake Bay as a system the way science tells us we must."*

**—Will Baker, President, Chesapeake Bay Foundation**

A visit to Deal Island, Maryland, on the Chesapeake Bay reveals a situation that is all too common in modern America. The community—once bustling with productive industries and growing populations—is suffering. Economic opportunities are few, and its populace is increasingly "graying" as more and more young people leave to find work elsewhere. In 1930, Deal Island had 1237 residents. In 2010, a mere 471 people lived here—and only 75 of them were under age 18.

Unlike other parts of the country with similar stories of economic decline, the demise of Deal Island and other bayside towns was not caused by the closing of a local factory, steel mill, or corporate headquarters. It resulted from the collapse of the Chesapeake Bay oyster fishery.

The Chesapeake Bay once was home to a thriving system of interacting plants, animals, and microbes. Blue crabs, scallops, and fish such as giant sturgeon, striped bass, and shad thrived in the bay. Nutrients carried by streams from the bay's 168,000 km$^2$ (64,000 mi$^2$) **watershed** (the land area that funnels water to the bay through rivers) nourished fields of underwater grasses that provided food and refuge to juvenile fish, shellfish, and crabs. Hundreds of millions of oysters kept the bay's water clear by filtering nutrients and phytoplankton (microscopic photosynthetic algae, protists, and cyanobacteria that drift near the surface) from the water column.

Although oysters had been eaten locally for some time, the intensive harvest of bay oysters for export began in the 1830s, and by the 1880s the bay boasted the world's largest oyster fishery. People flocked to the Chesapeake to work on oystering ships or in canneries, dockyards, and shipyards. Bayside towns prospered along with the oyster industry and developed a unique maritime culture that defined the region.

By 2010, however, the bay's oyster populations had been reduced to a mere 1% of their historical abundance, and the oyster industry was all but ruined. Perpetual overharvesting, habitat destruction, water pollution, and virulent oyster diseases had nearly eradicated this economically and ecologically important species from bay waters. The fishery collapse cost Maryland and Virginia an estimated $4 billion in lost economic activity from 1980 to 2010 alone.

One of the biggest impacts in recent decades has been pollution with high levels of nitrogen and phosphorus from agricultural fertilizers, animal manure, stormwater runoff, and atmospheric compounds produced by fossil fuel combustion. Oysters naturally filter nutrients from water, but with so few oysters today, elevated nutrient levels have caused phytoplankton populations in the bay to increase. When phytoplankton die, settle to the bay bottom, and are decomposed by bacteria, oxygen in the water is depleted. This condition, called **hypoxia,** creates "dead zones" in the bay. Grasses, oysters, and other immobile organisms perish in dead zones when deprived of oxygen. Crabs, fish, and other mobile organisms are forced to flee to areas where oxygen levels are higher, but they face smaller food supplies and increased predation pressure. Because of hypoxia, along with other human impacts, the Environmental Protection Agency (EPA) includes Chesapeake Bay on its list of highly polluted waters.

Fortunately, recent events have, at long last, given reason for hope for the recovery of the Chesapeake Bay system. In 2010, for the first time in the region, the EPA agreed to hold bay states to strict pollutant "budgets" that aim to substantially reduce inputs of nitrogen and phosphorus into the bay by 2025. Further, oyster restoration efforts are finally showing promise in the Chesapeake (see **THE SCIENCE BEHIND THE STORY,** pp. 36–37). If these initiatives can begin to restore the bay to health, then Deal Island and other communities may again enjoy the prosperity they once did on the scenic shores of the Chesapeake. ☐

# Earth's Environmental Systems

Understanding the rise and fall of the oyster industry in the Chesapeake Bay, as with many other human impacts on the environment, involves comprehending the complex, interlinked systems that comprise Earth's environment. A **system** is a network of relationships among parts, elements, or components that interact with and influence one another through the exchange of energy, matter, or information. Earth's natural systems include processes that shape the landscape, affect planetary climate, govern interactions between species and the nonliving entities around them, and cycle chemical elements vital to life. Because we depend on these systems and processes for our very survival, understanding how they function and how human activities affect them is an important aspect of environmental science.

There are many ways to delineate natural systems. For instance, scientists sometimes divide Earth's components into structural spheres to help make our planet's dazzling complexity comprehensible. The **lithosphere** (p. 227) is the rock and sediment beneath our feet, the planet's uppermost mantle and crust. The **atmosphere** (p. 280) is composed of the air surrounding our planet. The **hydrosphere** (p. 250) encompasses all water—salt or fresh; liquid, ice, or vapor—in surface bodies, underground, and in the atmosphere. The **biosphere** (p. 55) consists of all the planet's organisms and the abiotic (nonliving) portions of the environment with which they interact.

Natural systems seldom have well-defined boundaries, so deciding where one system ends and another begins can be difficult. As an analogy, consider a smartphone. It is certainly a system—a network of circuits and parts that interact and exchange energy and information—but where are its boundaries? Is the system merely the phone itself, or does it include the other phones you call, the websites you access on it, and the cellular networks that keep it connected? What about the energy grid that recharges the phone's battery? No matter how we attempt to isolate or define a system, we soon see that it has connections to systems larger and smaller than itself. Systems may exchange energy, matter, and information with other systems, and they may contain or be contained within other systems. Thus, where we draw boundaries may depend on the spatial (space) or temporal (time) scale we choose to focus on.

## Systems involve feedback loops

Earth's environmental systems receive inputs of energy, matter, or information; process these inputs; and produce outputs. The Chesapeake Bay system receives inputs of fresh water, sediments, nutrients, and pollutants from the rivers that empty into it. Oystermen, crabbers, and fishermen harvest some of the bay system's output: matter and energy in the form of seafood. This output subsequently becomes input to the nation's economic system and to the digestive systems of people who consume the seafood.

Sometimes a system's output can serve as input to that same system, a circular process described as a **feedback loop.** Feedback loops are of two types, negative and positive. In a **negative feedback loop** (**FIGURE 2.1a**), a system moving in one direction

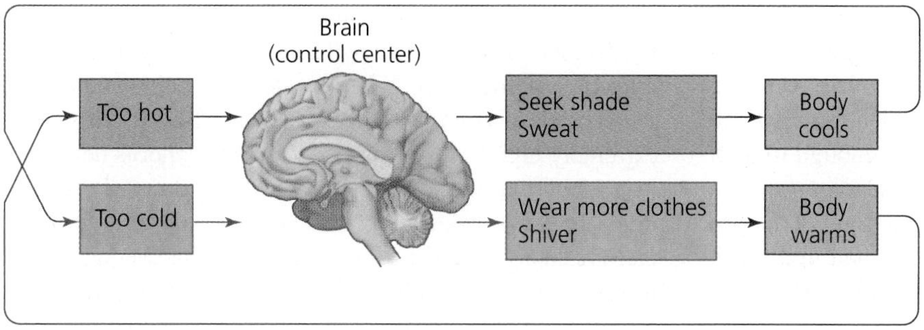


**(a) Negative feedback**

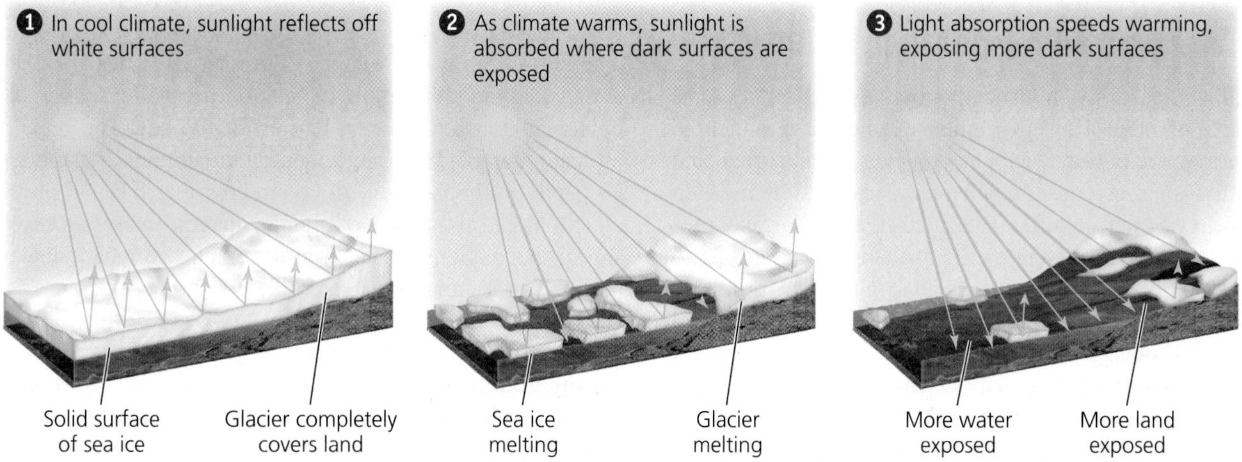


**(b) Positive feedback**

**FIGURE 2.1 Negative feedback loops exert a stabilizing influence on systems, whereas positive feedback loops have a destabilizing effect.** The human body's response to heat and cold **(a)** involves a negative feedback loop that keeps core body temperatures relatively stable. Positive feedback loops, such as the accelerated melting of Arctic ice caused by global warming **(b),** push systems away from equilibrium.

yields output that acts as an input that moves the system in the other direction. Input and output essentially neutralize one another's effects, stabilizing the system. For instance, a thermostat stabilizes a room's temperature by turning the furnace on when the room gets cold and shutting it off when the room gets hot. Similarly, negative feedback regulates our body temperature. If we get too hot, our sweat glands pump out moisture that evaporates to cool us down, or we move into the shade. If we get too cold, we shiver, creating heat, or we move into the sun.

In a system stabilized by negative feedback, processes that move in opposing directions at equivalent rates so that their effects balance out are said to be in **dynamic equilibrium.** Processes in dynamic equilibrium can contribute to **homeostasis,** the tendency of a system to maintain relatively constant or stable internal conditions. A system (such as an organism) in homeostasis keeps its internal conditions within a range that allows it to function. However, the steady state of a homeostatic system may itself change slowly over time. For instance, Earth has experienced gradual changes in atmospheric composition and ocean chemistry over its long history, yet life persists and our planet remains, by most definitions, a homeostatic system.

**Positive feedback loops** have the opposite effect. Rather than stabilizing a system, they drive it further toward an extreme. In positive feedback, increased output leads to increased input, leading to further increased output. Exponential growth in a population (p. 61) is one such example. The more individuals there are, the more offspring can be produced. Another example is the self-accelerating process of cancer cells dividing uncontrollably.

One positive feedback cycle of great concern to environmental scientists today involves the melting of glaciers and sea ice in the Arctic as a result of global warming (pp. 315–316). Ice and snow, being white, reflect sunlight and keep surfaces cool. But if the climate warms enough to melt the ice and snow, darker surfaces of land and water are exposed, and these darker surfaces absorb sunlight. This absorption warms the surface, causing further melting, which in turn exposes more dark surface area, leading to further warming (**FIGURE 2.1b**). Runaway cycles of positive feedback are rare in nature, but they are common in natural systems altered by human impact.

## Environmental systems interact

Assessing questions holistically by taking a "systems approach" is helpful in environmental science, in which so many issues are multifaceted and complex. Such a broad and integrative approach poses challenges, because systems often show behavior that is difficult to predict. However, environmental scientists are rising to the challenges, helping us to develop comprehensive solutions to complicated problems that result from the interactions of complex, diverse natural systems.

The Chesapeake Bay and the rivers that empty into it provide an example of interacting systems. The rivers form a branched and braided network of water channels surrounded by farms, cities, and forests (**FIGURE 2.2**). But where are the boundaries of this system? For a scientist interested in **runoff** (precipitation that flows over land and enters waterways) and the flow of water, sediment, or pollutants, it may make the most sense to view the bay's watershed as a system. However, for a scientist interested in the bay's dead zones, it may be best to view the watershed together with the bay as the system of interest, because their interaction is central to the problem. In environmental science, identifying the boundaries of systems depends on the questions being addressed.

The dead zones in the Chesapeake Bay result from the extremely high levels of nitrogen and phosphorus delivered to its waters from the 6 states in its watershed and the 15 states in its **airshed**—the geographic area that produces air pollutants that are likely to end up in a waterway. On average, the bay receives about 160 million kg (350 million lb) of nitrogen and 8 million kg (18 million lb) of phosphorus, with roughly one-third of nitrogen inputs from atmospheric sources. Agriculture is a major source of these nutrients, contributing 40% of the nitrogen (**FIGURE 2.3a**) and 45% of the phosphorus (**FIGURE 2.3b**) entering the bay.

Elevated nitrogen and phosphorus inputs cause phytoplankton in the bay's waters to flourish. High densities lead to elevated mortality in phytoplankton populations, and dead phytoplankton settle to the bottom of the bay. The remains of dead phytoplankton are joined on the bottom by the waste products of zooplankton, tiny creatures that feed on phytoplankton. The abundance of organic material causes an explosion in populations of bacterial decomposers, which deplete the oxygen in bottom waters while consuming this material. Deprived of oxygen, organisms will suffocate or flee. Oxygen replenishes slowly at the bottom because fresh water entering the bay from rivers remains naturally stratified in a layer at the surface and is slow to mix with the denser, saltier bay water. This limits the amount of oxygenated surface water that reaches the bottom-dwelling life that needs it. The process of

**FAQ** **But isn't positive feedback "good" and negative feedback "bad"?**

Understanding negative and positive feedback in systems can be difficult, because it goes against the way we use those terms in everyday language. In daily life, positive feedback (such as a compliment on a writing assignment) is something that makes us feel good, whereas negative feedback (such as criticism) may make us feel bad. In essence, we have been trained to view positive feedback as a stabilizing force ("Keep up the good work, and you'll succeed") and negative feedback as a destabilizing force ("You need to change your approach to succeed").

In environmental systems, it's the opposite! Negative feedback resists change in systems and thereby enhances stability, typically keeping conditions within ranges beneficial to life. Positive feedback exerts destabilizing effects that push conditions to extremes, threatening organisms adapted to the system's normal conditions. Thus, negative feedback in environmental systems typically aids living things, whereas positive feedback often harms them.

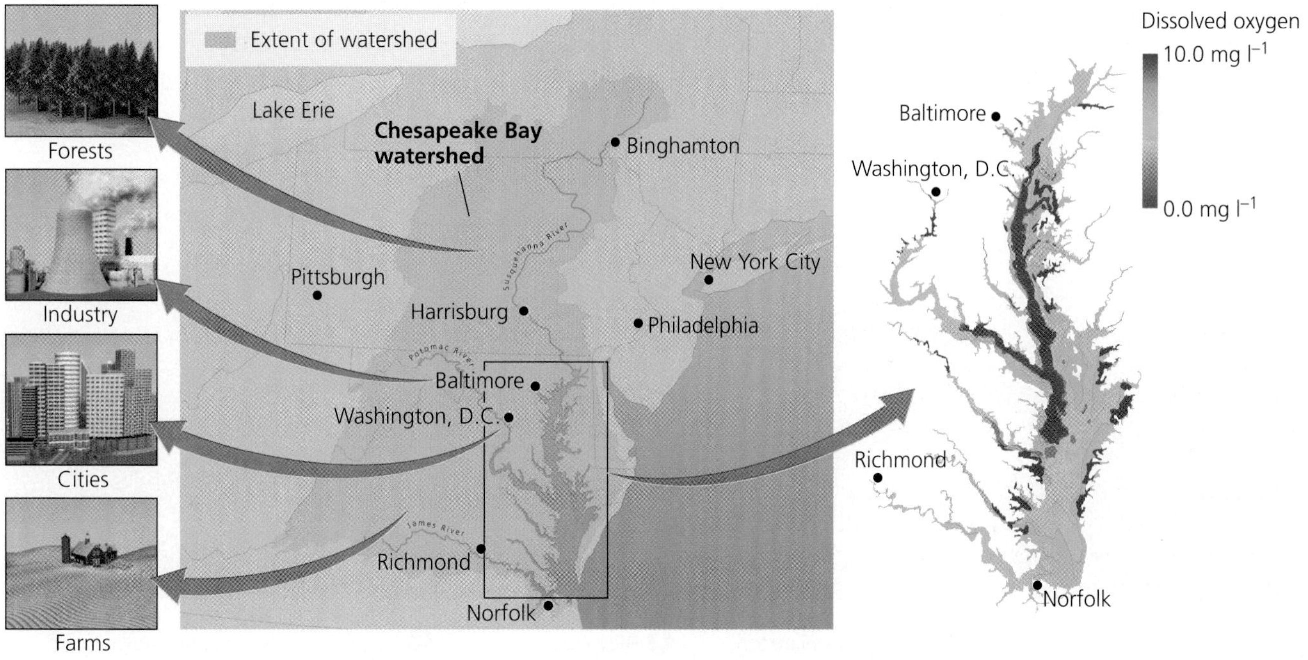


**FIGURE 2.2 The Chesapeake Bay watershed encompasses land in six states and the District of Columbia.** Nutrient pollution from its watershed has given rise to large areas of hypoxic waters in the bay. The zoomed-in map **(at right)** shows dissolved oxygen concentrations in the Chesapeake Bay in 2011. Oysters, crabs, and fish typically require a minimum of 3 mg/L of oxygen and are therefore excluded from large portions of the bay where oxygen levels are too low. *Source: Figure at right adapted from* Chesapeake Bay Record Dead Zone Map, *Chesapeake Bay Foundation.*

nutrient overenrichment, blooms of algae, increased production of organic matter, and subsequent ecosystem degradation is known as **eutrophication** (**FIGURE 2.4**, next page).

The Chesapeake Bay is not the only water body suffering from eutrophication. Nutrient pollution has led to more than 500 documented hypoxic dead zones (**FIGURE 2.5**, next page), including one that forms each year near the mouth of the Mississippi River (p. 267). The increase in the number of dead zones—there were 162 documented in the 1980s and only 49 in the 1960s—reflects how human activities are changing the chemistry of waters around the world. Let's now take a look at chemistry and its applications in environmental science.

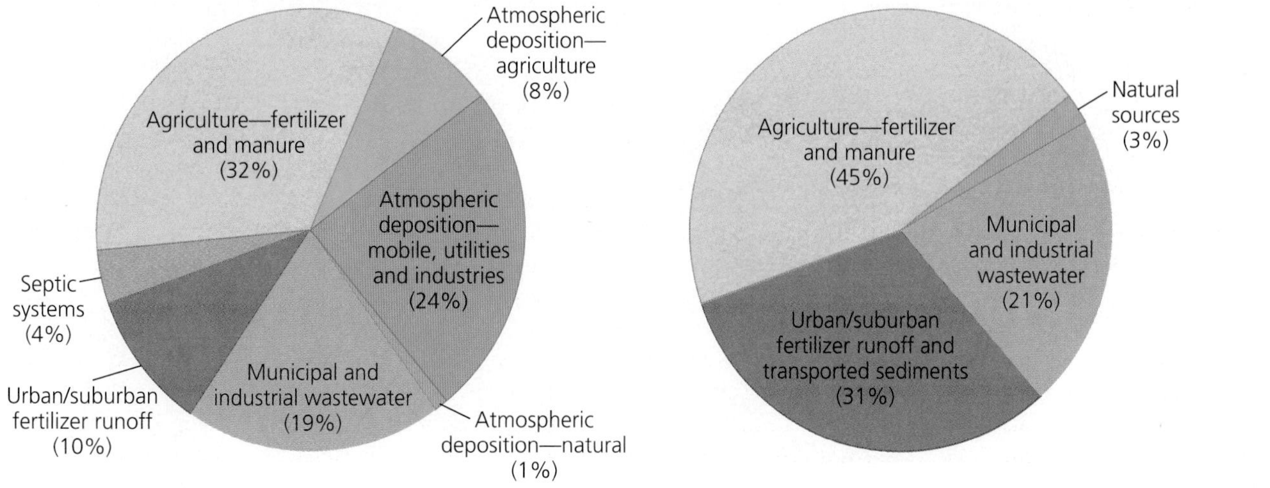


**(a) Sources of nitrogen entering the Chesapeake Bay**

**(b) Sources of phosphorus entering the Chesapeake Bay**

**FIGURE 2.3 The Chesapeake Bay receives inputs of nitrogen (a) and phosphorus (b) from many sources in its watershed.** *Data from* Chesapeake Bay Program Watershed Model Phase 4.3 *(Chesapeake Bay Program Office, 2009).*

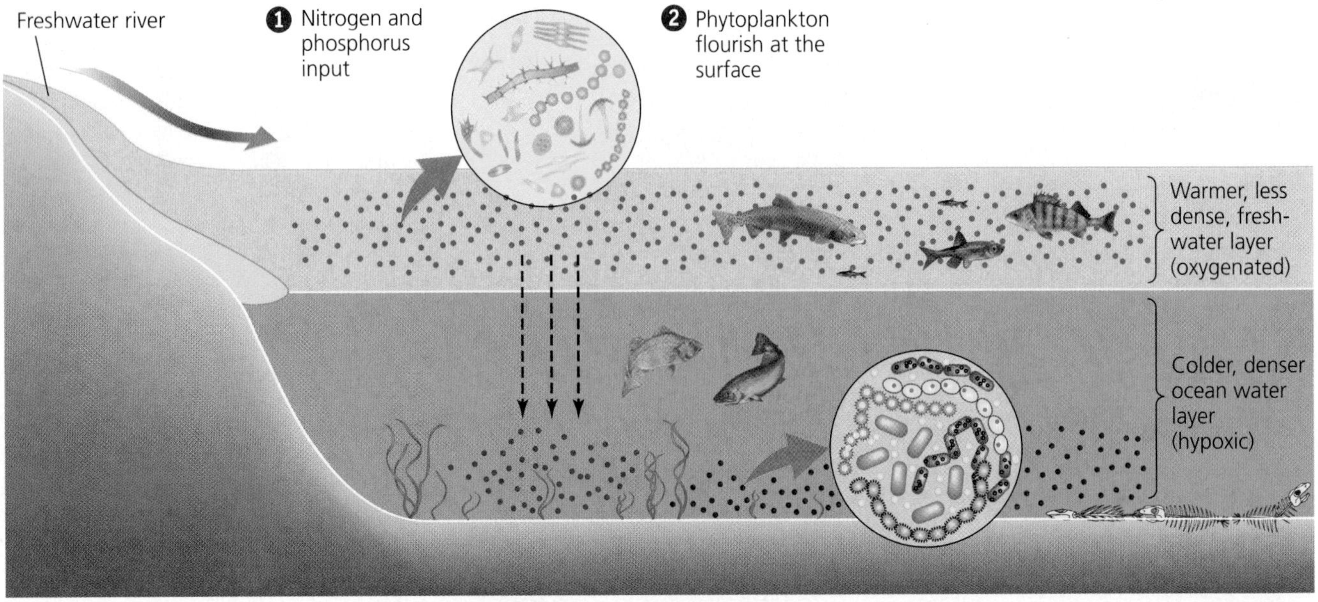
Freshwater river
1 Nitrogen and phosphorus input
2 Phytoplankton flourish at the surface
Warmer, less dense, fresh-water layer (oxygenated)
Colder, denser ocean water layer (hypoxic)
3 Dead phytoplankton and their waste drift to the bottom, providing more food for bacteria to decompose
4 Microbial decomposer population grows and consumes more oxygen
5 Insufficient oxygen suffocates oysters and grasses, fish and shrimp at the bottom; dead zone (hypoxic zone) forms

**FIGURE 2.4 Excess nitrogen and phosphorus cause eutrophication in aquatic systems such as the Chesapeake Bay.** Coupled with stratification (layering) of water, eutrophication can severely deplete dissolved oxygen. Nutrients from river water 1 boost growth of phytoplankton 2, which die and are decomposed at the bottom by bacteria 3. Stability of the surface layer prevents deeper water from absorbing oxygen to replace oxygen consumed by decomposers 4, and the oxygen depletion suffocates or drives away bottom-dwelling marine life 5. This process gives rise to hypoxic zones like those in the bay.

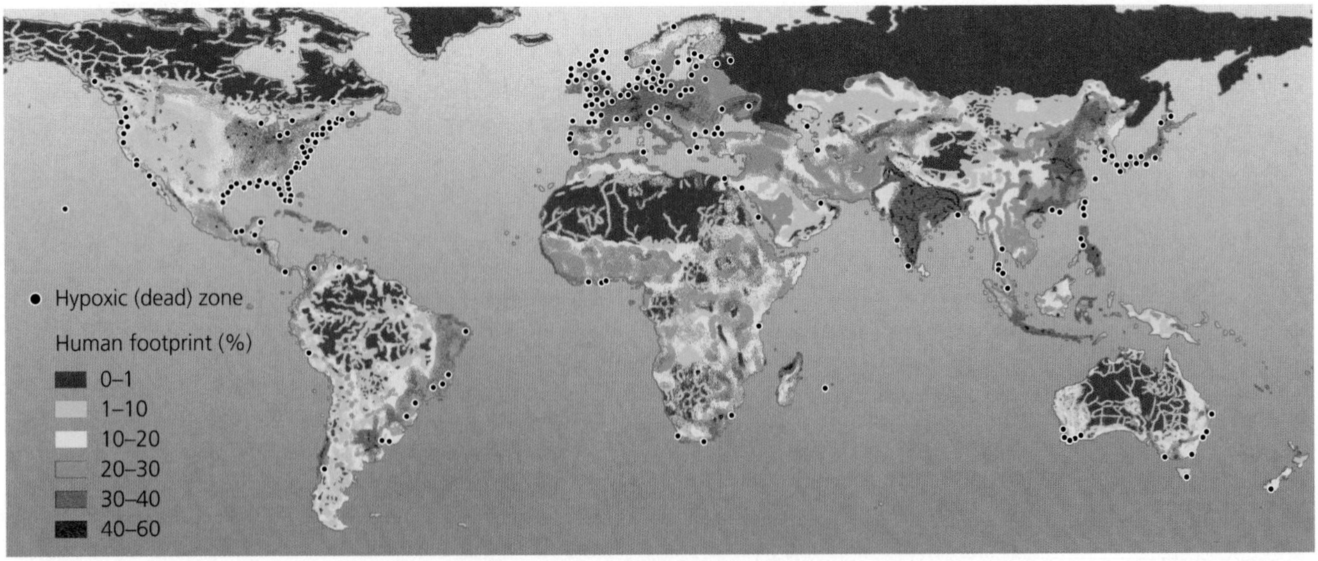
Hypoxic (dead) zone
Human footprint (%)
0–1
1–10
10–20
20–30
30–40
40–60

**FIGURE 2.5 Over 500 marine dead zones have been recorded across the world.** Dead zones (shown by dots on the map) occur mostly offshore from areas of land with the greatest human ecological footprints (here, expressed on a scale of 0 to 100, with higher numbers indicating bigger human footprints). *Data are as of 2010, from World Resources Institute, 2010, http://www.wri.org/project/eutrophication/map; and Diaz, R., and R. Rosenberg, 2008. Spreading dead zones and consequences for marine ecosystems.* Science *321: 926–929. Reprinted with permission from AAAS.*

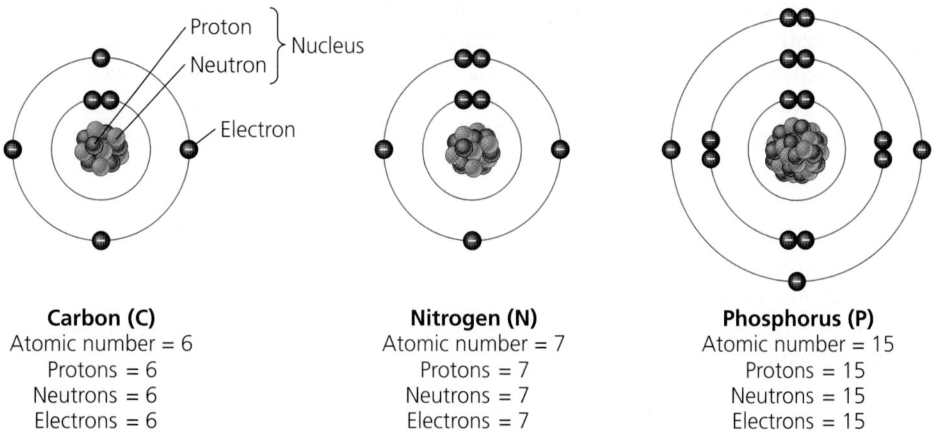


**FIGURE 2.6 In an atom, protons and neutrons stay in the nucleus, and electrons move around the nucleus.** Each chemical element has its own particular number of protons. Carbon possesses 6 protons, nitrogen 7, and phosphorus 15. These schematic diagrams are meant to clearly show and compare numbers of electrons for these three elements. In reality, however, electrons do *not* orbit the nucleus in rings as shown; they move through space in more complex ways.

# Matter, Chemistry, and the Environment

All material in the universe that has mass and occupies space—solid, liquid, and gas alike—is called **matter.** The study of types of matter and their interactions is called **chemistry.** Chemistry plays a central role in addressing the environmental challenges facing the Chesapeake Bay, as it helps us understand how too much nitrogen or phosphorus in one part of a system can lead to too little oxygen in another. Knowledge of chemistry is crucial for understanding how environmental chemicals affect the health of humans and wildlife, how greenhouse gases contribute to climate change, and how synthetic chemicals thin the ozone layer.

Matter may be transformed from one type of substance into others, but it cannot be created or destroyed. This principle is referred to as the **law of conservation of matter.** In environmental science, this principle helps us understand that the amount of matter stays constant as it is recycled in nutrient cycles and ecosystems (pp. 38–43). It also makes it clear that we cannot simply wish away "undesirable" matter, such as nuclear waste or toxic pollutants, and so must take prudent steps to lessen their impacts.

## Atoms and elements are chemical building blocks

An **element** is a fundamental type of matter, a chemical substance with a given set of properties, that cannot be chemically broken down into substances with other properties. Chemists currently recognize 98 elements occurring in nature, as well as about 20 others they have created in the lab.

An **atom** is the smallest unit that maintains the chemical properties of an element. Atoms of each element hold a defined number of **protons** (positively charged particles) in the atom's nucleus (its dense center), and this number is called the element's *atomic number*. (Elemental carbon, for instance, has six protons in its nucleus; thus, its atomic number is 6.) Most atoms also contain **neutrons** (particles lacking electric charge) in their nuclei. An atom's nucleus is surrounded by negatively charged particles known as **electrons,** which balance the positive charge of the protons (**FIGURE 2.6**).

Elements especially abundant on our planet include **hydrogen** (in water), **oxygen** (in the air), **silicon** (in Earth's crust), **nitrogen** (in the air), and **carbon** (in living organisms). Each element is assigned an abbreviation, or chemical symbol (for instance, H for hydrogen and O for oxygen). The *periodic table of the elements* (see **APPENDIX D**) organizes the elements according to their chemical properties and behavior.

Elements that organisms need for survival, such as carbon, nitrogen, calcium, and **phosphorus,** are called **nutrients.**

**Isotopes** Although all atoms of a given element contain the same number of protons, they do not necessarily contain the same number of neutrons. Atoms with differing numbers of neutrons are **isotopes** (**FIGURE 2.7a**). Isotopes are

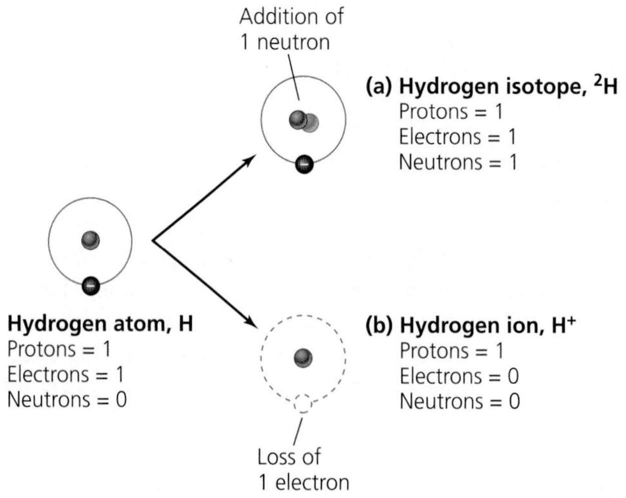


**FIGURE 2.7 Hydrogen has a mass number of 1 because a typical atom of this element contains one proton and no neutrons.** Deuterium (hydrogen-2 or $^2H$), an isotope of hydrogen **(a),** contains a neutron as well as a proton and thus has greater mass than a typical hydrogen atom; its mass number is 2. The hydrogen ion, $H^+$ **(b),** occurs when an electron is lost; it therefore has a positive charge.

denoted by their elemental symbol preceded by the *mass number,* or combined number of protons and neutrons in the atom. For example, $^{12}C$ (carbon-12), the most abundant carbon isotope, has six protons and six neutrons in the nucleus, whereas $^{14}C$ (carbon-14) has eight neutrons (and six protons). Because they differ slightly in mass, isotopes of an element differ slightly in their behavior.

Although elements cannot be broken down by chemical reactions, some isotopes are *radioactive* and "decay," changing their chemical identity as they shed subatomic particles and emit high-energy radiation. *Radioisotopes* decay into lighter and lighter radioisotopes, until they become *stable isotopes,* isotopes that are not radioactive. Each radioisotope decays at a rate determined by that isotope's *half-life,* the amount of time it takes for one-half the atoms in a given sample to give off radiation and decay. Different radioisotopes have very different half-lives, ranging from fractions of a second to billions of years. The radioisotope uranium-235 ($^{235}U$) is our society's source of energy for commercial nuclear power (pp. 356–358). It decays into a series of daughter isotopes, eventually forming lead-207 ($^{207}Pb$), and has a half-life of about 700 million years.

**Ions** Atoms may also gain or lose electrons to become **ions,** electrically charged atoms or combinations of atoms (**FIGURE 2.7b**). Ions are denoted by their elemental symbol followed by their ionic charge. For instance, a common ion used by mussels and clams to form shells is $Ca^{2+}$, a calcium atom that has lost two electrons and so has a charge of positive 2.

## Atoms bond to form molecules and compounds

Because of attractions between their electrons, atoms can bond together and form **molecules,** combinations of two or more atoms. Some molecules contain only a single element; common examples include oxygen gas ($O_2$) and nitrogen gas ($N_2$), both of which are abundant in air. As shown in these examples, scientists use a **chemical formula** (such as $O_2$ and $N_2$) as a shorthand way to indicate the type and number of atoms in the molecule. A molecule composed of atoms of two or more different elements is called a **compound. Water** is a compound; it is composed of two hydrogen atoms bonded to one oxygen atom and is denoted by the chemical formula $H_2O$. Another compound is **carbon dioxide,** consisting of one carbon atom bonded to two oxygen atoms; its chemical formula is $CO_2$.

Some compounds are made up of ions of differing charge that bind with one another to form **ionic bonds.** A crystal of table salt, sodium chloride (NaCl), is held together by ionic bonds between the positively charged sodium ions ($Na^+$) and the negatively charged chloride ions ($Cl^-$). Atoms that lack an electrical charge combine by "sharing" electrons. For example, two atoms of hydrogen bind together to form hydrogen gas ($H_2$) by sharing their electrons. This type of bond is called a **covalent bond.**

Elements, molecules, and compounds can also come together in **solutions** without chemically bonding. Air in the atmosphere is a solution formed of constituents such as nitrogen, oxygen, water vapor, carbon dioxide, **methane** ($CH_4$), and **ozone** ($O_3$). Human blood, ocean water, plant sap, and metal alloys (p. 235) such as brass are all solutions.

## The pH scale describes acids and bases

In any aqueous solution, a small number of water molecules split apart, each forming a hydrogen ion ($H^+$) and a hydroxide ion ($OH^-$). The product of hydrogen and hydroxide ion concentrations is always $10^{-14}$; as one increases, the other decreases. Pure water contains equal numbers of these ions, each at a concentration of $10^{-7}$, and we say that pure water is **neutral.** Solutions in which the $H^+$ concentration is greater than the $OH^-$ concentration are **acidic.** Solutions in which the $OH^-$ concentration is greater than the $H^+$ concentration are **basic.**

The pH scale (**FIGURE 2.8**) was devised to quantify the acidity or basicity of solutions. It ranges from 0 to 14. Pure water has a pH of 7 (neutral); solutions with pH less than 7 are acidic, and solutions with pH greater than 7 are basic. Each step on the scale represents a tenfold difference in hydrogen ion concentration. Thus, a substance with pH of 5 contains 10 times as many hydrogen ions as a substance with pH of 6, and 100 times as many hydrogen ions as a substance with pH of 7. Most biological solutions have a pH between 6 and 8, and substances that are strongly acidic (such as battery acid) or strongly basic (such as sodium hydroxide) are harmful to living things. The acidification of soils and water from acid rain (pp. 295–297) and acidic mine drainage (pp. 237–239) are examples of how pH changes caused by human activities can affect ecosystems.

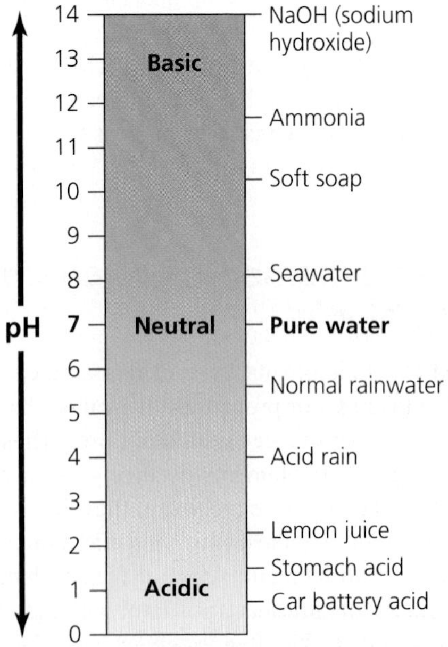


**FIGURE 2.8 The pH scale measures how acidic or basic (alkaline) a solution is.** The pH of pure water is 7, the midpoint of the scale. Acidic solutions have a pH less than 7, whereas basic solutions have a pH greater than 7.

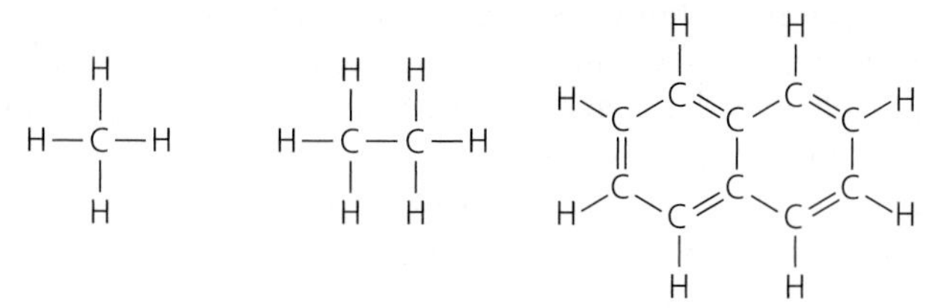

(a) Methane, $CH_4$ (b) Ethane, $C_2H_6$ (c) Naphthalene, $C_{10}H_8$

**FIGURE 2.9 Hydrocarbons have a diversity of chemical structures.** The simplest hydrocarbon is methane **(a).** Many hydrocarbons consist of linear chains of carbon atoms with hydrogen atoms attached; the shortest of these is ethane **(b).** The air pollutant naphthalene **(c)** is a ringed hydrocarbon.

## Matter is composed of organic and inorganic compounds

Beyond their need for water, living things also depend on organic compounds. **Organic compounds** consist of carbon atoms (and generally hydrogen atoms) joined by covalent bonds, and they may also include other elements, such as nitrogen, oxygen, sulfur, and phosphorus. *Inorganic compounds,* in contrast, lack carbon–carbon bonds.

Carbon's unusual ability to bond together in chains, rings, and other structures to build elaborate molecules has resulted in millions of different organic compounds. One class of such compounds that is important in environmental science is **hydrocarbons,** which consist solely of bonded atoms of carbon and hydrogen (although other elements may enter these compounds as impurities) (**FIGURE 2.9**). Fossil fuels and the many petroleum products we make from them (Chapter 15) consist largely of hydrocarbons.

## Macromolecules are building blocks of life

Just as carbon atoms in hydrocarbons may be strung together in chains, organic compounds sometimes combine to form long chains of repeated molecules. These chains are called **polymers.** Three types of polymers are essential to life: carbohydrates, proteins, and nucleic acids. Along with lipids (which are not polymers), these types of molecules are referred to as **macromolecules** because of their large sizes.

**Carbohydrates** include simple sugars and large molecules made up of chemically bonded simple sugars. Glucose ($C_6H_{12}O_6$) fuels living cells and serves as a building block for complex carbohydrates, such as starch. Plants use starch to store energy, and animals eat plants to acquire starch. Plants and animals also use complex carbohydrates to build structure. Insects and crustaceans form hard shells from the carbohydrate chitin. Cellulose, the most abundant organic compound on Earth, is a complex carbohydrate found in the cell walls of leaves, bark, stems, and roots.

**Proteins** are made up of long chains of organic molecules called *amino acids.* Organisms combine up to 20 types of amino acids into long chains to build proteins. Proteins comprise the majority of each organism's matter and serve many functions in living things. Some help produce tissues and provide structural support. In animals, for example, proteins are key components of skin, hair, muscles, and tendons. Some proteins help store energy, and others transport substances. Some function in the immune system, defending the organism against harmful viruses and bacteria. Others act as hormones, molecules that serve as chemical messengers within an organism. Proteins can also serve as enzymes, which are molecules that catalyze, or promote, certain chemical reactions.

**Nucleic acids** direct the production of proteins. The two nucleic acids—*deoxyribonucleic acid (DNA)* and *ribonucleic acid (RNA)*—carry hereditary information for organisms and are responsible for passing traits from parents to offspring. Nucleic acids are composed of series of nucleotides, each of which contains a sugar molecule, a phosphate group, and a nitrogenous base. DNA includes four types of nucleotides and can be pictured as a ladder twisted into a spiral, giving the molecule a shape called a double helix (**FIGURE 2.10**). Regions of DNA coding for particular proteins that perform particular functions are called **genes.**

**Lipids** are a chemically diverse group of compounds, classified together because they do not dissolve in water. Lipids include fats and oils (for energy storage), phospholipids (for cell membranes), waxes (for structure), and steroids (for hormone production). Organisms compartmentalize proteins, nucleic acids, carbohydrates, lipids, and other molecules in **cells.** Cells vary greatly in size, shape, and function. Some organisms are comprised of only a single cell while others are comprised of many cells. The human body, for example, contains tens of trillions of cells.

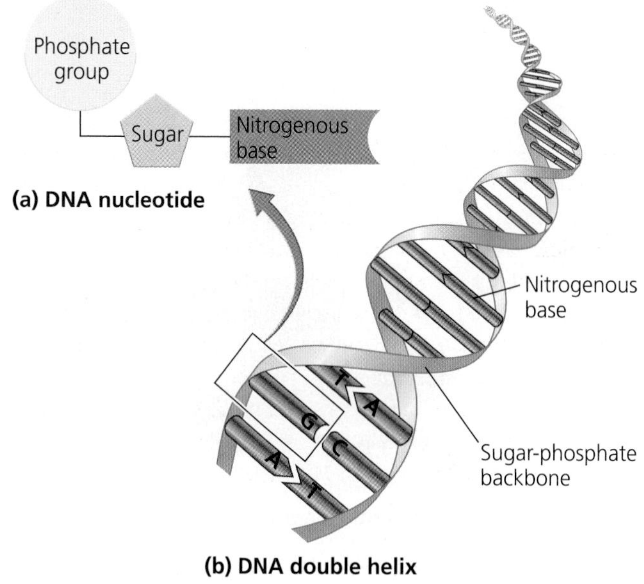


**FIGURE 2.10 Nucleic acids encode genetic information in the sequence of nucleotides, small molecules that pair together like rungs of a ladder.** DNA includes four types of nucleotides **(a),** each with a different nitrogenous base: adenine (A), guanine (G), cytosine (C), and thymine (T). Adenine (A) pairs with thymine (T), and cytosine (C) pairs with guanine (G). In RNA, thymine is replaced by uracil (U). DNA **(b)** twists into the shape of a double helix.

# Energy Fundamentals

Creating and maintaining organized complexity—of a cell, an organism, or an ecological system—requires energy. Energy is needed to organize matter into complex forms, to build and maintain cellular structure, to govern species' interactions, and to drive the geologic forces that shape our planet. Energy is somehow involved in nearly every chemical, biological, and physical phenomenon.

What, exactly, is energy? **Energy** is an intangible phenomenon that can change the position, physical composition, or temperature of matter. Scientists differentiate two types of energy: **potential energy,** or the energy of position; and **kinetic energy,** the energy of motion. Consider river water held behind a dam. Prevented from moving downstream, the water accumulates potential energy. When the dam gates are opened, this potential energy is converted to kinetic energy as the water rushes downstream.

Such energy transfers take place at the atomic level every time a chemical bond is broken or formed. **Chemical energy** is potential energy held in the bonds between atoms. Converting a molecule that has high-energy bonds (such as the carbon–carbon bonds of hydrocarbons in crude oil) into molecules with lower-energy bonds (such as the bonds in water or carbon dioxide) releases energy by changing potential energy into kinetic energy, and produces motion, action, or heat. Just as our automobile engines split the hydrocarbons of gasoline to release chemical energy and generate movement, our bodies split glucose molecules from our food for the same purpose (**FIGURE 2.11**).

Potential energy can also occur as nuclear energy, the energy that holds atomic nuclei together. Nuclear power plants utilize this energy when they break apart the nuclei of large atoms within their reactors. Mechanical energy, such as the energy stored in a compressed spring, is yet another type of potential energy. Kinetic energy can also express itself in different forms, including thermal energy, light energy, sound energy, and electrical energy—all of which involve the movement of atoms, subatomic particles, molecules, or objects.

## Energy is always conserved, but it changes in quality

Although energy can change from one form to another, it cannot be created or destroyed. Just as matter is conserved, the total energy in the universe remains constant and thus is said to be conserved. Scientists have dubbed this principle the **first law of thermodynamics.** For example, the potential energy of the water behind a dam will equal the kinetic energy of its eventual movement downstream. Likewise, we obtain energy from the food we eat and then expend it in exercise, apply it toward maintaining our body, or store it in fat. We do not somehow create additional energy or end up with less energy than the food gives us. Any particular system in nature can temporarily increase or decrease in energy, but the total amount in the universe always remains constant.

The total amount of energy is conserved in any energy transfer, but the **second law of thermodynamics** states that the nature of energy will change from a more-ordered state to a less-ordered state, if no force counteracts this tendency. That is, systems tend to move toward increasing disorder, or *entropy*. For instance, a log of firewood—the highly organized and structurally complex product of many years of tree growth—transforms in a campfire to carbon ash, smoke, and

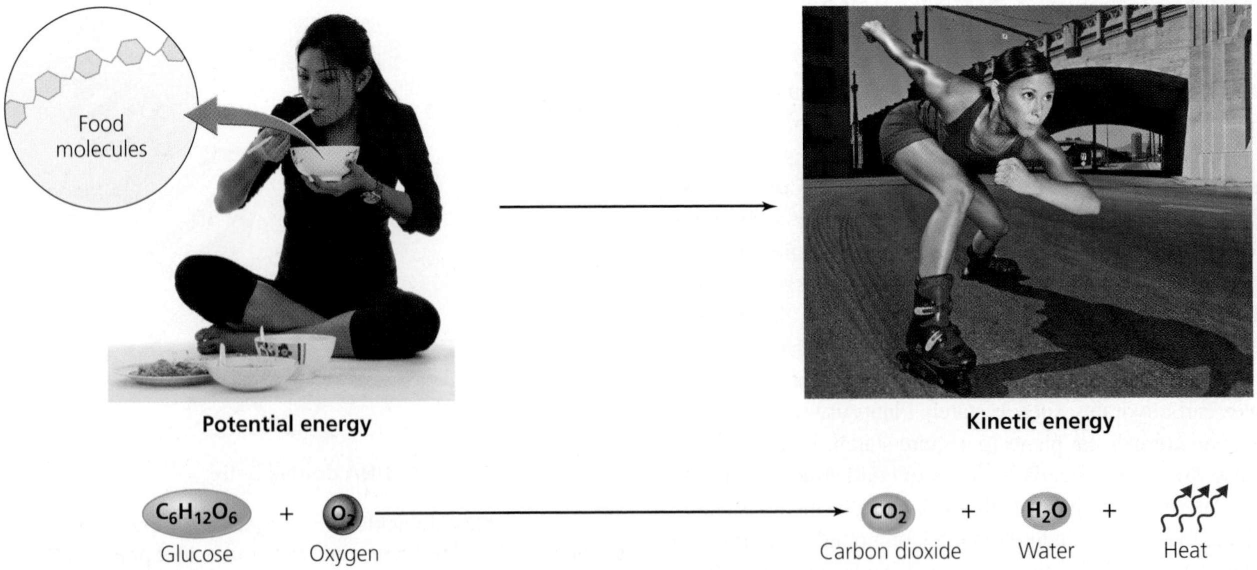


**FIGURE 2.11 Energy is released when potential energy is converted to kinetic energy.** Potential energy stored in sugars (such as glucose) in the food we eat, combined with oxygen, becomes kinetic energy when we exercise, releasing carbon dioxide, water, and heat as by-products.

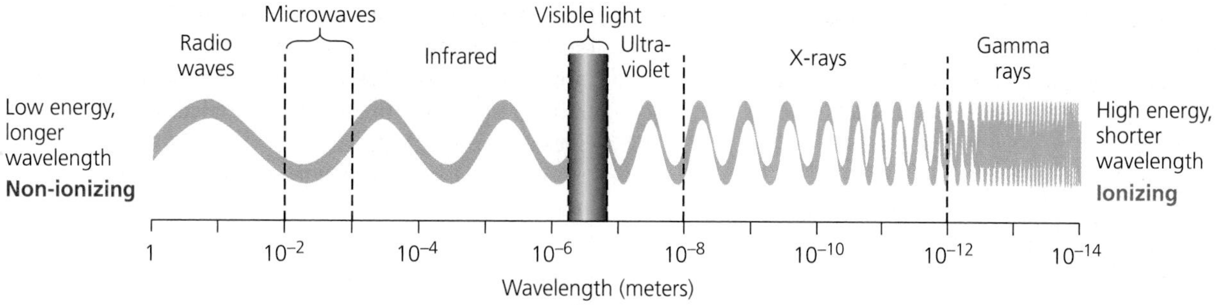


**FIGURE 2.12 The sun emits radiation from many portions of the electromagnetic spectrum.** Visible light makes up only a small proportion of this energy. Some radiation that reaches our planet is reflected back; some is absorbed by air, land, and water; and a small amount powers photosynthesis.

gases such as carbon dioxide and water vapor, as well as the light and heat of the flame. With the help of oxygen, the complex biological polymers making up the wood are converted into a disorganized assortment of rudimentary molecules and heat and light energy. When energy transforms from a more-ordered to less-ordered state, it cannot accomplish tasks as efficiently. For example, the energy available in ash (a less-ordered state of wood) is far lower than that available in a log of firewood (the more-ordered state of wood).

Living organisms are able to resist entropy through regular inputs of energy from food sources and photosynthesis. Once death occurs and those energy inputs cease, an organism undergoes decomposition and loses its highly organized structure.

## Light energy from the sun powers most living systems

The energy that powers Earth's environmental systems comes primarily from the sun. Solar energy drives our planet's winds, ocean currents, weather, and climate patterns. The sun releases radiation from large portions of the electromagnetic spectrum, although our atmosphere filters much of this out and we see only some of this radiation as visible light (**FIGURE 2.12**).

Some organisms use the sun's radiation directly to produce their own food. Such organisms, called **autotrophs** or **producers,** include green plants, algae, and cyanobacteria (a type of bacteria named for their cyan, or blue-green, color). Autotrophs turn light energy from the sun into chemical energy through a process called **photosynthesis** (**FIGURE 2.13**). In photosynthesis, sunlight powers a series of chemical reactions that convert carbon dioxide and water into sugars, transforming energy from the sun into high-quality chemical energy (sugars) the organism can use. This is an example of movement toward a state of lower entropy, and as such it requires a substantial input of outside energy.

Photosynthesis occurs within cell organelles called *chloroplasts,* where the light-absorbing pigment *chlorophyll* (the substance that makes plants green) uses solar energy to initiate a series of chemical reactions called *light reactions.* During these reactions, water molecules split and react to form hydrogen ions ($H^+$) and molecular oxygen ($O_2$), thus creating the oxygen that we breathe. The light reactions also produce small, high-energy molecules that are used to fuel reactions in the *Calvin cycle,* where carbon atoms from carbon dioxide are linked together to manufacture sugars.

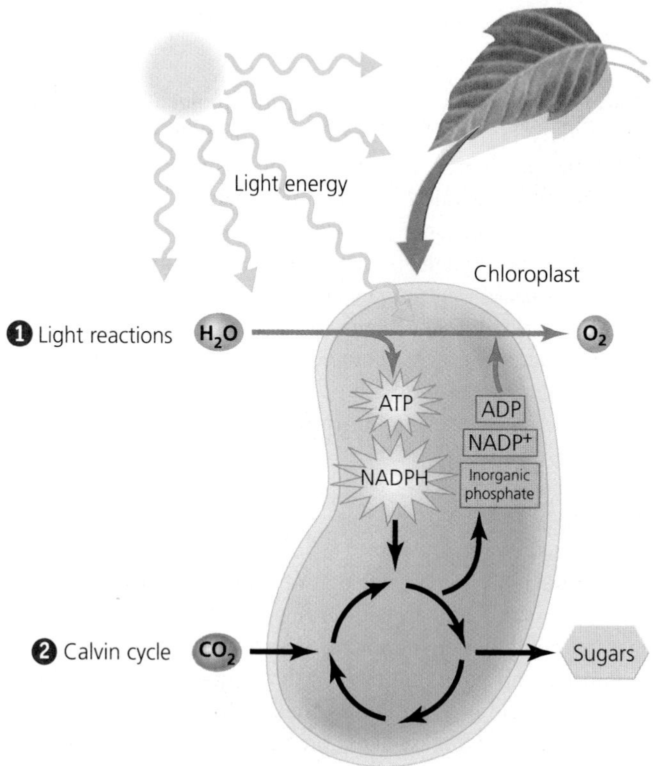


**FIGURE 2.13 In photosynthesis, autotrophs such as plants, algae, and cyanobacteria use sunlight to convert water and carbon dioxide into oxygen and sugar.** In the light reactions, water is converted to oxygen in the presence of sunlight, creating high-energy molecules (ATP and NADPH). These molecules help drive reactions in the Calvin cycle, in which carbon dioxide is used to produce sugars. Molecules of ADP, $NADP^+$, and inorganic phosphate created in the Calvin cycle, in turn, help power the light reactions, creating an endless loop.

Photosynthesis is a complex process, but the overall reaction can be summarized in the following equation:

$$6CO_2 + 6H_2O + \text{the sun's energy} \longrightarrow \underset{\text{(sugar)}}{C_6H_{12}O_6} + 6O_2$$

Thus in photosynthesis, green plants draw up water from the ground through their roots, absorb carbon dioxide from the air through their leaves, and harness the energy in sunlight to create sugars (such as $C_6H_{12}O_6$) for their growth and maintenance. As a by-product, they release the oxygen that we, and other organisms, breathe.

## Cellular respiration releases chemical energy

Organisms make use of the chemical energy created by photosynthesis in a process called **cellular respiration,** which is vital to life. To release the chemical energy of glucose, cells employ oxygen to convert glucose back into its original starting materials, water and carbon dioxide. The energy released during this process is used to form chemical bonds or to perform other tasks within cells. The net equation for cellular respiration is the opposite of that for photosynthesis:

$$\underset{\text{(sugar)}}{C_6H_{12}O_6} + 6O_2 \longrightarrow 6CO_2 + 6H_2O + \text{energy}$$

However, the energy gained per glucose molecule in respiration is only two-thirds of the amount of energy required to synthesize a glucose molecule in photosynthesis—a prime example of the second law of thermodynamics in action.

Respiration occurs in autotrophs and also in **heterotrophs,** or **consumers,** organisms that gain their energy by feeding on the tissues of other organisms. Heterotrophs include most animals, as well as the fungi and microbes that decompose organic matter. In most ecological systems, plants, algae, or cyanobacteria form the base of a food chain through which energy passes to heterotrophs (pp. 71–72).

# Ecosystems

Let's now apply our knowledge of chemistry and energy to see how energy, matter, and nutrients move through the living and nonliving environment. An **ecosystem** consists of all organisms and nonliving entities that occur and interact in a particular area at a particular time. Animals, plants, water, soil, nutrients—all these and more help comprise ecosystems.

## Energy flows and matter cycles through ecosystems

The ecosystem concept originated with scientists who recognized that biological entities are tightly intertwined with the chemical and physical aspects of their environment. For instance, in the Chesapeake Bay **estuary** (a water body where rivers flow into the ocean, mixing fresh water with salt water), aquatic organisms are intimately affected by the flow of water, sediment, and nutrients from the rivers that feed the bay and from the land that feeds those rivers. In turn, the photosynthesis, respiration, and decomposition of these organisms influence the chemical composition of the Chesapeake's waters.

Ecologists soon began analyzing ecosystems as an engineer might analyze the operation of a machine. In this view, ecosystems are systems that receive inputs of energy, process and transform that energy while cycling matter internally, and produce a variety of outputs (such as heat, water flow, and animal waste products) that can feed into other ecosystems.

Energy flows in one direction through ecosystems; it arrives mostly as radiation from the sun, powers the system, and exits in the form of heat (**FIGURE 2.14a**). Matter, in contrast, is generally recycled within ecosystems (**FIGURE 2.14b**). Energy and matter are passed among organisms (producers, consumers, and decomposers) through food web relationships (pp. 72–73). Matter is recycled because when organisms die and decay, their nutrients remain in the system. In contrast, most energy that organisms take in drives cellular respiration and is released as heat.

## Sunlight is converted to chemical energy in biomass

Energy flow in most ecosystems begins with the sun's radiation. Autotrophs convert solar energy to the energy of chemical bonds in sugars during photosynthesis, a process termed **primary production.** Specifically, the total amount of chemical energy produced by autotrophs is termed **gross primary production.** Autotrophs use most of this production to power their own metabolism by cellular respiration, however. The energy that remains and is used to generate *biomass* (such as leaves, stems, and roots) is called **net primary production.** Net primary production is equal to gross primary production minus cellular respiration.

Ecosystems vary in the rate at which autotrophs convert energy to biomass. This rate is termed **productivity,** and ecosystems whose producers convert solar energy to biomass rapidly are said to have high **net primary productivity.** Freshwater wetlands, tropical forests, and coral reefs tend to have the highest net primary productivities, whereas deserts, tundra, and open ocean tend to have the lowest (**FIGURE 2.15**). In terrestrial ecosystems, net primary productivity tends to increase with temperature and precipitation. In aquatic ecosystems, net primary productivity tends to rise with light and the availability of nutrients. The limiting nature of nutrients in waters such as the Chesapeake is why additions of nitrogen and phosphorus from people's activities have such significant effects on the system.

## Ecosystems interact across landscapes

Ecosystems vary widely in size. An ecosystem can be as small as a puddle of water containing algae and tadpoles or as large as a forest, lake, or bay that supports a diversity of

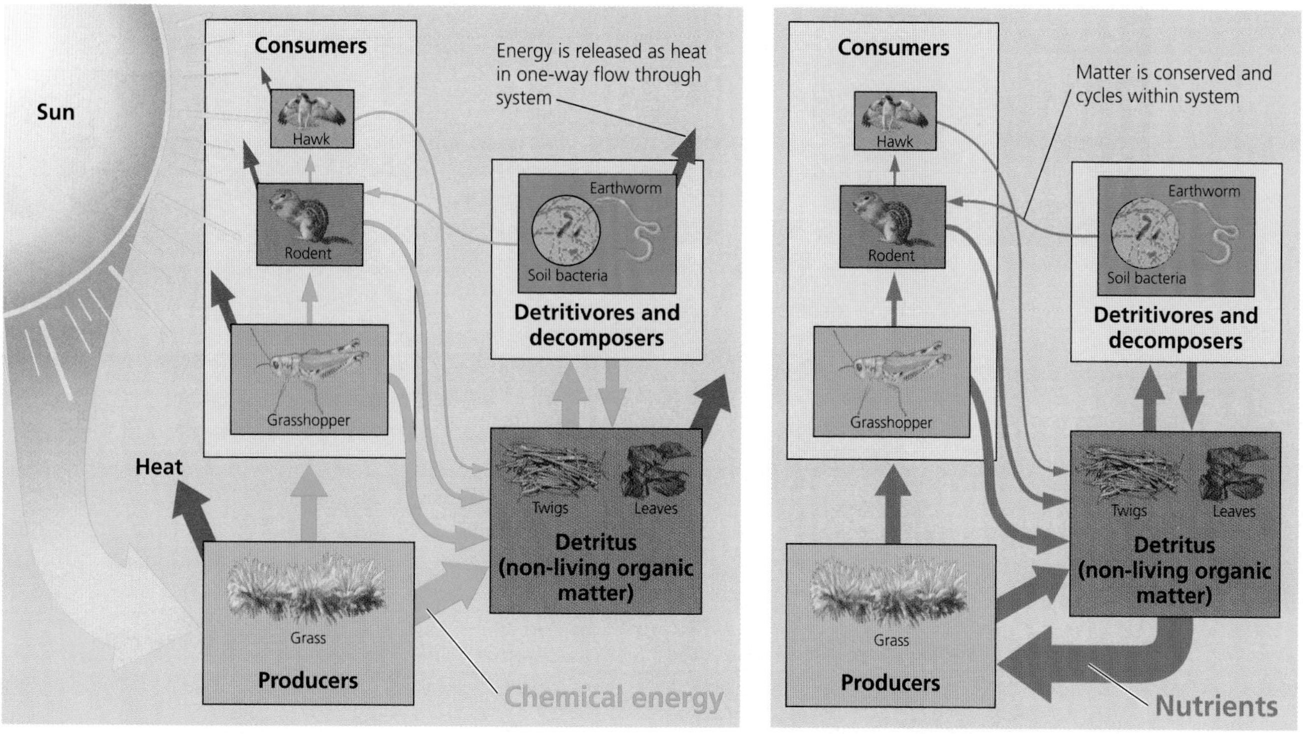


**(a) Energy flowing through an ecosystem**

**(b) Matter cycling within an ecosystem**

**FIGURE 2.14 In systems, energy flows in one direction, whereas matter is recycled.** In **(a),** light energy from the sun **(yellow arrow)** drives photosynthesis in producers, which begins the transfer of chemical energy **(green arrows)** among trophic levels (pp. 71–72) and detritus. Energy exits the system through respiration in the form of heat **(red arrows).** In **(b),** nutrients **(blue arrows)** move among trophic levels and detritus. In both diagrams, box sizes represent quantities of energy or matter content, and arrow widths represent relative magnitudes of energy or matter transfer. Such values may vary greatly among ecosystems. For simplicity, various abiotic components (such as water, air, and inorganic soil content) of ecosystems have been omitted.

**DATA Q** Based on the figure, which transfer of chemical energy is the largest in ecosystems? Which transfer is the largest for nutrient cycling in ecosystems?

GO TO **INTERPRETING GRAPHS & DATA** ON MasteringEnvironmentalScience®

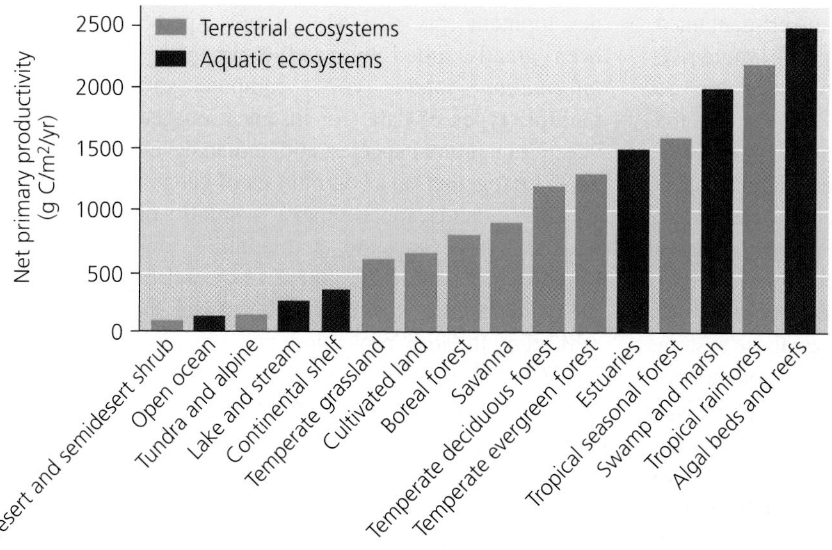


**FIGURE 2.15 Net primary productivity varies greatly between ecosystem types.** Freshwater wetlands, tropical forests, coral reefs, and algal beds show high values on average, whereas deserts, tundra, and the open ocean show low values. *Data from Whittaker, R.H., 1975.* Communities and ecosystems, *2nd ed. New York: MacMillan.*

**FIGURE 2.16 Landscape ecology deals with spatial patterns above the ecosystem level.** This generalized diagram of a landscape shows a mosaic of patches of five ecosystem types (three terrestrial types, a marsh, and a river). Thick red lines indicate ecotones. A stretch of lowland broadleaf forest running along the river serves as a corridor connecting the large region of forest on the left to the smaller patch of forest alongside the marsh. The inset shows a magnified view of the forest-grassland ecotone and how it consists of patches on a smaller scale.

habitats and species. In general, the term refers to a system of moderate geographic extent that is somewhat self-contained. For example, the tidal marshes in the Chesapeake where river water empties into the bay are an ecosystem, as are the sections of the bay dominated by oyster reefs. Ecosystems that border one another may interact extensively. For instance, rivers, tidal marshes, and open waters in estuaries all interact, as do forests and grasslands where they converge. Areas where ecosystems meet may consist of transitional zones called *ecotones,* in which elements of each ecosystem mix.

Because components of different ecosystems may intermix, ecologists often find it useful to view these systems on larger geographic scales that encompass multiple ecosystems. In such a broad-scale approach, called **landscape ecology,** scientists study how landscape structure affects the abundance, distribution, and interaction of organisms. Taking a view across the landscape is important in studying birds that migrate long distances, mammals that move seasonally between mountains and valleys, and fish such as salmon that swim upriver from the ocean to reproduce. A landscape-level approach is also useful for city planning and regional development (pp. 412–414). Landscape-level studies have been greatly aided by satellite imaging and *geographic information systems (GIS)*—computer software that takes multiple types of data (for instance, on geology, hydrology, vegetation, animal species, and human development) and layers them together on a common set of geographic coordinates.

For a landscape ecologist, a landscape is made up of **patches** (of ecosystems, communities, or habitat) arrayed spatially in a *mosaic* (**FIGURE 2.16**). Landscape ecology is of great interest to **conservation biologists** (p. 177), scientists who study the loss, protection, and restoration of biodiversity (see *The Science Behind the Story,* pp. 200–201). Populations of organisms have specific habitat requirements and so occupy suitable patches across the landscape. If habitat patches are highly fragmented and isolated (pp. 171, 199–203), the populations in those patches may perish. Accordingly, establishing corridors of habitat (p. 202; also see Figure 2.16) to link patches is one approach that conservation biologists use to preserve biodiversity.

## WEIGHING THE ISSUES

**Ecosystems Where You Live** Think about the area where you live, and briefly describe this region's ecosystems. How do these systems interact? For instance, does any water pass from one to another? Describe the boundaries of watersheds in your region. If one ecosystem were greatly modified (say, if a shopping mall were built atop a wetland or amid a forest), what impacts on nearby systems might result? (Note: If you live in a city, realize that urban areas can be thought of as ecosystems, too!)

## Modeling helps ecologists understand systems

Another way in which ecologists seek to make sense of the natural systems they study is by working with models. In science, a **model** is a simplified representation of a complicated natural process, designed to help us understand processes and make predictions.

**Ecological modeling** is the practice of constructing and testing models that aim to explain and predict how ecological systems function (**FIGURE 2.17**). These models are grounded in actual data and based on hypotheses about how system components interact. Ecological modeling is extremely useful in large, intricate systems that are difficult to isolate and study. For example, ecological models are useful for understanding the flow of nutrients into the Chesapeake Bay and in predicting the responses of fish, crabs, oysters and underwater grasses to changing water conditions. Modeling is a vital pursuit in the scientific study of Earth's changing climate (pp. 310–311).

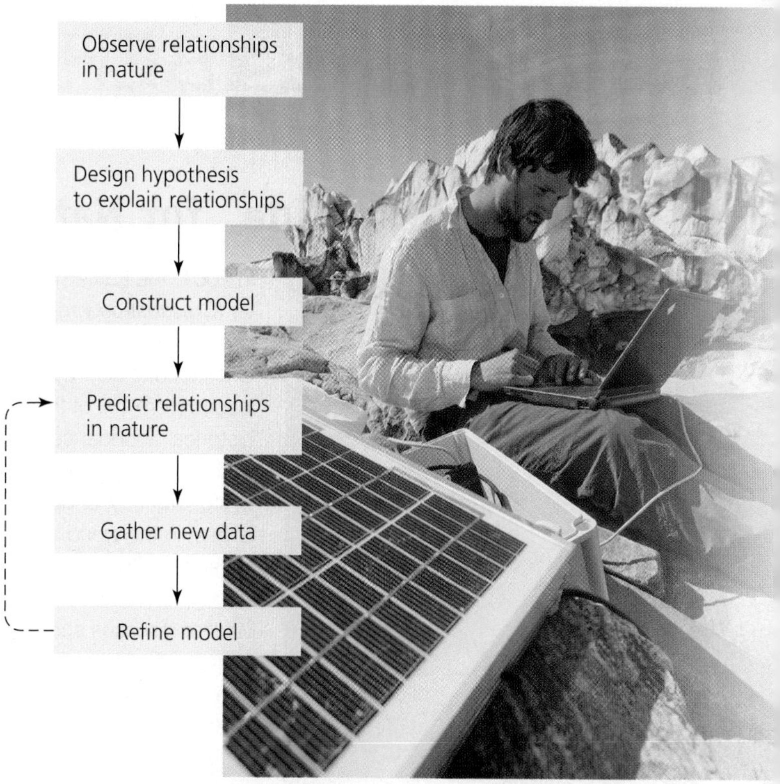


**FIGURE 2.17 Ecological modelers observe relationships among variables in nature and then construct models to explain those relationships and make predictions.** They test and refine the models by gathering new data from nature and seeing how well the models predict those data.

## Ecosystem services sustain our world

All life on our planet, including human life, depends on healthy, functioning ecosystems. When Earth's ecosystems function normally and undisturbed, they provide goods and services that we could not survive without. When human activities impair ecosystem functioning, we must devote resources in an attempt to provide these services ourselves, requiring extra investments of time and money.

Consider the **ecosystem services** of controlling agricultural pests, such as crop-eating insects. When we eliminate insect predators from agricultural fields by using intensive farming practices, we impair the ecological service of pest control that nature provides. Farmers must then apply synthetic pesticides to reduce pest populations, costing them resources and increasing human and wildlife exposure to these chemicals.

Functioning ecosystems provide many other ecosystem services, include regulating atmospheric gases; providing people with food and natural resources; pollinating plants; preventing soil erosion; filtering wastes and pollutants; and providing recreational opportunities (TABLE 2.1).

**TABLE 2.1 Ecosystem Services**

| Ecological processes do many things that benefit us: |
|---|
| • Regulate oxygen, carbon dioxide, stratospheric ozone, and other atmospheric gases |
| • Cycle carbon, nitrogen, phosphorus, sulfur, and other nutrients |
| • Regulate temperature and precipitation by means of ocean currents, cloud formation, and so on |
| • Provide habitat for organisms to breed, feed, rest, migrate, and winter |
| • Store and regulate fresh water supplies |
| • Protect against storms, floods, and droughts, mainly by means of vegetation |
| • Filter waste, recover nutrients, and control pollution |
| • Control crop pests with predators and parasites |
| • Produce fish, game, crops, nuts, and fruits that we eat |
| • Supply lumber, fuel, metals, fodder, and fiber |
| • Provide recreation such as ecotourism, fishing, hiking, birding, hunting, and kayaking |
| • Provide aesthetic, artistic, educational, spiritual, and scientific amenities |

# "Turning the Tide" for Native Oysters in Chesapeake Bay

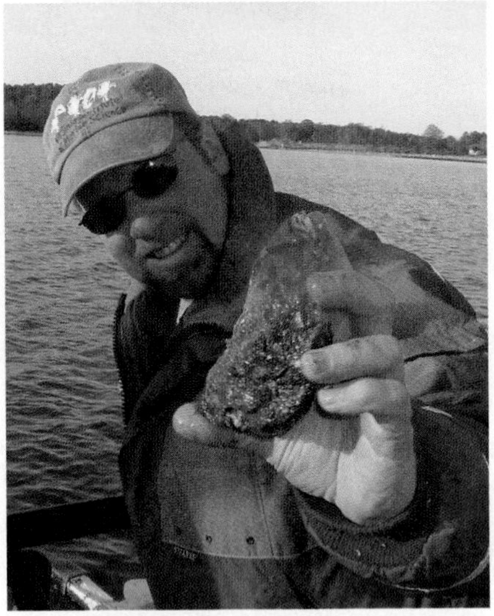

**David Schulte, U.S. Army Corps of Engineers.**

In 2001, the Eastern oyster (*Crassostrea virginica*) was in dire trouble in the Chesapeake Bay. Populations were 1% of their historical abundance, and the Chesapeake's oyster industry, once the largest in the world, had collapsed. Poor water quality, reef destruction, virulent diseases spread by transplanted oysters, and 200 years of overharvesting all contributed to the collapse.

Restoration efforts had largely failed. Moreover, when scientists or resource managers proposed to rebuild oyster populations by significantly restricting oyster harvests or establishing oyster reef "sanctuaries," these initiatives were typically defeated by the politically powerful oyster industry. All this had occurred in a place whose very name (derived from the Algonquin word *Chesepiook*) means "great shellfish bay."

With the collapse of the native oyster fishery and with political obstacles blocking restoration projects for native oysters, support grew among the oyster industry, state resource managers, and some scientists for the introduction of Suminoe oysters (*Crassostrea ariakensis*) from Asia. This species seemed well suited for conditions in the bay and showed resistance to the parasitic diseases that were ravaging native oysters. Proponents argued that introducing Suminoe oysters would reestablish thriving populations in the bay and revitalize the oyster fishery.

Introducing oysters would also improve the bay's water quality, proponents said, because as oysters feed, they filter phytoplankton and sediments from the water column. Filter-feeding by oysters is an important ecological service in the bay because it reduces phytoplankton densities, clarifies waters, and supports the growth of underwater grasses that provide food and refuge for waterfowl and young crabs.

However, introducing invasive species can have profound ecological impacts (pp. 78–80), so the Army Corps of Engineers was directed to coordinate an environmental impact statement (EIS, p. 106) on oyster restoration approaches in the Chesapeake.

It was in this politically charged, high-stakes environment that David Schulte, a scientist with the Corps and doctoral student at the College of William and Mary, set out to create a viable approach for restoring native oyster populations. The work he and his team began would help turn the tide in favor of native oysters in the bay's restoration efforts.

One of the biggest impacts on native oysters was the destruction of oyster reefs by a century of intensive harvesting. Oysters settle and grow best on the shells of other oysters, and over long periods this process forms reefs (underwater outcrops of living oysters and oyster shells) that solidify and become as hard as stone. Throughout the bay, massive reefs that at one time had jutted out of the water were reduced to rubble on the bottom from a century of repeated scouring by metal dredgers used by oyster-harvesting ships. The key to restoration, Schulte realized, was to construct artificial reefs like those that once existed, to get oysters off the bottom—away from smothering sediments and hypoxic waters—and up into the plankton-rich upper waters.

Armed with the resources available to the Corps, he used a landscape ecology approach and restored patches of reef habitat on nine complexes of reefs covering a total of 35.3 hectares (87 acres) in an oyster sanctuary near the mouth of the Great Wicomico River (**FIGURE 1**) in the lower Chesapeake Bay. This approach was very different from the smaller-scale restoration efforts of the past.

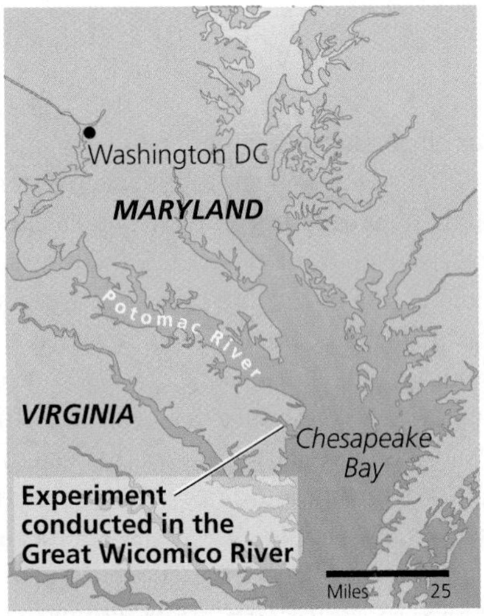


**FIGURE 1 Schulte's study was conducted in the Great Wicomico River in Virginia in the lower Chesapeake Bay.**

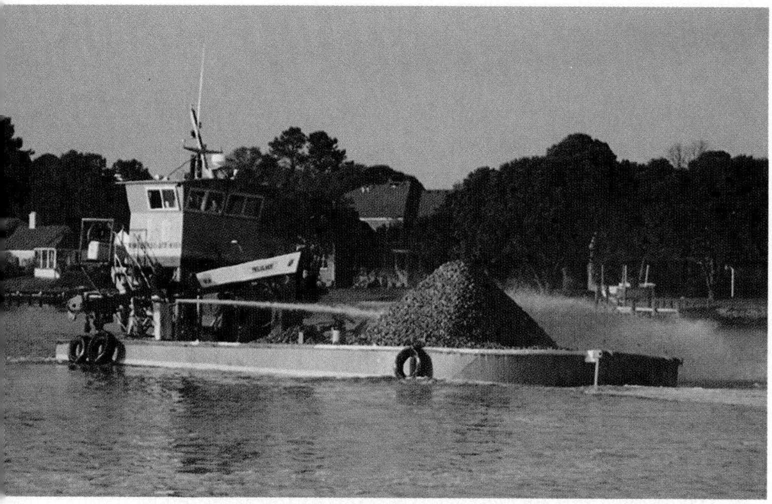

**FIGURE 2 A water cannon blows oyster shells off a barge and onto the river bottom to create an artificial oyster reef for the experiment.**

Artificial reefs of two heights were constructed in 2004 (**FIGURE 2**), and oysters were allowed to colonize the reefs, safe from harvesting. Oyster populations on the constructed reefs were then sampled in 2007, and the results were stunning. The reef complex supported an estimated 185 million oysters, a number nearly as large as the 200 million oysters estimated to live on the remaining degraded habitat in all of Maryland's waters.

Higher reefs supported an average of over 1000 oysters per square meter—four times more than the lower reefs and 170 times more than unrestored bottom (**FIGURE 3**). Like natural reefs, the constructed reefs began to solidify, providing a firm foundation for the settlement of new oysters. In 2009, Schulte's research team published its findings in the journal *Science*, bringing international attention to their study.

After reviewing eight alternative approaches to oyster restoration as part of its EIS, the Corps advocated an approach that avoided the introduction of non-native oysters. Instead it proposed a combination of native oyster restoration, a temporary moratorium on oyster harvests (accompanied by a compensation program for the oyster industry), and enhanced support for oyster aquaculture in the bay region.

Schulte's restoration project cost roughly $3 million and will require substantial investments if it is to be repeated elsewhere in the bay. This is particularly true in upper portions of the bay, where oyster reproduction levels are lower (requiring restored reefs to be "seeded" with oysters), water conditions are poorer, and oysters are less resistant to disease. Many scientists contend that expanded reef restoration efforts are worth the cost, however, because they enhance oyster populations and provide a vital service to the bay through water filtering. Some scientists also see value in promoting oyster farming (pp. 151, 318), in which restoration efforts would be supported by businesses instead of taxpayers.

These efforts are encouraged by the success of the project to date. By summer 2013, the majority of high-relief reef acreage was thriving, despite pressures from poachers and hypoxic conditions in several years. Moreover, many of the low-relief reefs had accumulated enough shell to now be classified as high-relief reef. Further, oyster reproduction in 2012 was some of the highest Schulte had seen during the project, and a follow-up study in 2013 found that spat from the sanctuary reefs were seeding other parts of the Great Wicomico River and increasing oyster populations outside protected areas.

Protected sites for oyster restoration efforts are now being established elsewhere the bay. Maryland recently designated 3640 hectares (9000 acres) of new oyster sanctuaries—25% of existing oyster reefs in state waters—and seeded these reefs with over a billion hatchery-raised spat. This movement toward increased protection for oyster populations, coupled with findings of increased disease resistance in bay oysters, has given new hope that native oysters may once again thrive in the bay that bears their name. ☐

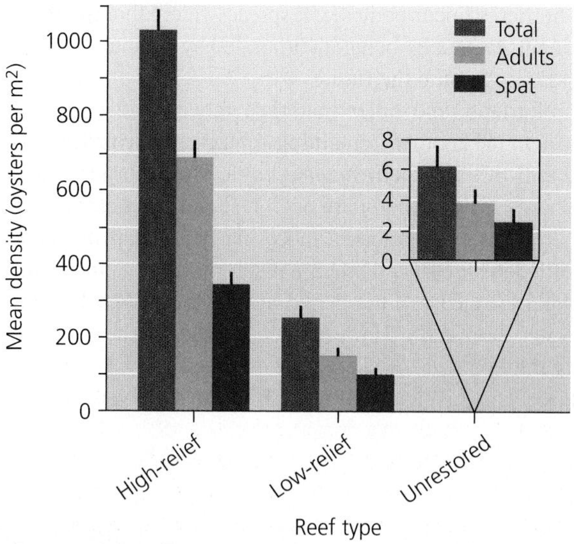


**FIGURE 3 Reef height had a profound effect on the density of adult oysters and spat (newly settled oysters).** Schulte's work suggested that native oyster populations could rebound in portions of the Chesapeake Bay if they were provided elevated reefs and were protected from harvest. *Data from Schulte, D.M., R.P. Burke, and R.N. Lipicus, 2009. Unprecedented restoration of a native oyster meta population. Science 325: 1124–1128.*

One of the most important ecosystem services is the cycling of nutrients. Through the processes that take place within and among ecosystems, the chemical elements and compounds that we need—water, carbon, nitrogen, phosphorus, and many more—cycle through our environment in intricate ways.

# Biogeochemical Cycles

Just as nitrogen and phosphorus from fertilizer on Pennsylvania cornfields end up in Chesapeake Bay oysters on our dinner plates, all nutrients move through the environment in complex ways. Whereas energy enters an ecosystem from the sun, flows from organism to organism, and dissipates to the atmosphere as heat, the physical matter of an ecosystem is circulated over and over again.

## Nutrients circulate through ecosystems in biogeochemical cycles

Nutrients move through ecosystems in **nutrient cycles** (or **biogeochemical cycles**) that circulate elements or molecules through the lithosphere, atmosphere, hydrosphere, and biosphere. A carbon atom in your fingernail today might have been part of the muscle of a cow a year ago, may have resided in a blade of grass a month before that, and may have been part of a dinosaur's tooth 100 million years ago. After we die, the nutrients in our bodies will spread widely through the environment, eventually being incorporated by an untold number of organisms far into the future.

Nutrients and other materials move from one *pool,* or *reservoir,* to another, remaining in each reservoir for varying amounts of time (the *residence time*). The dinosaur, the grass, the cow, and your body are each reservoirs for carbon atoms. When a reservoir releases more materials than it accepts, it is called a *source,* and when a reservoir accepts more materials than it releases, it is called a *sink.* **FIGURE 2.18** illustrates these concepts in a simple manner. The rate at which materials move between reservoirs is termed a *flux,* and the flux between any given pair of reservoirs can change over time. As we will see in the following sections, human activities affect the cycling of nutrients by altering fluxes, residence times, and the relative amounts of nutrients in reservoirs.

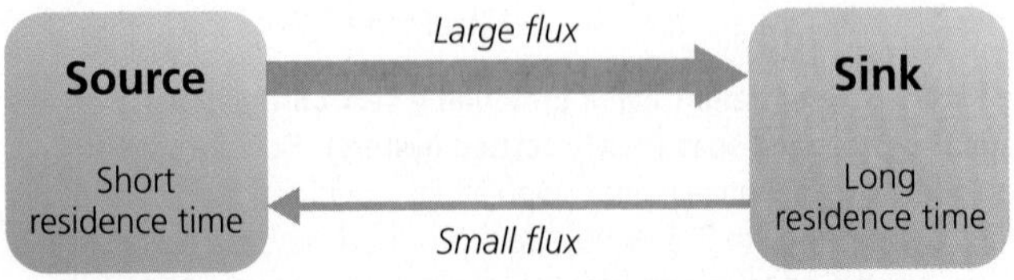


**FIGURE 2.18 The main components of a biogeochemical cycle are reservoirs and fluxes.** A source reservoir releases more materials than it accepts, and a sink reservoir accepts more materials than it releases.

## The water cycle affects all other cycles

Water is so integral to life that we frequently take it for granted. The essential medium for many biochemical reactions, water plays key roles in nearly every environmental system, including each nutrient cycle. Water transports nutrients, sediments, and pollutants from the continents to the oceans via rivers, streams, and surface runoff. Nutrients can then be carried thousands of miles on ocean currents. Water also brings atmospheric pollutants from the air back down to the surface when they dissolve in falling rain or snow. These activities make the water cycle, or **hydrologic cycle** (**FIGURE 2.19**), an integral part of nutrient cycling on Earth.

The oceans are the largest reservoir in the hydrologic cycle, holding more than 97% of all water on Earth. The fresh water we depend on for our survival accounts for the remaining water, and two-thirds of this small amount is tied up in glaciers, snowfields, and ice caps (p. 250). Thus, considerably less than 1% of the planet's water is in a form that we can readily use—groundwater, surface fresh water, and rain from atmospheric water vapor.

Water moves from oceans, lakes, ponds, rivers, and moist soil into the atmosphere by **evaporation,** the conversion of a liquid to gaseous form. Water also enters the atmosphere by **transpiration,** the release of water vapor by plants through their leaves. Water returns from the atmosphere to Earth's surface as **precipitation** when water vapor condenses and falls as rain or snow. Precipitation may be taken up by plants and used by animals, but much of it flows as runoff (p. 252) into streams, rivers, lakes, ponds, and oceans.

Some water soaks down through soil and rock through a process called **infiltration** to recharge underground reservoirs known as **aquifers.** Aquifers are porous regions of rock and soil that hold **groundwater,** water found underground beneath layers of soil. The uppermost level of groundwater held in an aquifer is referred to as the **water table.** Some aquifers hold groundwater for short periods of time, whereas others contain quite ancient water.

Human activity has affected nearly every aspect of the water cycle. By damming rivers and holding their water in reservoirs, we increase evaporation and slow the movement of water from land to sea. We have removed natural vegetation by clear-cutting and developing land, which increases surface runoff, decreases infiltration and transpiration, and promotes soil erosion. Our withdrawals of surface water and groundwater for agriculture, industry, and domestic uses have depleted rivers, lakes, and streams and have lowered water tables. (We will revisit the water cycle, water resources, and human impacts in more detail in Chapter 12.)

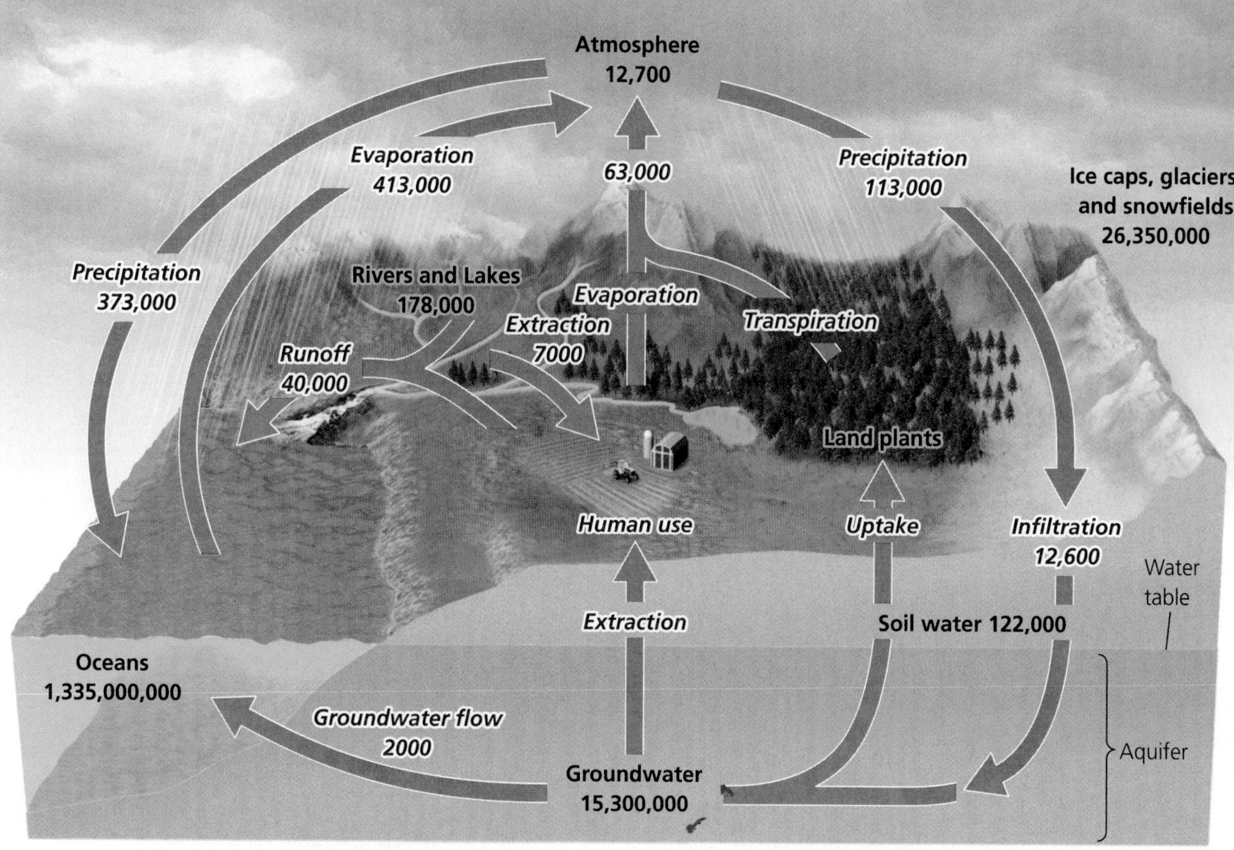


**FIGURE 2.19 The water cycle, or hydrologic cycle, summarizes the many routes that water molecules take as they move through the environment.** Gray arrows represent fluxes among reservoirs, or pools, for water. Oceans hold 97% of our planet's water, whereas most fresh water resides in groundwater and ice caps. Water vapor in the atmosphere condenses and falls to the surface as precipitation, then evaporates from land and transpires from plants to return to the atmosphere. Water flows downhill into rivers, eventually reaching the oceans. In the figure, pool names are printed in black type, and numbers in black type represent pool sizes expressed in units of cubic kilometers ($km^3$). Processes, printed in italic red type, give rise to fluxes, printed in italic red type and expressed in $km^3$ per year. *Data from Schlesinger, W.H.,* 2013. Biogeochemistry: An analysis of global change, *3rd ed. Academic Press, London.*

## The carbon cycle circulates a vital organic nutrient

The **carbon cycle** describes the routes that carbon atoms take through the environment (**FIGURE 2.20**, next page). Because carbon forms the backbone of biological molecules, its cycling is of great importance. Producers pull carbon dioxide out of the atmosphere and surface water to use in photosynthesis. Autotrophs and the heterotrophs that consume them use carbohydrates produced in photosynthesis to fuel cellular respiration and for structural growth, releasing carbon back into the atmosphere and waters as $CO_2$. Because plants are a sizable reservoir of carbon and $CO_2$ is a greenhouse gas (p. 305), researchers are attempting to measure the amount of $CO_2$ that plants sequester to evaluate the potential of vegetation in combating global climate change.

Oceans absorb carbon-containing compounds from the atmosphere, terrestrial runoff, undersea volcanoes, and the detritus of marine organisms. Ocean waters are the second largest reservoir of carbon on Earth.

The largest reservoir of carbon, sedimentary rock (p. 229), is formed in oceans and freshwater wetlands. When organisms in these habitats die, their remains can settle in sediments, and as layers of sediment accumulate, the older layers are buried more deeply and experience high pressure for long periods. Over time, these conditions can convert the soft tissues of dead organisms into fossil fuels—coal, oil, and natural gas—deep underground (p. 338). The high underground pressures convert the shells and skeletons of aquatic organisms, which are rich in calcium carbonate ($CaCO_3$), to sedimentary rock, such as limestone.

Although any given carbon atom spends a relatively short time in the atmosphere, carbon trapped in sedimentary rock may reside there for hundreds of millions of years. Carbon trapped in sedimentary rocks and fossil fuel deposits may eventually be released into the oceans or atmosphere by geologic

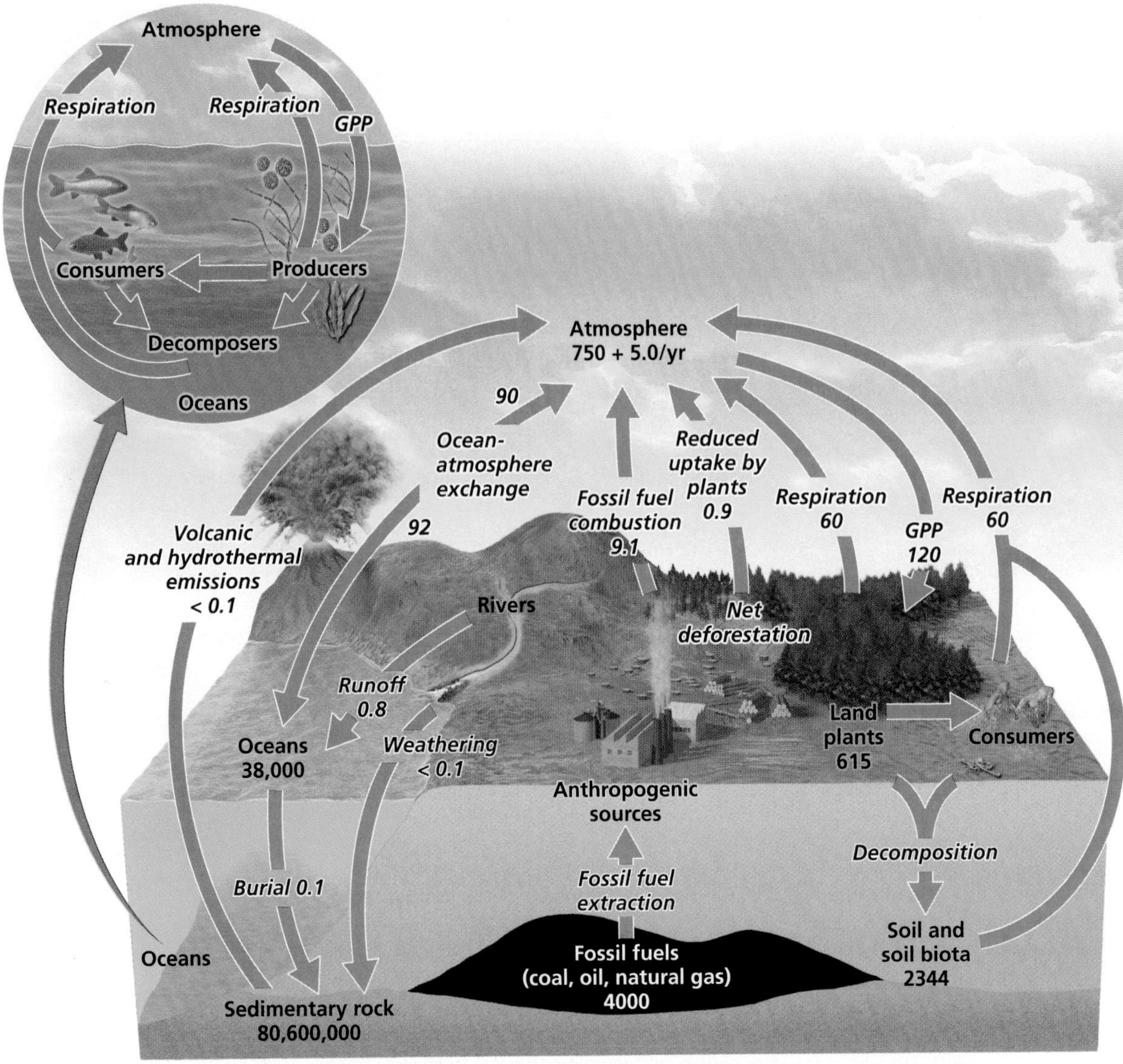


**FIGURE 2.20 The carbon cycle summarizes the many routes that carbon atoms take as they move through the environment.** Gray arrows represent fluxes among reservoirs, or pools, for carbon. In the carbon cycle, plants use carbon dioxide from the atmosphere for photosynthesis (gross primary production, or "GPP" in the figure). Carbon dioxide is returned to the atmosphere through cellular respiration by plants, their consumers, and decomposers. The oceans sequester carbon in their water and in deep sediments. The vast majority of the planet's carbon is stored in sedimentary rock. In the figure, pool names are printed in black type, and numbers in black type represent pool sizes expressed in petagrams (units of $10^{15}$ g) of carbon. Processes, printed in italic red type, give rise to fluxes, printed in italic red type and expressed in petagrams of carbon per year. *Data from Schlesinger, W.H., 2013.* Biogeochemistry: An analysis of global change, *3rd ed. Academic Press, London.*

processes such as uplift, erosion, and volcanic eruptions. It also reenters the atmosphere when we extract and burn fossil fuels.

The largest human impact on the carbon cycle is through our use of fossil fuels as an energy source. The burning of coal, oil, and natural gas has greatly increased the flux of carbon from the lithosphere to the atmosphere and has shortened the residence time of carbon in fossil fuel deposits.

Moreover, when people burn forests and fields to clear land for agriculture, the carbon in wood and leaves is released to the atmosphere. Because these cleared sites have less vegetation, photosynthesis removes less carbon dioxide from the atmosphere than before. As a result of vegetation removal and fossil fuel combustion, scientists estimate that today's atmospheric carbon dioxide reservoir is the largest that Earth has experienced in the past 800,000 years—likely in the past 20 million years (p. 306)—and is a driving force behind anthropogenic global climate change (Chapter 14). Some of the excess $CO_2$ in the atmosphere is now being absorbed by ocean water. This is causing ocean water to become more acidic, threatening many marine organisms (pp. 258–259).

Our understanding of the carbon cycle is not yet complete. Scientists remain baffled by the so-called "missing carbon sink." Of the carbon dioxide released by human activities, researchers have measured how much goes into the atmosphere and oceans, but roughly 2.3–2.6 billion metric tons remain unaccounted for. Many scientists think that plants or soils of the temperate and boreal forests (pp. 83, 84, 86–87) absorb this $CO_2$. If so, this would provide strong incentives to conserve these ecosystems.

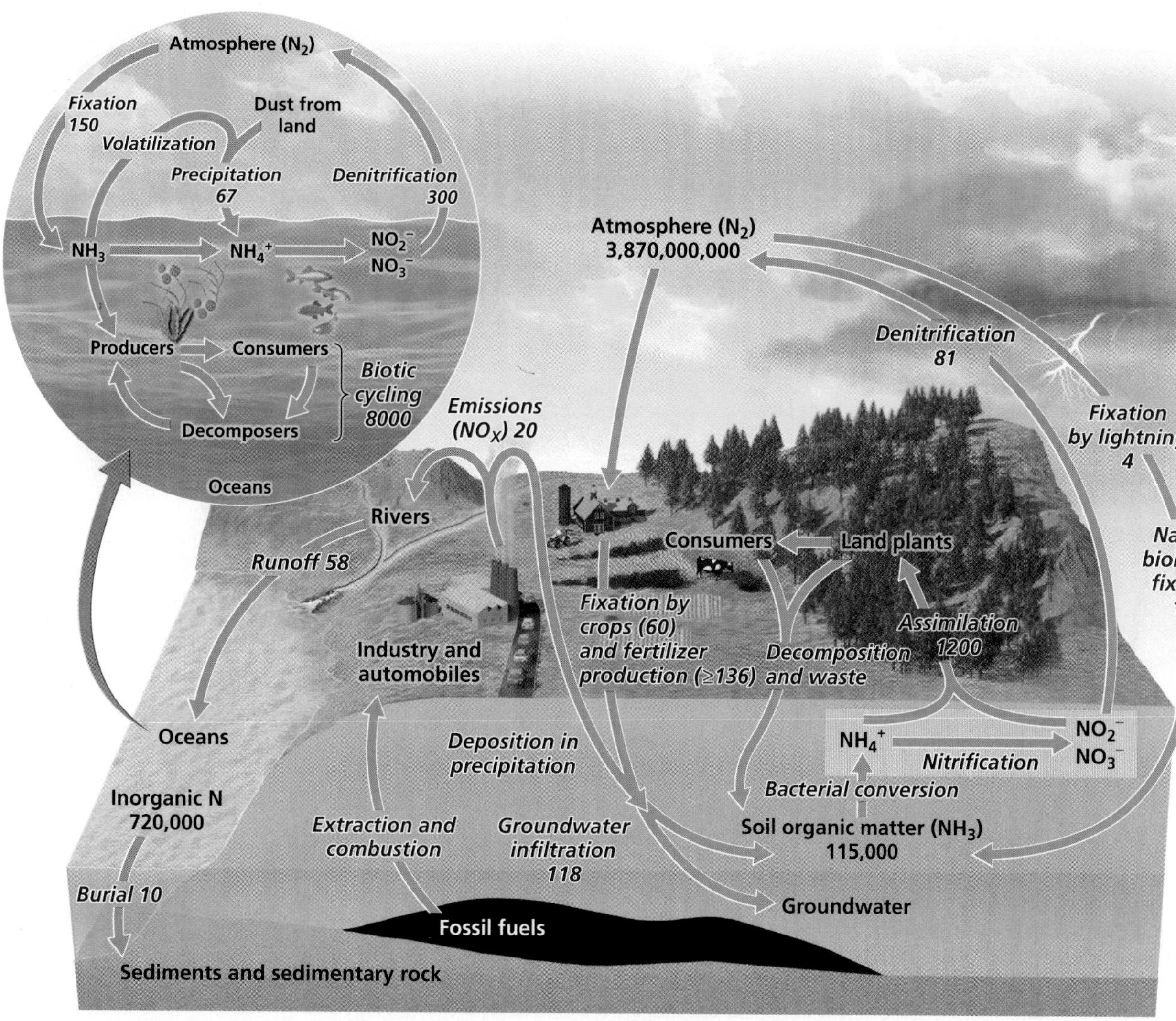


**FIGURE 2.21 The nitrogen cycle summarizes the many routes that nitrogen atoms take as they move through the environment.** Gray arrows represent fluxes among reservoirs, or pools, for nitrogen. In the nitrogen cycle, specialized bacteria play key roles in "fixing" atmospheric nitrogen and converting it to chemical forms that plants can use. Other types of bacteria convert nitrogen compounds back to the atmospheric gas, $N_2$. In the oceans, inorganic nitrogen is buried in sediments, whereas nitrogen compounds are cycled through food webs as they are on land. In the figure, pool names are printed in black type, and numbers in black type represent pool sizes expressed in teragrams (units of $10^{12}$ g) of nitrogen. Processes, printed in italic red type, give rise to fluxes, printed in italic red type and expressed in teragrams of nitrogen per year. *Data from Schlesinger, W.H., 2013.* Biogeochemistry: An analysis of global change, *3rd ed. Academic Press, London.*

## The nitrogen cycle involves specialized bacteria

Nitrogen makes up 78% of our atmosphere by mass and is the sixth most abundant element on Earth. It is an essential ingredient in proteins, DNA, and RNA and, like phosphorus, is an essential nutrient for plant growth. Thus the **nitrogen cycle** (**FIGURE 2.21**) is of vital importance to all organisms. Despite its abundance in the air, nitrogen gas ($N_2$) is chemically inert and cannot cycle out of the atmosphere and into living organisms without assistance from lightning, highly specialized bacteria, or human intervention. However, once nitrogen undergoes the right kind of chemical change, it becomes biologically active and available to the organisms that need it, and it can act as a potent fertilizer.

To become biologically available, inert nitrogen gas ($N_2$) must be "fixed," or combined with hydrogen in nature to form ammonia ($NH_3$), whose water-soluble ions of ammonium ($NH_4^+$) can be taken up by plants. **Nitrogen fixation** can be accomplished in two ways: by the intense energy of lightning strikes, or by particular types of **nitrogen-fixing bacteria** that inhabit the top layer of soil. These bacteria live in a mutualistic relationship (p. 70) with many types of plants, including soybeans and other legumes, providing them nutrients by converting nitrogen to a usable form. Other types of bacteria then perform a process known as **nitrification**, converting ammonium ions first into nitrite ions ($NO_2^-$), then into nitrate ions ($NO_3^-$). Plants can take up these ions, which also become available after atmospheric deposition on soils or in water or after application of nitrate-based fertilizer.

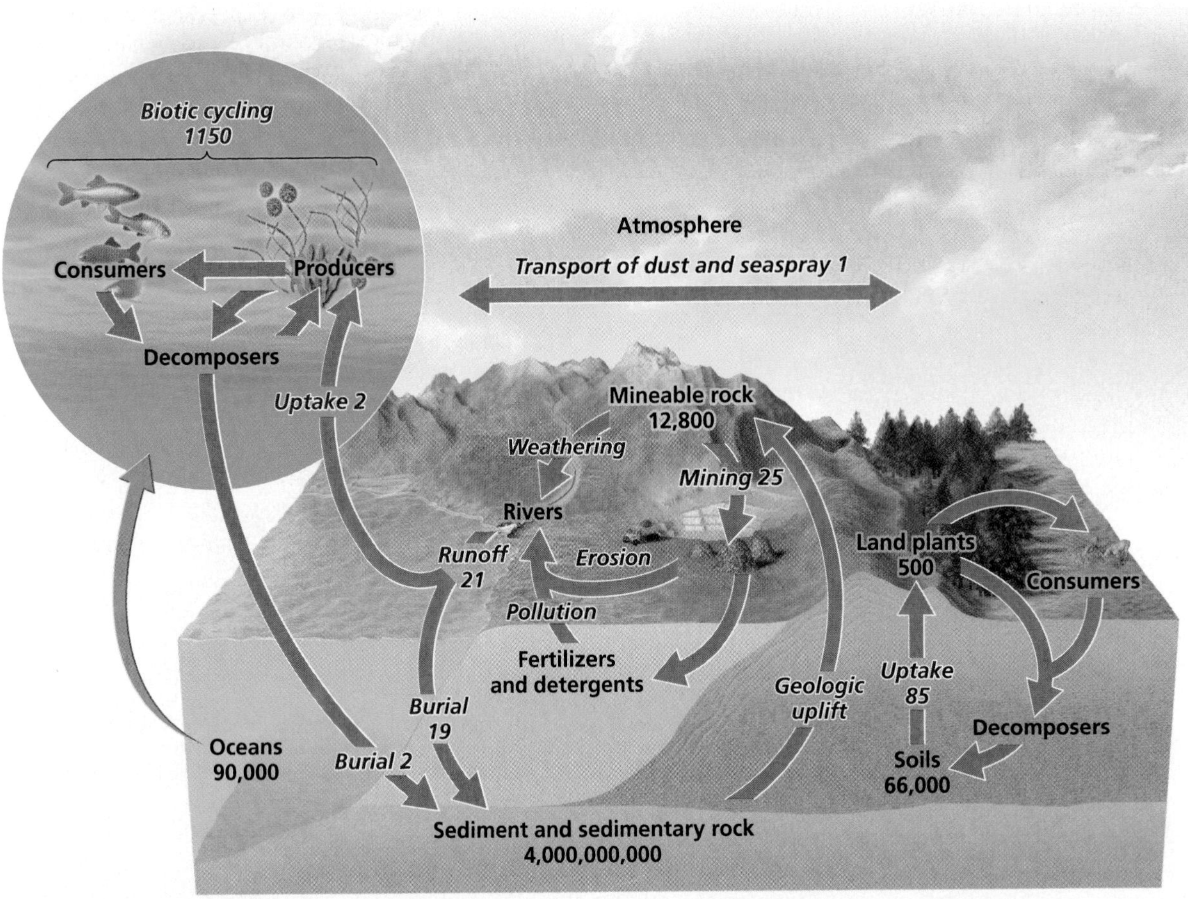


FIGURE 2.22 **The phosphorus cycle summarizes the many routes that phosphorus atoms take as they move through the environment.** Gray arrows represent fluxes among reservoirs, or pools, for phosphorus. Most phosphorus resides underground in rock and sediment. Rocks containing phosphorus are uplifted geologically and slowly weathered away. Small amounts of phosphorus cycle through food webs, where this nutrient is often a limiting factor for plant growth. In the figure, pool names are printed in black type, and numbers in black type represent pool sizes expressed in teragrams (units of $10^{12}$ g) of phosphorus. Processes, printed in italic red type, give rise to fluxes, printed in italic red type and expressed in teragrams of phosphorus per year.
*Data from Schlesinger, W.H., 2013.* Biogeochemistry: An analysis of global change, *3rd ed. Academic Press, London.*

Animals obtain the nitrogen they need by consuming plants or other animals. Decomposers obtain nitrogen from dead and decaying plant and animal matter and from the urine and feces of animals. Once decomposers process the nitrogen-rich compounds, they release ammonium ions, making these available to nitrifying bacteria to convert again to nitrates and nitrites. The next step in the nitrogen cycle occurs when **denitrifying bacteria** convert nitrates in soil or water to gaseous nitrogen. **Denitrification** thereby completes the cycle by releasing nitrogen back into the atmosphere as a gas.

Historically, nitrogen fixation was a *bottleneck,* a step that limited the flux of nitrogen out of the atmosphere into water-soluble forms. Once people discovered how to fix nitrogen on massive scales, a process called **industrial fixation,** we accelerated its flux into other reservoirs. Today, our species is fixing at least as much nitrogen artificially as is being fixed naturally, and we are overwhelming nature's denitrification abilities.

While the impacts of nitrogen runoff have become painfully evident to oystermen and scientists in the Chesapeake Bay, hypoxia in waters is by no means the only human impact on the nitrogen cycle. Oddly enough, the overapplication of nitrogen-based fertilizers can strip the soil of other vital nutrients, such as calcium and potassium, thereby reducing soil fertility. Additionally, burning fossil fuels, forests, or fields generates nitrogenous compounds in the atmosphere that act as greenhouse gases (p. 305), cause acid deposition (p. 295), promote eutrophication (p. 25), and contribute to photochemical smog (p. 291).

## The phosphorus cycle circulates a limited nutrient

Sedimentary rocks are the largest reservoir in the **phosphorus cycle** (**FIGURE 2.22**). Environmental concentrations of phosphorus available to organisms tend to be very low, because weathering (p. 140), which releases phosphate ions ($PO_4^{3-}$) into water, is the only process that makes phosphorus available for uptake. Phosphorus is a key component of cell membranes, DNA, RNA, and a number of vital biochemical compounds, so its rarity in the environment makes it a limiting factor for plant growth. Aquatic producers take up phosphates from surrounding waters, whereas terrestrial producers take up phosphorus from soil water through their roots. Phosphorus is incorporated into the tissues of producers, and consumers obtain their required phosphorus by eating the tissues of other organisms. Phosphorus from dead organisms or waste products in the oceans can precipitate into solid form and settle to the bottom in sediments, which are eventually compressed into sedimentary rock.

Human activities have increased concentrations of phosphorus in surface waters, most notably through runoff of the phosphorus-rich fertilizers we apply to lawns and farmlands. People also add phosphorus to waterways through releases of treated wastewater rich in phosphates from domestic use of phosphate detergents. These inputs promote eutrophication in waters such as the Chesapeake Bay.

## Tackling nutrient enrichment requires diverse approaches

Given our reliance on synthetic fertilizers for food production and fossil fuels for energy, nutrient enrichment of ecosystems will pose a challenge for many years to come. Fortunately, a number of approaches are available to control nutrient pollution in waterways:

- Reducing fertilizer use on farms and lawns and timing its application to reduce water runoff
- Planting and maintaining vegetation "buffers" around streams to trap nutrient and sediment runoff
- Using natural and constructed wetlands (pp. 252–253, 274) to filter stormwater and farm runoff
- Improving technologies in sewage treatment plants to enhance nitrogen and phosphorus capture
- Upgrading stormwater systems to capture runoff from roads and parking lots
- Reducing fossil fuel combustion to minimize atmospheric inputs of nitrogen to waterways

Some of these methods cost more than others for similar results. For example, planting vegetation buffers and restoring wetlands can reduce nutrient inputs into waterways at a fraction of the cost of some other approaches, such as upgrading wastewater treatment plants (**FIGURE 2.23**).

### WEIGHING THE ISSUES

**Nutrient Pollution and Its Financial Impacts** A sizable amount of the nitrogen and phosphorus that enters the Chesapeake Bay originates from farms and other faraway sources, yet it is people living near the bay, such as oystermen and crabbers, who bear many of the negative impacts. Who do you believe should be responsible for addressing the problems posed by nutrient pollution? Should policies on this issue be developed and enforced by state governments, the federal government, both, or neither? Explain the reasons for your answer.

## A systemic approach to restoration offers hope for the Chesapeake Bay

The federal government and the states around the Chesapeake Bay are now managing the bay as a holistic system—an approach that, at last, offers prospects for recovery. Arriving at this endpoint was not easy, though. After 25 years of failed pollution control agreements and nearly $6 billion spent on cleanup efforts, the Chesapeake Bay Foundation (CBF), a nonprofit organization dedicated to conserving the bay, sued the Environmental Protection Agency (EPA) in 2009 for failing to use its available powers under the Clean Water Act to clean up the bay. This spurred federal action, and in 2010 a comprehensive "pollution budget" was developed and implemented by the EPA with the assistance of the District of Columbia, Delaware, Maryland, New York, Pennsylvania, Virginia, and West Virginia.

Although the Chesapeake Bay remains highly degraded and much work is still needed, efforts to reduce nutrient and sediment inputs and to limit harvests of oysters, crabs, and fish are already leading to modest improvement in some aspects of

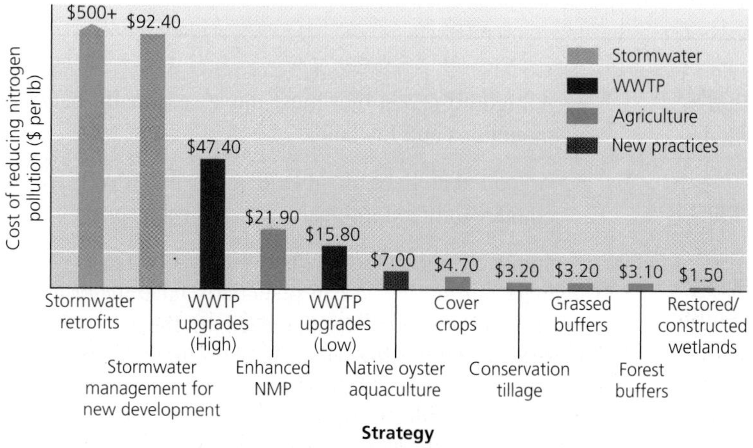


**FIGURE 2.23 The cost per pound of reducing nitrogen inputs into the Chesapeake Bay varies widely.** Approaches that involve slowing nitrogen-containing runoff cost a few dollars per pound. Upgrades to wastewater treatment plants (WWTP), enhanced nutrient management plans (NMP—careful regulation of nutrient applications), and stormwater upgrades can be considerably more expensive.

DATA Q For what it costs to remove 1 pound of nitrogen by using enhanced nutrient management programs (NMP), how many pounds of nitrogen could be kept out of waterways by planting forested buffers around streams instead?

GO TO **INTERPRETING GRAPHS & DATA** ON MasteringEnvironmentalScience®

the bay's health. This provides hope for the 17 million people living in its watershed that the Chesapeake Bay of tomorrow may be healthier than it is today, thanks to the collaborative efforts of advocacy organizations, the EPA, state governments, and concerned citizens.

## Conclusion

Earth hosts many interacting systems, and the way we perceive them depends on the questions we ask. Life interacts with its abiotic environment in ecosystems, systems through which energy flows and materials are recycled. Understanding the biogeochemical cycles that describe the movement of nutrients within and among ecosystems is crucial, because human activities are causing significant changes in the ways these cycles function.

Understanding energy, energy flow, and chemistry enhances our comprehension of how organisms interact with one another, how they relate to their nonliving environment, and how environmental systems function. Energy and chemistry are tied to nearly every environmental issue, and applications of chemistry can provide solutions to environmental problems involving agricultural practices, water resources, air quality, energy policy, and environmental health.

Thinking at the systems level is important in understanding how our planet works, so that we may learn to minimize and alleviate our disruptions of its processes. By studying the environment from a systems perspective and by integrating scientific findings with the policy process, people who care about the Chesapeake Bay and other environmental issues are working today to solve pressing problems.

# Testing Your Comprehension

1. Which type of feedback loop is more common in nature, and which more commonly results from human action? For either type of feedback loop, can you think of an example that was not mentioned in the text?
2. Describe how hypoxic conditions can develop in aquatic ecosystems such as the Chesapeake Bay.
3. Differentiate an ion from an isotope.
4. Describe the two major forms of energy, and give examples of each. Compare and contrast the first law of thermodynamics with the second law of thermodynamics.
5. What substances are produced by photosynthesis? By cellular respiration?
6. Describe the typical movement of energy through an ecosystem. Describe the typical movement of matter through an ecosystem.
7. List five ecosystem services provided by functioning ecosystems, and rank them according to your perceived value of each.
8. What role do each of the following play in the carbon cycle?
   - Cars
   - Photosynthesis
   - The oceans
   - Earth's crust
9. Distinguish the function performed by nitrogen-fixing bacteria from that performed by denitrifying bacteria.
10. How has human activity altered the hydrologic cycle? The carbon cycle? The phosphorus cycle? The nitrogen cycle? To what environmental problems have these changes given rise?

# Seeking Solutions

1. Can you think of an example of an environmental problem *not* mentioned in this chapter that a good knowledge of chemistry could help us solve? Explain your answer.
2. Consider the ecosystem(s) that surround(s) your campus. How is each affected by human activities?
3. For a conservation biologist interested in sustaining populations of the organisms below, why would it be helpful to take a landscape ecology perspective? Explain your answer in each case.
   - A forest-breeding warbler that suffers poor nesting success in small fragmented forest patches
   - A bighorn sheep that must move seasonally between mountains and lowlands
   - A toad that lives in upland areas but travels cross-country to breed in localized pools each spring
4. A simple change in the flux between just two reservoirs in a single nutrient cycle can potentially have major consequences for ecosystems and, indeed, for the globe. Explain how this can be, using one example from the carbon cycle and one example from the nitrogen cycle.

5. **THINK IT THROUGH** You are an oysterman in the Chesapeake Bay, and your income is decreasing because the dead zone is making it harder to harvest oysters. One day your senator comes to town, and you have a one-minute audience with her. What steps would you urge her to take in Washington, D.C., to try to help alleviate the dead zone and bring back the oyster fishery?

Now suppose you are a Pennsylvania farmer who has learned that the government is offering incentives to farmers to help reduce fertilizer runoff into the Chesapeake Bay. What types of approaches described in the text might you be willing to try, and why?

# Calculating Ecological Footprints

In ecological systems, a rough rule of thumb is that when energy is transferred from plants to plant-eaters or from prey to predator, the efficiency is only about 10% (p. 72). Much of this inefficiency is a consequence of the second law of thermodynamics. Another way to think of this is that eating 1 calorie of material from an animal is the ecological equivalent of eating 10 calories of plant material.

Humans are considered omnivores because we can eat both plants and animals. Our food choices have significant ecological impacts. With this in mind, calculate the ecological energy requirements for four different diets, each of which provides a total of 2000 dietary calories per day.

| Diet | Source of calories | Number of calories consumed | Ecologically equivalent calories | Total ecologically equivalent calories |
|---|---|---|---|---|
| 100% plant | Plant | | | |
| 0% animal | Animal | | | |
| 90% plant | Plant | 1800 | 1800 | 3800 |
| 10% animal | Animal | 200 | 2000 | |
| 50% plant | Plant | | | |
| 50% animal | Animal | | | |
| 0% plant | Plant | | | |
| 100% animal | Animal | | | |

1. How many ecologically equivalent calories would it take to support you for a year, for each of the four diets listed?
2. How does the ecological impact from a diet consisting strictly of animal products (meat, eggs, milk, and other dairy products) compare with that of a strictly vegetarian diet? How many additional ecologically equivalent calories do you consume each day by including as little as 10% of your calories from animal sources?
3. What percentages of the calories in your own diet do you think come from plant versus animal sources? Estimate the ecological impact of your diet, relative to a strictly vegetarian one.
4. Describe some challenges of providing food for the growing human population, especially as people in many poorer nations develop a taste for an American-style diet rich in animal protein and fat.

## MasteringEnvironmentalScience®

**STUDENTS**

Go to **MasteringEnvironmentalScience** for assignments, the etext, and the Study Area with practice tests, videos, current events, and activities.

**INSTRUCTORS**

Go to **MasteringEnvironmentalScience** for automatically graded activities, current events, videos, and reading questions that you can assign to your students, plus Instructor Resources.

# 3

# Evolution, Biodiversity, and Population Ecology

## Upon completing this chapter, you will be able to:

- ☐ Explain the process of natural selection and cite evidence for this process
- ☐ Describe how evolution influences biodiversity
- ☐ Discuss reasons for species extinction and mass extinction events
- ☐ List the levels of ecological organization
- ☐ Outline the characteristics of populations that help predict population growth
- ☐ Explain logistic growth, carrying capacity, limiting factors, and other fundamental concepts in population ecology

**Photo:** Native Hawaiian forest at Hakalau Forest National Wildlife Refuge, and the endangered ʻakiapōlāʻau

CENTRAL CASE STUDY

# Saving Hawaii's Native Forest Birds

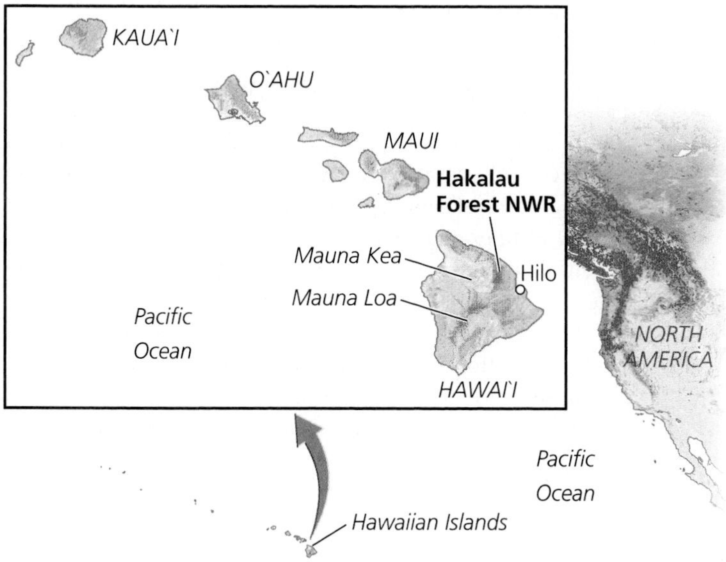


*"When an entire island avifauna ... is devastated almost overnight because of human meddling, it is, quite simply, a tragedy."*

**—H. Douglas Pratt, ornithologist and expert on Hawaiian birds**

*"To keep every cog and wheel is the first precaution of intelligent tinkering."*

**—Aldo Leopold**

Jack Jeffrey stopped in his tracks. "I hear one!" he said. "Over there in those trees!"

Jeffrey led his group of ecotourists through a misty woodland of ferns, grasses, koa trees, and red-flowering ʻōhiʻa-lehua trees toward an emphatic chirping sound. At last they spotted the bird that was calling—an ʻakiapōlāʻau, one of fewer than 1500 of its kind left in the world.

The ʻakiapōlāʻau (or "aki" for short) is a sparrow-sized wonder of nature—one of many exquisite birds that evolved on the Hawaiian Islands and exists only here. For millions of years, this chain of islands in the middle of the Pacific Ocean has acted as a cradle of evolution, generating new and unique species. Yet today many of these species are going from the cradle to the grave. Half of Hawaii's native bird species (70 of 140) have gone extinct in recent times, and many of those that remain—like the aki—teeter on the brink of extinction.

The aki is a type of Hawaiian honeycreeper. The Hawaiian honeycreepers include 18 living species (and at least 38 species recently extinct), all of which originated from an ancestral species that reached Hawaii several million years ago. As new volcanic islands emerged from the ocean and then eroded away, and as forests expanded and contracted over millennia, populations were split and new honeycreeper species evolved.

As honeycreeper species diverged from their common ancestor and from one another, they evolved different colors, sizes, body shapes, feeding behaviors, mating preferences, diets, and bill shapes. Bills in some species became short and straight, allowing birds to glean insects from leaves. In other species, bills became long and curved, enabling birds to probe into flowers to sip nectar. The bills of still other species became thick and strong for cracking seeds. The aki's bill is highly specialized: It uses the short, straight lower half to peck into dead branches to find beetle grubs, then uses the long, curved upper half to reach in and extract them.

Hawaii's honeycreepers long thrived in the islands' forests amid a unique community of plants. The ʻōhiʻa, a stately tree that can live for 1500 years, spreads twisting gnarled limbs covered with moss and lichens through the misty air, offering bright red flowers loaded with nectar and pollen. It grows alongside the koa, a fast-growing acacia tree. A multitude of shrubs, herbs, and vines found nowhere else in the world used to fill out the forest understory.

Today native Hawaiian forests are under siege. The crisis began 750 or more years ago as Polynesian settlers colonized the islands, cutting down trees and introducing non-native animals. Europeans arrived in the late 1700s and did more of the same. Pigs, goats, and cattle ate their way through the native plants, transforming lush forests into ragged grasslands. Rats, cats, dogs, and mongooses destroyed the eggs and young of native ground-nesting birds. Foreign plants from Asia, Europe, and America, whose seeds accompanied the people and animals, spread across the altered landscape.

Foreign diseases also arrived, including strains of pox and malaria that target birds. The native fauna were not adapted to resist pathogens they had never encountered. Avian malaria and avian pox, carried by introduced mosquitoes, killed off native birds everywhere except on high mountain slopes, where it gets too cold for the mosquitoes and the pathogens they carry to survive. Today few native forest birds exist anywhere on the Hawaiian Islands below 1500 m (4500 ft) in elevation.

The aki being watched by Jack Jeffrey's group inhabits the Hakalau Forest National Wildlife Refuge, high atop the slopes of Mauna Kea, a volcano on the island of Hawaiʻi, the largest island in the chain. At Hakalau, native birds find a rare remaining patch of disease-free native forest.

Refuge managers and conservation biologists have worked hard to protect Hakalau's forest. Jack Jeffrey was a biologist here for 20 years, leading innovative projects to save native plants and birds from extinction. Staff and volunteers at Hakalau fenced out pigs and planted thousands of native plants in areas deforested by cattle grazing. Young restored native forest is now regrowing on thousands of acres.

However, today global climate change is presenting new challenges. As temperatures climb, mosquitoes move upslope, and malaria and pox spread deeper into the forests, so that even protected areas such as Hakalau are no longer immune. The next generation of managers will need to innovate further strategies to fend off extinction for the island's native species.

Plenty of challenges remain, but the successes at Hakalau so far provide hope that through responsible management we can restore Hawaii's native flora and fauna and preserve the priceless bounty of millions of years of evolution on this extraordinary chain of islands. ◻

# Evolution: The Source of Earth's Biodiversity

The animals and plants native to the Hawaiian Islands help reveal how our world became populated with the remarkable diversity of life we see around us today—a lush cornucopia of millions of species (**FIGURE 3.1**).

A **species** is a particular type of organism. More precisely, it is a population or group of populations whose members share characteristics and can freely breed with one another to produce fertile offspring. A **population** is a group of individuals of a given species that live in a particular region at a particular time. Over vast spans of time, the process of evolution has shaped populations and species, giving us the vibrant abundance of life that enriches Earth today.

In its broad sense, the term *evolution* means change over time. In its biological sense, **evolution** consists of change in populations of organisms from generation to generation. Changes in genes (p. 29) often lead to modifications in appearance or behavior.

**FIGURE 3.1 Hawaii hosts a treasure trove of biodiversity, including ① happyface spider, ② i'iwi, ③ nene, and ④ Haleakala silversword.**

Evolution is one of the best-supported and most illuminating concepts in all of science, and it is the very foundation of modern biology. Perceiving how species adapt to their environments and change over time is crucial for comprehending ecology and learning the history of life. Evolutionary processes influence many aspects of environmental science, including agriculture, pesticide resistance, medicine, and environmental health.

## Natural selection shapes organisms

Natural selection is a primary mechanism of evolution. In the process of **natural selection,** inherited characteristics that enhance survival and reproduction are passed on more frequently to future generations than characteristics that do not, thereby altering the genetic makeup of populations through time.

Natural selection is a simple concept that offers a powerful explanation for patterns evident in nature. The idea of natural selection follows logically from a few straightforward facts that are readily apparent to anyone who observes the life around us:

- Organisms face a constant struggle to survive and reproduce.
- Organisms tend to produce more offspring than can survive to maturity.
- Individuals of a species vary in their attributes.

Variation is due to differences in genes, the environments in which genes are expressed, and the interactions between genes and environment. As a result of this variation, some individuals of a species will be better suited to their environment than others and will be better able to reproduce.

Attributes are passed from parent to offspring through the genes, and a parent that produces many offspring will pass on more genes to the next generation than a parent that produces few or no offspring. In the next generation, therefore, the genes of better-adapted individuals will outnumber those of individuals that are less well adapted. From one generation to another through time, characteristics, or traits, that lead to better and better reproductive success in a given environment will evolve in the population. This process is termed **adaptation,** and a trait that promotes success is also called an **adaptation** or an **adaptive trait.**

## Selection acts on genetic variation

For an organism to pass a trait along to future generations, genes in its DNA (p. 29) must code for the trait. In an organism's lifetime, its DNA will be copied millions of times by millions of cells. In all this copying, sometimes a mistake is made. Accidental changes in DNA, called **mutations,** give rise to genetic variation among individuals. If a mutation occurs in a sperm or egg cell, it may be passed on to the next generation. Most mutations have little effect, but some can be deadly, and others can be beneficial. Those that are not lethal provide the genetic variation on which natural selection acts.

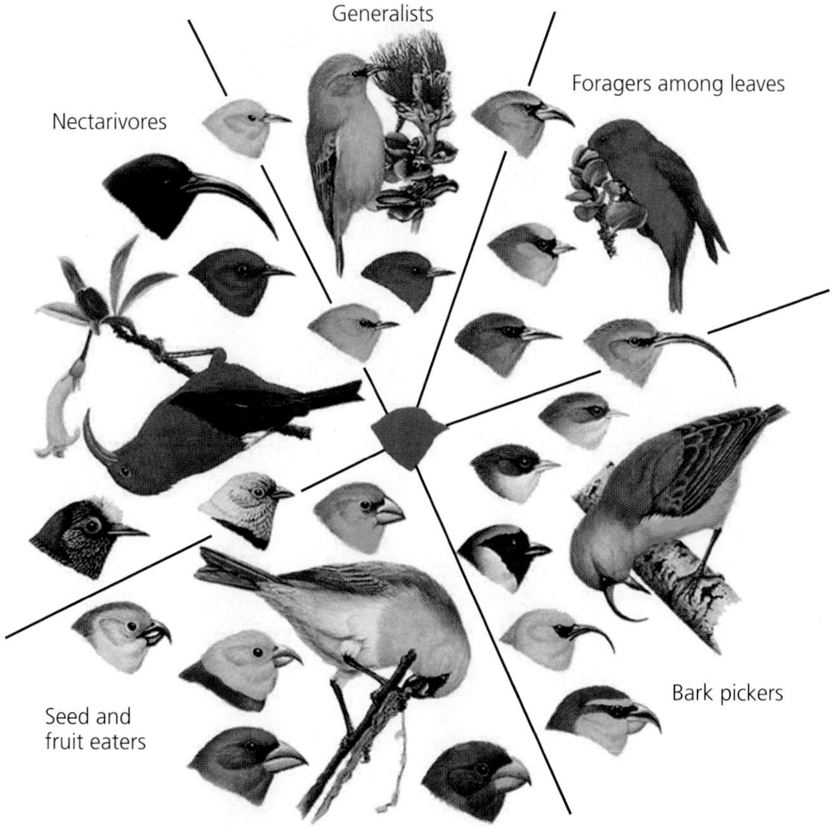


**(a) Divergent evolution of Hawaiian honeycreepers**

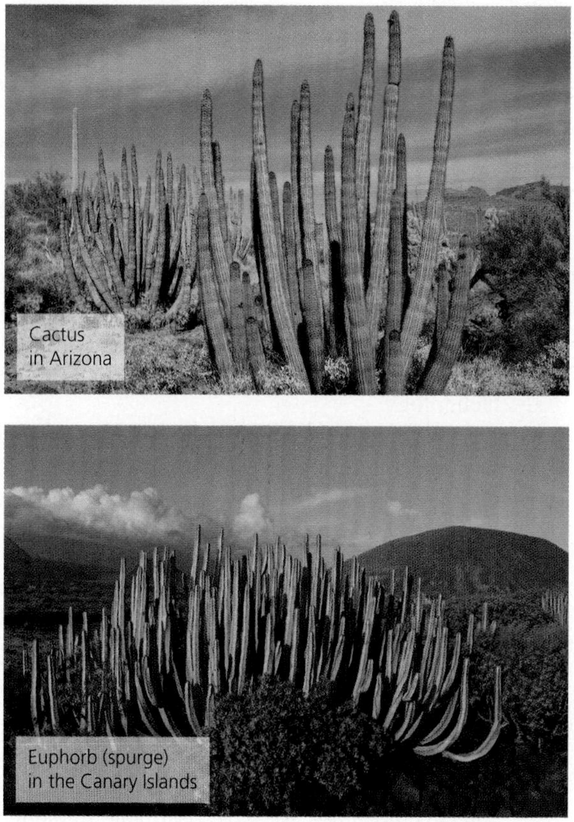


**(b) Convergent evolution of cactus and spurge**

**FIGURE 3.2 Natural selection can cause closely related species to diverge in appearance or distantly related species to converge in appearance.** Hawaiian honeycreepers **(a)** diversified as they adapted to different food resources and habitats, as indicated by the diversity of their plumage colors and bill shapes. In contrast, distantly related cacti of the Americas and euphorbs of Africa **(b)** became similar to one another as they independently adapted to arid environments. These plants each evolved succulent tissues to hold water, thorns to keep thirsty animals away, and photosynthetic stems without leaves to reduce surface area and water loss.

Genetic variation also results as organisms mix their genetic material through sexual reproduction. When organisms reproduce sexually, each parent contributes to the genes of the offspring, producing novel combinations of genes and generating variation among individuals.

Genetic variation can help populations adapt to changing environmental conditions. For example, one of the honeycreeper species of the Hakalau Forest, the 'amakihi, has in recent years been discovered in 'ōhi'a trees at low elevations where avian malaria has killed off all other honeycreepers. Researchers have determined that some of the 'amakihis living here when malaria arrived had genes that, by chance, gave them a natural resistance to the disease. These resistant birds survived malaria's onslaught, and their descendants reestablished a population that is growing today.

Environmental conditions determine what pressures natural selection will exert, and these selective pressures affect which members of a population will survive and reproduce. Over many generations, this results in the evolution of traits that enable success within a given environment. Closely related species that live in different environments tend to diverge in their traits as differing selective pressures drive the evolution of different adaptations (**FIGURE 3.2a**). Conversely, sometimes very unrelated species living in similar environments in separate locations may independently acquire similar traits as they adapt to selective pressures; this is called **convergent evolution** (**FIGURE 3.2b**).

## Evidence of selection is all around us

The concept of natural selection was first proposed in 1858 by **Charles Darwin** and **Alfred Russel Wallace,** two exceptionally keen British naturalists. By this time, scientists and laypeople had been observing nature, puzzling over fossils, and discussing the notion that populations might evolve, but no one could say how or why. After spending years studying and cataloguing an immense variety of natural phenomena from his English garden across the world to the Galápagos Islands, Darwin finally concluded that natural selection helped explain the world's great variety of living things. Yet he put off publishing his work, fearing the social disruption that could ensue if people felt their religious convictions threatened. Darwin was at last driven to go public when Wallace wrote to him from the Asian tropics, independently describing the idea of natural selection. The two men's shared ideas were presented together at a scientific meeting, and Darwin soon published his book, *On the Origin of Species*.

With natural selection, scientists finally gained a precise and feasible mechanism to explain how and why organisms change across generations. Once geneticists worked out how traits are inherited, modern evolutionary biology was born. In the century and a half since Darwin and Wallace, many thousands of scientists have refined our understanding of evolution, driving progress in biology and facilitating spectacular advances in agriculture, medicine, and technology.

The results of natural selection are all around us, visible in every adaptation of every organism. Moreover, scientists have demonstrated the rapid evolution of traits by selection in countless lab experiments, mostly with fast-reproducing organisms such as bacteria, yeast, and fruit flies.

The evidence for selection that may be most familiar to us is that which Darwin himself cited prominently in his work 150 years ago: our breeding of domesticated animals. In dogs, cats, and livestock, we have conducted selection under our own direction; that is, **artificial selection.** We have chosen animals possessing traits we like and bred them together, while culling out individuals with traits we do not like. Through such *selective breeding,* we have been able to augment particular traits we prefer.

Consider the great diversity of dog breeds (**FIGURE 3.3a**). Every dog alive today is descended from a single ancestral species; the differences among breeds are the result of human choices in selecting for particular desired traits as individuals were bred together. From Great Dane to Chihuahua, all dogs are able to interbreed and produce viable offspring, yet breeders maintain striking differences among them by allowing only like individuals to breed.

Artificial selection through selective breeding has also given us the many crop plants and livestock we depend on for food, all of which people domesticated from wild species and carefully bred over years, centuries, or millennia (**FIGURE 3.3b**). Through selective breeding, we have created corn with bigger, sweeter kernels; wheat and rice with larger and more numerous grains; and apples, pears, and oranges with better taste. We have diversified single types into many—for instance, breeding variants of the plant *Brassica oleracea* to create broccoli, cauliflower, cabbage, and brussels sprouts. Our entire agricultural system is based on artificial selection.

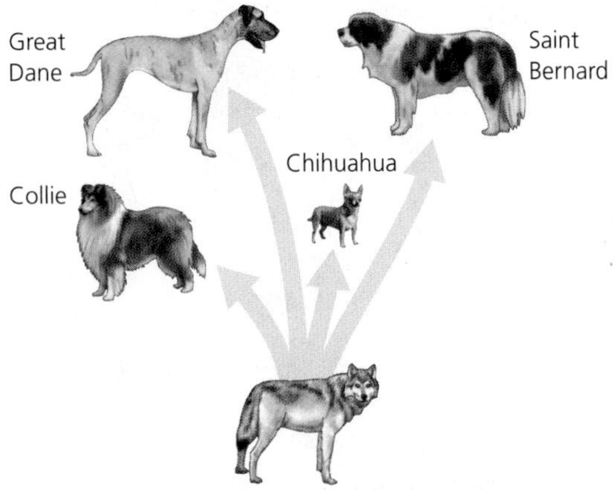


**(a) Ancestral wolf (*Canis lupus*) and derived dog breeds**

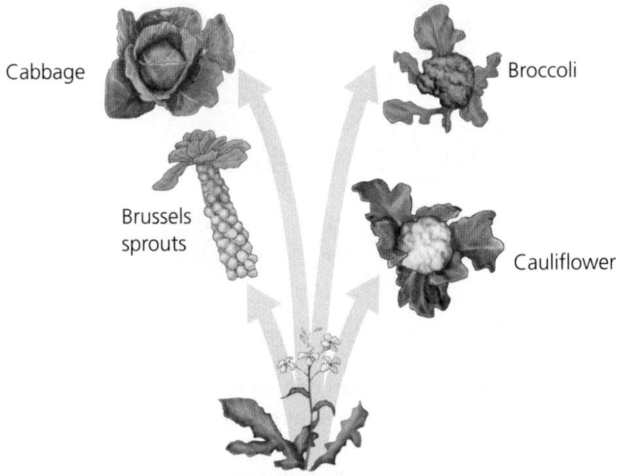


**(b) Ancestral *Brassica oleracea* and derived crops**

**FIGURE 3.3 Selective breeding, or artificial selection, has given us our many different breeds of dogs and varieties of crops.** The ancestral wild species for dogs **(a)** is the gray wolf (*Canis lupus*). By breeding like with like and selecting for traits we preferred, we have produced breeds as different as Great Danes and Chihuahuas. Using the same process, we have created an immense variety of crop plants **(b)**. Cabbage, brussels sprouts, broccoli, and cauliflower were all generated from a single ancestral species, *Brassica oleracea*.

## Understanding evolution is vital for modern society

Evolutionary processes play key roles in today's society and in our everyday lives. As we've just seen, the food we eat and the clothes we wear, each and every day, have been made possible through the selective breeding of crops and livestock.

Many medical advances have also resulted from our understanding of evolution. It teaches us how infectious diseases spread and how they gain or lose potency. It allows scientists to track the constantly evolving strains of influenza, HIV, and other pathogens. It enables biomedical experts to predict which flu strains will most likely spread in a given year and then design effective vaccines targeting them. Comprehending evolution also enables us to understand how dangerous bacteria develop antibiotic resistance so that we can detect it and respond.

In agriculture, our understanding of evolution helps us prevent antibiotic resistance in feedlots and pesticide resistance in crop-eating insects (pp. 147–148). It informs our technology as well; from studying how organisms adapt to challenges and evolve new abilities, we develop ideas on how to design technologies and engineer solutions.

## Evolution generates biodiversity

Just as selective breeding helps us create new types of pets, farm animals, and crop plants, evolution can elaborate and diversify traits in wild organisms, helping to form new species and whole new types of organisms. Life's complexity can be expressed as **biological diversity,** or **biodiversity.** These terms refer to the variety of life across all levels, including the diversity of species, genes, populations, and communities.

Scientists have described about 1.8 million species, but many more remain undiscovered or unnamed. Estimates for the actual number of species in the world range from 3 million up to 100 million. Hawaii's insect fauna provides one example of how much we have yet to learn. Scientists studying fruit flies in the Hawaiian Islands have described over 500 species of them, but they have also identified about 500 others that have not yet been formally named and described. Still more fruit fly species probably exist but have not yet been found.

Subtropical islands such as Hawai'i are by no means the only places rich in biodiversity. Step outside just about anywhere, and you will find many species within close reach. Plants poke up from cracks in asphalt in every city in the world, and even Antarctic ice harbors microbes. A handful of backyard soil may contain an entire miniature world of life, including insects, mites, millipedes, nematode worms, plant seeds, fungi, and millions of bacteria. (We will examine Earth's biodiversity in detail in Chapter 8.)

## Speciation produces new types of organisms

How did Earth come to have so many species? The process by which new species are generated is termed **speciation.** Speciation can occur in a number of ways, but the main mode is generally thought to be *allopatric speciation*, whereby species form from populations that become physically separated over some geographic distance. To understand allopatric speciation, begin by picturing a population of organisms. Individuals within the population possess many similarities that unify them as a species because they are able to breed with one another, sharing genetic information. However, if the population is broken up into two or more isolated areas, individuals from one area cannot reproduce with individuals from the others.

When a mutation arises in the DNA of an organism in one of these newly isolated populations, it cannot spread to the other populations. Over time, each population will independently accumulate its own set of mutations. Eventually, the populations may diverge, growing so different that their members can no longer mate with one another and produce viable offspring (because of changes in reproductive organs, hormones, courtship behavior, breeding timing, or other factors). Once this has happened, there is no turning back; populations that no longer exchange genetic information will embark on their own independent paths as separate species (**FIGURE 3.4**). As mutations accumulate, causing the populations to differ more and more, the populations will continue diverging in their traits.

1 Single population
2 Geographically isolated populations
3 Divergence due to long-term isolation
4 Isolated populations come together; they can no longer interbreed and are now two species

**FIGURE 3.4 In allopatric speciation, species form from populations that become physically separated by geographic distance.** This long, slow process begins when a geographic barrier splits a population—as when forest 1 is destroyed by lava flowing from a volcano, but isolated patches of forest 2 are left. In Hawaii, such forested patches are called kipukas. Hawaiian fruit flies are weak fliers and become isolated in kipukas. Over the centuries, each population accumulates its own independent set of genetic changes 3, until individuals become genetically distinct from and unable to breed with individuals from the other population. The two populations now represent separate species and will remain so even when the geographic barrier disappears 4, as new forest grows over the eroding lava rock, and the new species intermix.

Populations can undergo long-term geographic isolation in various ways. Lava flows can destroy forest, leaving small isolated patches intact (as shown in Figure 3.4). Glacial ice sheets may move across continents during ice ages and split populations in two. Major rivers may change course or mountain ranges may be uplifted, dividing regions and their organisms. Sea level may rise, flooding low-lying regions and isolating areas of higher ground as islands. Drying climate may partially evaporate lakes, subdividing them into smaller bodies of water. Warming or cooling climate may cause plant communities to shift, creating new patterns of plant and animal distribution.

Alternatively, sometimes organisms colonize newly created areas, establishing isolated populations. This can occur when a long string of islands, an *archipelago*, forms as a result of tectonic activity. (The Hawaiian Islands are an example; flip to Figure 11.7, p. 231, for an illustration of the process.) As each new island is formed, plants and animals that colonize it may undergo allopatric speciation if they are isolated enough from their source population.

## FAQ Are humans still evolving?

As our civilization has developed, it's become easy to fall under the illusion that we're above and separate from nature. This has given rise to the idea that human beings are no longer subject to natural selection, only to cultural forces. It's true that the "natural" environment of our hunter-gatherer days has given way to a largely human-constructed environment in our modern urban-techno-industrial society. But when environments change, new selective pressures come into play. For instance, in the past many people died prematurely from diseases, injuries, and infection. Today, medical advances save and extend our lives. Ironically, as a result, illnesses and disabilities that can be inherited get passed on to more children. At the same time, humanity's old foes continue to have an impact: In developing countries, malaria, measles, tuberculosis, and other infectious diseases remain major killers of young people, exerting strong selection on the human genome. Natural selection never stops, and all organisms—people included—are always evolving. Indeed, the pace of evolutionary change quickens as environments change.

## We can infer the history of life's diversification

Innumerable speciation events have generated complex patterns of diversity beyond the species level. Scientists represent histories of divergence by using branching diagrams called **phylogenetic trees.** Similar to family genealogies, these diagrams illustrate hypotheses proposing how divergence took place (**FIGURE 3.5**). Phylogenetic trees can show relationships among species, groups of species, populations, or genes. Scientists construct these trees by analyzing patterns of similarity among the genes or external traits of present-day organisms.

Evolutionary biologists are interested not only in how groups of organisms arose, but also in how they evolved the attributes they show. Once we have a phylogenetic tree, we can map traits onto the tree according to which organisms possess them, and we can thereby trace when in evolutionary history the traits appeared.

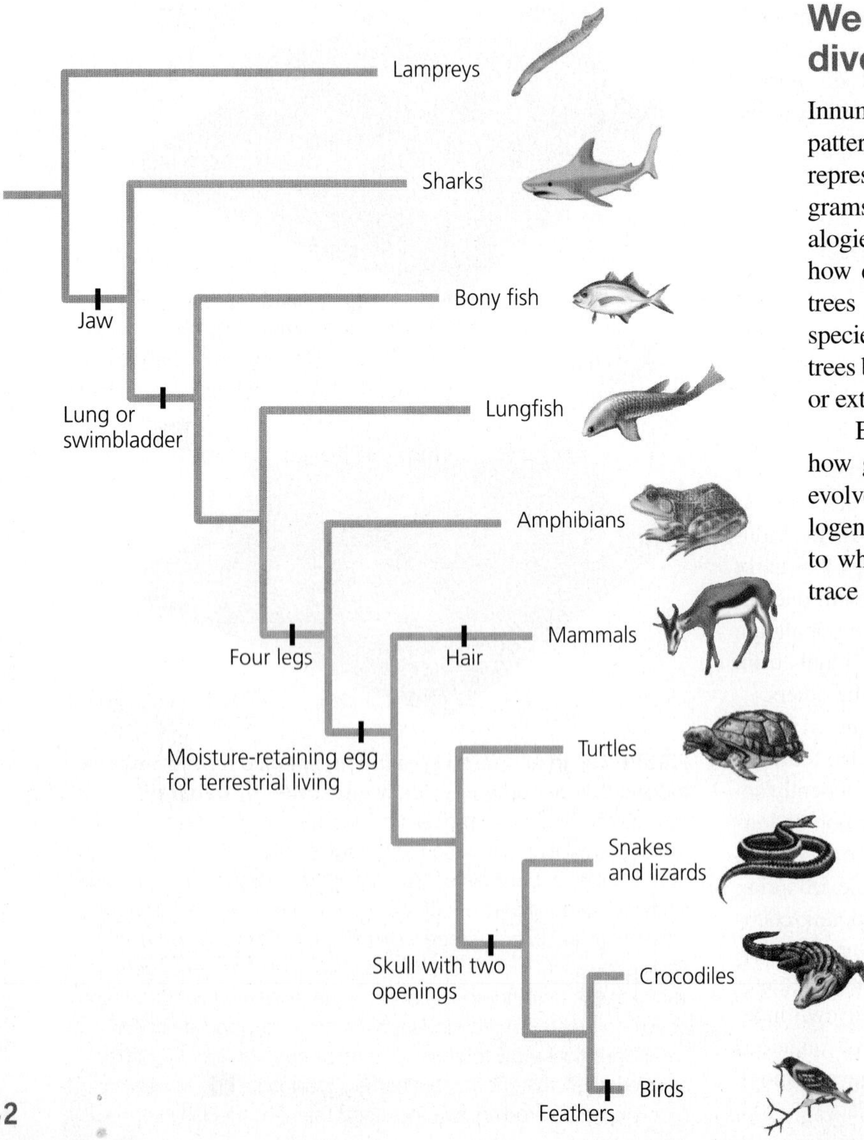


**FIGURE 3.5 Phylogenetic trees show the history of life's divergence.** The tree here illustrates relationships among groups of vertebrates—just one small portion of the huge and complex "tree of life." Each branch results from a speciation event; as you follow the tree left to right from its trunk to the tips of its branches, you proceed forward in time, tracing life's history. In this tree, major traits are "mapped" to indicate when they originated.

DATA Q Find the hash mark indicating the origin of jaws. Which groups of vertebrates possess jaws? Which group(s) diverged before jaws originated?

GO TO **INTERPRETING GRAPHS & DATA** ON MasteringEnvironmentalScience®

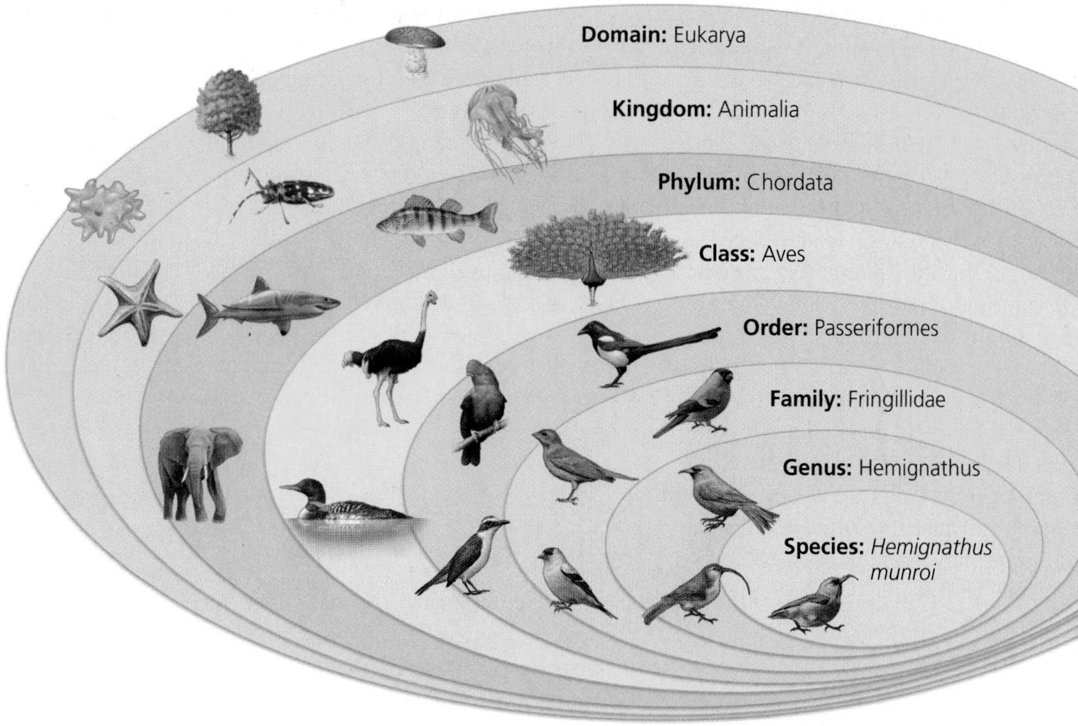


**FIGURE 3.6 Taxonomists classify organisms using a hierarchical system meant to reflect evolutionary relationships.** Species similar in appearance, behavior, and genetics (because they share recent common ancestry) are placed in the same genus. Organisms of similar genera are placed in the same family. Families are placed within orders, orders within classes, classes within phyla, phyla within kingdoms, and kingdoms within domains. For example, honeycreepers belong to the class Aves, along with peacocks, loons, and ostriches. However, the differences between these species, which have diverged across millions of years of evolution, are great enough that they are placed in different orders, families, and genera.

Knowing how organisms are related to one another also helps scientists classify them and name them. *Taxonomists* use an organism's genetic makeup and physical appearance to determine its species identity. These scientists then group species by their similarity into a hierarchy of categories meant to reflect evolutionary relationships. Related species are grouped together into *genera* (singular: *genus*), related genera are grouped into families, and so on (**FIGURE 3.6**). Each species is given a two-part Latin or Latinized scientific name denoting its genus and species.

For example, the ʻakiapōlāʻau, *Hemignathus munroi,* is similar to other Hawaiian honeycreepers in the genus *Hemignathus*. These species are closely related in evolutionary terms, as indicated by the genus name they share. They are more distantly related to honeycreepers in other genera, but all honeycreepers are classified together in the family Fringillidae.

## Fossils reveal life's history

Scientists also decipher life's history by studying fossils. As organisms die, some are buried by sediment. Under certain conditions, the hard parts of their bodies—such as bones, shells, and teeth—may be preserved as sediments are compressed into rock (p. 229). Minerals replace the organic material, leaving behind a **fossil,** an imprint in stone of the dead organism (**FIGURE 3.7**). Over millions of years, geologic processes have buried sediments and later brought sedimentary rock layers to the surface, revealing assemblages of fossilized plants and animals from different time periods. By dating the rock layers that contain fossils, scientists can determine when particular organisms lived. The cumulative body of fossils worldwide is known as the **fossil record.**

**FIGURE 3.7 The fossil record helps reveal the history of life on Earth.** Here, a paleontologist in India excavates a 50,000-year-old fossilized elephant skull.

The fossil record shows that the number of species has increased through time, but that the species on Earth today are just a small fraction of all species that ever existed. During the 3.5 billion years that life has existed on Earth, complex structures have evolved from simple ones, and large sizes from small ones. However, simplicity and small size have also evolved when favored by natural selection; it is easy to argue that Earth still belongs to the bacteria and other microbes, some of them little changed over eons.

Even enthusiasts of microbes, however, must marvel at some of the exquisite adaptations of animals, plants, and fungi: the heart that beats so reliably for an animal's entire lifetime that we take it for granted; the complex circulatory system to which the heart belongs; the stunning plumage of a peacock; the ability of plants to lift water and nutrients from the soil, gather light from the sun, and produce food; and the human brain and its ability to reason. All these adaptations and more have come about as evolution has generated new species and whole new branches on the tree of life.

Although speciation generates Earth's biodiversity, it is only part of the equation, because the fossil record teaches us that the vast majority of creatures that once lived are now gone. The disappearance of a species from Earth is called **extinction.** From studying fossils, paleontologists calculate that the average time a species spends on Earth is 1–10 million years. The number of species in existence at any one time is equal to the number added through speciation minus the number removed by extinction.

## Some species are especially vulnerable to extinction

In general, extinction occurs when environmental conditions change rapidly or drastically enough that a species cannot adapt genetically to the change; the slow process of natural selection simply does not have enough time to work. Small populations are especially vulnerable to extinction because fluctuations in their size could, by chance, bring the population size to zero. Small populations also sometimes lack enough genetic variation to buffer them against environmental change. Species narrowly specialized to some particular resource or role are also vulnerable, because environmental changes that make that resource or way of life unavailable can spell doom. Species that are **endemic** to a particular region occur nowhere else on Earth. Endemic species face elevated risks of extinction because when some impact influences their region, it may affect all members of the species.

Island-dwelling species are particularly vulnerable. Because islands are smaller than mainland areas and are isolated by water, fewer species inhabit islands. As a result, some of the pressures and challenges faced daily by organisms on a mainland simply don't exist on islands. For instance, only one land mammal—a bat—ever reached Hawaii naturally, so Hawaii's birds evolved for millions of years without needing to protect against the threat of predation by mammals. Likewise, Hawaii's plants did not need to invest in defenses (such as thick bark, spines, or chemical toxins) against mammals that might eat them. Because defenses are costly to invest in, most island birds and plants have lost any defenses their mainland ancestors may have had against mammals.

Eventually, people—first Polynesians and then Europeans—arrived in Hawaii, bringing cattle, goats, pigs, rats, dogs, cats, and mongooses. Hawaii's native organisms were completely unprepared (**FIGURE 3.8**). Rats, cats, and mongooses preyed voraciously on ground-nesting seabirds, ducks, geese, and flightless rails, driving many of these birds extinct. Livestock ate through the vegetation, turning luxuriant forests into desolate grasslands. All in all, half of Hawaii's native birds were driven extinct soon after human arrival.

## Earth has seen episodes of mass extinction

Most extinction occurs gradually, one species at a time, at a rate referred to as the *background extinction rate*. However, the fossil record reveals that Earth has seen five events of staggering proportions that killed off massive numbers of species at once. These episodes, called **mass extinction events,** have occurred at widely spaced intervals in our planet's history and have wiped out 50–95% of Earth's species each time (see Table 8.3, p. 169).

**(a) Hawaiian Petrel**

**(b) Mongoose**

**FIGURE 3.8 Island-dwelling species that have lost defenses are vulnerable to extinction when enemies are introduced.** The Hawaiian petrel **(a)**, a seabird that nests in the ground, is endangered as a result of predation by mammals introduced to Hawaii, such as **(b)** the Indian mongoose.

The best-known mass extinction occurred 65 million years ago and brought an abrupt end to the dinosaurs (although birds are modern representatives of dinosaurs). Evidence suggests that the collision of a gigantic asteroid with Earth caused this event. Still more catastrophic was the mass extinction 250 million years ago at the end of the Permian period (see **APPENDIX E** for Earth's geologic periods). Scientists estimate that 75–95% of all species perished during this event, described by one researcher as the "mother of all mass extinctions." Massive volcanism may have caused the end-Permian extinction event; other hypotheses include an asteroid impact, methane releases and global warming, or some combination of these factors.

## The sixth mass extinction is upon us

Many biologists have concluded that Earth is currently entering its sixth mass extinction event—and that we are the cause. Changes to our planet's natural systems set in motion by human population growth, development, and resource depletion have driven many species to extinction and are threatening countless more. As we alter and destroy natural habitats; overhunt and overharvest populations; pollute air, water, and soil; introduce invasive non-native species; and alter climate, we set in motion processes that combine to threaten Earth's biodiversity (pp. 173–178). Extinction may, in fact, be the single biggest problem we face, because the loss of a species is irreversible.

Biodiversity loss affects people directly, because organisms provide us with life's necessities—food, fiber, medicine, and vital ecosystem services (pp. 3, 35, 97). Because we depend on what nature has to offer, biodiversity loss ultimately threatens our own survival.

# Ecology and the Organism

Extinction, speciation, and other evolutionary forces play key roles in ecology. **Ecology** is the scientific study of the interactions among organisms and of the relationships between organisms and their environments. Ecology allows us to explain and predict the distribution and abundance of organisms in nature. It is often said that ecology provides the stage on which the play of evolution unfolds. The two are intertwined in many ways.

## We study ecology at several levels

Life exists in a hierarchy of levels, from atoms, molecules, and cells (pp. 27–29) up through the **biosphere,** the cumulative total of living things on Earth and the areas they inhabit. **Ecologists** are scientists who study relationships at the higher levels of this hierarchy (**FIGURE 3.9**), namely at the levels of the organism, population, community, ecosystem, landscape, and biosphere.

At the organismal level, ecology describes the relationships between an organism and its physical environment. For example, organismal ecology helps us understand what

Biosphere
The sum total of living things on Earth and the areas they inhabit
Landscape
A geographic region including an array of ecosystems
Ecosystem
A functional system consisting of a community, its nonliving environment, and the interactions between them
Community
A set of populations of different species living together in a particular area
Population
A group of individuals of a species that live in a particular area
Organism
An individual living thing

**FIGURE 3.9 Ecologists study questions on multiple levels.** Hawaii's green sea turtles are part of a coral reef community that inhabits reef ecosystems along many miles of Hawaii's coasts.

aspects of a Hawaiian honeycreeper's environment are important to it, and why. In contrast, **population ecology** examines the dynamics of population change and the factors that affect the distribution and abundance of members of a population. It helps us understand why populations of some species decline while populations of others increase.

In ecology, a **community** consists of an assemblage of interacting populations that inhabit the same area. A population of 'akiapōlā'au, a population of koa trees, a population of wood-boring grubs, and a population of ferns, together with all the other interacting plant, animal, fungal, and microbial populations in the Hakalau Forest, would be considered a community. **Community ecology** (Chapter 4) focuses on patterns of species diversity and on interactions among species, ranging from one-to-one interactions up to complex interrelationships involving the entire community.

**Ecosystems** (p. 32) encompass living biotic communities and the abiotic (nonliving) materials and forces with which community members interact. Hakalau's cloud-forest ecosystem consists of its community plus the air, water, soil, nutrients, and energy used by the community's organisms. **Ecosystem ecology** (Chapter 2) addresses the flow of energy and nutrients by studying living and nonliving components of systems in conjunction. Today's warming climate (Chapter 14) is having ecosystem-level consequences as it exerts impacts on Hakalau and other ecosystems across the world.

Today, concerns such as climate change and habitat loss, together with technologies such as satellite imagery, are invigorating the study of how ecosystems are arrayed across the landscape. **Landscape ecology** (p. 34) helps us understand how and why ecosystems, communities, and populations are distributed across geographic regions. Indeed, as new technologies help scientists study the complex dynamics of natural systems at a global scale, ecologists are expanding their horizons to the biosphere as a whole.

## Habitat, niche, and specialization are key concepts in ecology

At the level of the organism, each individual relates to its environment in ways that tend to maximize its survival and reproduction. One key relationship involves the specific environment in which an organism lives, its **habitat.** A species' habitat consists of the living and nonliving elements around it, including rock, soil, leaf litter, humidity, plant life, and more. The 'akiapōlā'au (**FIGURE 3.10**) lives in a habitat of cool, moist, montane forest of native koa and 'ōhi'a trees, where it is high enough in elevation to be safe from avian pox and malaria.

Each organism thrives in certain habitats and not in others, leading to nonrandom patterns of **habitat use.** Mobile organisms actively select habitats from among the range of options they encounter, a process called **habitat selection.** Each species assesses habitats differently because each species has different needs. Indeed, a species' needs may vary with the season; many migratory birds use distinct breeding, wintering, and migratory habitats. In the case of plants and of rooted animals (such as sea anemones), whose young disperse and settle passively, patterns of habitat use result from success in some habitats and failure in others.

**FIGURE 3.10 The 'akiapōlā'au fills a unique niche by virtue of its odd bill.** Its straight bottom half and long, curved top half allows the bird to specialize on digging grubs out from native trees in its montane forest habitat.

Habitat is a vital concept in environmental science. Because habitats provide everything an organism needs, including nutrition, shelter, breeding sites, and mates, the organism's survival depends on the availability of suitable habitats. Often this need results in conflict with people who want to alter a habitat for their own purposes.

Another way in which an organism relates to its environment is through its **niche,** its functional role in a community. A species' niche reflects its use of habitat and resources, its consumption of certain foods, its role in the flow of energy and matter, and its interactions with other organisms. The niche is a multidimensional concept, a kind of summary of everything an organism does. The pioneering ecologist Eugene Odum once wrote that "habitat is the organism's address, and the niche is its profession."

Organisms vary in the breadth of their niches. A species with narrow breadth (and thus very specific requirements) is said to be a **specialist.** One with broad tolerances, a "jack-of-all-trades" able to use a wide array of resources, is a **generalist.** A native Hawaiian honeycreeper like the 'akiapōlā'au (see Figure 3.10) is a specialist, because its unique bill is exquisitely adapted for feeding on grubs that tunnel through the wood of certain native trees. In contrast, the common myna (a bird introduced to Hawaii from Asia) is a generalist; its unremarkable bill allows it to eat many foods in many habitats. As a result, the common myna has spread throughout the Hawaiian Islands wherever human development has altered the landscape.

Generalists like the myna succeed by being able to live in many different places and withstand variable conditions, yet they do not thrive in any single situation as well as a specialist adapted for those specific conditions. (A jack-of-all-trades, as the saying goes, is a master of none.) Specialists succeed over evolutionary time by being extremely good at the things they do, yet they are vulnerable when conditions change and threaten the habitat or resource on which they have specialized. An organism's habitat preferences, niche, and degree of specialization each reflect adaptations of the species and are products of natural selection.

# Population Ecology

A population, as we have seen, consists of individuals of a species that inhabit a particular area. A species may consist of multiple populations that are geographically isolated. This is the case with Hawaii's state bird, the nēnē, or Hawaiian goose (see Figure 3.1b). Originally common throughout the Hawaiian Islands, the nēnē (pronounced "nay-nay") was nearly driven to extinction by human hunting; plants and livestock that people introduced, which destroyed and displaced the vegetation the nēnē fed on; and rats, cats, dogs, pigs, and mongooses that preyed on its eggs and young. Nēnēs disappeared from all islands except the island of Hawai'i. In recent decades, biologists and wildlife managers have labored to breed it in captivity and reintroduce it to protected areas. These efforts are succeeding, and today nēnēs can be found in at least seven populations on four of the Hawaiian Islands.

In contrast, the human species faces few threats from other animals, is a consummate generalist, and has spread into nearly every corner of the planet. As a result, it is difficult to define a distinct human population on anything less than the global scale. In the ecological sense of the word, all 7 billion of us comprise one population.

## Populations show features that help predict their dynamics

All populations—from humans to nēnēs—exhibit attributes that help population ecologists predict their dynamics. The ability to predict a population's growth or decline is useful for managing wildlife and fisheries and for monitoring threatened and endangered species (see **THE SCIENCE BEHIND THE STORY**, pp. 58–59). It is also crucial for understanding the dynamics of our human population (Chapter 6)—a central element of environmental science and one of the prime challenges for our society today.

**Population size** Expressed as the number of individual organisms present at a given time, **population size** may increase, decrease, undergo cyclical change, or remain stable over time. Populations generally grow when resources are abundant and natural enemies are few. Populations can decline in response to loss of resources, natural disasters, or impacts from other species. Researchers estimate that the nēnē population numbered over 25,000 birds before Europeans reached the Hawaiian Islands. By the 1950s, impacts from hunting, agriculture, non-native mammals, and invasive plants had brought the population size down to just 30 individuals. Since then, intensive conservation efforts have turned this decline around, and now over 2000 nēnēs live on the Hawaiian Islands.

The passenger pigeon, now extinct, illustrates extremes in population size (**FIGURE 3.11**). Not long ago it was the most abundant bird in North America; flocks of passenger pigeons literally darkened the skies. In the early 1800s, ornithologist Alexander Wilson watched a flock of 2 billion birds 390 km (240 mi) long that took 5 hours to fly over and sounded like a tornado. Passenger pigeons nested in gigantic colonies in the forests of the upper Midwest and southern Canada. Once settlers began cutting the forests, the birds were easy targets for market hunters, who gunned down thousands at a time and shipped them to market by the wagonload. By 1890, the population had declined to such low numbers that the birds could not form the large colonies they evidently needed to breed. In 1914, the last passenger pigeon on Earth died in the Cincinnati Zoo, bringing the continent's most numerous bird species to extinction in just a few decades.

**(a) Passenger pigeon**

**(b) 19th-century lithograph of pigeon hunting in Iowa**

**FIGURE 3.11 Flocks of passenger pigeons literally darkened the skies as billions of birds passed overhead.** However, hunting and deforestation drove the species to extinction within just a few decades.

# Monitoring Bird Populations at Hakalau Forest

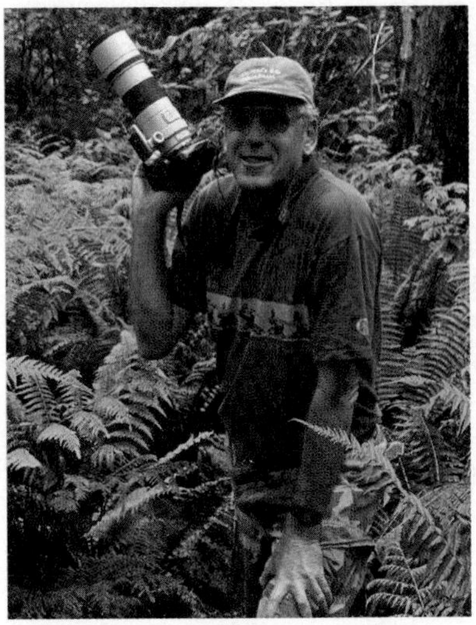

**Jack Jeffrey in the Hakalau Forest**

Are populations of honeycreepers increasing or decreasing? It's a high-stakes question. The answer could tell us whether efforts to save them are on the right track, or whether the birds may be headed for extinction.

At Hakalau Forest National Wildlife Refuge, on the rainy slope of Mauna Kea on the island of Hawai'i, biologists have been working for years to understand the dynamics of bird populations. But monitoring population trends is not easy, and there is still debate and uncertainty.

Established in 1985, the Hakalau Refuge is home to nine species of native forest birds, including four federally endangered ones: the Hawai'i 'ākepa, the Hawai'i creeper, the 'akiapōlā 'au, and the 'io, or Hawaiian hawk. A number of non-native birds now occur here as well. Much of the region's native forest had been cleared years earlier for cattle ranching, while free-roaming pigs and invasive plants had degraded the rest.

The biologists employed after 1985 to manage Hakalau built fences to keep pigs and cattle out of the forest. They labored to restore the area by removing invasive weeds and by planting half a million native plants. Gradually the forest recovered, and stands of young trees took root at higher elevations. But has the restored forest brought higher populations of native forest birds?

In 1987, federal biologists and trained observers began conducting periodic surveys of birds on the refuge. Following standard protocols, they performed "point counts" by walking transects, stopping for 8 minutes in predetermined spots, and counting numbers of each species seen or heard. With 343 points along 15 transects across the refuge, this sampling allowed them to estimate population densities for each bird species. Researchers then analyzed changes in these samples through time to make inferences about changes in population densities and population sizes.

After 21 years of point counts, refuge biologists summarized their analyses. At a 2008 workshop and in a 2009 technical report, they concluded that populations of most native birds were either stable or slowly increasing across most of the refuge.

Their report included a series of graphs similar to that shown in **FIGURE 1.** Densities of birds varied from year to year, which is a normal and expected result of varying conditions (such as weather). Because of such variation, long-term studies are necessary; scientists expect that over time, year-to-year variation will be overshadowed by long-term trends that reveal the actual growth or decline of the population.

To interpret trends, researchers use a statistical method called *regression*. Linear regression analyzes how values vary through time and determines the straight line that, when drawn through data points on a graph, most accurately represents their trend (see Figure 1). With the data from Hakalau, linear regression led managers to conclude that over the 21-year period:

- In high-elevation pasture that managers were restoring to forest, populations of birds that used the young trees were rising sharply.
- In middle-elevation open forest being managed, populations of all native birds were either stable or increasing.

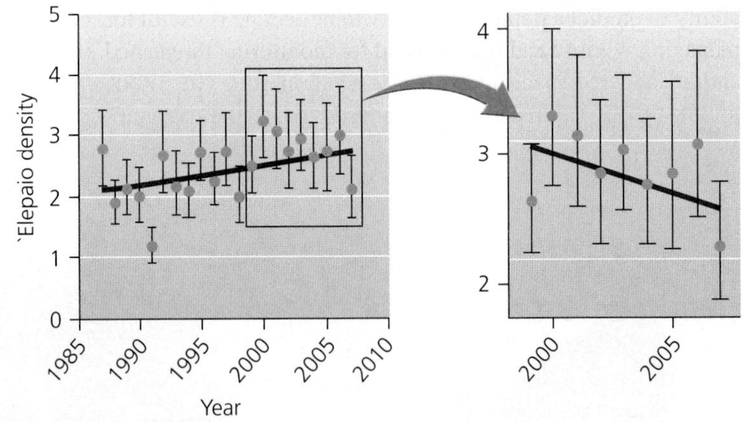


**FIGURE 1 Population density data for the Hawai'i'elepaio, a native forest bird, typify data gathered at Hakalau Forest NWR.** Over a 21-year period, regression shows an increase **(red line).** However, over the most recent 9 years, regression shows a decrease **(blue line in box).** Thin black bars indicate 95% confidence intervals. *Data from Camp, Richard J., et al. 2009.* Passerine bird trends at Hakalau Forest National Wildlife Refuge, Hawai'i. *Hawai'i Cooperative Studies Unit Technical Report HCSU-011.*

**DATA Q** What is the trend in the data during the first 12 years of the period shown? Draw a line through the first 12 data points that best represents this trend.

GO TO **INTERPRETING GRAPHS & DATA** ON MasteringEnvironmentalScience®

The long-term picture thus appeared bright.

However, during the most recent 9 years of the 21-year period, many of the populations had decreased. Regressions run on the last 9 years alone showed declines for many species (see Figure 1).

Biologist Leonard Freed of the University of Hawai'i at Manoa argued that federal biologists were overemphasizing the positive long-term trends and ignoring the negative near-term trends. For years, Freed and his colleagues had conducted research at Hakalau, focusing on the breeding biology of the Hawai'i 'ākepa (**FIGURE 2a**). Their research suggested that the 'ākepa began suffering competition for food once the non-native Japanese white-eye (**FIGURE 2b**) became abundant in the forest. Freed maintained that competition from the Japanese white-eye, along with attacks from parasitic lice in the nests, stunted the growth of the young birds and threatened the survival of the 'ākepa population. Freed urged that white-eyes be trapped and killed in order to save the 'ākepa. Refuge biologists called Freed's results controversial and said they needed validation.

**(a) Hawai`i `ākepa** **(b) Japanese white-eye**

**FIGURE 2 Work by biologist Leonard Freed suggests that the 'ākepa is suffering competition from the non-native Japanese white-eye.**

The debate came to a head in a pair of papers published back-to-back in the ornithological journal *Condor* in 2010. The team of federal biologists, led by Richard Camp and including Jack Jeffrey, presented their data and acknowledged that many populations showed downward trends in the most recent 9 years. Freed and his colleague Rebecca Cann reanalyzed the federal data using alternative methods and questioned whether the earlier apparent increases were reliable.

At stake is how best to manage the forest and its birds. If Camp and his colleagues are right, then management actions taken so far seem to have been effective, boosting populations or holding them stable in the face of dire threats. If Freed and Cann are right, then management strategies may need to be rethought. Researchers on all sides are debating the issues diligently because they care deeply about the forest and its birds and are trying their utmost to save them during a time of crisis.

Many factors could account for the apparent recent declines in populations of 'ākepas and other native honeycreepers, including the simple fact that thicker forest vegetation has made it harder for counters to detect birds. Of concern, though, is the possibility that challenges from outside the refuge—such as avian malaria and pox moving upslope with climate warming—might eventually overwhelm even the best management efforts.

To clarify the situation, researchers today continue to survey Hakalau's birds, adding to their valuable long-term database. But government budget cuts are threatening their ability to analyze the data—as well as their capacity to safeguard the refuge. Amid shortfalls in funding and staffing, pigs broke through fences and began degrading the newly restored forest. When budgets are cut, refuge staff must work even harder just to regain the progress previously made.

Despite the apparent recent declines in bird populations at Hakalau, populations there seem to be faring better than elsewhere on the island of Hawai'i (**FIGURE 3**). Moreover, the reforestation of Hakalau's upper zone is creating new habitat into which birds are moving. This success is a hopeful sign that research and careful management can help undo past damage and preserve endangered island species. ◻

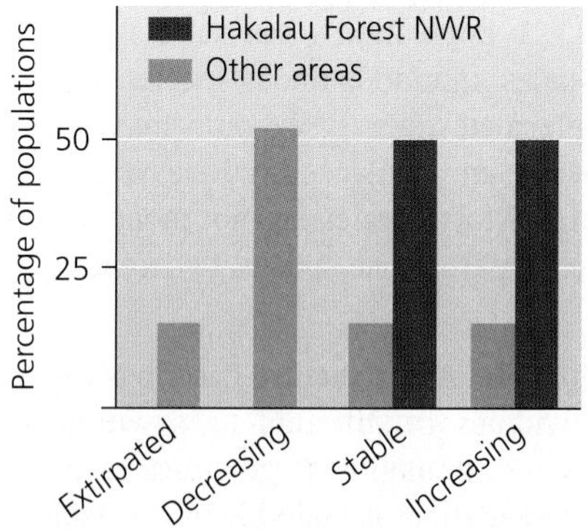


**FIGURE 3 At Hakalau Forest NWR, native forest bird populations were judged to be stable or increasing, whereas at four other protected areas on the island of Hawai'i, most populations were judged to be decreasing, and some have recently vanished.** *Data from Camp, Richard J., et al. 2009.* Passerine bird trends at Hakalau Forest National Wildlife Refuge, Hawai'i. Hawai'i *Cooperative Studies Unit Technical Report HCSU-011.*

**Population density** The flocks and breeding colonies of passenger pigeons showed high population density, another attribute that ecologists assess. **Population density** describes the number of individuals per unit area in a population. High population density makes it easier for organisms to group together and find mates, but it can also lead to competition and conflict if space, food, or mates are in limited supply. Close contact among overcrowded individuals may also increase the transmission of infectious disease. In contrast, at low population densities, individuals benefit from more space and resources but may find it harder to locate mates and companions.

**Population distribution** **Population distribution** describes the spatial arrangement of organisms in an area. Ecologists define three distribution types: random, uniform, and clumped (**FIGURE 3.12**). In a *random distribution*, individuals are located haphazardly in no particular pattern. This type of distribution can occur when resources are plentiful throughout an area and other organisms do not strongly influence where members of a population settle.

A *uniform distribution*, in which individuals are evenly spaced, can occur when organisms compete for space. Animals such as nesting birds (see Figure 3.12b) may hold and defend territories. Plants need space for their roots to gather moisture, and they may even exude chemicals that poison one another's roots as a means of competing for space. As a result, competing individuals may end up distributed at equal distances from one another.

A *clumped distribution* often results when organisms seek habitats or resources that are unevenly spaced. In arid areas, many plants grow in patches near isolated springs, ponds, or streambeds. Hawaiian honeycreepers tend to cluster near flowering trees that offer nectar. People frequently aggregate in villages, towns, and cities.

**Sex ratio** A population's **sex ratio** is its proportion of males to females. In monogamous species (in which each sex takes a single mate), a 1:1 sex ratio maximizes population growth, whereas an unbalanced ratio leaves many individuals without mates. Most species are not monogamous, however, so sex ratios vary from one species to another.

**Age structure** **Age structure** describes the relative numbers of individuals of different ages within a population. By combining a population's age structure with data on the reproductive potential of individuals in each age class, a population ecologist can predict whether the population will grow or shrink.

Many plants and animals grow steadily in size as they age, and in these species, older individuals often reproduce more. Older, larger trees in a population produce more seeds than smaller, younger trees. Larger, older fish produce more eggs than smaller, younger fish of the same species. Birds use the experience they gain with age to become more successful breeders at older ages. Human beings are unusual because we often survive past our reproductive years.

**(a) Random:** Distribution of organisms displays no pattern.

**(b) Uniform:** Individuals are spaced evenly.

**(c) Clumped:** Individuals concentrate in certain areas.

**FIGURE 3.12 Individuals in a population can spatially distribute themselves in three fundamental ways.**

A human population made up largely of older (post-reproductive) individuals will tend to decline over time, whereas one with many young people (of reproductive or pre-reproductive age) will tend to increase.

We will use diagrams to explore these ideas further in Chapter 6 (pp. 121–122) as we study human population growth. Understanding the fundamentals of population dynamics in plants and animals gives us a solid basis for studying our own population dynamics—and all the pressing social, cultural, and environmental issues related to human population growth.

## We can measure population growth

Let's now take a more quantitative approach by examining some simple mathematical concepts used by population ecologists and by **demographers** (scientists who study human populations). Population change is determined by four factors:

- *Natality* (births within the population)
- *Mortality* (deaths within the population)
- *Immigration* (arrival of individuals from outside the population)
- *Emigration* (departure of individuals from the population)

We can measure a population's **rate of natural increase** simply by subtracting the death rate from the birth rate:

$$(\text{birth rate}) - (\text{death rate}) = \text{rate of natural increase}$$

To obtain the **population growth rate,** the total rate of change in a population's size per unit time, we must also include the effects of immigration and emigration:

$$(\text{birth rate} - \text{death rate}) + (\text{immigration rate} - \text{emigration rate}) = \text{population growth rate}$$

The rates in these formulas are often expressed in numbers per 1000 individuals per year. So for example, a population with a birth rate of 18 per 1000/yr, a death rate of 10 per 1000/yr, an immigration rate of 5 per 1000/yr, and an emigration rate of 7 per 1000/yr would have a population growth rate of 6 per 1000/yr:

$$(18/1000 - 10/1000) + (5/1000 - 7/1000) = 6/1000$$

Given these rates, a population of 1000 in one year will reach 1006 in the next. If the population is 1,000,000, it will grow to 1,006,000 the next year. Such population increases are often expressed in percentages, as follows:

$$\text{population growth rate} \times 100\%$$

Thus, a growth rate of 6/1000 would be expressed as:

$$6/1000 \times 100\% = 0.6\%$$

By measuring population growth in percentages, we can compare changes among populations of far different sizes. We can also project changes into the future. Understanding such changes helps wildlife and fisheries managers to regulate hunting and fishing to ensure sustainable harvests, informs conservation biologists trying to protect rare and declining species, and assists policymakers planning for human population growth in cities, regions, and nations.

## Unregulated populations increase by exponential growth

When a population increases by a fixed percentage each year, it is said to undergo **exponential growth.** Imagine you put money in a savings account at a fixed interest rate and leave it untouched for years. Each time the principal accrues interest and grows, you earn still more interest the next time, and the sum grows by escalating amounts each year. The reason is that a fixed percentage of a small number makes for a small increase, but the same percentage of a larger number produces a larger increase. Thus, as savings accounts (or populations) grow larger, each incremental increase of the same percentage becomes larger in absolute terms. Such acceleration is characteristic of exponential growth.

We can visualize changes in population size by using population growth curves. The J-shaped curve in **FIGURE 3.13** shows exponential growth. Populations increase exponentially unless they meet constraints. Each organism reproduces by a certain amount, and as populations grow, there are more individuals reproducing by that amount. Thus, if there are adequate resources and no external limits, ecologists predict exponential growth.

Normally, exponential growth occurs in nature only when a population is small, competition is minimal, and environmental conditions are ideal for the organism in question. Most often, these circumstances hold when the organism arrives in a new environment containing abundant resources. Mold growing on a piece of fruit, or bacteria decomposing a dead animal, are cases in point. Plants colonizing regions during primary succession (p. 77) after glaciers recede or volcanoes erupt also show exponential growth. In Hawaii, many species that colonized the islands underwent exponential growth for a time after their arrival. One current example of exponential growth in North America is the Eurasian collared-dove (see Figure 3.13). Unlike its extinct relative the passenger pigeon, this species arrived here from Europe, thrives in human-disturbed areas, and has spread across the continent in a matter of years.

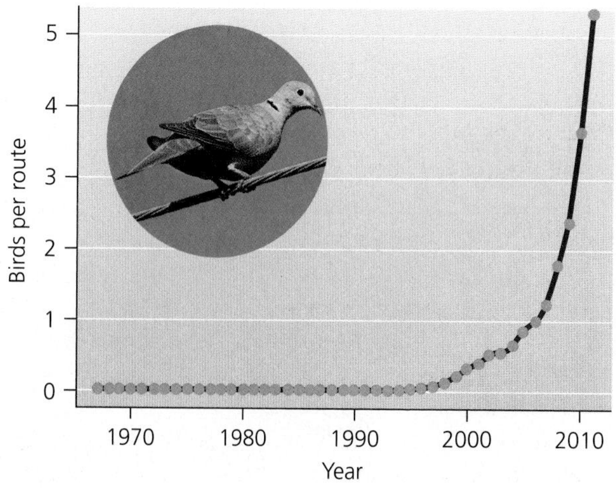


**FIGURE 3.13 A population may grow exponentially when colonizing an unoccupied environment or exploiting an unused resource.** The Eurasian collared-dove is spreading across North America, propelled by exponential growth. *Data from Sauer, J.R., et al., 2012.* The North American Breeding Bird Survey, results and analysis 1966–2011. *v. 12.13.2011. USGS Patuxent Wildlife Research Center, Laurel, MD.*

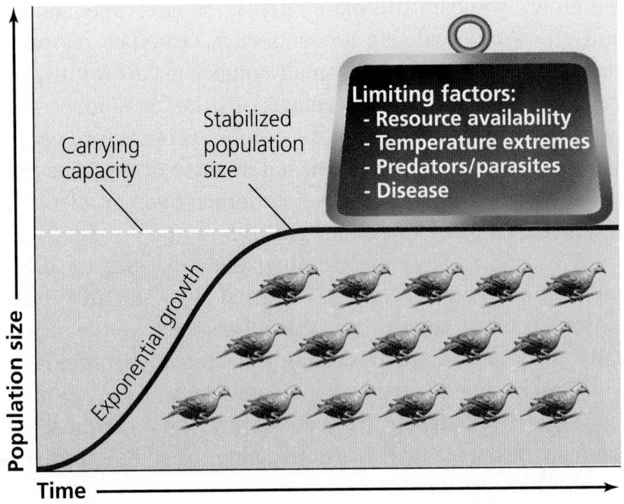


**FIGURE 3.14 The logistic growth curve shows how population size may increase rapidly at first, then slow, and finally stabilize at a carrying capacity.**

## Limiting factors restrain growth

Exponential growth rarely lasts long. If even a single species were to increase exponentially for very many generations, it would blanket the planet's surface! Instead, every population eventually is constrained by **limiting factors**—physical, chemical, and biological attributes of the environment that restrain population growth. These limiting factors determine the **carrying capacity,** the maximum population size of a species that a given environment can sustain.

Ecologists use the S-shaped curve in **FIGURE 3.14** to show how exponential growth is slowed and eventually brought to a standstill by limiting factors. Called the **logistic growth curve,** it rises sharply at first but then begins to level off as the effects of limiting factors become stronger. Eventually the collective force of these factors stabilizes the population size at its carrying capacity.

We can witness this process by taking a closer look at data for the Eurasian collared-dove, gathered by thousands of volunteer birders and analyzed by government biologists in the Breeding Bird Survey, a long-running citizen science project. The dove first appeared in Florida a few decades ago and then spread north and west. Today its numbers are growing fastest in western areas it has recently reached, and more slowly in eastern areas where it has been present for longer. In Florida, it has apparently reached carrying capacity (**FIGURE 3.15**). Populations of other European birds that spread across North America in the past, such as the house sparrow and European starling, have peaked and are today beginning to decline.

Many factors influence a population's growth rate and carrying capacity. For animals in terrestrial environments, limiting factors include temperature extremes; disease prevalence; predator abundance; and the availability of food, water, mates, shelter, and breeding sites. Plants are often limited by sunlight, moisture, soil chemistry, disease, and plant-eating animals. In aquatic systems, limiting factors include salinity, sunlight, temperature, dissolved oxygen, fertilizers, and pollutants.

A population's density can enhance or diminish the impact of certain limiting factors. Recall that high population density can help organisms find mates but may also worsen competition, predation, and disease. Such limiting factors are said to be **density-dependent** because their influence rises and falls with population density. The logistic growth curve in Figure 3.14 represents the effects of density dependence. The larger the population size, the stronger the effects of the limiting factors.

**Density-independent** factors are those whose influence is independent of population density. Temperature extremes and

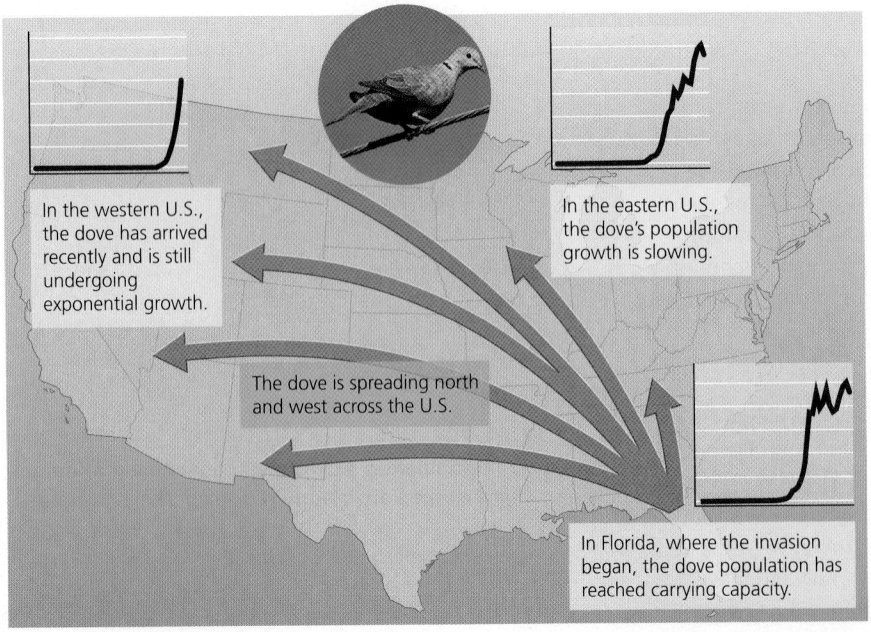


**FIGURE 3.15 Exponential growth slows over time and gives way to logistic growth.** By breaking down the continent-wide data for the Eurasian collared-dove from Figure 3.13, we can track its spread north and west from Florida, where it first arrived. Today its population growth is fastest in the west, slower in the east (where the species has been present longer), and stable in Florida (where it has apparently reached carrying capacity).
*Data from Sauer, J.R., et al., 2012.* The North American Breeding Bird Survey, results and analysis 1966–2011. *v. 12.13.2011. USGS Patuxent Wildlife Research Center, Laurel, MD.*

**DATA Q** Looking ahead several decades into the future, what do you predict the population growth graph for the Eurasian collared-dove in the western United States will look like?

GO TO **INTERPRETING GRAPHS & DATA** ON MasteringEnvironmentalScience®

catastrophic events such as floods, fires, and landslides are examples of density-independent factors, because they can eliminate large numbers of individuals without regard to their density.

The logistic curve is a simple model, and real populations in nature can behave differently. Some may cycle above and below the carrying capacity. Others may overshoot the carrying capacity and then crash, destined either for extinction or recovery.

## Carrying capacities can change

Because environments are complex and ever-changing, carrying capacities can vary. If a fire destroys a forest, the carrying capacities for most forest animals will decline, while carrying capacities for species that benefit from fire will increase. Our own species has proved capable of intentionally altering our environment so as to raise our carrying capacity. When our ancestors began to build shelters and use fire for heating and cooking, they eased the limiting factors of cold climates and were able to expand into new territory. As human civilization developed, we overcame limiting factors time and again by creating new technologies and cultural institutions. People have managed so far to increase the planet's carrying capacity for our species, but we have done so by appropriating immense proportions of the planet's natural resources. In the process, we have reduced carrying capacities for countless other organisms that rely on those same resources.

### WEIGHING THE ISSUES

**Carrying Capacity and Human Population Growth** The global human population recently surpassed 7 billion, far exceeding our population's size throughout our species' long history on Earth. Name some specific means by which we have apparently raised Earth's carrying capacity for our species. Do you think we can continue to raise our carrying capacity? How might we do so? Are the ways in which we might accommodate more people in the future the same as the ways in which we have expanded our population so far? What limiting factors exist for our population today? Might Earth's carrying capacity for us decrease? Why or why not?

# Conserving Biodiversity

Populations have always been affected by environmental change, but today human development, resource extraction, and population pressure are speeding the rate of change and bringing new types of impacts. Fortunately, committed people are taking action to safeguard biodiversity and to preserve and restore Earth's ecological and evolutionary processes (as we shall see more fully in Chapter 8).

**FIGURE 3.16 Hawaii protects some of its diverse natural areas, helping to stimulate its economy with ecotourism.** Here, a scuba diver observes raccoon butterflyfish at a coral reef along the Kona coast of Hawaii's Big Island.

## Innovative solutions are working

Amid all the challenges of Hawaii's extinction crisis, hard work is resulting in some inspirational success stories, and several species have been saved from imminent extinction. At Hakalau Forest, ranchland is being restored to forest, invasive plants are being removed, native ones are being planted, and nēnē are being protected while new populations of them are being established.

Early work at Hawai'i Volcanoes National Park inspired the work at Hakalau, as well as efforts by managers and volunteers from the Hawaii Division of Forestry and Wildlife, The Nature Conservancy of Hawai'i, Kamehameha Schools, and local watershed protection groups. Across Hawaii, people are protecting land, removing alien mammals and weeds, and restoring native habitats. Offshore, Hawaiians are striving to protect their fabulous coral reefs, seagrass beds, and beaches from pollution and overfishing. The northwesternmost Hawaiian Islands are now part of the largest federally declared marine reserve (p. 276) in the world.

Hawaii's citizens are reaping economic benefits from their conservation efforts. The islands' wildlife and natural areas draw visitors from around the world, a phenomenon called **ecotourism** (**FIGURE 3.16**). A large proportion of Hawaii's tourism is ecotourism, and altogether tourism draws over 7 million visitors to Hawaii each year, creates thousands of jobs, and pumps $12 billion annually into the state's economy.

## Climate change poses an extra challenge

Traditionally, people sought to conserve populations of threatened species by preserving and managing tracts of land (or areas of ocean) designated as protected areas. However, global climate change (Chapter 14) now threatens this strategy. As temperatures climb and rainfall patterns shift, conditions within protected areas may become unsuitable for the species they were meant to protect.

Hawaii's systems are especially vulnerable. At Hakalau Forest on the slopes of Mauna Kea, mosquitoes and malaria are moving upslope toward the refuge as temperatures rise, exposing more and more birds to disease. Some research suggests that climate change will also reduce rainfall here, pushing the upper limit of the forest downward. If this comes to pass, Hakalau's honeycreepers may become trapped within a shrinking band of forest by disease from below and drought from above.

On the Hawaiian island of Kaua'i, the outlook is worse: Forests there are closer to the mountaintops, and climate warming is expected to shift the forests upward until they vanish. Already mosquitoes have moved upslope and bird populations are diminishing. Two honeycreeper species, the 'akeke'e and the 'akikiki, were recently added to the Endangered Species List (p. 177).

The challenges posed by climate change mean that scientists and managers need to come up with new ways to help save declining populations. We will learn about the many efforts being made across the world in our exploration of conservation biology in Chapter 8. In Hawaii, management and ecotourism can help conserve natural systems, but resources to preserve habitat and protect endangered species will likely need to be stepped up. Restoring communities—as at Hakalau Forest—will also be necessary. The restoration of ecological communities is one phenomenon we will examine in our next chapter, as we shift from populations to communities.

## Conclusion

The honeycreepers of Hakalau Forest National Wildlife Refuge, along with many other Hawaiian species, have helped to illuminate the fundamentals of evolution and population ecology that are integral to environmental science. Earth's biodiversity is continually shaped by the evolutionary processes of natural selection, speciation, and extinction. Understanding how ecological processes function at the population level is crucial to protecting biodiversity threatened by the mass extinction event that many biologists maintain is now underway. Population ecology also informs the study of human populations (Chapter 6), another key endeavor in environmental science.

# Testing Your Comprehension

1. Define the concept of natural selection in your own words, and explain how it follows logically from a few common observations of nature.
2. Describe an example of evidence for natural selection and an example of evidence for artificial selection.
3. Describe the steps involved in allopatric speciation.
4. Name two organisms that have become extinct or are threatened with extinction. For each, give a probable reason for its decline.
5. Define the terms *species, population,* and *community*. How does a species differ from a population? How does a population differ from a community?
6. Define and contrast the concepts of habitat and niche.
7. List and describe each of the five major population characteristics discussed. Briefly explain how each shapes population dynamics.
8. Can a species undergo exponential growth forever? Explain your answer.
9. Describe how limiting factors affect carrying capacity.
10. What are some advantages of ecotourism for a state like Hawaii? Can you think of any potential disadvantages?

# Seeking Solutions

1. In what ways has artificial selection changed people's quality of life? Provide examples. How might artificial selection be used to improve our quality of life further? Can you envision a way it could be used to reduce our environmental impact?
2. In your region, what species are threatened with extinction? Why are they vulnerable? Suggest steps that could be taken to increase their populations.
3. Do you think the human species can continue raising its global carrying capacity? Why or why not? Do you think we *should* try to keep raising our carrying capacity? Why or why not?
4. Describe two of the threats facing native species in Hawaii, and two actions people have taken to address these threats. What new trend is now jeopardizing some of these efforts? What steps do you think might be required in the future if we are to safeguard native species in mountainous habitats in places like Hawaii?

5. **THINK IT THROUGH** You are a population ecologist studying animals in a national park, and park managers are asking for advice on how to focus their limited conservation funds. How might you rate the following three species, from most vulnerable (and thus most in need of conservation attention) to least vulnerable? Give reasons for your choices.
   - A bird that is a generalist in its use of habitats and resources
   - A salamander endemic to the park that lives in high-elevation forest
   - A fish that specializes on a few types of invertebrate prey and has a large population size

# Calculating Ecological Footprints

Professional demographers delve deeply into the latest statistics from cities, states, and nations to bring us updated estimates on human populations. The table shows population estimates for two consecutive years that take into account births, deaths, immigration, and emigration. Simply divide the 2013 data by the 2012 data, subtract 1.00, and multiply by 100, and you will obtain the population growth rate (p. 61) for each area.

Texas was one of the fastest-growing U.S. states during this particular one-year period, and Pennsylvania was one of the slowest-growing states. You can find data for your own state, city, or metropolitan area by exploring the Web pages of the U.S. Census Bureau.

| Region | 2012 population | 2013 population | Population growth rate |
|---|---|---|---|
| Hawaii | 1,390,090 | 1,404,054 | |
| Texas | 26,060,796 | 26,448,193 | |
| Pennsylvania | 12,764,475 | 12,773,801 | |
| Your state or city | | | |
| United States | 313,873,685 | 316,128,839 | 0.72% |
| World | 6,986,951,000 | 7,057,075,000 | |

*Data from U.S. Census Bureau and Population Reference Bureau.*

1. What would the population of the United States have been in 2014, assuming its population growth rate remained at the calculated rate?
2. The birth rate of the United States during this period was 13 births per 1,000 people, and the death rate was 8 deaths per 1,000 people. Using the formula on p. 61, what was the rate of natural increase for the United States between 2012 and 2013? Now subtract this rate from the population growth rate to obtain the net migration rate. Was the United States experiencing more immigration or more emigration during this year?
3. At the growth rate you calculated for Texas, the population of that state will double in just 47 years. What impacts would you expect this to have on (1) food supplies, (2) drinking water supplies, (3) forests and other natural areas, and (4) wildlife populations?
4. How does your own state, city, or metropolitan area compare in its growth rate with other regions in the table? What steps could your region take to lessen potential impacts of population growth on (1) food supplies, (2) drinking water supplies, (3) forests and other natural areas, and (4) wildlife populations?

## MasteringEnvironmentalScience®

**STUDENTS**

Go to **MasteringEnvironmentalScience** for assignments, the etext, and the Study Area with practice tests, videos, current events, and activities.

**INSTRUCTORS**

Go to **MasteringEnvironmentalScience** for automatically graded activities, current events, videos, and reading questions that you can assign to your students, plus Instructor Resources.

# 4

# Species Interactions and Community Ecology

**Upon completing this chapter, you will be able to:**

- ☐ Compare and contrast the major types of species interactions
- ☐ Characterize feeding relationships and energy flow, using them to identify trophic levels and construct food webs
- ☐ Distinguish characteristics of a keystone species
- ☐ Characterize disturbance, succession, and notions of community change
- ☐ Perceive the potential impacts of invasive species in communities, and offer solutions to biological invasions
- ☐ Explain the goals and methods of restoration ecology
- ☐ Identify and describe the major terrestrial biomes of the world



**Photo: Workers use a pressure hose to remove zebra mussels from the pump room of a power plant near Detroit.**

# CENTRAL CASE STUDY

# Black and White, and Spread All Over: Zebra Mussels Invade the Great Lakes

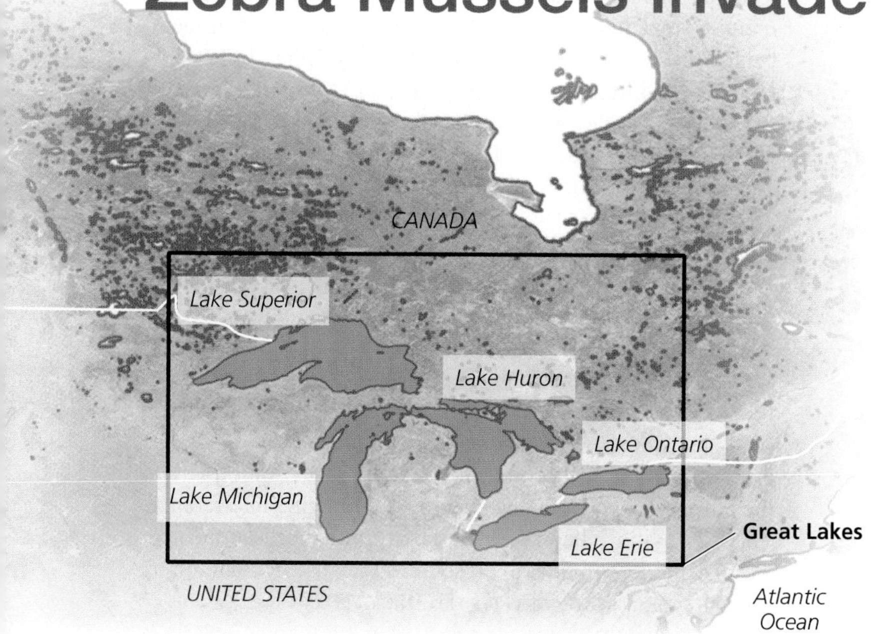


*"We are seeing changes in the Great Lakes that are more rapid and more destructive than any time in [their] history."*

**—Andy Buchsbaum, National Wildlife Federation**

*"When you tear away the bottom of the food chain, everything that is above it is going to be disrupted."*

**—Tom Nalepa, National Oceanic and Atmospheric Administration**

Things were looking up for the Great Lakes. Their waters were becoming cleaner and cleaner as regulation by the U.S. and Canadian governments brought industrial pollution under control starting in the 1970s. People started venturing back onto the lakes for recreation, and populations of fish began to rebound.

Then the zebra mussel arrived. Black-and-white shellfish the size of a dime, zebra mussels attach to hard surfaces and feed on algae by filtering water through their gills. This mollusk is native to the Caspian and Black Seas where Europe meets Asia, but in 1988 it was discovered in North America in Lake St. Clair, which connects Lake Erie with Lake Huron. People brought the shellfish to this continent by accident when ships from Europe discharged ballast water containing the mussels or their larvae.

Within just two years of their discovery, zebra mussels had multiplied and reached all five Great Lakes. The next year, they invaded New York's Hudson River to the east, and the Illinois River at Chicago to the west. From the Illinois River and its canals, they reached the Mississippi River, gaining access to a vast watershed covering 40% of the United States. In just three more years, they spread to 19 states and two provinces. By 2010, they had colonized waters in 30 U.S. states.

How could the zebra mussel spread so quickly? Its larval stage is well adapted for long-distance dispersal: The tiny larvae drift freely for several weeks, traveling as far as currents take them. Adults that adhere to boats and ships may be transported from place to place, even overland to isolated lakes and ponds. Once established, the mussels had free rein: In North America they were free of the predators, competitors, and parasites that had evolved with them in the Old World and limited their population growth there.

Why all the fuss? For one thing, zebra mussels clog water intake pipes at factories, power plants, municipal water supplies, and wastewater facilities (**FIGURE 4.1**). At one Michigan power plant, workers counted 700,000 mussels per square meter of pipe. Clusters of these organisms also damage boat engines, degrade docks, foul fishing gear, and sink buoys that ships use for navigation. In these ways, zebra mussels cost Great Lakes economies an estimated $5 billion in the first decade of the invasion and continue to impose costs of hundreds of millions of dollars each year.

Zebra mussels also exert severe ecological impacts. They eat primarily phytoplankton (p. 22), microscopic photosynthetic algae, protists, and cyanobacteria that drift in open water. Because each mussel filters a liter or more of water

**FIGURE 4.1 Zebra mussels clog water intake pipes of industrial facilities.**

every day, zebra mussels can consume enough phytoplankton to deplete populations. Phytoplankton is the foundation of the Great Lakes food web, so its depletion is bad news for zooplankton (p. 24), the tiny aquatic animals that eat phytoplankton—and for the fish that eat both. Researchers find that water bodies with zebra mussels contain fewer zooplankton and open-water fish than water bodies without them. And although zebra mussels benefit many bottom-feeding animals, they can suffocate native mollusks by attaching to their shells.

Zebra mussels are just one of many invasive species affecting ecological communities and human economies today—but the zebra mussel story has a few extra twists. For one, on the heels of its invasion came another: The quagga mussel, a close relative of the zebra mussel from Ukraine, is also spreading through the Great Lakes and beyond. This species, named after an extinct type of zebra, is replacing the zebra mussel in many locations.

In a second twist, today there are signs that the zebra mussel invasion may be losing steam. In a number of areas, zebra mussel populations have apparently peaked and begun to decline. In some cases they are being displaced by quagga mussels, but in others predation by fish, crabs, or ducks may be driving down their numbers as native predators develop a taste for the invader.

In areas where zebra mussels are declining, some of the affected native fish and invertebrates are now recovering. This is welcome news, especially since the Great Lakes today are being bombarded with a variety of other non-native species, including several voracious species of Eurasian carp that are eating their way through the lakes' native flora and fauna.

Ecologists are monitoring populations closely to see how things develop. No one expects zebra mussels to disappear, but there is now hope that some of their impacts may be reversed and that the ecological communities of the Great Lakes can begin to be restored. ◻

# Species Interactions

By interacting with many species in a variety of ways, zebra mussels and quagga mussels have set in motion an array of changes in the communities they have invaded. Interactions among species are the threads in the fabric of ecological communities. Ecologists organize species interactions into several main categories (**TABLE 4.1**).

**TABLE 4.1 Species Interactions: Effects on Each Participant**

| TYPE OF INTERACTION | EFFECT ON SPECIES 1 | EFFECT ON SPECIES 2 |
|---|---|---|
| Competition | – | – |
| Predation, parasitism, herbivory | + | – |
| Mutualism | + | + |

"+" denotes a positive effect; "–" denotes a negative effect.

## Competition can occur when resources are limited

When multiple organisms seek the same limited resource, their relationship is said to be one of **competition.** Competing organisms do not usually fight with one another directly and physically. Competition is generally more subtle and indirect, involving organisms vying with one another to procure resources. Such resources include food, water, space, shelter, mates, sunlight, and more. Competitive interactions can take place between members of the same species (*intraspecific competition*) or between members of different species (*interspecific competition*).

If one species is a very effective competitor, it may exclude another species from resource use entirely. This occurred in parts of the Great Lakes as zebra mussels displaced native mussels—and it is happening now as quagga mussels displace zebra mussels. Alternatively, species may be able to coexist, adapting over evolutionary time as natural selection (p. 48) favors individuals that use slightly different resources or that use shared resources in different ways. For example, if two bird species eat the same type of seeds, natural selection might drive one species to specialize on larger seeds and the other to specialize on smaller seeds. Or one bird species might become more active in the morning and the other more active in the evening, minimizing interference. This process is called **resource partitioning** because the species partition, or divide, the resources they use in common by specializing in different ways (**FIGURE 4.2**).

In competitive interactions, each participant exerts a negative effect on other participants by taking resources the others could have used. This is reflected in the two minus

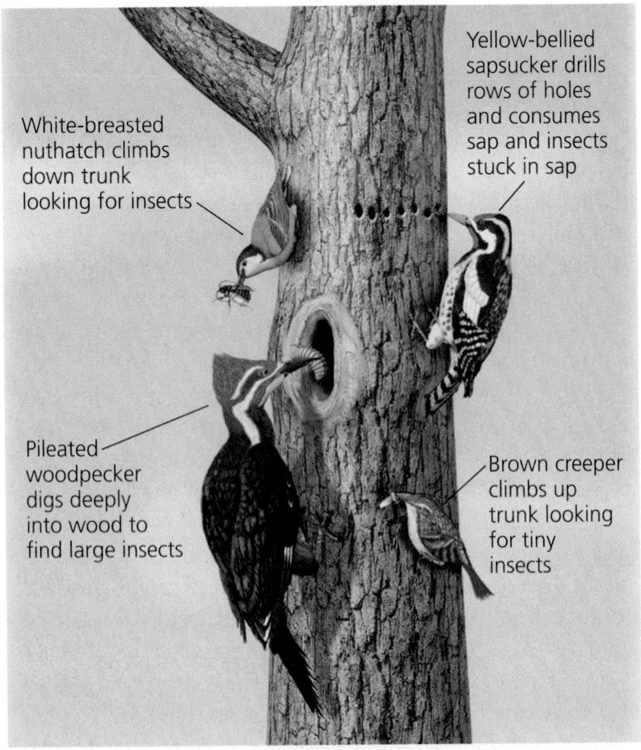
Yellow-bellied sapsucker drills rows of holes and consumes sap and insects stuck in sap
White-breasted nuthatch climbs down trunk looking for insects
Pileated woodpecker digs deeply into wood to find large insects
Brown creeper climbs up trunk looking for tiny insects

**FIGURE 4.2 When species compete, they may partition resources.** Birds that forage for insects on tree trunks use different portions of the trunk and seek different foods in different ways.

**FIGURE 4.3 Predation, parasitism, and herbivory are exploitative interactions in which one participant benefits at the expense of another.** In predation **(a)**, a predator kills and eats prey. In parasitism **(b)**, a parasite gains nourishment while harming its host. In herbivory **(c)**, an animal feeds on plants.

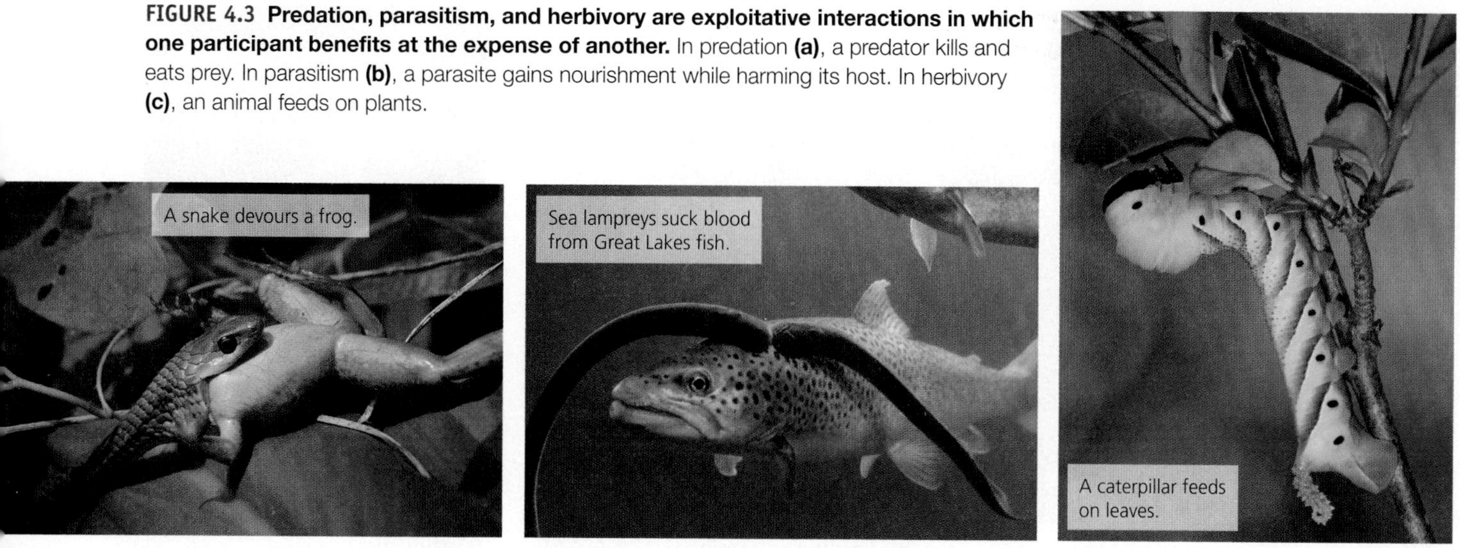

A snake devours a frog.
Sea lampreys suck blood from Great Lakes fish.
A caterpillar feeds on leaves.


**(a) Predation** **(b) Parasitism** **(c) Herbivory**

signs shown for competition in Table 4.1. In other types of interactions, some participants benefit while others are harmed; that is, one species exploits the other (note the +/– interactions in Table 4.1). Such exploitative interactions include predation, parasitism, and herbivory (**FIGURE 4.3**).

## Predators kill and consume prey

Every living thing needs to procure food, and for most animals, this means eating other organisms. **Predation** is the process by which individuals of one species—the *predator*—hunt, capture, kill, and consume individuals of another species, the *prey* (see Figure 4.3a). Interactions among predators and prey structure the food webs we will examine shortly, and they help shape community composition by influencing the relative numbers of predators and prey.

Zebra mussels consume the smaller types of zooplankton, and this predation has diminished zooplankton populations by up to 70% in Lake Erie and the Hudson River. Most predators are also prey, however. Zebra mussels have become a food source for muskrats, crayfish, and a variety of North American fish and ducks.

Predation can sometimes drive population dynamics. An increase in the population size of prey creates more food for predators, which may survive and reproduce more effectively as a result. As the predator population rises, intensified predation drives down the population of prey. Diminished numbers of prey in turn cause some predators to starve, and the predator population declines. This allows the prey population to rise again, starting the cycle anew (**FIGURE 4.4**).

Predation also has evolutionary ramifications. Individual predators that are more adept at capturing prey may live longer lives, reproduce more, and be better providers for their offspring. Natural selection (p. 48) will thereby lead to the evolution of adaptations (p. 48) that enhance hunting skills. Prey, however, face a stronger selective pressure—the risk of immediate death. As a result, predation pressure has driven the evolution of an elaborate array of defenses against being eaten (**FIGURE 4.5**, next page).

## Parasites exploit living hosts

Organisms can exploit other organisms without killing them. **Parasitism** is a relationship in which one organism, the *parasite*, depends on another, the *host*, for nourishment or some other benefit while doing the host harm (see Figure 4.3b). Unlike predation, parasitism usually does not result in an organism's immediate death.

Many types of parasites live inside their hosts. For example, tapeworms live in their hosts' digestive tracts, robbing them of nutrition. The larvae of parasitoid wasps burrow into the tissues of caterpillars and consume them from the inside. Other parasites are free-living. Cuckoos of Eurasia and cowbirds of the Americas lay their eggs in other birds' nests and let the host raise their young. Still other parasites live on the exterior of their hosts, such as fleas or ticks that suck

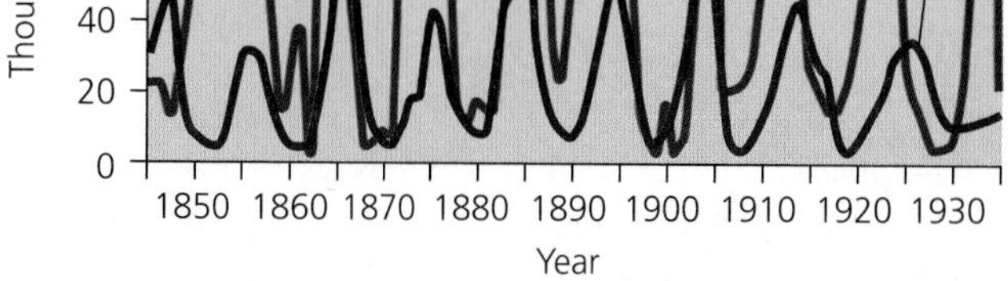

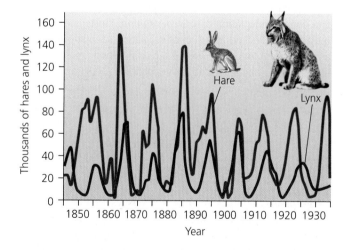

**FIGURE 4.4 Predator–prey systems sometimes show paired cycles.** A classic case is that of snowshoe hares and the lynx that prey on them in Canada. These data come from fur-trapping records of the Hudson Bay Company and represent numbers of each animal trapped. *Data from MacLulich, D.A., 1937. Fluctuation in the numbers of varying hare* (Lepus americanus)*. Univ. Toronto Stud. Biol. Ser. 43, Toronto: University of Toronto Press.*

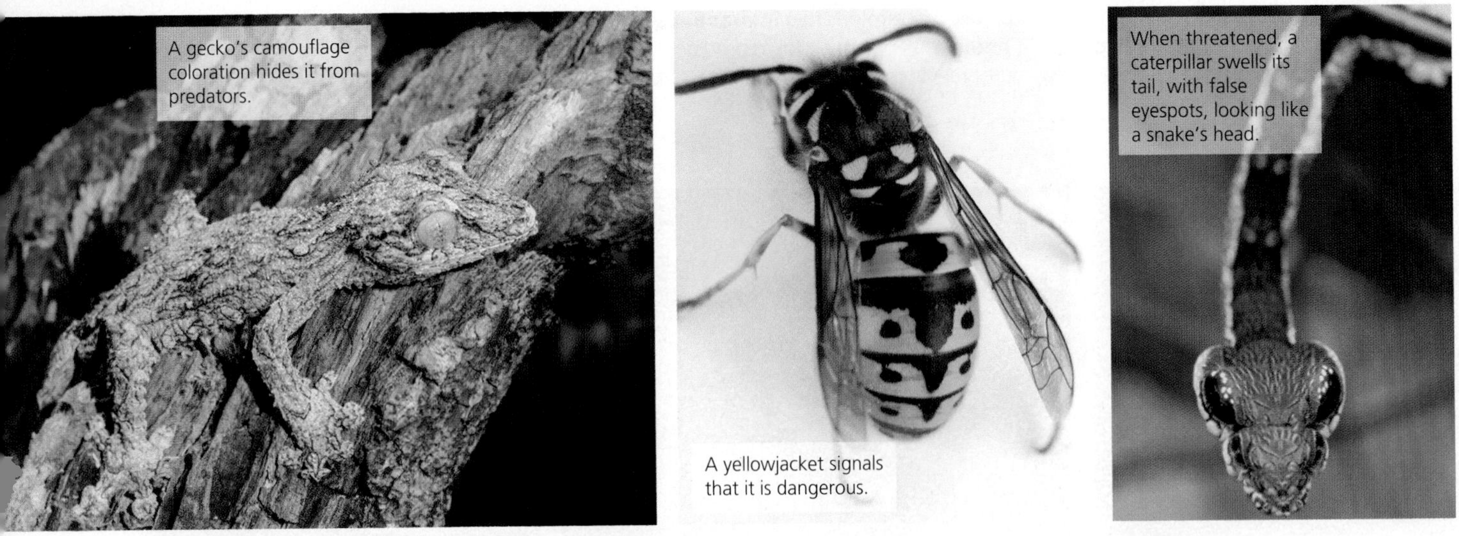

A gecko's camouflage coloration hides it from predators.
A yellowjacket signals that it is dangerous.
When threatened, a caterpillar swells its tail, with false eyespots, looking like a snake's head.
(a) Crypsis
(b) Warning coloration
(c) Mimicry


**FIGURE 4.5 Natural selection to avoid predation has resulted in fabulous adaptations.** Some prey species use cryptic coloration **(a)** to blend into their background. Others are brightly colored **(b)** to warn predators they are toxic, distasteful, or dangerous. Others use mimicry **(c)** to fool predators.

blood through the skin. The sea lamprey is a tube-shaped vertebrate that grasps fish with its suction-cup mouth and rasping tongue, sucking their blood for days or weeks (see Figure 4.3b). Sea lampreys invaded the Great Lakes from the Atlantic Ocean after people dug canals to connect the lakes for shipping, and the lampreys soon devastated economically important fisheries of chubs, lake herring, whitefish, and lake trout.

Parasites that cause disease in their hosts are called *pathogens*. Common human pathogens include the protists that cause malaria and amoebic dysentery, the bacteria that cause pneumonia and tuberculosis, and the viruses that cause hepatitis and AIDS.

Just as predators and prey evolve in response to one another, so do parasites and hosts, in a reciprocal process of adaptation and counter-adaptation called *coevolution*. Hosts and parasites may become locked in a duel of escalating adaptations, known as an *evolutionary arms race*. Like rival nations racing to stay ahead of one another in military technology, host and parasite repeatedly evolve new responses to the other's latest advance. In the long run, though, it may not be in a parasite's best interest to do its host too much harm. In many cases, a parasite may leave more offspring by allowing its host to live longer.

## Herbivores exploit plants

In **herbivory,** animals feed on the tissues of plants. Insects that feed on plants are the most common type of herbivore; nearly every plant in the world is attacked by insects (see Figure 4.3c). Herbivory generally does not kill a plant outright but may affect its growth and reproduction.

Like animal prey, plants have evolved an impressive arsenal of defenses against the animals that feed on them. Many plants produce chemicals that are toxic or distasteful to herbivores. Others arm themselves with thorns, spines, or irritating hairs. In response, herbivores evolve ways to overcome these defenses, and the plant and the animal may embark on an evolutionary arms race.

## Mutualists help one another

Unlike exploitative interactions, **mutualism** is a relationship in which two or more species benefit from interacting with one another. Generally each partner provides some resource or service that the other needs.

Many mutualistic relationships—like many parasitic ones—occur between organisms that live in close physical contact. Physically close association is called **symbiosis,** and symbiosis can be either mutualistic or parasitic. Most terrestrial plant species depend on mutualisms with fungi; plant roots and some fungi together form symbiotic associations called mycorrhizae. In these relationships, the plant provides energy and protection to the fungus while the fungus helps the plant absorb nutrients from the soil. In the ocean, coral polyps, the tiny animals that build coral reefs (p. 258), provide housing and nutrients for specialized algae in exchange for food the algae produce through photosynthesis (pp. 31–32).

You, too, are engaged in a symbiotic mutualism. Your digestive tract is filled with microbes that help you digest food and carry out other bodily functions—microbes that you are providing a place to live. Without these mutualistic microbes, none of us would survive for long.

Not all mutualists live in close proximity. **Pollination** (p. 149) involves free-living organisms that may encounter each other only once. Bees, birds, bats, and other creatures transfer pollen (containing male sex cells) from flower to flower, fertilizing ovaries (containing female sex cells) that grow into fruits with seeds. These pollinators consume pollen or nectar (a reward the plant uses to entice them), and the plants are pollinated and reproduce. Bees pollinate three-fourths of our crops—from soybeans to potatoes to tomatoes to beans to cabbage to oranges.

# Ecological Communities

A **community** is an assemblage of populations of organisms living in the same area at the same time (as we saw in Figure 3.9, p. 55). Members of a community interact in the ways discussed above, and these species interactions have indirect effects that ripple outward to affect other community members. The species interactions just discussed help determine the structure, function, and species composition of communities. Community ecology (p. 56) is the scientific study of species interactions and the dynamics of communities. Community ecologists study which species coexist, how they interact, how communities change through time, and why these patterns occur.

## Energy passes among trophic levels

Some of the most important interactions among community members involve who eats whom. As organisms feed on one another, matter and energy move through the community from one **trophic level,** or rank in the feeding hierarchy, to another (**FIGURE 4.6**; and compare Figure 2.14, p. 33).

**Producers** *Producers*, or *autotrophs* ("self-feeders," p. 31), comprise the first trophic level. Terrestrial green plants, cyanobacteria, and algae capture solar energy and use photosynthesis to produce sugars (p. 32). The chemosynthetic bacteria of hydrothermal vents use geothermal energy in a similar way to produce food (p. 259).

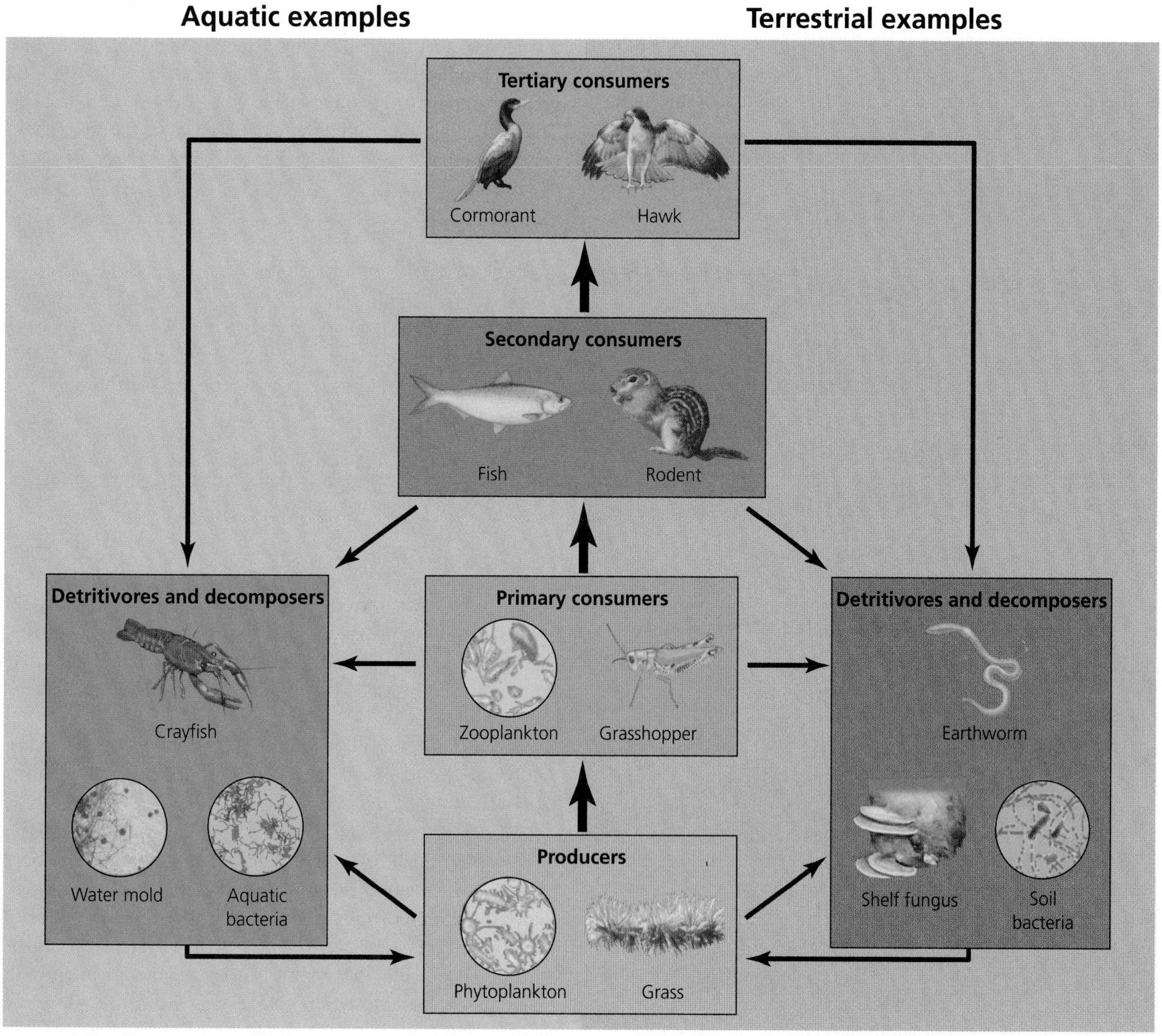


**FIGURE 4.6 Ecologists organize species hierarchically by trophic level.** The diagram shows aquatic **(left)** and terrestrial **(right)** examples at each level. Arrows indicate the direction of energy flow. Producers generate food by photosynthesis, primary consumers (herbivores) feed on producers, secondary consumers eat primary consumers, and tertiary consumers eat secondary consumers. Detritivores and decomposers feed on nonliving organic matter and "close the loop" by returning nutrients to the soil or the water column for use by producers.

**Consumers** Organisms that consume producers are known as *primary consumers* and comprise the second trophic level. Herbivorous grazing animals, such as deer and grasshoppers, are primary consumers. The third trophic level consists of *secondary consumers*, which prey on primary consumers. Wolves that prey on deer are secondary consumers, as are rodents and birds that prey on grasshoppers. Predators that feed at still higher trophic levels are known as *tertiary consumers*. Examples of tertiary consumers include hawks and owls that eat rodents that have eaten grasshoppers.

**Detritivores and decomposers** Detritivores and decomposers consume nonliving organic matter. *Detritivores*, such as millipedes and soil insects, scavenge the waste products or dead bodies of other community members. *Decomposers*, such as fungi and bacteria, break down leaf litter and other nonliving matter into simpler constituents that can be taken up and used by plants. These organisms enhance the topmost soil layers (pp. 139–140) and play essential roles as the community's recyclers, making nutrients from organic matter available for reuse by living members of the community.

In Great Lakes communities, phytoplankton are the main producers, conducting photosynthesis with sunlight that penetrates the water. Zooplankton are primary consumers, feeding on the phytoplankton. Fish that eat phytoplankton are primary consumers, and fish that eat zooplankton are secondary consumers. Tertiary consumers include birds and larger fish that feed on plankton-eating fish. (The left side of Figure 4.6 shows these relationships in very generalized form.) Zebra mussels and quagga mussels, by eating both phytoplankton and zooplankton, function on multiple trophic levels. When an organism dies and sinks to the bottom, detritivores scavenge its tissues and decomposers recycle its nutrients.

## Energy, numbers, and biomass decrease at higher trophic levels

At each trophic level, organisms use energy in cellular respiration (p. 32), and most of the energy ends up being given off as heat. Only a small portion of the energy is transferred to the next trophic level through predation, herbivory, or parasitism. A general rule of thumb is that each trophic level contains about 10% of the energy of the trophic level below it (although the actual proportion varies greatly). This pattern can be visualized as a pyramid (**FIGURE 4.7**).

This pattern also tends to hold for numbers of organisms; in general, fewer organisms exist at high trophic levels than at low ones. A grasshopper eats many plants in its lifetime, a rodent eats many grasshoppers, and a hawk eats many rodents. Thus, for every hawk there must be many rodents, still more grasshoppers, and an immense number of plants. Because the difference in numbers of organisms among trophic levels tends to be large, the same pyramid-like relationship often holds true for **biomass,** the collective mass of living matter.

This pyramid pattern illustrates why eating at lower trophic levels—being vegan or vegetarian, for instance—decreases a person's ecological footprint. Each amount of meat or other animal product we eat requires the input of a considerably greater amount of plant material (see Figure 7.15, p. 150). Thus, when we eat animal products, we use up far more energy per calorie that we gain than when we eat plant products.

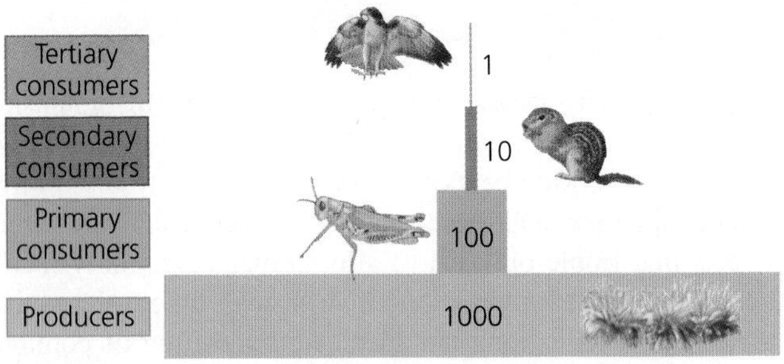


**FIGURE 4.7 Lower trophic levels generally contain more organisms, biomass, and energy content than higher trophic levels.** The 10:1 ratio shown here is typical, but varies greatly.

DATA Q Using the ratios shown in this example, let's suppose that a system has 3000 grasshoppers. How many rodents would you expect?

GO TO **INTERPRETING GRAPHS & DATA** ON MasteringEnvironmentalScience®

### WEIGHING THE ISSUES

**THE FOOTPRINTS OF OUR DIETS** What proportion of your diet would you estimate consists of meat, milk, eggs, or other animal products? Would you choose to decrease this proportion in order to reduce your ecological footprint? Describe other ways in which you could reduce your footprint through your food choices.

## Food webs show feeding relationships and energy flow

As energy is transferred from lower trophic levels to higher ones, it is said to pass up a **food chain,** a linear series of feeding relationships. Plant, grasshopper, rodent, and hawk make up a food chain—as do phytoplankton, zooplankton, fish, and fish-eating birds.

Thinking in terms of food chains is conceptually useful, but ecological systems are far more complex than simple linear chains. A more accurate representation of the feeding relationships in a community is a **food web**—a visual map of energy flow that shows the many paths along which energy passes as organisms consume one another.

**FIGURE 4.8** portrays a food web from a temperate deciduous forest of eastern North America. It is greatly simplified and leaves out the vast majority of species and interactions that occur. Note, however, that even within this simplified diagram we can pick out a number of food chains involving different sets of species. For instance, grasses are eaten by deer mice that may be consumed by black rat snakes—while in another food chain, blackberry leaves are consumed by caterpillars that provide food for spiders, which in turn may be eaten by American toads.

**FIGURE 4.8 Food webs represent feeding relationships in a community.** This food web shows organisms on several trophic levels in eastern North America's temperate deciduous forest. Arrows indicate the direction of energy flow as a result of predation, parasitism, or herbivory. Like most such diagrams, this one is simplified; the actual community contains many more species and interactions than can be shown.

A Great Lakes food web would include phytoplankton that photosynthesize, zooplankton that eat them, fish that eat phytoplankton and zooplankton, larger fish that eat the smaller fish, and lampreys that parasitize the fish. It would include a number of native mussels and clams and, since 1988, the zebra and quagga mussels that are displacing them. It would include diving ducks that formerly relied on native bivalves and now prey on the mussels. It would also show that crayfish and other bottom-dwelling invertebrate animals feed from the refuse of the mussels. Finally, the food web would include underwater plants and macroscopic algae, whose growth is enhanced as the non-native mussels filter out phytoplankton, allowing sunlight to penetrate deeply into the water column. (Jump ahead to Figure 4.11a, p. 79, for an illustration of these effects.)

Overall, zebra and quagga mussels alter Great Lakes food webs by shifting productivity from open-water regions to *benthic* (bottom) and *littoral* (nearshore) regions. In so doing, the mussels help benthic and littoral fishes and make life harder for open-water fishes (see **THE SCIENCE BEHIND THE STORY**, pp. 74–75).

## Some organisms play outsized roles

"Some animals are more equal than others," George Orwell wrote in his classic novel *Animal Farm*. Orwell was making wry sociopolitical commentary, but his remark hints at an ecological truth. In communities, ecologists have found, some species exert greater influence than do others. A species that

THE SCIENCE BEHIND THE STORY

# Determining Zebra Mussels' Impacts on Fish Communities

When zebra mussels appeared in the Great Lakes, people feared for sport fisheries and predicted that fish population declines could cost billions of dollars. The mussels would deplete the phytoplankton and zooplankton that fish relied on for food, people reasoned.

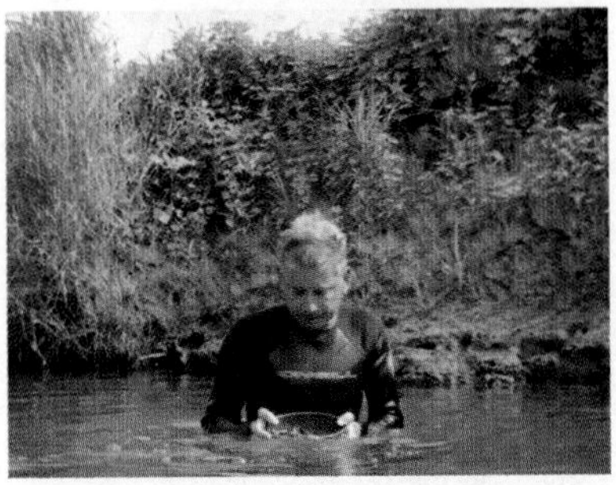

**Dr. David Strayer samples aquatic invertebrates**

However, food webs are complicated, and disentangling them to infer the impacts of any one species is difficult.

Thus, even 15 years after the arrival of zebra mussels, there was no solid evidence of widespread harm to fish populations.

So, aquatic ecologist David Strayer of the Institute of Ecosystem Studies in Millbrook, New York, joined Kathryn Hattala and Andrew Kahnle of New York State's Department of Environmental Conservation (DEC). They mined data sets on fish populations in the Hudson River, which zebra mussels had invaded in 1991.

Strayer and other scientists had been studying this community for years. Their data showed that after zebra mussels invaded the Hudson:

- Biomass of phytoplankton fell by 80%.
- Biomass of small zooplankton fell by 76%.
- Biomass of large zooplankton fell by 52%.

Zebra mussels increased filter-feeding in the community 30-fold, depleting phytoplankton and small zooplankton and leaving larger zooplankton with less to eat. Overall, zooplankton and invertebrate animals of the open water declined by 70%.

However, Strayer had also found that *benthic* (bottom-dwelling) invertebrates in shallow water (especially in the nearshore, or *littoral,* zone) had increased, because the mussels' shells provide habitat structure and their feces provide nutrients.

These contrasting trends in the benthic shallows and the open deep water led Strayer's team to hypothesize that zebra mussels would harm open-water fish that ate plankton but would help littoral-feeding fish. They predicted that larvae and juveniles of open-water fish species would decline in number, grow more slowly, and shift downriver toward saltier water, where mussels are absent. Conversely, they predicted that larvae and juveniles of littoral fish species would increase in number, grow more quickly, and shift upriver to regions of high zebra mussel density.

To test their predictions, the researchers analyzed data from fish surveys carried out by DEC scientists over 26 years. Strayer's team compared data on abundance, growth, and distribution of young fish before and after the zebra mussel's arrival in 1991.

The results supported their predictions. Larvae and juveniles of open-water fish, such as American shad, blueback herring, and alewife, declined in abundance after zebra mussels were introduced (**FIGURE 1a**). Those of littoral fish, such as tessellated darter, bluegill, and largemouth bass, increased (**FIGURE 1b**). Growth rates showed the same trend: Open-water fish grew more slowly after zebra mussels invaded, whereas littoral fish grew more quickly.

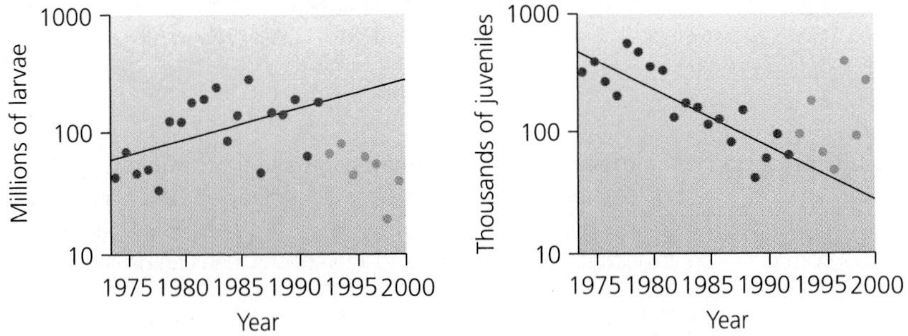

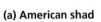

**(a) American shad** **(b) Tessellated darter**

**FIGURE 1 Zebra mussels harm open-water fish but help littoral fish.** Larvae of American shad **(a)**, an open-water fish, were increasing before zebra mussels invaded **(red points and trend line)**. After zebra mussels invaded, shad larvae decreased **(orange points)**. In contrast, juveniles of the tessellated darter **(b)**, a littoral fish, were decreasing **(red points and trend line)**, but increased after the mussels invaded **(orange points)**. *Source: Strayer, D., et al., 2004. Effects of an invasive bivalve* (Dreissena polymorpha) *on fish in the Hudson River estuary.* Canadian Journal of Fisheries and Aquatic Sciences *61: 924–941. © 2004. Reprinted by permission of NRC Research Press.*

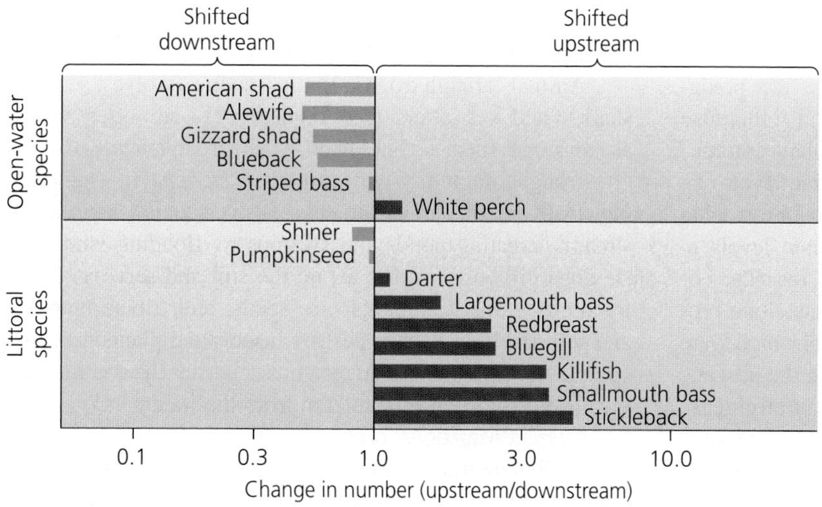


**FIGURE 2 Fish respond in different ways to zebra mussels.** Following the arrival of zebra mussels, the young of open-water fish tended to shift downstream toward areas with fewer zebra mussels. Young of littoral fish tended to shift upstream toward areas with more zebra mussels. *Source: Strayer, D., et al., 2004. Effects of an invasive bivalve* (Dreissena polymorpha) *on fish in the Hudson River estuary.* Canadian Journal of Fisheries and Aquatic Sciences *61: 924–941. © 2004. Reprinted by permission of NRC Research Press.*

growing to full size. Indeed, Strayer's teams determined that the zebra mussel's survival rate had fallen to less than 1% of what it was in the early years of the invasion! Predators such as the blue crab were eating more and more of the mussels, and large mussels were becoming rare. As the average size of mussels decreased, their filtering capacity fell by more than 80%, and more zooplankton began to survive.

Do these trends suggest that the zebra mussel might just fade away and prove harmless in the long run? Strayer feels it is too early to answer that question; much remains unknown, and he is continuing his research. He cautions that zebra mussels remain abundant, that phytoplankton levels have not yet bounced back, and that there is no guarantee that zebra mussel impacts will continue to diminish. Nonetheless, the apparent turnaround in the Hudson River is intriguing ecologists and providing hope that the Hudson's native systems may recover. ☐

And as predicted, along the 248-km (154-mi) stretch of river studied, open-water fish shifted downstream toward areas with fewer zebra mussels, whereas littoral fish shifted upstream toward areas with more zebra mussels (**FIGURE 2**). Overall, the data supported the hypothesis that the fish community would respond to changes in food resources caused by zebra mussels.

Yet then, surprisingly, some of the zebra mussel's impacts began to reverse. Populations of native mussels and clams in the Hudson that had crashed after the zebra mussel invaded (likely because of competition for food) began to stabilize and persist at about 4–22% of their pre-invasion population sizes. Crustaceans, flatworms, and other invertebrates also rebounded. Several types of zooplankton began to increase (**FIGURE 3**).

To determine why these changes were occurring, Strayer's research teams stepped up monitoring efforts and ran experiments placing cages (to keep out predators) around some areas of mussels. They found that mussels within the cages grew larger than mussels outside, indicating that predators were feeding on mussels and preventing most of them from

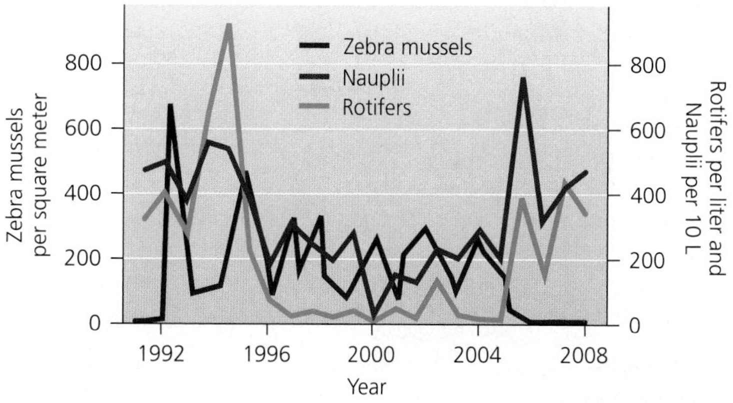


**FIGURE 3 Zooplankton bounced back once zebra mussels declined.** Large, mature zebra mussels have decreased in density since their initial sudden increase upon colonizing the Hudson River **(red line)**. Two types of zooplankton **(blue and green lines)** declined following the zebra mussel's introduction but recovered after 2005, once large mussels disappeared. *Source: Pace, M.L, et al., 2010. Recovery of native zooplankton associated with increased mortality of an invasive mussel.* Ecosphere *1(1): Article 3.*

DATA Q What were the densities of nauplii, rotifers, and zebra mussels in 1991? In 2000? In 2008? What pattern do you see in these data?

GO TO **INTERPRETING GRAPHS & DATA** ON MasteringEnvironmentalScience®

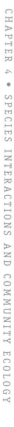

has strong or wide-reaching impact far out of proportion to its abundance is often called a **keystone species.** A keystone is the wedge-shaped stone at the top of an arch that holds the structure together; remove the keystone, and the arch will collapse (**FIGURE 4.9a**). In an ecological community, removal of a keystone species will likewise have major consequences.

Often, secondary or tertiary consumers near the tops of food chains are considered keystone species. Top predators control populations of herbivores, which otherwise would multiply and could greatly modify the plant community (**FIGURE 4.9b**). Thus, predators at high trophic levels can indirectly promote populations of organisms at low trophic levels by keeping species at intermediate trophic levels in check, a phenomenon referred to as a **trophic cascade.** For example, government bounties in the United States long promoted the hunting of wolves and mountain lions, which were largely exterminated by the mid-20th century. In the absence of these predators, deer populations grew unnaturally dense and have overgrazed forest-floor vegetation and eliminated tree seedlings, causing major changes in forest structure.

The removal of top predators in the United States was an uncontrolled large-scale experiment with unintended consequences, but ecologists have verified the keystone species concept in controlled scientific experiments. Classic research by biologist Robert Paine established that the predatory sea star *Pisaster ochraceus* shapes the community composition of intertidal organisms (p. 257) on North America's Pacific coast. When *Pisaster* is present in this community, species diversity is high, with various types of barnacles, mussels, and algae. When *Pisaster* is removed, the mussels it preys on become numerous and displace other species, suppressing species diversity.

Animals at high trophic levels—such as wolves, sea stars, sharks, and sea otters (see Figure 4.9)—are often viewed as keystone species, but other types of organisms also exert strong community-wide effects. "Ecosystem engineers" physically modify environments. Beavers build dams across streams, creating ponds and swamps by flooding land. Prairie dogs dig burrows that aerate the soil and serve as homes for other animals. Ants disperse seeds, redistribute nutrients, and selectively protect or destroy insects and plants near their colonies. And zebra and quagga mussels alter the communities they invade by filtering plankton from the water.

Less conspicuous organisms at low trophic levels can exert still more impact. Remove the fungi that decompose dead matter, or the insects that control plant growth, or the phytoplankton that support the marine food chain, and a community may change rapidly indeed. However, because there are usually more species at lower trophic levels, it is less likely that

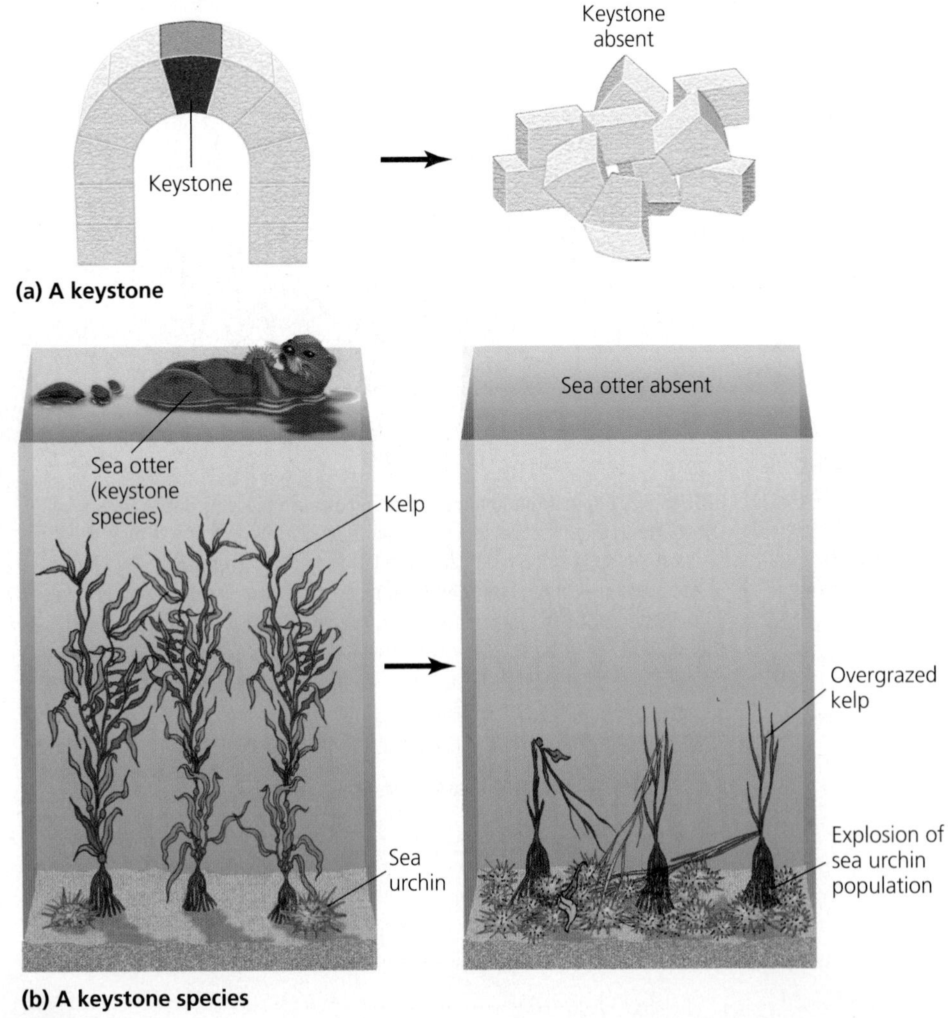


**FIGURE 4.9 Sea otters are a keystone species.** A keystone **(a)** is the wedge-shaped stone atop an arch that holds its structure together. A keystone species **(b)** is a species that exerts great influence on a community's composition and structure. Sea otters consume sea urchins that eat kelp in coastal waters of the Pacific. Otters keep urchin numbers down, allowing lush underwater forests of kelp to grow, providing habitat for many species. When otters are absent, urchins increase and devour the kelp, destroying habitat and depressing species diversity.

any single one of them alone has wide influence. Often if one species is removed, other species that remain may be able to perform many of its functions.

## Communities respond to disturbance in various ways

The removal of a keystone species is just one type of disturbance that can modify a community. In ecological terms, a **disturbance** is an event that has rapid and drastic impacts on environmental conditions, resulting in changes to the community and ecosystem. A disturbance can be localized, such as when a tree falls in a forest, creating a gap in the canopy that lets in additional sunlight. Or it can be as large and severe as a hurricane, tornado, or volcanic eruption. Some disturbances are sudden, such as landslides or floods, whereas others are more gradual. Some recur regularly and are considered normal aspects of a system (such as periodic fire, seasonal storms, or cyclic insect outbreaks). Today, human impacts are major sources of disturbance for ecological communities worldwide.

Communities are dynamic and may respond to disturbance in several ways. A community that resists change and remains stable despite disturbance is said to show **resistance** to the disturbance. Alternatively, a community may show **resilience**, meaning that it changes in response to disturbance but later returns to its original state. Or, a community may be modified by disturbance permanently and never return to its original state.

## Succession follows severe disturbance

If a disturbance is severe enough to eliminate all or most of the species in a community, the affected site may then undergo a predictable series of changes that ecologists have traditionally called **succession.** In the conventional view of this process, there are two types of succession (**FIGURE 4.10**). **Primary succession** follows a disturbance so severe that no vegetation or soil life remains from the community that had occupied the site. In primary succession, a community is built essentially from scratch. In contrast, **secondary succession** begins when a disturbance dramatically alters an existing community but does not destroy all life and organic matter. In secondary succession, vestiges of the previous community remain, and these building blocks help shape the process.

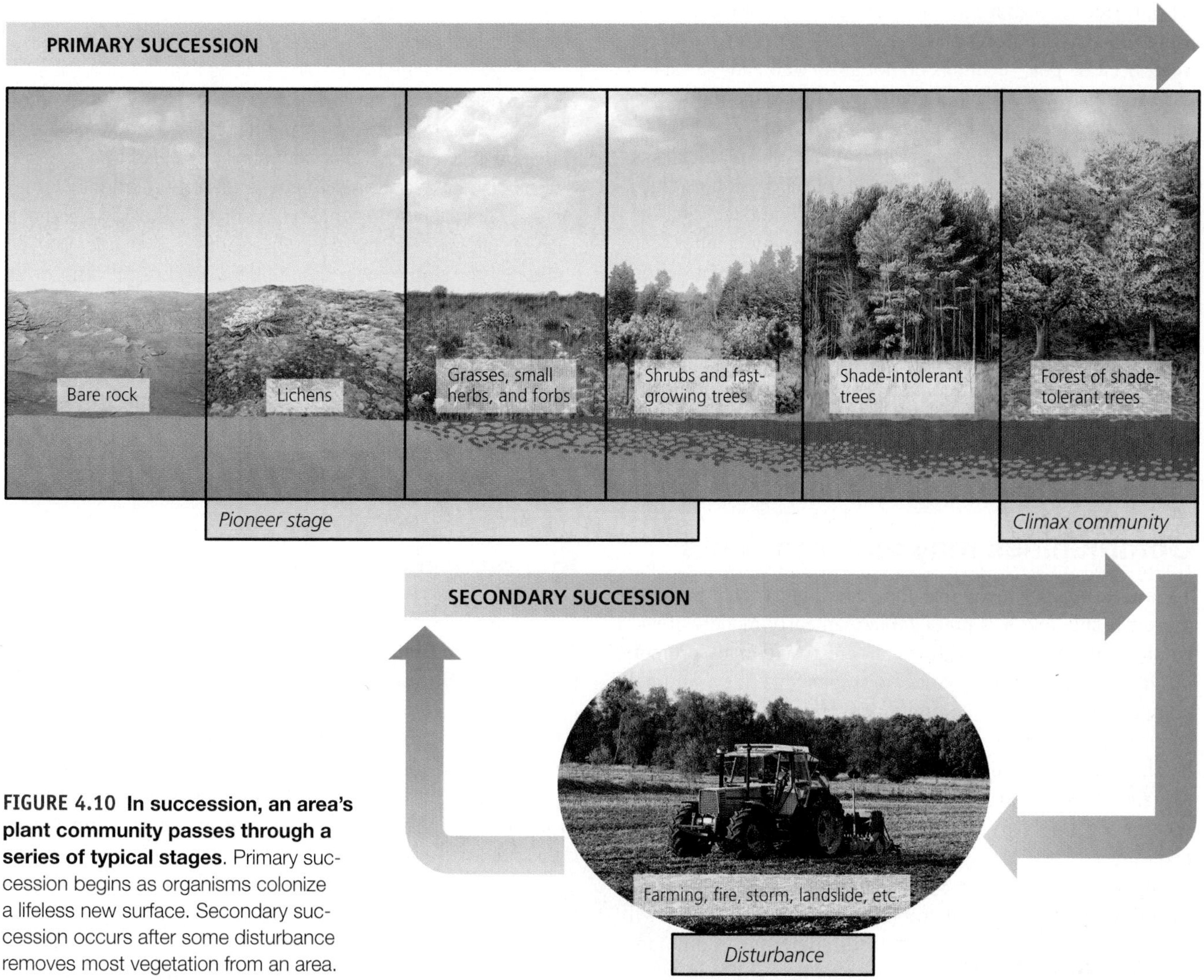


**FIGURE 4.10 In succession, an area's plant community passes through a series of typical stages**. Primary succession begins as organisms colonize a lifeless new surface. Secondary succession occurs after some disturbance removes most vegetation from an area.

At terrestrial sites, primary succession takes place after a bare expanse of rock, sand, or sediment becomes exposed to the atmosphere. This can occur when glaciers retreat, lakes dry up, or volcanic lava or ash covers a landscape. Species that arrive first and colonize the new substrate are referred to as **pioneer species.** Pioneer species are adapted for colonization, having traits such as spores or seeds that can travel long distances.

The pioneers best suited to colonizing bare rock are the mutualistic aggregates of fungi and algae known as lichens. In lichens, the algal component provides food and energy via photosynthesis while the fungal component grips the rock and captures moisture. As lichens grow, they secrete acids that break down the rock surface, beginning the process that forms soil. Small plants and insects arrive, providing more nutrients and habitat. As time passes, larger plants and animals establish themselves, vegetation increases, and species diversity rises.

Secondary succession begins when a fire, a storm, logging, or farming removes much of the biotic community. Consider a farmed field in eastern North America that has been abandoned. After farming ends, the site is colonized by pioneer species of grasses, herbs, and forbs that disperse well or were already in the vicinity. Soon, shrubs and fast-growing trees such as aspens and poplars begin to grow. As time passes, pine trees rise above the pioneer trees and shrubs, forming a pine forest. This forest attains an understory of hardwood trees, because pine seedlings do not grow well under a canopy but some hardwood seedlings do. Eventually the hardwoods outgrow the pines, creating a hardwood forest.

Processes of succession occur in many ecological systems. For instance, a pond may undergo succession as algae, microbes, plants, and zooplankton grow, reproduce, and die, gradually filling the water body with organic matter. The pond acquires further organic matter and sediments from streams and runoff, and eventually it may fill in, becoming a bog (p. 252) or even a terrestrial system.

In the traditional view of succession described here, the process leads to a *climax community*, which remains in place until some disturbance restarts succession. Early ecologists felt that each region had its own characteristic climax community, determined by climate.

## Communities may undergo shifts

Today, ecologists recognize that community change is far more variable and less predictable than early models of succession suggested. Conditions at one stage may promote progression to another stage, or organisms may, through competition, inhibit a community's progression to another stage. The trajectory of change can vary greatly according to chance factors, such as which particular species happen to gain an early foothold. And climax communities are not determined solely by climate, but vary with soil conditions and other factors from one time or place to another.

Once a community is disturbed and changes are set in motion, there is no guarantee that the community will return to its original state. Instead, sometimes communities may undergo a **phase shift,** or **regime shift,** in which the character of the community fundamentally changes. This can occur if some crucial climatic threshold is passed, a keystone species is lost, or a non-native species invades. For instance, many coral reef communities have undergone a phase shift and become dominated by algae after people overharvested fish or turtles that once ate the algae and kept their levels in check. Phase shifts show that we cannot count on being able to reverse damage caused by human disturbance, because some changes we set in motion may become permanent.

Many ecologists now think that human disturbance is creating wholly new communities that have not previously occurred on Earth. These **novel communities,** or **no-analog communities,** are composed of novel mixtures of plants and animals and have no analog or precedent. As we enter more deeply into an age of fast-changing climate, habitat alteration, species extinctions, and species invasions, scientists predict that we will see more and more novel communities.

**FAQ** **Once we disturb a community, won't it return to its original state if we just leave the area alone?**

Probably not, if the disturbance has been substantial. For example, if soil has become compacted or if water sources have dried up, then the plant species that grew at the site originally may no longer be able to grow. Different plant species may take their place—and among them, a different suite of animal species may find habitat. Because species interact and because they rely on particular habitat conditions, a change in one aspect of a community can lead to a cascade of other changes. Sometimes a whole new community may arise. For instance, in some grasslands, livestock grazing and fire suppression have led shrubs and trees to invade, forming shrublands. And removing sea otters can lead to the loss of kelp forests (p. 76). In the past, people didn't realize how permanent such changes could be, because we tended to view natural systems as static, predictable, and liable to return to equilibrium. Today ecologists recognize that systems are highly dynamic and can sometimes undergo rapid, extreme, and long-lasting change.

## Invasive species pose threats to community stability

Traditional concepts of communities involve species native to an area. But what if a species not native to the area (a non-native, alien, or exotic species) arrives from elsewhere? In our age of global mobility and trade, people have moved countless organisms from place to place, intentionally or by accident. As a result, today most non-native arrivals in a community are **introduced species,** species introduced by people.

Most introduced species fail to establish populations, but some turn invasive, spreading widely and coming to dominate communities. Such **invasive species** often thrive in disturbed communities and, in turn, disturb them further. By altering communities, invasive species are one of the central ecological forces in today's world.

Introduced species may become invasive when limiting factors (p. 62) that regulate their population growth are absent. Plants and animals brought to a new area may leave behind the predators, parasites, herbivores, and competitors that had exploited them in their native land. If few organisms in the new environment eat, parasitize, or compete with the introduced species, then it may thrive and spread. As the species proliferates, it may exert diverse influences on other community members (**FIGURE 4.11**).

Zebra and quagga mussels spread with global trade, inadvertently transported in the ballast water of cargo ships. (To maintain stability at sea, ships take water into their hulls as they begin their voyage and then discharge that water at their destination.) Countless species have been ferried across the oceans over decades of unregulated exchange of ballast water.

Examples abound of invasive species that have had major ecological impacts (pp. 174–176). The chestnut blight, an Asian fungus, killed nearly every mature American chestnut, a dominant tree species of eastern North American forests, between 1900 and 1930. Asian trees had evolved defenses against the fungus over millennia of coevolution, but the American chestnut had not. A different fungus caused Dutch elm disease, destroying most of the American elms that once gracefully lined the streets of many U.S. cities. Fish introduced into streams for sport compete with and exclude native fish. Grasses introduced in the American West for ranching have overrun entire regions, pushing out native vegetation. Hundreds of island-dwelling animals and plants worldwide

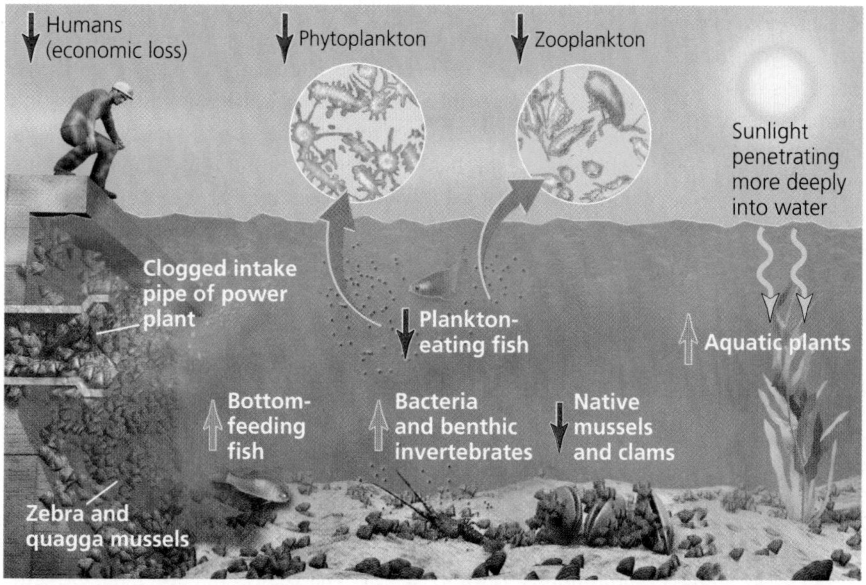


**(a) Impacts of zebra and quagga mussels on a Great Lakes nearshore community**

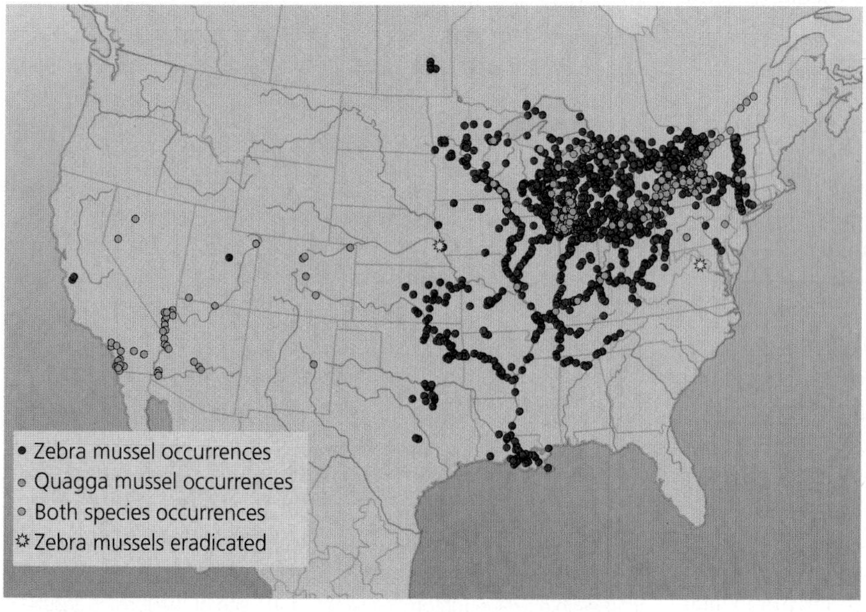


**(b) Occurrence of zebra and quagga mussels in North America**

**FIGURE 4.11 The zebra mussel and the quagga mussel modify ecological communities.** By filtering phytoplankton and small zooplankton from open water, they exert impacts **(a)** on other species, both negative (red downward arrows) and positive (green upward arrows). The map **(b)** shows known occurrences of zebra and quagga mussels as of 2014. In just two decades they spread across North America, assisted by accidental transport on boats. *Source (b): U.S. Geological Survey.*

DATA Q What differences do you see between the distribution of quagga mussels and the distribution of zebra mussels? Suggest two hypotheses for one of these differences.

GO TO **INTERPRETING GRAPHS & DATA** ON MasteringEnvironmentalScience®

have been driven extinct by goats, pigs, and rats introduced by human colonists (Chapter 3).

Ecologists tend to view the impacts of invasive species—and introduced species in general—in a negative light. Yet we enjoy the beauty of introduced ornamental plants in our gardens. Some organisms are introduced intentionally to control pests through biocontrol (p. 148). And some invasive species provide economic benefits, such as the European honeybee, which pollinates many of our crops (p. 149). Whatever view one takes, the impacts of introduced species on native populations and communities are significant, and they grow year by year with our increasing mobility and the globalization of our society.

### WEIGHING THE ISSUES

**Are Invasive Species All Bad?** Some ethicists question the notion that all invasive species should automatically be considered bad. If we introduce a non-native species to a community and it greatly modifies the community, do you think that is a bad thing? What if it drives another species extinct? What if the invasive species arrived on its own, rather than through human intervention? What if it provides economic services, as the European honeybee does? What ethical standard(s) (p. 12) would you apply to assess whether we should reject or welcome an invasive species?

## We can respond to invasive species with control, eradication, or prevention

Scientific research and media attention to zebra and quagga mussels helped put invasive species on the map as a major environmental and economic issue. In 1990, the U.S. Congress passed legislation that led to the National Invasive Species Act of 1996. Among other things, this law required ships to dump their freshwater ballast at sea and exchange it with salt water before entering the Great Lakes.

Since then, funding has become available for the control and eradication of invasive species. Eradication (total elimination of a population) is difficult, so managers usually aim merely to control populations; that is, to limit their growth, spread, and impact. Managers have tried to control zebra mussels by removing them manually; applying toxic chemicals; drying them out; depriving them of oxygen; introducing predators and diseases; and stressing them with heat, sound, electricity, carbon dioxide, and ultraviolet light. However, most of these are localized and short-term fixes. With one invasive species after another, managers find that control and eradication are so difficult and expensive that prevention strategies (such as ballast water regulations) represent a better investment.

To prevent invasions, it helps to be able to predict where a given species might spread. By analyzing the biology of the organism, scientists can try to model the environmental conditions in which it will thrive. To predict where zebra and quagga mussels might do best, researchers in 2007 applied knowledge of how these mussels use calcium from water to create their shells. The researchers mapped low-risk and high-risk regions across North America. One unanswered question is why quagga mussels have leapfrogged zebra mussels by spreading into some western states (see Figure 4.11b).

## Altered communities can be restored

Invasive species are adding to the transformations that people have forced on natural systems through habitat alteration, deforestation, pollution, overhunting, and climate change. Ecological systems support our civilization and all of life, so when systems are degraded and cease to function, our health and well-being are threatened.

This realization has given rise to the science of **restoration ecology.** Restoration ecologists research the historical conditions of ecological communities as they existed before our industrialized civilization altered them. They then devise ways to restore altered areas to an earlier condition. In some cases, the intent is to restore the functionality of a system—to reestablish a wetland's ability to filter pollutants and recharge groundwater, for example, or a forest's ability to cleanse the air, build soil, and provide wildlife habitat. In other cases, the aim is to return a community to its natural "pre-settlement" condition. Either way, the science of restoration ecology informs the practice of **ecological restoration,** the on-the-ground efforts to carry out these visions and restore communities.

For instance, nearly all the tallgrass prairie in the United States was converted to agriculture in the 1800s. Now, people are restoring patches of prairie by planting native vegetation, weeding out invaders and competitors, and introducing prescribed fire to mimic the fires that historically maintained this community (**FIGURE 4.12**). The region outside Chicago, Illinois, boasts several of the largest prairie restoration projects,

**FIGURE 4.12 Ecological restoration is used to restore prairies.** Ecologists at the Midewin National Tallgrass Prairie inspect native grasses in a prairie restoration area on the site of the former Joliet Arsenal in Illinois.

including over 400 ha (1000 acres) of restored prairie at the Fermilab nuclear accelerator in Batavia.

The world's largest restoration project is the ongoing effort to restore parts of the Florida Everglades, a vast ecosystem of marshes and seasonally flooded grasslands. This wetland system has been drying out for decades because the water that feeds it has been managed for flood control and overdrawn for irrigation and development. Economically important fisheries have suffered greatly as a result, and the region's famed populations of wading birds have dropped by 90–95%.

The 30-year, $7.8-billion restoration project intends to restore natural water flow by undoing damming and diversions of hundreds of miles of canals and levees. Because the Everglades provides drinking water for millions of Florida citizens, as well as considerable tourism revenue, restoring its ecosystem services (pp. 3, 35, 97) should prove economically beneficial as well as ecologically valuable. Unfortunately, the huge project has struggled against budgeting shortfalls and political interference, and most of its goals have not yet been met.

As our population grows and development spreads, ecological restoration is becoming an increasingly vital conservation strategy. However, restoration is difficult, time-consuming, and expensive, and it is not always successful. It is therefore best, whenever possible, to protect natural systems from degradation in the first place.

# Earth's Biomes

Across the world, each location is home to different sets of species, leading to endless variety in community composition. However, communities in far-flung places can share strong similarities, so we can classify communities into broad types. A **biome** is a major regional complex of similar communities—a large-scale ecological unit recognized primarily by its dominant plant type and vegetation structure. The world contains a number of biomes, each covering large geographic areas (**FIGURE 4.13**).

## Climate helps determine biomes

Because biomes are largely a function of climate, global climate patterns cause biomes to occur in large patches in different parts of the world. For instance, temperate deciduous forest occurs in Europe, China, and eastern North America. Note in Figure 4.13 that patches of any given biome tend to occur at similar latitudes. This is due to Earth's north–south gradients in temperature and to atmospheric circulation patterns (p. 282).

More specifically, which biome covers each portion of the planet depends on a variety of abiotic factors, including temperature, precipitation, soil conditions, and the circulation patterns of wind in the atmosphere and water in the

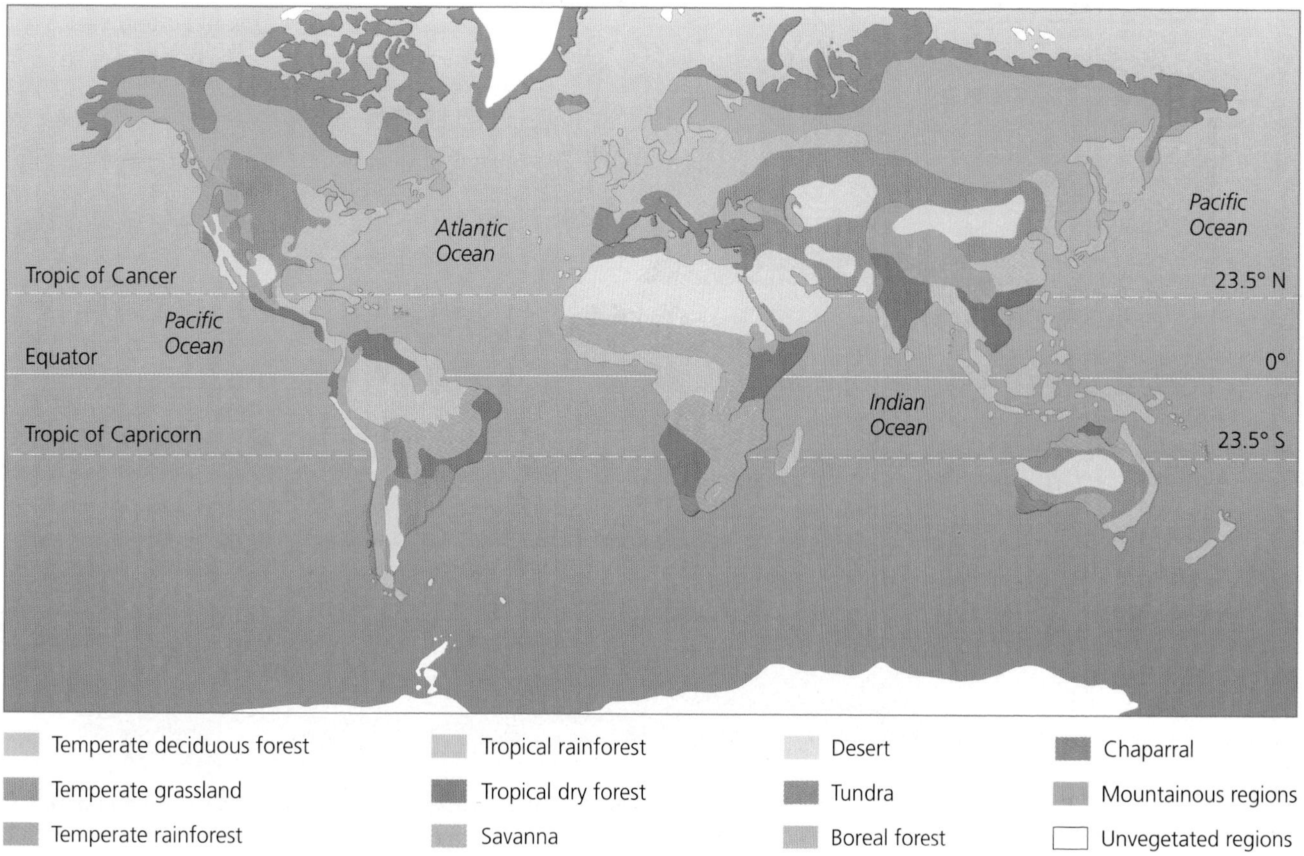


**FIGURE 4.13** **Biomes are distributed around the world, correlated roughly with latitude.**

oceans. Among these factors, temperature and precipitation exert the greatest influence (**FIGURE 4.14**). Scientists use **climate diagrams,** or **climatographs,** to depict information on temperature and precipitation. As we tour the world's biomes on the following pages, you will see climatographs from specific localities. The data in each graph is typical of the climate for the biome the locality lies within. In these graphs, the blue curve shows average monthly precipitation and the red curve shows average monthly temperature. Moist times of year are indicated in green and dry times of year in yellow.

## Aquatic and coastal systems resemble biomes

In our discussion of biomes, we focus exclusively on terrestrial systems because the biome concept, as traditionally developed and applied, has done so. However, areas equivalent to biomes also exist in the oceans, along coasts, and in freshwater systems. One might consider the shallows along the world's coastlines to represent one aquatic system, the continental shelves another, and the open ocean, the deep sea, coral reefs, and kelp forests as still others. Many coastal systems—such as salt marshes, rocky intertidal communities, mangrove forests, and estuaries—share both terrestrial and aquatic components. And freshwater systems such as those of the Great Lakes are widely distributed across the world.

Aquatic systems are shaped not by air temperature and precipitation, but by water temperature, salinity, dissolved nutrients, wave action, currents, depth, light levels, and type of substrate (e.g., sandy, muddy, or rocky bottom). Marine communities are also more clearly delineated by their animal life than by their plant life. We will examine freshwater, marine, and coastal systems in the greater detail they deserve in Chapter 12.

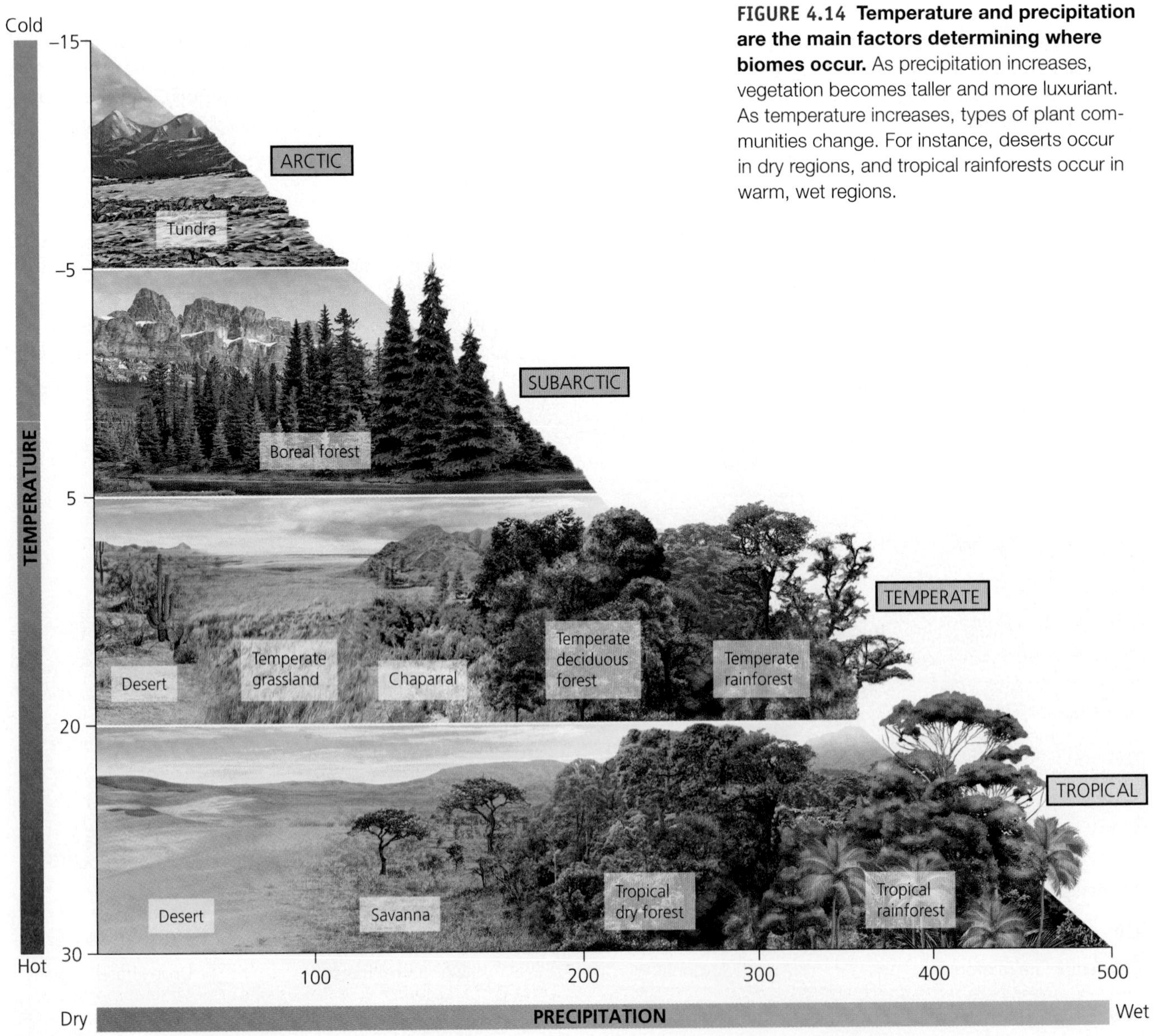


**FIGURE 4.14 Temperature and precipitation are the main factors determining where biomes occur.** As precipitation increases, vegetation becomes taller and more luxuriant. As temperature increases, types of plant communities change. For instance, deserts occur in dry regions, and tropical rainforests occur in warm, wet regions.

(a) Temperate deciduous forest

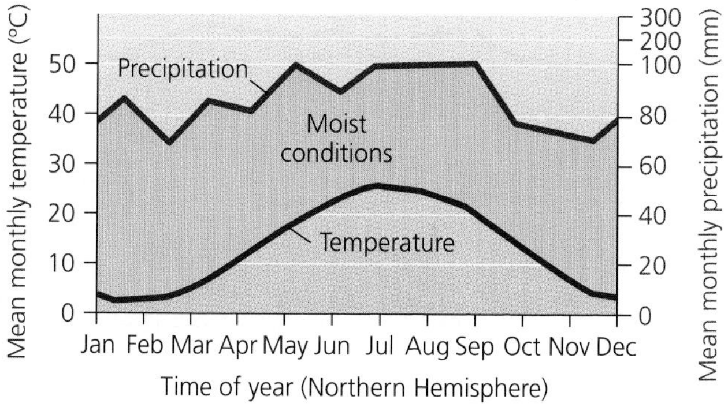


(b) Washington, D.C., USA

**FIGURE 4.15 Temperate deciduous forests (a) experience fairly stable seasonal precipitation but varied seasonal temperatures.** Scientists use climate diagrams **(b)** to illustrate average monthly precipitation and temperature. In these diagrams, the lines indicate precipitation **(blue)** and temperature **(red)** from month to month. *Climatograph here and in the following figures adapted from Breckle, S.W., and H. Walter, trans. by G. Lawlor. 2002.* Walter's vegetation of the Earth: The ecological systems of the geo-biosphere, *4th ed. Originally published by Eugen Ulmer KG, 1999, used by permission.*

(a) Temperate grassland

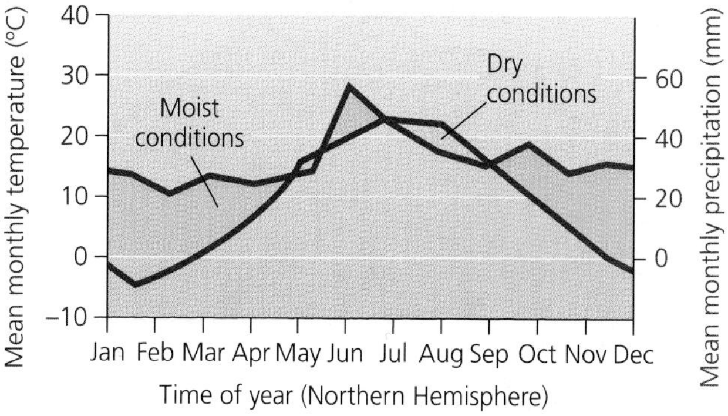


(b) Odessa, Ukraine

**FIGURE 4.16 Temperate grasslands experience seasonal temperature variation and too little precipitation for trees to grow.** *Climatograph adapted from Breckle, S.W., 2002.*

**DATA Q** How would you explain the dry conditions from July to September? Given the temperature and precipitation patterns shown, what role do you think evaporation might play, and why?

GO TO **INTERPRETING GRAPHS & DATA** ON MasteringEnvironmentalScience®

## We can divide the world into 10 terrestrial biomes

**Temperate deciduous forest** The **temperate deciduous forest** (**FIGURE 4.15**) that dominates the landscape around the southern Great Lakes is characterized by broad-leafed trees that are deciduous, meaning that they lose their leaves each fall and remain dormant during winter, when hard freezes would endanger leaves. These midlatitude forests occur in much of Europe and eastern China as well as in eastern North America—all areas where precipitation is spread relatively evenly throughout the year.

Soils of the temperate deciduous forest are fertile, but this biome contains far fewer tree species than do tropical rainforests. Oaks, beeches, and maples are a few of the most common trees in these forests. Some typical animals of the temperate deciduous forest of eastern North America are shown in Figure 4.8 (p. 73).

**Temperate grassland** Traveling westward from the Great Lakes, temperature differences between winter and summer become more extreme, rainfall diminishes, and we find **temperate grasslands** (**FIGURE 4.16**). This is because the limited precipitation in the Great Plains supports grasses more easily than trees. Also known as *steppe* or *prairie*, temperate grasslands were once widespread in much of North and South America and central Asia.

Vertebrate animals of North America's native grasslands include American bison, prairie dogs, pronghorn antelope, and ground-nesting birds such as meadowlarks and prairie chickens. People have converted most of the world's grasslands for farming and ranching, however, so most of these animals exist today at a tiny fraction of their historic population sizes.

(a) Temperate rainforest

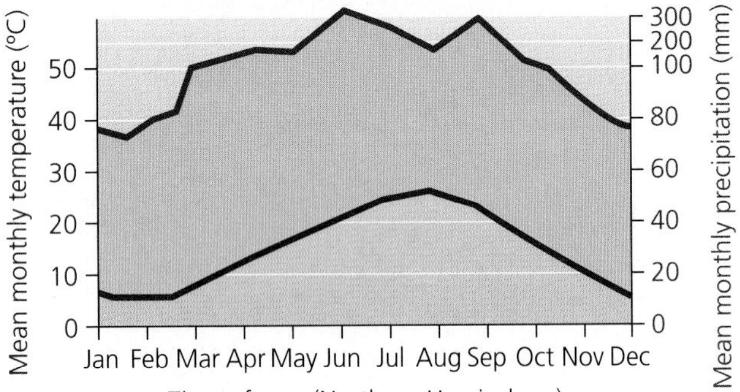


(b) Nagasaki, Japan

**FIGURE 4.17 Temperate rainforests receive a great deal of precipitation and have moist, mossy interiors.** *Climatograph adapted from Breckle, S.W., 2002.*

(a) Tropical rainforest

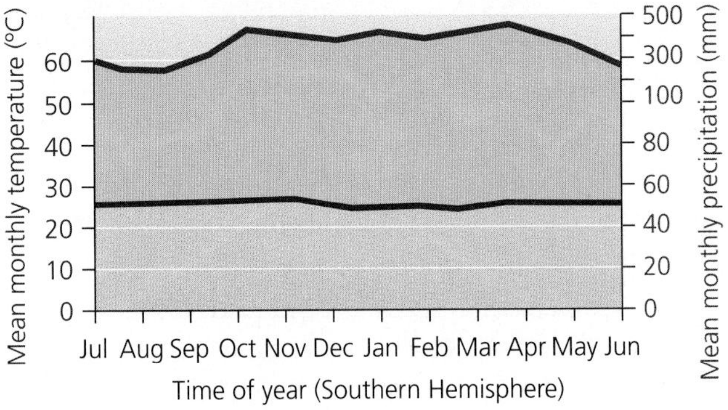


(b) Bogor, Java, Indonesia

**FIGURE 4.18 Tropical rainforests, famed for their biodiversity, grow under constant, warm temperatures and a great deal of rain.** *Climatograph adapted from Breckle, S.W., 2002.*

**Temperate rainforest** Further west in North America, the topography becomes varied, and biome types intermix. The coastal Pacific Northwest region, with its heavy rainfall, features **temperate rainforest** (**FIGURE 4.17**). Coniferous trees such as cedars, spruces, hemlocks, and Douglas fir grow very tall in the temperate rainforest, and the forest interior is shaded and damp. Moisture-loving animals such as the bright yellow banana slug are common.

The soils of temperate rainforests are fertile but are susceptible to landslides and erosion when forests are cleared. We extract commercially valuable products from temperate rainforests, but timber harvesting has eliminated most old-growth trees, driving species such as the spotted owl and marbled murrelet toward extinction.

**Tropical rainforest** In tropical regions we see the same pattern found in temperate regions: Areas of high rainfall grow rainforests, areas of intermediate rainfall support dry or deciduous forests, and areas of low rainfall are dominated by grasses. However, tropical biomes differ from their temperate counterparts in other ways because they are closer to the equator and therefore warmer on average year-round. For one thing, they hold far greater biodiversity.

**Tropical rainforest** (**FIGURE 4.18**)—found in Central America, South America, Southeast Asia, west Africa, and other tropical regions—is characterized by year-round rain and uniformly warm temperatures. Tropical rainforests have dark, damp interiors, lush vegetation, and highly diverse communities, with more species of insects, birds, amphibians, and other animals than any other biome. These forests consist of high numbers of tree species intermixed, each at a low density. A tree may be draped with vines and loaded with epiphytes (orchids and other plants that grow in trees). Indeed, trees sometimes collapse under the weight of all the life they support!

Despite this profusion of life, tropical rainforests have poor, acidic soils that are low in organic matter. Nearly all nutrients in this biome are contained in the plants, not in the soil. An unfortunate consequence is that once tropical rainforests are cleared, the nutrient-poor soil can support agriculture for only a short time (p. 141). As a result, farmed areas are abandoned quickly, and farmers move on and clear more forest.

**(a) Tropical dry forest**

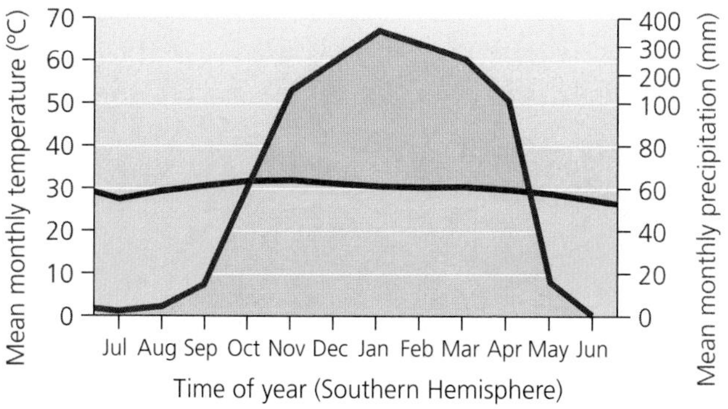


**(b) Darwin, Australia**

**FIGURE 4.19 Tropical dry forests experience significant seasonal variation in precipitation but relatively stable, warm temperatures.** *Climatograph adapted from Breckle, S.W., 2002.*

**(a) Savanna**

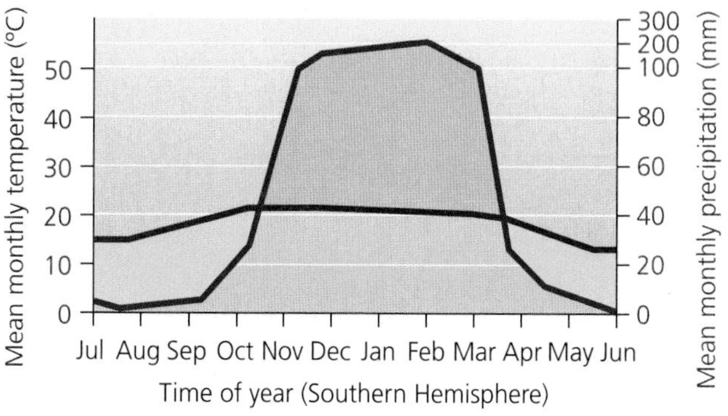


**(b) Harare, Zimbabwe**

**FIGURE 4.20 Savannas are grasslands with clusters of trees.** They experience slight seasonal variation in temperature but significant variation in rainfall. *Climatograph adapted from Breckle, S.W., 2002.*

**Tropical dry forest** Tropical areas that are warm year-round but where rainfall is lower overall and highly seasonal give rise to **tropical dry forest,** or *tropical deciduous forest* (**FIGURE 4.19**), a biome widespread in India, Africa, South America, and northern Australia. Wet and dry seasons each span about half a year in tropical dry forest. Organisms that inhabit tropical dry forest have adapted to seasonal fluctuations in precipitation and temperature. For instance, plants leaf out and grow profusely with the rains, then drop their leaves during dry times of year.

Rains during the wet season can be heavy and, coupled with erosion-prone soils, can lead to severe soil loss where people have cleared forest. Across the globe, we have converted a great deal of tropical dry forest to agriculture. Clearing is straightforward because vegetation is lower and canopies less dense than in tropical rainforest.

**Savanna** Drier tropical regions give rise to **savanna** (**FIGURE 4.20**), tropical grassland interspersed with clusters of acacias or other trees. The savanna biome is found across stretches of Africa, South America, Australia, India, and other dry tropical regions. Precipitation usually arrives during distinct rainy seasons, whereas in the dry season grazing animals concentrate near widely spaced water holes. Common herbivores on the African savanna include zebras, gazelles, and giraffes. Predators of these grazers include lions, hyenas, and other highly mobile carnivores.

**Desert** Where rainfall is very sparse, **desert** (**FIGURE 4.21**, next page) forms. The driest biome on Earth, most deserts receive well under 25 cm (10 in.) of precipitation per year, much of it during isolated storms months or years apart. Some deserts, such as Africa's Sahara, are mostly bare sand dunes. Others, such as the Sonoran Desert of Arizona and northwest Mexico, receive more rain and are more heavily vegetated.

Deserts are not always hot; the high desert of the western United States is positively cold in winter. Because deserts have low humidity and little vegetation to insulate them from temperature extremes, sunlight readily heats them in the daytime, but heat is quickly lost at night. As a result, temperatures vary greatly from day to night and from season to season. Desert soils can be saline and are sometimes known as lithosols, or stone soils, for their high mineral and low organic-matter content.

**(a) Desert**

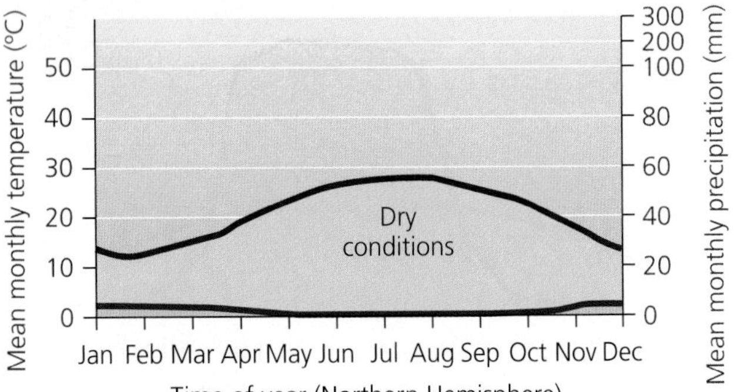


**(b) Cairo, Egypt**

**FIGURE 4.21 Deserts are dry year-round.** The precipitation curve is consistently below the temperature curve in this climatograph for Cairo, Egypt, indicating that the region experiences dry conditions all year. *Climatograph adapted from Breckle, S.W., 2002.*

**(a) Tundra**

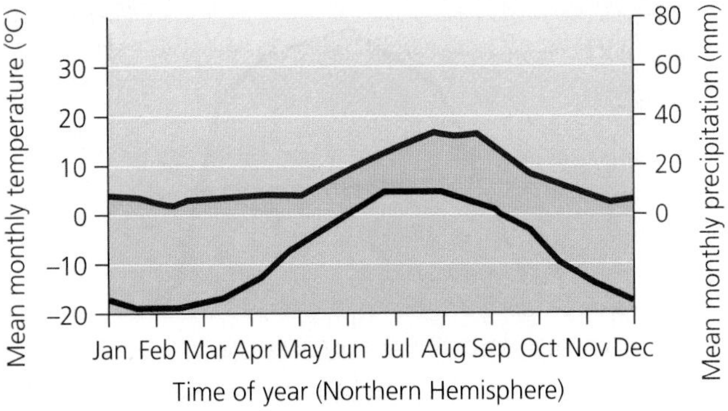


**(b) Vaigach, Russia**

**FIGURE 4.22 Tundra is a cold, dry biome found near the poles.** Alpine tundra occurs atop high mountains at lower latitudes. *Climatograph adapted from Breckle, S.W., 2002.*

Desert animals and plants show many adaptations to deal with a harsh climate. Most reptiles and mammals, such as rattlesnakes and kangaroo mice, are active in the cool of night. Many Australian desert birds are nomadic, wandering long distances to find areas of recent rainfall and plant growth. Desert plants tend to have thick, leathery leaves to reduce water loss, or green trunks so that the plant can photosynthesize without leaves. The spines of cacti and other desert plants guard them from being eaten by herbivores desperate for the precious water they hold. Such traits have evolved by convergent evolution in deserts across the world (see Figure 3.2b, p. 49).

**Tundra** Nearly as dry as desert, **tundra** (**FIGURE 4.22**) occurs at very high latitudes in northern Russia, Canada, and Scandinavia. Extremely cold winters with little daylight and summers with lengthy days characterize this landscape of lichens and low, scrubby vegetation without trees. The great seasonal variation in temperature and day length results from this biome's high-latitude location, angled toward the sun in summer and away from the sun in winter.

Because of the cold climate, underground soil remains permanently frozen and is called *permafrost.* During winter, surface soil freezes as well. When the weather warms, the soil melts and produces pools of surface water, forming ideal habitat for mosquitoes and other insects. The swarms of insects benefit bird species that migrate long distances to breed during the brief but productive summer. Caribou also migrate to the tundra to breed, then leave for the winter. Only a few animals, such as polar bears and musk oxen, can survive year-round here.

Tundra also occurs as alpine tundra at the tops of mountains in temperate and tropical regions. Here, high elevation creates conditions similar to those of high latitude.

**Boreal forest** The northern coniferous forest, or **boreal forest,** often called *taiga* (**FIGURE 4.23**), extends across much of Canada, Alaska, Russia, and Scandinavia. A few species of evergreen trees, such as black spruce, dominate large stretches of forest, interspersed with many bogs and lakes. Boreal forests occur in cooler, drier regions than do temperate forests, and they experience long, cold winters and short, cool summers.

(a) Boreal forest

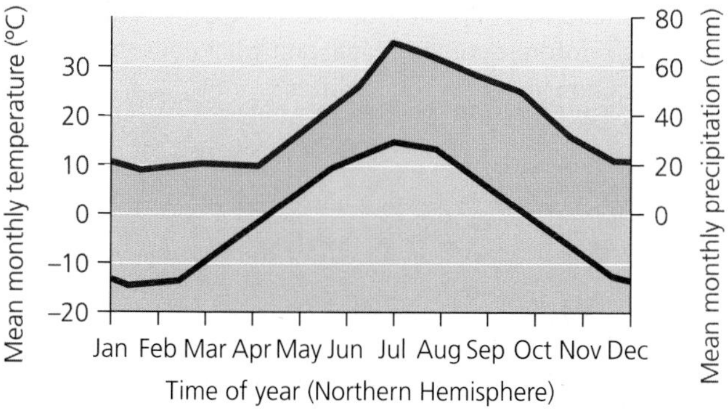


(b) Archangelsk, Russia

**FIGURE 4.23 Boreal forest experiences long, cold winters, cool summers, and moderate precipitation.** *Climatograph adapted from Breckle, S.W., 2002.*

(a) Chaparral

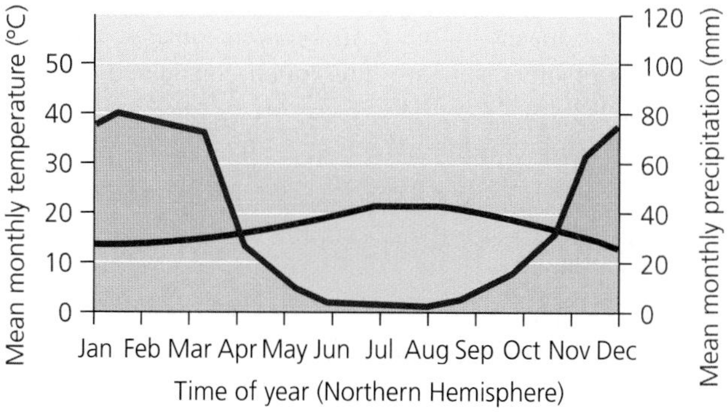


(b) Los Angeles, California, USA

**FIGURE 4.24 Chaparral is a seasonally variable biome dominated by shrubs, influenced by marine weather, and dependent on fire.** *Climatograph adapted from Breckle, S.W., 2002.*

Soils are typically nutrient-poor and somewhat acidic. As a result of strong seasonal variation in day length, temperature, and precipitation, many organisms compress a year's worth of feeding and breeding into a few warm, wet months. Year-round residents of boreal forest include mammals such as moose, wolves, bears, lynx, and rodents. Many insect-eating birds migrate here from the tropics to breed during the brief, intensely productive, summers.

**Chaparral** In contrast to the boreal forest's broad, continuous distribution, **chaparral** (**FIGURE 4.24**) is limited to small patches widely flung around the globe. Chaparral consists mostly of evergreen shrubs and is densely thicketed. This biome is highly seasonal, with mild, wet winters and warm, dry summers—a climate influenced by ocean waters and often termed "Mediterranean." Besides ringing the Mediterranean Sea, chaparral occurs along the coasts of California, Chile, and southern Australia. Chaparral communities experience frequent fire, and their plants are adapted to resist fire or even to depend on it for germination of their seeds.

## Conclusion

The natural world is so complex that we can visualize it in many ways and at various scales. Classifying the world's communities into major types, or biomes, is informative at the broadest geographic scales. Understanding how communities function at more local scales requires understanding how species interact. Species interactions such as competition, predation, parasitism, herbivory, and mutualism give rise to effects weak and strong, direct and indirect. We can represent feeding relationships with the concepts of trophic levels and food webs, and particularly influential species are sometimes called keystone species. People alter communities, in part by introducing non-native species that sometimes turn invasive. But more and more, through ecological restoration, we are also attempting to undo some of the changes we have set in motion.

# Testing Your Comprehension

1. Explain how competition promotes resource partitioning.
2. Compare and contrast the three main types of exploitative species interactions. How do predation, parasitism, and herbivory differ?
3. Give examples of symbiotic and nonsymbiotic mutualisms. Describe at least one way in which a mutualism affects your daily life.
4. Using the concepts of trophic levels and energy flow, explain why the ecological footprint of a vegetarian person is smaller than that of a meat-eater.
5. Differentiate a food chain from a food web. Which best represents the reality of communities, and why?
6. What is meant by the term *keystone species*, and what types of organisms are most often considered keystone species?
7. Describe the process of primary succession. How does it differ from secondary succession? Give an example of each.
8. What is restoration ecology? Why is it an important scientific pursuit in today's world?
9. What factors most strongly influence the type of biome that forms in a particular place on land? What factors determine the type of aquatic system that may form in a given location?
10. Draw a typical climate diagram for a tropical rainforest. Label all parts of the diagram, and describe what information an ecologist can glean from it. Now draw a climate diagram for a desert. How does it differ from your rainforest climatograph, and what does this tell you about how the two biomes differ?

# Seeking Solutions

1. Suppose you spot two species of birds feeding side by side, eating seeds from the same plant. You begin to wonder whether competition is at work. Describe how you might design scientific research to address this question. What observations would you try to make at the outset? Would you try to manipulate the system to test your hypothesis that the two birds are competing? If so, how?
2. Spend some time outside on your campus, in your yard, or in the nearest park or natural area. Find at least 10 species of organisms (plants, animals, or others), and observe each one long enough to watch it feed or to make an educated guess about how it derives its nutrition. Now, using Figure 4.8 as a model, draw a simple food web involving all the organisms you observed.
3. Can you think of one organism not mentioned in this chapter as a keystone species that you believe may be a keystone species? For what reasons do you suspect this? How could an ecologist experimentally test whether an organism is a keystone species?
4. Why do scientists consider invasive species to be a problem? What makes a species "invasive," and what impacts can invasive species have? Give examples.
5. **THINK IT THROUGH** A federal agency has put you in charge of devising responses to the invasion of zebra mussels and quagga mussels. Based on what you know from this chapter, how would you seek to control the spread of these species and reduce their impacts? What strategies would you consider pursuing immediately, and for which strategies would you commission further scientific research? For each of your ideas, name one benefit or advantage, and identify one obstacle it might face in being implemented.

# Calculating Ecological Footprints

Environmental scientists David Pimentel, Rodolfo Zuniga, and Doug Morrison of Cornell University reviewed scientific estimates for the economic and ecological costs imposed by introduced and invasive species in the United States. They found that, as of 2005, approximately 50,000 species had been introduced in the United States and that these accounted for over $120 billion in economic costs each year. These costs include direct losses and damage, as well as costs required to control the species. (The researchers did not quantify monetary estimates for losses of biodiversity, ecosystem services, and aesthetics, which they said would drive total costs several times higher.) Calculate values missing from the table to determine the number of introduced species of each type of organism and the annual cost that each imposes on our economy.

| Group of organism | Percentage of total introduced | Number of species introduced | Percentage of total annual costs | Annual economic costs |
|---|---|---|---|---|
| Plants | 50.0 | 25,000 | 27.2 | |
| Microbes | 40.0 | | 20.2 | |
| Arthropods | 9.0 | | 15.7 | |
| Fish | 0.28 | | 4.2 | |
| Birds | 0.19 | | 1.5 | $1.9 billion |
| Mollusks | 0.18 | | 1.7 | |
| Reptiles and amphibians | 0.11 | | 0.009 | |
| Mammals | 0.04 | 20 | 29.4 | $37.5 billion |
| TOTAL | 100 | 50,000 | 100 | $127.4 billion |

*Source: Pimentel, D., R. Zuniga, and D. Morrison, 2005. Update on the environmental and economic costs associated with alien-invasive species in the United States.* Ecological Economics 52: 273–288.

1. Of the 50,000 species introduced into the United States, half are plants. Describe two ways in which non-native plants might be brought to a new location. How might we help prevent non-native plants from establishing in new areas and altering native communities?
2. Organisms that damage crop plants are the most costly of introduced species. Weeds, pathogenic microbes, and arthropods that attack crops together account for half of the costs documented by Pimentel's team. What steps can we—farmers, policymakers, and all of us as a society—take to minimize the impacts of invasive species on crops?
3. How might your own behavior influence the influx and ecological impacts of non-native species such as those listed above? Name three things you could personally do to help reduce the impacts of invasive species.

## MasteringEnvironmentalScience®

**STUDENTS**

Go to **MasteringEnvironmentalScience** for assignments, the etext, and the Study Area with practice tests, videos, current events, and activities.

**INSTRUCTORS**

Go to **MasteringEnvironmentalScience** for automatically graded activities, current events, videos, and reading questions that you can assign to your students, plus Instructor Resources.

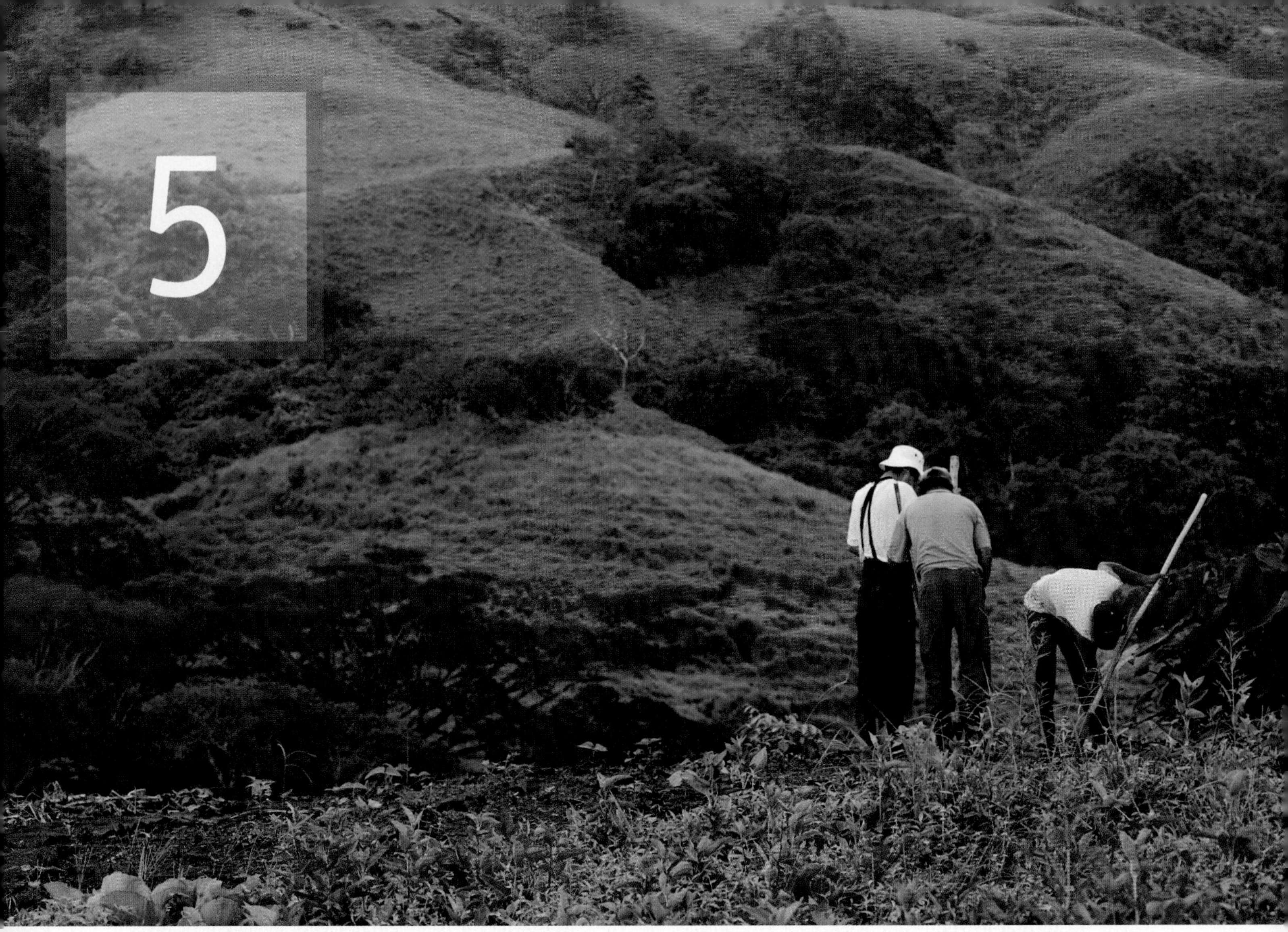

# 5 Economics, Policy, and Sustainable Development

## Upon completing this chapter, you will be able to:

- ☐ Explain how our economies exist within the environment and rely on ecosystem services
- ☐ Describe principles of classical and neoclassical economics, and summarize their implications for the environment
- ☐ Illustrate aspects of environmental economics and ecological economics, including valuation of ecosystem services and full-cost accounting
- ☐ Describe environmental policy and assess its societal context
- ☐ Explain the role of science in policymaking
- ☐ Discuss the history of U.S. environmental policy and summarize major U.S. environmental laws
- ☐ List institutions that influence international environmental policy and describe how nations handle transboundary issues
- ☐ Compare and contrast the different approaches to environmental policymaking
- ☐ Define sustainable development, explain the "triple bottom line," and describe how sustainable development is pursued worldwide.



**Photo: Costa Rican farmers tend crops where forest once stood.**

# CENTRAL CASE STUDY

# Costa Rica Values Its Ecosystem Services

*"Costa Rica's PSA program has been one of the conservation success stories of the last decade."*

**—Stefano Pagiola, The World Bank**

*"In the last 25 years, my home country has tripled its GDP while doubling the size of its forests."*

**—Carlos Manuel Rodriguez, former Minister of Energy and the Environment, Costa Rica**

Few nations have transformed their path of development in just a few decades—but Costa Rica has. In the 1980s, this small Central American country was losing its forests as fast as any place on Earth. Yet today this nation of 4.7 million people has regained much of its forest cover, boasts a world-class park system, and stands as a global model for sustainable resource management.

Costa Rica took many steps on this impressive road to success. One key step was to begin paying landholders to conserve forest on private land, in a novel government program called *Pago por Servicios Ambientales* (PSA)—Payment for Environmental Services.

Nature provides ecosystem services (pp. 3, 35, 97), such as air and water purification, climate regulation, and nutrient cycling. For example, forests in Costa Rica's mountains capture rainfall and provide clean drinking water for towns and cities below. Ecosystem services are vital for our lives and our society, but historically we have taken them for granted and have not paid for them in the marketplace. As a result, these services have diminished as we degrade the natural systems that provide them. For this reason, many economists want to create financial incentives for conserving ecosystem services.

In Costa Rica, which had lost over three-quarters of its forest, political leaders adopted this approach in Forest Law 7575, passed in 1996. Since then, the government has been paying farmers and ranchers to preserve forest on their land, replant cleared areas, allow forest to regenerate naturally, or establish sustainable forestry systems. Payments are designed to be competitive with potential profits from farming or cattle ranching.

The PSA program recognized four ecosystem services that forests provide:

1. Watershed protection: Forests cleanse water by filtering pollutants, and they conserve water and reduce soil erosion by slowing runoff.
2. Biodiversity: Tropical forests such as Costa Rica's are especially rich in life.
3. Scenic beauty: This encourages recreation and ecotourism, which bring money to the economy.
4. Carbon sequestration: By pulling carbon dioxide from the atmosphere, forests slow global warming.

To fund the PSA program, Costa Rica's government sought money from people and companies that benefit from these services. For watershed protection, irrigators, bottlers, municipal water suppliers, and utilities that generate hydropower all made voluntary payments into the program, and a tariff on water users was added in 2005. For biodiversity and scenery, the country targeted ecotourism, while international lending agencies provided loans and donations. Because carbon dioxide is emitted when fossil fuels are burned, the nation used a 3.5% tax on fossil fuels to help fund the program. It also sought to sell carbon offsets in global markets (p. 328).

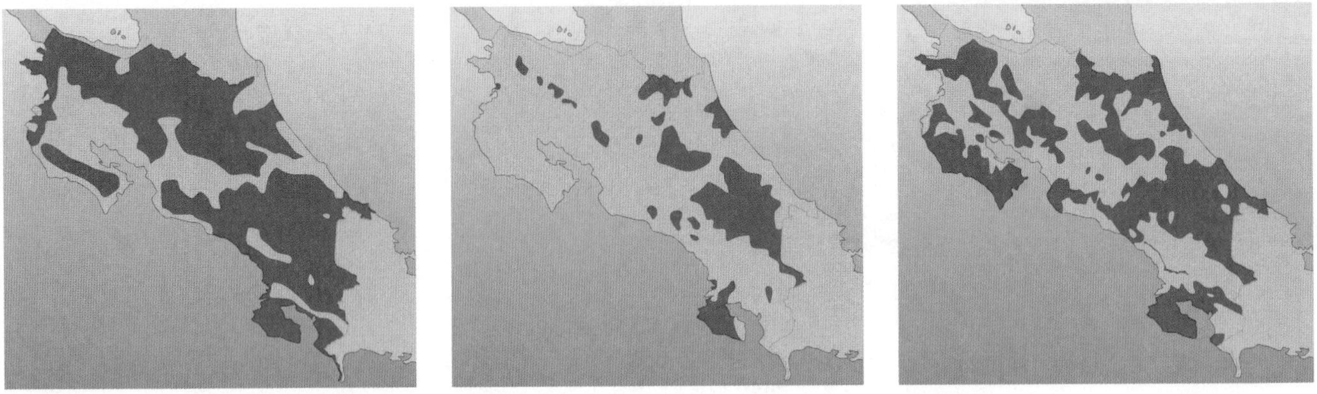

(a) 1940 (b) 1987 (c) 2005

**FIGURE 5.1 Forest cover in Costa Rica decreased between 1940 and 1987, but it increased by 2005.**
*Data from FONAFIFO.*

Costa Rican landholders rushed to sign up for the PSA program. The agency administering it, *Fondo Nacional de Financiamiento Forestal* (FONAFIFO), signed landowners to contracts and sent agents to advise them on forest conservation and to monitor compliance. By 2009, FONAFIFO had paid 57 billion colónes ($110 million) to over 8300 landholders and had registered over 670,000 ha (1.66 million acres)—13% of the nation's land area.

Deforestation began to slow in Costa Rica, and forest cover rose by 10% in the decade after 1996. Policymakers, economists, and environmental advocates cheered the PSA program's apparent success in safeguarding forests and the ecosystem services they provide.

However, some observers argued that forest loss had been slowing for other reasons and that the program itself was having little effect. They contended that payments were being wasted on people who had no plans to cut down their trees. Critics also lamented that large wealthy landowners utilized the program more than low-income small farmers. All these concerns were borne out by researchers (see **THE SCIENCE BEHIND THE STORY**, pp. 100–101).

In response, the government modified its policies, making the program more accessible to small farmers and targeting the payments to locations where forest is most at risk and environmental assets are greatest.

Today forest cover in Costa Rica has risen from a low of 17% in 1983 to over 52% (**FIGURE 5.1**). The nation has thrived economically while protecting its environment; since the PSA program began, Costa Ricans have enjoyed an increase in per capita income of over 50%—a rise in wealth surpassing the vast majority of nations.

Many factors have contributed to Costa Rica's success in building a wealthier society while protecting its ecological assets. Back in 1948, Costa Rica abolished its army and shifted funds from the military budget into health and education. With a stable democracy and a healthy and educated citizenry, the stage was set for well-managed development, including innovative advances in conservation. The nation created one of the world's finest systems of national parks, covering one-quarter of its territory. Ecotourism at the parks brings wealth to the country: Each year 2 million foreign tourists inject $2 billion into Costa Rica's economy.

As a result, Costa Ricans understand the economic value of protecting their natural capital. They see how innovative policies and economic incentives can help conserve resources while boosting the economy and enhancing the quality of people's everyday lives. By placing economic value on nature, Costa Rica is pointing the way toward truly sustainable development. ◻

# Economics and the Environment

An **economy** is a social system that converts resources into *goods* (material commodities made and bought by individuals and businesses) and *services* (work done for others as a form of business). **Economics** is the study of how people decide to use potentially scarce resources to provide goods and services that are in demand. The word *economics* and the word *ecology* come from the same Greek root, *oikos*, meaning "household." Economists traditionally have studied the household of human society, and ecologists the broader household of all life.

## Economies rely on goods and services from the environment

Our economies and our societies exist within the natural environment and depend on it in vital ways. Economies receive inputs (such as natural resources and ecosystem services) from the environment, process them, and discharge outputs (such as waste) into the environment (**FIGURE 5.2**).

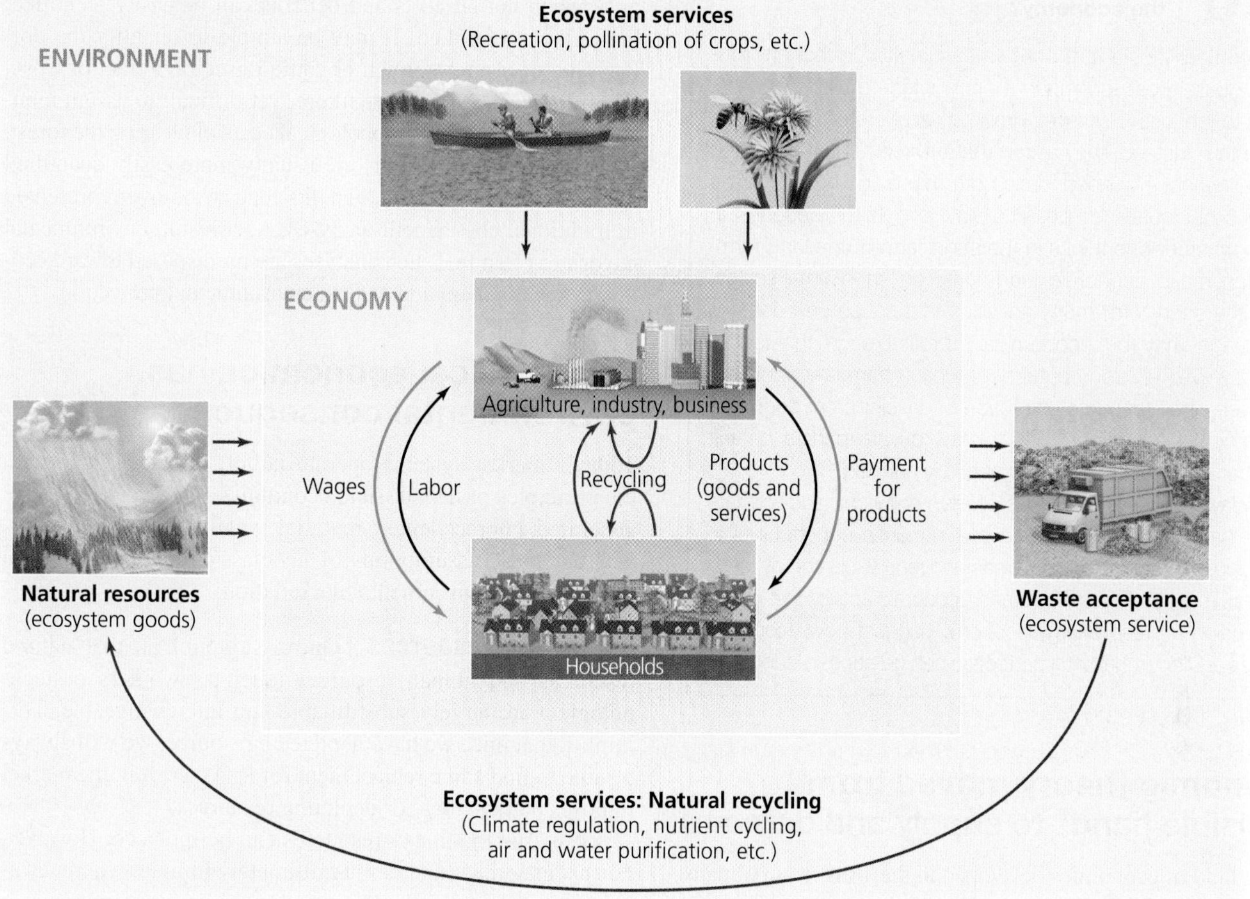


**FIGURE 5.2 Economies exist within the natural environment, receiving resources from it, discharging waste into it, and benefiting from ecosystem services.** Conventional neoclassical economics focuses only on processes of production and consumption among households and businesses **(yellow box in middle)** and views the environment merely as an external factor. In contrast, environmental and ecological economists emphasize that economies exist within the natural environment and depend on all that it offers.

These interactions are readily apparent, yet traditional economic schools of thought have long ignored them or overlooked their importance. Many mainstream economists still adhere to a worldview that largely ignores the environment (and instead considers only the yellow box in the middle of Figure 5.2). This conventional view, which continues to drive policy decisions, implies that natural resources are free and limitless and that wastes can be endlessly absorbed at no cost. In contrast, modern economists in the fast-growing fields of environmental economics and ecological economics (p. 95) emphasize that economies exist within the environment and depend on it for natural resources and ecosystem services.

Natural resources (pp. 2–3) are the substances and forces that sustain our society and our everyday lives: the fresh water we drink, the trees that provide our lumber, the rocks that provide our metals, and the energy from sun, wind, water, and fossil fuels. We can think of natural resources as "goods" produced by nature. Environmental systems also naturally function in a manner that supports our economies. Earth's ecological systems purify air and water, form soil, cycle nutrients, regulate climate, pollinate plants, and recycle waste. Such essential ecosystem services (pp. 3, 35, 97) support the very life that makes our economic activity possible. Together, nature's resources and services make up the natural capital (pp. 15–16) on which we depend.

When we deplete natural resources and generate pollution, we degrade the capacity of ecological systems to function. Scientists with the Millennium Ecosystem Assessment, a worldwide review undertaken in 2005, concluded that 15 of 24 ecosystem services they surveyed globally were being degraded or used unsustainably. The degradation of ecosystem services can weaken economies. In Costa Rica, rapid forest loss was causing soil erosion, water pollution, and biodiversity loss. Low-income small farmers were the first to feel these impacts. Indeed, across the world, ecological degradation is harming poor and marginalized people before wealthy ones, the Millennium Ecosystem Assessment found. As a result, restoring ecosystem services stands as a prime avenue for alleviating poverty.

**FAQ** **Doesn't environmental protection hurt the economy?**

We often hear it said that policy measures to protect environmental quality cost too much money, interfere with progress, or lead to job loss. However, growing numbers of economists dispute this. Instead, they assert that environmental protection tends to *enhance* our economic bottom line and nearly always improves our quality of life. The view one takes depends in part on whether one thinks in the short term or the long term. We often make economic judgments on short time scales, and in the short term many activities that cause environmental damage may be economically profitable. In the longer term, however, environmental degradation harms economies. Moreover, when resource extraction or development degrades environmental conditions, often a few private parties benefit economically while the broader public is harmed.

Today, concern over fossil fuels, pollution, and climate change has led many people to see immense opportunities in revamping our economies with clean and renewable energy technologies. The jobs, investment, and economic activity that come with building a green energy economy suggest how economic progress and environmental protection can go hand-in-hand.

## Economic theory moved from "invisible hand" to supply and demand

As the field of economics developed in the 18th century, many people argued that individuals acting in their own self-interest harm society. However, Scottish philosopher **Adam Smith** (1723–1790) argued that self-interested economic behavior can benefit society, as long as the behavior is constrained by the rule of law and private property rights in a competitive marketplace. A founder of **classical economics,** Smith famously wrote that when people pursue their economic self-interest under these conditions, the marketplace will behave as if guided by "an invisible hand" to benefit society as a whole.

Today, Smith's philosophy remains a pillar of free-market thought, which many credit for the tremendous gains in material wealth that capitalist market economies have achieved. Others contend that free-market policies tend to intensify environmental degradation and worsen inequalities between rich and poor.

Economists subsequently adopted more quantitative approaches. **Neoclassical economics** examines consumer choices and explains market prices in terms of our preferences for units of particular commodities. In neoclassical economics, buyers desire a low price, whereas sellers desire a high price. This conflict results in a compromise price being reached and the "right" quantity of commodities being bought and sold. This is phrased in terms of *supply*, the amount of a product offered for sale at a given price, and *demand*, the amount of a product people will buy at a given price if free to do so.

To evaluate an action or decision, economists use **cost-benefit analysis**, which compares the estimated costs of a proposed action with the estimated benefits. If benefits exceed costs, the action should be pursued; if costs exceed benefits, it should not. Given a choice of actions, the one with the greatest excess of benefits over costs should be chosen.

This reasoning seems eminently logical, but problems arise when not all costs and benefits can be easily identified, defined, or quantified. It may be simple to quantify the dollar value of bananas grown or cattle raised on a tract of Costa Rican land cleared for agriculture, yet difficult to assign monetary value to the various ecological costs of clearing the forest. Because monetary benefits are usually more easily quantified than environmental costs, benefits tend to be overrepresented in traditional cost-benefit analyses. As a result, environmental advocates often feel such analyses are predisposed toward economic development and against environmental protection.

## Neoclassical economics has environmental consequences

Today's market systems operate largely in accordance with the principles of neoclassical economics. These systems have generated unprecedented material wealth for our societies, yet four basic assumptions of neoclassical economics often contribute to environmental degradation.

**Replacing resources** One assumption is that natural resources and human resources (such as workers or technologies) are largely substitutable and interchangeable. This implies that once we have depleted a resource, we will always be able to find some replacement for it. As a result, the market imposes no penalties for depleting resources.

It is true that many resources can be replaced. However, Earth's material resources are ultimately limited. Nonrenewable resources can be depleted, and many renewable resources (such as soils, fish stocks, timber, and clean water) can be used up if we exploit them faster than they are replenished.

**External costs** A second assumption of neoclassical economics is that all costs and benefits associated with an exchange of goods or services are borne by individuals engaging directly in the transaction. In other words, it is assumed that the costs and benefits are "internal" to the transaction, experienced by the buyer and seller alone.

However, many transactions affect other members of society. When a landowner fells a forest, people nearby suffer poorer water quality, dirtier air, and less wildlife. When a factory, power plant, or mining operation pollutes the air or water, it harms the health of those who live nearby. In such cases, people who are not involved in degrading the environment end up paying the costs. Costs of a transaction that affect people other than the buyer or seller are known as **external costs** (**FIGURE 5.3**). Often whole communities suffer external costs while certain individuals enjoy private gain. External costs commonly include the following:

- Health problems, stress, or anxiety among people downstream or downwind from a pollution source
- Declines in resources, such as fewer fish in a stream
- Aesthetic damage, such as from air pollution, clear-cutting, or strip-mining
- Declining real estate values, lost tourism revenue, higher health care expenses, and more

**FIGURE 5.3 Air pollution is one common type of external cost.** Here, residents of an Indonesian town cycle through smoke from fires set to clear nearby forests for oil palm plantations.

If market prices do not take the social, ecological, or economic costs of environmental degradation into account, then taxpayers bear the burden of paying them. When economists ignore external costs, this creates a false impression of the consequences of our choices. External costs are one reason that governments develop environmental policy (p. 102).

**Discounting** Third, neoclassical economics grants an event in the future less value than one in the present. In economic terminology, future effects are "discounted." Discounting is meant to reflect the way people value things (we'd all rather have an ice cream cone today than be promised one next month). Yet giving more weight to current costs and benefits than to future costs and benefits encourages us to ignore the long-term consequences of our decisions.

Many environmental problems unfold gradually, and discounting causes us to downplay the impacts on future generations of pollution we create and resources we deplete today. Discounting has emerged as a flashpoint in the debate over how to respond to climate change. Economists agree that climate change will impose major costs on society, but they differ on how much to discount future effects—and so they differ on how much we should invest today to battle climate change.

**Growth** **Economic growth** can be defined as an increase in an economy's production and consumption of goods and services. Neoclassical economics assumes that economic growth is essential for maintaining social order, because a growing economy can alleviate the discontent of the poor by creating opportunities for poor people to become wealthier. A rising tide raises all boats, as the saying goes; if we make the overall economic pie larger, then each person's slice can become larger (even if some people still have much smaller slices than others). However, critics of the growth paradigm maintain that endless growth cannot be sustained, because resources to support growth are ultimately limited.

## How sustainable is economic growth?

Our global economy is seven times the size it was just half a century ago. All measures of economic activity are greater than ever before. Economic expansion has brought unprecedented material wealth to many people (although the rich have gained more than the poor, and gaps between haves and have-nots are wide and growing).

Economic growth can occur in two ways: (1) by an increase in inputs to the economy (such as more labor or natural resources) or (2) by improvements in the efficiency of production due to better methods or technologies (ideas or equipment that enable us to produce more goods with fewer inputs).

As our population grows and consumption rates rise, it is becoming clearer that we cannot sustain growth forever by using the first approach. Nonrenewable resources are finite in quantity, and renewable resources can also be exhausted if we overexploit them (as is happening with many fisheries today). As for the second approach to growth, we have used technological innovation to push back the limits on growth time and again. More-efficient technologies for extracting minerals, fossil fuels, and groundwater allow us to mine these resources more fully with less waste. Better machinery in our factories speeds manufacturing. We continue to make computer chips more powerful with less material input. In such ways, we are producing more goods and services with relatively fewer resources.

Can we conclude, then, that technology will allow us to overcome all environmental constraints and continue economic growth forever? We can certainly continue to innovate and achieve further efficiency. Yet ultimately, if our population and consumption continue to grow and we do not shift to full reuse and recycling, we will continue to diminish our natural capital, putting ever-greater demands on our capacity to innovate.

More and more economists recognize the challenges of achieving growth sustainably. Those in the field of **environmental economics** feel we can modify neoclassical economic principles to make resource use more efficient and thereby attain sustainability within our current economic system. Environmental economists were the first to develop methods to tackle the problems of external costs and discounting.

Economists in the field of **ecological economics** feel that sustainability requires more far-reaching changes. They stress that in nature, every population has a carrying capacity (p. 62) and systems generally operate in self-renewing cycles. Ecological economists maintain that societies, like natural populations, cannot surpass environmental limitations. Many of these economists advocate economies that neither grow nor shrink, but are stable. Such **steady-state economies** are intended to mirror natural systems. Critics of steady-state economies assert that to halt growth would dampen our quality of life. Proponents respond that technological advances would continue under a steady-state economy and wealth and happiness would rise.

Attaining sustainability will certainly require the reforms pioneered by environmental economists and may require the fundamental shifts advocated by ecological economists. One approach they each take is to assign monetary values to ecosystem goods and services, so as to better integrate them into traditional cost-benefit analyses.

## We can assign monetary value to ecosystem goods and services

Ecosystems provide us essential resources and life-support services, including arable soil, waste treatment, clean water, and clean air. Yet we often abuse the very ecological systems that sustain us. Why? From the economist's perspective, people overexploit natural resources and processes largely because the market assigns these entities no quantitative monetary value—or assigns values that underestimate their true worth.

Ecosystem services are said to have **nonmarket values,** values not usually included in the price of a good or service (**FIGURE 5.4**). For example, the aesthetic and recreational pleasure we obtain from natural landscapes is something of real value. Yet because we do not pay money for this, its value is hard to quantify and appears in no traditional measures of economic worth. Or consider Earth's water cycle (pp. 38–39): Rain fills our reservoirs with drinking water, rivers give us hydropower and flush away our waste, and water evaporates, purifying itself of contaminants and later falling as rain. This natural cycle is vital to our very existence, yet because we do not pay for it, markets impose no financial penalties when we disturb it.

For these reasons, economists have sought ways to assign market values to ecosystem services. They use surveys to determine how much people are willing to pay to protect or restore a resource. They measure the money, time, or effort people expend to travel to parks. They compare housing prices for similar homes in different settings to infer the dollar value of landscapes, views, or peace and quiet. They calculate how much it costs to restore natural systems that are damaged, to replace their functions with technology, or to clean up pollution.

For example, in Costa Rica, a team led by Taylor Ricketts of Stanford University studied native bees at a coffee plantation. By carefully measuring how bees pollinated the coffee plants and comparing the resulting coffee production in areas near forest and far from forest, Ricketts calculated that forests were providing the farm with pollination services worth $60,000 per year.

**FIGURE 5.4 Accounting for nonmarket values may help us make better environmental and economic decisions.**

**(a) Use value:** The worth of something we use directly

**(b) Existence value:** The worth of knowing that something exists, even if we never experience it ourselves

**(c) Option value:** The worth of something we might use later

**(d) Aesthetic value:** The worth of something's beauty or emotional appeal

**(e) Scientific value:** The worth of something for research

**(f) Educational value:** The worth of something for teaching and learning

**(g) Cultural value:** The worth of something that sustains or helps define a culture

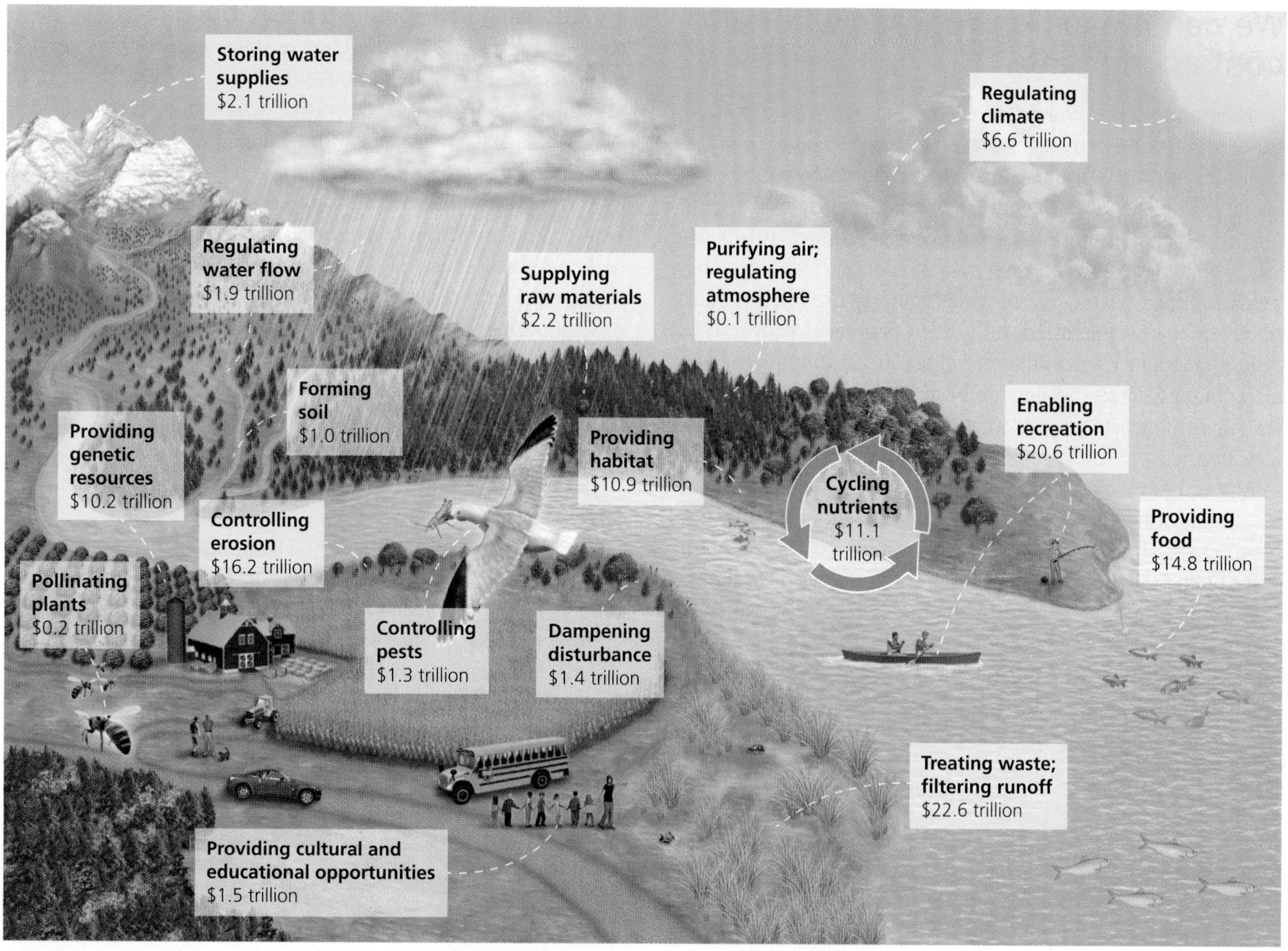


**FIGURE 5.5 Ecological economists have estimated the value of the world's ecosystem services at more than $143 trillion (in 2014 dollars).** This amount is an underestimate because it does not include ecosystems and services for which adequate data were unavailable. Shown are subtotals for each ecosystem service, in 2007 dollars.
*Data from Costanza, R., et al., 2014. Changes in the global value of ecosystem services.* Global Env. Change *26: 152–158.*

Researchers have even set out to calculate the total economic value of all the services that oceans, forests, wetlands, and other systems provide across the world. Teams headed by ecological economist Robert Costanza have combed the scientific literature and evaluated hundreds of studies that estimated dollar values for 17 major ecosystem services (**FIGURE 5.5**). The researchers reanalyzed the data using multiple valuation techniques to improve accuracy, then multiplied average estimates for each ecosystem by the global area it occupied. Their initial analysis in 1997 was groundbreaking, and in 2014 they updated their research. The 2014 study calculated that Earth's biosphere in total provides more than $125 trillion worth of ecosystem services each year, in 2007 dollars. This is equal to $143 trillion in 2014 dollars, an amount that exceeds the GDP of all nations combined!

Costanza also joined Andrew Balmford and 17 other colleagues to compare the benefits and costs of preserving natural systems intact versus converting wild lands for agriculture, logging, or fish farming. After reviewing many studies, they reported that a global network of nature reserves covering 15% of Earth's land surface and 30% of the ocean would be worth $4.4 to $5.2 trillion. This amount is 100 times greater than the value of those areas were they to be converted for direct exploitative human use.

Such research has sparked debate. Some ethicists argue that we should not put dollar figures on amenities such as clean air and water, because they are priceless and we would perish without them. Others say that arguing for conservation purely on economic grounds risks not being able to justify it whenever it fails to deliver clear economic benefits. However, backers of the research counter that valuation does not argue for making decisions on monetary grounds alone, but instead clarifies and quantifies values that we already hold implicitly.

In 2010, researchers wrapped up a large international effort to summarize and assess attempts to quantify the economic value of natural systems. *The Economics of Ecosystems and Biodiversity* study has published a number of fascinating reports that you can find online. This effort describes the valuation of nature's economic worth as "a tool to help recalibrate [our] faulty economic compass." It concludes that this is useful because "the invisibility of biodiversity values has often encouraged inefficient use or even destruction of the natural capital that is the foundation of our economies."

## We can measure progress with full cost accounting

If assigning market values to ecosystem services gives us a fuller and truer picture of costs and benefits, then we can take a similar approach in measuring our economic progress as a society. For decades, we have assessed each nation's economy by calculating its **Gross Domestic Product (GDP),** the total monetary value of final goods and services the nation produces each year. Governments regularly use GDP to make policy decisions that affect billions of people. However, GDP fails to account for nonmarket values. It also lumps together all economic activity, desirable and undesirable. GDP can rise in response to crime, war, pollution, and natural disasters, because we spend money to protect ourselves from these things and to recover from them.

Environmental economists have developed indicators meant to distinguish desirable from undesirable economic activity and to better reflect our well-being. One such alternative to the GDP is the **Genuine Progress Indicator (GPI).** To calculate GPI, we begin with conventional economic activity and add to it positive contributions not paid for with money, such as volunteer work and parenting. We then subtract negative impacts, such as crime and pollution (**FIGURE 5.6a**).

GPI can differ strikingly from GDP: **FIGURE 5.6b** compares these indices for the United States across 50 years. On a per-person basis, the nation's GDP rose greatly, but its GPI remained flat for 30 years. This discrepancy suggests that Americans have been spending more and more money but that their quality of life is not improving.

The GPI is an example of **full cost accounting,** also called **true cost accounting,** because it aims to account for all costs and benefits. Several U.S. states are beginning to use the GPI to measure progress and help guide policy. Since Maryland's governor embraced the approach in 2010, that state's GPI has grown faster than its GDP.

Critics of full cost accounting argue that the approach is subjective and too easily driven by ideology. Proponents respond that making a subjective attempt to measure progress is better than using a more objective indicator such as the GDP in a task for which it wasn't intended.

Today, attempts are gaining ground to measure happiness (rather than economic output) as the prime goal of national policy. The small Asian nation of Bhutan has pioneered this approach with its measure of Gross National Happiness. Another indicator is the Happy Planet Index, which measures how much happiness we gain per amount of resources we consume. By this measure, Costa Rica was calculated to be the top nation in the world.

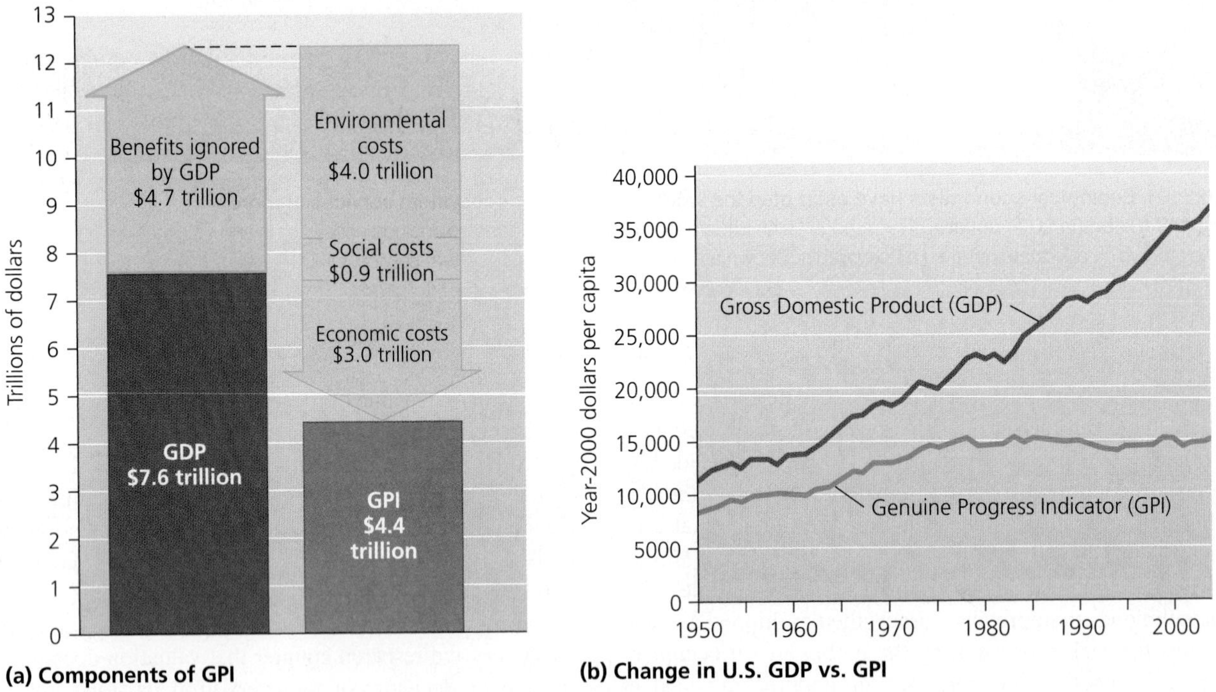


**FIGURE 5.6 Full cost accounting indicators such as the GPI attempt to measure progress and well-being more effectively than pure economic indicators such as the GDP.** We see in **(a)** how the Genuine Progress Indicator **(orange bar at right)** adds to the Gross Domestic Product **(red bar at left)** benefits such as volunteering and parenting **(upward gold arrow).** The GPI then subtracts external environmental costs such as pollution, social costs such as divorce and crime, and economic costs such as borrowing and the gap between rich and poor **(downward gold arrow).** Shown are values for the United States in 2004. We see in **(b)** that per capita U.S. GDP has increased dramatically since 1950, yet per capita U.S. GPI leveled off after 1975. *Data from Talberth, J., C. Cobb, and N. Slattery, 2007.* The Genuine Progress Indicator 2006: A tool for sustainable development. *Redefining Progress, Oakland, CA. By permission of John Talberth, Ph.D. All data are adjusted for inflation by using year-2000 dollars.*

**DATA Q** What was the ratio of GDP to GPI in 1950? (Divide GDP by GPI.) What was this ratio in the year you were born? What was this ratio in 2004? What do the changes in these values indicate to you?



GO TO **INTERPRETING GRAPHS & DATA** ON MasteringEnvironmentalScience®

Costa Rica is also one of five nations working with the World Bank (p. 108) in a program to implement full cost accounting methods. Together they are addressing questions such as how much economic benefit the nation's forests, national parks, and other natural amenities generate through tourism and watershed protection.

**FAQ** **Does having more money make a person happier?**

This age-old question has long been debated in the realm of philosophy. In recent years, though, social scientists have conducted serious research on the issue. So far, studies have found a surprising degree of consensus: In general, we become happier as we get wealthier, but once we gain a moderate level of wealth (roughly $50,000–$90,000 in yearly income), attaining further money no longer increases our happiness. Apparently reaching a basic level of financial security alleviates day-to-day economic worries, but once those worries are taken care of, our happiness revolves around other aspects of our lives (such as family, friends, and the satisfaction of helping others). Research on happiness can help us guide our personal life decisions. It also suggests that enhancing a society's happiness might best be achieved by raising many people's incomes up by a little, rather than by raising some people's incomes up by a lot.

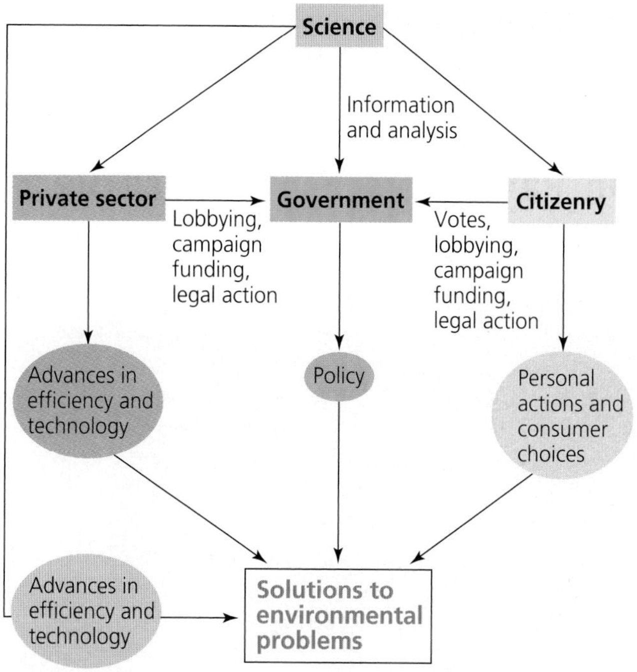


**FIGURE 5.7 Policy plays a central role in addressing environmental problems.**

## Markets can fail

When markets do not take into account the positive outside effects on economies (such as ecosystem services) or the negative side effects of economic activity (external costs), economists call this **market failure.** Traditionally, we have tried to counteract market failure by using government intervention. Government can restrain individual and corporate behavior through laws and regulations. It can tax harmful activities. It can also design economic incentives that use market mechanisms to promote fairness, resource conservation, and economic sustainability. Paying for the conservation of ecosystem services, as Costa Rica does, is one way of deploying economic incentives toward policy goals. We will now examine these approaches in our discussion of environmental policy.

# Environmental Policy: An Overview

When a society comes to feel that a problem exists, its leaders may try to resolve the problem using **policy,** a formal set of general plans and principles intended to guide decision making. **Public policy** is policy made by governments. **Environmental policy** pertains to our interactions with our environment. Environmental policy generally aims to regulate resource use or reduce pollution to promote human welfare and protect natural systems.

Forging effective policy requires input from science, ethics, and economics. Science provides information and analyses needed to identify and understand problems and devise solutions. Ethics and economics offer criteria by which to assess problems and help clarify how society might address them. Government interacts with citizens, organizations, and the private sector to formulate policy (**FIGURE 5.7**).

## Environmental policy addresses issues of fairness and resource use

Because market capitalism is driven by incentives for short-term economic gain rather than long-term social and environmental stability, it provides businesses and individuals little motivation to minimize environmental impact or to equalize costs and benefits among parties. As we noted, such *market failure* has traditionally been viewed as justification for government involvement. Governments typically intervene in the marketplace for several reasons:

- To provide social services, such as national defense, healthcare, and education
- To provide "safety nets" (for the elderly, the poor, victims of natural disasters, and so on)
- To eliminate unfair advantages held by single buyers or sellers
- To manage publicly held resources
- To minimize pollution and other threats to health and quality of life

Environmental policy aims to protect people's health and well-being, to safeguard environmental quality and conserve natural resources, and to promote equity or fairness in people's use of resources.

## Do Payments Help Preserve Forest?

**Workers on a Costa Rican farm plant native trees for Arbofilia, an organization helping farmers to restore forest.**

Costa Rica's program to pay for ecosystem services has garnered international praise and inspired other nations to implement similar policies. But have Costa Rica's payments actually been effective in preventing forest loss? A number of research teams have sought to answer this surprisingly difficult question by analyzing data from the PSA program.

Some early studies were quick to credit the PSA program for saving forests. A 2006 study conducted for FONAFIFO, the agency administering the program, concluded that PSA payments in the central region of the country had prevented 108,000 ha (267,000 acres) of deforestation—38% of the area under contract. Indeed, deforestation rates fell as the program proceeded; rates of forest clearance in 1997–2000 were half what they were in the preceding decade.

However, some researchers hypothesized that PSA payments were not responsible for this decline and that forest loss would have slowed anyway because of other factors. To test this hypothesis, a team led by G. Arturo Sanchez-Azofeifa of the University of Alberta and Alexander Pfaff of Duke University worked with FONAFIFO's payment data, as well as data on land use and forest cover from satellite surveys. They layered these data onto maps using a geographic information system (GIS) (p. 34) and explored the patterns revealed.

In 2007 in the journal *Conservation Biology,* they reported that only 7.7% of PSA contracts were located within 1 km of regions where forest was at greatest risk of clearance. PSA contracts were only slightly more likely to be near such a region than far from it. This meant, they argued, that PSA contracts were not being targeted to regions where they could have the most impact.

Moreover, since enrollment was voluntary, most landowners applying for payments likely had land unprofitable for agriculture and were not actually planning to clear forest (**FIGURE 1**). In a 2008 paper, these researchers compared lands under PSA contracts with similar lands not under contracts. PSA lands experienced no forest loss, whereas the deforestation rate on non-PSA lands was 0.21%/yr. However, their analyses indicated that PSA lands had only a 0.08%/yr likelihood of being cleared in the first place, suggesting that the program prevented only 0.08%/yr of forest loss, not 0.21%/yr. Other research was bearing this out; at least two studies found that many PSA participants, when interviewed, said they would have retained their forest even without the PSA program.

These researchers argued that Costa Rica's success in halting forest loss was likely due to other factors. In particular, Forest Law 7575, which had established the PSA system, had also banned forest clearing nationwide. This top-down government mandate, assuming it was enforceable, in theory made the PSA payments unnecessary. However, the PSA program

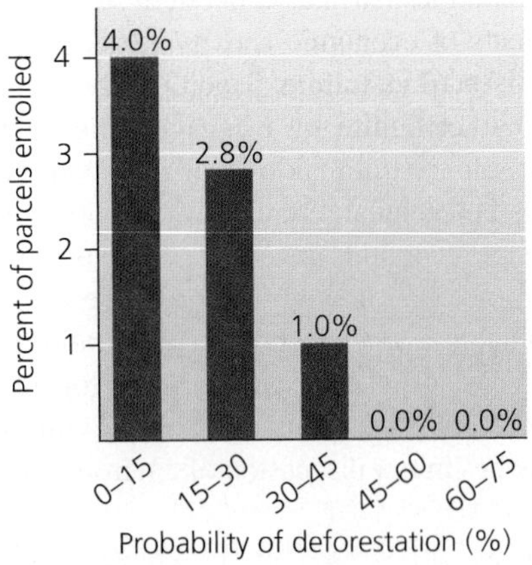


**FIGURE 1 In areas at greater risk of deforestation, lower percentages of land parcels were enrolled in the PSA program.** This is because land more profitable for agriculture was less often enrolled. *Data: Pfaff, A., et al. 2008.* Payments for environmental services: Empirical analysis for Costa Rica. *Working Papers Series SAN08-05, Terry Sanford Institute of Public Policy, Duke University.*

**The tragedy of the commons** When publicly accessible resources are open to unregulated exploitation, they tend to become overused, damaged, or depleted. So argued environmental scientist Garrett Hardin in his 1968 essay "The Tragedy of the Commons." Basing his argument on an age-old scenario, Hardin explained how in a public pasture (or "common") open to unregulated grazing, each person who grazes animals will be motivated by self-interest to increase the number of his or her animals in the pasture. Because no single person owns the pasture, no one has

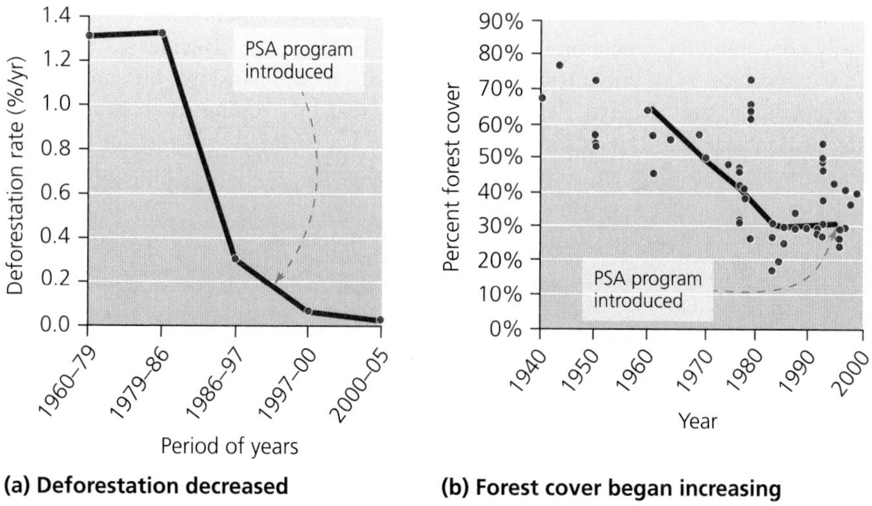


**FIGURE 2 Forest recovery was underway in Costa Rica before the PSA program began.** Deforestation rates **(a)** had already dropped steeply, indicating that other factors were responsible. Forest cover **(b)** also began increasing shortly before the program's initiation (the graph gathers together estimates from many studies). *Data: From Daniels, Amy E., et al. 2010. Understanding the impacts of Costa Rica's PES: Are we asking the right questions?* Ecological Economics *69: 2116–2126.*

made the mandate far more palatable to legislators, and Forest Law 7575 might never have passed had it not included the PSA payments.

Despite the PSA program's questionable impact in preserving existing forest, scientific studies show that it has been effective in regenerating new forest. In Costa Rica's Osa Peninsula, Rodrigo Sierra and Eric Russman of the University of Texas at Austin found in 2006 that PSA farms had five times more regrowing forest than did non-PSA farms. Interviews with farmers indicated that the program encouraged them to let land grow back into forest if they did not soon need it for production.

In the nation's northern Caribbean plain, a team led by Wayde Morse of the University of Idaho combined satellite data with on-the-ground interviews, finding that PSA payments plus the clearance ban reduced deforestation rates from 1.43%/year to 0.10%/year and that the program encouraged forest regrowth still more. Meanwhile, dissertation work by Rodrigo Arriagada indicated that the regeneration of new forest seemed to be the PSA program's major effect at the national level as well.

Most researchers today hold that Costa Rica's forest recovery results from a long history of conservation policies and economic developments. Data show that deforestation rates had been dropping before the PSA program was initiated (**FIGURE 2**). There are several major reasons:

- Earlier policies (tax rebates and tax credits for timber production) encouraged forest cover.
- The creation of national parks fed a boom in ecotourism, so Costa Ricans saw how conserving natural areas could bring economic benefits.
- Falling market prices for meat discouraged ranching.
- After an economic crisis roiled Latin America in the 1980s, Costa Rica ended subsidies that had encouraged ranchers and farmers to expand into forested areas.

To help the PSA program make better use of its money, most scientists today feel that PSA payments should be targeted. Instead of paying equal amounts to anyone who applies, FONAFIFO should prioritize applicants, or pay more money, in regions that are ecologically most valuable or that are at greatest risk of deforestation. In a 2008 study, Tobias Wünscher of Bonn, Germany, and colleagues modeled and tested seven possible ways to target the payments, using data from Costa Rica's Nicoya Peninsula. Wünscher's team suggested using auctions, in which applicants for PSA funds put in bids stating how much they were requesting. Because applicants have outnumbered available contracts 3 to 1, FONAFIFO could favor the lower bids to keep costs down, while the auction system could make differential payments politically acceptable.

Costa Rica's government is responding to suggestions from researchers by aiming payments toward regions of greater environmental value and by making the program more accessible to low-income farmers in undeveloped regions. The government has also raised the payment amounts considerably. Researchers—and other nations—are watching closely to see how the program develops. ■

incentive to expend effort taking care of it. Instead, each person takes what he or she can until overgrazing causes the pasture's food production to collapse, hurting everyone. This scenario, known as the **tragedy of the commons,** pertains to many resources held and used in common by the public: forests, fisheries, clean air, clean water—even the global climate.

When shared resources are being depleted or degraded, it is in society's interest to develop guidelines for their use. In Hardin's example, guidelines might limit the number of animals each person can graze or might require pasture users to help restore and manage the resource. These two concepts—restriction of use, and management—are central to environmental policy today.

Public oversight through government is a standard way to alleviate the tragedy of the commons, but we can also address it in other ways. Resource users can voluntarily cooperate to prevent overexploitation. This may work if the resource is localized and enforcement is simple, but these conditions are rare. Alternatively, the resource can be subdivided and allotments sold into private ownership, so that each owner gains incentive to manage his or her portion. Privatization may be effective if property rights can be clearly assigned (as with land), but it tends not to work with resources such as air or water. Privatization also opens the door to short-term profit-taking at the long-term expense of the resource.

**Free riders** A second reason we develop policy for publicly held resources is the **free rider** predicament. Let's say a community on a river suffers from water pollution that emanates from 10 different factories. The problem could in theory be solved if every factory voluntarily agreed to reduce its own pollution. However, once they all begin reducing their pollution, it becomes tempting for any one of them to stop doing so. A factory that avoids the efforts others are making would in essence get a "free ride." If enough factories take a free ride, the whole collective endeavor will collapse. Because of the free rider problem, private voluntary efforts are often less effective than efforts mandated by public policy.

**External costs** Environmental policy also aims to promote fairness by dealing with external costs (pp. 94–95). For example, a factory that discharges waste into a river imposes external costs (water pollution, health impacts, reduced fish populations, aesthetic impacts) on downstream users of the river. If the government forces the company to clean up its pollution, pay fees, or reimburse affected residents, this helps to "internalize" costs. The costs are paid by the company, which will likely pass them on to consumers by raising the prices of its products. Higher market prices may reduce demand for the products, and consumers may instead favor less-expensive products whose production imposes fewer costs on society.

### WEIGHING THE ISSUES

**Internalizing External Costs** Imagine that we were to use policy to internalize all the external costs of gasoline (pollution, health impacts, climate change, damage from oil drilling and transport, etc.), and that as a result, gas prices rise to $13 per gallon. What effects do you think this would have on the choices we make as consumers (such as driving behavior and types of vehicles purchased)? What influence might it have on the types of vehicles produced and the types of energy sources developed? What effects might it have on our taxes and our health insurance premiums? In the long run, do you think that internalizing external costs in this way would end up costing society more money, or saving society money? What factors might be important in determining the outcome?

## Various factors can obstruct environmental policy

If the goals of environmental policy are so noble, why are environmental laws and regulations often challenged? One reason is the perception that environmental protection means economic sacrifice (see FAQ, p. 94). Businesses and individuals often view regulations as restrictive, bureaucratic, or costly. Landowners may fear that zoning (p. 413) or protections for endangered species (p. 179) will restrict how they can use their land. Developers complain of time and money lost in obtaining permits; reviews by government agencies; and required environmental controls, monitoring, and mitigation.

Another hurdle for environmental policy stems from the nature of environmental problems, which often develop gradually over long time periods. Human behavior is geared toward addressing short-term needs, and this is reflected in our social institutions. Businesses usually opt for short-term financial gain. The news media give more coverage to new and sudden events than gradual long-term trends. Politicians often act in their short-term interest because they depend on reelection every few years. For all these reasons, environmental policy may be obstructed.

Policy in general can be held up for many reasons, even if a majority of people favor it. Checks and balances in a constitutional democracy seek to ensure that new policy is implemented only after extensive review and debate. However, less desirable factors can also hinder policy. In democracies such as the United States, each person has a political voice and can make a difference—yet money wields influence. People, organizations, industries, or corporations with enough wealth to buy access to power exert disproportionate influence over policymakers.

## Science informs policy but is sometimes disregarded

Policy that is effective is generally informed by science. For instance, Costa Rica's PSA program was inspired by scientific research into the importance of ecosystem services. Once researchers diagnosed shortcomings in the way the program was being run, policymakers responded with remedies. When deciding whether to regulate a substance that may pose a public health risk, government agencies may comb the scientific literature for information or commission new studies to resolve outstanding questions. When crafting a bill to reduce pollution, a legislator may consult scientific data that quantifies impacts of the pollution or that predicts benefits from its reduction. In today's world, a nation's strength depends on its commitment to science. This is why governments devote a portion of our taxes to fund scientific research.

Alas, sometimes policymakers allow political ideology, rather than science, to determine policy on scientific matters. Politicians may ignore scientific consensus on well-established matters such as evolution, vaccination, or climate change if it suits their political needs. Some may reject or distort scientific advice if this helps to please campaign contributors or powerful constituencies. Whenever taxpayer-funded science is suppressed or distorted for political ends, society loses. We cannot take for granted that science will play a role in policy. As scientifically literate citizens of a democracy, we need to stay vigilant and help ensure that our representatives in government are making proper use of the tremendous scientific assets we have at our disposal.

# U.S. Environmental Law and Policy

The United States provides a good focus for understanding environmental policy in constitutional democracies worldwide for several reasons. First, the United States has pioneered innovative environmental policy. Second, U.S. policies serve as models—of both success and failure—for other nations and for international bodies. Third, the United States exerts a great deal of influence on the affairs of other nations. Finally, understanding U.S. policy at the federal level helps us to understand policy at local, state, and international levels.

## The federal government's three branches shape policy

Federal policy in the United States results from actions of the legislative, executive, and judicial branches of government. Congress creates laws, or **legislation,** by crafting bills that can become law with the signature of the head of the executive branch, the president. Once a law is enacted, its implementation and enforcement is assigned to an administrative agency in the executive branch. Administrative agencies create **regulations,** specific rules intended to achieve objectives of a law. These agencies also monitor compliance with laws and regulations. Several dozen administrative agencies influence U.S. environmental policy, ranging from the Environmental Protection Agency to the Forest Service to the Food and Drug Administration to the Bureau of Land Management.

The judiciary, consisting of the Supreme Court and various lower courts, is charged with interpreting law and is an important arena for environmental policy. Grassroots environmental advocates and organizations use lawsuits to help level the playing field with large corporations and agencies. Conversely, the courts hear complaints from businesses and individuals challenging the constitutional validity of environmental laws they feel to be infringing on their rights. Individuals and organizations also lodge suits against government agencies when they feel the agencies are failing to enforce their own regulations.

The structure of the federal government is mirrored at the state level with governors, legislatures, judiciaries, and agencies. States, counties, and municipalities all generate policy of their own. They can act as laboratories experimenting with novel ideas, so that policies that succeed may be adopted elsewhere. In the "cooperative federalism" approach, a federal agency sets national standards and then works with state agencies to achieve them in each state.

## Early U.S. environmental policy promoted development

Environmental policy in the United States was created in three periods. Laws enacted during the first period, from the 1780s to the late 1800s, accompanied the westward expansion of the nation and were intended mainly to promote settlement and the extraction and use of the continent's abundant natural resources (**FIGURE 5.8**).

Among these early laws were the *General Land Ordinances of 1785 and 1787,* by which the new federal government gave itself the right to manage the lands it was expropriating from Native Americans. These laws created a grid system for surveying these lands and readying them for private ownership.

**(a) Settlers in Nebraska, circa 1860**

**(b) Loggers felling an old-growth tree, Washington**

**FIGURE 5.8 Early U.S. environmental policy promoted settlement and natural resource extraction.** The Homestead Act of 1862 allowed settlers **(a)** to claim, for a $16 fee, 160 acres of land by living there for 5 years and farming or building a home. The timber industry was allowed to clear-cut the nation's ancient forests **(b)** with little policy to encourage conservation. The General Mining Act of 1872 legalized and promoted mining by private individuals on public land for just $5 per acre, with no government oversight.

From 1785 onward, the government promoted settlement in the Midwest and West and doled out millions of acres to its citizens and to railroad companies, encouraging settlers, entrepreneurs, and land speculators to move west.

Western settlement provided U.S. citizens with means to achieve prosperity while relieving crowding in Eastern cities. It expanded the geographical reach of the United States at a time when the young nation was still jostling with European powers for control of the continent. It also wholly displaced the millions of Native Americans who had inhabited these lands for millennia. U.S. environmental policy of this era reflected a perception that the vast western lands were inexhaustible in natural resources.

## The second wave of U.S. environmental policy encouraged conservation

In the late 1800s, as the continent became more populated and its resources were increasingly exploited, public perception and government policy toward natural resources began to shift. Reflecting the emerging conservation and preservation ethics (p. 13) in American society, laws of this period aimed to alleviate some of the environmental impacts of westward expansion.

In 1872, Congress designated Yellowstone the world's first national park. In 1891, Congress authorized the president to create forest reserves to prevent overharvesting and protect forested watersheds. In 1903, President Theodore Roosevelt created the first national wildlife refuge. These acts launched the creation of a national park system, national forest system, and national wildlife refuge system that still stand as global models (pp. 194, 199). These developments reflected a new understanding that the continent's resources were exhaustible and required protection.

Land management policies continued through the 20th century, targeting soil conservation in the wake of the Dust Bowl (p. 142) and wilderness preservation with the Wilderness Act of 1964 (p. 199).

## The third wave responded to pollution

Further social changes in the 20th century gave rise to the third major period of U.S. environmental policy. In a more densely populated nation driven by technology, industry, and intensive resource consumption, Americans found themselves better off economically but living amid dirtier air, dirtier water, and more waste and toxic chemicals. Events in the 1960s and 1970s triggered greater awareness of environmental problems, bringing about a profound shift in public policy.

A landmark event was the 1962 publication of *Silent Spring,* a book by American scientist and writer Rachel Carson (**FIGURE 5.9**). *Silent Spring* awakened the public to the ecological and health impacts of pesticides and industrial chemicals (pp. 210–211). The book's title refers to Carson's warning that pesticides might kill so many birds that few would be left to sing in springtime.

Ohio's Cuyahoga River (**FIGURE 5.10**) also drew attention to pollution hazards. The Cuyahoga was so polluted with oil and industrial waste that the river actually caught fire near Cleveland a number of times in the 1950s and 1960s. This spectacle, coupled with an oil spill offshore from Santa Barbara, California, in 1969, moved the public to prompt Congress and the president to better safeguard water quality and public health. The first Earth Day event in 1970 helped to galvanize public support for action to address pollution problems.

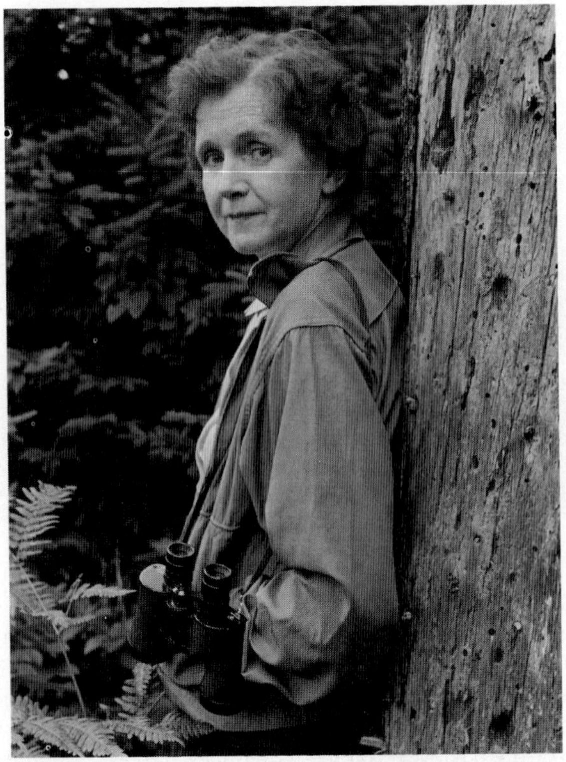

**FIGURE 5.9 Scientist and writer Rachel Carson revealed the effects of DDT and other pesticides in her 1962 book, *Silent Spring.***

**FIGURE 5.10 Ohio's Cuyahoga River was so polluted with oil and waste that the river caught fire multiple times in the 1950s and 1960s and would burn for days at a time.**

**TABLE 5.1** Major U.S. Environmental Protection Laws, 1963–1980

**Clean Air Act**
1963;
amended 1970 and 1990

Sets standards for air quality, restricts emissions from new sources, enables citizens to sue violators, funds research on pollution control, and established an emissions trading program for sulfur dioxide. As a result, the air we breathe today is far cleaner (pp. 284–287).

**Resource Conservation and Recovery Act**
1976

Sets standards and permitting procedures for the disposal of solid waste and hazardous waste (p. 392). Requires that the generation, transport, and disposal of hazardous waste be tracked "from cradle to grave."

**Endangered Species Act**
1973

Seeks to protect species threatened with extinction. Forbids destruction of individuals of listed species or their critical habitat on public and private land, provides funding for recovery efforts, and allows negotiation with private landholders (pp. 177–179).

**Clean Water Act**
1977

Regulates the discharge of wastes, especially from industry, into rivers and streams (p. 272). Aims to protect wildlife and human health, and has helped to clean up U.S. waterways.

**Safe Drinking Water Act**
1974

Authorizes the EPA to set quality standards for tap water provided by public water systems, and to work with states to protect drinking water sources from contamination.

**Soil and Water Conservation Act**
1977

Directs the U.S. Department of Agriculture to survey and assess soil and water conditions across the nation and prepare conservation plans. Responded to worsening soil erosion and water pollution on farms and rangeland as production intensified.

**Toxic Substances Control Act**
1976

Directs the EPA to monitor thousands of industrial chemicals and gives it power to ban those found to pose too much health risk (p. 221). However, the number of chemicals continues to increase far too quickly for adequate testing.

**CERCLA ("Superfund")**
1980

Funds the Superfund program to clean up hazardous waste at the nation's most polluted sites (p. 404). Costs were initially charged to polluters but most are now borne by taxpayers. The EPA continues to progress through many sites that remain. Full name is the Comprehensive Environmental Response Compensation and Liability Act.

Public demand for a cleaner environment during this period inspired a number of major laws that underpin modern U.S. environmental policy (TABLE 5.1). You will encounter many of them again later in this book, and they have already helped to shape the quality of your life.

Historians suggest that major advances in environmental policy occurred in the 1960s and 1970s because (1) environmental problems became readily apparent and were directly affecting people's lives, (2) people could visualize policies to deal with the problems, and (3) citizens were politically active and leaders were willing to act. In addition, photographs from NASA's space program allowed humanity to see, for the first time ever, images of Earth from space (see photos on pp. 1 and 425). It is hard for us to comprehend the power those images had at the time, but they revolutionized many people's worldviews by making us aware of the finite nature of our planet.

Today, largely because of policies enacted since the 1960s, our health is better protected and the nation's air and water are considerably cleaner. Thanks to the many citizens who worked tirelessly in grassroots efforts, and to the policymakers who listened and chose to make a difference in people's lives, we now enjoy a cleaner environment where industrial chemicals, waste disposal, and resource extraction are more carefully regulated. Much remains to be done, but all of us alive today owe a great deal to the dedicated people who inspired policy to tackle pollution during this period.

## Passage of NEPA and creation of the EPA were milestones

One of the foremost U.S. environmental laws is the **National Environmental Policy Act (NEPA),** signed into law by Republican President Richard Nixon in 1970. NEPA created an agency called the Council on Environmental Quality and required that an **environmental impact statement (EIS)** be prepared for any major federal action that might significantly affect environmental quality. An EIS summarizes results from studies that assess environmental impacts that could result from development projects undertaken or funded by the federal government.

The EIS process forces government agencies and businesses that contract with them to evaluate impacts in a cost-benefit approach (p. 94) before proceeding with a new dam, highway, building project, or similar action. The EIS process rarely halts development projects, but it serves as an incentive to lessen environmental damage. NEPA grants ordinary citizens input in the policy process by requiring that EISs be made publicly available and that policymakers solicit and consider public comment on them.

In 1970 policymakers also created the **Environmental Protection Agency (EPA).** The EPA was charged with conducting and evaluating research, monitoring environmental quality, setting and enforcing standards for pollution levels, assisting the states in meeting the standards, and educating the public.

## The social context for policy evolves

In the 1980s Congress strengthened, broadened, and elaborated upon the laws of the 1970s. But the political climate in the United States soon changed, and advocates of environmental protection watched their hard-won gains begin to erode. Although public support for the goals of environmental protection remained high, many people began to feel that the regulatory means used to achieve these goals too often imposed economic burdens on businesses or individuals. Attempts were made to roll back environmental policy, beginning with the administration of Ronald Reagan and culminating in an array of efforts by the George W. Bush administration and by the Congresses in power from 1994 through 2006.

Today in the United States, legal protections for public health and environmental quality remain strong in some areas but have been weakened in others. Past policies restricting toxic substances such as lead and DDT have improved public health, but scientists and regulators cannot keep up with the flood of new chemicals being introduced by industry. As demand for energy rises while concern over climate change intensifies, today's policy debates focus on oil exploration, coal-burning power plants, hydraulic fracturing ("fracking") to extract oil and gas, piping petroleum from oil sands, exporting coal to other nations, taxpayer subsidies, and mandates for renewable sources.

## Environmental policy advances today on the international stage

Amid the heightened partisanship of U.S. politics today, environmental policy has gotten caught in the political crosshairs. Despite the fact that some of the greatest early conservationists were Republicans, and even though the words *conservative* and *conservation* share the same root meaning, environmental issues have today become identified as a predominantly Democratic concern. As a result, significant bipartisan advances rarely occur, and the United States now wields less clout internationally on environmental policy.

Meanwhile, other nations have forged ahead with innovative policy. Germany has used policy to make impressive strides with solar energy (pp. 365–366). Sweden maintains a thriving society while promoting progressive environmental policies. Small developing nations such as Costa Rica and Bhutan are bettering their citizens' lives while protecting and restoring their natural capital. Even China, despite becoming the world's biggest polluter, is taking the world's biggest steps toward renewable energy, reforestation, and pollution control.

Worldwide, we have embarked on a fourth wave of environmental policy, one focused on sustainability and sustainable development (pp. 18, 111–112) to safeguard natural systems while raising living standards for the world's people. In addition, global climate change (Chapter 14) has come to dominate discussion of environmental policy (**FIGURE 5.11**). A series of international conferences (pp. 325–326) has brought together the world's nations to grapple with issues of how to reduce the greenhouse gas emissions that drive climate change.

**FIGURE 5.11 Concerns over climate change are driving environmental policy in all nations.** Here, college students and activists urge U.S. leaders to enact policies to help bring the atmosphere's carbon dioxide concentration back down to 350 parts per million. Many scientists feel this is the level needed to avert catastrophic climate change.

# International Environmental Policy

Environmental systems pay no heed to political boundaries, and neither do environmental problems. Climate change is a global issue because carbon pollution from any one nation spreads through the atmosphere and oceans, affecting all nations. Because one nation's laws have no authority in other nations, international policy is vital to solving "transboundary" problems in our globalizing world.

## Globalization makes international institutions vital

We live in an era of rapid and profound change. **Globalization** describes the process by which the world's societies have become more interconnected, linked by trade and communication technologies in countless ways. Globalization has brought us many benefits by facilitating the spread of ideas and technologies that empower individuals and enhance our lives. Billions of people enjoy a degree of access to news, education, arts, and science that we could barely have imagined in the past, and billions also now live under more democratic governments. A better awareness of other cultures promotes peace and understanding, and warfare is on the decline.

Yet as globalization proceeds, societies and ecological systems are altered at unprecedented rates and scales. People move organisms from one continent to another, allowing invasive species to affect ecosystems everywhere. Multinational corporations operate outside the reach of national laws and rarely have incentive to conserve resources or limit pollution while moving from nation to nation. Our biggest environmental challenges are now global in scale (such as climate change, ozone depletion, overfishing, and biodiversity loss)—yet we lack an adequate global legal framework to address these issues effectively. For all these reasons, in today's globalizing world the institutions that do shape international law and policy play increasingly vital roles.

## International law includes customary law and conventional law

International law known as **customary law** arises from long-standing practices, or customs, held in common by most cultures. International law known as **conventional law** arises from conventions, or treaties (written contracts), into which nations enter. One example is the United Nations Framework Convention on Climate Change, a 1994 treaty that established a framework for agreements to reduce greenhouse gas emissions that contribute to global climate change. The Kyoto Protocol (a *protocol* is an amendment or addition to a convention) later specified the agreed-upon details of the emissions limits (p. 325). TABLE 5.2 shows a selection of major environmental treaties ratified by most of the world's nations.

Treaties are also signed among pairs or groups of nations. The United States, Mexico, and Canada entered into the **North American Free Trade Agreement (NAFTA)** in 1994 (**FIGURE 5.12**). NAFTA eliminated trade barriers such as tariffs on imports and exports, making goods cheaper to buy. Yet NAFTA also threatened to undermine protections for workers and the environment by steering economic activity to areas where regulations were most lax. Side agreements aimed to address these concerns, and NAFTA's impacts on jobs and on environmental quality in the three nations have been complex. Some U.S. jobs moved to Mexico, but fears that pollution would soar and regulations would be gutted largely did not come to pass, while some sustainable products and practices spread from nation to nation. The debates over such issues recur with each proposed free trade agreement among nations, as we try to find ways to gain the benefits of free trade while avoiding environmental damage and harm to working people.

**TABLE 5.2 Major International Environmental Treaties**

| CONVENTION OR PROTOCOL | YEAR IT CAME INTO FORCE | NATIONS THAT HAVE RATIFIED IT | U.S. STATUS |
|---|---|---|---|
| **CITES:** Convention on International Trade in Endangered Species of Wild Fauna and Flora (p. 179) | 1975 | 175 | Ratified |
| **Ramsar Convention** on Wetlands of International Importance | 1975 | 159 | Ratified |
| **Montreal Protocol,** of the Vienna Convention for the Protection of the Ozone Layer (p. 294) | 1989 | 196 | Ratified |
| **Basel Convention** on the Control of Transboundary Movements of Hazardous Wastes and Their Disposal (p. 403) | 1992 | 172 | Signed but has not ratified |
| **Convention on Biological Diversity** (p. 179) | 1993 | 168 | Signed but has not ratified |
| **Stockholm Convention** on Persistent Organic Pollutants (p. 222) | 2004 | 152 | Signed but has not ratified |
| **Kyoto Protocol,** of the UN Framework Convention on Climate Change (p. 325) | 2005 | 184 | Signed but has not ratified |

**FIGURE 5.12 The North American Free Trade Agreement (NAFTA) eliminated trade barriers to make goods cheaper.** However, some U.S. manufacturing jobs moved to Mexico (such as to this garment factory in Tehuacan), where wages are lower and health and environmental regulations are more lax.

## Several organizations shape international environmental policy

In this age of globalization, a number of international institutions act to influence the policy and behavior of nations by providing funding, applying political or economic pressure, and directing media attention.

**The United Nations** Founded in 1945 and including representatives from all nations of the world, the **United Nations (U.N.)** seeks "to maintain international peace and security; to develop friendly relations among nations; to cooperate in solving ... problems and in promoting respect for human rights and fundamental freedoms; and to be a centre for harmonizing the actions of nations in attaining these ends." Headquartered in New York City, the United Nations plays an active role in environmental policy by sponsoring conferences, coordinating treaties, and publishing research.

**The World Bank** Established in 1944 and based in Washington, D.C., the **World Bank** is one of the largest funding sources for economic development. It shapes policy by funding dams, irrigation systems, and other major projects. In fiscal year 2013, the World Bank provided $31.5 billion in loans and support for projects designed to benefit low-income people in developing countries.

Despite its admirable mission, the World Bank is often criticized for funding projects that cause environmental impacts, such as dams that flood forests and farmland to provide electricity. Providing for the needs of growing human populations in poor nations while minimizing damage to the ecological systems people rely on can be a tough balancing act. Environmental scientists agree that the concept of sustainable development must be the guiding principle for such efforts.

**The World Trade Organization** Based in Geneva, Switzerland, the **World Trade Organization (WTO)** represents multinational corporations and promotes free trade by reducing obstacles to international commerce and enforcing fairness among nations in trading practices. The WTO has authority to impose financial penalties on nations that do not comply with its directives.

The WTO has interpreted some national environmental laws as unfair barriers to trade. In 1995, the U.S. EPA issued regulations requiring cleaner-burning gasoline in U.S. cities. Brazil and Venezuela filed a complaint with the WTO, saying the new rules discriminated against the dirtier-burning petroleum they exported to the United States. The WTO agreed, ruling that even though the dirty gasoline posed a threat to human health in the United States, the EPA rules were an illegal trade barrier. The ruling forced the United States to weaken its regulations.

### WEIGHING THE ISSUES

**Trade Barriers and Environmental Protection** If Canada has stricter laws for environmental protection than Mexico, and if these laws limit Mexico's ability to export its goods to Canada, then by WTO policy Canada's laws could be overruled in the name of free trade. Do you think this is fair? Now consider that Canada is wealthier than Mexico and that Mexico could use an economic boost. Does this affect how you would respond?

**Nongovernmental organizations** A number of **nongovernmental organizations (NGOs)**—nonprofit, mission-driven organizations not overseen by any government—have become international in scope and exert influence over policy. Groups such as the Nature Conservancy focus on conservation objectives on the ground (such as purchasing and managing land and habitat for rare species) without becoming politically involved. Other groups, such as Greenpeace, Conservation International, and Population Connection, attempt to shape policy through research, education, lobbying, or protest.

# Approaches to Environmental Policy

When most of us think of environmental policy, what comes to mind are major laws or regulations. However, environmental policy is diverse.

## Policy can follow three approaches

Environmental policy can utilize a variety of strategies within three major approaches (**FIGURE 5.13**).

**Lawsuits in the courts** Prior to the legislative push of the 1960s and 1970s, most environmental policy questions were addressed with lawsuits in the courts. Individuals suffering external costs from pollution would sue polluters, one case at a time. The courts sometimes punished polluters by ordering them to stop their operations or pay damages to the affected parties. However, as industrialization proceeded and population grew, pollution became harder to avoid, and judges became reluctant to hinder industry. People began to view legislation and regulation as more effective means of protecting public health and safety.

**Command-and-control policy** Most environmental laws and regulations of recent decades use a **command-and-control** approach, in which a regulating agency prohibits certain actions—or sets rules, standards, or limits—and threatens punishment for violations. This simple and direct approach to policymaking has brought citizens of the United States and other nations cleaner air, cleaner water, safer workplaces, healthier neighborhoods, and many other advances. The relatively safe, healthy, comfortable lives most of us enjoy today owe much to the command-and-control environmental policy of the past several decades.

Even in plain financial terms, command-and-control policy has been effective. Each year the White House Office of Management and Budget analyzes U.S. regulatory policy to calculate the economic costs and benefits of regulations. These analyses have consistently revealed that benefits far outweigh costs and that environmental regulations are most beneficial of all. You can explore some of these data in *Calculating Ecological Footprints* (p. 114).

**Economic policy tools** Despite the successes of command-and-control policy, many people dislike the top-down nature of an approach that dictates particular solutions to problems. As an alternative approach, we can aim to channel the innovation and economic efficiency of market capitalism in ways that benefit the public. Economic policy tools use financial incentives to promote desired outcomes by encouraging private entities competing in a marketplace to innovate and generate new or better solutions at lower cost.

All three of these approaches aim to "internalize" external costs suffered by the public by building these costs into

## Problem
**Pollution from factory harms people's health**

## Solutions
**Three policy approaches**

❶ People can sue factory in court.

❷ Government can regulate emissions.

❸ Economic policy tools can create incentives: A factory that pollutes less (right) will outcompete one that pollutes more (left) through permit trading, avoiding green taxes, or selling ecolabeled products.

**FIGURE 5.13 Three major policy approaches exist to resolve environmental problems.** To address pollution from a factory, we might ❶ seek damages through lawsuits, ❷ limit pollution through legislation and regulation, or ❸ reduce pollution using market-based strategies.

market prices. Each approach has strengths and weaknesses, and each is best suited to different conditions. The approaches may also be used together. For instance, Costa Rica's Forest Law 7575 was a command-and-control law that banned forest clearing, but it also established the PSA program as an economic policy tool to help the policy succeed. Government regulation is often needed to frame market-based efforts, and citizens can use the courts to ensure that regulations are enforced. Let's now explore several types of economic policy tools: taxes, subsidies, ecolabeling, and emissions trading.

## Green taxes discourage undesirable activities

In taxation, money passes from private parties to the government, which uses it to pay for services to benefit the public. Taxing undesirable activities helps to internalize external costs by making these costs part of the normal expense of doing business. Taxes on environmentally harmful activities and products are called **green taxes.** When a business pays a green tax, it is essentially reimbursing the public for environmental damage it causes.

Under green taxation, a corporation owning a polluting factory pays taxes on the pollution it discharges—the more pollution, the higher the tax payment. This gives factory owners a financial incentive to reduce pollution while allowing them the freedom to decide how to do so. One polluter might invest in pollution control technology if this is more affordable than paying the tax. Another polluter might instead choose to pay the tax—funds the government can use to reduce pollution in some other way.

Costa Rica uses a green tax to help fund its PSA program. It applies a tax of 3.5% to sales of fossil fuels and then uses the revenue to pay for conserving forests, which soak up carbon emissions from fossil fuel combustion. In the United States, similar "sin taxes" on cigarettes and alcohol are long-accepted tools of U.S. social policy. Taxes on pollution are more common in Europe, where many nations have adopted the **polluter-pays principle,** which specifies that the party creating pollution be held responsible for covering the costs of its impacts. Today there is wide debate over carbon taxes—taxes on gasoline, coal-based electricity, and fossil-fuel-intensive products—to fight climate change (p. 327).

## Subsidies promote certain activities

Another economic policy tool is the **subsidy,** a government giveaway of money or resources that is intended to support or promote an industry or activity. Subsidies take many forms, and one is the *tax break*, which relieves the tax burden on an industry, firm, or individual. Costa Rica's PSA program subsidizes the conservation and restoration of forests by transferring public money to landowners who conserve and restore forests. Ironically, much of the nation's deforestation had resulted from ranching and farming that the government had previously been subsidizing.

Subsidies like Costa Rica's payments for ecological services promote environmentally sustainable activities—but all too often subsidies are used to prop up unsustainable ones. In the United States, subsidies for grazing (p. 145), timber extraction (p. 194), and mineral extraction (p. 243) on public lands all benefit private parties while often degrading publicly held resources.

Fossil fuels have been a major recipient of subsidies over the years. From 1950 to 2010, the U.S. government gave $594 billion of its citizens' money to oil, gas, and coal corporations (most of this in tax breaks), according to one recent compilation (**FIGURE 5.14**). In comparison, just $171 billion was granted to renewable energy, and most of these subsidies went to hydropower and to corn ethanol, which is not widely viewed as a sustainable fuel (p. 382). Globally in 2007–2010, fossil fuel subsidies outpaced renewable energy subsidies by nearly 8 to 1, according the International Energy Agency.

In 2009, President Obama and other leaders of the Group of 20 (G-20) nations resolved to gradually phase out their collective $300 billion of annual fossil fuel subsidies. Doing so would hasten a shift to cleaner renewable energy sources and accomplish half the greenhouse gas emissions cuts needed to hold global warming to 2°C. However, since that time, fossil fuel subsidies have *grown*, not shrunk. A prime reason is that consumers are accustomed to artificially low subsidized prices for gasoline and electricity and might punish policymakers who let these prices rise.

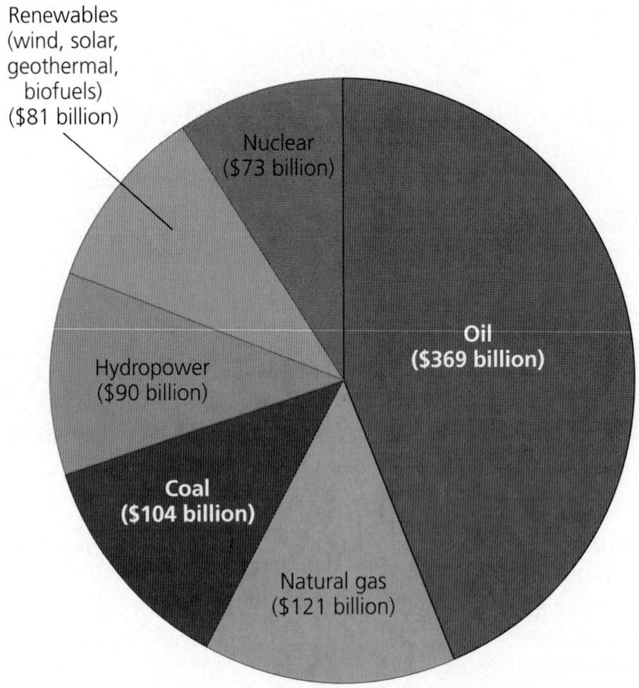


**FIGURE 5.14 The well-established fossil fuel industries receive more subsidies than the young renewable energy industries.** Cumulative data for the United States from 1950 to 2010 are shown. Apportionments remain similar today. *Data from: Management Information Services, Inc. 2011.* 60 years of energy incentives: Analysis of federal expenditures for energy development. *Washington, D.C.*

DATA **Q** How many dollars in subsidies have gone to fossil fuels (oil, coal, and natural gas) for every dollar that has gone to renewable energy (excluding hydropower)?

GO TO **INTERPRETING GRAPHS & DATA** ON MasteringEnvironmentalScience®

## Ecolabeling empowers consumers

With subsidies and green taxes, policymakers deploy financial incentives in direct and selective ways. However, we may also pursue policy goals by establishing financial incentives and then letting marketplace dynamics run their course. With **ecolabeling,** sellers who use sustainable practices in growing, harvesting, or manufacturing products advertise this fact on their labels, hoping to win approval from buyers (**FIGURE 5.15**). Examples include labeling recycled paper (pp. 395–396), organic foods (pp. 155–158), dolphin-safe tuna, and sustainably harvested lumber (p. 197).

In many cases, ecolabeling grew from initial steps taken by government to require the disclosure of information to consumers. Once established, ecolabeling can spread in a free market as businesses seek to win consumer confidence and to outcompete less sustainable brands. Of course, some businesses may try to mislead us into thinking their products are more sustainable than they actually are—a phenomenon called *greenwashing*. Independent certification by an outside party helps ensure that consumers get accurate information. When labeling information is accurate, each of us as consumers can provide businesses a powerful incentive to switch to more sustainable processes when we buy ecolabeled products.

**FIGURE 5.15 Ecolabeling enables all of us to promote sustainable business practices through our purchasing decisions.** Organic, fair-trade, and shade-grown coffee varieties are examples of the many ecolabeled products now widely available.

## Emissions trading can produce cost-effective results

Another approach that employs market dynamics to achieve policy goals is **emissions trading.** In an emissions trading system, a government creates a market in permits for the emission of pollutants, and companies, utilities, or industries then buy and sell the permits among themselves. In a **cap-and-trade** emissions trading system, the government first caps the overall amount of pollution it will allow, then grants or auctions off permits to polluters that allow them each to emit a certain fraction of that amount. As polluters trade these permits, the government progressively lowers the cap of overall emissions allowed (see Figure 14.29, p. 327).

Suppose you own an industrial plant with permits to release 10 units of pollution, but you find that you can make your plant more efficient and release only 5 units instead. You now have a surplus of permits, which you can sell to some other plant owner who needs them. Doing so generates income for you and meets the needs of the other plant, while the total amount of pollution does not rise. By providing firms an economic incentive to reduce pollution, emissions trading can lower expenses for both industry and the public relative to a conventional regulatory system.

The United States pioneered the cap-and-trade approach with its program to reduce sulfur dioxide emissions, established by the 1990 Clean Air Act amendments. Since then, sulfur dioxide emissions have declined by 67%, acid rain has been reduced, and air quality has improved (see Figure 13.21, p. 297). Similar cap-and-trade programs have shown success with smog in the Los Angeles basin and with nitrogen oxides in northeastern states.

To address climate change, European nations are operating a market in greenhouse gas emissions (p. 327). In the United States, carbon trading markets are running in California and among northeastern states (pp. 326–327), while other states and nations are considering programs. We will assess some of these efforts in Chapter 14.

## Market incentives are diverse at the local level

You may have already taken part in transactions involving financial incentives as policy tools. Many municipalities charge residents for waste disposal according to the amount of waste they generate. Some cities place taxes or fees on items whose safe disposal is costly, such as tires and motor oil. Others give rebates to residents who buy water-efficient toilets and appliances, because rebates can cost a city less than upgrading its wastewater treatment system. Likewise, power utilities may offer discounts to customers who buy high-efficiency appliances, because doing so is less costly than expanding the generating capacity of their plants.

The creative use of economic policy tools is growing, while command-and-control regulation and legal action in the courts continue to play vital roles in environmental policy. As a result, we have a variety of strategies available as we seek sustainable solutions to our society's challenges.

# Sustainable Development

Today's vital search for sustainable solutions centers on **sustainable development,** economic progress that maintains resources for the future. The United Nations defines sustainable development as development that "meets the needs of the present without sacrificing the ability of future generations to meet their own needs." Sustainable development is an economic pursuit shaped by policy and informed by science. It is also an ethical pursuit because it asks us to manage our resource use so that future generations can enjoy similar access to resources.

## Sustainable development involves environmental protection, economic well-being, and social equity

*Development* involves making purposeful changes intended to improve our quality of life. Construction of homes, schools, hospitals, power plants, factories, and transportation networks are all examples of development. In the past, "sustainable development" might have been viewed as an oxymoron—a phrase that contradicts itself. Advocates of development felt that protecting the environment threatened people's economic needs, whereas advocates of environmental protection held that development degrades the environment, jeopardizing the very improvements for our lives that were intended. Today, however, people increasingly perceive how we all depend on a healthy and functional natural environment.

We also now recognize that society's poorer people tend to suffer the most from environmental degradation. As a result, advocates of environmental protection, economic development, and social justice began working together toward common goals. This cooperation gave rise to the modern drive for sustainable development, which seeks ways to promote social justice, economic well-being, and environmental quality at the same time (**FIGURE 5.16**). Governments, businesses, industries, and organizations pursuing sustainable development aim to satisfy a **triple bottom line,** a trio of goals including economic advancement, environmental protection, and social equity.

Programs that pay for ecosystem services are one example of a sustainable development approach that seeks to satisfy a triple bottom line. Costa Rica's PSA program aims to enhance its citizens' well-being by conserving the country's natural assets while compensating affected landholders for any economic losses. The intention is to achieve a win-win result that pays off in economic, social, and environmental dimensions.

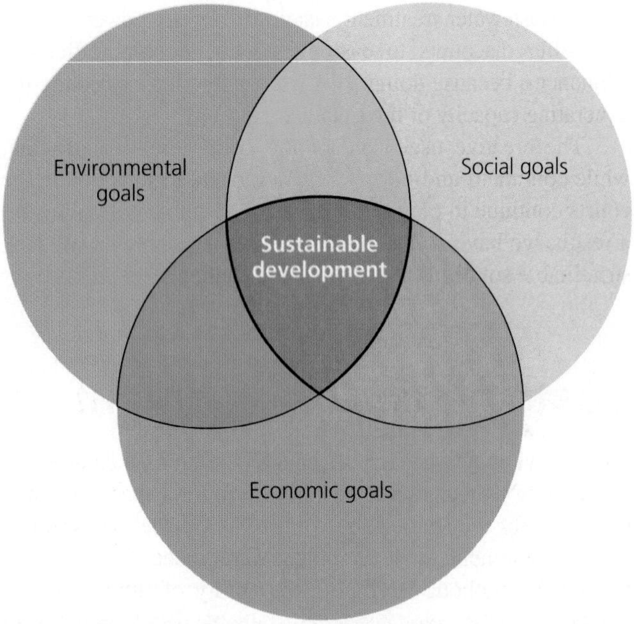


**FIGURE 5.16 Sustainable development occurs where three sets of goals overlap: social, economic, and environmental goals.**

**TABLE 5.3 U.N. Millennium Development Goals for 2015**

| |
|---|
| • Eradicate extreme poverty and hunger |
| • Achieve universal primary education |
| • Promote gender equality and empower women |
| • Reduce child mortality |
| • Improve maternal health |
| • Improve maternal health |
| • Combat HIV/AIDS, malaria, and other diseases |
| • Ensure environmental sustainability |
| • Develop a global partnership for development |

***Source:*** United Nations. *End poverty 2015: Millennium Development Goals.* © United Nations. Reproduced with permission.

## Sustainable development is global

Sustainable development has blossomed as an international movement. The United Nations, the World Bank, and other global organizations sponsor conferences, fund projects, publish research, and facilitate collaboration across borders among governments, businesses, and nonprofit organizations.

The Earth Summit at Río de Janeiro, Brazil, in 1992 was the world's first major gathering focused on sustainable development. With representatives from over 200 nations, this conference gave rise to notable achievements, including the Convention on Biological Diversity (p. 179) and the Framework Convention on Climate Change (p. 325). Ten years later, nations met in Johannesburg, South Africa, at the 2002 World Summit on Sustainable Development. Then in 2012, the world returned to Río de Janeiro for the Rio-Plus-20 conference and explored the latest strategies for promoting economic vitality, social equity, and environmental quality.

Meanwhile, in 2000, world leaders adopted the United Nations Millennium Declaration, which set forth eight *Millennium Development Goals* for humanity (**TABLE 5.3**). Each broad goal for sustainable development has several specific underlying targets to be met by implementing concrete strategies. For instance, strategies to "develop a global partnership for development" include working with governments and corporations of wealthy nations to provide poor nations with more financial aid, more debt relief, freer access to global markets, and better access to inexpensive drugs and to technologies such as cell phones and Internet access.

Many of the Millennium Development Goals were given a 2015 target date, and we have made better progress on some than on others. We still have a long way to go to resolve the many challenges facing our world. Pursuing solutions that meet a triple bottom line of environmental, economic, and social goals can help pave the way for a truly sustainable global society.

## Conclusion

Environmental policy is a problem-solving tool that makes use of science, ethics, and economics. Command-and-control legislation and regulation remain the most common policy approaches, but innovative market-based policy tools are also being deployed. Environmental and ecological economists are quantifying the value of ecosystem services and devising alternative means of measuring progress, thereby helping to show how economic progress is tied to environmental protection and resource conservation. In pursuing sustainable development, we recognize that economic, social, and environmental well-being depend on one another and can be mutually reinforcing. If we can enhance our economic and social well-being without depleting natural resources, then truly sustainable solutions will be within reach.

# Testing Your Comprehension

1. Name and describe two key contributions that the natural environment makes to our economies.
2. Describe four ways in which neoclassical economic approaches can contribute to environmental problems.
3. Compare and contrast the views of neoclassical economists, environmental economists, and ecological economists, particularly regarding the issue of economic growth.
4. What are ecosystem services? Give several examples. Describe how some economists have tried to assign monetary values to ecosystem services.
5. Describe at least one major goal of and justification for environmental policy. Now articulate three problems that environmental policy commonly seeks to address.
6. Summarize how the first, second, and third waves of environmental policy in U.S. history differed from one another. Describe two current priorities in international environmental policy.
7. What did the National Environmental Policy Act accomplish? Briefly describe the origin and mission of the U.S. Environmental Protection Agency.
8. Compare and contrast the three major approaches to environmental policy: lawsuits, command-and-control, and economic policy tools. Describe an advantage and disadvantage of each.
9. Explain how each of the following work: a green tax, a subsidy, and emissions permits.
10. How can *sustainable development* be defined? What is meant by the triple bottom line? Why is it important to pursue sustainable development?

# Seeking Solutions

1. Do you think that a steady-state economy is a practical alternative to our current approach that prioritizes economic growth? Why or why not?
2. Do you think we should attempt to quantify and assign market values to ecosystem services? Why or why not? What consequences might this have?
3. Reflect on causes for the transitions in U.S. history from one type of environmental policy to another. Now peer into the future, and consider how life and society might be different in 25, 50, or 100 years. What would you predict about the environmental policy of the future, and why? What issues might it address? Do you predict we will have more or less environmental policy?
4. Compare the roles of the United Nations, the World Bank, the World Trade Organization, and nongovernmental organizations. If you could gain the support of just one of these institutions for a policy you favored, which would you choose? Why?
5. **THINK IT THROUGH** You have just returned from serving in the U.S. Peace Corps in Costa Rica, where you worked closely with farmers, foresters, ecologists, and policymakers on issues related to Costa Rica's PSA program. You have now been hired as an adviser on natural resource issues to the governor of your state. Think about the condition of the forests, land and soil, water supplies, and other natural resources in your state. Given what you learned in Costa Rica, would you advise your governor to institute some kind of program to pay citizens to conserve ecosystem services? Why or why not? Describe what policies you would advocate to best conserve your state's resources and ecosystem services while advancing the economic and social condition of its citizens.

# Calculating Ecological Footprints

Critics of command-and-control policy often argue that regulations are costly to business and industry, yet cost-benefit analyses (p. 94) repeatedly show that regulations bring citizens more benefits than costs, overall. Each year the U.S. Office of Management and Budget assesses costs and benefits of major federal regulations of administrative agencies. Results from the most recent report, covering the decade from 2003 to 2013, are presented in the table (shown are averages from ranges of estimates). Subtract costs from benefits, and enter these values for each agency in the third column. Divide benefits by costs, and enter these values in the fourth column.

**Costs and benefits of major U.S. federal regulations, 2003–2013**
(average values from ranges of estimates, in billions of dollars)

| Agency | Benefits | Costs | Benefits minus costs | Benefit : Cost ratio |
|---|---|---|---|---|
| Department of Energy | 15.6 | 5.9 | 9.7 | 2.6 |
| Department of Health and Human Services | 32.4 | 4.6 | | |
| Department of Transportation | 25.4 | 11.6 | | |
| Environmental Protection Agency (EPA) | 507.2 | 42.2 | | |
| Other departments | 25.1 | 8.1 | | |
| **Total (billions of dollars)** | **605.7** | **72.4** | | |

*Data from U.S. Office of Management and Budget, 2014.* 2014 draft report to Congress on the benefits and costs of federal regulations and unfunded mandates on state, local, and tribal entities. *OMB, Washington, D.C.*

1. For how many of the agencies shown do regulations exert more costs than benefits? For how many do regulations provide more benefits than costs?
2. Which agency's regulations have the greatest excess of benefits over costs? Which agency's regulations have the greatest ratio of benefits to costs?
3. What percentage of total benefits from regulations comes from EPA regulations? Most of the benefits and costs from EPA regulations are from air pollution rules resulting from the Clean Air Act and its amendments. Judging solely by these data, would you say that Clean Air Act legislation has been a success or a failure for U.S. citizens? Why?

## MasteringEnvironmentalScience®

**STUDENTS**

Go to **MasteringEnvironmentalScience** for assignments, the etext, and the Study Area with practice tests, videos, current events, and activities.

**INSTRUCTORS**

Go to **MasteringEnvironmentalScience** for automatically graded activities, current events, videos, and reading questions that you can assign to your students, plus Instructor Resources.

# 6

# Human Population

## Upon completing this chapter, you will be able to:

- ☐ Describe the scope of human population growth
- ☐ Evaluate how human population, affluence, and technology affect the environment
- ☐ Explain and apply the fundamentals of demography
- ☐ Describe the concept of demographic transition
- ☐ Relate family planning, the status of women, and affluence to population growth

**Photo:** Crowded street in Shanghai, one of China's largest cities

# CENTRAL CASE STUDY

# China's One-Child Policy: Is It a Population "Time Bomb"?

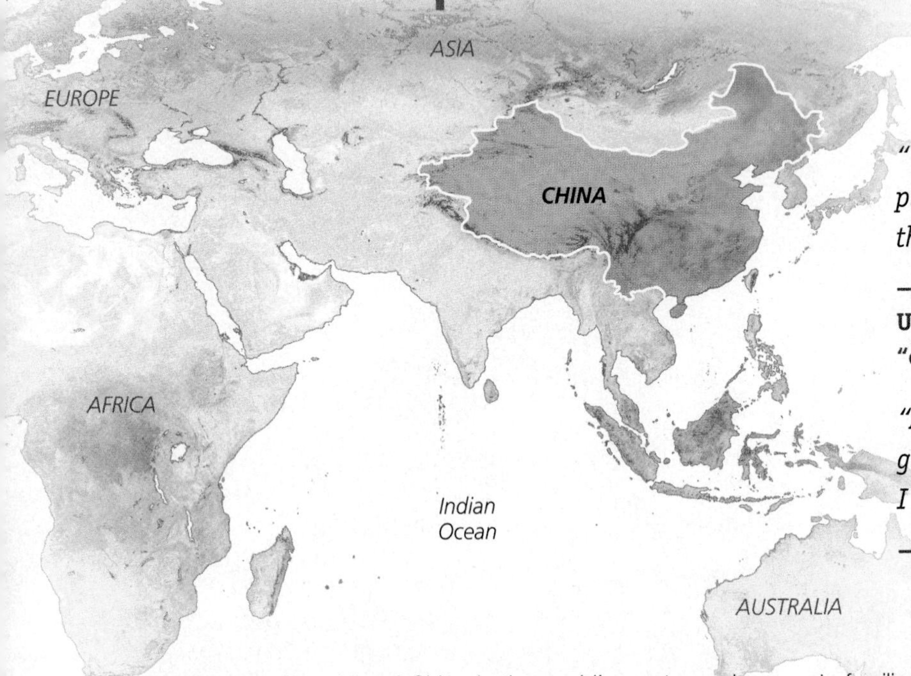


*"We don't need adjustments to the family-planning policy. What we need is a phaseout of the whole system."*

**—Gu Baochang, Chinese demographer at People's University, Beijing, referring to the nation's "one-child" policy**

*"As you improve health in a society, population growth goes down.... Before I learned about it, I thought it was paradoxical."*

**—Bill Gates, Chair, Microsoft Corporation**

The People's Republic of China is the world's most populous nation, home to one-fifth of the more than 7 billion people living on Earth. It is also the site of one of the most controversial social experiments in history.

When Mao Zedong founded the country's current regime six decades ago, roughly 540 million people lived in a mostly rural, war-torn, impoverished nation. Mao's policies encouraged population growth, and by 1970 improvements in food production, food distribution, and public health allowed China's population to swell to 790 million people. At that time, the average Chinese woman gave birth to 5.8 children in her lifetime.

However, the country's burgeoning population and its industrial and agricultural development were eroding the nation's soils, depleting its water, and polluting its air. Realizing that the nation might not be able to continue to feed its people, Chinese leaders decided in 1970 to institute a population control program that prohibited most Chinese couples from having more than one child.

The program began with outreach efforts encouraging people to marry later and have fewer children. Along with these efforts, the Chinese government increased the accessibility of contraceptives and abortion. Fertility declined with these initiatives, and by 1975 China's annual population growth rate had dropped from 2.8% to 1.8%.

In 1979, the government began rewarding one-child families with government jobs and better housing, medical care, and access to schools. Families with more than one child, meanwhile, were subjected to costly monetary fines, employment discrimination, and social scorn. The one-child program applied mostly to families in urban areas. Many farmers and ethnic minorities in rural areas were exempted, because success on the farm often depends on having multiple children. The experiment has been a success in slowing population growth: The nation's growth rate is now down to 0.5%, and the average Chinese woman now has only 1.5 children in her lifetime.

However, the one-child policy has also produced a population with a shrinking labor force, increasing numbers of elderly people, and too few women. These unintended consequences have led some demographers to question whether China's one-child policy simply traded one population problem—overpopulation—for other population problems.

The rapid reduction in fertility that resulted from this policy has drastically changed China's age structure. Once consisting predominantly of young people, China's population has shifted, such that the numbers of children and elderly are more even (**FIGURE 6.1**). This means there will be relatively fewer workers for China's growing economy and proportionately larger numbers of elderly people relying on governmental support and services.

The shrinking workforce caused by the one-child policy may now slow the growth of the thriving Chinese economy it helped to produce. Many employers in China are struggling to find workers and must pay wages up to 35% higher than they did a few years ago to keep their employees. Although this is a welcome development for workers, it threatens China's ability to produce goods as cheaply as it once did, inducing companies to relocate from China to nations where labor costs are lower.

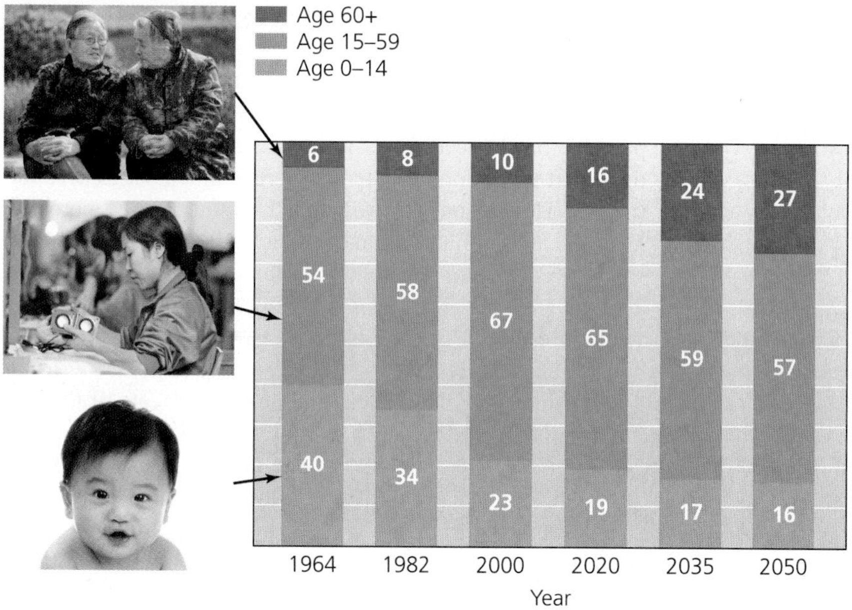


**FIGURE 6.1 China's one-child policy is leading to a shrinking workforce and rising numbers of elderly citizens.** Values in the figure represent the percentage of the Chinese population in each age group. *Figure from Population Reference Bureau, 2004.* China's Population: New Trends and Challenges.

The growing number of elderly Chinese poses problems because the Chinese government lacks the resources to fully support them. And because tradition dictates that Chinese sons care for their parents and grandparents in old age, this puts a heavy economic burden on the millions of "only" children produced under the one-child policy.

Modern China also has too few women. Chinese culture has traditionally valued sons because they carry on the family name, assist with farm labor in rural areas, and care for aging parents. Daughters, in contrast, will most likely marry and leave their parents, as the culture dictates. Thus, faced with being limited to just one child, many Chinese couples prefer a son to a daughter. Tragically, this has led to selective abortion and the killing of female infants. This has caused a highly unbalanced ratio of young men and women in China, leading to the social instability that arises when large numbers of young men are unable to find brides and remain longtime bachelors.

Chinese authorities have attempted to address these population issues by loosening the one-child policy. The government has tested initiatives where people, in certain provinces, are given greater control of their reproductive choices. Chinese authorities also announced in 2013 that if either member of a married couple is an only child, the couple will be allowed to have a second child—expanding a previous exemption that allowed a second child to couples whose parents were *both* only children.

China's reproductive policies have long elicited intense criticism worldwide from people who oppose government intrusion into personal reproductive choices, but it has proven highly effective in slowing growth rates. As other nations become more crowded and wish to emulate China's economic growth, might their governments also feel forced to turn to drastic policies that restrict individual freedoms? In this chapter, we examine human population dynamics worldwide, consider their causes, and assess their consequences for the environment and our society. ☐

# Our World at Seven Billion

China receives a great deal of attention with regard to population issues because of its unique reproductive policies and its status as the world's most populous nation. But China is not alone in struggling with population matters. India soon will surpass China in possessing the world's largest population (**FIGURE 6.2**). India was the first nation to implement comprehensive population control policies—but when India's policymakers introduced forced sterilization in the 1970s, the resulting outcry forced the government to change its policies. Since then, India's efforts have been more modest and far less coercive, focusing on family planning and reproductive health care.

Like India, many of the world's poorer nations continue to experience substantial population growth. These nations are often ill equipped to handle such growth, and this leads to stresses on society, the environment, and people's well-being. In our world of now more than 7 *billion* people, one of our greatest challenges is finding ways to slow the growth of the human population without coercive measures such as those used in China, but rather by establishing conditions that lead people to desire to have fewer children.

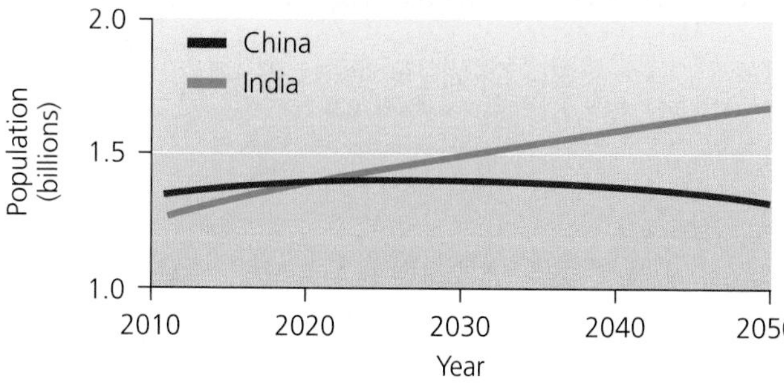


**FIGURE 6.2 India will likely soon surpass China as the most populous nation.** China's rate of growth is now lower than India's as a result of China's aggressive population policies. *Data from Population Division of the Department of Economic and Social Affairs of the United Nations Secretariat, 2013.* World population prospects: The 2012 revision, http://esa.un.org/wpp. © *United Nations, 2013.*

**FAQ** **How big is a billion?**

Human beings have trouble conceptualizing huge numbers. As a result, we often fail to recognize the true magnitude of a number such as 7 billion. Although we know that a billion is bigger than a million, we tend to view both numbers as impossibly large and therefore similar in size. For example, guess (without calculating) how long it would take a banker to count out $1 million if she did it at a rate of a dollar a second for 8 hours a day, 7 days a week. Now guess how long it would take to count $1 billion at the same rate. The difference between your estimate and the answer may surprise you. Counting $1 million would take a mere 35 days, whereas counting $1 billion would take 95 years! Living 1 million seconds takes only 12 days, while living for 1 billion seconds requires more than 31 years. You couldn't live for 7 billion seconds if you tried, because that would take 221 years. Examples like these can help us appreciate the "b" in "billion."

## The human population is growing rapidly

Our global population grows by over 85 million people each year. This is the equivalent of adding all the people of California, Texas, and New Jersey to the world annually—and it means that we add more than two people to the planet *every second*. Take a look at **FIGURE 6.3** and note just how recent and sudden our rapid increase has been. It took until after 1800, virtually all of human history, for our population to reach 1 billion. Yet by 1930 we had reached 2 billion, and 3 billion in just 30 more years. Our population added its next billion in just 15 years, and it has taken only 12 years to add each of the next three installments of a billion people.

What accounts for our unprecedented growth? Exponential growth—the increase in a quantity by a fixed percentage per unit time—accelerates increase in population size over time, just as compound interest accrues in a savings account (p. 61). The reason, you will recall, is that a fixed percentage of a small number makes for a small increase, but that same percentage of a large number produces a large increase. Thus, even if the growth *rate* remains steady, population *size* will increase by greater increments with each successive generation.

For much of the 20th century, the growth rate of the human population rose from year to year. This rate peaked at 2.1% during the 1960s and has declined to 1.2% since then. Although 1.2% may sound small, a hypothetical population starting with one man and one woman that grows at 1.2% gives rise to a population of 2,939 after only 40 generations and 112,695 after 60 generations. Also, although the global growth rate is 1.2%, rates vary widely from region to region and are highest in developing nations (**FIGURE 6.4**).

At a 1.2% annual growth rate, a population doubles in size in only 58 years. We can roughly estimate doubling times with a handy rule of thumb. Just take the number 70 (which is 100 times 0.7, the natural logarithm of 2) and divide it by the annual percentage growth rate: 70/1.2 = 58.3. China's current growth rate of 0.5% means it would take roughly 140 years (70/0.5 = 140) for its current population to double, but India's current growth rate of 1.5% predicts a doubling in only about 47 years (70/1.5 = 46.7). Had China not instituted its one-child policy and had its growth rate remained at 2.8%, it would have taken only 25 years (70/2.8 = 25) to double in size.

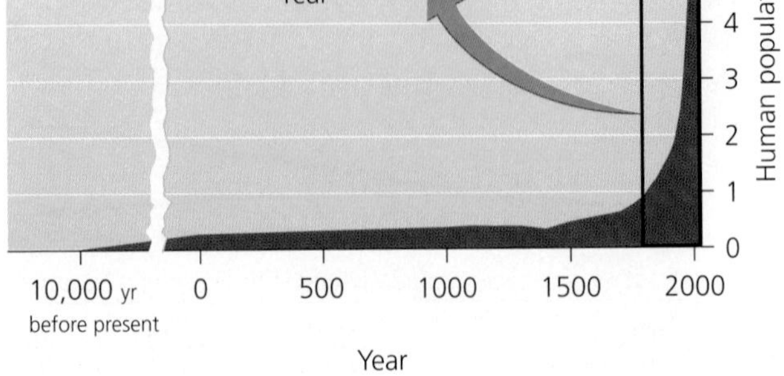

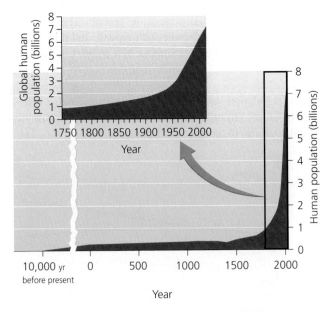

**FIGURE 6.3 We have risen from fewer than 1 billion people in 1800 to more than 7 billion today.** Viewing global human population size over a long time scale **(bottom graph)** and growth since the industrial revolution **(inset top graph)** shows that nearly all growth has occurred in just the past 200 years. *Data from U.S. Bureau of the Census.*

## Is there a limit to human population growth?

Our spectacular growth in numbers has resulted largely from technological innovations, improved sanitation, better medical care, increased agricultural output, and other factors that have brought down death rates. These improvements have been particularly successful in reducing **infant mortality rate,** the frequency of children dying in infancy. Birth rates have not declined as much, so births have outpaced deaths for many years now, leading to population growth. But can the human population continue to grow indefinitely?

Environmental factors set limits on the growth of populations (p. 62), but environmental scientists who have tried to pin a number to the human carrying capacity (p. 62) have come up with wildly differing estimates. The most rigorous estimates range from 1–2 billion people living prosperously in a healthy environment to 33 billion people living in extreme poverty in a degraded world of intensive cultivation without natural areas.

The difficulty in estimating the carrying capacity for humans is that we are a particularly successful species, and we have repeatedly overcome predicted limits on growth by developing new technologies and ways of securing resources.

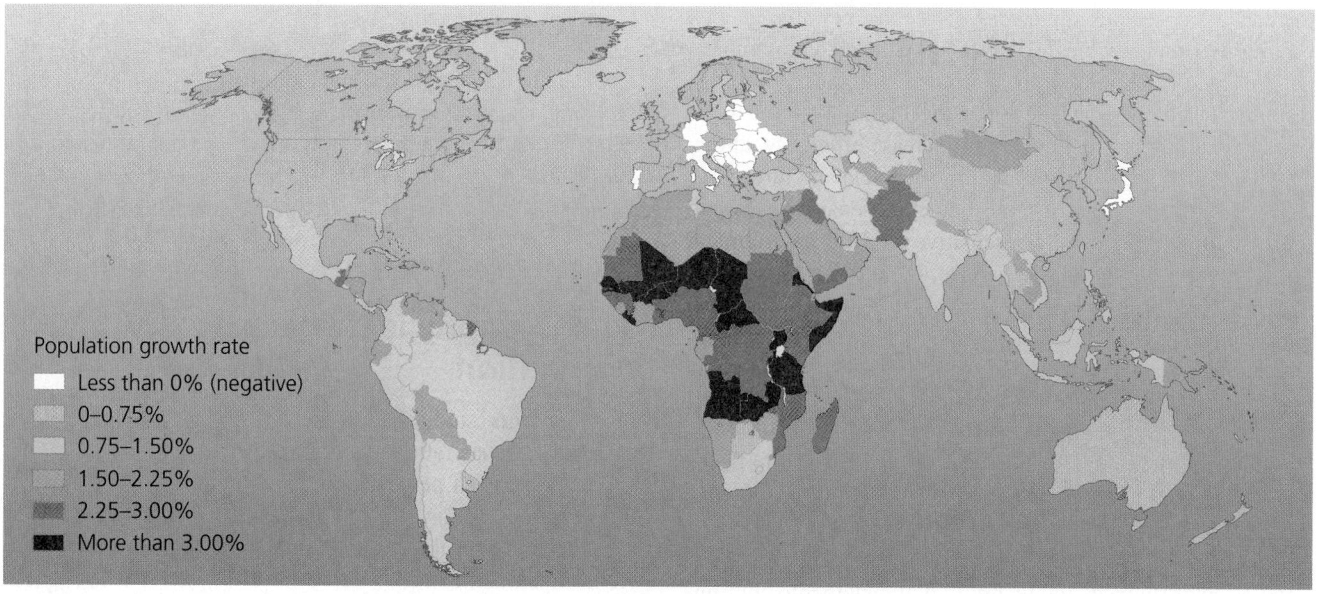


**FIGURE 6.4 Population growth rates vary greatly from place to place.** Populations are growing fastest in poorer nations while populations are beginning to decrease in some industrialized nations. Shown are rates of natural increase (p. 61) as of 2013. *Data from Population Reference Bureau, 2013.* 2013 World population data sheet.

DATA **Q** Which world region has the highest population growth rates? Which world region has the lowest population growth rates?

GO TO **INTERPRETING GRAPHS & DATA** ON MasteringEnvironmentalScience®

For example, British economist Thomas Malthus (1766–1834) argued in his influential work *An Essay on the Principle of Population* (1798) that if society did not reduce its birth rate, then rising death rates would reduce the population through war, disease, and starvation. Although his contention was reasonable at the time, agricultural improvements in the 19th century increased food supplies, and his prediction did not come to pass. Similarly, biologist Paul Ehrlich predicted in his 1968 book *The Population Bomb* that human population growth would soon outpace food production and unleash massive famine and conflict in the latter 20th century. However, thanks to the way the "Green Revolution" (p. 138) increased food production in developing regions in the decades after his book, Ehrlich's dire forecasts did not fully materialize.

Does this mean human innovation will always find a way to support our population? Under the view that many economists hold, resource depletion due to population increase is not a problem if new resources can be found or created to replace depleted ones (p. 94). In contrast, environmental scientists recognize that not all resources can be replaced once they are depleted. For example, once a species is extinct, we cannot replicate its exact functions in an ecosystem or know what benefits it may have provided us. As the human population continues to climb, we may yet continue to find ways to raise our carrying capacity. Given our knowledge of population ecology and logistic growth (pp. 57–63), however, we cannot presume that human numbers can increase forever.

So, is continued human population growth a problem? To answer this, we must ask whether we could maintain the quality of life we desire even if resource substitution could (hypothetically) enable our population to grow indefinitely. Unless the availability and quality of all resources keeps pace forever with population growth, the average person in the future will have less space in which to live, less food to eat, and less material wealth than the average person does today. Thus population growth is indeed a problem if it depletes resources, stresses social systems, and degrades the natural environment, such that our quality of life declines.

## Population is one of several factors that affect the environment

One widely used formula gives us a handy way to think about population and other factors that affect environmental quality. Nicknamed the **IPAT model,** it is a variation of a formula proposed in 1974 by Paul Ehrlich and John Holdren. The IPAT model represents how our total impact ($I$) on the environment results from the interaction among population ($P$), affluence ($A$), and technology ($T$):

$$I = P \times A \times T$$

We can interpret impact in various ways, but we can generally boil it down either to unsustainable resource consumption or to the degradation of ecosystems by pollution. Increased population intensifies impact on the environment as more individuals take up space, use natural resources, and generate waste. Increased affluence magnifies environmental impact through greater per capita resource consumption, which generally has accompanied enhanced wealth. Technology that enhances our abilities to exploit minerals, fossil fuels, old-growth forests, or fisheries generally increases impact, but technology to reduce smokestack emissions, harness renewable energy, or improve manufacturing efficiency can decrease impact. One reason our population has kept growing, despite limited resources, is that we have developed technology—the *T* in the IPAT equation—time and again to increase efficiency, alleviate our strain on resources, and allow us to expand further.

We might also add a fourth factor, sensitivity (*S*), to the equation to denote how sensitive a given environment is to human pressures. For instance, the arid lands of western China are more sensitive to human disturbance than the moist regions of southeastern China. Plants grow more slowly in the arid west, making the land more vulnerable to deforestation and soil degradation. Thus, adding an additional person to western China has more environmental impact than adding one to southeastern China. We could refine the IPAT equation further by adding terms for the influence of social factors such as education, laws, ethical standards, and social stability and cohesion. Such factors all affect how population, affluence, and technology translate into environmental impact.

Modern-day China shows how all elements of the IPAT formula can combine to cause tremendous environmental impact in little time. Although China boasts one of the world's fastest growing economies, the country is battling unprecedented environmental challenges brought about by this rapid economic development. Intensive agriculture has expanded westward out of the nation's moist rice-growing regions, causing farmland to erode and blow away, much like the Dust Bowl tragedy that befell the U.S. heartland in the 1930s (p. 142). China has overpumped aquifers and has drawn so much water for irrigation from the Yellow River that this once-mighty waterway now dries up in many stretches (p. 260). The nation also faces severe, chronic urban air pollution (p. 287) and massive traffic jams from rapidly rising numbers of automobiles. As the world's developing countries try to attain the material prosperity that industrialized nations enjoy, China is a window into what much of the rest of the world could soon become.

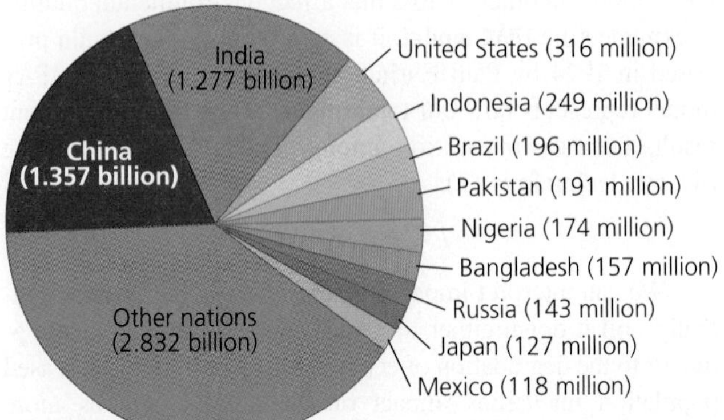


**FIGURE 6.5 Almost one in five people in the world lives in China, and more than one of every six live in India.** Three of every five people live in one of the 11 nations that have populations above 100 million. *Data from Population Reference Bureau, 2013.* 2013 World population data sheet.

# Demography

People do not exist outside of nature. We exist within our environment as one species of many. As such, all the principles of population ecology that drive biological change in the natural world (Chapter 3) apply to humans as well. The application of principles from population ecology to the study of statistical change in human populations is the focus of **demography.**

## Demography is the study of human population

**Demographers** study population size, density, distribution, age structure, sex ratio, and rates of birth, death, immigration, and emigration of people, just as population ecologists study these characteristics in other organisms. Each of these characteristics is useful for predicting population dynamics and environmental impacts.

**Population size** Our global human population of more than 7.1 billion is spread among more than 200 nations with populations ranging up to China's 1.36 billion, India's 1.28 billion, and the 316 million of the United States (**FIGURE 6.5**). The United Nations Population Division estimates that by the year 2050, the global population will surpass 9 billion (**FIGURE 6.6**). However, population size alone—the absolute number of individuals—doesn't tell the whole story. Rather,

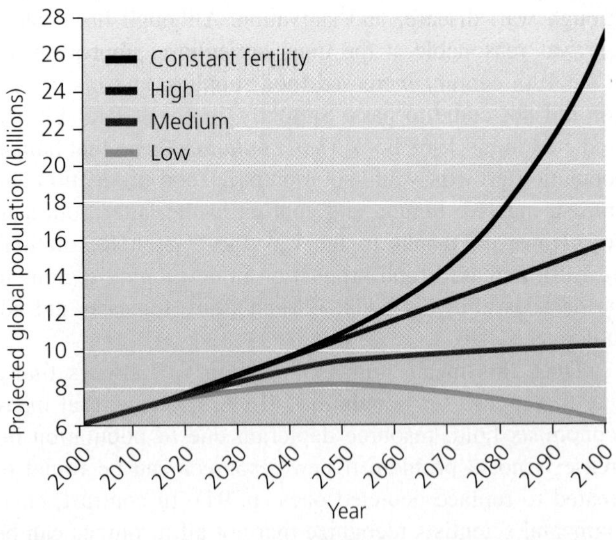


**FIGURE 6.6 The United Nations predicts world population growth.** In the latest projection, population is estimated to reach 11.0 billion in the year 2050 if fertility rates remain constant at 2005–2010 levels (top line in graph). However, U.N. demographers expect fertility rates to continue falling, so they arrived at a best guess (medium scenario) of 9.3 billion for 2050 and around 10.1 billion in 2100. In the high scenario, women on average have 0.5 child more than in the medium scenario. In the low scenario, women have 0.5 child fewer than in the medium scenario. *Adapted by permission from Population Division of the Department of Economic and Social Affairs of the United Nations Secretariat, 2013.* World population prospects: The 2012 revision. http://esa.un.org/wpp, *Fig 1. © United Nations, 2013.*

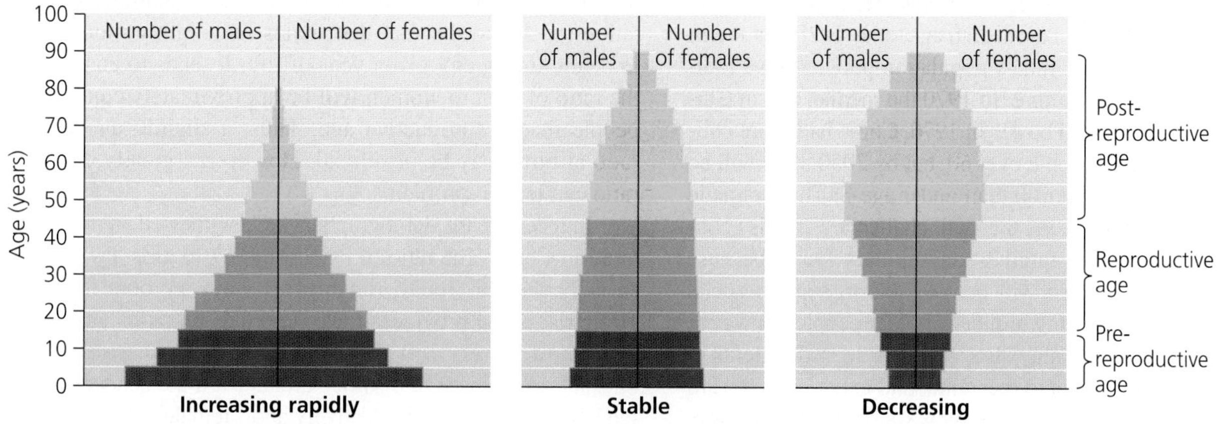


**FIGURE 6.7 Age structure diagrams show numbers of individuals of different age classes in a population.** A diagram like that on the left is weighted toward younger age classes, indicating a population that will grow quickly. A diagram like that on the right is weighted toward older age classes, indicating a population that will decline. Populations with balanced age structures, as shown in the middle diagram, will remain relatively stable in size.

a population's environmental impact depends on its density, distribution, and composition (as well as on affluence, technology, and other factors outlined earlier).

**Population density and distribution** People are distributed unevenly across our planet. In ecological terms, our distribution is clumped (p. 60) at all spatial scales. At the global scale, population density is highest in regions with temperate, subtropical, and tropical climates and lowest in regions with extreme-climate biomes, such as desert, rainforest, and tundra. Human population is dense along seacoasts and rivers, and less dense farther away from water. At more local scales, we cluster together in cities and towns.

This uneven distribution means that certain areas bear more environmental impact than others. Just as the Yellow River experiences pressure from Chinese farmers, the world's other major rivers all receive more than their share of human impact. At the same time, some areas with low population density are sensitive (a high *S* value in our revised IPAT model) and thus vulnerable to impact. Deserts and arid grasslands, for instance, are easily degraded by agriculture and ranching that commandeer too much water.

**Age structure** Age structure describes the relative numbers of individuals of each age class within a population (p. 60). Data on age structure are especially valuable to demographers trying to predict future dynamics of human populations. A population made up mostly of individuals past reproductive age will tend to decline over time. In contrast, a population with many individuals of reproductive age or pre-reproductive age is likely to increase. A population with an even age distribution will likely remain stable as births keep pace with deaths.

**Age structure diagrams,** often called *population pyramids*, are visual tools scientists use to illustrate age structure (**FIGURE 6.7**). The width of each horizontal bar represents the number of people in each age class. A pyramid with a wide base denotes a large proportion of people who have not yet reached reproductive age—and this indicates a population soon capable of rapid growth.

As an example, compare age structures for Canada and Nigeria (**FIGURE 6.8**). Nigeria's large concentration of individuals in younger age classes predicts a great deal of future reproduction. Not surprisingly, Nigeria has a higher population growth rate than Canada.

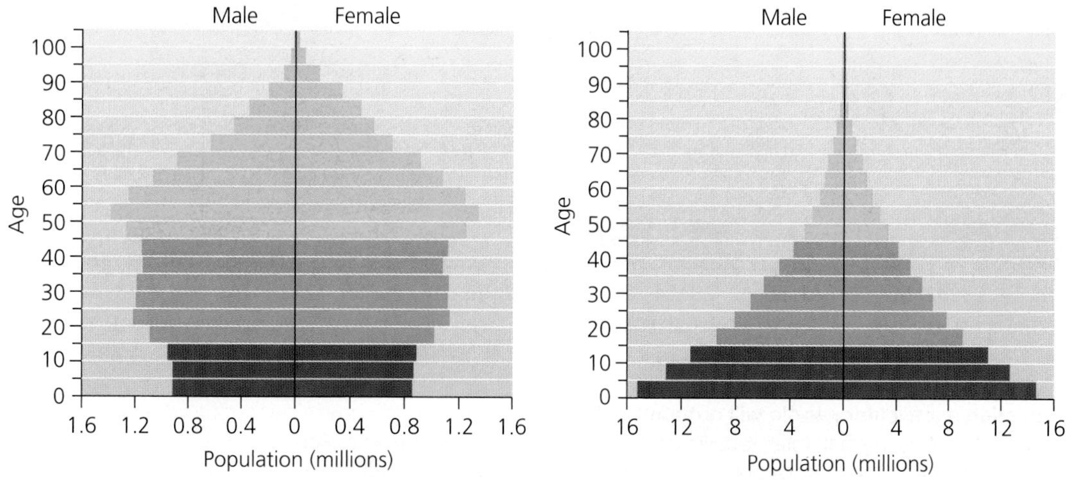


**(a) Age structure diagram of Canada**

**(b) Age structure diagram of Nigeria**

**FIGURE 6.8 Canada (a) shows a fairly balanced age structure, whereas Nigeria (b) shows an age distribution heavily weighted toward young people.** Nigeria's population growth rate (2.8%) is seven times greater than Canada's (0.4%). *Data are for 2013, from United States Census Bureau International Database.* http://www.census.gov/population/international/data/idb/.

Today, populations are aging in many nations, including the United States. The global median age today is 28, but it will be 38 in the year 2050. China's population policies have radically changed its age structure. In 1970 the median age in China was 20; by 2050 it will be 45. In 1970, China had more children under age 5 than people over 60, but by 2050 there will be 12 times more people over 60 than under age 5! This dramatic shift in age structure (**FIGURE 6.9**) will challenge China's economy, health care system, families, and military forces because fewer working-age people will be available to support social programs to assist the rising number of older people. However, a changing age structure may also lead to increased benefits to society as retirees contribute their time and experience toward productive volunteer efforts. In many ways both good and bad, "graying" populations will affect societies in China, the United States, and elsewhere throughout our lifetimes.

**Sex ratios** The ratio of males to females also can affect population dynamics. The naturally occurring sex ratio at birth in human populations features a slight preponderance of males; for every 100 female infants born, about 106 male infants are born. This phenomenon is an evolutionary adaptation (p. 48) to the fact that males are slightly more prone to death during any given year of life. It tends to ensure that the ratio of men to women will be approximately equal when people reach reproductive age. Thus, a slightly uneven sex ratio at birth may be beneficial. However, a greatly distorted ratio can lead to problems.

In recent years, demographers have witnessed an unsettling trend in China: The ratio of newborn boys to girls has become strongly skewed. Today, roughly 120 boys are born for every 100 girls. Some provinces have reported sex ratios as high as 138 boys for every 100 girls. The leading hypothesis for these unusual sex ratios is that some parents learn the gender of their fetus by ultrasound and selectively abort female fetuses.

China's skewed sex ratio may lower population growth rates further. Already the scarcity of young women has led to intense competition among young men, and in parts of rural China, teenaged girls are being kidnapped and sold to wealthy families as brides for single men. Many other Chinese men are being left single. Some of these men find employment as migrant workers and tend to engage in more risky sexual activity than their married counterparts. Researchers speculate that this could lead to a higher incidence of HIV in China in coming decades, as tens of millions of bachelors adopt such a lifestyle.

## WEIGHING THE ISSUES

**China's Reproductive Policy**

Describe what benefits may come from a reproductive policy such as China's. Now describe several problems that may result. Do you think a government should be able to enforce strict penalties for citizens who fail to abide by such a policy? Why or why not? What alternatives to China's policy can you suggest for dealing with the resource demands of a rapidly growing population?

**(a) Billboard promoting China's "one child" policy**

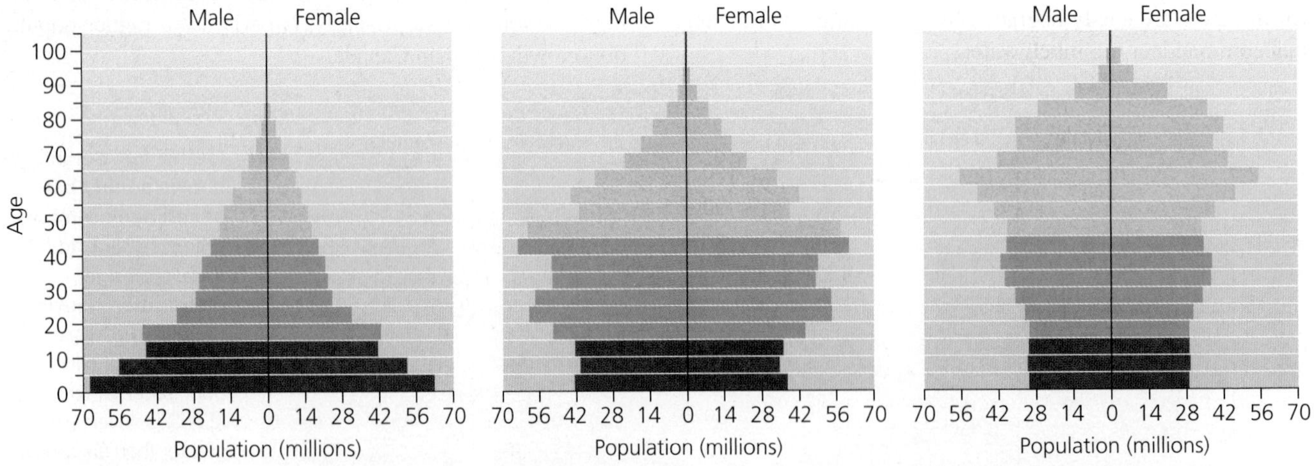


**(b) China in 1970** **(c) China in 2013** **(d) China in 2050 (projected)**

**FIGURE 6.9 As China's population ages, older people will outnumber the young.** China's "one child" policy **(a)** has been highly successful in reducing birth rates, significantly changing China's age structure. Population pyramids show the predicted graying of the Chinese population from **(b)** 1970 to **(c)** 2013 to **(d)** 2050.
*Data from United States Census Bureau International Database.* http://www.census.gov/population/international/data/idb.

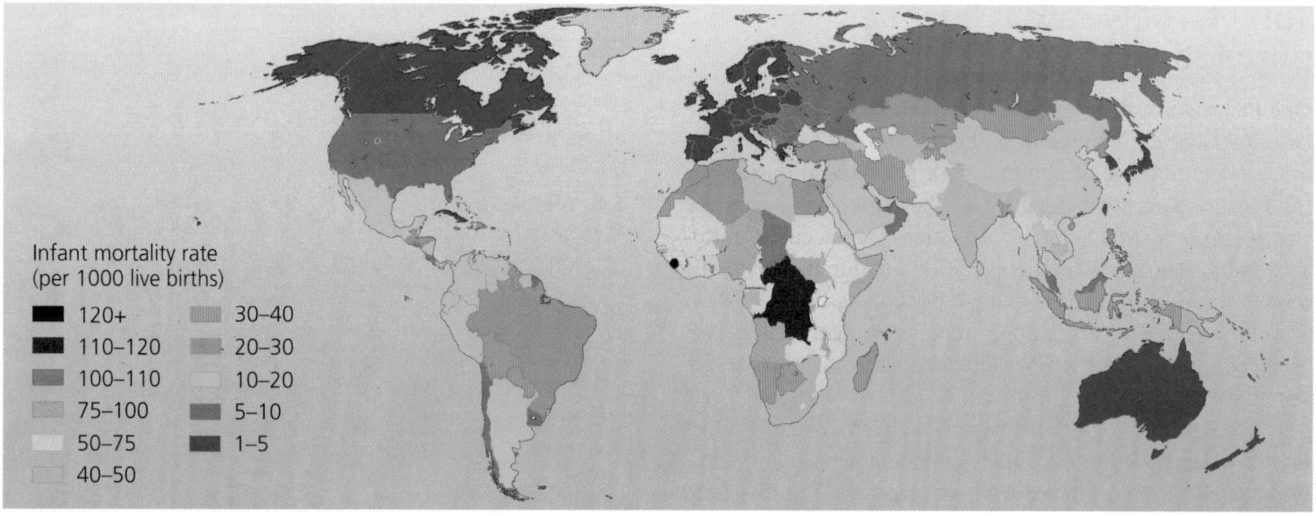


**FIGURE 6.10** Infant mortality rates are highest in poorer nations, such as in sub-Saharan Africa, and lowest in wealthier nations. Industrialization brings better nutrition and medical care, which greatly reduce the number of children dying in their first year of life. *Data from Population Reference Bureau, 2013.* 2013 World population data sheet.

## Population change results from birth, death, immigration, and emigration

Rates of birth, death, immigration, and emigration determine whether a population grows, shrinks, or remains stable. Birth and immigration add individuals to a population, whereas death and emigration remove individuals (see formula on p. 61). Technological advances during the last 200 years led to a dramatic decline in human death rates, widening the gap between birth rates and death rates and resulting in the rapid growth of the global human population.

Falling infant mortality rates have played an especially large role. Throughout much of human history, parents needed to have larger families as insurance against the likelihood that one or more of their children would die during infancy. Poor nutrition, disease, exposure to hostile elements, and limited medical care claimed the lives of many infants in their first year of life. As societies have industrialized and become more affluent, infant mortality rates have plummeted as a result of better nutrition, prenatal care, and the presence of medically trained practitioners during birth.

As shown in **FIGURE 6.10**, infant mortality rates vary widely around the world and are closely tied to a nation's level of industrialization. China, for example, saw its infant mortality rate drop from 47 children per 1000 live births in 1980 to 16 children per 1000 live births in 2013 as the nation industrialized and prospered. Many other industrializing nations enjoyed similar success in reducing infant mortality during this time period.

In recent decades, reductions in birth rates around the world have led to an overall decline in the global growth rate (**FIGURE 6.11**). Note, however, that although the rate of growth is slowing, the absolute size of our population continues to increase. Our growth rate is getting smaller, but tens of millions of people are added to the planet each year.

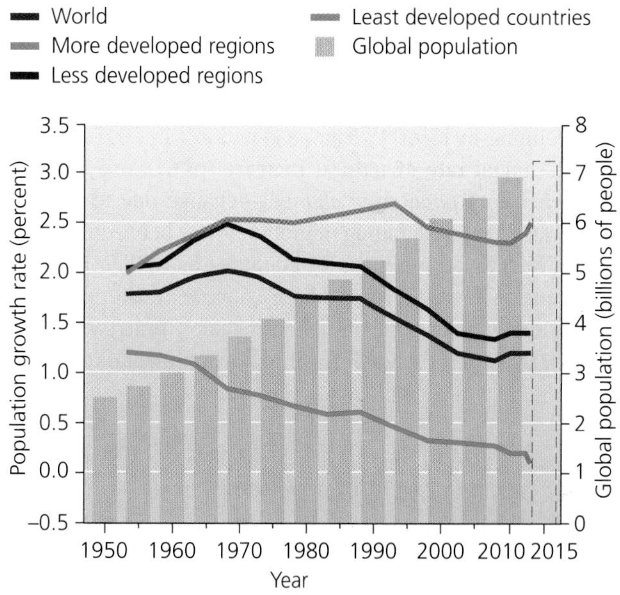


**FIGURE 6.11 The annual growth rate of the global human population peaked in the late 1960s and has declined since then.** Growth rates of developed nations have fallen since 1950, whereas those of developing nations have fallen since the global peak in the late 1960s. For the world's least developed nations, growth rates began to fall in the 1990s. Although growth rates are declining, global population size is still growing about the same amount each year, because smaller percentage increases of ever-larger numbers produce roughly equivalent additional amounts. *Data from Population Division of the Department of Economic and Social Affairs of the United Nations Secretariat, 2013.* World population prospects: The 2012 revision. http://esa.un.org/wpp. *© United Nations, 2013. Data updates for 2011–2013 from Population Reference Bureau, 2011–2013 World population data sheets. Global population in 2015 is projected.*

## Total fertility rate influences population growth

One key statistic demographers calculate to examine a population's potential for growth is the **total fertility rate (TFR),** the average number of children born per woman during her lifetime. **Replacement fertility** is the TFR that keeps the size of a population stable. For humans, replacement fertility roughly equals a TFR of 2.1. (Two children replace the mother and father, and the extra 0.1 accounts for the risk of a child dying before reaching reproductive age.) If the TFR drops below 2.1 and remains there, population size (in the absence of immigration) will shrink.

Reduced infant mortality rates have led to lower TFRs, as more children now survive infancy than in the past. Urbanization has also driven TFRs down; whereas rural families need children to contribute to farm labor, in urban areas children are usually excluded from the labor market, are required to go to school, and impose economic costs on their families. Moreover, if a government provides some form of social security for retirees, parents need fewer children to support them in their old age. Finally, with greater educational opportunities and changing roles in society, women tend to shift into the labor force, putting less emphasis on child rearing.

Factors such as industrialization, improved women's rights, and quality health care have driven TFR downward in many nations in recent years. All these factors have come together in Europe, where TFR has dropped from 2.6 to 1.6 in the past half-century. Nearly every European nation now has a fertility rate below the replacement level, and populations are declining in 18 of 45 European nations. In 2013, Europe's overall annual **rate of natural increase** (p. 61; also called the *natural rate of population change*)—change due to birth and death rates alone, excluding migration—was between 0.0% and 0.1%. Worldwide by 2013, 77 countries had fallen below the replacement fertility of 2.1. These countries make up roughly half of the world's population and include China (with a TFR of 1.5). TABLE 6.1 shows TFRs of major continental regions.

**TABLE 6.1 Total Fertility Rates for Major Regions**

| REGION | TOTAL FERTILITY RATE (TFR) |
|---|---|
| Africa | 4.8 |
| Australia and the South Pacific | 2.4 |
| Latin America and the Caribbean | 2.2 |
| Asia | 2.2 |
| North America | 1.9 |
| Europe | 1.6 |

*Data from Population Reference Bureau, 2013.* 2013 World population data sheet.

## Many nations have experienced the demographic transition

Many nations with lowered birth rates and TFRs are experiencing a common set of interrelated changes. In countries with reliable food supplies, good public sanitation, and effective health care, more people than ever before are living long lives. As a result, over the past 50 years the life expectancy (the time a person can expect to live) for the average person worldwide has increased from 46 to 70 years, with much of this increase attributed to reductions in infant mortality rates. Societies going through these changes are generally those that have undergone urbanization and industrialization and have generated personal wealth for their citizens.

Demographers have summarized these trends in a concept called the **demographic transition.** This is a model of economic and cultural change first proposed in the 1940s and 1950s by demographer Frank Notestein to explain the declining death rates and birth rates that have occurred in Western nations as they industrialized. Notestein believed nations move from a stable pre-industrial state of high birth and death rates to a stable post-industrial state of low birth and death rates (**FIGURE 6.12**). Industrialization, he proposed, causes these rates to fall by first decreasing mortality and then lessening the need for large families. Parents thereafter choose to invest in quality of life rather than quantity of children. Because death rates fall before birth rates fall, a period of net population growth results. Thus, under the demographic transition model, population growth is seen as a temporary phenomenon that occurs as societies move from one stage of development to another.

**The pre-industrial stage** The first stage of the demographic transition model is the **pre-industrial stage,** characterized by conditions that have defined most of human history. In pre-industrial societies, both death rates and birth rates are high. Death rates are high because disease is widespread, medical care rudimentary, and food supplies unreliable and difficult to obtain. Birth rates are high because people must compensate for infant mortality by having many children and because reliable methods of birth control are not available. In this stage, children are valuable as workers who can help meet a family's basic needs. Populations in the pre-industrial stage are not likely to experience much growth, which is why the human population grew very slowly until the industrial revolution.

**Industrialization and falling death rates** Industrialization initiates the second stage of the demographic transition, known as the **transitional stage.** This transition from the pre-industrial stage to the industrial stage is generally characterized by declining death rates due to increased food production and improved medical care. Birth rates in the transitional stage remain high, however, because people have not yet grown used to the new economic and social conditions. As a result, population growth surges.

**The industrial stage and falling birth rates** The third stage in the demographic transition is the **industrial stage.** Industrialization increases opportunities for employment outside the home, particularly for women. Children become less valuable, in economic terms, because they do not

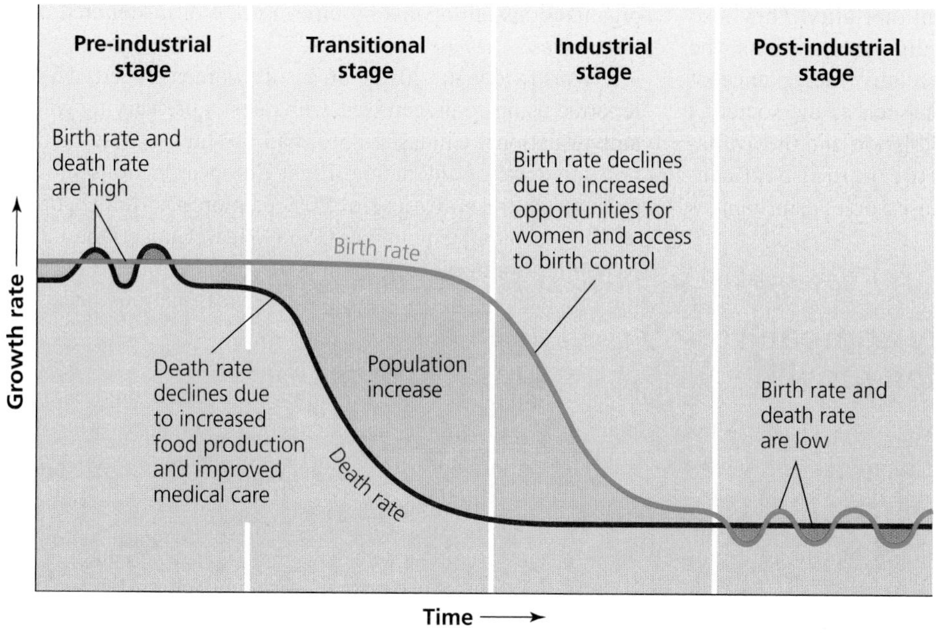


**FIGURE 6.12 The demographic transition is a process in which a population moves from a pre-industrial stage of high birth rates and high death rates to a post-industrial stage of low birth rates and low death rates.** In this diagram, the wide green area between the two curves illustrates the gap between birth and death rates. This gap is the source of the rapid population growth that occurs during the intermediate stages of this process. *Adapted from Kent, M. and K. Crews, 1990.* World population: Fundamentals of growth. *By permission of the Population Reference Bureau.*

DATA Q In which stage of the demographic transition does population increase the most? Is growth greatest at the beginning or end of this stage?

GO TO **INTERPRETING GRAPHS & DATA** ON MasteringEnvironmentalScience®

help meet family food needs as they did in the pre-industrial stage. If couples are aware of this, and if they have access to birth control, they may choose to have fewer children. Birth rates fall, closing the gap with death rates and reducing population growth.

**The post-industrial stage** In the final stage, called the **post-industrial stage,** both birth and death rates have fallen to low and stable levels. Population sizes stabilize or decline slightly. The society enjoys the fruits of industrialization without the threat of runaway population growth.

**FAQ**

**Why hasn't AIDS drastically lowered Africa's fertility rates?**

Africa's TFR is the highest of all world regions (see Table 6.1) despite rates of HIV/AIDS infection six times higher than in the next closest region. It seems contradictory that the region with the highest incidence of HIV/AIDS also have the world's highest fertility. This seeming paradox is related to the mode of transmission of HIV/AIDS and the time it takes for the disease to affect its victims' health. Because the disease is typically spread in Africa through sexual behavior, it tends to afflict individuals who are already in the reproductive age group. Hence, many people already have children by the time they contract the disease. Moreover, infected people often live for several years before succumbing to AIDS, and they continue having children even after becoming HIV-positive. As a result, nations with high rates of HIV infection still have extremely high fertility. The southern African nation of Swaziland in 2013, for example, had one of Africa's highest HIV infection rates (26% of its citizens infected) but also one of southern Africa's highest total fertility rates (3.5 children per woman).

## Is the demographic transition a universal process?

The demographic transition has occurred in many European countries, the United States, Canada, Japan, and a number of other developed nations over the past 200 to 300 years. It is a model that may or may not apply to all developing nations as they industrialize now and in the future. On the one hand, note in Figure 6.11 that growth rates fell first for industrialized nations, then for less developed nations, and finally for least developed nations. This pattern suggests that it may merely be a matter of time before all nations experience the transition. On the other hand, some developing nations may already be suffering too much from the impacts of large populations to replicate the developed world's transition. And some demographers assert that the transition will fail in cultures that place greater value on childbirth or that grant women fewer freedoms.

Moreover, natural scientists estimate that for people of all nations to attain the material standard of living that North Americans now enjoy, we would need the natural resources of three additional planet Earths. Whether developing nations (which include the vast majority of the planet's people) pass through the demographic transition is one of the most important and far-reaching questions for the future of our civilization and Earth's environment.

# Population and Society

Demographic transition theory links the quantitative study of how populations change with the societal factors that influence (and are influenced by) population dynamics. Many factors affect fertility in a given society. These factors include public health issues, such as people's access to medical care and the rate of infant mortality. They also include cultural factors—such as religious traditions, the degree of gender equality, and the relative acceptance of contraceptive use. Economic factors, such as the society's level of affluence, the degree of child labor, and the availability of governmental support for retirees are also influential. Let's now examine a few of these societal influences on fertility more closely.

## Family planning is a key approach for controlling population growth

Perhaps the greatest single factor enabling a society to slow its population growth is the ability of women and couples to engage in **family planning,** the effort to plan the number and spacing of one's children. Family-planning programs and clinics offer information and counseling to potential parents on reproductive issues.

An important component of family planning is **birth control,** the effort to control the number of children one bears, particularly by reducing the frequency of pregnancy. Birth control relies on **contraception,** the deliberate attempt to prevent pregnancy despite sexual intercourse. Common methods of modern contraception include condoms, spermicide, hormonal treatments (birth control pill/hormone injection), intrauterine devices (IUDs), and permanent sterilization through tubal ligation or vasectomy. Many family-planning organizations aid clients by offering free or discounted contraceptives.

Worldwide in 2013, 56% of women (aged 15–49) reported using contraceptives, with rates of use varying widely among nations. China, at 85%, had the highest rate of contraceptive use of any nation. Eight European nations showed rates of contraceptive use of 70% or more, as did Australia, Brazil, Canada, Colombia, Costa Rica, Cuba, Dominican Republic, South Korea, Thailand, and the United States. At the other end of the spectrum, 11 African nations had rates below 10%.

In some societies, rates of contraceptive use are low because religious doctrine or cultural influences hinder family planning. When counseling and contraceptives are unavailable to people who might otherwise use them, people have larger families than they may desire, and rates of population growth remain elevated.

In a physiological sense, access to family planning (and the civil rights to use it) gives women control over their **reproductive window.** This is the period of a woman's life, beginning with sexual maturity and ending with menopause, in which she may become pregnant. A woman can bear up to 25 children within this window (**FIGURE 6.13a**), but she may choose to delay the birth of her first child to pursue education and employment. She may also use contraception to delay her first child, space births within the window, and "close" her reproductive window after achieving her desired family size (**FIGURE 6.13b**).

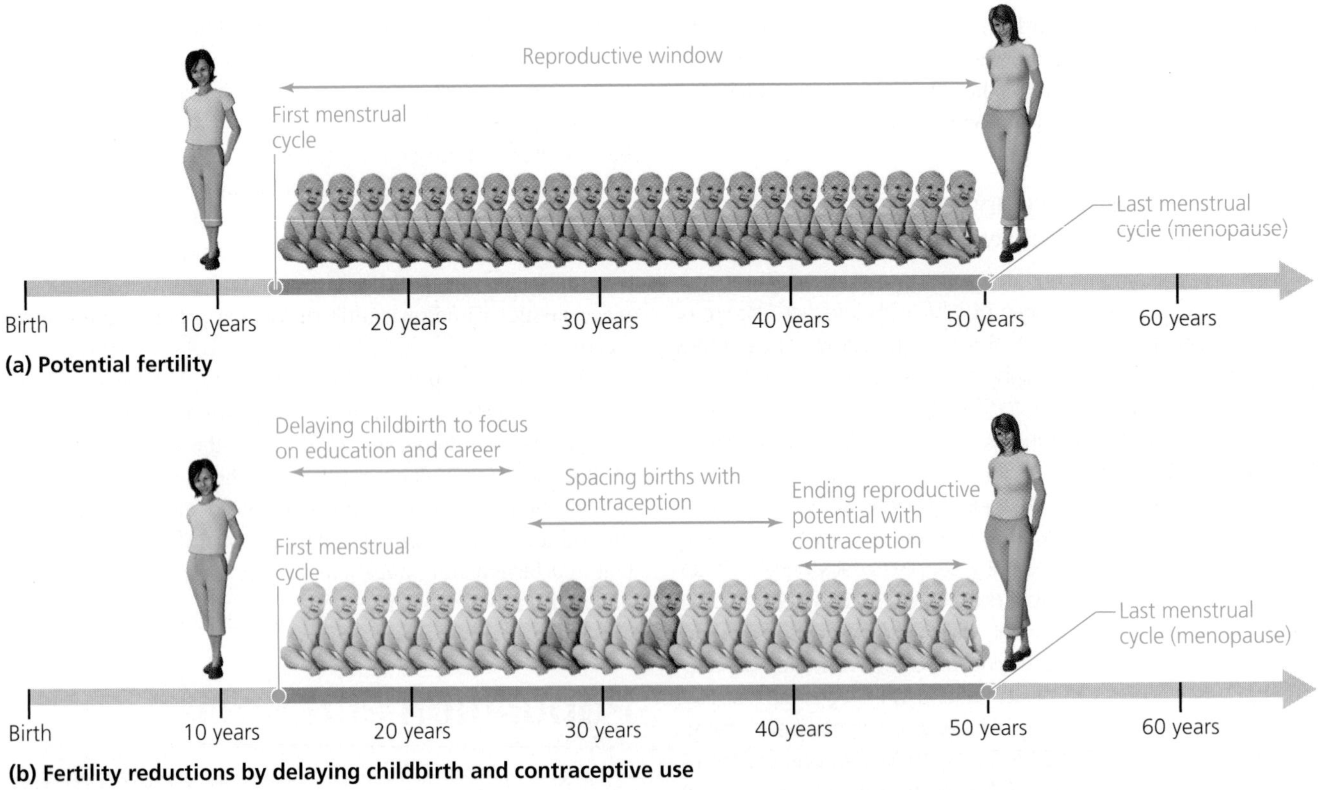


**FIGURE 6.13 Women potentially have very high fertility within their "reproductive window" (a) but can choose to reduce the number of children they bear (b).** They may do this by delaying the birth of their first child or by using contraception to space pregnancies or to end their reproductive window at the time of their choosing.

## Family-planning programs are working around the world

Data show that policies that encourage family planning can lower population growth rates in all types of nations, even those that are least industrialized. No nation has pursued a sustained population control program as intrusive as China's, but other rapidly growing nations have implemented programs that are less restrictive but nonetheless effective at reducing population growth.

The government of Thailand, for example, has reduced birth rates and slowed population growth. In the 1960s, Thailand's growth rate was 2.3%, but today it stands at 0.4%. This decline was achieved without a one-child policy, resulting instead from an education-based approach to family planning and the increased availability of contraceptives. Brazil, Mexico, Iran, Cuba, and many other developing countries have instituted active programs that entail setting targets and providing incentives, education, contraception, and reproductive health care.

In 1994, the United Nations hosted a milestone conference on population and development in Cairo, Egypt, at which 179 nations endorsed a platform calling on all governments to offer their citizens universal access to reproductive health care. The conference marked a change from older notions of top-down population policy geared toward pushing contraception and lowering population to preset targets. Instead, it urged governments to offer better education and health care and to address social needs that affect population from the bottom up (such as alleviating poverty, disease, and sexism). Such bottom-up approaches have now become the norm for many nations' population initiatives. They act to lower population growth rates while avoiding the social disruptions that can accompany top-down approaches such as China's.

### WEIGHING THE ISSUES

**Abstaining from International Family Planning?** Over the years, the United States has joined 180 other nations in providing millions of dollars to the United Nations Population Fund (UNFPA), which advises governments on family planning, sustainable development, poverty reduction, reproductive health, and AIDS prevention in many nations, including China. In 2001, the George W. Bush administration withheld funds from UNFPA, saying that U.S. law prohibits funding any organization that "supports or participates in the management of a program of coercive abortion or involuntary sterilization," and maintaining that the Chinese government has been implicated in both. Many nations criticized the U.S. decision, and the European Union offered UNFPA additional funding to offset the loss of U.S. contributions. Once President Obama came to office, he reinstated funding to the program. What do you think U.S. policy should be? Should the United States fund family-planning efforts in other nations? What conditions, if any, should it place on the use of such funds?

## Empowering women reduces fertility rates

Today, many social scientists and policymakers recognize that for population growth to slow and stabilize, women must be granted equal power to men in societies worldwide. This would have many benefits: Studies show that where women are freer to decide whether and when to have children, fertility rates fall, and the resulting children are better cared for, healthier, and better educated.

For women, one benefit of equal rights is the ability to make reproductive decisions. In some societies, men restrict women's decision-making abilities, including decisions as to how many children they will bear. Birth rates have dropped the most in nations where women have gained reliable access to contraceptives and to family planning. This trend indicates that giving women the right to control their reproduction reduces fertility rates.

Efforts to improve women's rights are enhanced when young women are exposed to empowered women they can emulate. These women can be relatives, friends, social workers, politicians, or even characters on television programs (see **THE SCIENCE BEHIND THE STORY**, pp. 128–129).

Equality for women also involves expanding educational opportunities, because in many nations girls are discouraged from pursuing an education or are kept out of school altogether. Over two-thirds of the world's people who cannot read are women. Data clearly show that as women receive educational opportunities, fertility rates decline (**FIGURE 6.14**). Education helps more women pursue careers, delay childbirth, and have a greater say in reproductive decisions.

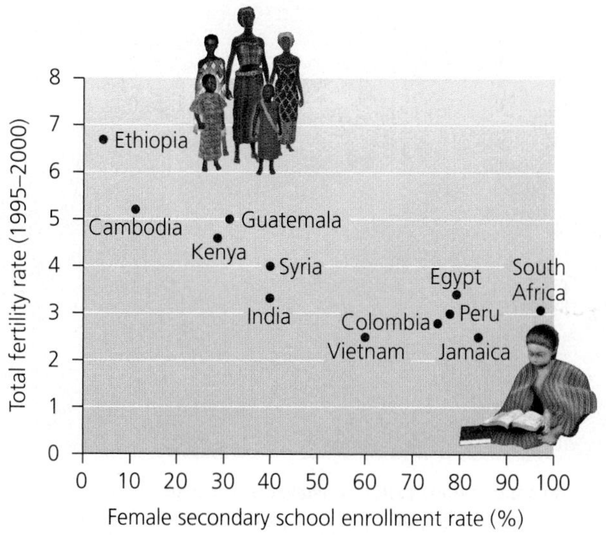


**FIGURE 6.14 Increasing female literacy is strongly associated with reduced birth rates in many nations.** *Data from McDonald, M., and D. Nierenberg, 2003.* Linking population, women, and biodiversity. *State of the world 2003. Washington, D.C.: Worldwatch Institute.*

**DATA Q** What is the relationship between total fertility rate and the rate of enrollment of girls in secondary school? Is it positive (as variable 1 increases, so does variable 2), negative (as variable 1 increases, variable 2 decreases), or is there no obvious relationship (increases in variable 1 are not related to changes in variable 2)?

GO TO **INTERPRETING GRAPHS & DATA** ON MasteringEnvironmentalScience®

## Did Soap Operas Reduce Fertility in Brazil?

Over the past 50 years, the South American nation of Brazil experienced the second-largest drop in fertility among developing nations with large populations—second only to China. In the 1960s, the average woman in Brazil had six children. Today, Brazil's total fertility rate is 1.9 children per woman, which is lower than that of the United States. Brazil's drastic decrease in fertility is interesting because, unlike China, it occurred without intrusive governmental policies to control its citizens' reproduction.

Brazil accomplished this, in part, by providing women equal access to education and opportunities to pursue careers outside the home. Women now comprise 40% of the workforce in Brazil and graduate from college in greater numbers than men. In 2010, Brazilians elected a woman, Dilma Rousseff, as their nation's president.

The Brazilian government also provides family planning and contraception to its citizens free of charge. Eighty percent of married women of childbearing age in Brazil currently utilize contraception, a rate higher than that in the United States or Canada. Universal access to family planning has given women control over their desired family size and has helped reduce fertility across all economic groups, from the very rich to the very poor. Induced abortion is not utilized in Brazil as it is in China; the procedure is illegal except in rare circumstances.

**Brazilian soap operas, called *telenovelas*, are a surprising cultural force for promoting lower fertility.** Here, residents gather outside a cafe in Rio de Janeiro to watch the popular program *Avenida Brasil*.

As Brazil's economy grew with industrialization, people's nutrition and access to health care improved, greatly reducing infant mortality rates. Increasing personal wealth promoted materialism and greater emphasis on career and possessions over family and children. The nation also urbanized as people flocked to growing cities such as Río de Janeiro and São Paolo. This brought about the fertility reductions that typically occur when people leave the farm for the city.

It turns out, however, that Brazil may also have had a rather unique influence affecting its fertility rates over the past several decades—soap operas. Brazilian soap operas, called *telenovelas* or *novelas*, are a cultural phenomenon and are watched religiously by people of all ages, races, and incomes. Each *novela* follows the activities of several fictional families, and these TV shows are wildly popular because they have characters, settings, and plot lines with which everyday Brazilians can identify.

*Telenovelas* do not overtly address fertility issues, but they do promote a vision of the "ideal" Brazilian family. This family is typically middle- or upper-class, materialistic, individualistic, and full of empowered women. By challenging existing cultural and religious values through their characters, *novelas* had, and continue to have, a profound impact on Brazilian society. In essence, these programs provided a model family for Brazilians to emulate—with small family sizes being a key characteristic.

In a 2012 paper in the *American Economic Journal: Applied Economics*, a team of researchers from Bocconi University in Italy, George Washington University, and the Inter-American Development Bank (based in Washington, D.C.) analyzed various parameters to investigate statistical relationships between *telenovelas* and fertility patterns in Brazil from 1965 to 2000. Rede Globo, the network that has a virtual monopoly on the most popular *novelas*, increased the number of areas that received its signal in Brazil over those 35 years

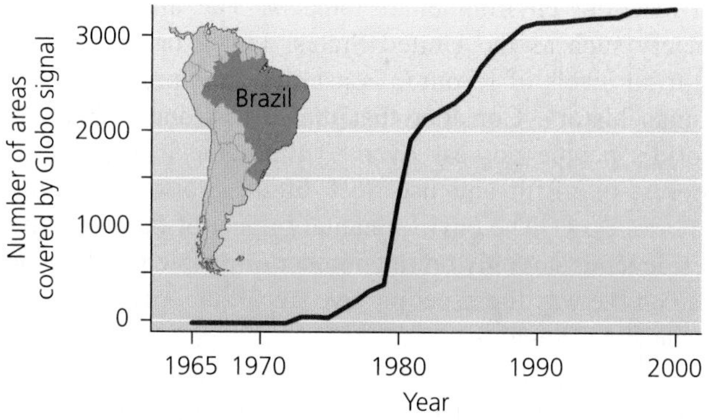


**FIGURE 1 The Globo television network expanded over time and now reaches nearly all households in Brazil.** Fertility declines were correlated with the availability of Globo, and its *novelas*, over the time periods in the study. *Source: La Ferrara, E., et al., 2012. Soap operas and fertility: Evidence from Brazil.* Am. Econ. J. Appl. Econ. *4: 1–31.*

(**FIGURE 1**), and it now reaches 98% of Brazilian households. By combining data on Rede Globo broadcast range with demographic data, the researchers were able to compare changes in fertility patterns over time in areas of Brazil that received access to *novelas* with areas of Brazil that did not.

The team found that women in areas that received the Globo signal had significantly lower fertility than those in areas not served by Rede Globo. They also found that fertility declines were age-related, with substantial reductions in fertility occurring in women aged 25–44, but not in younger women (**FIGURE 2**). The authors hypothesized that this effect was likely due to the fact that women between 25 and 44 were closer in age to the main female characters in *novelas*, who typically had no children or only a single child. The depressive effect on fertility among women in areas served by Globo was therefore attributed to wider spacing of births and earlier ending of reproduction by women over 25, rather than to younger women delaying the birth of their first child.

Further evidence for the influence of *novelas* on fertility was found in school records. The researchers found that fifth-graders living in areas reached by the Globo network were four times more likely to be named after characters in *novelas* than were children in areas not served by Globo. These results were compelling because most *novela* characters have relatively unusual names.

The researchers determined that access to television alone did not depress fertility. For example, comparisons of fertility rate in areas with access to a different television network, Sistema Brasileiro de Televisão, found no relationship. The study authors concluded that this was likely due to the network's reliance on programming imported from other nations, with which everyday Brazilians did not connect as they did with *novelas* from Rede Globo.

Television's ability to influence fertility is not limited to Brazil. A 2014 study found that in the United States, tweets and Google searches for terms such as "birth control" increased significantly the day following the airing of new episodes of "MTV's 16 and Pregnant." By correlating geographic patterns in viewership with fertility data, the study authors concluded that MTV's "Teen Mom" series may have been responsible for reducing teenage births by up to 20,000 per year.

The factors that affect human fertility can be complex and vary greatly from one society to another. Although this correlative study (p. 10) does not prove causation between watching *telenovelas* and reduced fertility, it does show that the factors that influence fertility can sometimes come from unexpected sources. ☐

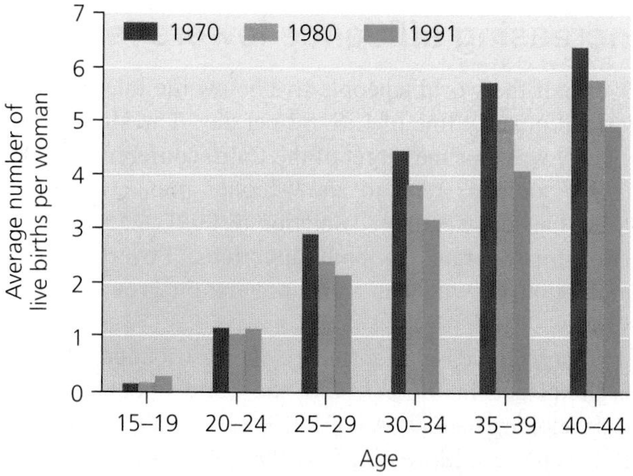


**FIGURE 2 Fertility declines among Brazilian women between 1970 and 1991 were most pronounced in later age classes.** Researchers attribute some of this decline to women in those age classes emulating the low fertility of lead female characters in *novelas*. *Source: La Ferrara, E., et al., 2012. Soap operas and fertility: Evidence from Brazil.* Am. Econ. J. Appl. Econ. *4: 1–31.*

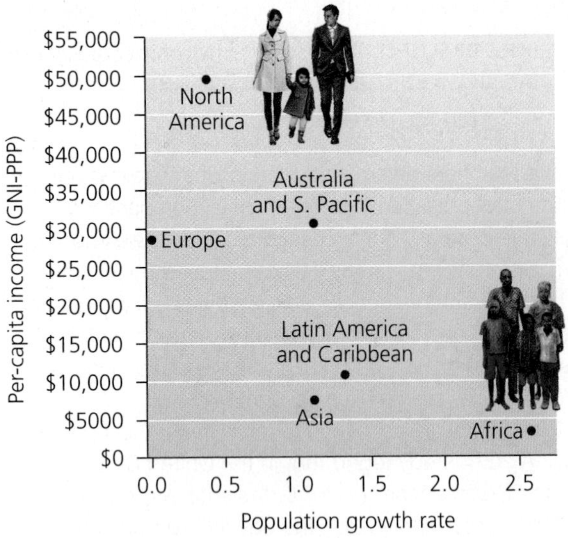


**FIGURE 6.15 Poverty and population growth show a fairly strong correlation, despite the influence of many other factors.** Regions with the lowest per capita incomes tend to have the most rapid population growth. Per capita income is here measured in GNI PPP, or "gross national income in purchasing power parity." GNI PPP is a measure that standardizes income among nations by converting it to "international" dollars, which indicate the amount of goods and services one could buy in the United States with a given amount of money. *Data from Population Reference Bureau, 2013.* 2013 World population data sheet.

## Increasing affluence lowers fertility

Over half the world's people live below the internationally defined poverty line of U.S. $2 per day. The alleviation of poverty was a prime target of the Cairo conference because poorer societies tend to show higher population growth rates than do wealthier societies (**FIGURE 6.15**). This relationship operates in both directions: Poverty worsens population growth, and rapid population growth worsens poverty.

These trends also influence the distribution of people on the planet. In 1960, 70% of the world's population lived in developing nations. As of 2013, 83% of all people live in these countries. Moreover, fully 99% of the next billion people to be added to the global population will be born into developing nations (**FIGURE 6.16**).

This is unfortunate from a social standpoint, because these people will be added to nations that are often unable to adequately provide for them. It is also unfortunate from an environmental standpoint, because poverty often results in environmental degradation. People who depend on agriculture in areas of poor farmland, for instance, may need to try to farm even if doing so degrades the soil and is not sustainable. Poverty also drives people to cut forests and to deplete biodiversity. For example, impoverished settlers and miners hunt large mammals for "bush meat" in Africa's forests, including the great apes that are now heading toward extinction.

## Expanding wealth can increase the environmental impact per person

Poverty can lead people into environmentally destructive behavior, but wealth can produce even more severe and far-reaching environmental impacts. The affluence of a society such as the United States, Japan, or Germany is built on levels of resource consumption unprecedented in human history. Consider that the richest one-fifth of the world's people possess over 80 times the income of the poorest one-fifth and use 86% of the world's resources (**FIGURE 6.17**). The environmental impact of human activities depends not only on the number of people involved but also on the way those people live (recall the *A* for *affluence* in the IPAT equation).

An *ecological footprint* represents the cumulative amount of Earth's surface area required to provide the raw materials a person or population consumes and to dispose of or recycle the waste produced (p. 4). Individuals from affluent societies leave considerably larger per capita ecological footprints (see Figure 1.17, p. 17). In this sense, the addition of one American to the world has as much environmental impact as the addition of 3.4 Chinese, 8 Indians, or 14 Afghans. This fact reminds us that the "population problem" does not lie solely with the developing world!

Indeed, just as population is rising, so is consumption. Researchers have found that humanity's global ecological footprint surpassed Earth's capacity to support us in the 1970s (p. 4), and that our species is now living 50% beyond

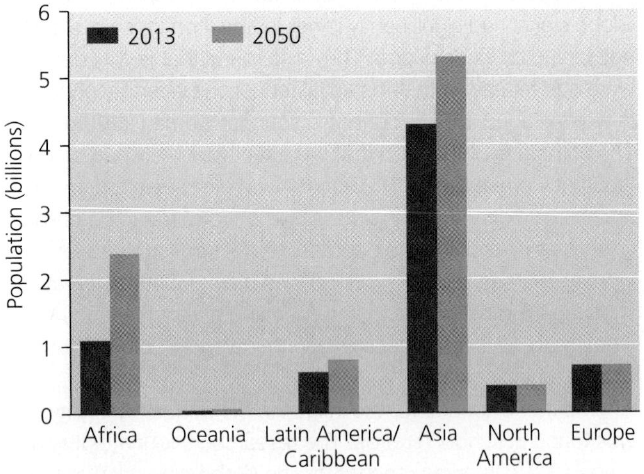


**FIGURE 6.16 Africa will experience the greatest population growth of any region in coming decades.** The vast majority of future population growth will occur in developing regions. The highly industrialized regions of Europe and North America are predicted to experience only minor population change. *Figure from Population Reference Bureau, 2013.* 2013 World population data sheet.

**DATA Q** Which region will add the most people between 2013 and 2050? Which will increase by the greatest percentage during this time period. Which regions will experience very little population growth during this period?

GO TO **INTERPRETING GRAPHS & DATA** ON MasteringEnvironmentalScience®

(a) A family living in the United States

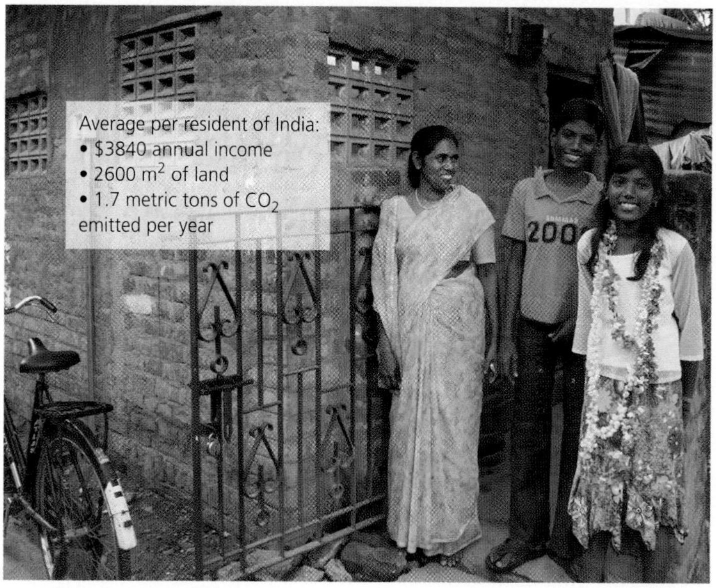


(b) A family living in India

**FIGURE 6.17 Material wealth varies widely from nation to nation.** A typical U.S. family **(a)** may own a large house with a wealth of material possessions. A typical family in a developing nation such as India **(b)** may live in a much smaller home with far fewer material possessions. Compared with the average resident of India, the average U.S. resident enjoys 13 times more income, uses 11 times more land, and emits 10 times more carbon dioxide emissions. Average annual incomes were calculated by dividing the gross national income (GNI) of each country by its population. *Data from Population Reference Bureau, 2013.* 2013 World population data sheet *and World Bank, 2013,* http://data.worldbank.org.

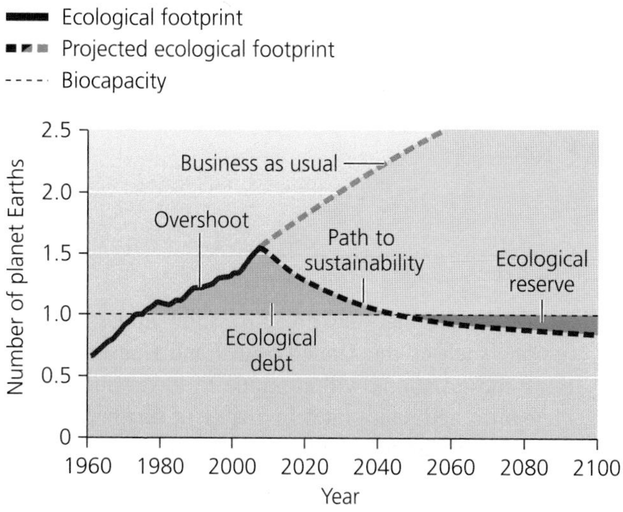


**FIGURE 6.18 The global ecological footprint of the human population is estimated to be 50% greater than what Earth can bear.** If population and consumption continue to rise **(orange dashed line),** we will increase our ecological deficit, or degree of overshoot, until systems give out and populations crash. If, instead, we pursue a path to sustainability **(red dashed line),** we can eventually repay our ecological debt and sustain our civilization. *Adapted from WWF International. 2012.* Living planet report 2012. *Published by WWF–World Wide Fund for Nature. © 2012 WWF* (panda.org)*, Zoological Society of London, and Global Footprint Network.*

its means (**FIGURE 6.18**). The rising consumption that is accompanying the rapid industrialization of China, India, and other populous nations makes it all the more urgent for us to find a path to global sustainability.

If humanity's overarching goal is to generate a high standard of living and quality of life for all people, then developing nations must find ways to slow their population growth. However, those of us living in the developed world must also be willing to reduce our consumption. Earth does not hold enough resources to sustain all 7 billion of us at the current North American standard of living, nor can we venture out and bring home extra planets. We must make the best of the one place that supports us all.

# Conclusion

Today's human population is larger than at any time in the past. Our growing population and our growing consumption affect the environment and our ability to meet the needs of all the world's people. The great majority of children born today are likely to live their lives in conditions far less healthy and prosperous than most of us in the industrialized world are accustomed to.

However, there are at least two major reasons to feel encouraged. First, although global population is still rising, the rate of growth has decreased everywhere. Some countries are even seeing population declines. Most developed nations have passed through the demographic transition, reducing death rates while stabilizing population and creating more prosperous societies. Second, progress has been made in expanding rights for women worldwide. Although there is still a long way to go, women are obtaining better education, more economic independence, and more ability to control their reproductive decisions.

The human population cannot continue to rise forever. The question is how it will stop rising: Will it be through the gentle and benign process of the demographic transition, through restrictive governmental intervention such as China's one-child policy, or through the Malthusian checks of disease and social conflict caused by overcrowding and competition for scarce resources? How we answer this question today will determine the quality of our own lives and the quality of the world we leave to our children and grandchildren.

# Testing Your Comprehension

1. What is the approximate current human global population? How many people are being added to the population each day?
2. Why has the human population continued to grow despite environmental limitations? Give one example of an innovation that increased the carrying capacity for humans.
3. Contrast the views of environmental scientists with those of economists regarding whether population growth is a problem. Name several reasons why population growth is viewed as a problem.
4. Explain the IPAT model. How can technology increase environmental impact? Provide at least one example. How can technology decrease environmental impact? Provide at least one example.
5. How do the size, density, distribution, age structure, and sex ratio of a population help determine the impact of human populations on the environment?
6. What is the total fertility rate (TFR)? Why is the replacement fertility for humans approximately 2.1? How is Europe's TFR affecting its rate of natural increase?
7. Why have fertility rates fallen in many countries?
8. How does the demographic transition model explain the increase in population growth rates in recent centuries? How does it explain the recent decrease in population growth rates in many countries?
9. Why are the empowerment of women and the pursuit of gender equality viewed as important to controlling population growth? Describe the aim of family-planning programs.
10. Why do poorer societies have higher population growth rates than wealthier societies? How does poverty affect the environment? How does affluence affect the environment?

# Seeking Solutions

1. China's reduction in birth rates is changing the nation's age structure. Review Figure 6.9, which shows that the population is growing older, leading to the top-heavy population pyramid for the year 2050. What effects might this ultimately have on Chinese society? What steps could be taken in response?
2. Apply the IPAT model to the example of China provided in the chapter. How do population, affluence, technology, and ecological sensitivity each affect China's environment? Now consider your own country or your own state. How do population, affluence, technology, and ecological sensitivity each affect your environment? How can we minimize the environmental impacts of growth in the human population?
3. Do you think that all of today's developing nations will complete the demographic transition and come to enjoy a permanent state of low birth and death rates? Why or why not? What steps might we as a global society take to help ensure that they do? Now think about developed nations such as the United States and Canada. Do you think these nations will continue to lower and stabilize their birth and death rates in a state of prosperity? What factors might affect whether they do so?
4. **THINK IT THROUGH** India's prime minister puts you in charge of that nation's population policy. India has a population growth rate of 1.5% per year, a TFR of 2.4, a 47% rate of contraceptive use, and a population that is 69% rural. What policy steps would you recommend to reduce growth rates, and why?
5. **THINK IT THROUGH** Now suppose that you have been tapped to design population policy for Germany. Germany is losing population at an annual rate of 0.2%, has a TFR of 1.4, a 66% rate of contraceptive use, and a population that is 73% urban. What policy steps would you recommend, and why?

# Calculating Ecological Footprints

A nation's population size and the affluence of its citizens each influence its resource consumption and environmental impact. As of 2013, the world's population passed 7.1 billion, average per capita income was $11,690 per year, and the latest estimate for the world's average ecological footprint was 2.7 hectares (ha) per person. The sampling of data in the table will allow you to explore patterns in how population, affluence, and environmental impact are related.

| Nation | Population (millions of people) | Affluence (per capita income, in GNI PPP)[1] | Personal impact (per capita footprint, in ha/person) | Total impact (national footprint, in millions of ha) |
|---|---|---|---|---|
| Brazil | 195.5 | $ 11,720 | 2.9 | 567 |
| China | 1357.4 | $ 9,210 | 2.1 | |
| Ethiopia | 89.2 | $ 1,140 | 1.1 | |
| India | 1276.5 | $ 3,840 | 0.9 | |
| Japan | 127.3 | $ 36,320 | 4.2 | |
| Mexico | 117.6 | $ 16,630 | 3.3 | |
| Russia | 143.5 | $ 22,760 | 4.4 | |
| Sweden | 9.6 | $ 43,160 | 5.7 | |
| United States | 316.2 | $ 50,610 | 7.1 | 2245 |

[1]*GNI PPP is "gross national income in purchasing power parity," a measure that standardizes income among nations by converting it to "international" dollars, which indicate the amount of goods and services one could buy in the United States with a given amount of money.*

*Data Sources: Population and affluence data are from Population Reference Bureau, 2013.* World population data sheet 2013. *Footprint data are for 2008, from WWF International, Zoological Society of London, and Global Footprint Network.* Living Planet Report 2012.

1. Calculate the total impact (national ecological footprint) for each country.
2. Draw a graph illustrating per capita impact (on the *y* axis) versus affluence (on the *x* axis). What do the results show? Explain why the data look the way they do.
3. Draw a graph illustrating total impact (on the *y* axis) in relation to population (on the *x* axis). What do the results suggest to you?
4. Draw a graph illustrating total impact (on the *y* axis) in relation to affluence (on the *x* axis). What do the results suggest to you?
5. You have just used three of the four variables in the IPAT equation. Now give one example of how the T (technology) variable could potentially increase the total impact of the United States, and one example of how it could potentially decrease the U.S. impact.

# 7

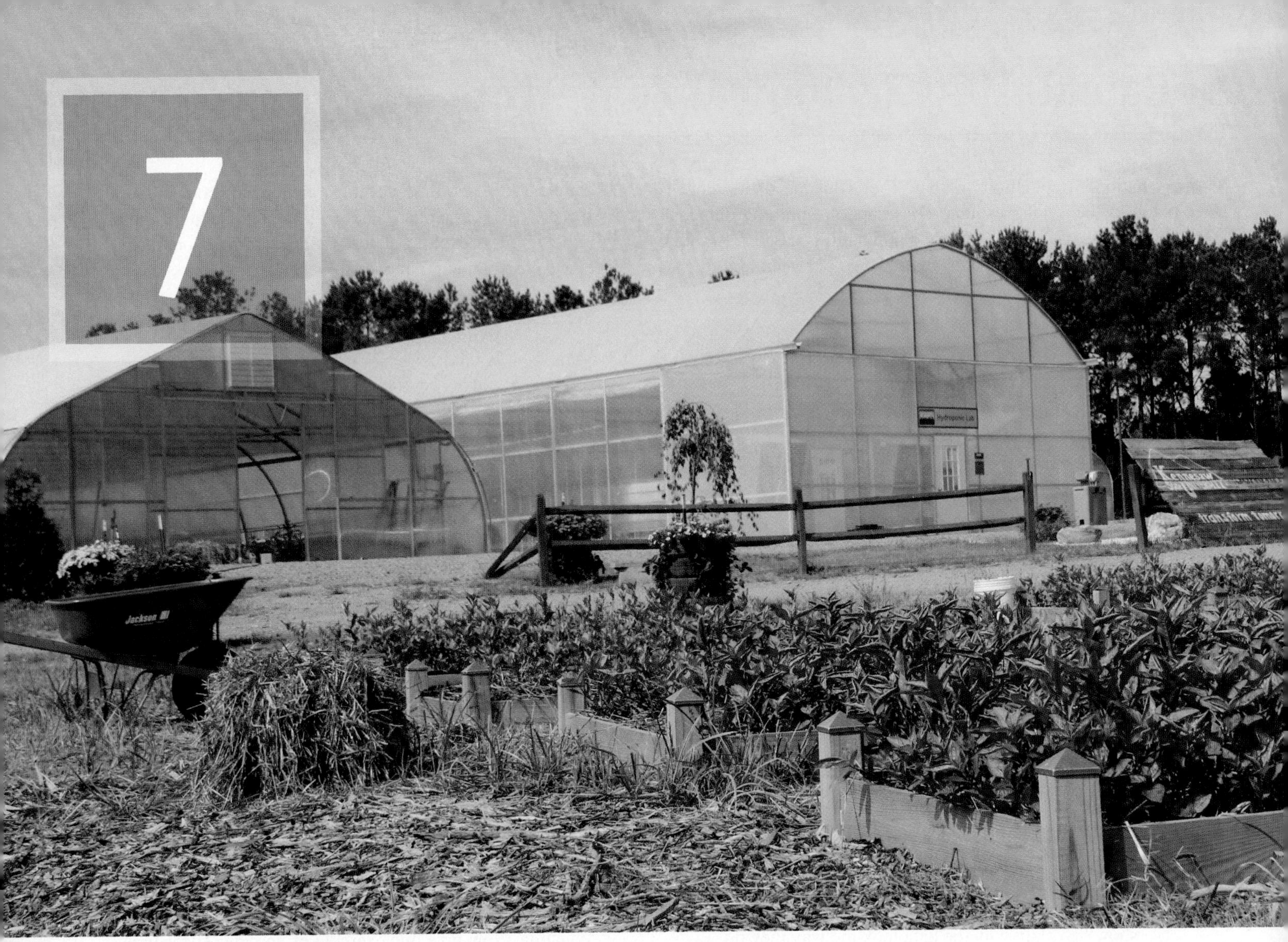

# Soil, Agriculture, and the Future of Food

## Upon completing this chapter, you will be able to:

- ☐ Explain the challenges of feeding a growing human population
- ☐ Compare and contrast traditional, industrial, and sustainable agricultural approaches
- ☐ Identify the goals, methods, and consequences of the Green Revolution
- ☐ Explain the importance of soils to agriculture
- ☐ Analyze the causes and impacts of soil erosion and land degradation
- ☐ Explain the principles of soil conservation and provide solutions to soil erosion and land degradation
- ☐ Compare and contrast approaches to irrigation, fertilization, and pest management in industrial and sustainable agriculture
- ☐ Describe the science behind genetic engineering and evaluate the public debate over its use
- ☐ Assess how we raise animals for food
- ☐ Analyze the nature, growth, and potential of organic agriculture



**Photo:** Kennesaw State University's Hickory Grove farm.

CENTRAL CASE STUDY

# Farm to Table—And Back Again: The Commons at Kennesaw State University

*"What we eat has changed more in the last 40 years than in the previous 40,000."*

**—Eric Schlosser, *Fast Food Nation* (2005)**

*"There are two spiritual dangers in not owning a farm. One is the danger of supposing that breakfast comes from the grocery, and the other that heat comes from the furnace."*

**—Aldo Leopold, conservationist and philosopher**

It's not surprising to see phrases such as "think global, eat local" and "farm to table" when you're dining at a trendy restaurant, but would you expect to see the same phrases used at your campus dining hall?

Believe it or not, campus dining services around the country are among the industry leaders in culinary sustainability, which embraces the use of sustainably produced, local foods to provide diners delicious, nutritious meals. One leader in sustainable dining is Kennesaw State University (KSU) in suburban Atlanta. In 2009, the university opened The Commons, a dining facility that serves up to 5000 students daily. The Commons offers nine themed food stations and a rotating menu of 200–300 items, but it's the facility's commitment to sustainability that make it noteworthy.

This commitment began when the university made sustainability a prime consideration in all aspects of the facility's construction and operation. Gary Coltek, a former executive chef and current Director of Culinary and Hospitality Services at KSU, used his extensive professional experience to design a facility that minimized energy use, water consumption, and waste generation while serving thousands of students a day.

As one of the nation's largest LEED-Gold certified (p. 416) campus dining halls, The Commons embraces efficiency in all parts of its operation. It is illuminated with floor-to-ceiling windows and highly efficient LED lights. A "trayless" approach to food service reduces the amount of water used in dishwashing, as diners without trays tend to eat less and use fewer plates. The Commons even generates biodiesel from its used cooking oil (p. 383) for use in university vehicles.

The Commons' commitment to sustainability extends far beyond the confines of the dining hall. The facility takes a farm-to-table approach by growing 20% of its produce on 27 hectares (67 acres) of land on three nearby campus farms using practices that minimize the use of pesticides and synthetic fertilizer. Herbs are grown in a greenhouse behind the dining hall and in rainwater-fed hydroponic stations scattered throughout the dining area. The university has also cultivated relationships with local farms and meat producers, sourcing locally produced food whenever possible.

Foods are prepared to-order or in small batches according to demand. This means that food is prepared only when it is needed, drastically reducing the amount of leftover food. Food waste is fed into a digester behind the facility and is broken down to generate a nutrient-rich liquid. This liquid is then used on the campus farms as a fertilizer to grow new produce. As Coltek puts it, "We go beyond farm-to-campus. We embrace farm-to-campus and back-to-farm operations."

The impacts of The Commons go beyond mealtime. In 2013, Kennesaw State opened its Institute for Culinary Sustainability and Hospitality, the first such degree program in the United States. In it students learn the culinary, business, and scientific skills required to implement sustainable food practices in restaurants, hotels, hospitals, and schools.

The Commons has received wide recognition as a notable campus dining destination, and in 2013 it became the first educational institution to win the National Restaurant Association's Innovator of the Year award, beating out competitors such as Walt Disney Parks and Resorts and the U.S. Air Force.

Kennesaw State is not the only university integrating sustainability into its campus dining operations. Boston University's newly constructed, LEED-Gold certified Marciano Commons also reduces waste by composting food scraps and using compostable packaging, and it commits one-fourth of its annual food budget to purchasing locally produced food. Michigan State University's recently remodeled The Vista at Shaw (LEED-Silver certification pending) offers diners a view of the Red Cedar River through floor-to-ceiling windows while they enjoy prepared-to-order food items from three distinct restaurants. This facility is reducing water consumption by 20% by embracing water-conserving appliances and practices in its kitchens.

Not all collegiate dining operations have the benefits of a newly constructed or remodeled facility, but food service operations on campuses across the country are embracing culinary sustainability in ways both big and small. Collectively, these efforts are decreasing the ecological footprint of American colleges and universities and helping to chart a path to a more sustainable food future. Most important, The Commons and other modern dining halls demonstrate that your taste buds don't have to suffer for the Earth to benefit.

Throughout this chapter, we'll see how principles of sustainability embraced at The Commons are being replicated at ever-larger scales, helping us chart a path toward ensuring adequate food for all people while minimizing the environmental impacts of agriculture. □

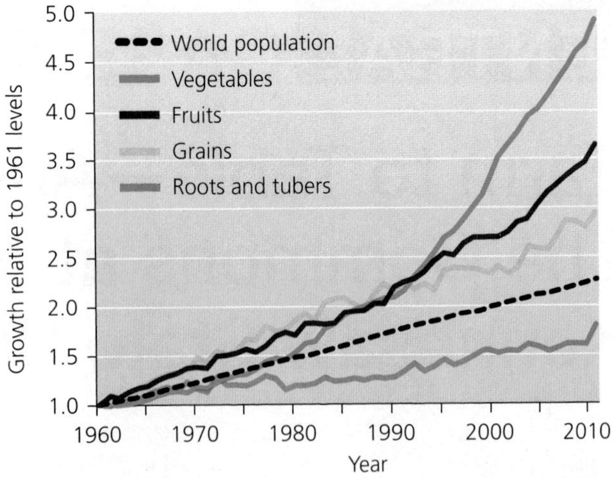


**FIGURE 7.1 Global production of most foods has risen more quickly than world population.** This means that we have produced more food per person each year. Trend lines show cumulative increases relative to 1961 levels (for example, a value of 2.0 means twice the 1961 amount). Food is measured by weight. *Data from U.N. Food and Agriculture Organization (FAO).*

# The Race to Feed the World

Someone dining at The Commons, surrounded by a diversity of food options, might find it hard to imagine that many people in the world today struggle on a daily basis to get enough to eat. Unfortunately, such is the case, and as the human population continues to grow, we can expect our numbers to swell to over 9 billion by the middle of this century. Feeding 2 billion more people than we do today while protecting the integrity of soil, water, and ecosystems will require the large-scale embrace of farming practices that are more sustainable than those we currently use. Providing **food security,** the guarantee of an adequate, safe, nutritious, and reliable food supply available to all people at all times, will be one of our greatest challenges in coming decades.

There is good news, however. Over the past half-century, our ability to produce food has grown even faster than the global population (**FIGURE 7.1**). Improving people's quality of life by producing more food per person is a monumental achievement of which humanity can be proud.

## We face undernutrition, overnutrition, and malnutrition

Despite our rising food production, roughly 870 million people worldwide suffer from **undernutrition,** receiving fewer calories than the minimum dietary energy requirement. As a result, every 5 seconds, somewhere in the world, a child dies because he or she did not get enough to eat.

Many people are undernourished because they are too poor to purchase the food they need, but political obstacles, conflict, and inefficiencies in distribution contribute to hunger as well. Even our energy choices affect food supplies. For example, sizable amounts of cropland are devoted to growing crops for the production of biofuels (pp. 382–384). Biofuels are derived from organic materials and used in internal combustion engines as replacements for petroleum. In the United States, ethanol produced from corn (p. 382) is the primary biofuel.

Globally, the number of people suffering from undernourishment has been falling since the 1960s, as has the percentage of the global population that is undernourished (**FIGURE 7.2**). We have managed to reduce hunger in part because prices for food have become lower through the years, making food more accessible for poor families. (Some households in developing nations spend as much as 70% of their budget on food, compared to 6–15% in developed nations.) However, food prices have been rising since 2002, so careful monitoring and management of food prices will be needed to guarantee food security.

Although nearly 1 billion people lack access to nutritious foods, many others consume too many calories each day. **Overnutrition** causes unhealthy weight gain, which leads to cardiovascular disease, diabetes, and other health problems. In the United States, more than one in three adults is obese; worldwide, 1.5 billion adults are overweight, and among these, at least 500 million are obese. The growing availability of highly processed foods (which are often calorie-rich, nutrient-poor, and affordable to people of all incomes) suggests that overnutrition will remain a global nutritional problem, along with undernutrition, for the foreseeable future.

Just as the *quantity* of food a person eats is important for health, so is the *quality* of food. **Malnutrition,** a shortage of nutrients the body needs, occurs when a person fails to obtain a complete complement of proteins (p. 29), essential lipids (p. 29), vitamins, and minerals. Malnutrition can lead to disease (**FIGURE 7.3**). For example, people who eat a diet that is

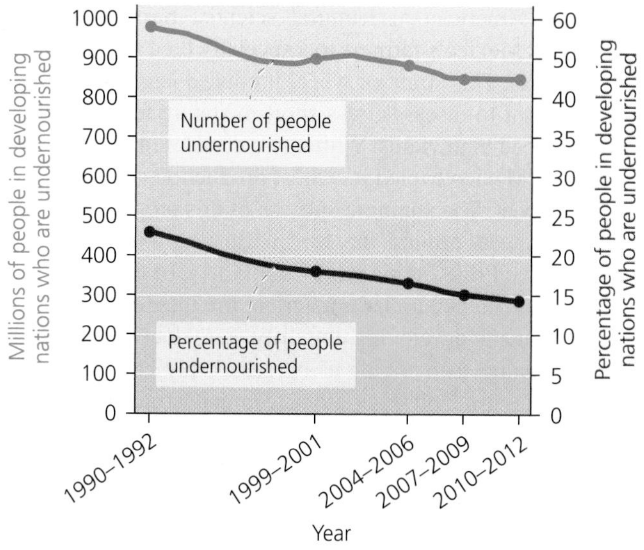


**FIGURE 7.2 The number and the percentage of people in the developing world who suffer undernutrition have each been declining.** *Data from Food and Agriculture Organization of the United Nations, 2012.* The state of food insecurity in the world, 2012. *FAO, Rome.*

DATA Q Explain how the percentage of undernourished people decreased from 2007–2009 to 2010–2012 while the number of undernourished people stayed roughly the same.

GO TO **INTERPRETING GRAPHS & DATA** ON MasteringEnvironmentalScience®

high in starch but deficient in protein can develop *kwashiorkor.* Children who have recently stopped breast-feeding and are no longer getting protein from breast milk are most at risk for developing kwashiorkor, which causes bloating of the abdomen, deterioration and discoloration of hair, mental disability, immune suppression, developmental delays, and reduced growth. Protein deficiency together with a lack of calories can lead to *marasmus,* which causes wasting or shriveling among millions of children in the developing world. Dietary deficiencies are also alarmingly prevalent. For example, iron deficiency can lead to anemia, which causes fatigue and developmental disabilities; iodine deficiency can cause swelling of the thyroid gland and brain damage; and vitamin A deficiency can lead to blindness.

**FIGURE 7.3 Millions of children suffer from forms of malnutrition, such as kwashiorkor and marasmus.**

# The Changing Face of Agriculture

If we are to enhance global food security, we will need to examine the ways we produce our food and how we can make them more sustainable. We can define **agriculture** as the practice of raising crops and livestock for human use and consumption. We obtain most of our food and fiber from **cropland,** land used to raise plants for human use, and from **rangeland,** land used for grazing livestock.

As the human population has grown, so have the amounts of land and resources we devote to agriculture. Agriculture is currently practiced on 38% of Earth's land surface, and uses more land area than any other human activity. Of this land, 26% is rangeland and 12% is cropland.

## Industrial agriculture is a recent human invention

During most of the human species' 200,000-year existence, we were hunter-gatherers, depending on wild plants and animals for our food and fiber. Then about 10,000 years ago, as glaciers retreated and the climate warmed, people in some cultures began to raise plants from seed and to domesticate animals.

For thousands of years, the work of cultivating, harvesting, storing, and distributing crops was performed by human and animal muscle power, along with hand tools and simple machines—an approach known as **traditional agriculture.** Traditional farmers typically plant **polycultures** ("many types"), mixtures of different crops in small plots of farmland, such as the Native American farming systems that mixed maize, beans, squash, and peppers. Traditional agriculture is still practiced today but has rapidly been overtaken by newer methods of farming that increase crop yields.

Thousands of years after humans began practicing traditional agriculture, the industrial revolution (p. 3) introduced large-scale mechanization and fossil fuel combustion to agriculture, just as it did to industry. Farmers replaced horses and oxen with machinery that provided faster and more powerful means of cultivating, harvesting, transporting, and processing crops. Such **industrial agriculture** also boosted yields by intensifying irrigation and introducing synthetic fertilizers, while the advent of chemical pesticides reduced herbivory by crop pests and competition from weeds. Today, industrial agriculture is practiced on over 25% of the world's cropland.

The use of machinery created a need for highly organized approaches to farming, and this led to the planting of vast areas with single crops in orderly, straight rows. Such **monocultures** ("one type") make farming more efficient, but they reduce biodiversity by eliminating habitats used by organisms in and around traditional farm fields. Moreover, when all plants in a field are genetically similar, as in monocultures, all are equally susceptible to bacterial and viral diseases, fungal pathogens, or insect pests that can spread quickly from plant to plant.

Industrial agriculture's reliance on genetically similar crop varieties is a source of efficiency, but also an Achilles heel. Concern over potential crop failure has led to coordinated efforts to conserve the wild relatives of crop plants and crop varieties indigenous to various regions, because they contain genes we may one day need to introduce into our commercial crops. Seeds of these species are stored in some 1400 **seed banks,** institutions that preserve some 1–2 million different seed types in locations around the world. Many agricultural scientists feel that we also need to protect the genetic integrity of wild relatives of crop plants by preventing gene exchange between these species and crop plants that have been genetically modified with genes from other species (p. 154). These scientists want to avoid genetic "contamination" of wild populations so that we preserve the natural gene combinations of plants that are well adapted to their environments.

Our use of monocultures also contributes to a narrowing of the human diet. Globally, 90% of the food we consume now comes from just 15 crop species and eight livestock species—a drastic reduction in the diversity of food humans have historically eaten. For example, only 30% of the maize varieties that grew in Mexico as recently as the 1930s still exist today. In the United States, the varieties of some fruits and vegetables cultivated by farmers has decreased by 90% in less than a century.

## The Green Revolution boosted production—and exported industrial agriculture

The desire for greater quantity and quality of food for our growing population led in the mid- and late 20th century to the **Green Revolution,** which introduced new technology, crop varieties, and farming practices to the developing world and drastically increased food production in these nations. The transfer of technology and knowledge to the developing world that marked the Green Revolution began in the 1940s, when the American agricultural scientist **Norman Borlaug** introduced Mexico's farmers to a specially bred type of wheat (**FIGURE 7.4**). This strain of wheat produced large seed heads, was resistant to diseases, was short in stature to resist wind, and produced high yields. Within two decades of planting this new crop, Mexico tripled its wheat production and began exporting wheat. The stunning success of this program inspired similar projects around the world. Borlaug—who won the Nobel Peace Prize for his work—took his wheat to India and Pakistan and helped transform agriculture there.

Soon many developing countries were doubling, tripling, or quadrupling their yields using selectively bred strains of wheat, rice, corn, and other crops from industrialized nations. These crops dramatically increased yields and helped millions avoid starvation. When Borlaug died in 2009 at age 95, he was widely celebrated as having "saved more lives than anyone in history."

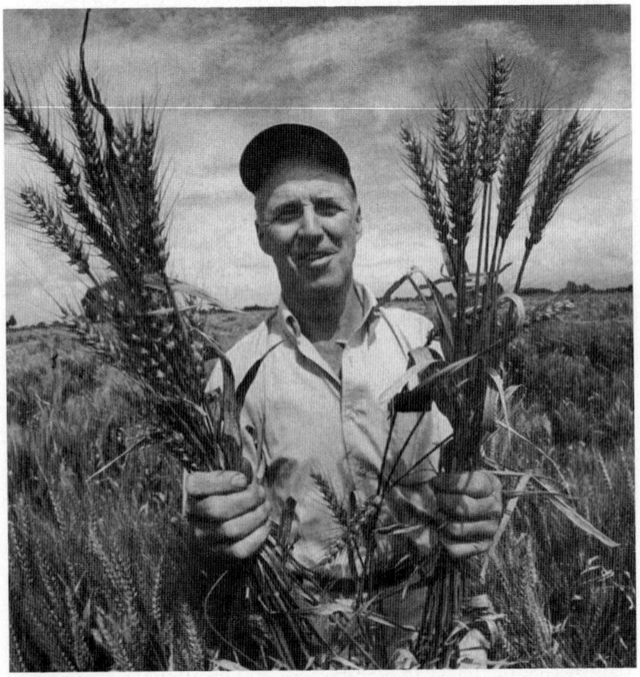

**FIGURE 7.4 Norman Borlaug helped launch the Green Revolution.** The high-yielding, disease-resistant wheat that he bred helped boost agricultural productivity in many developing countries.

## The effects of industrial agriculture have been mixed

Industrial agriculture has allowed food production to keep pace with our growing population, but it has had many adverse environmental and social impacts. On the positive side, high-input industrial agriculture has succeeded dramatically in producing higher crop yields from each hectare of land and reducing pressure to develop natural areas for new farmland. Between 1961 and 2010, food production more than tripled and per-person food production rose 48%, while the amount of land converted for agriculture increased by only 10%.

On the negative side, the intensive application of water, fossil fuels, inorganic fertilizers, and synthetic pesticides has worsened pollution, topsoil losses, and soil quality. Industrial agriculture also requires far more energy than traditional agriculture. From 1900 to 2000, as industrial agriculture expanded around the world, humanity increased energy inputs into agriculture by 80 times while expanding the world's cultivated area by just 33%. Industrial agriculture also displaces low-income farmers who cannot afford the advanced technologies required to compete with industrial-scale farms. This can force people to quit farming and to move to cities, adding to the immense migration of poor rural people to urban areas of the developing world (p. 409).

## Sustainable agriculture reduces environmental impacts

We have achieved our impressive growth in food production by devoting more fossil fuel energy to agriculture; intensifying our use of irrigation, fertilizers, and pesticides; cultivating more land; planting and harvesting more frequently; and developing (through crossbreeding and genetic engineering) more productive varieties of crops and livestock. However, many of these practices have degraded soils, polluted waters, and affected biodiversity.

We cannot simply keep expanding agriculture into new areas, because land suitable and available for farming is running out. Instead, we must find ways to improve the efficiency of food production in areas already under cultivation. Industrial agriculture in some form seems necessary to feed our planet's more than 7 billion people, but many experts feel we will be better off in the long run by raising animals and crops in ways that are less polluting, are less resource-intensive, and cause less impact on natural systems. To this end, many farmers and agricultural scientists are creating agricultural systems that better mimic the way a natural ecosystem functions.

**Sustainable agriculture** describes agriculture that maintains the healthy soil, clean water, and genetic diversity essential to long-term crop and livestock production. It is agriculture that can be practiced in the same way far into the future while maintaining high yields. Treating agricultural systems as ecosystems is a key aspect of sustainable agriculture, and this general principle applies regardless of location, scale, or crop.

One key component of making agriculture sustainable is reducing the fossil-fuel inputs we devote to agriculture and decreasing the pollution these inputs cause. Fossil fuels power the machines that plant and harvest crops and the pumps that draw irrigation water. Petroleum is also used as a raw material in the production of some synthetic fertilizers and pesticides.

*Low-input agriculture* describes approaches to agriculture that use lesser amounts of fossil fuel energy, water, pesticides, fertilizers, growth hormones, and antibiotics than are used in industrial agriculture. Low-input approaches also seek to reduce food production costs by allowing nature to provide valuable ecosystem services, such as fertilizing crops and pest control, that farmers using industrial agricultural methods must pay for themselves.

In the sections that follow, we will examine the basic elements of agriculture—maintaining soil fertility, watering and fertilizing crops, and controlling agricultural pests—and how more sustainable methods can reduce these impacts while maintaining high crop yields. We will begin with soils, the foundation of agriculture.

### WEIGHING THE ISSUES

**The Green Revolution and Population Growth** In the 1960s, India's population was skyrocketing, and its traditional agriculture was not producing enough food to support this growth. India had long been visited by famine, but by adopting Green Revolution practices, the nation avoided mass starvation. Still, Norman Borlaug called his Green Revolution methods "a temporary success in man's war against hunger and deprivation," something to give us breathing room in which to deal with what he called the "Population Monster." Indeed, in the years since intensifying its agriculture, India has added several hundred million more people and continues to suffer widespread poverty and hunger.

Do you think the Green Revolution has solved problems, deferred problems, or created new ones? Which aspects of the Green Revolution do you think help in the quest for sustainability, which do not, and why? Have the benefits of the Green Revolution outweighed its costs?

## Soils

**Soil** is not merely lifeless dirt; it is a complex system consisting of disintegrated rock, organic matter, water, gases, nutrients, and microorganisms. Healthy soil is vital for agriculture, for forests (Chapter 9), and for the functioning of Earth's natural systems. If we abuse soil through careless or uninformed practices, we can greatly reduce its ability to sustain agriculture.

By volume, soil consists very roughly of 50% mineral matter and up to 5% organic matter. The rest consists of the space between soil particles (pore space) taken up by air or water. The organic matter in soil includes living and dead microorganisms as well as decaying material derived from plants and animals. The soil ecosystem supports a diverse collection of bacteria, fungi, protists, worms, insects, and burrowing animals (**FIGURE 7.5**). The composition of a region's soil can have as much influence on its ecosystems as do climate, latitude, and elevation.

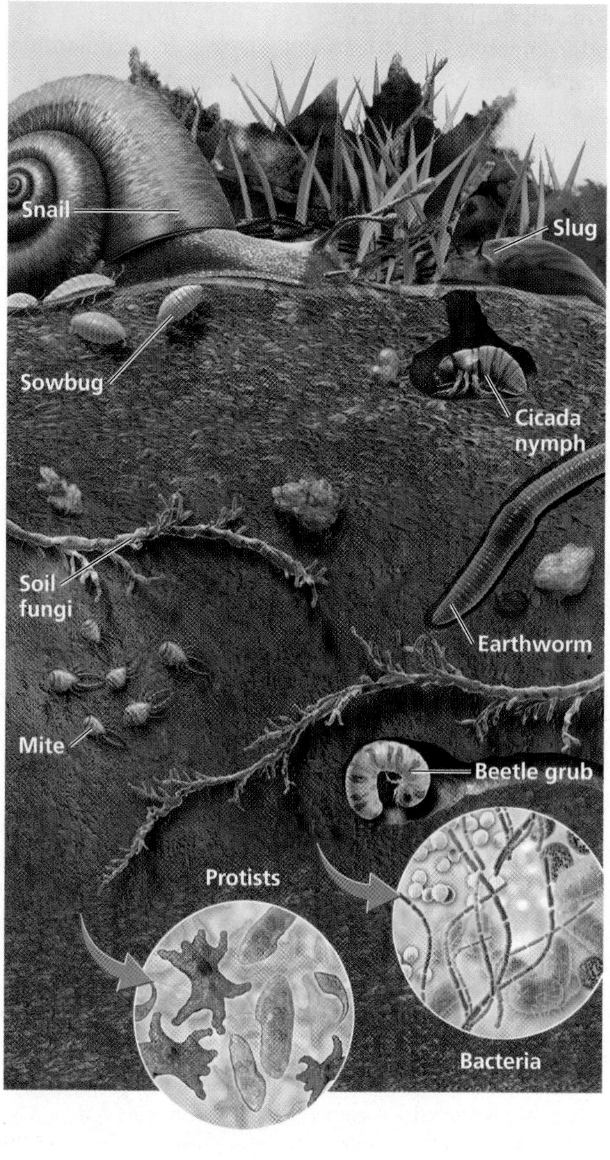


**FIGURE 7.5 Soil is a complex mixture of organic and inorganic components and is full of living organisms.** Entire ecosystems exist in soil. Most soil organisms decompose organic matter. Some, such as earthworms, also help to aerate soil.

## Soil forms slowly

Soil formation begins when the lithosphere's parent material is exposed to the effects of the atmosphere, hydrosphere, and biosphere (p. 23). **Parent material** is the base geologic material in a particular location. It can be hardened lava or volcanic ash; rock or sediment deposited by glaciers; sediments deposited in riverbeds, floodplains, lakes, and the ocean; or **bedrock,** the continuous mass of solid rock that makes up Earth's crust. Parent material is broken down by **weathering,** the physical, chemical, and biological processes that convert large rock particles into smaller particles.

Once weathering has produced fine particles, biological activity contributes to soil formation through the deposition, decomposition, and accumulation of organic matter. As plants, animals, and microbes die or deposit waste, this material is incorporated amid the weathered rock particles, mixing with minerals. For example, the deciduous trees of temperate forests drop their leaves each fall, making leaf litter available to the detritivores and decomposers (p. 72) that break it down and incorporate its nutrients into the soil. In decomposition, complex organic molecules are broken down into simpler ones, which plants can take up through their roots. Partial decomposition of organic matter creates *humus,* a dark, spongy, crumbly mass of material made up of complex organic compounds. Soils with high humus content hold moisture well and are productive for plant life. Soils on the Kennesaw State University farms have well-developed quantities of humus, because they are regularly resupplied with organic material collected from dining hall operations.

Forming just 1 inch of soil can easily require hundreds or thousands of years, so we would be wise to conserve the soil we have. Soil is a renewable resource, but it forms so slowly that for all practical purposes we cannot regain fertile soil once it has been lost.

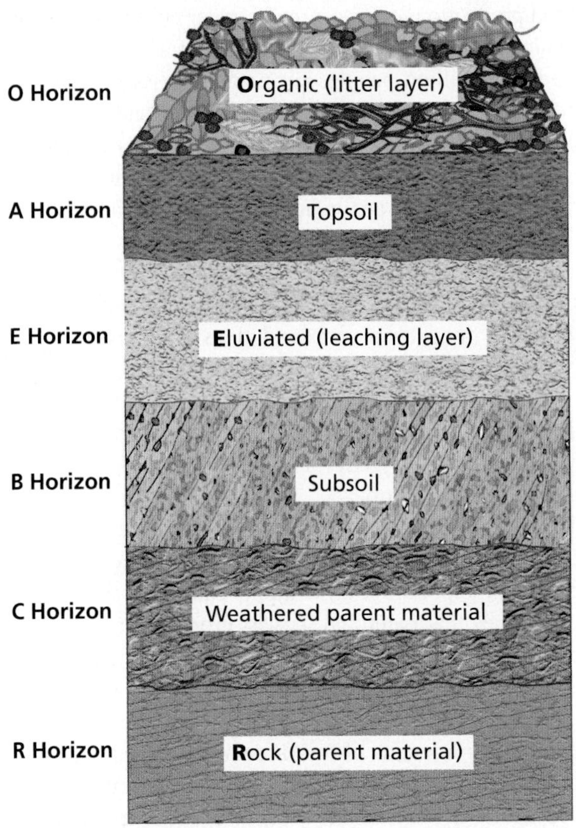


**FIGURE 7.6 Mature soil consists of layers, or horizons, that have different compositions and characteristics.** Uppermost is the O horizon, or litter layer (O = organic), consisting of organic matter deposited by organisms. Below it lies the A horizon, or topsoil, consisting of some organic material mixed with mineral components. Minerals and organic matter tend to leach out of the E horizon (E = eluviation, or leaching) into the B horizon, or subsoil, where they accumulate. The C horizon of weathered parent material overlies an R horizon (R = rock) of pure parent material.

## A soil profile consists of layers known as horizons

As wind, water, and organisms move and sort the fine particles that weathering creates, distinct layers eventually develop. Each layer of soil is known as a **horizon,** and the cross-section as a whole, from surface to bedrock, is known as a **soil profile.**

The simplest way to categorize soil horizons is to recognize A, B, and C horizons corresponding respectively to topsoil, subsoil, and parent material. However, soil scientists often recognize at least three additional horizons, including an O horizon (litter layer) that consists primarily of organic matter (**FIGURE 7.6**). Soils from different locations vary, and few soil profiles contain all six of these horizons, but any given soil contains at least some of them.

Generally, the degree of weathering and the concentration of organic matter decrease as one moves downward in a soil profile. Minerals are generally transported downward as a result of **leaching,** the process whereby solid particles suspended or dissolved in liquid are transported to another location. In some soils, minerals may be leached so rapidly that plants are deprived of nutrients. Leached minerals may enter groundwater, and some can pose human health risks when the water is extracted for drinking.

A crucial horizon for agriculture and ecosystems is the A horizon, or **topsoil.** Topsoil consists mostly of inorganic mineral components such as weathered substrate, with organic matter and humus from above mixed in. Topsoil is the portion of the soil that is most nutritive for plants, and it takes its loose texture, dark coloration, and strong water-holding capacity from its humus content. The O and A horizons are home to most of the countless organisms that give life to soil. Topsoil is vital for agriculture, but agriculture practiced unsustainably over time will deplete organic matter, reducing the soil's fertility and ability to hold water.

## Regional differences in soil traits affect agriculture

Soil characteristics vary from place to place, and they are deeply affected by climate and other variables. In regions of tropical rainforest (such as Indonesia, central Africa, or

(a) Swidden agriculture on nutrient-poor soil in Indonesia

(b) Industrial agriculture on Iowa's rich topsoil

**FIGURE 7.7 Regional soil differences affect how people farm.** In tropical forested areas such as Indonesia **(a),** farmers pursue swidden agriculture by the slash-and-burn method because tropical rainforest soils **(inset)** are nutrient-poor and easily depleted. On the Iowa prairie **(b),** less rainfall means less leaching of nutrients from the topsoil and more accumulation of organic matter, forming a thick, dark topsoil layer **(inset).**

South America's Amazon Basin), heavy and sustained rainfall quickly leaches minerals and nutrients out of the topsoil and E horizon. Warm temperatures speed the decomposition of leaf litter and the uptake of nutrients by plants, so only small amounts of humus remain in the thin topsoil layer (**FIGURE 7.7a**).

Conversely, in the agricultural U.S. Midwest, there is less rainfall, and therefore less leaching, which keeps nutrients within reach of plants' roots. Plants return nutrients to the topsoil when they die, maintaining its fertility. This creates the thick, rich topsoil typical of temperate grasslands, or prairies, in the central United States (**FIGURE 7.7b**).

Because of the low nutrient content of tropical soils, people farming them have traditionally employed *swidden* agriculture, in which the farmer clears a plot of forest and cultivates the land for only one to a few years. Once the soil's fertility is exhausted, the farmer moves on to clear another plot, leaving the first to grow back to forest. Many abandoned sites do not regrow forest, however, as cleared plots are often used as pastureland for grazing livestock.

**FAQ** **What is "slash-and-burn" agriculture?**

Soils in tropical forests are not well suited for cultivating crops because they contain relatively low levels of plant nutrients. But although the soil is nutrient-poor, there are large amounts of nutrients tied up in the forest's lush vegetation. When farmers develop plots of tropical rainforest for agriculture, they enrich the soil by cutting down and burning the plants on the site. The nutrient-rich ash is then tilled into the soil, providing sufficient fertility to grow crops. The practice is called *slash-and-burn* agriculture.

The nutrients from the ash, however, are usually depleted in only a few years. Lacking the resources to purchase synthetic fertilizer, many poor farmers simply move deeper into the forest, repeat the process on a new plot of forested land, and cause further impacts on these productive and biologically diverse ecosystems.

# Maintaining Healthy Soils

Throughout the world, especially in drier regions, it has gotten more difficult to raise crops and graze livestock. Soils have deteriorated in quality and declined in productivity—a process termed **soil degradation.** Each year, our planet gains around 85 million people yet loses 5–7 million ha (12–17 million acres, about the size of West Virginia) of productive cropland to degradation. The common causes include soil erosion, nutrient depletion, water scarcity, salinization (p. 145), waterlogging (p. 145), chemical pollution, changes in soil structure and pH, and loss of organic matter from the soil. Over the past 50 years, scientists estimate that soil degradation has reduced potential rates of global grain production on cropland by 13%. This is a dangerous trend, given that we will need to maximize the productivity of our farmland if we are feed an additional 2 billion people by mid-century.

## Erosion can degrade soil ecosystems

**Erosion** is the transfer of material from one place to another by the action of wind or water. *Deposition* occurs when eroded material is deposited at a new location. Erosion and deposition are natural processes that can help create soil. Flowing water can deposit freshly eroded nutrient-rich sediment across river valleys and deltas, producing fertile soils. However, erosion can be a problem locally because it generally occurs much more quickly than soil is formed. Erosion also tends to thin the biologically important topsoil layer. In general, steeper slopes, greater precipitation intensities, and sparser vegetation all lead to greater water erosion.

People have made land more vulnerable to erosion through three widespread practices: overcultivating fields through poor planning or excessive tilling (plowing); overgrazing rangeland with more livestock than the land can support; and clearing forests on steep slopes or with large clear-cuts (p. 195).

## Soil erosion is a global problem

In today's world, humans are the primary cause of soil erosion, and we have accelerated it to unnaturally high rates. A 2004 study concluded that human activities move over 10 times more soil than all other natural processes on the surface of the planet combined. More than 19 billion ha (47 billion acres) of the world's croplands suffer from erosion and other forms of soil degradation. Farmlands in the United States, for example, lose roughly 5 metric tons of soil for every ton of grain harvested. A 2007 study found an even greater degree of human impact, but also pointed toward a solution, revealing that land farmed with sustainable approaches erodes at slower rates than land under industrial farming practices.

## Desertification reduces productivity of arid lands

Much of the world's population lives and farms in *drylands*, arid and semi-arid environments that cover about 40% of Earth's land surface. These areas are prone to **desertification,** a form of land degradation in which more than 10% of productivity is lost as a result of erosion, soil compaction, forest removal, overgrazing, drought, salinization, climate change, water depletion, and other factors. Most such degradation results from wind and water erosion.

By some estimates, desertification endangers the food supply or well-being of more than 1 billion people in over 100 countries and costs tens of billions of dollars in income each year through reduced productivity.

## The Dust Bowl prompted the United States to fight erosion

Prior to the large-scale cultivation of North America's Great Plains, native prairie grasses of this temperate grassland region held soils in place. In the late 19th and early 20th centuries, many homesteading settlers arrived in Oklahoma, Texas, Kansas, New Mexico, and Colorado with hopes of making a living there as farmers. Farmers grew abundant wheat, and ranchers grazed many thousands of cattle, contributing to erosion by removing native grasses and altering soil structure.

In the early 1930s, a drought exacerbated the ongoing human impacts, and the region's strong winds began to erode millions of tons of topsoil. Dust storms traveled up to 2000 km (1250 mi), blackening rain and snow as far away as New York and Washington, D.C. Some areas lost 10 cm (4 in.) of topsoil in a few years. The most-affected region in the southern Great Plains became known as the **Dust Bowl,** a term now also used for the historical event itself (**FIGURE 7.8**). The "black blizzards" of the Dust Bowl forced thousands of farmers off their land.

In response, the U.S. government, along with state and local governments, increased its support for research on soil conservation practices. The U.S. Congress passed the Soil Conservation Act of 1935, establishing the Soil Conservation Service (SCS). This new agency worked closely with farmers to develop conservation plans for individual farms. The SCS (now renamed the *Natural Resources Conservation Service*) served as a model for other nations that established their own soil conservation agencies to aid farmers in fighting soil erosion.

## Sustainable agriculture begins with soil management

A number of farming techniques can reduce the impacts of conventional cultivation on soils and combat soil degradation (**FIGURE 7.9**). Not all of these approaches are new, as some have been practiced since the dawn of agriculture.

**(a) Kansas dust storm, 1930s**

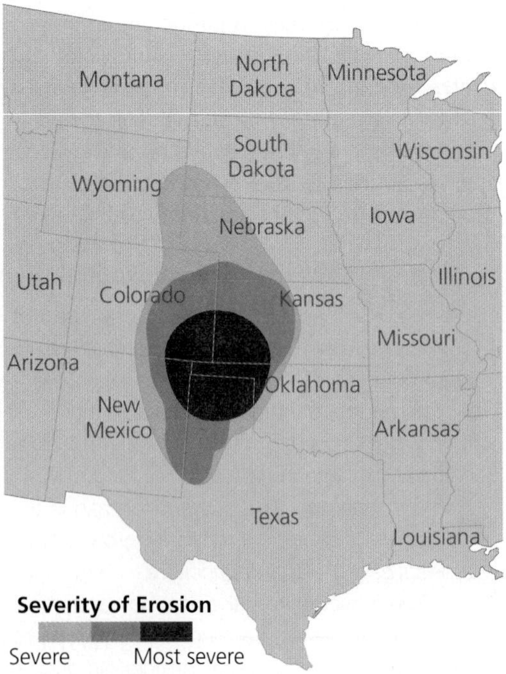


**(b) Dust Bowl region**

**FIGURE 7.8 Drought and poor agricultural practices devastated millions of U.S. farmers in the 1930s in the Dust Bowl.** The photo **(a)** shows towering clouds of dust approaching homes near Dodge City, Kansas. The map **(b)** shows the Dust Bowl region. Eroded soil from this region blew eastward all the way to the Atlantic Ocean.

(a) Crop rotation
(b) Contour farming
(c) Terracing
(d) Intercropping
(e) Shelterbelts
(f) No-till farming

**FIGURE 7.9 Farmers have adopted various strategies to conserve soil.** Rotating crops **(a)** such as soybeans and corn helps restore soil nutrients and reduce impacts of pests. Contour farming **(b)** reduces erosion on hillsides. Terracing **(c)** minimizes erosion in mountainous areas. Intercropping **(d)** reduces soil loss and maintains soil fertility. Shelterbelts **(e)** protect against wind erosion. In **(f),** corn grows through the remnants of a cover crop used in no-till agriculture.

**Crop rotation** In **crop rotation,** farmers alternate the type of crop grown in a given field from one season or year to the next (**FIGURE 7.9a**). Rotating crops can return nutrients to the soil, break cycles of disease associated with continuous cropping, and minimize the erosion that can come from letting fields lie uncultivated. Many U.S. farmers rotate their fields between wheat or corn and soybeans from one year to the next. Soybeans are legumes, plants that have specialized bacteria on their roots that fix nitrogen (p. 41), revitalizing soil that the previous crop had partially depleted of nutrients. Crop rotation also reduces insect pests; if an insect is adapted to feed and lay eggs on one crop, planting a different type of crop will leave its offspring with nothing to eat.

In a practice similar to crop rotation, many farmers plant cover crops, such as nitrogen-replenishing clover, to prevent erosion after harvesting crops.

**Contour farming** Water running down a hillside with little vegetative cover can easily carry soil away, so farmers have developed several methods for cultivating slopes. **Contour farming** (**FIGURE 7.9b**) consists of plowing furrows sideways across a hillside, perpendicular to its slope and following the natural contours of the land. In contour farming, the side of each furrow acts as a small dam that slows runoff and captures eroding soil. Farmers also plant buffer strips of vegetation along the borders of their fields and along nearby streams, which further protect against erosion and water pollution.

**Terracing** On extremely steep terrain, the most effective method for preventing erosion is **terracing** (**FIGURE 7.9c**). Terraces are level platforms, sometimes with raised edges, that are cut into steep hillsides to contain water from irrigation and precipitation. Terracing transforms slopes into series of steps like a staircase, enabling farmers to cultivate hilly land without losing huge amounts of soil to water erosion.

**Intercropping** Farmers may also minimize erosion by **intercropping,** planting different types of crops in alternating bands (**FIGURE 7.9d**). Intercropping helps slow erosion by providing more ground cover than does a single crop. Like crop rotation, intercropping reduces vulnerability to insects and disease and, when a nitrogen-fixing legume is planted, replenishes the soil with plant nutrients.

**Shelterbelts** A widespread technique to reduce erosion from wind is to establish **shelterbelts,** or *windbreaks* (**FIGURE 7.9e**). These are rows of trees or other tall plants that are planted along the edges of fields to slow the wind. On the Great Plains, fast-growing species such as poplars are often used. Shelterbelts can be combined with intercropping; mixed crops are planted in rows surrounded by or interspersed with rows of trees that provide fruit, wood, and wildlife habitat, as well as protection from wind.

**FIGURE 7.10 Farmers practice no-till farming with a no-till drill.** The drill 1 cuts a furrow through the soil surface, 2 drops in a seed, and 3 closes the furrow over the seed. This disturbs far less soil than does conventional tilling, reducing erosion rates on farm fields.

**Conservation tillage** *Conservation tillage* describes an array of approaches that reduce the amount of tilling relative to conventional farming (**FIGURE 7.9f**). Turning the earth by tilling (plowing, disking, harrowing, or chiseling) aerates the soil and works weeds and old crop residue into the soil to nourish it, but tilling also leaves the surface bare, allowing wind and water to erode away precious topsoil.

**No-till farming** is the ultimate form of conservation tillage. Rather than plowing after each harvest, farmers leave crop residues atop their fields, keeping the soil covered with plant material at all times. To plant the next crop, they cut a thin, shallow groove into the soil surface, drop in seeds, and cover them, using a machine called a no-till drill (**FIGURE 7.10**). By planting seeds of the new crop through the residue of the old, less soil erodes away, organic material accumulates, and the soil soaks up more water—all of which encourage better plant growth.

By increasing organic matter and soil biota while reducing erosion, no-till farming and conservation tillage can improve soil quality and combat global climate change by storing carbon in soils. In the United States today, nearly one-quarter of farmland is under no-till cultivation, and over 40% under conservation tillage.

No-till and conservation tillage methods have also become widespread in subtropical and temperate South America. In Brazil, Argentina, and Paraguay, over half of all cropland is now under no-till cultivation. In this part of the world, heavy rainfall promotes erosion, causing tilled soils to rapidly lose organic matter and nutrients, and hot weather can overheat tilled soil. Thus, the no-till approach is especially helpful here, and results have exceeded those in the United States: Crop yields have increased (in some cases they nearly doubled), erosion has been reduced, soil quality has been enhanced, and pollution has declined, all while costs to have farmers have dropped by roughly 50%.

Critics of no-till farming in the United States have noted that this approach often requires substantial use of chemical herbicides (because weeds are not physically removed from fields) and synthetic fertilizer (because non-crop plants take up some of the soil's nutrients). In many industrialized countries, this has indeed been the case.

Proponents of no-till farming, however, point out that in developing regions of South America, farmers have departed from the industrialized model by relying more heavily on *green manures* (dead plants as fertilizer) and by rotating fields with cover crops, including nitrogen-fixing legumes. The manures and legumes nourish the soil, and cover crops reduce weeds by taking up space that weeds would otherwise occupy. Although this approach is often not practical for large-scale intensive agriculture, farmers are educating themselves on the available approaches and choosing those that are best for their farms.

## Grazing practices can contribute to soil degradation

We have focused in this chapter largely on the cultivation of crops as a source of impacts on soils and ecosystems, but raising livestock also has impacts. Humans keep more than 3 billion cattle, sheep, and goats that graze primarily on grasses on the open range. As long as livestock populations do not exceed a range's carrying capacity (p. 62) and do not consume grass faster than it can regrow, grazing may be sustainable. However, grazing too many livestock that eat too much of the plant cover impedes plant regrowth. Without the regeneration of plant biomass, the result is **overgrazing.**

When livestock remove too much plant cover or churn up the soil with their hooves, soil is exposed and made vulnerable to erosion. In a positive feedback cycle (p. 23–24), soil erosion makes it difficult for vegetation to regrow, a problem that perpetuates the lack of cover and gives rise to even more erosion. Moreover, non-native weedy plants that are unpalatable or poisonous to livestock may outcompete native vegetation in the new, modified environment.

Too many livestock trampling the ground can also compact soils and alter their structure. Soil compaction makes it more difficult for water to infiltrate, for soils to be aerated, for plants' roots to expand, and for roots to conduct cellular respiration (p. 32). All of these effects further decrease plant growth and survival.

Private cattle grazing has affected millions of acres of public land across the western United States, where ranchers are allowed to graze for minimal fees. As a cause of soil degradation worldwide, overgrazing is equal to cropland agriculture, and it is a greater cause of desertification. Fully 70% of the world's rangeland is classified as degraded, and lost productivity on these lands is estimated at $23 billion per year.

## Agricultural subsidies affect soil degradation

Many nations spend billions of dollars in government subsidies to support agriculture. In the United States, roughly one-fifth of the income of the average farmer comes from subsidies.

Proponents of such subsidies stress that the uncertainties of weather make profits and losses from farming unpredictable from year to year. To persist, these proponents say, an agricultural system needs some way to compensate farmers for bad years. This may be the case, but subsidies can encourage people to cultivate land that would otherwise not be farmed; to produce more food than is needed, driving down prices for other producers; and to practice unsustainable farming methods that further degrade the land. Thus, opponents of subsidies argue that subsidizing environmentally destructive practices is unsustainable. They suggest that a better model is for farmers to buy insurance to protect against short-term production failures, an approach increasingly embraced by the U.S. government in its support for agriculture.

## A number of U.S. and international programs promote soil conservation

While government subsidies may promote soil degradation, the U.S. Congress has also enacted provisions promoting soil conservation through the farm bills it passes every five to six years. Many of these provisions require farmers to adopt soil conservation plans and practices before they can receive government subsidies.

The **Conservation Reserve Program,** established in the 1985 farm bill, pays farmers to stop cultivating highly erodible cropland and instead place it in conservation reserves planted with grasses and trees. Lands under the Conservation Reserve Program now cover an area nearly the size of Iowa, and the United States Department of Agriculture (USDA) estimates that each dollar invested in this program saves nearly 1 metric ton of topsoil. Besides reducing erosion, the Conservation Reserve Program generates income for farmers, improves water quality, and provides habitat for wildlife. Recently, the federal government paid farmers about $1.8 billion per year for the conservation of 11 million ha (27 million acres) of land. The farm bill passed in 2014 continues this program, but caps the amount at 10 million ha (24 million acres) of land. Internationally, the United Nations promotes soil conservation and sustainable agriculture through a variety of programs led by its Food and Agriculture Organization (FAO).

# Watering and Fertilizing Crops

Soil degradation is not the only impact of agriculture on the environment. When we supply plants with unsustainable levels of supplemental water and nutrients to boost crop yields, we affect natural systems, including those that are far away from farm fields.

## Irrigation boosts productivity but can damage soil

Plants require water for optimum growth, and people have long supplemented the water that crops receive from rainfall. The artificial provision of water to support agriculture is known as **irrigation.** By irrigating crops, people maintain high yields in times of drought and turn previously dry and unproductive regions into fertile farmland.

Worldwide, irrigated acreage has increased along with the adoption of industrial farming methods, and 70% of the fresh water withdrawn by people is applied to crops. In some cases, withdrawing water for irrigation has depleted aquifers and dried up rivers and lakes (pp. 260–262).

In some locations, excessive irrigation has degraded soils. **Waterlogging** occurs when over-irrigation causes the water table to rise to the point that water drowns plant roots, depriving them of access to gases and essentially suffocating them. A more frequent problem is **salinization,** the buildup of salts in surface soil layers. In dryland areas where precipitation is minimal and evaporation rates are high, the evaporation of water from the soil's A horizon may pull water with dissolved salts up from lower horizons. When the water evaporates at the surface, those salts remain on the soil, often turning the soil surface white. Salinization now reduces productivity on one-fifth of all irrigated cropland, costing more than $11 billion each year.

## Sustainable approaches to irrigation maximize efficiency

One of the most effective ways to reduce water use in agriculture is to better match crops and climate. Many arid regions have been converted into productive farmland through extensive irrigation, often with the support of government subsidies that make irrigation water artificially inexpensive. Some farmers in these areas cultivate crops that require large amounts of water, such as rice and cotton. This wastes a great deal of water when it readily evaporates in the arid climate. Choosing other crops that require far less water, such as beans or wheat, could enable these areas to remain agriculturally productive and yet greatly reduce water use.

Another approach is to embrace new technologies that improve water use efficiency in irrigation. Currently, irrigation efficiency worldwide is low. Plants end up using only about 40% of the water that we apply—the rest evaporates or soaks into the soil away from plant roots (**FIGURE 7.11a**). New drip irrigation systems that deliver water directly to plant roots can increase efficiencies to over 90% (**FIGURE 7.11b**). Drip irrigation systems are utilized on the Kennesaw State University campus farm that supplies The Commons with much of its produce, greatly reducing water use.

Such systems were once quite expensive to install and were used largely by farmers in wealthier nations. They are becoming less expensive, though, making them more available to farmers in developing countries.

## Fertilizers boost crop yields but can be overapplied

Plants require nitrogen, phosphorus, and potassium to grow, as well as smaller amounts of more than a dozen other nutrients. Leaching and uptake by plants removes these nutrients from soil, and if soils come to contain too few nutrients, crop yields decline. Therefore, we go to great lengths to enhance nutrient-limited soils by adding **fertilizer,** substances that contain essential nutrients for plant growth.

There are two main types of fertilizers. **Inorganic fertilizers** are mined or synthetically manufactured nutrient supplements. **Organic fertilizers** consist of the remains or wastes of organisms and include animal manure, crop residues, fresh vegetation (*green manure*), and *compost*, a mixture produced when decomposers break down organic matter, including food and crop waste, in a controlled environment.

One of the highlights of The Commons' sustainability efforts at KSU is its closed-loop system for recycling wastes. Uneaten food and scraps from food preparation are placed in a large "digester" tank outside the dining hall. Over time, the food items break down inside the digester, generating roughly 1900 L (500 gal) of nutrient-rich water daily. This liquid is trucked to the nearby campus farms where it is used as an organic fertilizer.

Historically, people relied on organic fertilizers to replenish soil nutrients. But during the latter half of the 20th century, farmers in industrialized and Green Revolution regions widely embraced the use of inorganic fertilizers (**FIGURE 7.12**). This use has greatly boosted our global food production, but the overapplication of inorganic fertilizers is causing increasingly severe pollution problems. Inorganic fertilizers are generally more susceptible to leaching than organic fertilizers and more readily contaminate groundwater supplies. Nutrients from these fertilizers can also have impacts far beyond the boundaries of the fields. For instance, nitrogen and phosphorus runoff from farms and other sources spurs

**(a) Flood-and-furrow irrigation**

**(b) Drip irrigation**

**FIGURE 7.11 Irrigation methods vary in their water use efficiency.** Flood-and-furrow irrigation **(a),** shown here in a California cotton field, wastes much of the water used. In drip irrigation systems **(b),** hoses drip water directly onto the plants, so that much less is wasted.

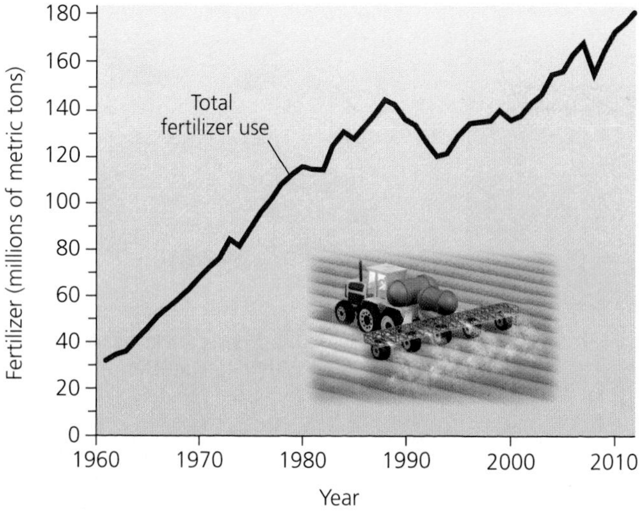


**FIGURE 7.12 Use of synthetic, inorganic fertilizers has risen sharply over the past half-century.** Today, usage stands at over 180 million metric tons annually. *Data from International Fertilizer Industry Association.*

phytoplankton blooms in the Chesapeake Bay and creates an oxygen-depleted "dead zone" that kills animal and plant life (Chapter 2). Such eutrophication (pp. 24–26, 267–268) occurs at countless river mouths, lakes, and ponds throughout the world. Air pollution can also arise from components of some nitrogen fertilizers that evaporate into the air, contributing to the formation of photochemical smog (p. 291) and acid deposition (pp. 295–297).

### Sustainable fertilizer use involves monitoring and targeting nutrients

Sustainable approaches to fertilizing crops target the delivery of nutrients to plant roots and avoid the overapplication of fertilizer. This is accomplished in many ways. Farmers using drip irrigation systems can add fertilizer to irrigation water, thereby releasing it above plant roots. No-till or conservation tillage systems often inject fertilizer along with seeds, concentrating it near the developing plant. Farmers can also avoid overapplication by regularly monitoring soil nutrient content and applying fertilizer only when soil nutrient levels are too low. Strips of vegetation planted at the edges of fields and along streams can capture nutrient runoff along with eroded soils.

Sustainable agriculture also embraces the use of organic fertilizers because they can provide some benefits that inorganic fertilizers cannot. Organic fertilizers provide not only nutrients, but also organic matter, which improves the soil's structure, nutrient retention, and water-retaining capacity.

The use of organic fertilizers is not without cost, though. For instance, when manure is applied in amounts needed to supply sufficient nitrogen for a crop, it may introduce excess phosphorus that can run off into waterways. Accordingly, some sustainable agriculture approaches do not rely on organic fertilizers alone, but rather integrate them with the targeted delivery of nutrients using inorganic fertilizer.

## Controlling Pests, Preserving Pollinators

Throughout the history of agriculture, the insects, fungi, viruses, rodents, and weeds that eat, infect, or compete with our crop plants have taken advantage of the ways we cluster food plants into agricultural fields. Pests pose an especially great threat to monocultures, where a pest adapted to specialize on the crop can move easily from plant to plant.

What people term a *pest* is any organism that damages crops that are valuable to us. What we term a *weed* is any plant that competes with our crops. There is nothing inherently malevolent in the behavior of a pest or a weed. These organisms are simply trying to survive and reproduce, like all living things, but they affect farm productivity when doing so.

### We have developed thousands of chemical pesticides

To suppress pests and weeds, people have developed thousands of chemicals to kill insects (*insecticides*), plants (*herbicides*), rodents (*rodenticides*), and fungi (*fungicides*). Such poisons are collectively termed **pesticides.** The highly modified ecosystems of industrial farming limit the ability of natural mechanisms to control pest populations. Hence, as industrial agriculture came into practice, farmers turned to chemical means to control agricultural pests.

Roughly 400 million kg (900 million lb) of active ingredients from conventional pesticides are used in the United States every year. Three-quarters of this total is applied on agricultural land. Since 1960, the worldwide use of pesticides has risen fourfold. Usage in industrialized nations has leveled off in the past two decades, but it continues to rise in the developing world. Exposure to synthetic pesticides can have health consequences for people and other organisms (Chapter 10), so their use in food production can have far-reaching effects.

### Pests evolve resistance to pesticides

Despite the toxicity of pesticides, their effectiveness tends to decline with time as pests evolve resistance to them. Recall from our discussion of natural selection (pp. 48–50) that individuals within populations vary in their genetic makeup. Because most insects, weeds, and microbes can exist in huge numbers, it is likely that a small fraction of individuals may by chance already have genes that enable them to metabolize and detoxify a pesticide (p. 214). These individuals will survive exposure to the pesticide, whereas individuals without these genes will not. If an insect that is genetically resistant to an insecticide survives and mates with another resistant individual, the genes for pesticide resistance will be

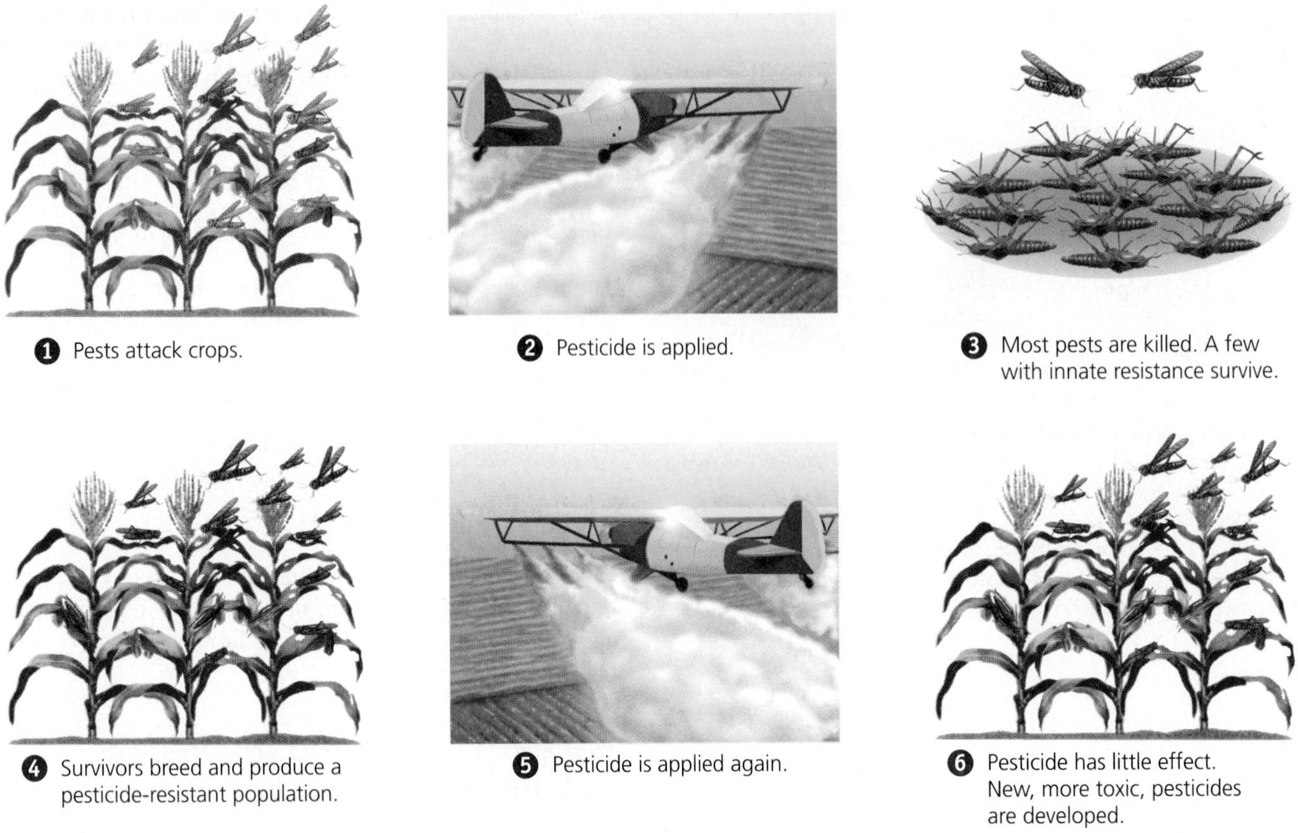

1 Pests attack crops.
2 Pesticide is applied.
3 Most pests are killed. A few with innate resistance survive.
4 Survivors breed and produce a pesticide-resistant population.
5 Pesticide is applied again.
6 Pesticide has little effect. New, more toxic, pesticides are developed.


**FIGURE 7.13 Through the process of natural selection, crop pests often evolve resistance to the toxic chemicals we apply to kill them.**

passed to their offspring. As resistant individuals become more prevalent in the pest population, insecticide applications will cease to be effective, and the population will increase in size (**FIGURE 7.13**).

In many cases, industrial chemists are caught up in an evolutionary arms race (p. 70) with the pests they battle, racing to increase or retarget the toxicity of their chemicals while the armies of pests evolve ever-stronger resistance to their efforts. Because we seem to be stuck in this cyclical process, it has been nicknamed the "pesticide treadmill." Currently, among arthropods (insects and their relatives) alone, there are more than 9000 known cases of resistance by 586 species to over 330 insecticides. Hundreds more weed species and plant diseases have evolved resistance to herbicides and other pesticides. Many species, including insects such as the green peach aphid, Colorado potato beetle, and diamondback moth, have evolved resistance to multiple chemicals.

## Biological control pits one organism against another

Because of pesticide resistance, toxicity to nontarget organisms, and human health risks from some synthetic chemicals, agricultural scientists increasingly battle pests and weeds with organisms that eat or infect them. This more sustainable pest control strategy, called **biological control** or **biocontrol,** operates on the principle that "the enemy of one's enemy is one's friend." For example, parasitoid wasps (p. 69) are natural enemies of many caterpillars. These wasps lay eggs on a caterpillar, and the larvae that hatch from the eggs feed on the caterpillar, eventually killing it. Parasitoid wasps are frequently used as biocontrol agents and have often succeeded in controlling pests and reducing chemical pesticide use.

One classic case of successful biological control is the introduction of the cactus moth, *Cactoblastis cactorum*, from Argentina to Australia in the 1920s to control the invasive prickly pear cactus that was overrunning rangeland. Within just a few years, the moth managed to free millions of hectares of Australian rangeland from the cactus.

However, biocontrol approaches entail risks. Biocontrol organisms are sometimes more difficult to manage than chemical controls, because they cannot be "turned off" once they are initiated. Further, biocontrol organisms have in some cases become invasive and harmed nontarget organisms. Following the cactus moth's success in Australia, for example, it was introduced in other countries to control non-native prickly pear. Moths introduced to Caribbean islands spread to Florida on their own and are now eating their way through rare native cacti in the southeastern United States. If these moths reach Mexico and the southwestern United States, they could decimate many native and economically important species of prickly pear cacti. Because of concerns about unintended impacts, researchers study biocontrol proposals carefully before putting them into action.

## Integrated pest management combines varied approaches to pest control

As it became clear that both chemical and biocontrol approaches pose risks, agricultural scientists and farmers began developing more sophisticated strategies, trying to combine the best attributes of each approach. **Integrated pest management (IPM)** incorporates numerous techniques, including close monitoring of pest populations, biocontrol approaches, use of synthetic chemicals when needed, habitat alteration, crop rotation, transgenic crops, alternative tillage methods, and mechanical pest removal.

IPM has become popular in many parts of the world that are embracing sustainable agriculture techniques. Indonesia stands as an exemplary case. This nation had subsidized pesticide use heavily for years, but its scientists came to understand that pesticides were actually making pest problems worse. They were killing the natural enemies of the brown planthopper, which began to devastate rice fields as its populations exploded. Concluding that pesticide subsidies were costing money, causing pollution, and apparently decreasing yields, the Indonesian government in 1986 banned the import of 57 pesticides, slashed pesticide subsidies, and promoted IPM. Within just four years, pesticide production fell by half, imports fell by two-thirds, and the government saved money by eliminating subsidy payments. Rice yields rose 13% with IPM, and since then the approach has spread to dozens of other nations, particularly in rice-growing regions in Asia.

## Pollinators are beneficial "bugs" worth preserving

Managing insect pests is such a major issue in agriculture that it is easy to fall into a habit of thinking of all insects as somehow bad or threatening. But in fact, most insects are harmless to agriculture—and some are absolutely essential. The insects that pollinate crops are among the most vital factors in our food production. Pollinators are the unsung heroes of agriculture.

**Pollination** (p. 70) is the process by which male sex cells of a plant (pollen) fertilize female sex cells of a plant; it is the botanical version of sexual intercourse. Pollinators are animals that move pollen from one flower to another. Flowers are, in fact, evolutionary adaptations that function to attract pollinators. The sugary nectar and protein-rich pollen in flowers serve as rewards to lure these sexual intermediaries, and the sweet smells and bright colors of flowers are signals to advertise these rewards.

Our staple grain crops are derived from grasses and are wind-pollinated, but 800 types of cultivated plants rely on bees, wasps, beetles, moths, butterflies and other insects for pollination. Preserving the biodiversity of native pollinators is especially important today because the domesticated workhorse of pollination, the honeybee (*Apis mellifera*), is in decline. In recent years, two introduced parasitic mites have decimated honeybee hives, pushing many beekeepers toward financial ruin. On top of this, entire hives inexplicably began dying off in the mid 2000s. These dieoffs continue today, and up to one-third of all honeybees in the United States have vanished from what is being called *colony collapse disorder*. Scientists are racing to discover the cause of this mysterious syndrome. Hypotheses include exposure to insecticides (particularly a widely used group of pesticides called neonicotinoids), an unknown new parasite, or a combination of stresses that weaken bees' immune systems and destroy social communication within the hive.

As this example clearly shows, when trying to control "bad" bugs, we must be very careful not to kill the "good" insects that aid us in agriculture. Honeybees, for example, pollinate over 100 crops that comprise one-third of the U.S. diet, contributing an estimated $15 billion in services annually.

# Raising Animals for Food

Food from cropland agriculture makes up a large portion of the human diet, but most of us also eat animal products. Just as farming methods have changed over time, so have the ways we raise animals for food.

As wealth and global commerce have increased, so has humanity's production of meat, milk, eggs, and other animal products (**FIGURE 7.14**). The world population of domesticated animals raised for food rose from 7.3 billion animals in 1961 to over 27 billion animals today. Most of these animals are chickens. Global meat production has increased fivefold since 1950, and per capita meat consumption has doubled. The United Nations Food and Agriculture Organization (FAO) estimates that as more developing nations go through the demographic transition (pp. 124–125) and become wealthier, total meat consumption will nearly double by the year 2050.

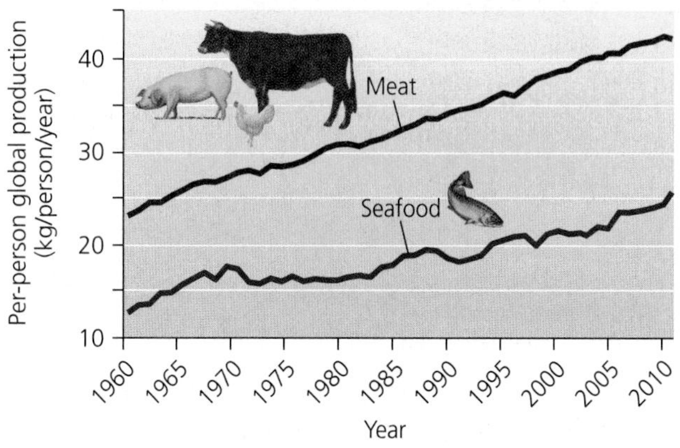


**FIGURE 7.14 Per-person production of meat from farmed animals and of seafood has risen steadily worldwide.** *Data from U.N. Food and Agriculture Organization (FAO).*

## Food choices are energy choices

What we choose to eat has ramifications for how we use energy and the land that supports agriculture. Recall our discussions of trophic levels and pyramids of energy (p. 72). Every time that one organism consumes another, only about 10% of the energy moves from one trophic level up to the next; the great majority of energy is used up in cellular respiration. For this reason, eating meat is far less energy-efficient than relying on a vegetarian diet and leaves a far greater ecological footprint.

Some animals convert grain feed into milk, eggs, or meat more efficiently than others (**FIGURE 7.15**). Scientists have calculated relative energy-conversion efficiencies for different types of animals. Such energy efficiencies have ramifications for land use: Land and water are required to raise food for the animals, and some animals require more than others. **FIGURE 7.16** shows the area of land and weight of water required to produce 1 kg (2.2 lb) of food protein for milk, eggs, chicken, pork, and beef. Producing eggs and chicken meat requires the least space and water, whereas producing beef requires the most. Such differences make clear that when we choose what to eat, we are also indirectly choosing how to make use of resources such as land and water.

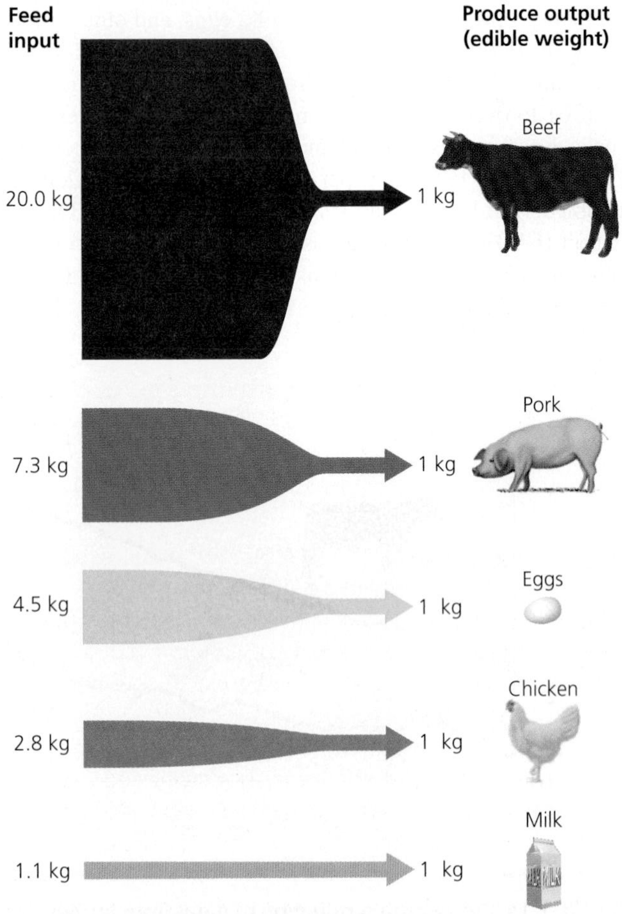


**FIGURE 7.15 Producing different animal food products requires different amounts of animal feed.** To produce 1 kg of beef, 20 kg of feed must be provided to cattle. *Data from Smil, V., 2001.* Feeding the world: A challenge for the twenty-first century. *Cambridge, MA: MIT Press.*

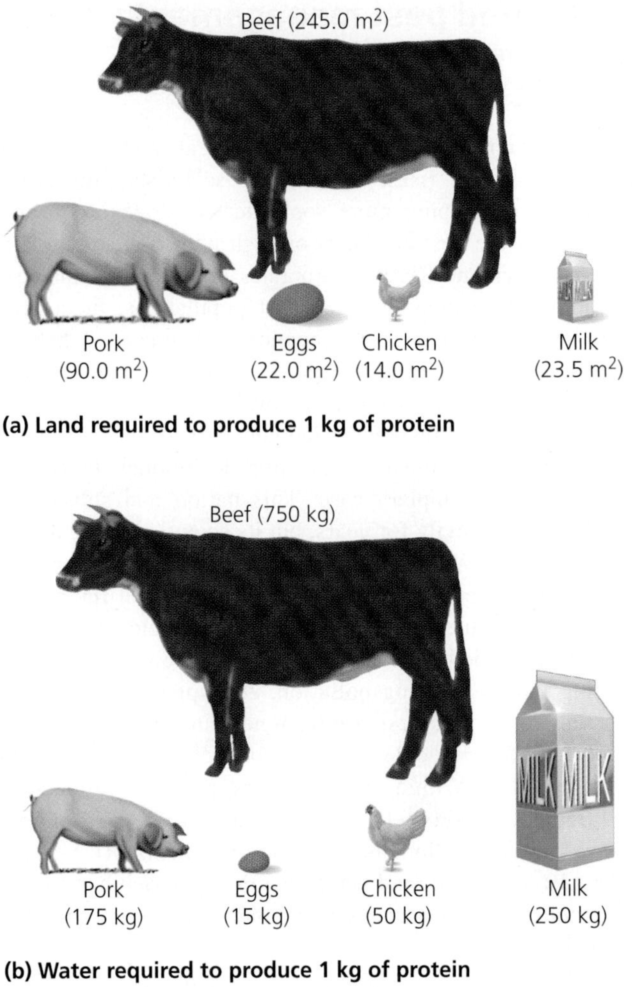


**FIGURE 7.16 Producing different types of animal products requires different amounts of (a) land and (b) water.** Raising cattle for beef requires by far the most land and water. *Data from Smil, V., 2001.* Feeding the world: A challenge for the twenty-first century. *Cambridge, MA: MIT Press.*

**DATA Q** In terms of protein, how many times more land does it take to produce beef than chicken? How many times more water does it take?

GO TO **INTERPRETING GRAPHS & DATA** ON MasteringEnvironmentalScience®

## Feedlots have benefits and costs

In traditional agriculture, livestock are kept by farming families near their homes or are grazed on open grasslands by nomadic herders or sedentary ranchers. These traditions survive, but the advent of industrial agriculture has brought a new method. **Feedlots,** also known as *factory farms* or *concentrated animal feeding operations (CAFOs)*, are essentially huge warehouses or pens designed to deliver energy-rich food to animals living at extremely high densities. Today nearly half the world's pork and most of its poultry come from feedlots.

Feedlot operations allow for economic efficiency and increased production, which make meat affordable to more people. Concentrating cattle and other livestock in feedlots also removes them from rangeland, thereby reducing the grazing impacts these animals would exert across large portions of the landscape.

Intensified animal production through the industrial feedlot model has some negative consequences, though. Forty-five percent of our global grain production goes to feed livestock and poultry. This elevates the price of staple grains and endangers food security for the very poor. Livestock produce prodigious amounts of manure and urine, and their waste can pollute surface water and groundwater near feedlots. The crowded conditions under which animals are often kept necessitate heavy use of antibiotics to control disease. The overuse of antibiotics can cause microbes to become resistant to the antibiotics (just as pests become resistant to pesticides; pp. 147–148), making these drugs less effective.

Livestock are also a major source of greenhouse gases, such as methane, that lead to climate change (p. 305). The FAO reported in 2013 that livestock agriculture contributes over 14% of greenhouse gas emissions worldwide. On a brighter note, the FAO judged that greenhouse gas emissions from livestock could be reduced by as much as 30% with the widespread adoption of best practices in livestock production.

Not all livestock are raised in intensive facilities, however. Meat, eggs, and dairy products are also produced from "free-range" animals that are not always confined indoors or in pens. The certification of an operation as free range is quite broad. For example, the only requirement the USDA imposes for certifying poultry operations as "free range" is that the birds be given access to the outdoors. The Kennesaw State University farms maintain some 60 free-range chickens, and these hens supply The Commons with 300 or more eggs weekly. Raising animals in this manner can be more costly than in intensive facilities, so prices for free-range meat and eggs are often higher than for conventionally raised animals.

## We raise seafood with aquaculture

Besides growing plants as crops and raising animals on rangelands and in feedlots, we rely on aquatic organisms for food. Increased demand and new technologies have led us to overharvest most marine fisheries (pp. 274–275); as a result, wild fish populations are plummeting throughout the world's oceans. This means that raising fish and shellfish on "fish farms" may be the only way to meet our growing demand for these foods.

The cultivation of aquatic organisms for food in controlled environments, called **aquaculture,** is now being pursued with over 220 freshwater and marine species (**FIGURE 7.17**). Many aquatic species are grown in open water in large, floating net-pens. Others are raised in ponds or holding tanks. Aquaculture is the fastest-growing type of food production; in the past 20 years, global output has increased fivefold. Most widespread in Asia, aquaculture today produces $125 billion worth of food and provides three-quarters of the freshwater fish and two-thirds of the shellfish that we eat.

Aquaculture helps reduce fishing pressure on overharvested and declining wild stocks. Furthermore, aquaculture consumes fewer fossil fuels and provides a safer work environment than does commercial fishing. Fish farming can also be remarkably energy-efficient, producing as much as 10 times more fish per unit area than is harvested from waters of the continental shelf and up to 1000 times more than is harvested from the open ocean.

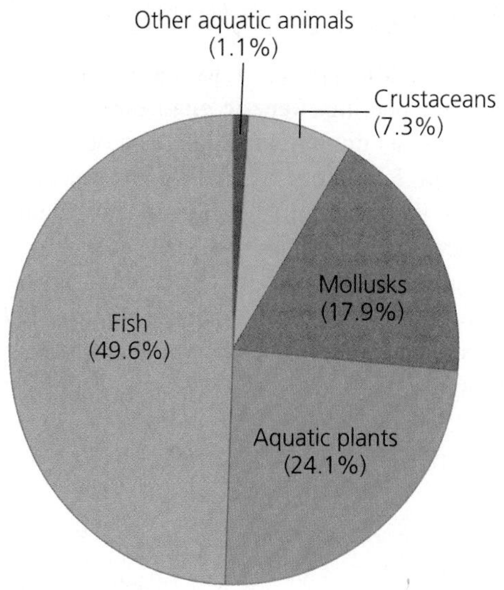


**FIGURE 7.17 Aquaculture involves many types of fish and a wide diversity of other marine and freshwater organisms.**
*Data from U.N. Food and Agriculture Organization (FAO).*

Along with its benefits, aquaculture has disadvantages. Aquaculture can produce prodigious amounts of waste, both from the fish and shellfish and from the feed that goes uneaten and decomposes in the water. Like feedlot animals, commercially farmed fish often are fed grain, and this affects food supplies for people. In other cases, farmed fish are fed fish meal made from wild ocean fish such as herring and anchovies, whose harvest from oceans may place additional stress on wild fish populations.

If farmed aquatic organisms escape into ecosystems where they are not native (as several carp species have done in U.S. waters), they may spread disease to native stocks or may outcompete native organisms for food or habitat. For example, if salmon that have been genetically engineered for rapid growth were to escape from aquaculture facilities, they could spread disease or introduce genes into wild salmon populations.

# Genetically Modified Food

The Green Revolution enabled us to feed a greater number and proportion of the world's people, but relentless population growth is demanding still more innovation. A new set of potential solutions began to arise in the 1980s and 1990s as advances in genetics enabled scientists to directly alter the genes of organisms, including crop plants and livestock. The genetic modification of organisms that provide us food holds promise to enhance nutrition and the efficiency of agriculture while lessening impacts on the planet's environmental systems. However, genetic modification may also pose risks that are not yet well understood. This possibility has given rise to anxiety and protest by consumer advocates, small farmers, environmental activists, and critics of big business.

## Foods can be genetically modified

The genetic modification of crops and livestock is one type of genetic engineering. **Genetic engineering** is any process whereby scientists directly manipulate an organism's genetic material in the laboratory by adding, deleting, or changing segments of its DNA (p. 29). **Genetically modified (GM) organisms** are organisms that have been genetically engineered using **recombinant DNA,** which is DNA that has been patched together from the DNA of multiple organisms (**FIGURE 7.18**). The goal is to place genes that code for certain desirable traits (such as rapid growth, disease resistance, or high nutritional content) into organisms lacking those traits. An organism that contains DNA from another species is called a **transgenic** organism, and the genes that have moved between them are called **transgenes.**

The creation of transgenic organisms is one type of **biotechnology,** the material application of biological science to create products derived from organisms. Biotechnology has helped us develop medicines, clean up pollution, understand the causes of cancer, dissolve blood clots, and make better beer and cheese. TABLE 7.1 shows several notable developments in GM foods. The stories behind them illustrate both the promises and pitfalls of food biotechnology.

The genetic alteration of plants and animals by people is nothing new; through artificial selection (p. 50), we have influenced the genetic makeup of our livestock and crop plants for thousands of years. However, as critics are quick to point out, the techniques geneticists use to create GM organisms differ from traditional selective breeding in several ways. For one, selective breeding mixes genes from individuals of the same or similar species, whereas scientists creating recombinant DNA routinely mix genes of organisms as different as bacteria and plants, or spiders and goats. For another, selective breeding deals with whole organisms living in the field, whereas genetic engineering works with genetic material in the lab. Third, traditional breeding selects from combinations of genes that come together naturally, whereas genetic engineering creates the novel combinations directly.

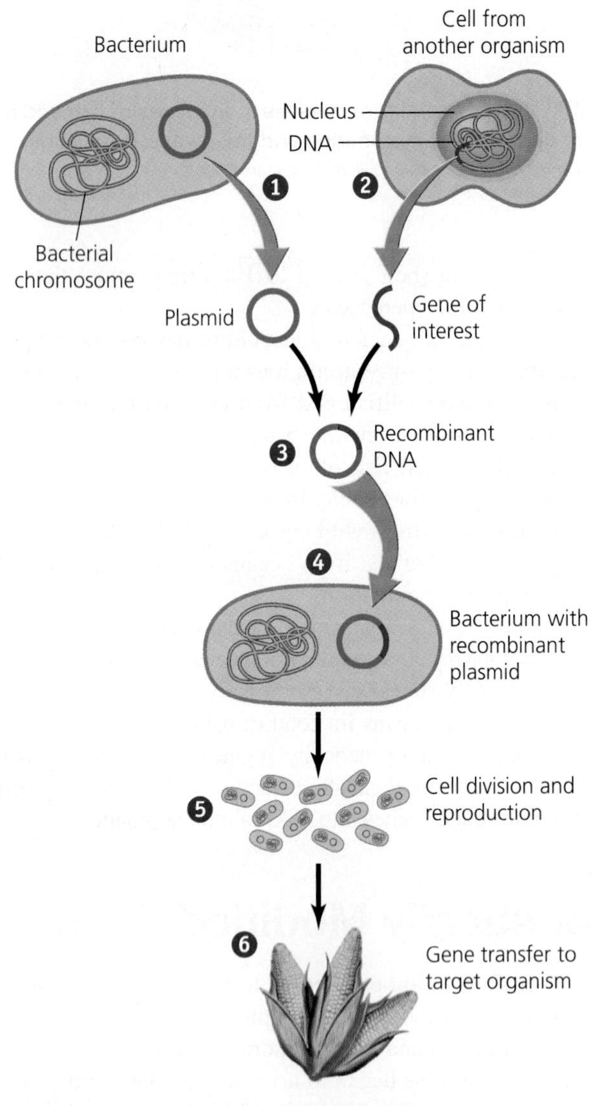


**FIGURE 7.18 To create recombinant DNA, scientists follow several steps.** First they isolate plasmids ❶, small circular DNA molecules, from a bacterial culture. DNA containing a gene of interest ❷ is then removed from another organism. Scientists insert this gene into the plasmid to form recombinant DNA ❸. This recombinant DNA enters new bacteria ❹, which reproduce ❺, generating many copies of the desired gene. The gene is then transferred to individuals of the target plant or animal ❻. It will be expressed in the genetically modified organism as a desirable trait, such as rapid growth or high nutritional content in a food crop.

## Biotechnology is transforming the products around us

In just three decades, GM foods have gone from science fiction to mainstream agriculture (**FIGURE 7.19**). Most GM crops

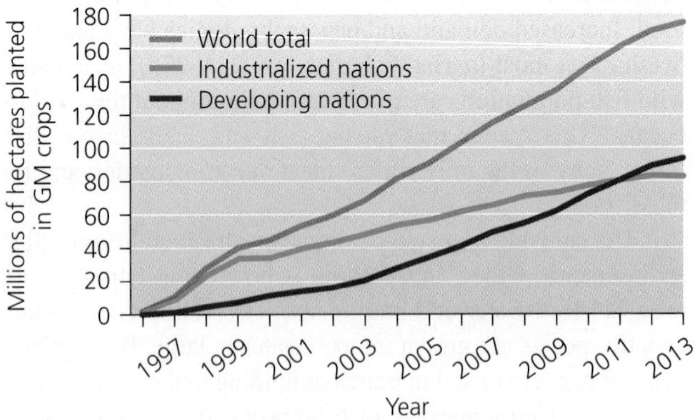


**FIGURE 7.19 GM crops have spread with remarkable speed since their commercial introduction in 1996.** They now are planted on over 10% of the world's cropland. *Data from the International Service for the Acquisition of Agri-Biotech Applications.*

**DATA Q** In the last 5 years, have GM crops been growing faster in industrialized nations or in developing nations? If current trends continue, which group of nations will have more GM crops in 2017? Can you estimate how much more they might have?

GO TO **INTERPRETING GRAPHS & DATA** ON MasteringEnvironmentalScience®

**TABLE 7.1 Several Notable Examples of Genetically Modified Food Technologies**

| CROP | DESCRIPTION AND STATUS | CROP | DESCRIPTION AND STATUS |
|---|---|---|---|
| **Golden rice**  | Engineered to produce beta-carotene to fight vitamin A deficiency in Asia and the developing world. May offer only moderate nutritional enhancement despite years of work. Still undergoing research and development. | **Bt cotton**  | Engineered with genes from bacterium *Bacillus thuringiensis* (Bt), which kills insects. Has increased yield, decreased insecticide use, and boosted income for 14 million small farmers in India, China, and other nations. |
| **Virus-resistant papaya**  | Resistant to ringspot virus and grown in Hawaii. In 2011 became the first biotech crop approved for consumption in Japan. | **Roundup-Ready alfalfa**  | One of many crops engineered to tolerate Monsanto's Roundup herbicide (glyphosate). Because the crop can withstand it, the chemical can be applied in great quantities to kill weeds. Unfortunately, many weeds are evolving resistance to glyphosate as a result. Planted in the United States from 2005 to 2007, GM alfalfa was then banned because a lawsuit forced the USDA to better assess its environmental impact. Reapproved in 2011. |
| **GM salmon**  | Engineered for fast growth and large size. Would be the first GM animal approved for sale as food. To prevent fish from breeding with wild salmon and spreading disease to them, company AquaBounty promised to make their fish sterile and raise them in inland pens. | **Roundup-Ready sugar beet** 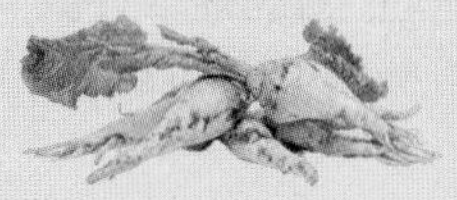 | Tolerant of Monsanto's Roundup herbicide (glyphosate). Swept to dominance (95% of U.S. crop) in just two years. As with alfalfa, a lawsuit forced more environmental review, after it had already become widespread. Reapproved in 2012. |
| **Biotech potato**  | Resistant to late blight, the pathogen that caused the 1845 Irish Potato Famine and that still destoys $7.5 billion of potatoes each year. Being developed by European scientists, but struggling with European Union (EU) regulations on research. | **Biotech soybean**  | The most common GM crop in the world, covering nearly half the cropland devoted to biotech crops. Engineered for herbicide tolerance, insecticidal properties, or both. Like other crops, soybeans may be "stacked" with more than one engineered trait. |
| **Bt corn**  | Engineered with genes from bacterium *Bacillus thuringiensis* (Bt), which kills insects. One of many Bt crops developed. | **Sunflowers and superweeds**  | Research on Bt sunflowers suggests that transgenes might spread to their wild relatives and turn them into vigorous "superweeds" that compete with the crop or invade ecosystems. This is most likely to occur with squash, canola, and sunflowers, which can breed with their wild relatives. |

today are engineered to resist herbicides, so that farmers can apply herbicides to kill weeds without having to worry about killing their crops. Other crops are engineered to resist insect attack. Some are modified for both types of resistance. Resistance to herbicides and insect pests enables large-scale commercial farmers to grow crops more efficiently. As a result, sales of GM seeds to these farmers in the United States and other countries have risen quickly.

Globally in 2013, more than 17 million farmers grew GM crops on 175 million ha (430 million acres) of farmland—11% of all cropland in the world. In the United States today, roughly 90% of corn, soybeans, cotton, and canola consist of genetically modified strains. This is incredible growth, considering that GM crop varieties have been commercially planted only since 1996. Worldwide, four of every five soybean and cotton plants are now transgenic, as are one of every three corn and canola plants. It is conservatively estimated that over 70% of processed foods in U.S. stores contain GM ingredients. Thus, it is highly likely that you consume GM foods on a daily basis.

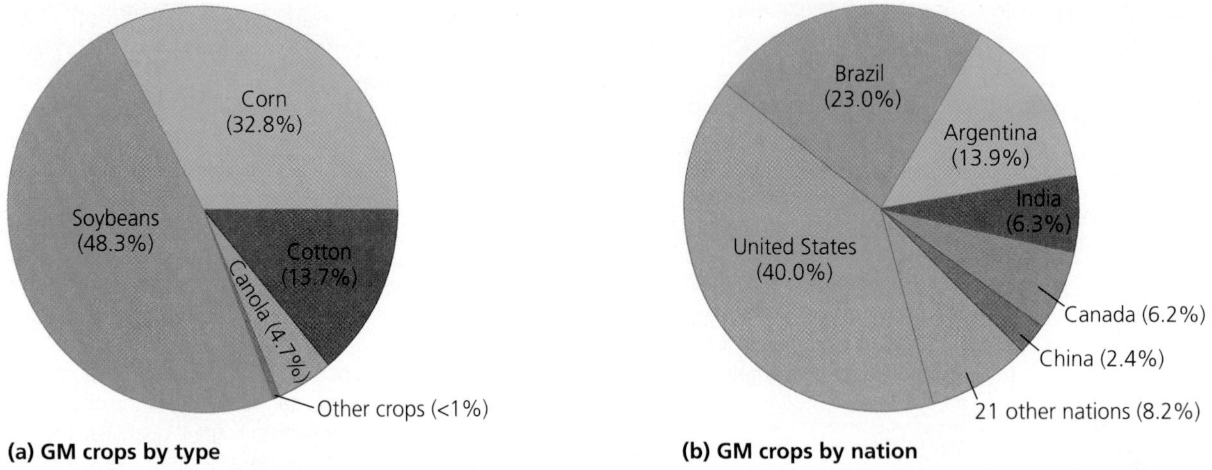


**FIGURE 7.20 So far, genetic engineering has mainly involved common crops grown in industrialized nations.** Of the world's GM crops **(a),** soybeans are the most common. Of global acreage planted in GM crops **(b),** the United States devotes the most area. *Data are for 2013, from the International Service for the Acquisition of Agri-Biotech Applications.*

Soybeans account for most of the world's GM crops (**FIGURE 7.20a**). Of the 27 nations growing GM crops in 2013, six (the United States, Brazil, Argentina, India, Canada, and China) accounted for over 90% of production, with the United States alone growing 40% of the global total (**FIGURE 7.20b**). Half of all GM crops worldwide, however, are grown in developing nations.

## What are the impacts of GM foods?

Genetic modification has the potential to advance agriculture by engineering crops with high drought tolerance for use in arid regions, and by developing high-yield crops that can feed our growing population on existing cropland. However, most of these noble intentions have not yet come to pass. This is largely because the corporations that develop GM varieties cannot easily profit from selling seed to small farmers in developing nations. Instead, most biotech crops have been engineered for insect resistance and herbicide tolerance, which improve efficiency for large-scale industrial farmers who can afford GM seeds.

Regardless, proponents of GM foods maintain that these foods bring environmental and social benefits and promote sustainable agriculture in several ways. Increased crop yields enhance food security and reduce the need for new farmland, conserving natural areas. Herbicide-resistant crops promote no-till farming. Planting insect-resistant GM crops, proponents maintain, also reduces pesticide applications.

Although GM crops do appear to result in lower levels of insecticide use, studies find that the cultivation of these crops tends to result in *more* herbicide use. As weeds evolve resistance to herbicides, farmers apply ever-larger quantities of herbicide. Worldwide, over 200 weed species have evolved resistance to herbicides, and resistance to the weed-killer glyphosate (p. 153) is being documented more widely than in the past (**FIGURE 7.21**).

Most scientists feel that ecological impacts of GM foods pose the greatest threat. Many conventional crops can interbreed with their wild relatives (crop varieties of rice can breed with wild rice, for example), so there seems little reason to believe that transgenic crops would not do the same. In the first confirmed case, GM oilseed rape was found hybridizing with wild mustard. In another case, creeping bentgrass engineered for use on golf courses—a GM plant not yet approved by the USDA—pollinated wild grass up to 21 km (13 mi) away from its experimental growing site. Most scientists think transgenes will inevitably make their way from GM crops into wild plants, but the ecological impacts of this are open to debate.

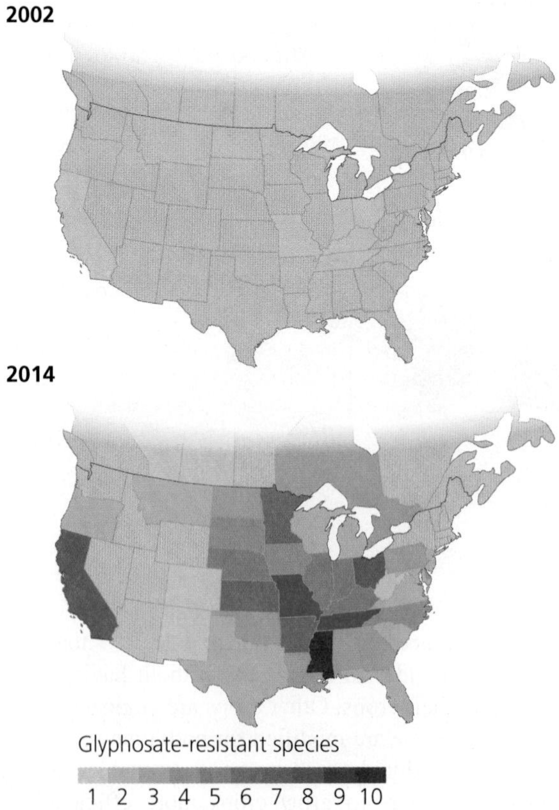


**FIGURE 7.21 In just over a decade, weed resistance to glyphosate spread across North America.** *Data from Heap, I. International Survey of Herbicide Resistant Weeds. Aug., 2014.* www.weed-science.com.

Because biotechnology is rapidly changing and the large-scale introduction of GM organisms into our environment is recent, there is much we don't yet know about the consequences. As a result, it is still too early to dismiss concerns about environmental impacts without further scientific research. Many experts feel we should proceed with caution, adopting the **precautionary principle,** the idea that one should not undertake a new action until the ramifications of that action are well understood.

## Public debate over GM foods continues

Science helps inform us about genetic engineering, but ethical and economic concerns have largely driven the public debate. For many people, the idea of "tinkering" with the food supply seems dangerous or morally wrong. Others fear that the global food supply is being dominated by a handful of large corporations that develop GM technologies, among them Monsanto, Syngenta, Bayer CropScience, Dow, DuPont, and BASF. Critics say these multinational corporations threaten family farmers, and recent legal actions by these corporations against small farmers have stoked such fears. Agrobiotech corporations have taken out patents on the transgenes they've developed for use in GM crops—and some have sued farmers who are found to have GM crops on their land, even if the farmers contend that GM crops from neighboring fields contaminated their non-GM crops. Court cases have largely been decided in the companies' favor. This has resulted in farmers being ordered to pay damages of tens of millions of dollars to agrobiotech companies.

Although American consumers have largely accepted GM foods, consumers in Europe, Japan, and other nations have expressed widespread unease about genetic engineering. For example, opposition to GM foods in Europe blocked the import of hundreds of millions of dollars in U.S. agricultural products from 1998 to 2003 until the United States brought a successful case before the World Trade Organization (p. 108) to force their import.

More than 60 nations require that GM foods be labeled so that consumers know what they are buying, but the United States is not one of them. Proponents of labeling argue that consumers have a right to know what's in the food they buy. Opponents argue that labeling implies the foods are dangerous and may result in lower sales.

# The Growth of Sustainable Agriculture

Industrial agriculture has allowed food production to keep pace with our growing population, but it involves many adverse environmental and social impacts. These range from the degradation of soils to reliance on fossil fuels to problems arising from pesticide use, genetic modification, and intensive feedlot and aquaculture operations. Although intensive commercial agriculture may help alleviate certain environmental pressures, it often worsens others. Throughout this chapter, we have seen examples of sustainable approaches to agriculture that maintain high crop yields, minimize resource inputs into food production, and lessen the environmental impacts of farming. Let's now take a closer look at the growth of sustainable agriculture and its adoption around the world.

## Organic agriculture is booming

One type of sustainable agriculture is **organic agriculture,** which uses no synthetic fertilizers, insecticides, fungicides, or herbicides. In 1990, the U.S. Congress passed the Organic Food Production Act to establish national standards for organic products and facilitate their sale. Under this law, in 2000 the USDA issued criteria by which crops and livestock could be officially certified as organic, and these standards went into effect in 2002 as part of the National Organic Program. California, Washington, and Texas established stricter state guidelines for labeling foods organic, and today many U.S. states and over 80 nations have laws spelling out organic standards.

For farmers, organic farming can bring a number of benefits: lower input costs, enhanced income from higher-value produce, and reduced chemical pollution and soil degradation (see **THE SCIENCE BEHIND THE STORY**, pp. 156–157). Transitioning to organic agriculture does involve some risk, however, because farmers must use organic approaches for three years before their products can be certified as organic and sold at the higher prices commanded by organic foods. One of Kennesaw State University's campus farms that supplies The Commons with fruits and vegetables, for example, had to undergo such a waiting period before being able to label its produce as "organic."

The main obstacle for consumers to organic foods is price. Organic products tend to be 10–30% more expensive than conventional ones, and some (such as milk) can cost

**FAQ** **Is it safe to eat genetically modified foods?**

In principle, there is nothing about the process of genetic engineering that should make genetically modified food any less safe to eat than food produced by conventional methods. The fact that technology is used to move a gene does not make the gene unsafe. Thus, to determine whether GM foods pose any health risks, studies must be done comparing those foods with conventional versions, one by one—just as researchers would study any other substance for health risks (Chapter 10). Thus far, no study has shown undeniable evidence of human health impacts on humans from any GM food, but this lack of evidence does not guarantee that such foods pose no risk. A great deal of research is controlled by the companies that develop GM foods, and we will never be able to test all GM foods. Hence, the effects on humans of consuming GM foods will largely be examined with correlative studies (p. 10) over long time periods in coming decades.

## THE SCIENCE BEHIND THE STORY

# How Productive Is Organic Farming?

**Swiss scientist Dr. Paul Mäder (center) visits colleagues at Pennsylvania's Rodale Institute.**

Organic farming puts fewer synthetic chemicals into the soil, air, and water than conventional industrial farming does. But can organic farming produce large enough crop yields to feed the human population? The world's two longest-running field experiments on the topic—in Switzerland and in Pennsylvania—suggest that the answer is yes.

Let's first visit Switzerland, where one in every nine hectares of agricultural land is managed organically (the fifth-highest rate in the world). Back in 1977, Swiss researchers established experimental farms at Therwil, near the city of Basel. At the research site, wheat, potatoes, and other crops are grown in plots cultivated in different treatments:

- Conventional farming using chemical pesticides, herbicides, and inorganic fertilizers
- Conventional farming that also uses organic fertilizer (cattle manure)
- Organic farming using only manure, mechanical weeding, and plant extracts to control pests
- Organic farming that also adds natural boosts, such as herbal extracts in compost

Researchers record crop yields at harvest each year. They analyze the soil regularly, measuring nutrient content, pH, structure, and other variables. They also measure the biological diversity and activity of microbes and invertebrates in the soil. Such indicators of soil quality help researchers assess the potential for long-term productivity.

In 2002, Paul Mäder and colleagues from two Swiss research institutes reported in the journal *Science* results from 21 years of data. Long-term studies are rare in agriculture and in ecology. They are highly valuable, because they can reveal slow processes or subtle effects that get swamped out by year-to-year variations in shorter-term studies. Over the study's 21 years, the organic fields yielded 80% of what the conventional fields produced. Organic potato crops averaged just 58–66% of conventional yields because of nutrient deficiency and disease, but organic crops of winter wheat produced 90% of conventional yields.

Although the organic plots produced 20% less on average, they received 35–50% less fertilizer than the conventional fields and 97% fewer pesticides. Thus, Mäder's team concluded, the organic plots were highly efficient and represent "a realistic alternative to conventional farming."

How can organic fields produce decent yields without relying on synthetic chemicals? The answer lies in the soil. Mäder's team found that soil in the organic plots had better structure, better supplies of some nutrients, and much more microbial activity and invertebrate biodiversity (**FIGURE 1**).

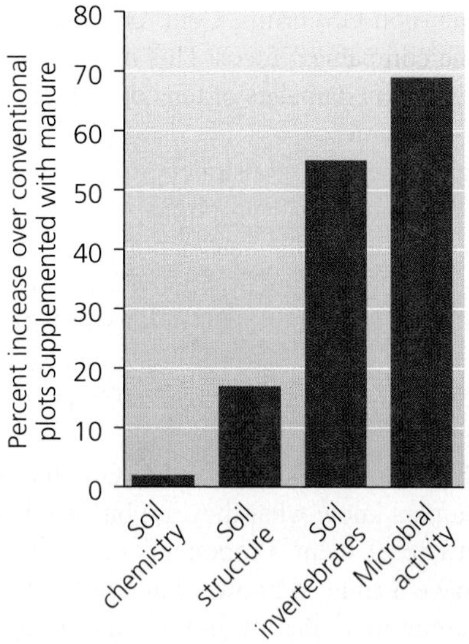


**FIGURE 1 Organic fields developed better soil quality than conventional fields supplemented with manure.** Values for soil chemistry (six variables), structure (three variables), invertebrates (five variables), and microbial activity (six variables) were compared. Organic fields outperformed conventional fields without manure (not shown) still more. *Data from Mäder, P., et al., 2002. Soil fertility and biodiversity in organic farming.* Science *296: 1694–1697.*

Studies are continuing at the Swiss plots today, producing new research results. As one example, Jens Leifeld and two colleagues at a Zürich research institute analyzed soil carbon content after 27 years. They found that soil carbon had decreased in all treatments but that conventional plots had suffered the greatest decline. Conventional plots supplemented with manure, however, did just as well as the organic plots in retaining soil carbon.

Organic farming has proved even more successful in Pennsylvania, where the Rodale Institute has compared organic and conventional fields of corn and soybeans in a large-scale experiment running since 1981 on its 330-acre farm. In 2011 it released results from 30 years' worth of data (**FIGURE 2**).

Averaged across the 30 years, yields of organically grown crops equaled yields of conventionally grown crops. Moreover, the organic crops required 30% less energy input, and raising them released 35% fewer greenhouse gas emissions. Because of lower energy inputs and higher crop prices for organic produce, farming organically created three times the profit for the farmer.

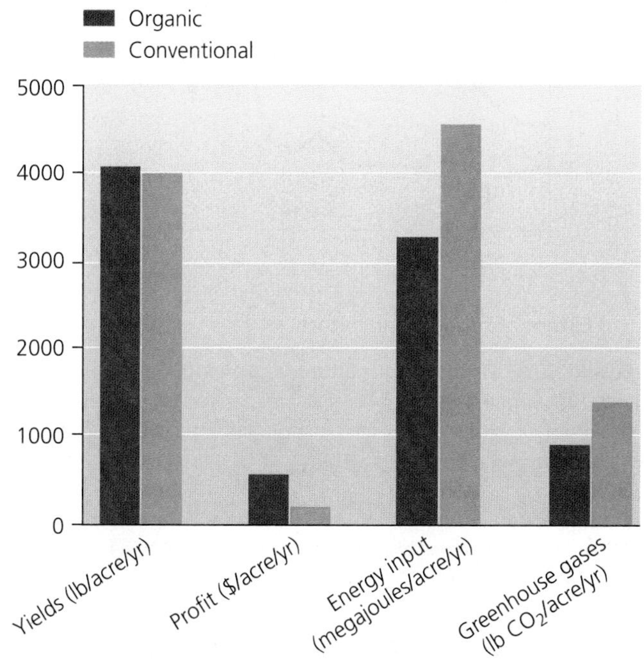


**FIGURE 2 Organic crops equaled conventional crops in yields while producing more profit, requiring less energy input, and releasing fewer greenhouse gas emissions.** *Data from Rodale Institute, 2011.* The farming systems trial: Celebrating 30 years. *Rodale Institute, Kutztown, Pennsylvania.*

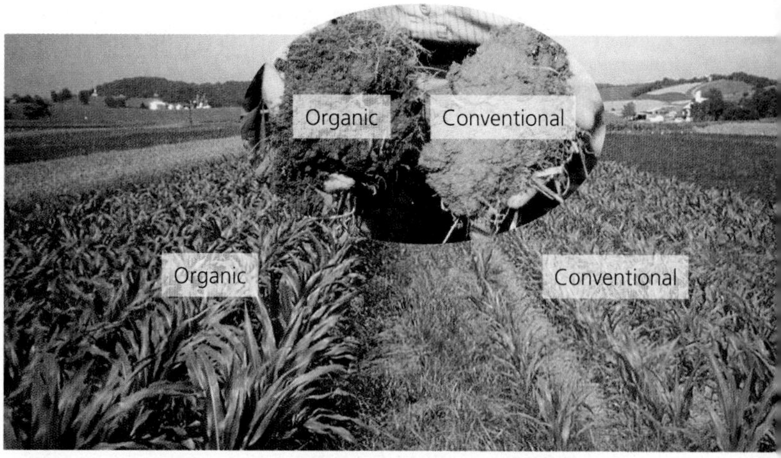


**FIGURE 3 Organically grown corn did better than conventionally grown corn during periods of drought in the Rodale Institute experiment.** Better-quality soil was part of the reason.

As with the Swiss experiment, the secret lies in the soil. At the Rodale Institute's farm over the years, the soil of the organic fields became visibly darker and better textured than the conventional fields' soil. This helped corn crops in the organic fields to outperform those in the conventional fields during times of drought (**FIGURE 3**).

Shorter-term experiments elsewhere have shown similar results. Researchers comparing organic and conventional farms in North Dakota and Nebraska have found that organic farming produces soils with more microbial life, earthworm activity, water-holding capacity, topsoil depth, and naturally occurring nutrients.

Given such differences in soil quality, researchers expect that organic fields should perform better and better relative to conventional fields as time goes by—in other words, that they are more sustainable. Moreover, the striking yield results of the Rodale experiment suggest that perhaps organic agriculture can feed the world's people every bit as reliably as today's conventional industrial agriculture.

As more long-term data are published and new studies begin, we are learning more and more about the benefits of organic agriculture for soil quality—and about how we might improve conventional methods to maximize crop yields while protecting the long-term sustainability of agriculture. ☐

twice as much. However, because there are many consumers willing to pay more for organic products, grocers and other businesses are making them more widely available.

Today, three of four Americans buy organic food at least occasionally, and more than four of five retail groceries offer it. U.S. consumers spent $30.5 billion on organic food in 2013, amounting to 4.3% of all food sales (**FIGURE 7.22a**). Worldwide, sales of organic food tripled between 2000 and 2013, when sales reached $63 billion.

Production of organic food is increasing along with demand (**FIGURE 7.22b**). Although organic agriculture takes up less than 1% of agricultural land worldwide, this area is rapidly expanding. Two-thirds of organic acreage is in developed nations. Farmers in all 50 U.S. states and about 2 million farmers in more than 160 nations practice organic farming commercially to some extent. Contrary to common perceptions, organic farming can be performed on both big and small farms, as long as the criteria for organic certification are satisfied.

Government policies have aided organic farming. In the United States, the 2014 Farm Bill had a number of provisions that directly aid organic agriculture, including funds to defray certification expenses. The European Union supports farmers financially during conversion to organic agriculture. Once conversion is complete, studies suggest that reduced inputs and higher market prices can make organic farming at least as profitable for the farmer as conventional methods.

## Locally supported agriculture is growing

Apart from organic methods, another component of the move toward sustainable agriculture is an attempt to reduce the use of fossil fuels for the long-distance transport of food. The average food product sold in a U.S. supermarket travels at least 1600 km (1000 mi) between the farm and the grocery. Because of the travel time, supermarket produce is often chemically treated to preserve freshness and color.

In response, increasing numbers of farmers and consumers in developed nations are supporting local small-scale agriculture and adopting the motto "think global, eat local." Farmers' markets are springing up throughout North America as people rediscover the joys of fresh, locally grown produce. At **farmers' markets,** consumers buy meats and fresh fruits and vegetables in season from local producers. These markets generally offer a wide choice of organic items and unique local varieties not found in supermarkets. At Kennesaw State University, a weekly, on-campus farmer's market provides students, faculty, and staff with an opportunity to purchase fresh produce while supporting local agriculture.

Some consumers are even partnering with local farmers in a phenomenon called **community-supported agriculture (CSA).** In a CSA program, consumers pay farmers in advance for a share of their yield, usually a weekly delivery of produce. Consumers get fresh seasonal produce, and farmers get a guaranteed income stream up front to invest in their crops—a welcome alternative to taking out loans and being at the mercy of the weather.

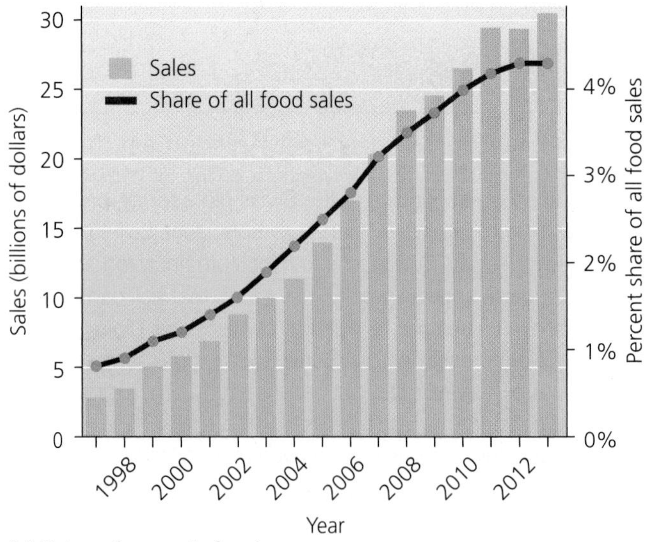


**(a) Sales of organic food**

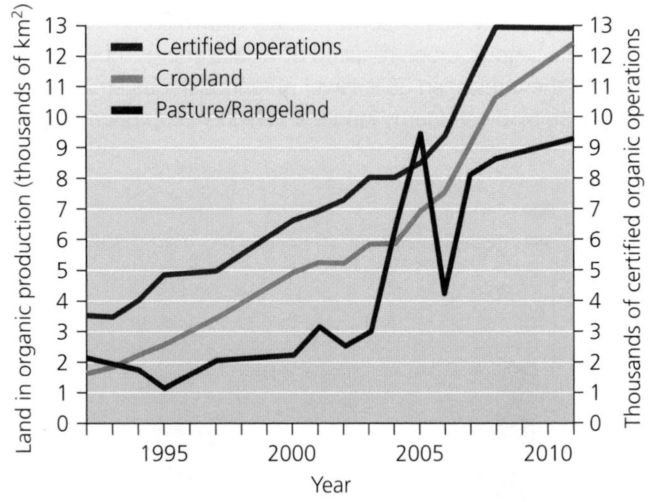


**(b) Extent of organic agriculture**

**FIGURE 7.22 Organic agriculture is growing.** Sales of organic food in the United States **(a)** have increased rapidly, both in total dollar amounts (bars) and as a percentage of the overall food market (line). Since the 1990s **(b),** U.S. acreage devoted to organic crops and livestock has quadrupled, and certified operations have more than tripled. *Sources: (a) Adapted by permission from Willer, Helga (2014).* Organic agriculture worldwide: Current statistics. *FiBL-IFOAM Report. IFOAM, Bonn; FiBL, Frick; ITC, Geneva. Data used with permission from Organic Trade Association OTA: Manufacturer Survey 2012–2013. (b) Data from USDA Economic Research Service.*

### WEIGHING THE ISSUES

**Do You Want Food Labeled?** The USDA issues labels to certify that products claiming to be organic have met the government's organic standards. Critics of genetically modified food want GM products to be labeled as well. Do you want your food to be labeled to indicate whether it is organic or genetically modified? Would you choose among foods based on such labeling? How might your food choices and purchasing decisions have environmental impacts (good or bad)?

## Sustainable agriculture provides a roadmap for the future

The best approach for making an agricultural system sustainable is to mimic the way a natural ecosystem functions. Ecosystems are sustainable because they operate in cycles and are internally stabilized with negative feedback loops (pp. 23–24). In this way, they provide a useful model for agriculture.

The Commons and other campus dining halls exemplify how food production can be integrated with farms and nearby natural ecosystems, reducing the environmental impacts from agriculture while providing delicious, nutritious, and affordable food. Efforts like these, together with other approaches taken by farmers around the world, are making agriculture sustainable. This will be crucial for all of us as we progress through the coming century.

# Conclusion

Our species has a 10,000-year history with agriculture, and over this time methods of food production have changed substantially. Many practices of intensive industrial agriculture exert substantial negative environmental impacts. At the same time, these practices have boosted food supplies and helped to relieve pressure to convert natural areas to new farmland.

If our planet is to support over 9 billion people by mid-century without further degrading the soil, water, pollinators, and other resources and ecosystem services that support our food production, we must find ways to shift to sustainable agriculture. Biological pest control, organic agriculture, pollinator conservation, preservation of native crop diversity, sustainable aquaculture, and likely some degree of careful and responsible genetic modification of food may all be parts of the game plan we will need to achieve a sustainable future.

# Testing Your Comprehension

1. Describe patterns in global food security from 1970 to the present. Name two nutritional deficiencies commonly seen in the modern world.
2. Compare and contrast the methods used in traditional and industrial agriculture. How does sustainable agriculture differ from industrial agriculture?
3. How are soil horizons created? List and describe the major horizons in a typical soil profile. How is organic matter distributed in a typical soil profile?
4. Name three human activities that can promote soil erosion. Describe several farming techniques (such as terracing and no-till farming) that can help reduce the risk of erosion.
5. Explain how overirrigation can damage soils and reduce crop yields.
6. How do fertilizers boost crop growth? How can large amounts of fertilizer added to soil also end up in water supplies and the atmosphere? How do sustainable agriculture approaches reduce fertilizer runoff?
7. Explain how pesticide resistance occurs.
8. What are some economic benefits of aquaculture? What are some negative environmental impacts?
9. How is a transgenic organism created? How is genetic engineering different from traditional agricultural breeding? How is it similar?
10. Describe the recent growth of organic food sales and land under organic management in the United States.

# Seeking Solutions

1. How do you think a farmer can best help to conserve soil? How do you think a scientist can best help to conserve soil? How do you think a national government can best help to conserve soil?
2. Describe and assess several ways in which high-input industrial agriculture can be beneficial for the environment and several ways in which it can be detrimental. Now suggest several ways in which we might modify industrial agriculture to reduce its environmental impacts.
3. What factors make for an effective biological control strategy of pest management? What risks are involved in biocontrol? If you had to decide whether to use biocontrol against a particular pest, what questions would you want to have answered before you decide?
4. **THINK IT THROUGH** You are the head of an international granting agency that assists farmers with soil conservation and sustainable agriculture. You have $10 million to disburse. Your agency's staff has decided that the funding should go to:

   (1) farmers in an arid area of Africa prone to salinization,

   (2) farmers in a fast-growing area of Indonesia where swidden agriculture is practiced,

   (3) farmers in Argentina practicing no-till agriculture, and

   (4) farmers in a dryland area of Mongolia undergoing desertification.

   What types of projects would you recommend funding in each of these areas, how would you apportion your funding among them, and why?

5. **THINK IT THROUGH** You are a USDA official and must decide whether to allow the planting of a new genetically modified strain of cabbage that produces its own pesticide and has twice the vitamin content of regular cabbage. What questions would you ask of scientists before deciding whether to approve the new crop? What scientific data would you want to see? Would you also consult nonscientists or consider ethical, economic, and social factors?

# Calculating Ecological Footprints

Many people who want to reduce their ecological footprint have focused on how much energy is expended (and how many climate-warming greenhouse gases are emitted) in transporting food from its place of production to its place of sale. The typical grocery store item is shipped by truck, air, and/or sea for many hundreds of miles before reaching the shelves, and this transport consumes oil. This concern over "food-miles" has helped drive the "locavore" movement to buy and eat locally sourced food.

However, food's transport from producer to retailer, as measured by food-miles, is just one source of carbon emissions in the overall process of producing and delivering food. In 2008, environmental scientists Christopher Weber and H. Scott Mathews conducted a thorough life-cycle analysis (p. 400) of U.S. food production and delivery. By filling in the table below, you will get a better idea of how our dietary choices contribute to climate change.

| Food type | Total emissions[1] across life cycle | Emissions[1] from delivery[2] | Percent emissions from delivery[2, 3] |
|---|---|---|---|
| Fruits and vegetables | 0.85 | 0.10 | 11.8 |
| Cereals and carbohydrates | 0.90 | 0.07 | |
| Dairy products | 1.45 | 0.03 | |
| Chicken/fish/eggs | 0.75 | 0.03 | |
| Red meat | 2.45 | 0.03 | |
| Beverages | 0.50 | 0.04 | |

[1] *Emissions are measured in metric tons of carbon-dioxide-equivalents, per household per year.*

[2] *"Delivery" means transport from producer to retailer.*

[3] *"Percent emissions from delivery" is calculated by dividing emissions from delivery/total emissions across life cycle and multiplying by 100 (to convert proportion to percentage).*

*Weber, C.L., and H.S. Mathews, 2008. Food-miles and the relative climate impacts of food choices in the United States.* Environmental Science and Technology *42: 3508–3513.*

1. Which type of food is responsible for the most greenhouse gas emissions across its whole life cycle? Which type is responsible for the least emissions?
2. What is the range of values for percent of emissions from transport of food to retailers (delivery)? Weber and Mathews found that 83% of food's total emissions came from its production process on the farm or feedlot. What do these numbers tell you about how you might best reduce your own footprint with regard to food?
3. After measuring mass, energy content, and dollar value for each food type, the researchers calculated emissions per kilogram, calorie, and dollar. In every case, red meat produced the most emissions, followed by dairy products and chicken, fish, and eggs. They then calculated that shifting one's diet from meat and dairy to fruits, vegetables, and grains for just one day per week would reduce emissions as much as eating 100% locally (cutting food-miles to zero) all the time. Knowing all this, how would you choose to reduce your own food footprint? By how much do you think you could reduce it?

## MasteringEnvironmentalScience®

**STUDENTS**

Go to **MasteringEnvironmentalScience** for assignments, the etext, and the Study Area with practice tests, videos, current events, and activities.

**INSTRUCTORS**

Go to **MasteringEnvironmentalScience** for automatically graded activities, current events, videos, and reading questions that you can assign to your students, plus Instructor Resources.

# 8

# Biodiversity and Conservation Biology

## Upon completing this chapter, you will be able to:

- Characterize the scope of biodiversity on Earth
- Specify the benefits that biodiversity brings us
- Understand today's extinction crisis in geologic context
- Evaluate the primary causes of biodiversity loss
- Assess the science and practice of conservation biology
- Analyze efforts to conserve threatened and endangered species
- Compare and contrast conservation efforts above the species level

**Photo:** Wildebeest crossing the vast plains of the Serengeti

CENTRAL CASE STUDY

# Will We Slice through the Serengeti?

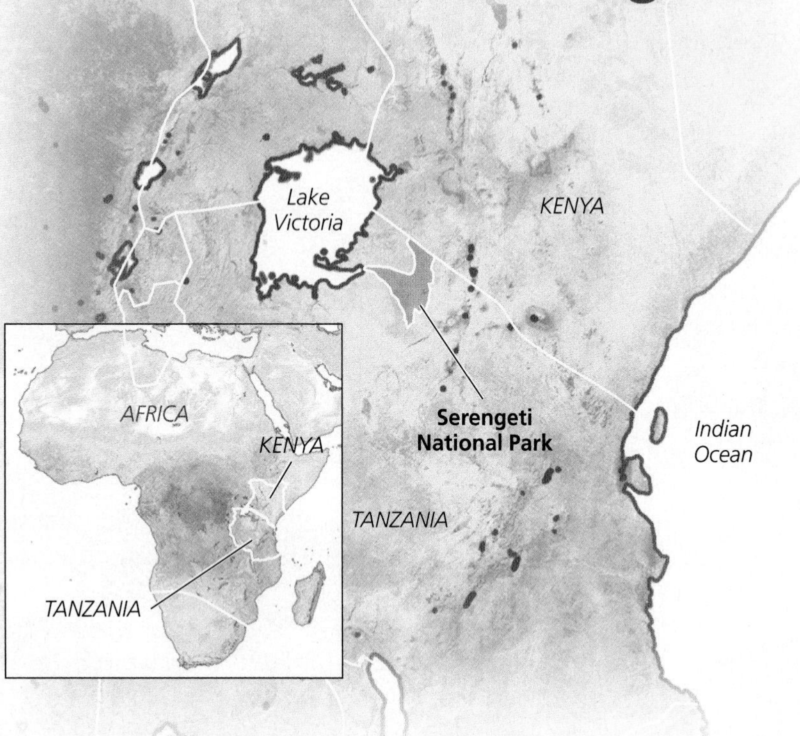


*"Construction of the road will be a huge relief for us. We will sell . . . maize and horticultural products to our colleagues in Arusha and they will bring us cows and goats."*

**—Bizare Mzazi, a farmer outside Serengeti National Park**

*"If we construct this road, all our rhinos will disappear. . . . We should strive to conserve our heritage for future generations."*

**—Sirili Akko, executive officer of the Tanzania Association of Tour Operators**

It's been called the greatest wildlife spectacle on Earth. Each year over 1.2 million wildebeest migrate across the vast plains of the Serengeti in East Africa, along with more than 700,000 zebras and hundreds of thousands of antelope. The herds can stretch as far as the eye can see. Packs of lions track the procession and pick off the weak and unwary, while hungry crocodiles wait in ambush at river crossings. After bearing their calves in the wet season, the wildebeest journey north to find fresh grass. The great herds spend the dry season at the northern end of the Serengeti ecosystem before turning back southward to complete their cyclical annual journey.

This epic migration, with its dramatic interplay of predators and prey, has cycled on for millennia. Yet today, the entire phenomenon may be threatened. Scientists and conservationists are alarmed by a proposal to build a commercial highway across the Serengeti, slicing straight across the animals' migratory route.

Before examining the highway proposal, let's step back for a broad view of the Serengeti. The people native to this region, the Maasai, are semi-nomadic herders who have long raised cattle on the grasslands and savannas. Because the Maasai subsist on their cattle and have lived at low population densities, wildlife thrived here long after it had declined in other parts of Africa.

When East Africa was under colonial rule, the British created game reserves to conserve wildlife for their own hunting. After Tanzania, Kenya, and other African nations gained independence in the mid-20th century, the British reserves became the basis for today's national protected areas. Serengeti National Park was established in 1951, and the Maasai Mara National Reserve was later created just across the border in Kenya. These two protected areas, together with several adjacent ones, encompass the Serengeti ecosystem. This 30,000-km$^2$ (11,500-mi$^2$) region is one of the last places on the planet where an ecosystem remains nearly intact and functional over a vast area.

Today 2 million people visit Tanzania and Kenya each year, most of them ecotourists who visit the parks and protected areas. Serengeti National Park alone receives 800,000 annual visitors. Tourism injects close to $3 billion into these nations' economies and creates jobs for tens of thousands of local people. Because the region's people see that functional ecosystems full of wildlife bring foreign dollars into their communities, many support the parks. Indeed, East Africa has been at the forefront of community-based conservation (p. 182), in which local people act as stewards managing their natural resources in collaboration with international conservationists.

However, most people living in northern Tanzania remain desperately poor. Farmers and townspeople on the shores of Lake Victoria feel isolated by a poor road system. Walled off by Serengeti National Park to their east (which does not allow commercial truck traffic on its few dirt roads), these people have little access to outside markets to buy and sell goods. In response, Tanzania's president Jakaya Kikwete promised them (as well as mining interests and contributors to his election campaign) that he would build a paved highway across the Serengeti. The highway would connect Lake Victoria with cities to the east and ports on the Indian Ocean. The World Bank and the German government offered to finance the $480-million project, and Chinese contractors stood ready to build it.

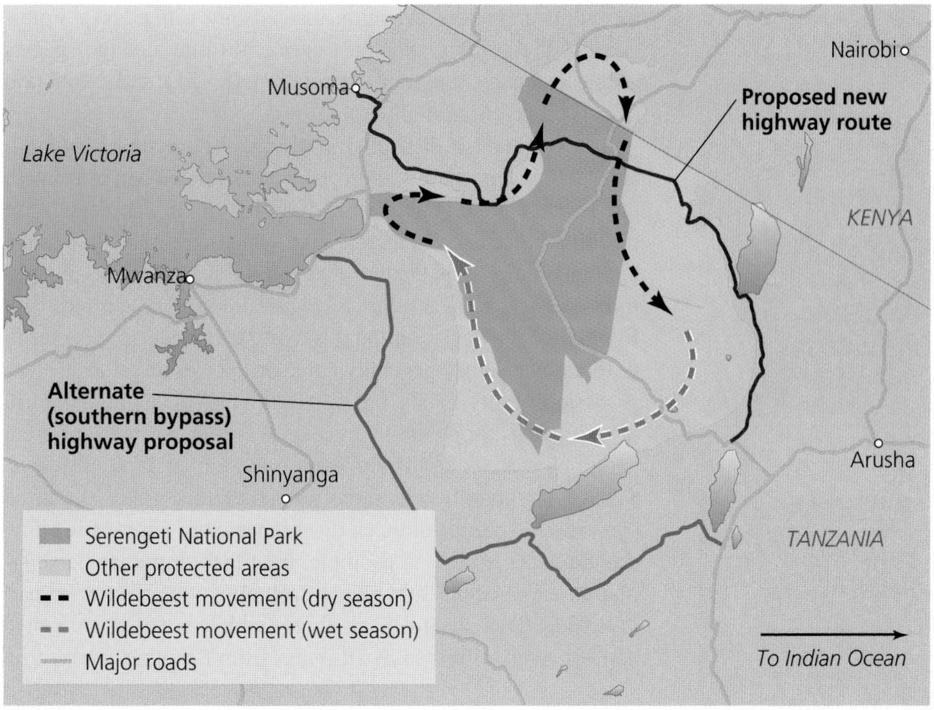


**FIGURE 8.1 A proposed highway would slice through Serengeti National Park.** It would increase commerce and connect Tanzanian people on each side, but it would also cut across the migration route for wildebeest and other animals. Highway opponents suggest an alternate route around the park's southern edge.

Around the world, conservationists reacted with alarm. The proposed highway would slice through the middle of the wildebeest migration path (**FIGURE 8.1**). Scientists predicted that the road would block migration and that vehicles would kill countless animals in collisions. A highway would also provide access for **poaching,** the illegal killing of wildlife for meat or body parts. Likewise, it would allow an entry corridor for exotic plant species that could invade the ecosystem. A highway would encourage human settlement right up to the park boundary, making the park an island of habitat hemmed in by agriculture, housing, and commerce. Moreover, development might encourage the towns on Lake Victoria to grow into large cities, creating demand for larger transportation corridors in the future. For all these reasons, experts predicted that the highway could destroy the migration spectacle.

Such an outcome could devastate tourism, so the region's tourism operators opposed the highway. So did most Kenyans, who feared that the highway would prevent migratory animals from reaching Kenya's Maasai Mara Reserve. In 2010 a Kenyan nongovernmental organization, the African Network for Animal Welfare, sought to stop the highway with a lawsuit in the East Africa Court of Justice, a body set up to adjudicate international matters in the region. This court temporarily halted the road project, and in 2013 heard testimony from both sides.

Meanwhile, international pressure rose on Tanzania to abandon its plans. Highway opponents proposed an alternative route wrapping around the Serengeti's southern end. Although longer, this southern route would pass through more towns, serving five times as many people. The World Bank and the German government offered to help fund this alternative route.

As of 2014, the Tanzanian government under President Kikwete continued to push for a highway directly across the Serengeti and had ordered the route paved right up to the park's borders, leaving unpaved only the 54-km (34-mi) portion that runs through the park. Unlike Tanzania's first president, Julius Nyerere, who left a rich conservation legacy, Kikwete has encouraged a number of development projects that critics say put the nation's ecological assets at risk for little gain. Many of these projects are inspired by the promise of greater trade with China, which is eager to obtain resources from Africa to fuel its red-hot economy.

Today in Tanzania, poaching is rising, and animal populations are falling. The Serengeti is one of our planet's last intact large ecosystems, so impacts here have global ramifications. We would all be impoverished if the Serengeti's biodiversity were lost, so we must hope that Africans can find ways to improve their standard of living while conserving their wildlife and natural systems. East Africa has helped to pioneer win-win solutions in conservation thus far, so perhaps it will show the way yet again. ◻

# Our Planet of Life

Rising human population and resource consumption are putting ever-greater pressure on the flora and fauna of our planet. We are diminishing the ultimate source of our civilization's wealth and happiness: Earth's diversity of life, the very quality that makes our planet unique in the known universe. Thankfully, many people around the world are working tirelessly to save threatened animals, plants, and ecosystems in efforts to stem the loss of our planet's priceless biological diversity.

## Biodiversity encompasses multiple levels

**Biological diversity,** or **biodiversity,** is the variety of life across all levels of biological organization, and includes diversity in species, genes, populations, communities, and ecosystems (p. 51). Biodiversity is a concept as multifaceted as life itself, and

**FIGURE 8.2 The concept of biodiversity encompasses multiple levels in the hierarchy of life.**

biologists employ different working definitions. Yet scientists agree that the concept applies across major levels in the organization of life (**FIGURE 8.2**). The level that people find easiest to visualize and that we use most commonly is species diversity.

**Species diversity** We can express **species diversity** in terms of the number or variety of species found in a particular region. One component of species diversity is *species richness,* the number of species. Another is *evenness* or *relative abundance,* the degree to which species differ in numbers of individuals (greater evenness means they differ less). Immigration, emigration, and local extinction may change species richness locally, but only speciation (p. 51) and extinction (p. 54) change it globally.

A **species** is a distinct type of organism, a set of individuals that uniquely share certain characteristics and can breed with one another and produce fertile offspring (p. 48). Biologists use differing criteria to distinguish one species from another. Some biologists emphasize features that species share because of common ancestry, whereas others emphasize the ability to interbreed. In practice, however, although scientists may define species in different ways, they generally agree on species identities.

Biodiversity exists below the species level in the form of *subspecies*, populations of a species that occur in different geographic areas and differ from one another in slight ways. Subspecies arise by the same processes that drive speciation (pp. 51– 52) but result when divergence stops short of forming separate species. As an example, the black rhinoceros diversified into about eight subspecies, each inhabiting a different part of Africa. The eastern black rhino, which is native to Kenya and Tanzania, differs in its attributes from the other subspecies.

**Genetic diversity** Scientists designate subspecies when they recognize substantial genetically based differences among individuals from different populations. However, all species consist of individuals that vary genetically to some degree, and this variation is an important component of biodiversity. **Genetic diversity** encompasses the differences in DNA composition (p. 29) among individuals, and this provides the raw material for adaptation to local conditions. In the long term, populations with more genetic diversity may be more likely to persist, because their variation better enables them to cope with environmental change.

Populations with little genetic diversity are vulnerable to environmental change if they lack genetic variants to help them adapt to changing conditions. Populations with low genetic diversity may also show less vigor, be more vulnerable to disease, and suffer *inbreeding depression,* which occurs when genetically similar parents mate and produce weak or defective offspring. Scientists have sounded warnings over low genetic diversity in species that have dropped to low population sizes, including American bison, elephant seals, and the cheetahs of the East African plains. Diminished genetic diversity in our crop plants is a prime concern to humanity (p. 138).

**Ecosystem diversity** Above the species level, **ecosystem diversity** refers to the number and variety of ecosystems (pp. 32, 56), but biologists may also refer to the diversity of communities (pp. 56, 71) or habitats (p. 56) within some specified area. Scientists may also consider the geographic arrangement of habitats, communities, or ecosystems at the landscape level, including the sizes and shapes of patches and the connections among them (p. 34). Under any of these concepts, a seashore of beaches, forested cliffs, offshore coral reefs, and ocean waters would hold more biodiversity than the same acreage of a monocultural cornfield. A mountain slope whose vegetation changes with elevation from desert to forest to alpine meadow would hold more biodiversity than a flat area the same size consisting of only desert, forest, or meadow.

The Serengeti's open plains are vast, but the region holds a diversity of habitats, including savanna (p. 85), grassland (p. 83), hilly woodlands, seasonal wetlands, and rock outcroppings. This habitat diversity contributes to the rich diversity of species in the region.

## Biodiversity is unevenly distributed

In numbers of species, insects show a staggering predominance over all other forms of life (**FIGURE 8.3**). Among insects, about 40% are beetles, and beetles alone outnumber all non-insect animals and all plants. No wonder the British biologist J.B.S. Haldane famously quipped that God must have had "an inordinate fondness for beetles."

Biodiversity is also greater in some places than others. Near the equator, greater amounts of solar energy, heat, and humidity spur plant growth, making tropical regions more productive than temperate regions and able to support larger numbers of organisms. Species diversity generally is higher near the equator, likely because the steady amount of sunlight year-round and the relatively stable climates of tropical regions allow numerous species to coexist. Whereas variable environmental conditions favor generalists (species that can tolerate a wide range of circumstances), stable conditions favor specialists (species highly adapted to particular circumstances).

Structurally diverse habitats tend to create more ecological niches (p. 56) and support greater species diversity. For instance, forests generally support more diversity than grasslands. For any given area, species diversity tends to increase with diversity of habitats, because each habitat supports a different mix of organisms.

Human disturbance often creates patchwork combinations of habitats. This increases habitat diversity locally; so in moderately disturbed areas, species diversity can rise. However, at larger scales, human disturbance decreases diversity because it replaces regionally unique habitats with homogenized disturbed habitats, causing many specialist species to disappear while a relative few generalist species thrive. Moreover, species that rely on large expanses of habitat disappear when those habitats are fragmented by human disturbance.

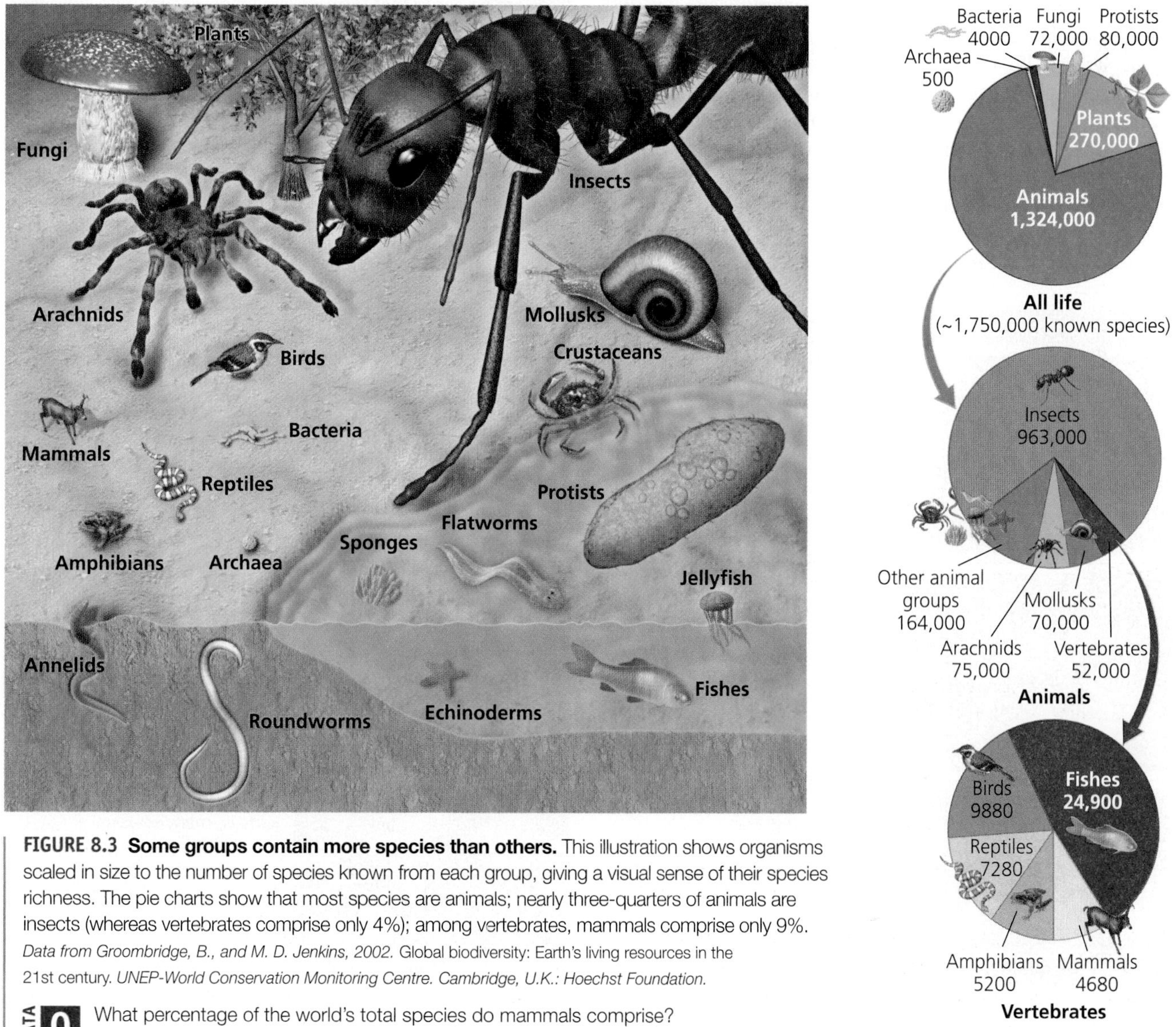


**FIGURE 8.3 Some groups contain more species than others.** This illustration shows organisms scaled in size to the number of species known from each group, giving a visual sense of their species richness. The pie charts show that most species are animals; nearly three-quarters of animals are insects (whereas vertebrates comprise only 4%); among vertebrates, mammals comprise only 9%. *Data from Groombridge, B., and M. D. Jenkins, 2002.* Global biodiversity: Earth's living resources in the 21st century. *UNEP-World Conservation Monitoring Centre. Cambridge, U.K.: Hoechst Foundation.*

DATA Q What percentage of the world's total species do mammals comprise?

GO TO **INTERPRETING GRAPHS & DATA** ON MasteringEnvironmentalScience®

## Many species await discovery

We still are profoundly ignorant of the number of species that exist. So far, scientists have identified and described about 1.8 million species of plants, animals, fungi, and microorganisms. However, estimates for the total number that actually exist range from 3 million to 100 million, with the most widely accepted estimates in the neighborhood of 14 million.

Our knowledge of species numbers is incomplete for several reasons. First, many species are tiny and easily overlooked. These include bacteria, nematodes, fungi, protists, and soil-dwelling arthropods. Second, many organisms are difficult to identify; sometimes, organisms that are thought to be of the same species turn out to be different species once biologists examine them more closely. Third, some areas of Earth remain little explored. We have barely sampled the ocean depths, hydrothermal vents (p. 259), or the tree canopies and soils of tropical forests. There remain many frontiers to explore!

# Benefits of Biodiversity

These days most of us live in cities and suburbs, spend nearly all our time indoors, and pass hours each day staring at electronic screens. It's no wonder that we often fail to appreciate how biodiversity relates to our lives! Yet our cities, homes, and technology could simply not exist without the resources and services that Earth's living species provide us—and neither could we. But what can we say more specifically about the direct ways that biodiversity helps us? Let's survey some of the major tangible and pragmatic ways in which biodiversity benefits people and supports our society.

## Biodiversity enhances food security

Biodiversity provides the food we eat. Throughout our history, human beings have used at least 7000 plant species and several thousand animal species for food. Today industrial agriculture has narrowed our diet. Globally, we now get 90% of our food from just 15 crop species and eight livestock species, and this lack of diversity leaves us vulnerable to crop failures. In a world where nearly 1 billion people go hungry, we can improve food security (the guarantee of an adequate, safe, nutritious, and reliable food supply; p. 136) by finding sustainable ways to harvest or farm wild species and rare crop varieties.

TABLE 8.1 shows a selection of promising food resources from just one region of the world—Central and South America. Plenty more exist here and elsewhere worldwide. The babassu palm of the Amazon produces more vegetable oil than any other plant. The serendipity berry generates a sweetener 3000 times sweeter than table sugar. Some salt-tolerant grasses and trees are so hardy that farmers can irrigate them with salt water to produce animal feed and other products.

Moreover, the wild relatives of our crops hold reservoirs of genetic diversity that can help protect the crops we grow in monocultures by providing helpful genes for crossbreeding or genetic engineering (p. 138). We have already received tens of billions of dollars' worth of disease resistance from the wild relatives of potatoes, wheat, corn, barley, and other crops.

**TABLE 8.1 Potential New Food Sources**

**Amaranths**
(three species of *Amaranthus*)

Grain and leafy vegetable; livestock feed; rapid growth, drought resistant

**Capybara**
(*Hydrochoeris hydrochaeris*)

World's largest rodent; meat esteemed; easily ranched in open habitats near water

**Buriti palm**
(*Mauritia flexuosa*)

"Tree of life" to Amerindians; vitamin-rich fruit; pith as source for bread; palm heart from shoots

**Vicuna**
(*Lama vicugna*)

Threatened species related to llama; source of meat, fur, and hides; can be profitably ranched

**Maca**
(*Lepidium meyenii*)

Cold-resistant root vegetable resembling radish, with distinctive flavor; near extinction

**Chachalacas**
(*Ortalis*, many species)

Tropical birds; adaptable to human habitations; fast-growing

*The wild species shown here are just some of the many plants and animals that could supplement our food supply.*
*Adapted from Wilson, E.O., 1992.* The diversity of life. *Cambridge, MA: Belknap Press.*

**TABLE 8.2 Natural Plant Sources of Pharmaceuticals**

| | | | |
|---|---|---|---|
|  | **Pineapple** (*Ananas comosus*)<br>**Drug:** Bromelain<br>**Application:** Controls tissue inflammation |  | **Pacific yew** (*Taxus brevifolia*)<br>**Drug:** Taxol<br>**Application:** Anticancer agent (especially ovarian cancer) |
|  | **Autumn crocus** (*Colchicum autumnale*)<br>**Drug:** Colchicine<br>**Application:** Anticancer agent |  | **Velvet bean** (*Mucuna deeringiana*)<br>**Drug:** L-Dopa<br>**Application:** Parkinson's disease suppressant |
|  | **Yellow cinchona** (several species of *Cinchona*)<br>**Drug:** Quinine<br>**Application:** Antimalarial agent |  | **Common foxglove** (*Digitalis purpurea*)<br>**Drug:** Digitoxin<br>**Application:** Cardiac stimulant |

*Shown are just a few of the many plants that provide chemical compounds of medical benefit.*
*Adapted from Wilson, E. O., 1992.* The diversity of life. *Cambridge, MA: Belknap Press.*

## Organisms provide drugs and medicines

People have made medicines from plants and animals for centuries, and many of today's pharmaceuticals are derived from chemical compounds from wild plants (TABLE 8.2). The rosy periwinkle produces compounds that treat Hodgkin's disease and a deadly form of leukemia. Had this plant from Madagascar become extinct, these two fatal diseases would have claimed far more victims. The Pacific yew of the Pacific Northwest produces a compound that forms the basis for the anti-cancer drug taxol. Aspirin was derived from chemicals found in willows and meadowsweet.

Each year, pharmaceutical products owing their origin to wild species generate up to $150 billion in sales and save thousands of human lives. The world's biodiversity holds a still-greater treasure chest of medicines still to be discovered. Yet with every species that goes extinct, we lose one more opportunity to find cures and treatments.

## Biodiversity provides ecosystem services

Contrary to popular opinion, some things in life can indeed be free—as long as we protect the ecological systems that provide them. Forests provide clean air and water, and they buffer hydrologic systems against flooding and drought. Native crop varieties provide insurance against disease and drought. Wildlife can attract tourism that boosts economies. Intact ecosystems provide these and other valuable processes, known as ecosystem services (pp. 3, 35), for all of us, free of charge.

According to scientists, biodiversity:

- Provides food, fuel, fiber, and shelter
- Purifies air and water
- Detoxifies and decomposes wastes
- Stabilizes Earth's climate
- Moderates floods, droughts, and temperatures
- Cycles nutrients and renews soil fertility
- Pollinates plants, including many crops
- Controls pests and diseases
- Maintains genetic resources for crop varieties, livestock breeds, and medicines
- Provides cultural and aesthetic benefits
- Gives us the means to adapt to change

In these ways, organisms and ecosystems support vital processes that people cannot replicate or would need to pay for if nature did not provide them. The economic value of just 17 of these ecosystem services has been estimated at over $143 trillion per year (p. 97).

## Biodiversity helps maintain ecosystem function

Ecological research demonstrates that biodiversity tends to enhance the stability of communities and ecosystems. Research has also found that biodiversity tends to increase the resilience (p. 77) of ecological systems—their ability to withstand disturbance, recover from stress, or adapt to change. Thus, the loss of biodiversity can diminish a natural system's ability to function and to provide services to our society.

Will the loss of a few species really make much difference in an ecosystem's ability to function? Consider a metaphor first offered by Paul and Anne Ehrlich (p. 119): The loss of one rivet from an airplane's wing—or two, or three—may not cause the plane to crash. But as rivets are removed the structure will be compromised, and eventually the loss of just one more rivet will cause it to fail.

Research shows that removing a keystone species (p. 76) will significantly alter an ecological system, because other species may disappear in response. Predators at the tops of food chains—such as lions, leopards, and cheetahs on the Serengeti—prey on herbivores, each of which consume many plants. The removal of a top predator can have consequences that multiply as they cascade down the food chain.

The loss of "ecosystem engineers" (such as ants or earthworms; p. 76) from a system can likewise set major changes in motion. A decline in wildebeest on the Serengeti would have consequences for grass growth, shrub invasion of grasslands, fire regime, soil quality, and populations of other animals. Scientists have already documented how African savannas can morph into scrub forests when poaching removes elephants. Ecosystems are complex, and it is difficult to predict which species may be most influential. Thus, many people prefer to apply the precautionary principle (p. 155) in the spirit of Aldo Leopold (p. 14), who advised, "To keep every cog and wheel is the first precaution of intelligent tinkering."

## Biodiversity boosts economies through tourism and recreation

When people travel to observe wildlife and explore natural areas, they create economic opportunities for area residents. Visitors spend money at local businesses, hire local people as guides, and support parks that employ the region's residents. The parks and wildlife of Kenya and Tanzania are prime examples. Ecotourism (p. 63) brings in fully a quarter of all foreign money entering Tanzania's economy each year. Leaders and citizens in both nations who recognize biodiversity's economic benefits have managed their parks and reserves diligently.

Ecotourism is a vital source of income for many nations, including Costa Rica, with its rainforests; Australia, with its Great Barrier Reef; and Belize, with its caves and coral reefs. The United States, too, benefits from ecotourism; its national parks draw millions of visitors from around the world. Although excessive development for ecotourism can damage the natural assets that draw people, ecotourism can serve as a powerful financial incentive for nations and local communities to preserve natural areas and reduce impacts on their wildlife and landscapes.

## People value connections with nature

Not all of biodiversity's benefits to people can be expressed in the hard numbers of economics or the practicalities of food and medicine. Some scientists and philosophers argue that people find a deeper value in biodiversity. Harvard University biologist Edward O. Wilson has popularized the notion of **biophilia**, asserting that human beings share an instinctive love for nature and feel an emotional bond with other living things (**FIGURE 8.4**). Wilson and others cite as evidence of biophilia our affinity for parks and wildlife, our love for pets, the high value of real estate with a view of natural landscapes, and our interest in hiking, bird-watching, fishing, hunting, backpacking, and similar outdoor pursuits.

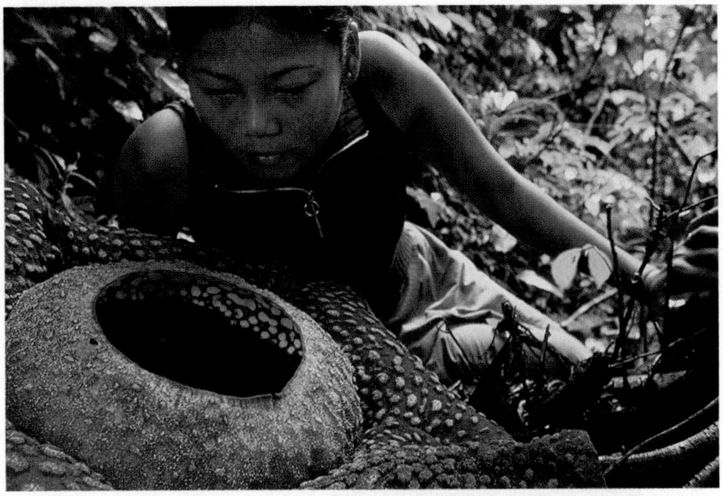

**FIGURE 8.4 An Indonesian girl gazes into a flower of *Rafflesia arnoldii*, the largest flower in the world.** Biophilia holds that human beings have an instinctive love and fascination for nature and a deep-seated desire to affiliate with other living things.

In a 2005 book, writer Richard Louv added that as today's children are increasingly deprived of outdoor experiences and direct contact with wild organisms, they suffer what he calls "nature-deficit disorder." Louv argues that this alienation from biodiversity and nature damages childhood development and may lie behind many of the emotional and psychological challenges young people in developed nations face today.

## Do we have ethical obligations toward other species?

Aside from all of biodiversity's pragmatic benefits, many people feel that living organisms simply have an inherent right to exist. Human beings are part of nature, and like any other animal we need to use resources and consume other organisms to survive. However, we also have conscious reasoning ability and can make deliberate decisions. Our ethical sense has developed from this intelligence and ability to choose. As our society's sphere of ethical consideration has widened, and as more people take up biocentric or ecocentric worldviews (pp. 12–13), more have come to feel that other organisms have intrinsic value. In this view, the conservation of biodiversity is justified on ethical grounds alone.

# Biodiversity Loss and Extinction

Despite our expanding ethical convictions and the many benefits biodiversity brings us, the future of life remains far from secure. Biological diversity is being rapidly lost to human

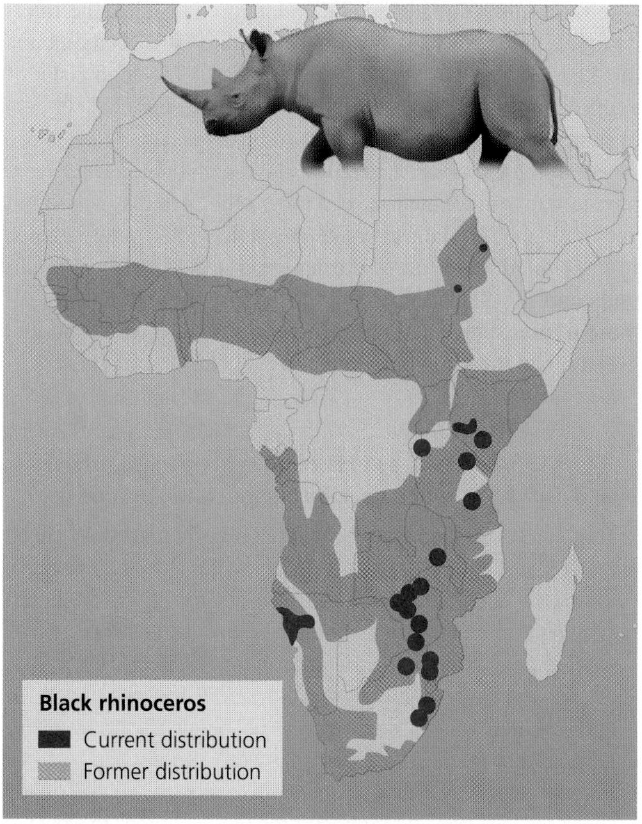


**FIGURE 8.5 The black rhinoceros has disappeared from most of its range across Africa.** *Based on data from Deon Furstenburg, wildliferanching.com, and other sources.*

impact, and the losses are irretrievable when species go extinct. **Extinction** (p. 54) occurs when the last member of a species dies and the species ceases to exist.

The disappearance of a population from an area, but not the entire species globally, is referred to as *extirpation* or *local extinction*. The black rhinoceros has been extirpated from most of its historic range across Africa (**FIGURE 8.5**), but as a species it is not yet extinct. However, at least three of its subspecies are extinct. For example, no individuals of the western black rhino have been seen since 2006, and this subspecies is now thought to be extinct. Extirpation is an erosive process that can, over time, lead to extinction.

Human impact is responsible for most extirpation and extinction today, but these processes also occur naturally at a much slower rate. If species did not naturally go extinct, our world would be filled with dinosaurs, trilobites, and millions of other creatures that vanished from Earth long before humans appeared. Paleontologists estimate that roughly 99% of all species that ever lived are now extinct. Thus, the wealth of species gracing our planet today represents just 1% of the species that have ever existed!

Most extinctions preceding the appearance of human beings occurred singularly for independent reasons, at a pace referred to as the *background extinction rate* (p. 54). By studying traces of organisms preserved in the fossil record (pp. 53–54), scientists infer that for mammals and marine animals, each year, on average, 1 species out of every 1–10 million has vanished.

## Earth has experienced five mass extinction events

Extinction rates rose far above this background rate at several points in Earth's history. In the past 440 million years, our planet experienced five major **mass extinction events** (pp. 54–55). Each event eliminated more than one-fifth of life's families and at least half its species (**TABLE 8.3**). The most severe episode occurred at the end of the Permian period (see **APPENDIX E**). At this time, 250 million years ago, close to 90% of all species went extinct. The best-known episode occurred 66 million years ago at the end of the Cretaceous period, when an asteroid impact (and possibly volcanism) brought an end to the dinosaurs and many other groups.

If current trends continue, the modern era, known as the Quaternary period, may see the extinction of more than half of all species. Although similar in scale to previous mass extinctions, today's ongoing mass extinction is different in two primary respects. First, we are causing it. Second, we will suffer as a result.

**TABLE 8.3 Mass Extinctions**

| EVENT | DATE (MILLIONS OF YEARS AGO [MYA]) | CAUSE | TYPES OF LIFE MOST AFFECTED | PERCENTAGE OF LIFE DEPLETED |
|---|---|---|---|---|
| Ordovician | 440 mya | Unknown | Marine organisms; terrestrial record is unknown | >20% of families |
| Devonian | 360 mya | Unknown | Marine organisms; terrestrial record is unknown | >20% of families |
| Permo-Triassic | 250 mya | Possibly volcanism | Marine organisms; terrestrial record is less known | >50% of families; 80–95% of species |
| End-Triassic | 200 mya | Unknown | Marine organisms; terrestrial record is less known | 20% of families; 50% of genera |
| Cretaceous-Paleogene | 66 mya | Likely asteroid impact | Marine and terrestrial organisms, including dinosaurs | 5% of families; >50% of species |
| Current | Beginning 0.01 mya | Human impacts | Large animals, specialized organisms, island organisms, organisms harvested by people | Ongoing |

**FIGURE 8.6 The ivory-billed woodpecker was one of North America's most majestic birds.** It lived in old-growth forests of the southeastern United States. Forest clearing and timber harvesting eliminated the mature trees it needed for food, shelter, and nesting, and this symbol of the South appeared to go extinct. Recent fleeting, controversial observations raised hopes that the species persists, but proof has been elusive.

## We are setting the sixth mass extinction in motion

In the past few centuries alone, history has recorded hundreds of instances of species extinction caused by humans. Among North American birds in the past two centuries, we have driven into extinction the Carolina parakeet, great auk, Labrador duck, passenger pigeon (p. 57), almost certainly the Bachman's warbler and Eskimo curlew, and likely the ivory-billed woodpecker (**FIGURE 8.6**). Several more species, including the whooping crane, Kirtland's warbler, and California condor (p. 180), teeter on the brink of extinction.

However, people may have been hunting species to extinction for thousands of years. Archaeological evidence shows that in case after case, a wave of extinction followed close on the heels of human arrival on islands and continents. After Polynesians reached Hawaii, half its birds went extinct. Birds, mammals, and reptiles vanished following human arrival on many other oceanic islands, including large island masses such as New Zealand and Madagascar. Dozens of species of large vertebrates died off in Australia after people arrived roughly 50,000 years ago. North America lost 33 genera of large mammals (such as camels, lions, horses, and giant ground sloths) after people arrived more than 10,000 years ago.

Today, species loss is accelerating as our population growth and resource consumption put increasing strain on habitats and wildlife. In 2005, scientists with the Millennium Ecosystem Assessment calculated that the current global extinction rate is 100 to 1000 times greater than the background extinction rate, and rising.

To monitor threatened and endangered species, the International Union for Conservation of Nature (IUCN) maintains the **Red List,** an updated list of species facing high risks of extinction. As of 2014, the Red List reported that 22% (1194) of mammal species, 13% (1308) of bird species, 31% (1961) of amphibian species, and 20% (1975) of fish species were threatened with extinction. For most other groups, we do not yet have enough data to make accurate global assessments. In the United States alone during the past 500 years, 236 animals and 30 plants are known to have gone extinct. For all these figures, the actual numbers of species are without doubt greater than the known numbers.

**FAQ** **If a mass extinction is happening, why don't I notice species disappearing around me?**

There are two reasons that most of us don't personally sense the scale of biodiversity loss. First, if you live in a city or suburb, the plants and animals you see from day to day are generalist species that thrive in disturbed areas. In contrast, the species most in trouble are those that rely on less-disturbed habitats.

Second, a human lifetime is very short! The loss of populations and species may seem slow to us, but on Earth's timescale it is sudden. Because each of us is born into a world that has already lost many species, we don't recognize what's already vanished. Likewise, our grandchildren won't appreciate what we lose in our lifetimes. Each human generation experiences just a portion of the overall phenomenon, so we have difficulty sensing the big picture. Nonetheless, researchers and naturalists who spend their time outdoors observing nature see biodiversity loss around them all the time—and that's precisely why they feel so passionate about preventing it.

## Biodiversity loss involves population declines

Extinction is only part of the story of biodiversity loss. The larger part involves declining population sizes. As a population shrinks, it loses genetic diversity, which may make it vulnerable to further declines. Moreover, as a population shrinks, its geographic range often gets smaller as the species disappears from parts of its range. Thus, many species today are less numerous and occupy less area than they once did. This is true of nearly all the large mammal species in Africa, and scientific studies have documented significant population declines among large mammals of the Serengeti in recent years (see **THE SCIENCE BEHIND THE STORY**, pp. 172–173).

To quantify and summarize such change globally, scientists at the World Wildlife Fund and the United Nations Environment Programme (UNEP) developed the *Living Planet Index*. This index expresses how large the average population size of a species is now, relative to its size in the baseline year of 1970. The most recent compilation summarized trends from populations of 1432 terrestrial species, 737 freshwater species, and 675 marine species that are sufficiently monitored. Between 1970 and 2008, the Living Planet Index fell by 28%—meaning that on average, population sizes became 28%

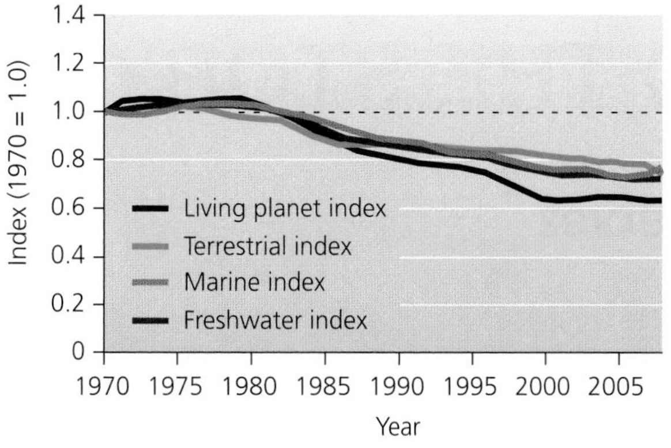


**FIGURE 8.7 The Living Planet Index is an indicator of the state of global biodiversity.** Index values summarize trends for 9014 populations of 2688 vertebrate species. Between 1970 and 2008, the Living Planet Index fell by 28%. The index for terrestrial species fell by 25%; for freshwater species, by 37%; and for marine species, by 22%. *Data from WWF, 2012.* Living planet report 2012. *WWF International, Gland, Switzerland.*

smaller than they were just four decades earlier (**FIGURE 8.7**). This decline was driven primarily by losses in tropical regions. In the temperate zones (where forests are regrowing, pollution is being controlled, and ecological restoration is taking hold in industrialized countries) the index rose by 31%. In tropical regions (where deforestation is rife), it decreased by 61%.

## Several major causes of biodiversity loss stand out

Scientists have identified four primary causes of population decline and species extinction: habitat loss, pollution, overharvesting, and invasive species. Global climate change (Chapter 14) now is becoming the fifth. Each cause is intensified by human population growth and by our increasing per capita consumption of resources.

**Habitat loss** Habitat loss is the single greatest cause of biodiversity decline today. Habitat is lost not only when it is destroyed outright, but also when it is fragmented or degraded. Because each organism has adapted over thousands or millions of years to the habitat in which it lives, any major change in the habitat is likely to render the habitat less suitable.

Many human activities alter, degrade, or destroy habitat. Farming replaces diverse natural communities with simplified ones of just a few plant species. Grazing modifies grasslands and can lead to desertification (p. 142). Clearing forests removes the food and shelter that forest-dwelling organisms need to survive. Dams turn rivers into reservoirs upstream while affecting water conditions and floodplain communities downstream. Urban sprawl supplants natural ecosystems, driving many species from their homes.

Habitat loss occurs most commonly by gradual, piecemeal degradation, such as **habitat fragmentation** (**FIGURE 8.8**). When farming, logging, road building, or development intrudes into an unbroken expanse of forest or grassland, this breaks up a continuous area of habitat into fragments, or patches. As habitat fragmentation proceeds across a landscape, animals and plants requiring the habitat disappear from one fragment after another. Fragmentation can also block the movement of animals from place to place; this is the concern of opponents of the proposed highway through the Serengeti. In response to habitat fragmentation, conservationists try to link fragments with corridors of habitat along which animals can travel. In Chapter 9 we will learn more about the fragmentation of forests, the effects on wildlife, and potential solutions (pp. 199–203).

Habitat loss affects almost all the world's biomes. Over half of the world's temperate forests, grasslands, and shrublands had been converted by 1950 (mostly for agriculture). Today habitat is being lost most rapidly in tropical rainforests, tropical dry forests, and savannas. Habitat loss is the primary source of population declines in 83% of threatened mammals and 85% of threatened birds, according to UNEP data. For

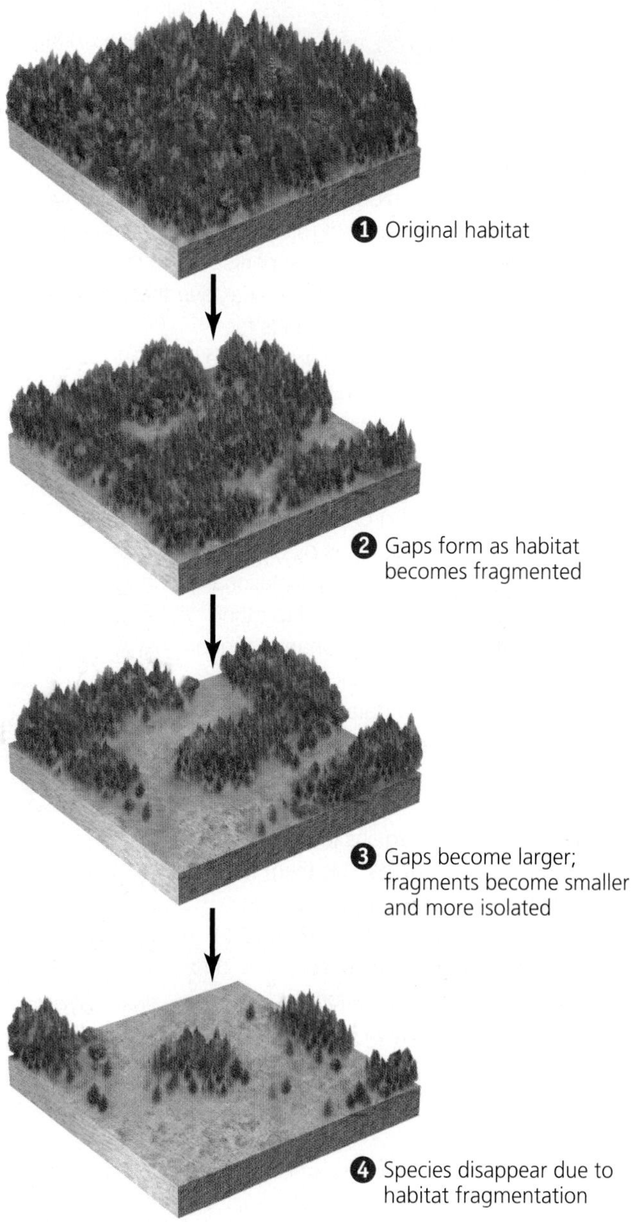


**FIGURE 8.8 Habitat fragmentation occurs as human impact creates gaps that expand and eventually come to dominate the landscape, stranding islands of habitat.** As habitat becomes fragmented, fewer populations can persist, and numbers of species in the fragments decline.

THE SCIENCE BEHIND THE STORY

## Wildlife Declines in African Reserves

**Zebras in Serengeti National Park**

Tanzania and Kenya have some of the world's most famous parks and protected areas, with the greatest variety and density of large mammals to be found anywhere. The parks are sizeable, well managed, and well funded. Yet even these places of refuge are not immune to pressures from rising human population, development, and resource extraction.

For several decades, biologists and park managers have censused wildlife in and around the parks and reserves. In recent years, researchers have been analyzing these long-term data sets to assess population trends for large mammals of the East African savanna. These studies are finding that most animals are declining in number—inside the parks and reserves as well as outside them.

In Tanzania, several government agencies and nonprofit groups have collaborated to census mammals by airplane. Aerial surveys began in Serengeti National Park in the 1970s and expanded to other parks in the 1980s. In 2006, Chantal Stoner and six Tanzanian and American colleagues compiled and analyzed data on 25 species over a 10-year period (roughly 1990–2000) from eight regions, each centered on a major park.

Across the eight regions in wet and dry seasons for all species, this team found that population declines outnumbered increases by more than 10 to 1 (**FIGURE 1**). In the Serengeti region, five species declined while two increased. Population declines were greater outside parks and in areas that received less protection. However, even within the boundaries of well-protected reserves, many species decreased in number.

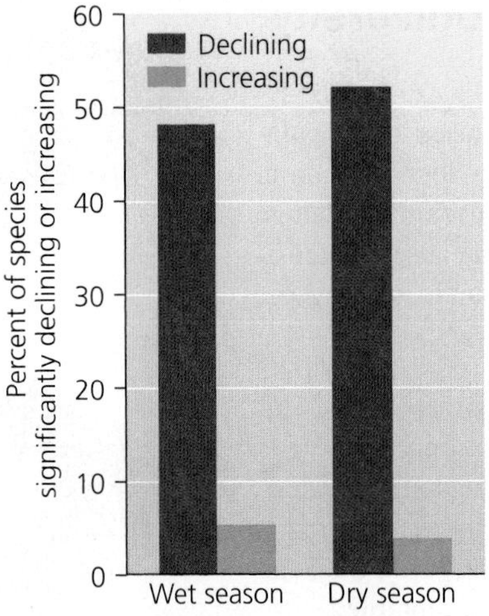


**FIGURE 1 Across protected areas in Tanzania, more animals have decreased than have increased.** *Data are combined from six locations, from Stoner, C., et al., 2006. Changes in large herbivore populations across large areas of Tanzania.* African Journal of Ecology *45: 202–215.*

Why are animals declining? For decades East Africa has had one of the world's fastest rates of human population growth, and this has intensified pressures on wildlife and ecosystems:

- Settlements increase as nomadic Maasai herders become sedentary and as people from elsewhere arrive.
- Farmers convert grasslands to crops (especially wheat). This destroys habitat for antelope, wildebeest, and the predators that follow them.
- Livestock compete with wild grazing animals for food on the grasslands.
- Local residents kill animals for food for their own subsistence, while criminal gangs poach animals and export "bush meat," elephant tusks, and rhino horns to rich consumers abroad.

In Kenya, researchers are seeing patterns similar to those in Tanzania. In 2009, conservation biologist David Western and two colleagues reviewed 30 years of data from aerial surveys conducted from 1977 to 2007 across Kenya's rangelands, which comprise three-quarters of the nation's land area. In most locations, Western's group found downward trends in the populations of most species. Cycles of rainfall and drought affect grazing mammals, but statistical analysis of the data indicated that the long-term declines transcended short-term effects due to rain or drought.

Moreover, the populations surveyed by Western's team were trending downward both inside and outside the parks (**FIGURE 2**). Many species, including wildebeest, zebras, and some antelope, migrate in and out of parks, so factors that affect their populations

example, the prairies of North America's Great Plains are today almost entirely converted to agriculture. Less than 1% of original prairie habitat remains. As a result, grassland bird populations have declined by an estimated 82–99%.

Of course, our habitat alteration benefits some species. Animals such as house sparrows, pigeons, starlings, raccoons, gray squirrels, rats, and cockroaches thrive in cities and towns. However, the species that benefit from our modification of natural habitats are relatively few; for every species that wins, more lose. Moreover, the species that do well in our midst tend to be weedy generalists that are in little danger of disappearing.

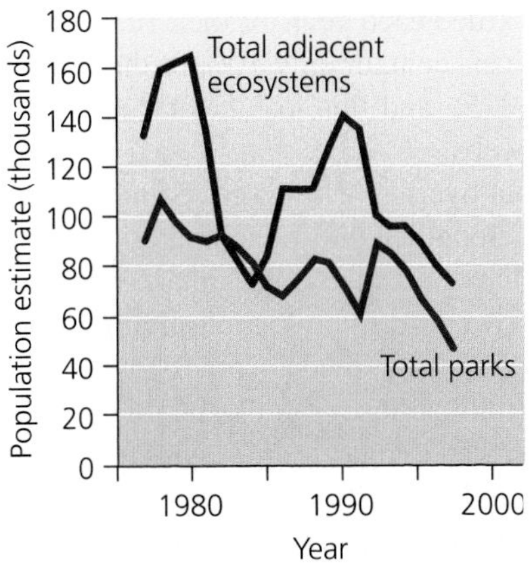


**FIGURE 2 Wildlife populations have declined across Kenya, both inside and outside parks.** *Data from Western, D., et al., 2009. The status of wildlife in protected areas compared to non-protected areas of Kenya.* PLoS ONE *4(7): e6140.*

outside the parks will affect numbers within park boundaries as well.

Researchers have given particular attention to the Maasai Mara National Reserve, on the Kenyan side of the border adjacent to Serengeti National Park. A decade ago Dutch scientist Wilber Ottichilo analyzed aerial survey data and concluded that nonmigratory mammals declined by 58% from 1977 to 1997 within the reserve and by a similar amount outside the reserve. Resident wildebeest fared even worse, showing an 81% decline.

More recently, Joseph Ogutu of the International Livestock Research Institute in Nairobi, Kenya, and his colleagues extended the analysis to 2009 and found that most species had continued declining, both on and off the reserve, and were now at just one-third of their 1977 population sizes (**FIGURE 3**). As elsewhere, drought accounted only for short-term fluctuations, while the long-term declines were due to habitat loss to farms, intensified human settlement near the parks, poaching, and competition with livestock.

As a result of studies like these, researchers have concluded that merely setting aside parks is not adequate to conserve wildlife and ecosystems. Instead, we need to view the big picture and consider animals' needs across the landscape, as well as how human impacts might spill into reserves. Conservation success thus requires linking reserves with corridors of habitat that animals can use, and it requires working with local people living near protected areas. Community-based conservation (p. 182) encourages people to be stewards, managing natural resources in the areas they live, and this approach shows promise for sustaining wildlife populations. □

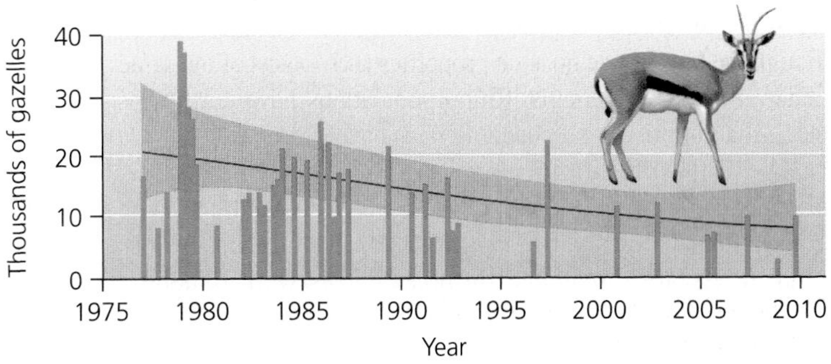


**(a) Inside the reserve**

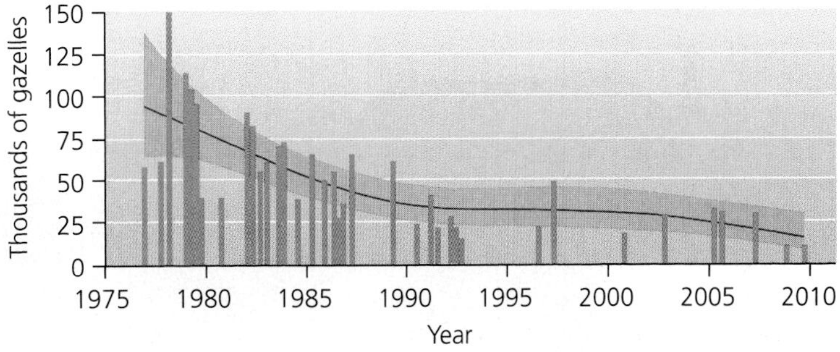


**(b) Outside the reserve**

**FIGURE 3 Thomson's gazelles, the most abundant species at Kenya's Maasai Mara National Reserve, decreased by 59% (a) inside the reserve, and by 77% (b) outside the reserve.** *Data from Ogutu, Joseph, et al., 2011. Continuing wildlife population declines and range contraction in the Mara region of Kenya during 1977–2009.* Journal of Zoology *285: 99–109.*

**Pollution** Pollution can harm organisms in many ways. Air pollution degrades forest ecosystems and affects the atmosphere and climate. Noise and light pollution impinge on the behavior and habitat use of animals. Water pollution impairs fish and amphibians. Agricultural runoff containing fertilizers, pesticides, and sediments harms many terrestrial and aquatic species. Heavy metals, endocrine-disrupting compounds, and other toxic chemicals poison people and wildlife. Plastic garbage in the ocean can strangle, drown, or choke marine creatures. The effects of oil spills on wildlife are dramatic and well known. We examine all these impacts in other chapters of this book. However, although pollution is a substantial threat, it tends to be less significant than public perception holds it to be, and it is far less influential than habitat loss.

**Overharvesting** For most species, being hunted or harvested will not in itself pose a threat of extinction. However, species that are long-lived and slow to reproduce—such as elephants, rhinoceroses, and other large mammals of the African savanna—are vulnerable to overhunting. People have long killed elephants to extract their tusks for ivory (**FIGURE 8.9**). By 1989, 7% of African elephants were being killed each year, so the world's nations enacted a global ban on the commercial trade of ivory. Elephant numbers recovered following the ban, but since 2007 poaching has increased to all-time highs, driven by high black-market prices for ivory paid by wealthy overseas buyers. From 2011 through 2013, well over 75,000 African elephants were killed—enough to send populations downward and threaten the species' future.

Poachers kill rhinoceroses for their horns, which are ground into powder and sold illegally to ultra-wealthy Asian consumers as cancer cures, party drugs, and hangover treatments, even though the horns have no such properties and consist of the same material as our fingernails. Rhino populations have crashed as black-market prices and poaching rates have skyrocketed.

With the illegal global trade in wildlife products surpassing $20 billion per year, poaching has led to steep population declines for many large-bodied animals. Across Asia, tigers are threatened by poaching as well as habitat loss; body parts from one tiger can fetch a poacher $15,000 on the black market, where they are sold as aphrodisiacs in Asian countries. Today half the world's tiger subspecies are extinct, and most of the remaining animals are crowded onto just 1% of the land they occupied historically.

In central Africa, gorillas and other primates are killed for their "bush meat" and could soon face extinction. In the oceans, many fish stocks today are overharvested (pp. 274–275). Whaling drove the Atlantic gray whale extinct and has left several other types of whales threatened or endangered. Thousands of sharks are killed each year for their fins, which are used in soup. The oceans contain only 10% of the large animals they once did (p. 275), and this loss has far-reaching impacts on marine food webs.

To combat overharvesting, governments pass laws, sign treaties, and strengthen anti-poaching efforts. Still, when demand from rich consumers raises market prices for wildlife products, and with so many people living in poverty, many individuals will be tempted to break laws to make money, and officials may choose to look the other way. In much of Africa today, protecting wildlife is a dangerous job. Poaching is conducted with brutal efficiency by organized crime syndicates using helicopters, night-vision goggles, and automatic weapons. Park rangers are heavily armed, yet are routinely outgunned in firefights with poachers, and many have lost their lives. In this way the demand for luxury goods by wealthy foreign consumers half a world away has grave consequences for Africans living in regions like the Serengeti.

**Invasive species** When non-native species are introduced to new environments, some may become invasive (pp. 78–80) and push native species toward extinction (**TABLE 8.4**). Some introductions are accidental, such as weeds whose seeds cling to our socks as we travel from place to place, and aquatic organisms such as zebra mussels, transported in the ballast water of ships (Chapter 4). Other introductions are intentional. People have long brought food crops and animals with

**FIGURE 8.9 Poachers slaughter elephants to sell their tusks for ivory.** Despite the ivory trade ban, poaching is at an all-time high. Here Kenyan officials at Maasai Mara National Reserve prepare to set fire to tusks confiscated from poachers, in an effort to discourage the trade.

**TABLE 8.4 Invasive Species**

**European gypsy moth**
(*Lymantria dispar*)

Introduced to Massachusetts in the hope it could produce silk. The moth failed to do so, and instead spread across the eastern United States, where its outbreaks defoliate trees over large regions every few years.

**Asian long-horned beetle**
(*Anoplophora glabripennis*)

Since the 1990s, has repeatedly arrived in North America in imported lumber. These insects burrow into wood and can kill the majority of trees in an area. Chicago, Seattle, Toronto, New York, and other cities have cleared thousands of trees to eradicate these invaders.

**European starling**
(*Sturnus vulgaris*)

Introduced to New York City in the 1800s by Shakespeare devotees intent on bringing every bird mentioned in Shakespeare's plays to America. Outcompeting native birds for nest holes, within 75 years starlings became one of North America's most abundant birds.

**Emerald ash borer**
(*Agrilus planipennis*)

Discovered in Michigan in 2002, this wood-boring insect reached 12 U.S. states and Canada by 2010, killing millions of ash trees in the upper Midwest. Billions of dollars will be spent in trying to control its spread.

**Cheatgrass**
(*Bromus tectorum*)

After introduction to Washington state in the 1890s, cheatgrass spread across the western United States. It crowds out other plants, uses up the soil's nitrogen, and burns readily. Fire kills many native plants, but not cheatgrass, which grows back stronger without competition.

**Sudden oak death**
(*Phytophthora ramorum*)

This disease has killed over 1 million oak trees in California since the 1990s. The pathogen (a water mold) was likely introduced via infected nursery plants. Scientists are concerned about damage to eastern U.S. forests if it spreads to oaks there.

**Brown tree snake**
(*Boiga irregularis*)

Nearly every native forest bird on the South Pacific island of Guam has disappeared, eaten by these snakes, which arrived from Asia as stowaways on ships and planes after World War II. Guam's birds had not evolved with snakes, and had no defenses against them.

**Nile perch**
(*Lates niloticus*)

A large fish from the Nile River. Introduced to Lake Victoria in the 1950s, it proceeded to eat its way through hundreds of species of native cichlid fish, driving a number of them extinct. People value the perch as food, but it has radically altered the lake's ecology.

**Kudzu**
(*Pueraria montana*)

A Japanese vine that can grow 30 m (100 ft) in a single season, the U.S. Soil Conservation Service introduced kudzu in the 1930s to help control erosion. Kudzu took over forests, fields, and roadsides throughout the southeastern United States.

**Polynesian rat**
(*Rattus exulans*)

One of several rat species that have followed human migrations across the world. Polynesians transported this rat to islands across the Pacific, including Easter Island (pp. 6–7). On each island it caused ecological havoc, and has driven extinct birds, plants, and mammals.

**FIGURE 8.10 The polar bear became the first species listed under the Endangered Species Act as a result of climate change.** As Arctic warming melts the sea ice from which they hunt seals, polar bears must swim farther and farther for food.

them as they colonized new places, and today we continue global trade in exotic pets and ornamental plants.

Most organisms introduced to new areas perish, but the few types that survive may do very well, especially if they are freed from the predators and parasites that attacked them back home or from the competitors that had limited their access to resources. Once released from such limiting factors (p. 62), an introduced species may proliferate and displace native species. Invasive species cause billions of dollars in economic damage each year.

In Lake Victoria near the Serengeti, the Nile perch was introduced as a food fish. It soon spread throughout the vast lake, preying on and driving extinct dozens of native species of cichlid fish from one of the world's most spectacular evolutionary radiations of animals. The Nile perch is providing people food, but at significant ecological cost.

If a highway is built through the Serengeti, ecologists fear that it would introduce weed seeds from passing vehicles. Park managers already are concerned about several American plants, such as datura, parthenium, and prickly poppy, that are toxic to native herbivores and that have spread rapidly in other African grasslands.

**Climate change** The preceding four types of human impacts affect biodiversity in discrete places and times. In contrast, our manipulation of Earth's climate (Chapter 14) is having global impacts. As we warm the atmosphere with emissions of greenhouse gases from fossil fuel combustion, we modify climate patterns and increase the frequency of extreme weather events.

Extreme weather events such as droughts and storms increase stress on populations. In the Arctic, melting sea ice and other impacts are threatening polar bears and people alike (**FIGURE 8.10** and p. 321). Across the world, warming temperatures are forcing organisms to shift their geographic ranges toward the poles and upward in altitude. Some species will not be able to adapt. Mountaintop organisms cannot move further upslope, so many may perish. Trees may not move toward the poles fast enough. As ranges shift, animals and plants encounter new communities of prey, predators, and parasites to which they are not adapted. All in all, scientists predict that a rise in global temperature of 1.5–2.5°C (2.7–4.5°F) could put 20–30% of the world's plants and animals at increased risk of extinction.

## A mix of causes threatens many species

For many species, multiple factors are conspiring to cause declines. The monarch butterfly, familiar to every American schoolchild, is today in precipitous decline (**FIGURE 8.11**). On its breeding grounds in the United States and Canada, industrial agriculture has eliminated most of the milkweed plants monarchs depend on. Our highly efficient monocultures leave no natural habitat remaining, while pesticides intended for crop pests also kill monarchs and other beneficial insects. In winter the entire monarch population migrates south and funnels into a single valley in Mexico, where the butterflies cluster by the millions in groves of tall trees. Here some people are illegally logging these forests while others fight to save the trees, the butterflies, and the ecotourism dollars they bring to the community.

Often, reasons for declines can be complex and difficult to determine. The worldwide collapse of amphibians provides an apparent example of a "perfect storm" of bewildering factors. Today entire populations of frogs, toads, and salamanders are vanishing without a trace. Over 40% of the

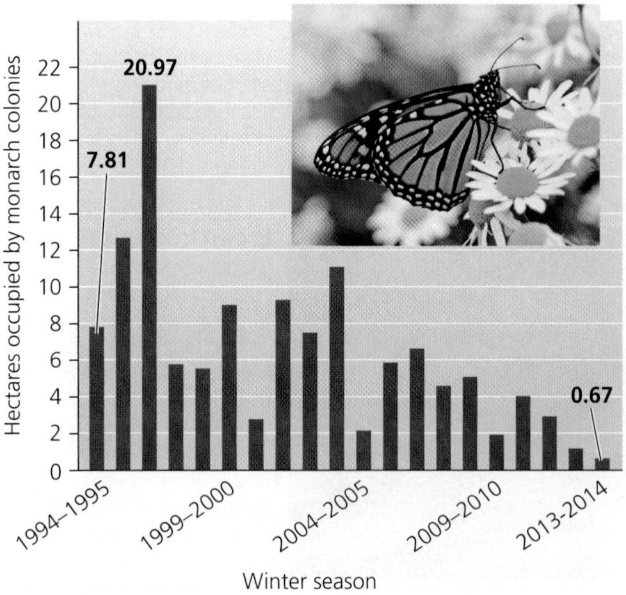


**FIGURE 8.11 The widespread and abundant monarch butterfly faces alarming declines due to pesticides and habitat loss in both summer and winter.** Annual surveys on its Mexican wintering grounds show the population occupying fewer and fewer hectares of forest. *Data from MonarchWatch, collected by the Monarch Butterfly Biosphere Reserve and World Wildlife Fund Mexico.*

**DATA Q** How much area did monarchs occupy in 2012–2013, as a proportion of the area occupied in 1994–1995? As a proportion of the area occupied in 1996–1997?

GO TO **INTERPRETING GRAPHS & DATA** ON MasteringEnvironmentalScience®

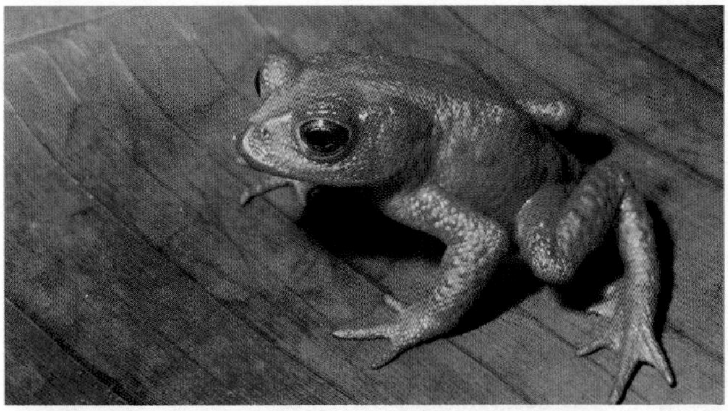

(a) Male golden toad from Monteverde, Costa Rica

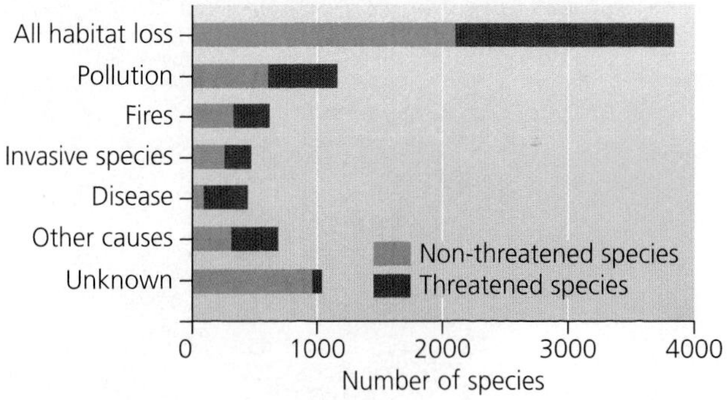


(b) Causes of amphibian declines

**FIGURE 8.12 The world's amphibians are declining.** The golden toad **(a)** is one of at least 170 species of amphibians that have suddenly gone extinct in recent years. This brilliant orange toad of Costa Rican cloud forests disappeared due to drought, climate change, and/or disease. Habitat loss **(b)** is the main reason for amphibian declines, but many declines remain unexplained. *Data from IUCN, 2008.* Global amphibian assessment.

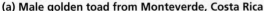

DATA Q What is the second greatest known cause of amphibian declines, after habitat loss? What is the greatest cause for threatened species?

GO TO **INTERPRETING GRAPHS & DATA** ON MasteringEnvironmentalScience®

7200 known species of amphibians are in decline, 30% are threatened, and at least 170 species studied just years or decades ago are thought to be extinct (**FIGURE 8.12a**). As these creatures disappear before our eyes, scientists are racing to discover the reasons, and studies implicate a wide array of factors (**FIGURE 8.12b**). These include habitat destruction, chemical pollution, invasive species, climate change, and a disease called chytridiomycosis caused by a fungal pathogen. Biologists suspect that multiple factors are interacting and multiplying one another's effects.

As scientists learn more, they are designing responses to amphibian declines. A conservation action plan published by the IUCN recommends that we protect and restore habitat, crack down on illegal harvesting, enhance disease monitoring, and establish captive breeding programs.

Today many people are striving to save vanishing species. The search for solutions to our biodiversity crisis is dynamic and inspiring, and scientists are developing innovative approaches to sustain Earth's diversity of life.

# Conservation Biology: The Search for Solutions

Today, more and more scientists and citizens perceive a need to stop the loss of biodiversity. In his 1994 autobiography, *Naturalist*, E.O. Wilson wrote:

> When the [20th] century began, people still thought of the planet as infinite in its bounty. [Yet] in one lifetime, exploding human populations have reduced wildernesses to threatened nature reserves. Ecosystems and species are vanishing at the fastest rate in 65 million years. Troubled by what we have wrought, we have begun to turn in our role from local conqueror to global steward.

## Conservation biology responds to biodiversity loss

The urge to act as responsible stewards of natural systems, and to use science as a tool in this endeavor, sparked the rise of **conservation biology.** This scientific discipline is devoted to understanding the factors, forces, and processes that influence the loss, protection, and restoration of biological diversity. Conservation biologists aim to develop solutions to such problems as habitat degradation and species loss (**FIGURE 8.13**). Conservation biology is thus an applied and goal-oriented science, with implicit values and ethical standards. Conservation biologists integrate an understanding of evolution and ecology as they use field data, lab data, theory, and experiments to study our impacts on other organisms. They also design, test, and implement responses to these impacts.

At the genetic level, *conservation geneticists* ask how small a population can become and how much genetic variation it can lose before running into problems such as inbreeding depression (p. 164). By determining a *minimum viable population size*, conservation geneticists help wildlife managers decide how vital it may be to increase a population. Studies of genes, populations, and species inform conservation efforts with habitats, communities, ecosystems, and landscapes. By examining how organisms disperse from one habitat patch to another, and how their genes flow among subpopulations, conservation biologists try to learn how likely a population is to persist or succumb in the face of environmental change.

## Endangered species are a focus of conservation efforts

The primary legislation for protecting biodiversity in the United States is the **Endangered Species Act.** Enacted in 1973, the Endangered Species Act (ESA) offers protection

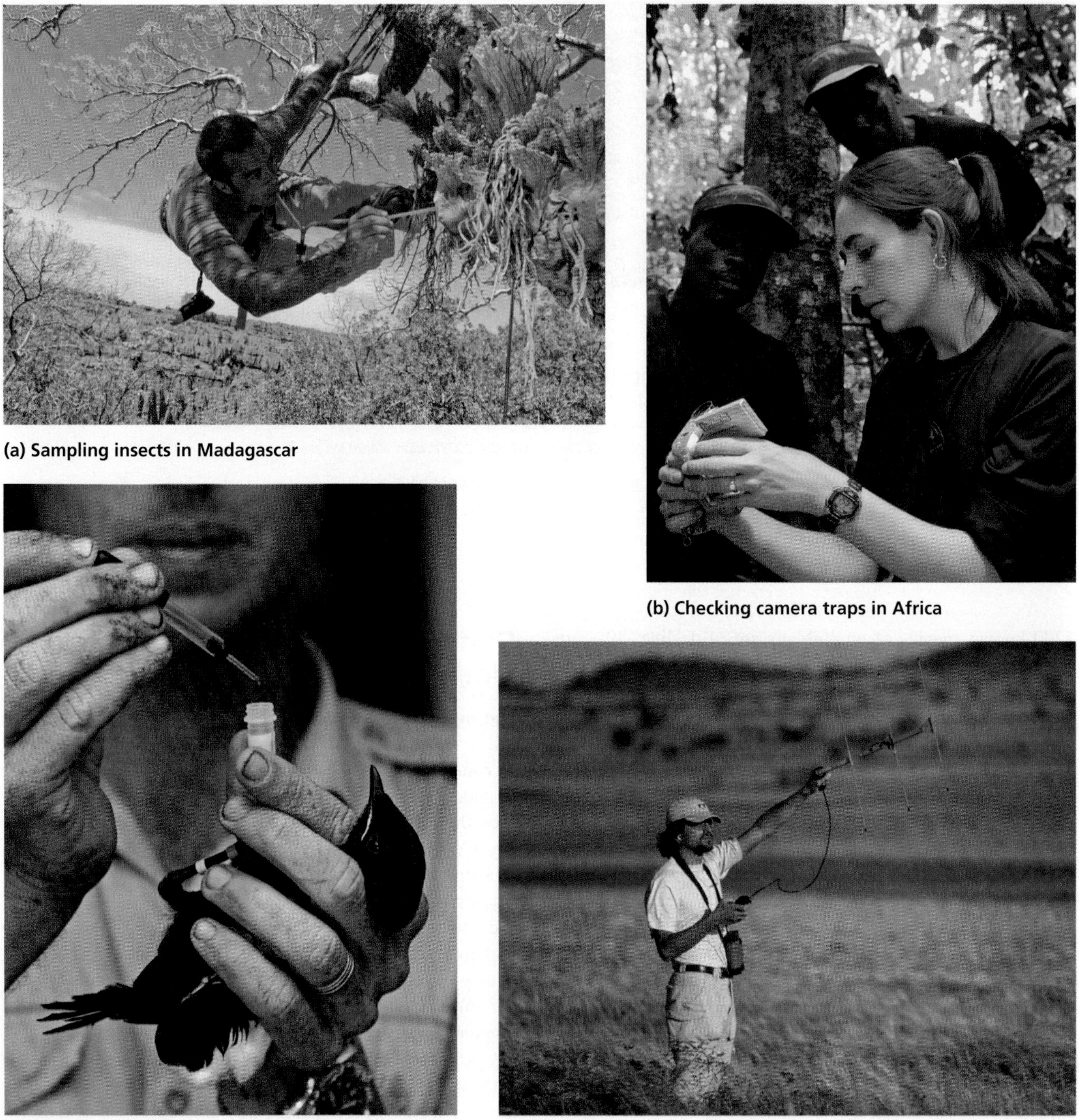

(a) Sampling insects in Madagascar

(b) Checking camera traps in Africa

(c) Drawing blood from a Seychelles Magpie Robin

(d) Radiotracking birds in Spain

**FIGURE 8.13 Conservation biologists use many approaches to study the loss, protection, and restoration of biodiversity, seeking to develop scientifically sound solutions.**

to species that are *endangered* (in danger of extinction) or *threatened* (likely to become endangered in the near future). It forbids the government and private citizens from taking actions that destroy individuals of these species or the habitats that are critical to their survival. The ESA also forbids trade in products made from threatened and endangered species. The aim is to prevent extinctions, stabilize declining populations, and enable populations to recover. As of 2014, there were 1189 species in the United States listed as endangered and 328 more listed as threatened. For about 75% of these species, government agencies are running recovery plans to protect them and stabilize or increase their populations.

The ESA has had a number of successes. Following the 1973 ban on the pesticide DDT (p. 210) and years of effort by wildlife managers, the bald eagle, peregrine falcon, brown pelican, and other birds have recovered and are no longer listed as endangered (**FIGURE 8.14**). Intensive management programs with species such as the red-cockaded woodpecker have held populations steady in the face of continued pressure on habitat. Overall, roughly 40% of declining populations have been stabilized.

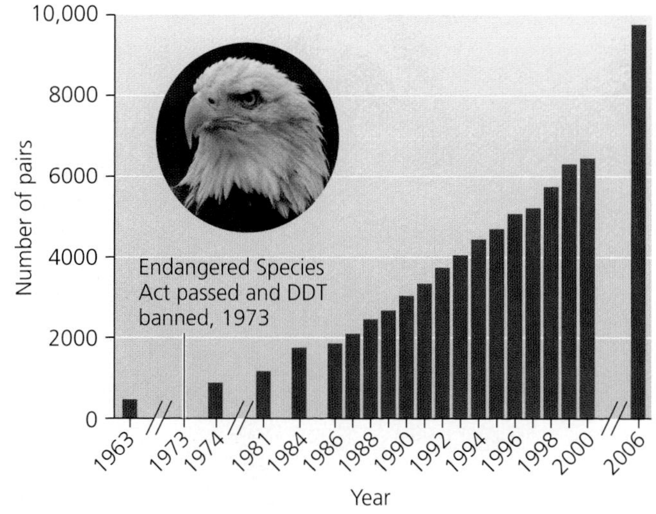


**FIGURE 8.14 The bald eagle's recovery is a success story.** The U.S. national symbol had been close to extinction in the Lower 48 states. Following protection under the Endangered Species Act and a ban on DDT, its numbers began to rebound. In 2007, the bald eagle was declared recovered and was removed from the Endangered Species List. *Data from U.S. Fish and Wildlife Service, based on annual volunteer surveys. Data after 2000 are scarce because surveys were discontinued once it became clear that eagles were recovering.*

**FIGURE 8.15 The greater sage grouse is a species whose addition to the Endangered Species List is warranted by science yet precluded by a lack of funding.** Sage grouse, like this displaying male, have lived in sagebrush habitats throughout the western United States. Their listing could complicate efforts to drill for oil and gas on these lands.

This success comes despite the fact that the U.S. Fish and Wildlife Service and the National Marine Fisheries Service, the agencies that administer the ESA, are perennially underfunded. Reauthorization of the ESA faced opposition from Republican-led Congresses in power from 1994 to 2006. Efforts in 2006 to weaken the ESA by stripping it of its ability to safeguard habitat were narrowly averted after 5700 scientists sent Congress a letter of protest.

Today, some species are judged by scientists to need protection yet have not been added to the endangered species list because the government has not supplied funding to protect them. Such species are said to be "warranted but precluded"—their listing is warranted by science, but precluded by lack of resources (**FIGURE 8.15**). This has led environmental advocacy groups to sue the government for failing to enforce the law. Well-meaning Fish and Wildlife Service staff are frequently caught in a no-win situation, battling lawsuits from the political left and budget cuts from the political right.

Polls repeatedly show that most Americans support protecting endangered species. Yet some opponents feel that the ESA imperils people's livelihoods. This has been a common perception in the Pacific Northwest, where protection for the northern spotted owl and the marbled murrelet—birds that rely on old-growth forest—have slowed timber harvesting, causing loggers to fear for their jobs. In addition, many landowners worry that federal officials will restrict the use of private land on which threatened or endangered species are found. This has led to a practice described as "shoot, shovel, and shut up," among landowners who want to conceal the presence of such species on their land.

In fact, however, the ESA has stopped very few development projects—and a number of its provisions and amendments promote cooperation with landowners. *Habitat conservation plans* and *safe harbor agreements* are arrangements that allow landowners to harm species in some ways if they improve habitat for them in others.

Today many nations have laws protecting species, although they are not always effective. When Canada enacted its *Species at Risk Act* in 2002, the Canadian government was careful to stress cooperation with landowners and provincial governments, and not to present the law as a decree from the national government. Environmental advocates and scientists protested that the law was weak and failed to protect habitat adequately.

## International treaties promote conservation

The United Nations has facilitated several international treaties to protect biodiversity. The 1973 **Convention on International Trade in Endangered Species of Wild Fauna and Flora (CITES)** protects endangered species by banning the international transport of their body parts. The 1990 global ban on the ivory trade was enacted under CITES and may be the treaty's biggest accomplishment so far. When nations enforce its provisions, CITES can protect rhinos, elephants, tigers, and other rare species whose body parts are traded internationally.

In 1992, leaders of many nations agreed to the **Convention on Biological Diversity.** This treaty aims to help nations conserve biodiversity, use it in a sustainable manner, and ensure the fair distribution of its benefits. The treaty helped Africans profit from ecotourism with their wildlife preserves. It also prompted nations to protect more areas, enhanced global markets for sustainable crops such as shade-grown coffee, and moved some rice-growing Asian nations away from pesticide-intensive farming practices. Yet the treaty's overall goal—"to achieve, by 2010, a significant reduction of the current rate of biodiversity loss"—was not met.

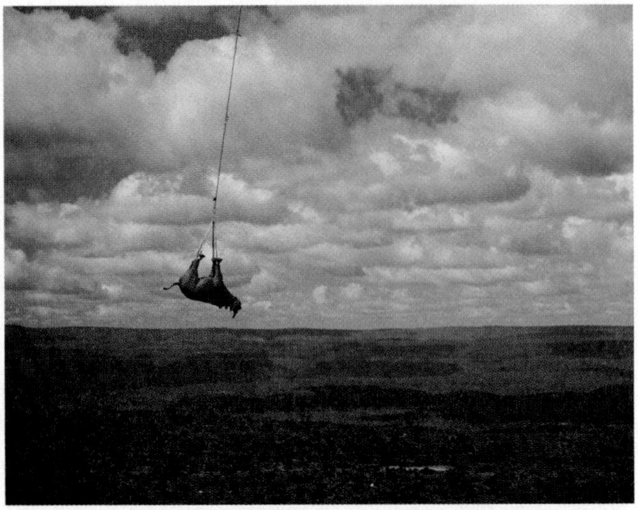

(a) A black rhino is air-lifted into Serengeti National Park

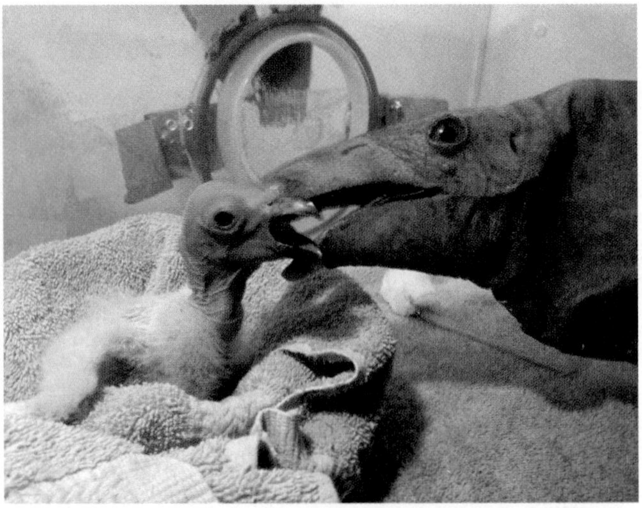

(b) Biologists use hand puppets to nurse condor chicks

**FIGURE 8.16 We can re-establish populations and rescue species by reintroducing them to areas where they were extirpated.** Black rhinos **(a)** have been helicoptered into Serengeti National Park from other areas where populations are increasing. To save the California condor **(b)** from extinction, biologists raise chicks in captivity with hand puppets that mimic the heads of adult condors.

## Captive breeding, reintroduction, and cloning are being pursued

In the effort to save species at risk, zoos and botanical gardens have become centers for **captive breeding,** in which individuals are bred and raised in controlled conditions with the intent of reintroducing their progeny into the wild. Many institutions are now linked in a program that trades animals among them to promote gene flow. The IUCN counts 65 plant and animal species that now exist *only* in captivity or cultivation.

Reintroducing species into areas they used to occur is resource-intensive, but it can pay big dividends. In 2010 the first of 32 black rhinos were translocated from South Africa to Serengeti National Park to help restore a former population (**FIGURE 8.16a**). This followed similar reintroduction projects elsewhere in Africa.

A prime example of captive breeding and reintroduction is the program to save the California condor, North America's largest bird (**FIGURE 8.16b**). Although condors are harmless scavengers of dead animals, people in the early 20th century used to shoot them. Condors also collided with electrical wires and succumbed to lead poisoning after scavenging carcasses of animals killed with lead shot. By 1982, only 22 condors remained, and biologists made the wrenching decision to take all the birds into captivity.

Today the collaborative program between the Fish and Wildlife Service and several zoos has boosted condor numbers. As of 2014 there were 198 birds in captivity and 237 birds living in the wild. Condors have been released in California, Arizona, and Baja California (Mexico), where they thrill people lucky enough to spot the huge birds soaring through the skies. Yet these birds are still dying from lead poisoning, and wild populations will likely not become sustainable until hunters convert from lead shot to nontoxic shot made of copper or steel. In 2013 California banned lead shot for all hunting, effective in 2019. This may give condors a fighting chance while helping to remove a highly toxic substance from the environment.

One new idea for saving species from extinction is to create individuals by cloning them. In this technique, DNA from an endangered species is inserted into a cultured egg without a nucleus, and the egg is implanted into a female of a closely related species that acts as a surrogate mother. Several mammals have been cloned in this way, with mixed results. Some scientists even talk of recreating extinct species from DNA recovered from preserved body parts, and an effort is underway to attempt this with the passenger pigeon. In 2009 a subspecies of Pyrenean ibex (a type of mountain goat) was cloned from cells taken from the last surviving individual, which had died in 2000. The cloned baby ibex died shortly after birth. Even if cloning can succeed from a technical standpoint, however, such efforts are not an adequate response to biodiversity loss. Without ample habitat and protection in the wild, having cloned animals in a zoo does little good.

## Forensics can help to protect species

**Forensic science**, or **forensics**, involves the scientific analysis of evidence to make an identification or answer a question relating to a crime or an accident. Conservation biologists are now employing forensics to protect species at risk from poaching and illegal trade. By analyzing DNA from organisms or their tissues sold at market, researchers can often determine the species or subspecies—and sometimes the geographic origin. This can help detect illegal activity and enhance law enforcement. Forensic analysis has helped researchers trace the origins of meat from whales sold in Asian markets, providing valuable data used to set policy for whaling and whale conservation.

Forensics is now being used to trace the origin of elephant tusks sold for ivory. To enforce the global ivory trade ban under CITES, officials seize illegal shipments when they find them (see Figure 8.9). Samples from seized tusks are sent to scientific labs, where researchers sequence DNA and compare it to a reference library of DNA that varies from one elephant population to another. For instance, after customs agents seized 3.9 tons of tusks in Hong Kong in 2006, they sent samples to researchers at the University of Washington, who found that their DNA matched that from forest elephants from Gabon in west-central Africa. Four years earlier, a 6.5-ton shipment had been intercepted in Singapore, and samples were sent to the same research lab. The DNA matched known samples from Zambian elephants, indicating that more elephants were being killed there than Zambia's leaders had realized. In response, the Zambian government replaced its wildlife director and began imposing harsher sentences on poachers and ivory smugglers.

The latest forensic research is showing that the vast majority of ivory comes from a relatively small number of locations. This suggests that by focusing enforcement on hotspots where poachers are targeting elephants, we may achieve more effective protection. Researchers are also helping to track elephants and ivory by putting radio collars on animals, satellite-tracking tusks with microchips, and flying drones (unmanned surveillance aircraft) over parks to capture real-time video of poachers.

## Biodiversity hotspots pinpoint regions of high diversity

To prioritize regions for conservation, scientists have mapped biodiversity hotspots (**FIGURE 8.17a**). A **biodiversity hotspot** is a region that supports an especially great number of species that are *endemic,* found nowhere else in the world (**FIGURE 8.17b**). To qualify as a hotspot, a region must harbor at least 1500 endemic plant species (0.5% of the world's total plant species). In addition, a hotspot must have already lost 70% of its habitat to human impact and be at risk of losing more.

The ecosystems of the world's biodiversity hotspots together once covered 15.7% of the planet's land surface. Today, because of habitat loss, they cover only 2.3%. This small amount of land is the exclusive home for half the world's plant species and 42% of terrestrial vertebrate species. The hotspot concept motivates us to focus on these areas, where the greatest number of unique species can be protected.

## Parks and protected areas conserve biodiversity at the ecosystem level

Scientists know that protecting species does little good if the larger systems they rely on are not also sustained. Although conservation biologists can restore degraded natural systems through ecological restoration (pp. 80–81), it is more effective to safeguard biodiversity by protecting natural areas before they become degraded. A prime way to conserve habitats, communities, ecosystems, and landscapes is to set aside areas of undeveloped land in parks and preserves. Currently people have set aside 13% of the world's land area in national parks, state parks, provincial parks, wilderness areas, biosphere reserves, and other protected areas. Many of these lands are managed for recreation, water quality protection, or other purposes (rather than for biodiversity), and many suffer from illegal logging, poaching, and resource extraction. Yet these areas offer animals and plants a degree of protection from human persecution, and some are large enough to preserve entire natural systems that otherwise would be fragmented, degraded, or destroyed. Areas of ocean are now also being protected in marine reserves and marine protected areas (p. 275).

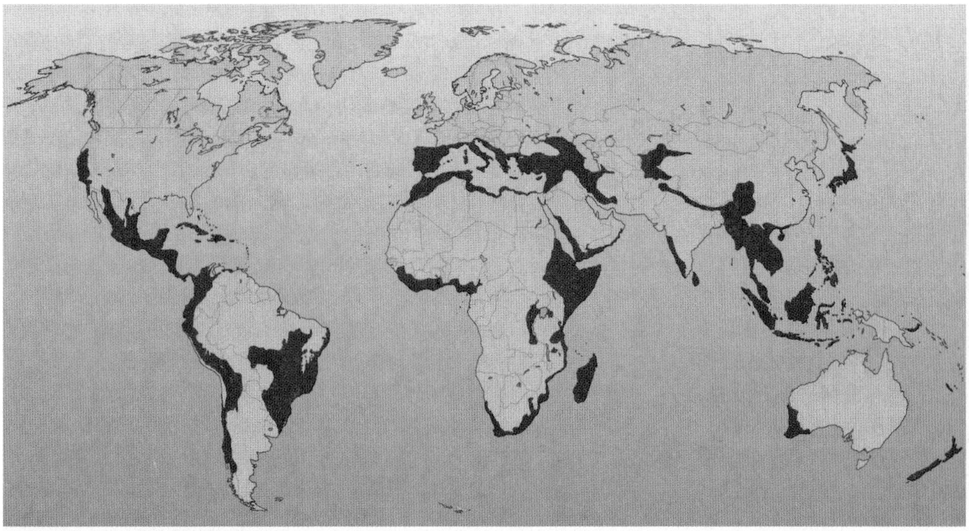

**(a) The world's biodiversity hotspots**

**(b) Ring-tailed lemur**

**FIGURE 8.17 Biodiversity hotspots are priority regions for habitat preservation.** In color **(a)** are the 34 hotspots mapped by Conservation International, a nongovernmental organization. (Only 15% of the colored area is actually habitat; most is developed.) These regions are home to species such as the ring-tailed lemur **(b),** a primate endemic to Madagascar that has lost over 90% of its forest habitat as a result of human population growth and resource extraction. *Data from Conservation International.*

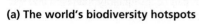

Serengeti National Park and the adjacent Maasai Mara National Reserve are two of the world's largest and most celebrated parks, but Tanzania and Kenya have each set aside a number of other protected areas. Some of the best known include (in Kenya) Amboseli, Tsavo, Mount Kenya, and Lake Nakuru National Parks; and (in Tanzania) Ngorongoro Conservation Area, Mafia Island Marine Park, Kilimanjaro National Park, and Gombe Stream National Park. Altogether roughly 25% of Tanzania's land area and 12% of Kenya's land area is protected.

Unfortunately, protecting land may not be enough to ensure effective conservation. Pressures from outside the reserves in Kenya and Tanzania are causing declines in wildlife within the reserves (see *The Science Behind the Story,* pp. 172–173). Similar situations occur in North America: Despite the large size of Yellowstone National Park, animals such as elk, bears, bison, and wolves roam seasonally in and out of the park, sometimes coming into conflict with ranchers. As a result, conservationists have tried to find ways to protect animals and habitats across the Greater Yellowstone Ecosystem, the larger region over which the animals roam.

As global warming (Chapter 14) drives species toward the poles and upward in elevation, this may force them out of protected areas. A major challenge today is to link protected areas across the landscape with corridors (pp. 202–203) of habitat so that species like wildebeest can move in response to climate change. We will explore parks and protected areas and the issues they face more fully in Chapter 9 (pp. 198–203).

### WEIGHING THE ISSUES

**Single-Species Conservation?**

What would you say are some advantages of focusing on conserving single species, versus trying to conserve broader communities, ecosystems, or landscapes? What might be some of the disadvantages? Which do you think is the better approach, or should we use both, and why?

## Community-based conservation is growing

Helping people, wildlife, and ecosystems all at the same time is the focus of many current conservation efforts. In the past, conservationists from industrialized nations, in their zeal to preserve ecosystems in developing nations, often neglected the needs of people in the areas they wanted to protect. Developing nations came to view this as a kind of neocolonialism. Today, in contrast, many conservation biologists actively engage local people in efforts to protect land and wildlife—a cooperative approach called **community-based conservation**. A quarter of the world's protected areas are now being managed using community-based conservation. In several African nations, the African Wildlife Foundation funds community-based conservation programs to help communities conserve elephants, lions, rhinos, gorillas, and other animals.

In East Africa, conservationists and scientists began working with the Maasai and other people of the region years ago, understanding that to conserve animals and ecosystems, local people need to be stewards of the land and feel invested

**FIGURE 8.18 Scientists and conservation advocates work cooperatively with local people to conserve wildlife.** Biologist Alayne Cotterill of the conservation research group Living with Lions works with Maasai warriors to monitor lion populations near the Serengeti.

in conservation (**FIGURE 8.18**). This has proven challenging because the parks and reserves were created on land historically used by local people. Residents were forcibly relocated; by some estimates 50,000 Maasai were evicted to create Serengeti National Park. In the view of many local people, the parks were a government land grab, and laws against poaching deprive them of a right to kill wildlife. As human population grew in the region, conflicts between people and wildlife increased. Ranchers worried that wildebeest and buffaloes might spread disease to their cattle. Farmers lost produce when elephants ate their crops. And the economic benefits of ecotourism were not being shared with all people in the region.

In response, proponents of conservation have tried to reallocate tourist dollars to local villages and to transfer some authority over wildlife management to local people. In the regions around the Maasai Mara Reserve, the Kenya Wildlife Service and international nonprofits have helped farmers and ranchers build electric fences to keep wildlife away from their crops and livestock. These efforts are reducing conflicts between people and wildlife and fostering more favorable attitudes, but they also raise new challenges. For instance, in some cases elephants are thriving on reserves, but because they are now hemmed in by fences, they become overcrowded and overgraze the vegetation, destroying trees and endangered plants the reserves were meant to protect.

Working cooperatively to make conservation beneficial for local people requires patience, investment, and trust on all sides. Setting aside land for preservation may deprive local people of access to exploitable resources, but it also helps ensure that those resources can be sustainably managed and will not be exhausted or sold to foreign corporations. If tourism revenues are adequately distributed, people gain direct economic benefits of conserving wildlife. Community-based conservation has not always been successful, but in a world of rising human population, sustaining biodiversity will require locally based management that sustainably meets people's needs.

## Conclusion

Data from scientists worldwide confirm what any naturalist who has watched the habitat change in his or her hometown already knows: From amphibians to zebras, biological diversity is being lost rapidly within our lifetimes. This erosion of biodiversity threatens to result in a mass extinction event equivalent to those of the geologic past. Habitat alteration, pollution, overharvesting, invasive species, and climate change are the primary causes of biodiversity loss. This loss matters, because society cannot function without biodiversity's pragmatic benefits.

Conservation biologists today are conducting research that guides efforts to save endangered species, protect their habitats, recover populations, and preserve and restore natural ecosystems. The innovative strategies these scientists are pursuing hold promise to slow, and perhaps reverse, the loss of biodiversity on Earth.

# Testing Your Comprehension

1. What is biodiversity? Describe three levels of biodiversity.
2. Define the term *ecosystem services*. Give three examples of ecosystem services that people would have a hard time replacing if these were lost.
3. What is the relationship between biodiversity and food security? Between biodiversity and pharmaceuticals? Give three examples of benefits of biodiversity conservation for food supplies and medicine.
4. List three reasons why people suggest that biodiversity conservation is important.
5. What are the five primary causes of biodiversity loss? Give one specific example of each.
6. List three invasive species, and describe their impacts.
7. Describe one successful accomplishment of the U.S. Endangered Species Act. Now describe one reason some people have criticized it.
8. Explain how captive breeding can help with endangered species recovery, and give an example. Now explain why cloning could never be, in itself, an effective response to species loss.
9. Name two reasons that a large national park like Serengeti or Yellowstone might not be adequate to effectively conserve a population of a threatened species. What solutions exist to address these reasons?
10. Explain the notion of community-based conservation. Why have conservation advocates been turning to this approach? What challenges exist in implementing it?

# Seeking Solutions

1. Many arguments have been advanced for the importance of preserving biodiversity. Which argument do you think is most compelling, and why? Which argument do you think is least compelling, and why?
2. Some people say we shouldn't worry about endangered species because extinction has always occurred. How would you respond to this view?
3. Conservation advocates from industrialized nations have long pushed to set aside land in biodiversity-rich developing nations. Leaders of developing nations have accused them of neocolonialism. "Your nations attained prosperity and power by overexploiting your environments decades or centuries ago," these leaders ask, "so why should we now sacrifice our development by setting aside our land and resources?" What would you say to these leaders? What would you say to the conservation advocates? Do you see ways that both preservation and development goals might be reached?
4. **THINK IT THROUGH** You are a legislator in a nation with no endangered species act. You want to introduce a law to protect your nation's biodiversity. Consider the U.S. Endangered Species Act, Canada's Species at Risk Act, and international efforts such as CITES and the Convention on Biological Diversity. What strategies would you write into your legislation? How would your law be similar to and different from each of these efforts?
5. **THINK IT THROUGH** As a resident of your community and a parent of two young children, you attend a town meeting called to discuss the proposed development of a shopping mall and condominium complex. The development would eliminate a 100-acre stand of forest, the last sizeable forest stand in your town. The developers say the forest loss will not matter because plenty of 1-acre stands still exist scattered throughout the area. Consider the development's possible impacts on the community's biodiversity, children, and quality of life. What will you choose to tell your fellow citizens and the town's decision-makers at this meeting, and why?

# Calculating Ecological Footprints

Research shows that much of humanity's footprint on biodiversity comes from our use of grasslands for grazing livestock and forests for timber and other resources. Grasslands and forests contribute different amounts to each nation's biocapacity (a region's natural capacity to provide resources and absorb our wastes), depending on how much of these habitats each nation has. Likewise, the per capita biocapacity and per capita ecological footprints of each nation vary further according to their populations. When footprints are equal to or below biocapacity, then resources are being used sustainably. When footprints surpass biocapacity, then resources are being used unsustainably.

In the table, fill in the proportion of each nation's per capita footprint accounted for by use of grazing land and forestland. Then fill in the proportion of each nation's per capita biocapacity provided by grazing land and forestland.

| Footprints and biocapacities (hectares per person) | Kenya | Tanzania | United States | Canada |
|---|---|---|---|---|
| Footprint from grazing land | 0.27 | 0.36 | 0.19 | 0.42 |
| Footprint from forestland | 0.28 | 0.24 | 0.86 | 0.74 |
| Total ecological footprint | 0.95 | 1.19 | 7.19 | 6.43 |
| Percent footprint from grazing and forest | 58 | | | |
| Biocapacity of grazing land | 0.27 | 0.39 | 0.26 | 0.23 |
| Biocapacity of forestland | 0.02 | 0.13 | 1.56 | 8.27 |
| Total biocapacity | 0.53 | 1.02 | 3.86 | 14.92 |
| Percent biocapacity from grazing and forest | | | | 57 |

*Data from WWF, 2012.* Living Planet Report 2012. *WWF International, Gland, Switzerland.*

1. In which nations is grazing land being used sustainably? In which nations is forestland being used sustainably?
2. Do forest use and grazing comprise a larger part of the footprint for temperate-zone industrialized nations such as Canada and the United States, or for tropical developing nations such as Kenya and Tanzania? What do you think accounts for this difference between these two types of nations? What else besides use of forests and grasslands contributes to an ecological footprint?
3. The Living Planet Index has declined 28% since 1970 (p. 171), but its temperate and tropical components differ. During this period, the index for temperate regions increased by 31%, whereas the index for tropical regions declined by 61%. Based on this information, do you predict that biodiversity loss has been steepest in Kenya and Tanzania or in Canada and the United States? Explain your answer.

## MasteringEnvironmentalScience®

**STUDENTS**

Go to **MasteringEnvironmentalScience** for assignments, the etext, and the Study Area with practice tests, videos, current events, and activities.

**INSTRUCTORS**

Go to **MasteringEnvironmentalScience** for automatically graded activities, current events, videos, and reading questions that you can assign to your students, plus Instructor Resources.

# Forests, Forest Management, and Protected Areas

## Upon completing this chapter, you will be able to:

- Summarize the ecological and economic contributions of forests
- Outline the history and current scale of deforestation
- Assess approaches to resource management, describe methods of harvesting timber, and appraise aspects of forest management
- Identify federal land management agencies and the lands they manage
- Recognize types of parks and protected areas and evaluate issues involved in their design

Photo: **Paper being produced at the mill in Escanaba, Michigan**

CENTRAL CASE STUDY

# Certified Sustainable Paper in Your Textbook

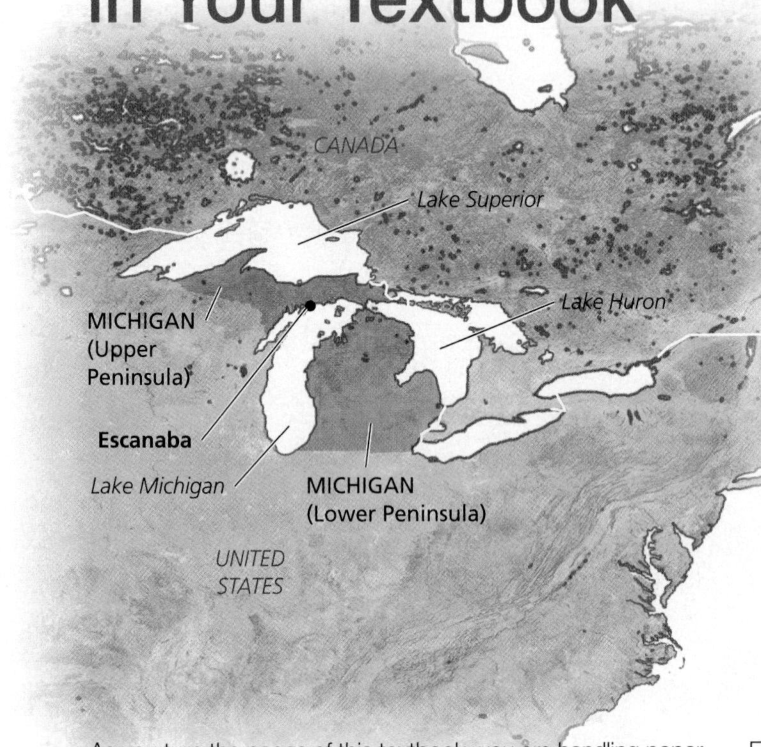


*"FSC is the high bar in forest certification, because its standards protect forests of significant conservation value and consistently deliver meaningful improvements in forest management on the ground."*

**—Kerry Cesareo, deputy director, WWF-US Forests Program**

*"As a company and as individuals, we must be able to look ourselves in the mirror and know that we are truly doing what is right and good for the environment."*

**—Rick Willett, president and CEO, NewPage Corporation**

As you turn the pages of this textbook, you are handling paper made from trees that were grown, managed, harvested, and processed using certified sustainable practices.

If you were to trace the paper in this edition's first printing back to its origin, you would find yourself standing in a diverse mixed forest of aspen, birch, beech, maple, spruce, and pine in the Upper Peninsula of Michigan. This sparsely populated region near the Canadian border, flanked by Wisconsin, Lake Michigan, and Lake Superior, remains heavily forested, even though it has supplied valuable timber to our society for nearly 200 years.

The trees cut to make this book's paper were selected for harvest based on a sustainable management plan designed to avoid depleting the forest of its mature trees or degrading the ecological functions the forest performs. The logs were then transported to a nearby pulp and paper mill at Escanaba, a community of 13,000 people on the shore of Lake Michigan. The mill is Escanaba's largest employer, providing jobs to about 1000 residents.

At the Escanaba mill, the wood is chipped and then fed into a digester, where the chips are cooked with chemicals to break down the wood's molecular bonds. Millworkers bleach, wash, and screen the cellulose fibers and mix them with water and dye. This mixture is poured onto a moving mat, where heavy rollers press it into thin sheets. The sheets of newly formed paper are then dried on heated rollers and made ready to receive a variety of coatings. The paper is wound into immense reels, later to be cut into sheets. In this way, the Escanaba mill produces about 785,000 tons of paper each year. It recycles chemicals and water used in the process, and it combusts discarded waste to help power the mill.

Each stage in this process has been examined by independent third-party inspectors accredited by the Forest Stewardship Council® (FSC®) to ensure that practices meet the FSC's criteria for sustainable forest management and paper production. The Forest Stewardship Council is an organization that certifies forests, companies, and products that meet sustainability standards. FSC-certified timber harvesting operations in Michigan's Upper Peninsula are required to protect rare species and sensitive habitats, safeguard water sources, control erosion, minimize pesticide use, and maintain the diversity of the forest and its ability to regenerate after harvesting. FSC certification is the best way for consumers of forest products to know that they are supporting sustainable practices that protect forests.

The Escanaba mill is the largest of eight mills run by the NewPage Corporation, based in Ohio. NewPage's mills (in Kentucky, Maine, Maryland, Michigan, Minnesota, and Wisconsin) together produce 3.5 million tons of coated paper each year for books, magazines, newspapers, food packaging, and more. Only some of this paper is FSC-certified, but NewPage seeks out suppliers of wood who are certified at least to the less-rigorous standards of the Sustainable Forestry Initiative®. NewPage states that it does not use wood from old-growth forests, rainforests, or forests of exceptional conservation value.

Your textbook's paper gained FSC certification because Pearson Education, the publisher of this book, is striving to follow sustainable practices. Pearson supported FSC paper for this book at the request of your authors and editors because our entire team feels that an environmental science textbook should walk its talk. So, as you flip through this book, you can feel satisfied that you are doing a small part to help safeguard the world's forests by supporting sustainable forestry practices. ◻

# Forest Ecosystems and Forest Resources

A **forest** is any ecosystem with a high density of trees. Forests provide habitat for countless organisms; help maintain the quality of soil, air, and water; and play key roles in our planet's biogeochemical cycles (pp. 38–43). Forests also provide humanity with wood for fuel, construction, paper production, and more.

## Many kinds of forests exist

Most of the world's forests occur as boreal forest (pp. 86–87), a biome that stretches across much of Canada, Scandinavia, and Russia; or as tropical rainforest (p. 84), a biome in Latin America, equatorial Africa, Indonesia, and Southeast Asia. Temperate deciduous forests (p. 83), temperate rainforests (p. 84), and tropical dry forests (p. 85) also cover large regions.

Within each forest biome, the plant community varies from region to region because of differences in soil and climate. As a result, ecologists and forest managers classify forests into **forest types**, categories defined by their predominant tree species (**FIGURE 9.1**). The eastern United States contains 10 forest types, ranging from spruce-fir to oak-hickory to longleaf–slash pine. The western United States holds 13 forest types, ranging from Douglas fir and hemlock–sitka spruce forests of the moist Pacific Northwest to ponderosa pine and pinyon-juniper woodlands of the dry interior. Michigan's Upper Peninsula holds a diverse mix of forest types. Near Escanaba, spruce-fir forest intermixes with forests of aspen, birch, maple, and beech, and with areas of white, red, and jack pine.

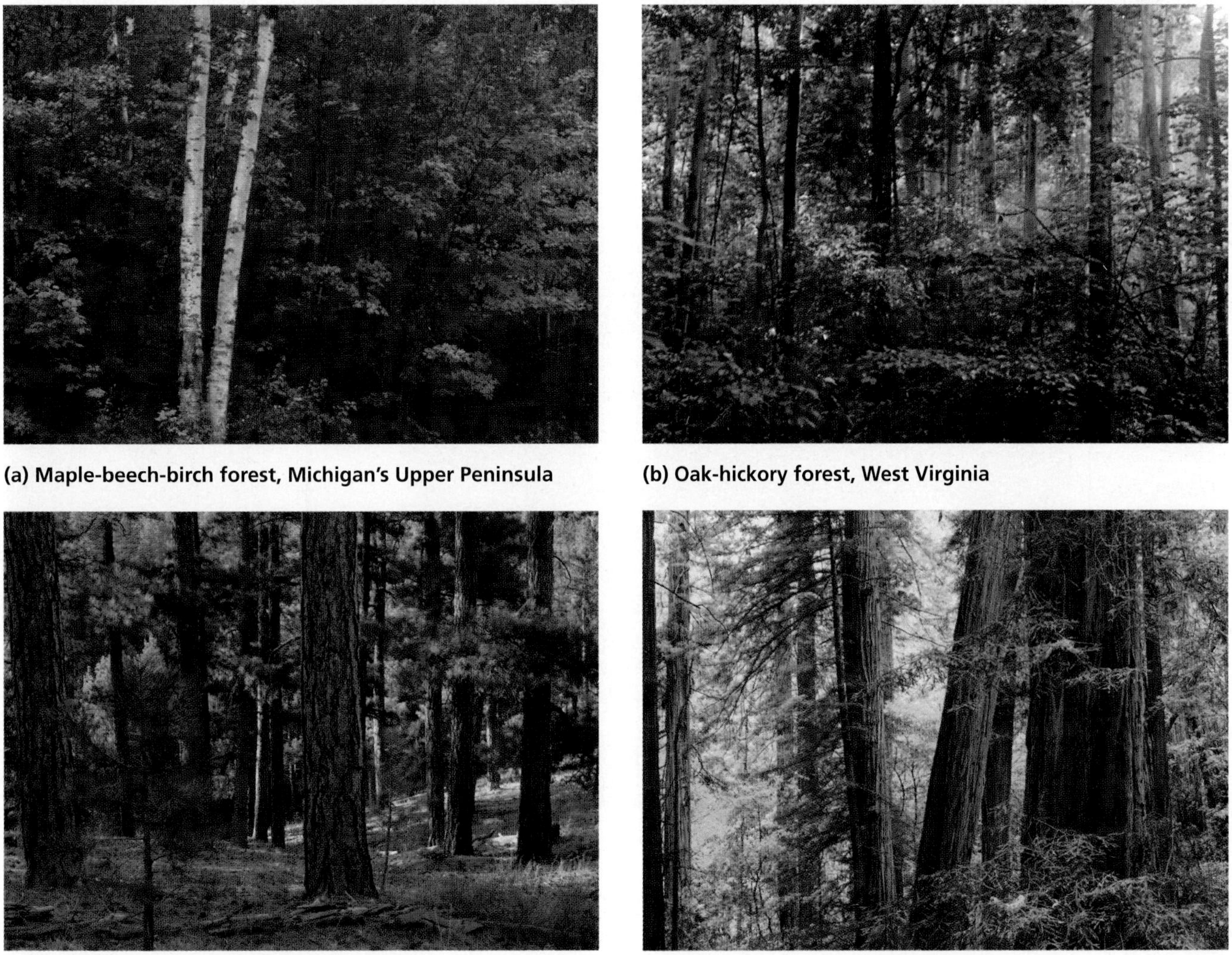

**(a) Maple-beech-birch forest, Michigan's Upper Peninsula**

**(b) Oak-hickory forest, West Virginia**

**(c) Ponderosa pine forest, northern Arizona**

**(d) Redwood forest, coastal northern California**

**FIGURE 9.1 Shown are four of the 23 forest types found in the continental United States.**

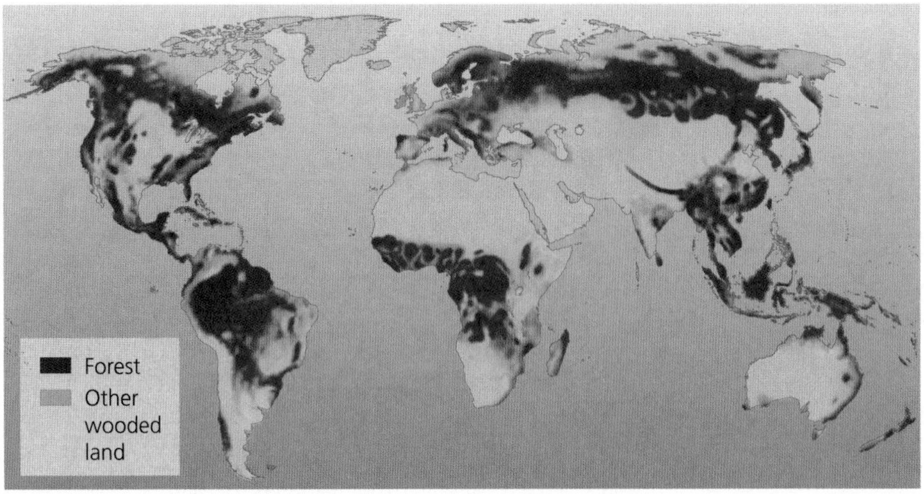


**FIGURE 9.2 Forests cover 31% of Earth's land surface.** Most widespread are boreal forests in the north and tropical forests near the equator. "Wooded land" supports trees at sparser densities. *Data from Food and Agriculture Organization of the United Nations, 2010.* Global forest resources assessment 2010.

Altogether, forests currently cover 31% of Earth's land surface (**FIGURE 9.2**). Forests occur on all continents except Antarctica.

## Forests are ecologically complex

Because of their structural complexity and their capacity to provide many niches for organisms, forests comprise some of the richest ecosystems for biodiversity (**FIGURE 9.3**). Insects, birds, mammals, and other animals subsist on the leaves, fruits, and seeds that trees produce, or find shelter in the cavities of tree trunks. Tree canopies are full of life, and understory shrubs and groundcover plants provide food and shelter for yet more organisms. Plants are colonized by an extensive array of fungi and microbes, in both parasitic and mutualistic relationships (pp. 69–70). And much of a forest's biodiversity resides on the forest floor, where the fallen leaves and branches of the leaf litter nourish the soil. A multitude of soil organisms helps to decompose plant material and cycle nutrients (p. 139).

Forests with a greater diversity of plants tend to host a greater diversity of organisms overall. As forests change by the process of succession (p. 77), their species composition changes

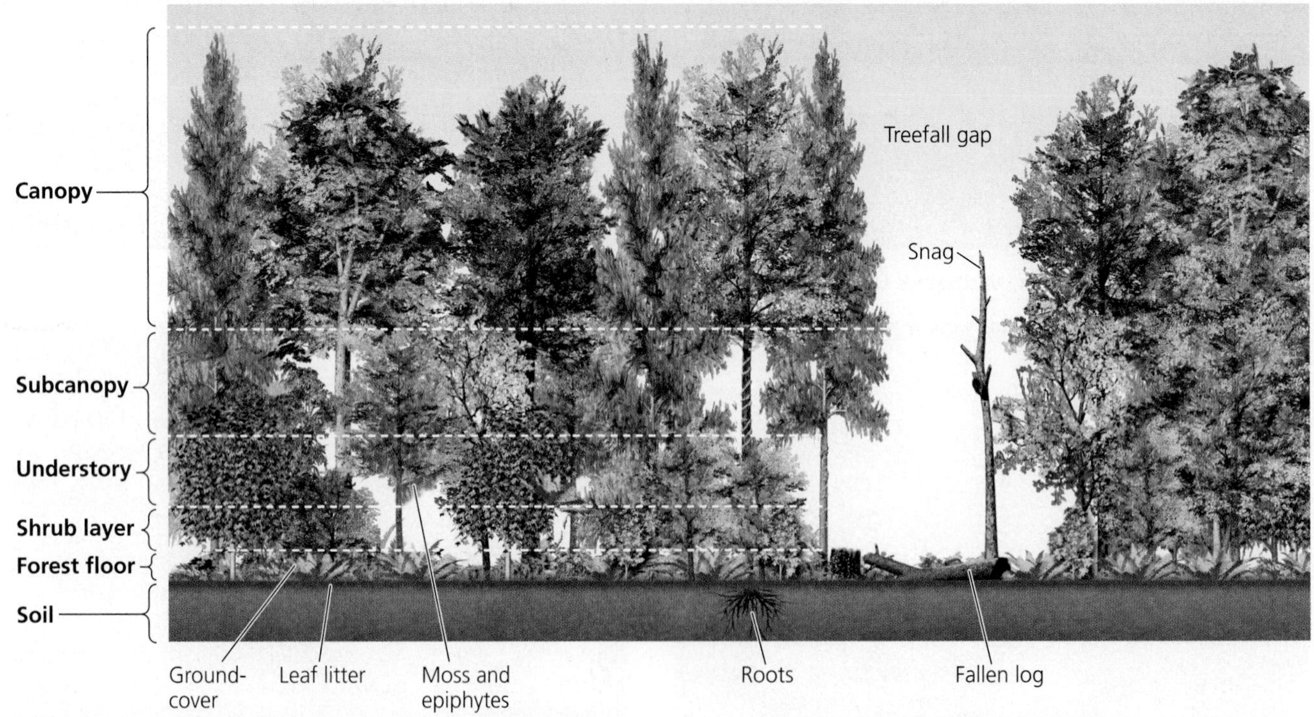


**FIGURE 9.3 A mature forest is complex in its structure.** Crowns of tall trees form the canopy, and trees beneath them form the subcanopy and understory. Shrubs and groundcover grow just above the forest floor, and vines, mosses, lichens, and epiphytes cover portions of trees and the forest floor. Snags (dead trees) provide food and nesting sites for woodpeckers and other animals, and logs nourish the soil. Fallen trees create openings called treefall gaps, letting light through and allowing early successional plants to grow.

along with their structure. In general, old-growth forests host more biodiversity than young forests, because older forests contain more structural diversity, microhabitats, and resources for more species. Old-growth forests also are home to more species that today are threatened, endangered, or declining, because old forests have become rare relative to young forests.

## Forests provide ecosystem services

Besides hosting biodiversity, forests supply us with vital ecosystem services (pp. 3, 35, 97, 167; **FIGURE 9.4**). As plants grow, their roots stabilize the soil and help to prevent erosion. When rain falls, leaves and leaf litter intercept water, slowing runoff. This helps water soak into the ground to nourish roots and recharge aquifers, thereby preventing flooding, reducing soil erosion, and helping keep streams and rivers clean.

Forest plants also filter pollutants and purify water as they take it up from the soil and release it to the atmosphere in transpiration (p. 38). Plants release the oxygen that we breathe, regulate moisture and precipitation, and moderate climate. Trees' roots draw minerals up from deep soil layers and deliver them to surface layers where other plants can use them. Plants also return organic material to the topsoil when they die or drop their leaves.

By performing all these ecological functions, forests are indispensable for our survival. Forests also enhance our health and quality of life by providing us with cultural, aesthetic, and recreational values (p. 96). People seek out forests for adventure and for spiritual solace alike—to admire beautiful trees, to observe wildlife, to enjoy clean air, and for many other reasons.

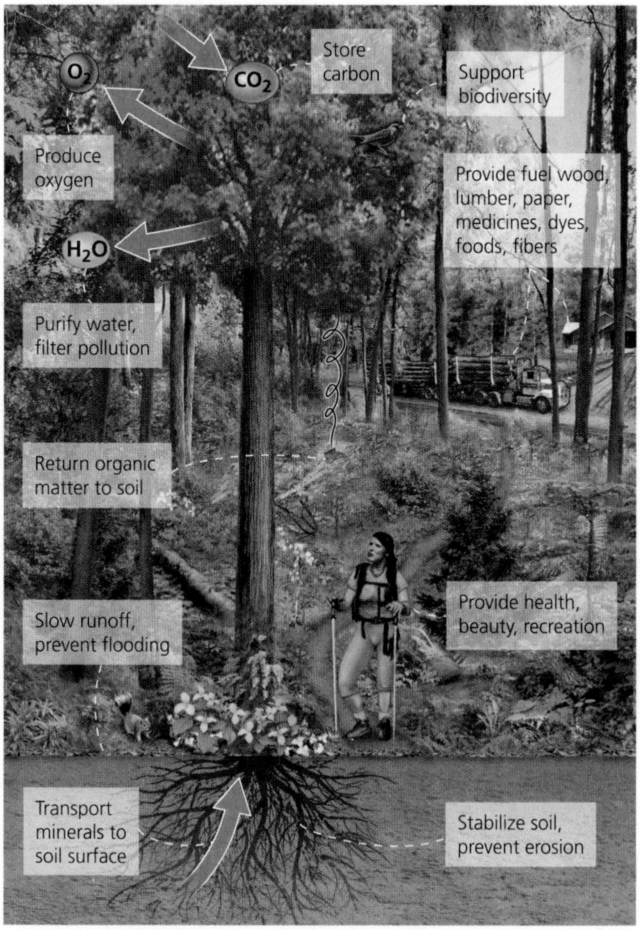


**FIGURE 9.4 Forests provide us a diversity of ecosystem services, as well as resources that we can harvest.**

## Carbon storage limits climate change

Of all the services that forests provide, their storage of carbon has elicited great interest as nations debate how to control global climate change (Chapter 14). Because trees absorb carbon dioxide from the air during photosynthesis (p. 31) and then store carbon in their tissues, forests serve as a major reservoir for carbon. Scientists estimate that the world's forests store over 280 billion metric tons of carbon in living tissue, which is more than the atmosphere contains. When plant matter is burned or when plants die and decompose, carbon dioxide is released—and thereafter less vegetation remains to soak it up. Carbon dioxide is the primary greenhouse gas contributing to global climate change (p. 305). Therefore, when we cut forests, we worsen climate change. The more forests we preserve or restore, the more carbon we keep out of the atmosphere, and the better we can address climate change.

## Forests provide us valuable resources

Carbon storage and other ecosystem services alone make forests priceless to our society, but forests also provide many economically valuable resources. Among these are plants for medicines, dyes, and fibers; animals, plants, and fungi for food; and, of course, wood from trees. For millennia, wood has fueled our fires, keeping us warm and well fed. It has built the homes that keep us sheltered. It built the ships that carried people and cultures between continents. And it gave us paper, the medium of the first information revolution. Forest resources have helped our society achieve the standard of living we now enjoy.

In recent decades, industrial harvesting has allowed us to extract more timber than ever before. Most commercial timber extraction today takes place in Canada, Russia, and other nations with large expanses of boreal forest; and in tropical nations with large areas of rainforest, such as Brazil and Indonesia. In the United States, most logging takes place in pine plantations of the South and conifer forests of the West.

# Forest Loss

When trees are removed more quickly than they can regrow, the result is **deforestation,** the clearing and loss of forests. Deforestation has altered landscapes across much of our planet. In the time it takes you to read this sentence, 2 hectares (5 acres) of tropical forest will have been cleared. As we alter, fragment, and eliminate forests, we lose biodiversity, worsen climate change, and disrupt the ecosystem services that support our societies.

## Agriculture and demand for wood put pressure on forests

We all use wood and other forest products, and we all rely on food and fiber grown on cropland and rangeland—much of which occupies land where forests once grew. To make way for agriculture and to extract wood products, people have been clearing forests for millennia. This has fed our civilization's growth, but the loss of forests also exerts damaging impacts as our population grows. Deforestation leads to biodiversity loss, soil degradation, and desertification (p. 142). It also adds carbon dioxide to the atmosphere, contributing to climate change.

In 2010, the U.N. Food and Agriculture Organization (FAO) released its latest *Global Forest Resources Assessment*, a periodic report based on remote sensing data from satellites, analysis from forest experts, questionnaire responses, and statistical modeling. The assessment concluded that we are eliminating 13 million hectares (32 million acres) of forest each year. Subtracting annual regrowth from this amount makes for an annual net loss of 5.2 million hectares (12.8 million acres)—an area about half the size of Kentucky or twice that of Massachusetts. The good news is that this rate (for the decade 2000–2010) is lower than in the 1990s, when 8.3 million ha (20.5 million acres) were lost worldwide each year.

However, in 2013, a large research team analyzed satellite data, published their work in the journal *Science*, and worked with staff from Google to create interactive maps of forest loss and gain across the world. They found more than twice as much deforestation as the FAO had: From 2000 to 2012, this team calculated annual losses of 19.2 million hectares and annual gains of 6.7 million hectares, for a net loss per year of 12.5 million hectares (30.9 million acres, larger than New York State).

## We deforested much of North America

Deforestation for timber and farmland propelled the expansion of the United States and Canada westward across the North American continent. The vast deciduous forests of the East were cleared by the mid-1800s, making way for countless small farms. Timber from these forests built the cities of the Atlantic seaboard and the upper Midwest.

As a farming economy shifted to an industrial one, wood was used to stoke the furnaces of industry. Logging operations moved south to the Ozarks of Missouri and Arkansas, and then to the pine woodlands and bottomland hardwood forests of the South, which were logged and converted to pine plantations. Once mature trees were removed from these areas, timber companies moved west, cutting the continent's biggest trees in the Rocky Mountains, the Sierra Nevada, the Cascade Mountains, and the Pacific Coast ranges. Exploiting forest resources helped American society to develop, but we were not harvesting forests sustainably. Instead, we were depleting our store of renewable resources for the future.

By the 20th century, very little **primary forest**—natural forest uncut by people—remained in the lower 48 U.S. states, and today even less is left (**FIGURE 9.5**). Nearly all the large oaks and maples found in eastern North America today, and even most redwoods of the California coast, are *second-growth* trees: trees that sprouted after old-growth trees were cut. Second-growth trees characterize **secondary forest,** which contains smaller, younger trees than does primary forest. The species composition, structure, and nutrient balance of a secondary forest may differ markedly from the primary forest that it replaced.

## Forests are being cleared most rapidly in developing nations

Uncut primary forests still remain in many developing countries. These nations are in the position the United States and Canada enjoyed a century or two ago: having a resource-rich frontier that they can develop. Today's powerful industrial technologies allow these nations to exploit their resources and push back their frontiers even faster than occurred in North America. As a result, deforestation is most rapid today in

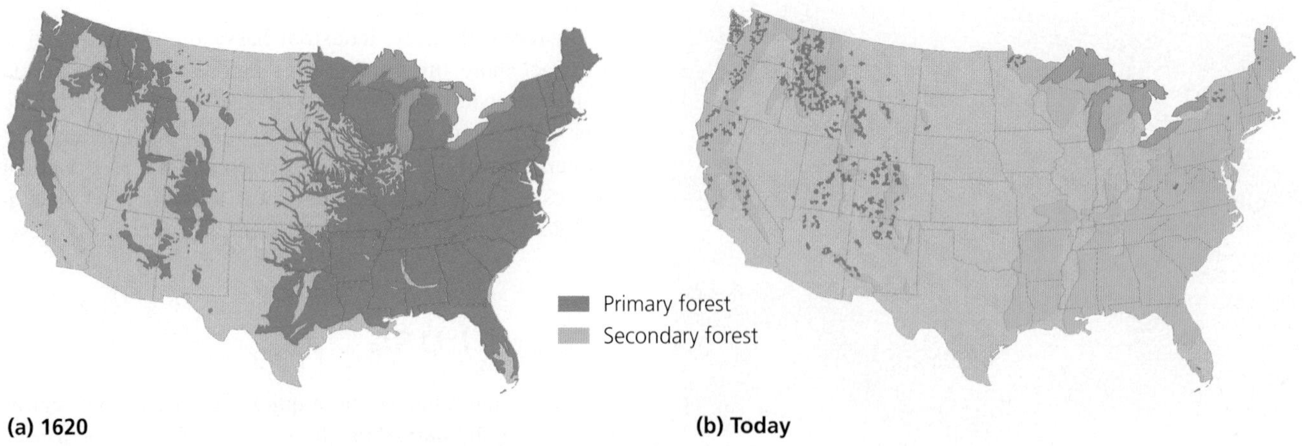


**FIGURE 9.5 Areas of primary (uncut) forest have been dramatically reduced.** When Europeans first colonized North America **(a)**, much of what is now the United States was covered in primary forest **(dark green)**. Today, nearly all this primary forest is gone **(b)**, cut for timber and to make way for agriculture. Much of the landscape has become reforested with secondary forest **(pale green)**. *Sources: Adapted from (a) U.S. Forest Service; and (b) maps by George Draffan, www.endgame.org, and Hansen, M. C. et al., 2013. High-resolution global maps of 21st-century forest cover change.* Science *342: 850–853.*

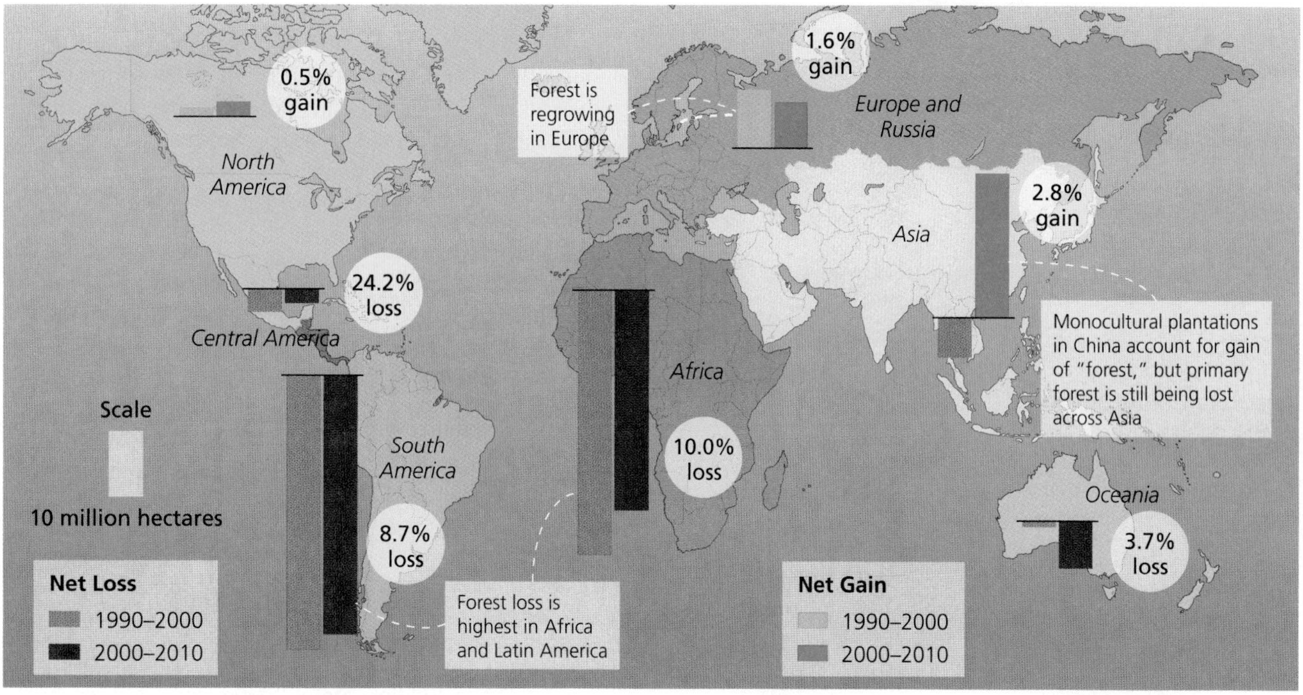


**FIGURE 9.6 Tropical forests are being lost.** Africa and Latin America are losing forests at rapid rates, whereas Europe and North America are slowly gaining secondary forest. In Asia, tree plantations are increasing, but natural primary forests are still being lost. *Data from Food and Agriculture Organization of the United Nations,* Global forest resources assessment 2010. *By permission.*

Indonesia, West Africa, and Latin America (**FIGURE 9.6**). In these regions, developing nations are striving to expand settlement for their burgeoning populations and to boost their economies by extracting natural resources. In contrast, parts of Europe and North America are gaining forest as they recover from past deforestation. This does not compensate for the loss of tropical forests in developing nations, however, because tropical forests are home to far more biodiversity than the temperate forests of North America and Europe.

The South American nation of Brazil, home to most of the vast Amazon rainforest, illustrates success in reducing deforestation but also the continuing pressures on tropical forests. Not long ago, Brazil was losing forests faster than any other country (**FIGURE 9.7**). Its government was promoting settlement on the forested frontier and was subsidizing an expansion of large-scale cattle ranching and soybean farming to meet demand from consumers in the United States and Europe. Since then, Brazil has reduced deforestation

**FIGURE 9.7 Deforestation of Amazonian rainforest has been rapid in recent decades.** Satellite images of the state of Rondonia in Brazil show extensive clearing resulting from settlement in the region.

**FIGURE 9.8 Oil palm plantations are replacing primary forest across Southeast Asia and Indonesia.** Since 1950, the immense island of Borneo (maps at bottom) has lost most of its forest.
*Data from Radday, M., WWF-Germany, 2007. Designed by Hugo Ahlenius, UNEP/GRID-Arendal.* Extent of deforestation in Borneo 1950–2001, and projection towards 2020. http://maps.grida.no/go/graphic/extent-of-deforestation-in-borneo-1950-2005-and-projection-towards-2020.

significantly. Today 80% of Brazil's forest remains intact, while the nation advances economically and politically as a stable democracy. However, in 2012 Brazil's legislature weakened the nation's Forest Code, which had helped to slow deforestation. President Dilma Rousseff vetoed some aspects of the legislation in an effort to balance conservation and development interests, but deforestation rose 28% over the following year.

Developing nations often are desperate enough for economic development and foreign capital that they impose few restrictions on logging. Often they allow their timber to be extracted by foreign multinational corporations, which pay fees for a **concession,** or right to extract the resource. Once a concession is granted, the corporation has little or no incentive to manage resources sustainably. Local people may receive temporary employment from the corporation, but once the timber is gone they no longer have the forest and the ecosystem services it provided. As a result, most economic benefits are short term and are reaped not by local residents but by the foreign corporation. Much of the wood extracted in developing nations is exported to Europe and North America. In this way, our consumption of high-end furniture and other wood products in industrialized nations can drive forest destruction in poorer nations.

Throughout Southeast Asia and Indonesia today, vast swaths of tropical rainforest are being cut to establish plantations of oil palms (**FIGURE 9.8**). Oil palm fruit produces palm oil, which we use in snack foods, soaps, cosmetics, and now as a biofuel. In Indonesia, the world's largest palm oil producer, oil palm plantations have displaced over 6 million ha (15 million acres) of rainforest. Clearing for plantations encourages further development and eases access for people to enter the forest and conduct logging illegally. The palm oil boom represents a conundrum for environmental advocates. Many people eager to fight climate change had urged the development of biofuels (pp. 382–384) to replace fossil fuels. Yet grown at the large scale that our society is demanding, monocultural plantations of biofuel crops such as oil palms are causing severe environmental impacts by displacing natural forests.

## WEIGHING THE ISSUES

**Logging Here or There** Suppose you are an activist protesting the logging of old-growth trees near your hometown. Now let's say you know that if the protest is successful, the company will move to a developing country and cut its primary forest instead. Would you still protest the logging in your hometown? Would you pursue any other approaches?

## Solutions are emerging

New solutions are being proposed to address deforestation in developing nations. Some conservation proponents are pursuing community-based conservation projects (p. 182) that empower local people to act as stewards of their forest resources. In other cases, conservation organizations are buying concessions and using them to preserve forest rather than to cut it down.

In Indonesia, NewPage Corporation, the supplier of this textbook's paper, has funded a project called POTICO (Palm Oil, Timber and Carbon Offsets) that aims to reduce deforestation and illegal logging. In this evolving project, the nonprofit World Resources Institute (WRI) is working with palm oil companies that own concessions to clear primary rainforest and is steering them instead to plant plantations on land that is already logged and degraded. WRI will then protect the primary forests that were slated for conversion or allow the forests to become certified to the FSC forest management standard. Because primary forests store more carbon than oil palm plantations, these land swaps can reduce Indonesia's greenhouse gas emissions and qualify for credit via carbon offsets (p. 328).

Carbon offsets are central to emerging international plans to curb deforestation and climate change. Forest loss accounts for at least 12% of the world's greenhouse gas emissions—nearly as much as all the world's vehicles emit. At recent international climate conferences (pp. 325–326), negotiators have outlined a program called *Reducing Emissions from Deforestation and Forest Degradation (REDD),* whereby wealthy industrialized nations pay poorer developing nations to conserve forest. The aim is to make forests more valuable when saved than when cut down. Under this plan, poor nations gain income while rich nations receive carbon credits to offset their emissions in an international cap-and-trade system (pp. 111, 326–327). Although the REDD plan has not been formally agreed to, leaders of rich nations have proposed to transfer $100 billion per year to poor nations by 2020. If this occurs, much of the funding could end up going toward REDD.

# Forest Management

Our demand for forest resources and amenities is rising, so we need to take care in managing forests. *Foresters* are professionals who manage forests through the practice of **forestry**. Foresters must balance our society's demand for forest products against the central importance of forests as ecosystems. Today, sustainable forest management practices are spreading as informed consumers demand sustainably produced products. Just as your textbook uses FSC-certified paper from sustainably managed forests, more and more paper, lumber, and other forest products are now made using certified sustainable practices.

Debates over how to manage forest resources reflect broader questions about how to manage natural resources in general. Resources such as fossil fuels and many minerals are nonrenewable, whereas resources such as the sun's energy are perpetually renewable (pp. 2–3). Between these extremes lie resources that are renewable if they are not exploited too rapidly. These include timber, as well as soils, fresh water, rangeland, wildlife, and fisheries.

**Resource management** describes our use of strategies to manage and regulate the harvest of renewable resources. Sustainable resource management involves harvesting these resources in ways that do not deplete them. Resource managers are guided by research in the natural sciences and by social, political, and economic factors.

## Resource managers follow several strategies

A key question in managing resources is whether to focus strictly on the resource of interest or to look more broadly at the environmental system of which it is a part. Taking a broader view often helps avoid degrading the system and thereby helps to sustain the resource.

**Maximum sustainable yield** A guiding principle in resource management has been **maximum sustainable yield**. Its aim is to achieve the maximum amount of resource extraction without depleting the resource from one harvest to the next. Recall the logistic growth curve (see Figure 3.14, p. 62), which shows how a population grows most quickly when it is at an intermediate size—specifically, at one-half of carrying capacity. A fisheries manager aiming for maximum sustainable yield will therefore prefer to keep fish populations at intermediate levels so that they rebound quickly after each harvest. Doing so should result in the greatest amount of fish harvested over time while sustaining the population indefinitely (**FIGURE 9.9**).

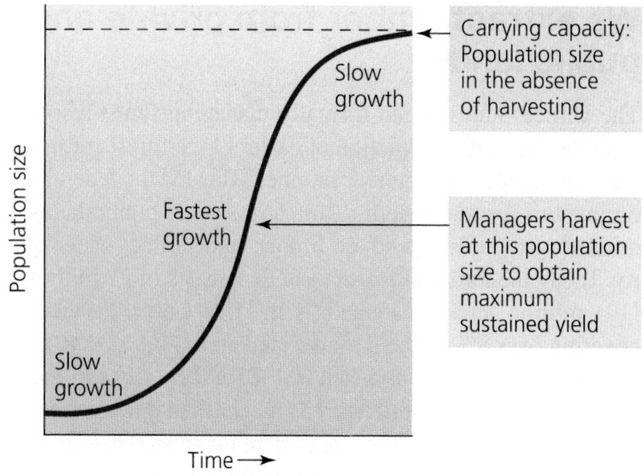


**FIGURE 9.9 Maximum sustainable yield maximizes the amount of resource harvested while sustaining the harvest in perpetuity.** For a wildlife population or fisheries stock that grows according to logistic growth, managers aim to keep the population at half the carrying capacity, because populations grow fastest at intermediate sizes.

This management approach, however, keeps the fish population at only half its carrying capacity—well below the size it would attain in the absence of fishing. Suppressing population size in this way will affect other species and alter the food web dynamics of the community. From an ecological point of view, managing for maximum sustainable yield may set in motion significant changes.

In forestry, maximum sustainable yield argues for cutting a stand of trees shortly after it goes through its fastest stage of growth. Because a stand of trees often increases in biomass most quickly at an intermediate age, trees are generally cut long before they grow as large as they would in the absence of harvesting. This practice maximizes timber production over time, but it also alters forest ecology and eliminates habitat for species that depend on mature trees.

**Ecosystem-based management** Because of these dilemmas, more and more managers espouse **ecosystem-based management**, which aims to minimize impact on the ecological processes that provide the resource. Many certified sustainable forestry plans protect certain forested areas, restore ecologically important habitats, and consider patterns at the landscape level (p. 34), allowing timber harvesting while preserving the functional integrity of the forest ecosystem. Ecosystems are complex, however, so it can be challenging to determine how best to implement this type of management. As a result, ecosystem-based management has come to mean different things to different people.

**Adaptive management** Some management actions will succeed, and some will fail. **Adaptive management** involves testing different approaches and trying to improve methods through time. For managers, it entails monitoring the results of one's practices and adjusting them as needed, based on what is learned. Adaptive management is intended to be a fusion of science and management, because it explicitly tests hypotheses about how best to manage resources.

## We extract timber from private and public lands

The United States began formally managing forest resources a century ago, after depletion of eastern U.S. forests prompted widespread fear of a "timber famine." This led the federal government to form a system of forest reserves: public lands set aside to grow trees, produce timber, and protect water quality. Today the **national forest** system consists of 77 million ha (191 million acres), managed by the U.S. Forest Service and covering over 8% of the nation's land area (**FIGURE 9.10**). The Forest Service was established in 1905 under Gifford Pinchot, whose conservation ethic (p. 13) meant managing forests for "the greatest good of the greatest number in the long run." Pinchot believed the nation should extract and use resources from its public lands, but that wise and careful management of timber resources was imperative.

Today most timber harvesting in the United States takes place on private land owned by the timber industry or by small landowners (**FIGURE 9.11**). Timber companies pursue maximum sustainable yield on their land, so as to obtain maximal profits year after year. Timber companies also extract timber from public forests. On the national forests, U.S. Forest Service employees conduct timber sales and build roads to provide access for logging companies, which sell the timber they harvest for profit. In this way, taxpayers subsidize private harvesting on public land (p. 110).

On the U.S. national forests, timber extraction increased in the 1950s as the nation underwent a postwar economic boom, paper consumption rose, and the growing population moved into newly built suburban homes. Harvests began to decrease in the 1980s as economic trends shifted, public concern over clear-cutting grew, and management evolved. By 2006, regrowth was outpacing removal on national forests by 11 to 1 (see Figure 9.11). At present, in an average year, about 2.1% of U.S. forest acreage is cut for timber. Across public and private land together, timber harvesting in the United States and other developed nations has remained stable for the past half-century, while it has more than doubled in developing nations.

Note, however, that even when the regrowth of trees outpaces their removal, the character of a forest may change. Once primary forest is replaced by secondary forest or single-species plantations, the resulting community may be very different and generally less ecologically valuable.

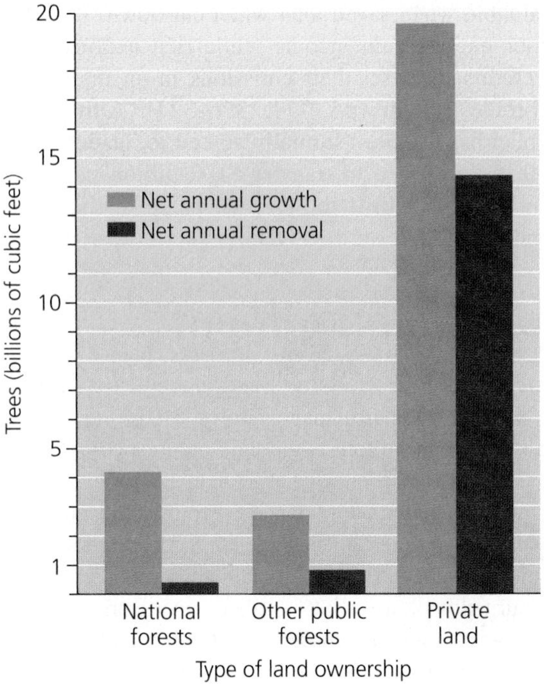


**FIGURE 9.11 In the United States today, trees are growing faster than they are being removed.** However, forests that regrow after logging often differ substantially from the forests that were removed. *"Private land" here combines land owned by the timber industry and by small landholders. Data are for 2006, from USDA Forest Service, 2008.* Forest resources of the United States, 2007. *U.S. Department of Agriculture, Washington, D.C.*

**DATA Q** For which of the three types of land is the ratio of growth to removal greatest?

GO TO **INTERPRETING GRAPHS & DATA** ON MasteringEnvironmentalScience®

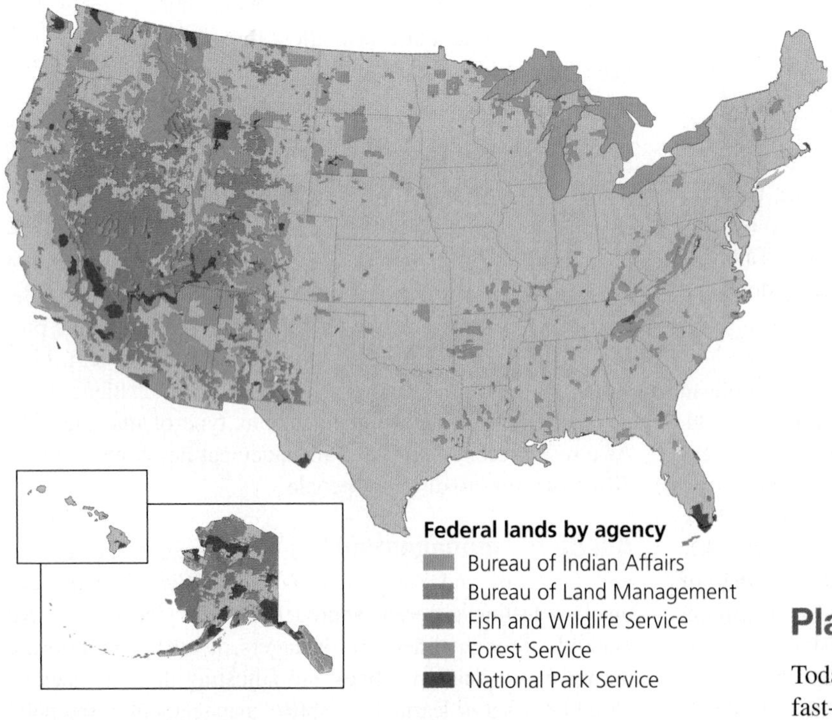


**FIGURE 9.10 U.S. citizens enjoy over 250 million ha (600 million acres) of public lands.** *Data from United States Geological Survey.*

## Plantation forestry has grown

Today's timber industry focuses on production of fast-growing tree species planted in single-species monocultures (p. 137). All trees in a stand are planted at the same time, so the stands are *even-aged*;

**FIGURE 9.12 Even-aged management, with all trees of equal age, is seen in the foreground in a plantation that is regrowing after having been clear-cut.** Uneven-aged management maintains a mix of tree ages, as seen in the more mature forest.

**FIGURE 9.13 Clear-cutting is cost-efficient for timber companies but has ecological consequences.** These include soil erosion, water pollution, and altered community composition.

that is, all trees are the same age (**FIGURE 9.12**). Stands are cut after a certain number of years (the *rotation time*), and the land is replanted with seedlings. Plantation forestry is growing, and 7% of the world's forests are now plantations. One-quarter of these feature non-native tree species.

Ecologists and foresters view plantations as akin to crop agriculture. Because there are few tree species and little variation in tree age, plantations do not offer habitat to many forest organisms. For instance, stands of red pine planted near Escanaba, Michigan, host far less biodiversity than the more-diverse forests that surround them. Plantations lack the structural complexity that characterizes a mature natural forest shown in Figure 9.3 (p. 188). Plantations are also vulnerable to outbreaks of pest species such as bark beetles, as we shall soon see. For all these reasons, some harvesting methods aim to maintain *uneven-aged* stands, where a mix of ages (and often a mix of tree species) makes the stand more similar to a natural forest.

## We harvest timber by several methods

Timber companies choose from several methods to harvest trees. In the simplest method, **clear-cutting**, all trees in an area are cut at once (**FIGURE 9.13**). Clear-cutting is cost-efficient, and to some extent it can mimic natural disturbance events such as fires, tornadoes, or windstorms. However, the ecological impacts of clear-cutting are considerable. An entire ecological community is removed, soil erodes away, and sunlight penetrates to ground level, changing microclimatic conditions. As a result, new types of plants replace those of the original forest.

Concerns about clear-cutting led foresters and the timber industry to develop alternative harvesting methods. In seed-tree systems and shelterwood systems, a few large trees are left standing in clear-cuts to reseed the area or to provide shelter for seedlings. In selection systems, a minority of trees is removed at any one time, while most are left standing. Selection systems preserve much of a forest's structural diversity, but are less cost-efficient for the industry.

All harvesting methods disturb soil, alter habitat, and affect plants and animals. All methods modify forest structure and composition. Most methods speed runoff, raise flooding risk, and increase soil erosion, thereby degrading water quality. When steep hillsides are clear-cut, landslides can result. Finding ways to minimize these impacts is important, because timber harvesting is necessary to obtain the wood products that all of us use.

## Forest management has evolved over time

As people became more aware of the impacts of clear-cutting, many citizens began to urge that public forests be managed for recreation, wildlife, and ecological integrity, as well as for timber. In 1976 the U.S. Congress passed the **National Forest Management Act**, mandating that every national forest draw up plans for renewable resource management, subject to public input under the National Environmental Policy Act (p. 106). Guidelines specified that these plans assess the ecological impacts of logging. As a result, timber-harvesting methods were integrated with ecosystem-based management goals, and the Forest Service developed programs to manage wildlife and restore degraded ecosystems.

The Hiawatha National Forest on Michigan's Upper Peninsula provides an example of management under the National Forest Management Act. Its most recent management plan seeks a balance of uses, aiming to:

- Carefully manage timber harvesting.
- Monitor fish and wildlife populations.
- Allow diverse recreation—hiking, fishing, boating, and more—in specified areas.
- Prohibit livestock grazing, protect historic and archaeological sites, and build no new roads.
- Maintain habitats for threatened plants and animals; restore wetlands, streams, forests, and soils; and protect stands of old-growth forest.

## Fire can hurt or help forests

One key area of policy debate involves how to handle wildfire. Smokey Bear, the Forest Service's beloved cartoon bear in a ranger's hat, advises us to fight forest fires—and for over a century, the Forest Service and other agencies suppressed fire whenever and wherever it broke out. Yet scientific research shows that many species and ecological communities depend on fire. Some plants have seeds that germinate only in response to fire, and researchers studying tree rings have documented that North America's grasslands and pine woodlands burned frequently. (Burn marks in a tree's growth rings reveal past fires, giving scientists an accurate history of fire events extending back hundreds or even thousands of years.)

In the long term, suppressing frequent, low-intensity fires leads to adverse effects on ecosystems that depend on fire: Shrubs invade grasslands, and pine woodlands become cluttered with hardwood understory. Invasive plants move in, and animal diversity and abundance decline. In addition, limbs, logs, sticks, and leaf litter accumulate on the forest floor, producing kindling for a catastrophic fire—the type that damages forests, destroys property, and threatens human lives. Severe fires have become more numerous in recent years (**FIGURE 9.14**). At the same time, residential development alongside forested land—in the *wildland-urban interface*—is placing more homes in fire-prone situations (**FIGURE 9.15**).

To reduce fuel loads, protect property, and improve the condition of forests, land management agencies now burn areas of forest intentionally with low-intensity fires under carefully controlled conditions (see Figure 1.7b, p. 8). These **prescribed burns** clear away fuel loads, nourish the soil with ash, and encourage the vigorous growth of new vegetation.

Once a fire burns a forest, it may leave many dead trees. The removal of dead trees, or snags, following a natural disturbance (such as a fire, windstorm, insect damage, or disease) is called **salvage logging.** From a short-term economic standpoint, salvage logging may seem to make good sense. However, ecologically, snags have immense value; the insects that decay them provide food for wildlife, and many animals depend on holes in snags for nesting and roosting. Removing timber from recently burned land can also cause soil erosion, impede forest regeneration, and promote further wildfire.

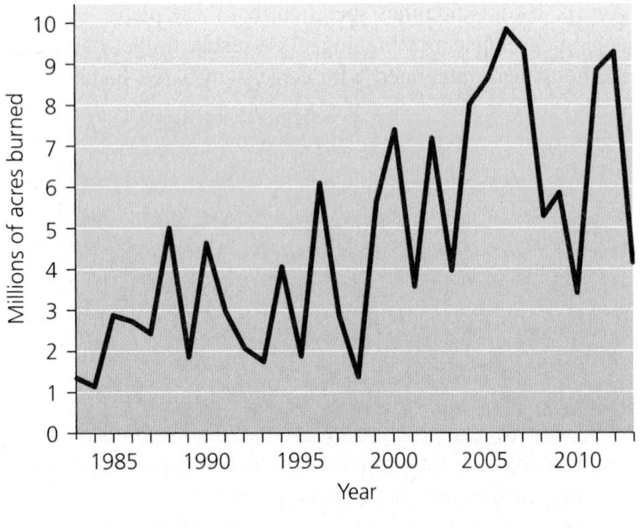


**FIGURE 9.14 Wildfires have become larger and more numerous in the United States.** Fuel buildup from decades of fire suppression has contributed to this trend. *Data from National Interagency Fire Center.*

**FIGURE 9.15 Habitual suppression of fire has led to catastrophic wildfires that damage forests and threaten homes.** To avoid these unnaturally severe fires, ecologists suggest we allow natural fires to burn when we can and conduct prescribed burns to reduce fuel loads and restore forest ecosystems.

### FAQ Aren't all forest fires bad?

No. Fire is a natural process that helps to maintain the health of many forests and grasslands. When allowed to occur naturally, low-intensity fires generally burn moderate amounts of material, return nutrients to the soil, and promote lush growth of new vegetation. When we suppress fire, we allow unnaturally large amounts of dead wood, dried grass, and leaf litter to accumulate. This material becomes kindling that eventually can feed a truly damaging fire that grows too big and too hot to control. This is why many land managers today conduct carefully controlled prescribed burns and also allow some natural fires to run their course. By doing so, they aim to help return our fire-dependent ecosystems to a healthier and safer condition.

### WEIGHING THE ISSUES

**How to Handle Fire?**

A century of fire suppression has left vast areas of North American forests vulnerable to catastrophic wildfires. Prescribed burning helps to alleviate this risk, yet we do not have adequate resources to conduct careful prescribed burning over all these lands. Can you suggest solutions to help protect people's homes in the wildland-urban interface while improving the ecological condition of forests? Do you think people should be allowed to develop homes in fire-prone areas?

## Climate change and pest outbreaks are altering forests

Global climate change (Chapter 14) is now worsening wildfire risk by bringing warmer weather to North America and drier weather to the American West. Scientific climate models predict further warming and drying (p. 320).

Climate change also is promoting pest insects such as bark beetles, which feed within the bark of conifer trees. These beetles attract one another to weakened trees and attack en masse, eating tissue, laying eggs, and bringing with them a small army of fungi, bacteria, and other pathogens. Bark beetle infestations can wipe out vast areas of trees with amazing speed (**FIGURE 9.16**). Since the 1990s, beetle outbreaks have devastated more than 11 million ha (27 million acres) of forest in western North America, killing an estimated 30 *billion* conifer trees.

Scientists find two primary reasons for today's unprecedented outbreaks. One is that past forest management has resulted in even-aged forests across large regions, and many trees in these forests are now at a prime age for beetle infestation. Most at risk are plantation forests dominated by single species that the beetles prefer. The second reason is climate change. Milder winters allow beetles to overwinter further north, and warmer summers speed up their feeding and reproduction. In Alaska, beetles have switched from a two-year life cycle to a one-year cycle. In parts of the Rocky Mountains, they now produce two broods per year instead of one. Meanwhile, droughts like those that have plagued the western and southern United States in recent years have stressed and weakened trees, making them vulnerable to attack.

As climate change interacts with pests, diseases, and management strategies, our forest systems could be altered in profound ways. Already, many dense, moist forests devastated by beetles have been replaced by drier woodlands, shrublands, or grasslands.

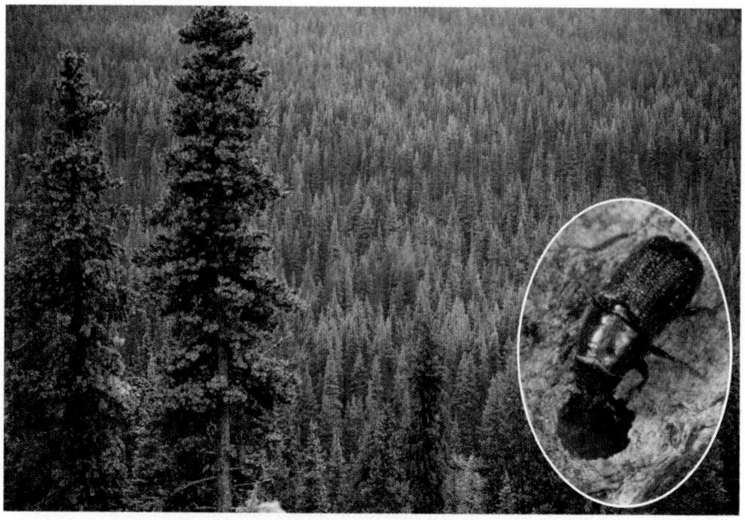

**FIGURE 9.16 Forests face more threats as temperatures rise.** Warmer summers and milder winters favor bark beetles, which are killing vast areas of trees throughout the West, especially in even-aged plantations.

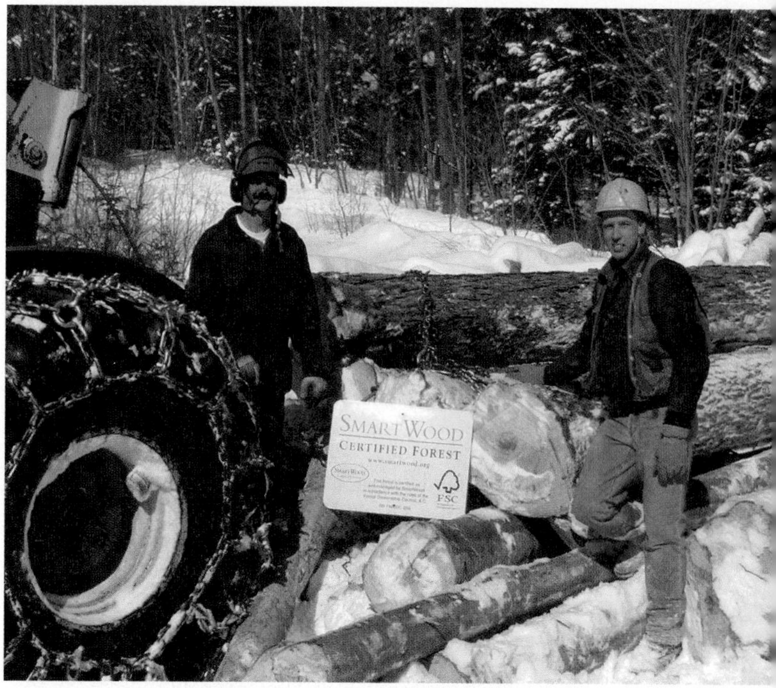


**FIGURE 9.17 Inspectors work with loggers in Michigan's Upper Peninsula to confirm that timber is harvested in accordance with FSC criteria for sustainable harvesting.**

## Sustainable forestry is gaining ground

We can help address all these challenges through sustainable forestry practices. Although the world overall is losing forested land, the FAO's 2010 global assessment showed that more and more forests are being managed for conservation of soil, water, and biodiversity and that we are increasingly getting our timber efficiently from plantations.

Any company can claim that its timber harvesting practices are sustainable, but how is the purchaser of wood products to know whether they really are? Organizations such as the Forest Stewardship Council (FSC) examine practices and rate them against criteria for sustainability. They grant **sustainable forest certification** to forests, companies, and products produced using methods they judge to be sustainable (**FIGURE 9.17**).

Several certification organizations exist, but the Forest Stewardship Council is considered to have the strictest standards. The paper for this textbook is "chain-of-custody certified" by the FSC, meaning that all steps in the life cycle of the paper's production—from timber harvest to transport to pulping to production—have met strict standards. FSC certification rests on 10 general principles and 56 more-detailed criteria.

The number of FSC-certified forests, companies, and products is growing quickly. More than 8% of the world's forests managed for timber production are now FSC certified—over 184 million ha (450 million acres) in 80 nations as of 2014. The growth of certification is tied to the increasing construction of green buildings (pp. 416–417).

Pursuing sustainable forestry practices is often more costly for producers, but producers recoup these costs when consumers pay more for certified products. Consumer demand has led Home Depot and other major retailers to carry sustainable wood, and these retailers' purchasing decisions are influencing timber harvesting practices around the world. You can look for the logos of certifying organizations on forest products where they are sold. If certification standards are kept strong, then we as consumers can exercise choice in the marketplace and thereby help to promote sustainable forestry practices.

# Parks and Protected Areas

As our world fills with more people consuming more resources, the sustainable management of resources from forests and other ecosystems becomes ever more important. So does our need to preserve functional ecosystems by setting aside tracts of undisturbed land to remain forever undeveloped. Today about 12–13% of the world's land area is designated for preservation in various types of parks, reserves, and protected areas.

## Why create parks and reserves?

People establish parks and protected areas for several reasons:

- Enormous or unusual scenic features inspire people to preserve them (**FIGURE 9.18**).
- Protected areas offer hiking, fishing, hunting, kayaking, bird-watching, and other recreation.
- Parks generate revenue from ecotourism (pp. 63, 168).
- Undeveloped land offers us peace of mind, health, exploration, wonder, and spiritual solace.
- Protected areas offer utilitarian benefits and ecosystem services. For example, undeveloped watersheds provide cities clean drinking water and a buffer against floods.
- Reserves protect biodiversity. These islands of habitat maintain species, communities, and ecosystems.

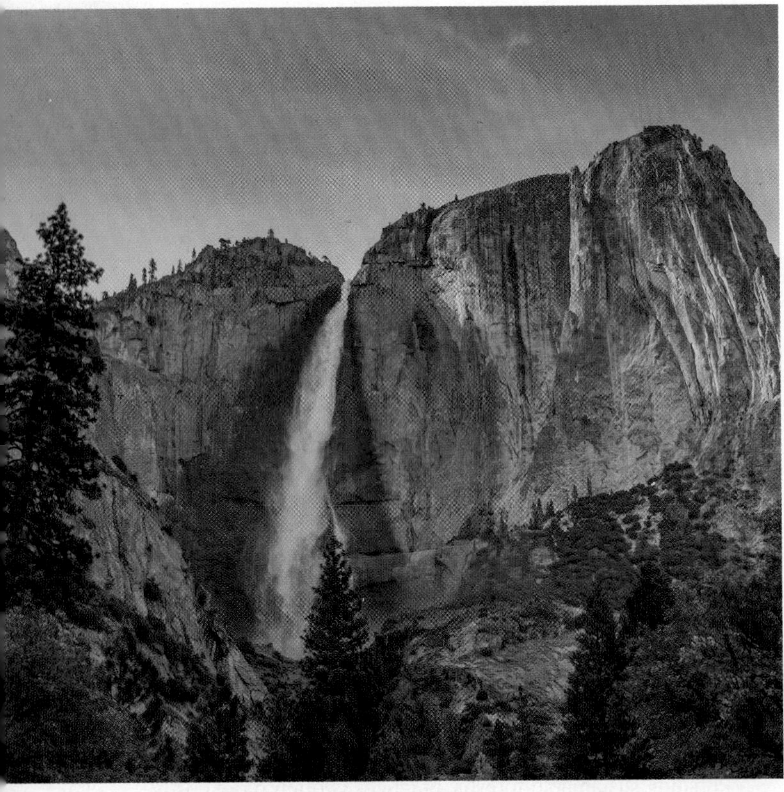

**FIGURE 9.18 The splendor of places such as Yosemite draws millions of people to America's national parks.**

## Federal parks and reserves began in the United States

The striking scenery of the American West persuaded U.S. leaders to create the world's first **national parks,** public lands protected from resource extraction and development but open to nature appreciation and recreation. Yellowstone National Park was established in 1872, followed by Sequoia, General Grant (now Kings Canyon), Yosemite, Mount Rainier, and Crater Lake National Parks.

The National Park Service was created in 1916 to administer the growing system of parks and monuments, which today numbers 401 sites totaling 34 million ha (84 million acres) and includes national historic sites, national recreation areas, national seashores, and other areas (see Figure 9.10).

Because America's national parks are open to everyone and showcase the nation's natural beauty in a democratic way, writer Wallace Stegner famously called them "the best idea we ever had." A mill worker in Escanaba can take his or her family for the weekend to Pictured Rocks National Lakeshore, where they can camp along sandstone cliffs on the shore of Lake Superior. They can head across the lake to remote Isle Royale National Park to hike and canoe through lands that are home to moose and wolves. Or they can cross Lake Michigan and enjoy climbing immense sand dunes at Sleeping Bear Dunes National Lakeshore.

Another type of federal protected area in the United States is the **national wildlife refuge.** The national wildlife refuge system, begun in 1903 by President Theodore Roosevelt, now totals over 560 sites comprising 61 million ha (151 million acres; see Figure 9.10). The U.S. Fish and Wildlife Service administers these refuges, which serve as havens for wildlife and encourage hunting, fishing, wildlife observation, photography, environmental education, and other public uses.

In response to the public's desire for undeveloped areas of land, in 1964 the U.S. Congress passed the Wilderness Act, which allowed some areas of existing federal lands to be designated as **wilderness areas.** These areas are off-limits to development but are open to hiking, nature study, and other low-impact public recreation. Overall the nation has more than 750 wilderness areas totaling 44 million ha (110 million acres), covering 5% of U.S. land area (2.7% outside Alaska).

## Many agencies and groups protect land

Efforts to set aside land at the national level are paralleled at regional and local levels. Each U.S. state has agencies that manage resources on public lands, as do many counties and municipalities. When Mackinac Island, the nation's second national park, was transferred to the state of Michigan, it became the first officially designated state park in the nation. Today it is one of nearly 7000 state parks across the United States, along with regional parks, county parks, and others.

Private nonprofit groups also preserve land. **Land trusts** are local or regional organizations that purchase land to preserve in its natural condition. The Nature Conservancy is the world's largest land trust, but nearly 1700 local land trusts in the United States together own 870,000 ha (2.1 million acres) and have helped preserve an additional 5.6 million ha (13.9 million acres), including scenic areas such as Big Sur on the California coast, Jackson Hole in Wyoming, and Maine's Mount Desert Island. Moreover, thousands of local volunteer groups (often named "Friends of" an area) have organized from the grassroots to help care for protected lands.

## Parks and reserves are increasing internationally

Many nations have established protected areas and are benefiting from ecotourism as a result—from Kenya and Tanzania (Chapter 8) to Costa Rica (Chapter 5) to Belize to Ecuador to India to Australia. Worldwide area in parks and reserves has increased nearly sevenfold since 1970, and today the world's 158,000 protected areas cover about 12.7% of the planet's land area. However, parks do not always receive funding adequate to manage resources, provide recreation, and protect wildlife from poaching and trees from logging. As a result, many of the world's protected areas are merely "paper parks"—protected on paper but not in reality.

Some types of protected areas are designated or partly managed by the United Nations. **Biosphere reserves** are tracts of land with exceptional biodiversity that couple preservation with sustainable development (pp. 111–112) to benefit local people. Each biosphere reserve consists of (1) a core area that preserves biodiversity; (2) a buffer zone that allows local activities and limited development; and (3) an outer transition zone where agriculture, settlement, and other land uses are pursued sustainably (**FIGURE 9.19**).

**World heritage sites** are another type of international protected area. Nearly 1000 sites in more than 150 countries are listed for their natural or cultural value. One such site is a reserve for mountain gorillas shared by three African countries. This reserve, which integrates national parklands of Rwanda, Uganda, and the Democratic Republic of Congo, is also an example of a *transboundary park*, a protected area overlapping national borders. Some transboundary reserves function as "peace parks," acting as buffers between nations that quarrel over boundaries.

Beyond all these efforts on land, the importance of conserving the oceans' natural resources is leading us to establish protected areas and reserves in marine waters (p. 275). Today 1.6% of the world's ocean area and 7.2% of coastal waters fall within designated protected areas.

## Habitat fragmentation makes preserves more vital

Protecting large areas of land has taken on new urgency now that scientists understand the risks posed by habitat fragmentation (see **THE SCIENCE BEHIND THE STORY**, pp. 200–201). Expanding agriculture, spreading cities, highways, logging, and other impacts routinely divide large expanses of habitat into small, disconnected ones (see Figure 8.8, p. 171). Forests are being fragmented everywhere these days. Even where

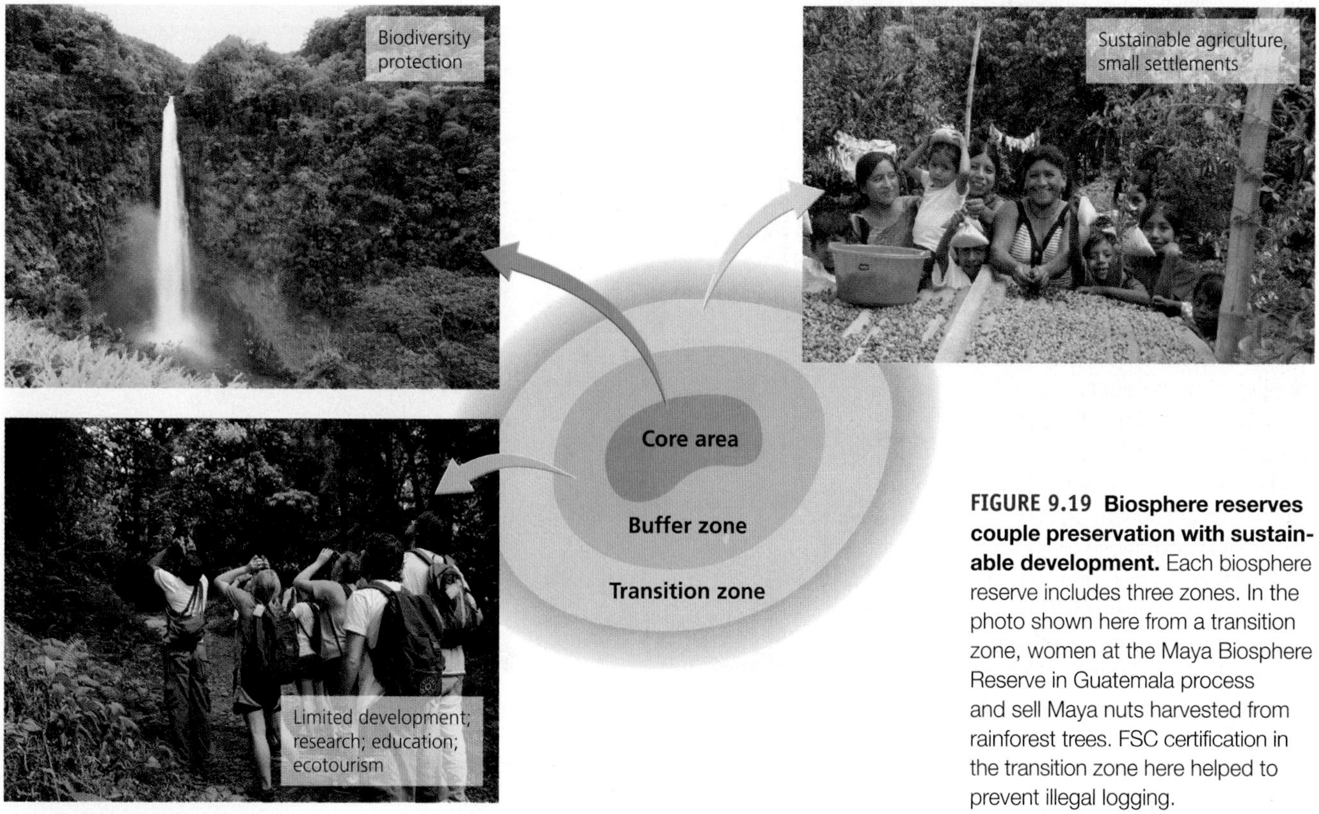


**FIGURE 9.19 Biosphere reserves couple preservation with sustainable development.** Each biosphere reserve includes three zones. In the photo shown here from a transition zone, women at the Maya Biosphere Reserve in Guatemala process and sell Maya nuts harvested from rainforest trees. FSC certification in the transition zone here helped to prevent illegal logging.

# Forest Fragmentation in the Amazon

**Dr. Thomas Lovejoy, founder of the BDFFP**

What happens to animals, plants, and ecosystems when we fragment a forest? A massive research experiment in the middle of the Amazon rainforest is helping scientists learn the answers.

Stretching across 1000 $km^2$ (386 $mi^2$), the Biological Dynamics of Forest Fragments Project (BDFFP) is the world's largest and longest-running experiment on forest fragmentation. For over 30 years, hundreds of researchers have muddied their boots here, publishing over 700 scientific research papers, graduate theses, and books.

The story begins back in the 1970s as conservation biologists debated how to apply ecological theory from ocean islands to forested landscapes that were being fragmented into islands of habitat. Biologist Thomas Lovejoy decided some good hard data were needed. He conceived a huge experiment to test ideas about forest fragmentation and established it in the heart of the biggest primary rainforest on the planet—South America's Amazon rainforest.

Farmers, ranchers, loggers, and miners were streaming into the Amazon then, and deforestation was rife. If scientists could learn how large fragments had to be in order to retain their species, this information would help them work with policymakers to preserve forests in the face of development pressures.

Lovejoy's team of Brazilians and Americans worked out a deal with Brazil's government: Ranchers could clear some forest within the study area if they left square plots of forest standing as fragments within those clearings. By this process, 11 fragments of three sizes (1, 10, and 100 ha [2.5, 25, and 250 acres]) were left standing, isolated as "islands" of forest, surrounded by "seas" of cattle pasture (**FIGURE 1**). Each fragment was fenced to keep cattle out. Then, 12 study plots (of 1, 10, 100, and 1000 ha) were established within the large expanses of continuous forest still surrounding the pastures; these would serve as control plots to compare the fragments against.

Besides comparing treatments (fragments) and controls (continuous forest), the project also surveyed populations in fragments before and after they were isolated. These data on trees, birds, mammals, amphibians, and invertebrates showed declines in the diversity of most groups (**FIGURE 2**).

As researchers studied the plots over the years, they found that small fragments lost more species, and lost them faster, than large fragments—just as theory predicts. To slow down species loss by 10 times, researchers found that a fragment needs to be 1000 times bigger. Even 100-ha fragments were not large enough for some animals, and they lost half their species in less than 15 years. Monkeys died out because they need large ranges. So did colonies of army ants and the birds that follow them to eat insects scared up as the ants swarm across the forest floor.

Fragments distant from continuous forest lost more species, but data revealed that even very small openings can stop organisms from dispersing to recolonize fragments. Many understory birds adapted to deep interior forest would not traverse cleared areas of only 30–80 m (100–260 ft). Distances of just 15–100 m (50–330 ft) were insurmountable for some bees, beetles, and tree-dwelling mammals.

Soon, a complication ensued: Ranchers abandoned many of the pastures because the soil was unproductive, and these

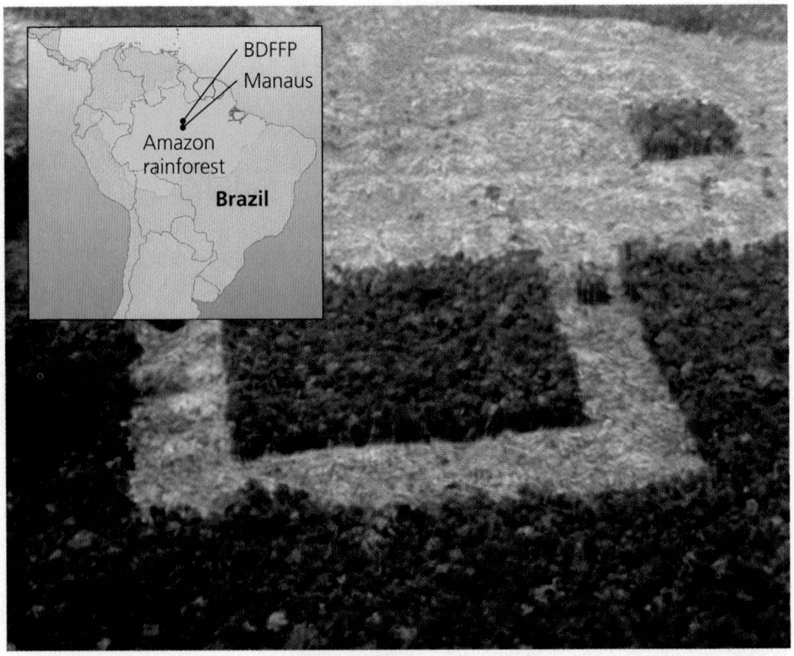


**FIGURE 1 Experimental forest fragments of 1, 10, and 100 ha were created in the BDFFP.**

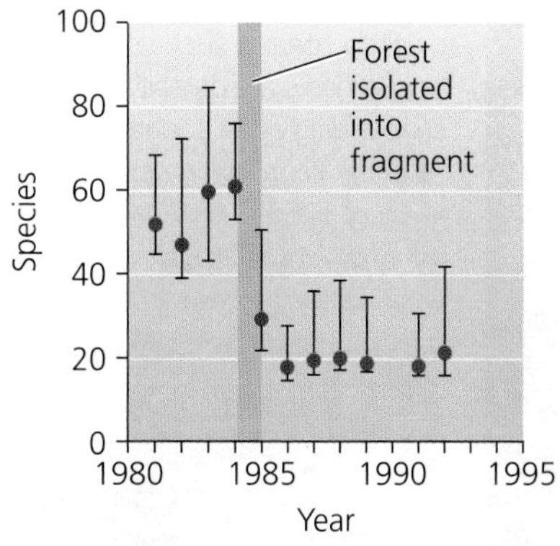


**FIGURE 2 Species richness of understory birds declined in this 1-ha forest plot after it was isolated as a fragment in 1984.** *Error bars show statistical uncertainty around mean estimates. Data from Ferraz, G., et al., 2003. Rates of species loss from Amazonian forest fragments.* Proc. Natl. Acad. Sci. *100: 14069–14073. © 2003 National Academy of Sciences. By permission.*

**DATA Q** On average, about how many bird species were present in this forest plot before fragmentation? How many were present after fragmentation?
GO TO **INTERPRETING GRAPHS & DATA** ON MasteringEnvironmentalScience®

areas began filling in with young secondary forest, making the fragments less like islands. However, this led to new insights. Researchers learned that secondary forest can act as a corridor for some species, allowing them to disperse from mature forest and recolonize fragments where they'd disappeared. By documenting which species did this, scientists learned which may be more resilient to fragmentation.

The secondary forest habitat also introduced new species—generalists adapted to disturbed areas. Frogs, leaf-cutter ants, and small mammals and birds that thrive in second growth soon became common in adjacent fragments. Open-country butterfly species moved in, displacing interior-forest butterflies.

The invasion of open-country species illustrated one type of edge effect (p. 202). There were more: Edges receive more sunlight, heat, and wind than interior forest, which can kill trees adapted to the dark, moist interior. Tree death releases carbon dioxide, worsening climate change. Sunlight also promotes growth of vines and shrubs that create a thick, tangled understory along edges. As BDFFP researchers documented these impacts, they found that many edge effects extended deep into the forest (**FIGURE 3**). Small fragments essentially became "all edge."

The results on edge effects are relevant for all of Amazonia, because forest clearance and road construction create an immense amount of edge. A satellite image study in the 1990s estimated that for every 2 acres of land deforested, 3 acres were brought within 1 km of a road or pasture edge.

BDFFP scientists emphasize that impacts across the Amazon will be more severe than the impacts revealed by their experiments. This is because most real-life fragments (1) are not protected from hunting, logging, mining, and fires; (2) do not have secondary forest to provide connectivity; (3) are not near large tracts of continuous forest that provide recolonizing species and maintain humidity and rainfall; and (4) are not square in shape, and thus feature more edge. Scientists say their years of data argue for preserving numerous tracts of Amazonian forest that are as large as possible.

The BDFFP has inspired another large-scale, long-term study, this one located in Borneo. The Stability of Altered Forest Ecosystems (SAFE) project is documenting changes that take place in tropical forests as they are logged and converted to oil palm plantations. This project should soon begin providing data and insights that may help us conserve biodiversity amid the rush to oil palm plantations in Southeast Asia.

Ironically, today the BDFFP study site is itself threatened by forest fragmentation. A Brazilian government agency has settled colonists just outside the site and has proposed settling 180 families inside it. Development is proceeding up a new highway from Manaus, a city of 1.7 million people, to Venezuela. Wherever roads are built and people settle, logging, hunting, mining, and burning follow. BDFFP researchers can now hear chainsaws and shotgun blasts from their study plots. "It would be tragic," says BDFFP scientist William Laurance, "to see a site that's given us so much information be lost so easily." ☐

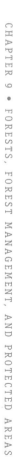

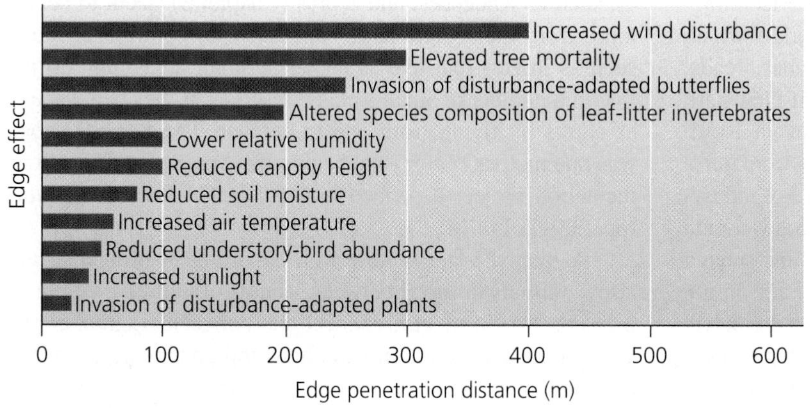


**FIGURE 3 Edge effects can extend far into the interior of forest fragments.** *Data from studies summarized in Laurance, W. F., et al., 2002. Ecosystem decay of Amazonian forest fragments: A 22-year investigation.* Conservation Biology *16: 605–618, Fig. 3, adapted. Reprinted by permission of John Wiley & Sons, Inc.*

**DATA Q** Would a tree inside a forest fragment, 275 meters in from the edge, be susceptible to edge effects? If so, which ones?
GO TO **INTERPRETING GRAPHS & DATA** ON MasteringEnvironmentalScience®

**(a) Fragmentation from clear-cuts in Mount Hood National Forest, Oregon**

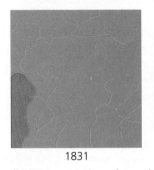

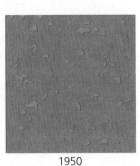

**(b) Fragmentation of wooded area (green) in Cadiz Township, Wisconsin**

**(c) Wood thrush**

**FIGURE 9.20 Forests are being fragmented, with ecological consequences.** Fragmentation results from clear-cutting **(a)**, agriculture, and residential development. Shown in **(b)** are historical changes in forested area in a region of Wisconsin. Fragmentation affects forest-dwelling species such as the wood thrush **(c)**, whose nests are parasitized by cowbirds from surrounding open country. *Source (b): Curtis, J. T., 1956. The modification of mid-latitude grasslands and forests by man. In Thomas, W. L. Jr., Ed.,* Man's role in changing the face of the earth. *©1956. Used by permission of the publisher, University of Chicago Press.*

forest cover is increasing, regrowing forests are becoming fragmented into ever-smaller parcels (**FIGURE 9.20a, b**).

When forests are fragmented, many species suffer. Bears, mountain lions, and other animals that need large areas of habitat may disappear. Birds that thrive in the interior of forests may fail to reproduce near the edge of a fragment (**FIGURE 9.20c**). Their nests are attacked by predators and parasites that favor open habitats or that travel along habitat edges. Because of such **edge effects**, avian ecologists judge forest fragmentation to be a main reason why populations of many North American songbirds are declining.

Fragmentation is affecting our national parks, which are islands of habitat surrounded by farms, ranches, roads, and cities. In 1983 conservation biologist William Newmark examined historical records of mammal sightings in North American national parks. He found that many parks were missing a few species they had held previously. The red fox and river otter had vanished from Sequoia and Kings Canyon National Parks, for example, and the white-tailed jackrabbit and spotted skunk no longer lived in Bryce Canyon National Park. In all, 42 species had disappeared. As ecological theory predicted, smaller parks lost more species than larger parks. Species were disappearing because the parks were too small to sustain their populations, Newmark concluded, and because the parks had become too isolated to be recolonized by new arrivals.

## Reserve design has consequences for biodiversity

Because habitat fragmentation is a central issue in biodiversity conservation, and because there are practical limits on how much land can feasibly be set aside, conservation biologists have argued heatedly about whether it is better to make reserves large in size and few in number, or many in number but small in size. This so-called **SLOSS** debate (for "**s**ingle **l**arge **o**r **s**everal **s**mall") is complex, but large species that roam great distances, such as wildebeest and zebras (Chapter 8), seem to benefit more from the "single large" approach. In contrast, creatures such as insects that live as larvae in small areas may do just fine in a number of small isolated reserves, if they can disperse as adults by flying from one to another. The SLOSS debate was one motivation for establishing the Amazonian forest fragmentation project described in our *Science Behind the Story* (pp. 200–201).

A related issue is how well **corridors** of protected land allow animals to travel between islands of habitat (see Figure 2.16, p. 34). In theory, connections between fragments provide animals access to more habitat and encourage gene flow to maintain populations in the long term. For these reasons, many land managers now try to establish corridors to join new reserves to existing reserves.

## Climate change threatens protected areas

Today global climate change (Chapter 14) threatens our investments in protected areas. As temperatures become warmer, species ranges shift toward the poles and upward in elevation (pp. 318–319). In a landscape of fragmented habitat, some organisms may be unable to move from one fragment to another. Species we had hoped to protect in parks may, in a warming world, become trapped in them. High-elevation species face the most risk from climate change, because they have nowhere to go once a mountaintop becomes too warm or dry. For this reason, corridors to allow movement from place to place become still more important. In response to these challenges, conservation biologists are now looking beyond parks and protected areas as they explore strategies for saving biodiversity.

# Conclusion

Forests are ecologically vital and economically valuable, yet we continue to lose them around the world. In forest management in North America, early emphasis on resource extraction evolved into policies of sustainable yield as land and resource availability began to decline. Public forests today are managed not only for timber production, but also for recreation, wildlife habitat, and ecosystem integrity. Sustainable forest certification now provides economic incentives for conservation on forested lands.

Public support for the preservation of natural lands has led to the establishment of parks and protected areas worldwide. As development spreads across the landscape, fragmenting habitats and subdividing populations, scientists trying to conserve species, communities, and ecosystems are thinking and working at the landscape level.

# Testing Your Comprehension

1. Name at least two reasons why natural primary forests contain more biodiversity than even-aged, single-species forestry plantations.
2. Describe three ecosystem services that forests provide.
3. Name several major causes of deforestation. Where is deforestation most severe today?
4. Compare and contrast maximum sustainable yield, ecosystem-based management, and adaptive management. How may pursuing maximum sustainable yield sometimes affect populations and communities?
5. Compare and contrast the major methods of timber harvesting.
6. Describe several ecological impacts of logging. How has the U.S. Forest Service responded to public concern over these impacts?
7. Are forest fires a bad thing? Explain your answer.
8. Name at least four reasons that people have created parks and reserves. How do national parks differ from national wildlife refuges? What is a wilderness area?
9. What percentage of Earth's land is protected? Describe one type of protected area that has been established outside North America.
10. Give two examples of how forest fragmentation affects animals. How might research like that of the Biological Dynamics of Forest Fragments Project help us design reserves?

# Seeking Solutions

1. People in industrialized nations are fond of warning people in industrializing nations to stop destroying rainforest. People of industrializing nations often respond that this is hypocritical, because the industrialized nations became wealthy by deforesting their land and exploiting its resources in the past. What would you say to the president of an industrializing nation, such as Indonesia or Brazil, in which a great deal of forest is being cleared?
2. Do you think maximum sustainable yield represents a desirable policy for resource managers to follow? Why or why not?
3. Consider the impacts that climate change may have on species' ranges. If you were trying to preserve an endangered mammal that occurs in a small area and you had generous funding to acquire land to help restore its population, how would you design a protected area for it? Would you use corridors? Would you include a diversity of elevations? Would you design a few large reserves or many small ones? Explain your answers.

4. **THINK IT THROUGH** You have just become supervisor of a national forest. Timber companies are asking to cut as many trees as you will let them, but environmental advocates want no logging at all. Ten percent of your forest is old-growth primary forest, and 90% is secondary forest. Your forest managers are split among preferring maximum sustainable yield, ecosystem-based management, and adaptive management. What management approach(es) will you take? Will you allow logging of all old-growth trees, some, or none? Will you allow logging of secondary forest? If so, what harvesting strategies will you encourage? What would you ask your scientists before deciding on policies regarding fire management and salvage logging?

5. **THINK IT THROUGH** You run a nonprofit environmental advocacy organization and are trying to save a tract of tropical forest in a poor industrializing nation. You have worked in this region for years and care for the local people, who want to save the forest and its animals but also need to make a living and use the forest's resources. The nation's government plans to sell a concession to a multinational timber corporation to log the entire forest unless your group can work out some other solution. Describe what solution(s) you would you try to arrange. Consider the range of options discussed in this chapter, including government protected areas, private protected areas, biosphere reserves, forest management techniques, carbon offsets, FSC-certified sustainable forestry, and more. Explain reasons for your choice(s).

# Calculating Ecological Footprints

We all rely on forest resources. The average North American consumes 225 kg (500 lb) of paper and paperboard each year. Using the estimates of paper and paperboard consumption for each region, calculate the per capita consumption for each using the population data in the table. Note: 1 metric ton = 2205 pounds.

| Nation | Population (millions) | Total paper consumed (millions of metric tons) | Per capita paper consumed (pounds) |
|---|---|---|---|
| Africa | 1072 | 7 | 14 |
| Asia | 4260 | 190 | |
| Europe | 740 | 93 | |
| Latin America | 599 | 26 | |
| North America | 349 | 78 | |
| Oceania | 37 | 4 | |
| World | 7058 | 400 | 125 |

*Data are for 2012, from Population Reference Bureau and U.N. Food and Agriculture Organization (FAO).*

1. How much paper would be consumed if everyone in the world used as much paper as the average North American?
2. How much paper would North Americans save each year if they consumed paper at the rate of Europeans?
3. North Americans have been reducing their per-person consumption of paper and paperboard by nearly 5% annually in recent years by recycling, shifting to online activity, and reducing packaging of some products. Name three specific things you personally could do to reduce your own consumption of paper products.
4. Describe three ways in which consuming FSC-certified paper rather than conventional paper can reduce the environmental impacts of paper consumption.

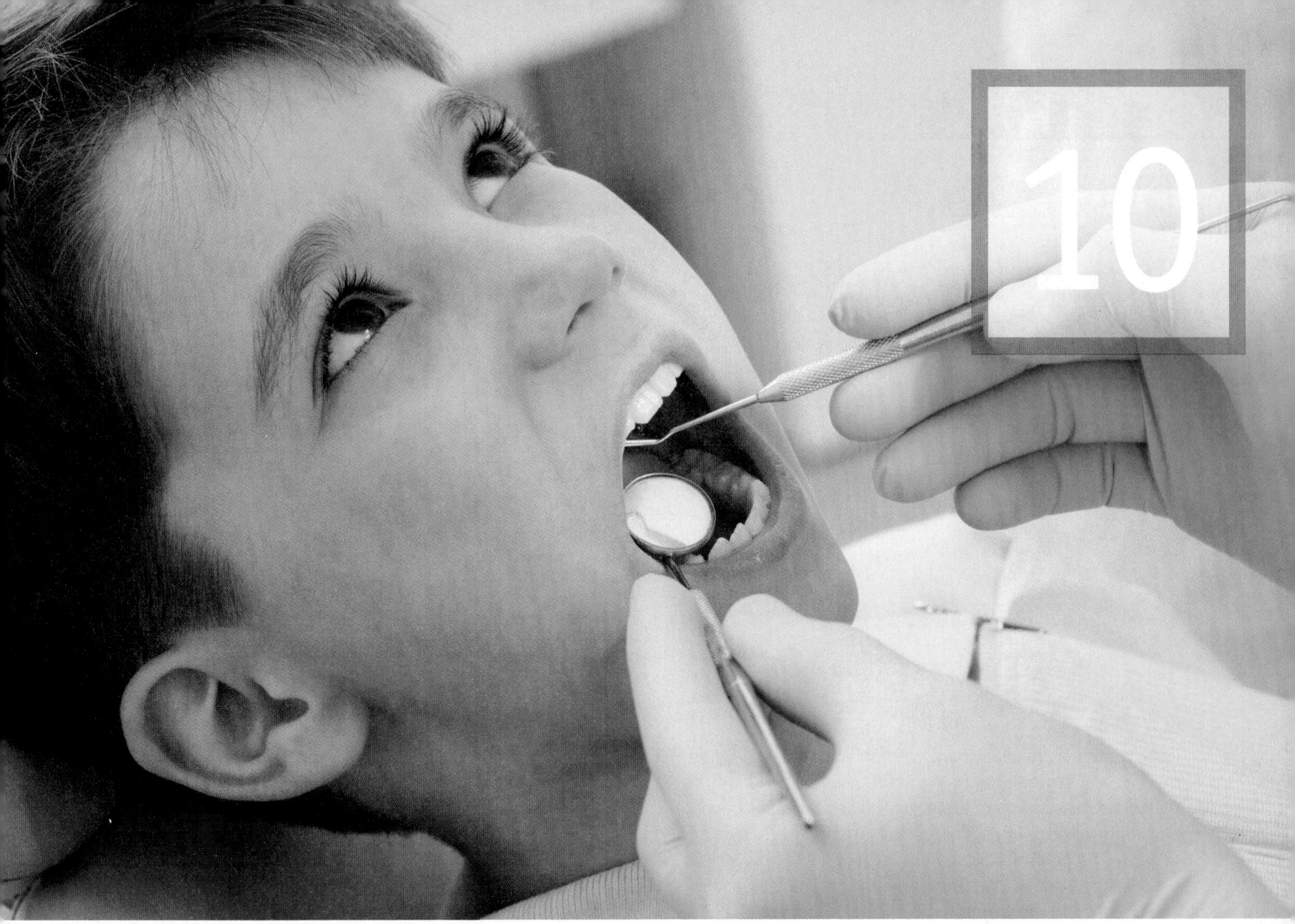

# 10

# Environmental Health and Toxicology

## Upon completing this chapter, you will be able to:

- ☐ Explain the goals of environmental health and identify major environmental health hazards
- ☐ Describe the types of toxic substances in the environment, the factors that affect their toxicity, and the defenses that organisms possess against them
- ☐ Explain the movements of toxic substances and how they affect organisms and ecosystems
- ☐ Discuss the study of chemical hazards and their effects, including wildlife toxicology, epidemiology, animal testing, and dose-response analysis
- ☐ Evaluate risk assessment and risk management
- ☐ Compare philosophical approaches to risk and how they relate to regulatory policy

**Photo:** **Dental sealants prevent cavities, but could they also make us sick?**

# CENTRAL CASE STUDY

# Bisphenol A Is Everywhere—But Is It Safe?

*"Babies in the U.S. are born pre-polluted with BPA. What more evidence do we need to act?"*

**—Dr. Janet Gray, Director of the Environmental Risks and Breast Cancer Project, Vassar College**

*"There is no basis for human health concerns from exposure to BPA."*

**—The American Chemistry Council**

How is it that a chemical found to alter reproductive development in animals gets used in baby bottles? How can it be that a substance linked to breast cancer, prostate cancer, and heart disease is routinely used in food and drink containers? The chemical bisphenol A (BPA for short) has been associated with everything from neurological effects to miscarriages. Yet there's a better than 9 in 10 chance that it is coursing through your body right now.

To understand how chemicals that may pose health risks come to be widespread in our society, we need to explore how scientists study toxic substances and other environmental health risks—and the vexing challenges these pursuits entail.

Chemists first synthesized bisphenol A, an organic compound (p. 29) with the chemical formula $C_{15}H_{16}O_2$, in 1891. As they began producing plastics in the 1950s, chemists found BPA to be useful in creating epoxy resins used in lacquers and coatings. Epoxy resins containing BPA were soon being used to line the insides of metal food and drink cans and the insides of pipes for our water supply, as well as in enamels, varnishes, adhesives, and even dental sealants for our teeth.

Chemists also found that linking BPA molecules into polymers (p. 29) helped create polycarbonate plastic, a hard, clear type of plastic that soon found use in water bottles, food containers, eyeglass lenses, CDs and DVDs, laptops and other electronics, auto parts, sports equipment, baby bottles, and children's toys. With so many uses, BPA has become one of the world's most-produced chemicals; each year we make about half a kilogram (1 lb) of BPA for each person on the planet, and 9 kg (20 lb) per person in the United States!

Unfortunately, BPA leaches out of its many products and into our food, water, air, and bodies. Fully 93% of Americans carry detectable concentrations in their urine, according to the Centers for Disease Control and Prevention (CDC). Because most BPA passes through the body within hours, these data suggest that we are receiving almost continuous exposure.

What, if anything, is BPA doing to us? To address such questions, scientists run experiments on laboratory animals, administering known doses of the substance and measuring the health impacts that result. Hundreds of studies with rats, mice, and other animals have shown many apparent health impacts from BPA, including a wide range of reproductive abnormalities (see **THE SCIENCE BEHIND THE STORY**, p. 213).

Many of these effects occur at extremely low doses—much lower than the exposure levels set so far by regulatory agencies for human safety. Scientists say this is because BPA mimics certain hormones, such as the female sex hormone estrogen; that is, it is structurally similar to estrogen and can induce some of its effects in animals (see Figure 10.4, p. 212). Hormones such as estrogen function at minute concentrations, so when a synthetic chemical that is similar to estrogen reaches the body in a similarly low concentration, it can fool the body into responding.

In reaction to research involving animals, a growing number of researchers, doctors, and consumer advocates are calling on governments to regulate bisphenol A and for manufacturers to stop using it. The chemical industry has long insisted that BPA is safe, pointing to industry-sponsored research that finds no direct health impacts on humans exposed to BPA.

But studies suggesting that BPA may be affecting human health are now emerging. A 2013 review found 91 studies that examined the relationship between the level of BPA in subjects' urine or blood and a variety of health problems. Although these studies are correlative (p. 10), they collectively suggest that exposure to elevated BPA levels is associated with health problems such as asthma, behavioral issues, and genital abnormalities in young males.

Nations vary widely in how they have responded to the growing body of scientific studies on BPA when considering regulations on its use in products. In some nations, such as Canada, BPA has been banned completely. In many nations, including the United States, its use in products for babies and small children is restricted.

Today, concerned parents can more easily find BPA-free items for their infants and children. The rest of us remain exposed through most food cans, many drink containers, and thousands of other products.

In the face of mounting public concern about the safety of BPA, many companies are voluntarily choosing to remove it from their products, even in the absence of regulation by the U.S. government. For example, Campbell's has announced that it is transitioning away from the use of BPA in its soup can liners, and food giants ConAgra, Nestlé, and Heinz have also pledged to remove BPA from their food packaging. There is precedent for such efforts, as BPA was voluntarily phased out of can liners in Japan starting in the late 1990s.

Bisphenol A is by no means one of our greatest environmental health threats. However, it provides a timely example of how we as a society assess health risks and decide how to manage them. As scientists and government regulators assess BPA's potential risks, their efforts give us a window on how hormone-disrupting chemicals are challenging the way we appraise and control the environmental health risks we face. ☐

# Environmental Health

Examining the impacts of human-made chemicals such as BPA is just one aspect of the broad field **environmental health**, which assesses environmental factors that influence our health and quality of life. These factors include both natural and anthropogenic (human-caused) factors. Practitioners of environmental health seek to prevent adverse effects on human health and on the ecological systems that are essential to our well-being.

## We face four types of environmental hazards

We can categorize environmental health hazards into four main types: physical, chemical, biological, and cultural. Although some amount of risk is unavoidable, much of environmental health focuses on taking steps to minimize the risks of encountering hazards and to lessen the impacts of the hazards we do encounter.

**Physical hazards** *Physical hazards* arise from processes that occur naturally in our environment and pose risks to human life or health. Some are ongoing natural phenomena, such as excessive exposure to ultraviolet (UV) radiation from sunlight, which damages DNA and has been tied to skin cancer, cataracts, and immune suppression (**FIGURE 10.1a**). We can reduce these risks by shielding our skin from intense sunlight with clothing and sunscreen.

Other physical hazards include discrete events such as earthquakes, volcanic eruptions, fires, floods, landslides,

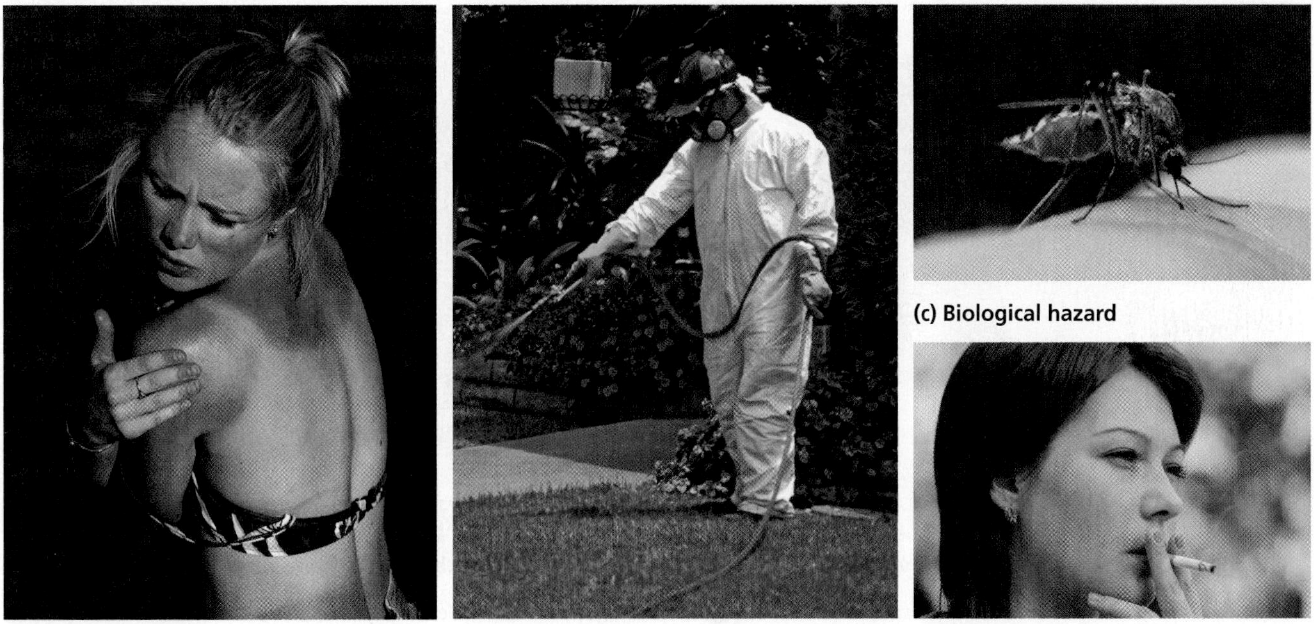

(a) Physical hazard (b) Chemical hazard (c) Biological hazard (d) Cultural hazard

**FIGURE 10.1 Environmental health hazards come in four types.** The sun's ultraviolet radiation is an example of a physical hazard **(a)**. Chemical hazards **(b)** include both synthetic and natural chemicals. Biological hazards **(c)** include diseases and the organisms that transmit them. Cultural or lifestyle hazards **(d)** include the behavioral decisions we make, such as smoking, as well as the socioeconomic constraints forced on us.

hurricanes, blizzards and droughts. We cannot prevent many of these hazards, but we can minimize risk by preparing ourselves with emergency plans and avoiding practices that make us vulnerable to certain physical hazards. For example, scientists can map geologic faults to determine areas at risk of earthquakes, engineers can design buildings to resist damage, and governments and individuals can create emergency plans to prepare for a quake's aftermath.

**Chemical hazards** *Chemical hazards* include many of the synthetic chemicals that our society manufactures, such as pharmaceuticals, disinfectants, and pesticides (**FIGURE 10.1b**). Some natural substances that we process for our use (such as hydrocarbons, lead, and asbestos) are also harmful to human health. Following our overview of environmental health, much of this chapter will focus on chemical health hazards and the ways we study and regulate them.

**Biological hazards** *Biological hazards* result from ecological interactions among organisms (**FIGURE 10.1c**). When we become sick from a virus, bacterial infection, or other pathogen, we are suffering parasitism (pp. 69–70). Some pathogenic viruses, bacteria, and protists attack us directly; others cause infection through a *vector*, an organism (such as a mosquito or a rat) that transfers the pathogen to the host. This is what we call **infectious disease**. Infectious diseases such as malaria, cholera, tuberculosis, and influenza (flu) are major environmental health hazards, especially in developing nations with widespread poverty and limited health care. As with physical and chemical hazards, it is impossible for us to avoid risk from biological agents completely, but through monitoring, sanitation, and medical treatment we can reduce the likelihood and impacts of infection.

**Cultural hazards** Hazards that result from our place of residence, our socioeconomic status, our occupation, or our behavioral choices can be thought of as *cultural hazards* or *lifestyle hazards*. We can minimize or prevent some of these hazards, whereas others may be beyond our control. For instance, people can choose whether or not to smoke cigarettes (**FIGURE 10.1d**), but exposure to secondhand smoke at home or work may be beyond one's control. Much the same might be said for other cultural hazards such as diet and nutrition, criminal behavior, and workplace hazards. Environmental justice advocates (pp. 14–15) argue that "forced" risks from cultural hazards, such as living near a hazardous waste site, are often higher for people with fewer economic resources or less political clout.

## Disease is a major focus of environmental health

Despite all our technological advances, disease causes the vast majority of human deaths worldwide (**FIGURE 10.2a**). Over half the world's deaths result from **noninfectious diseases**, such as cancer and heart disease. These diseases are not spread from one person to another, but rather are influenced by genetics, environmental factors, and lifestyle

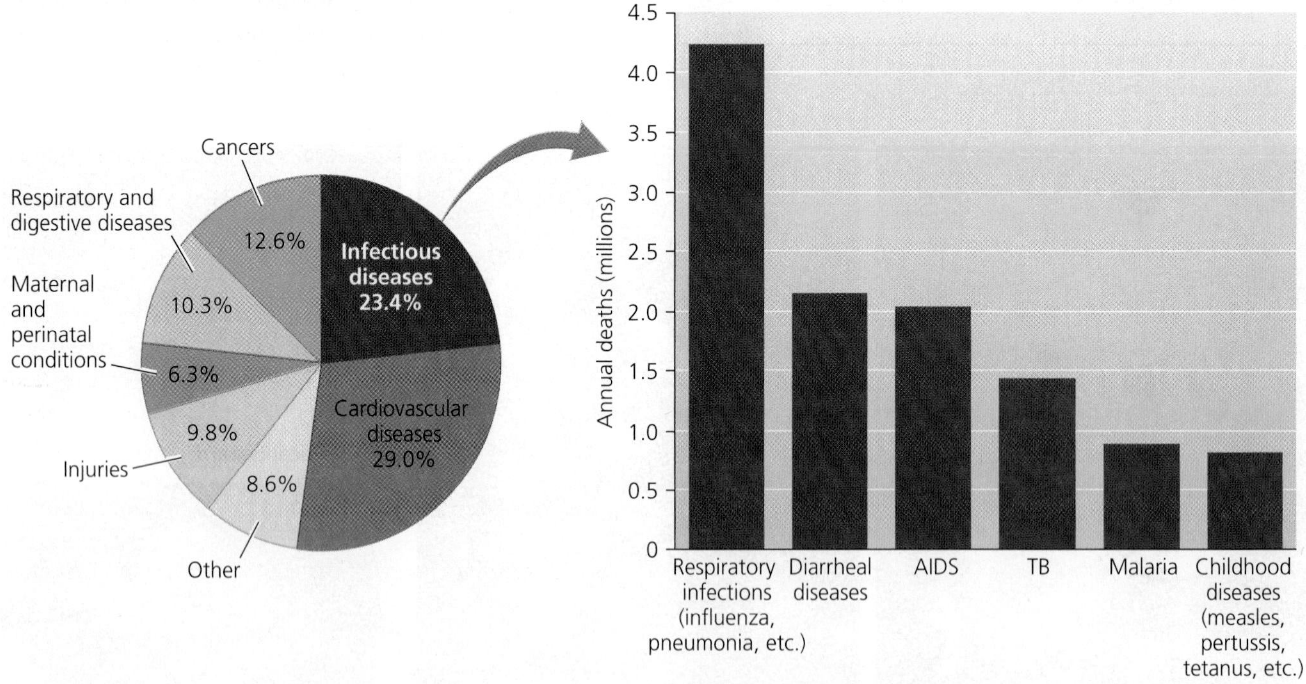


**(a) Leading causes of death across the world** **(b) Leading causes of death by infectious diseases**

**FIGURE 10.2 Infectious diseases are the second-leading cause of death worldwide.** Six types of diseases account for 80% of all deaths from infectious disease. *Data from World Health Organization, 2009.* World health statistics 2009. *WHO, Geneva, Switzerland.*

DATA Q AIDS is a well-known infectious disease, but respiratory and diarrheal diseases claim far more lives every year than AIDS. According to the figure, how many times more lives were lost to respiratory infections and diarrheal diseases than to AIDS?

GO TO **INTERPRETING GRAPHS & DATA** ON MasteringEnvironmentalScience®

choices. For instance, whether a person develops heart disease depends not only on his or her genes, but also on lifestyle choices such as diet and exercise.

In the United States, lifestyle trends are altering the prevalence of noninfectious disease. Over the past 20 years, the percentage of Americans who smoke cigarettes has declined significantly while obesity rates have increased. Obesity is typically due to low levels of physical activity coupled with a diet high in calories and fat, but other factors are involved. For example, studies have found that exposure to bisphenol A elevates the production of fat cells in mouse embryos and leads to increased fat storage in adult mice.

Although infectious disease accounts for fewer deaths than noninfectious disease, infectious disease robs society of more years of human life, because it strikes people at all ages, including the very young. Infectious diseases account for almost one of every four deaths that occur each year (**FIGURE 10.2b**). Infectious disease is a greater problem in developing countries, where it accounts for close to half of all deaths. Infectious disease causes many fewer deaths in developed nations because their wealth allows for better public sanitation and hygiene (which reduce the chance of contracting an infectious disease) and better access to medicine (which enables treatment if a disease is contracted).

Decades of public health efforts have lessened the impacts of infectious disease and even have eradicated some diseases—yet other diseases are posing new challenges. Some, such as acquired immunodeficiency syndrome (AIDS), continue to spread globally despite efforts to stop them. Others, such as tuberculosis and strains of malaria, are evolving resistance to antibiotics. Additionally, human-induced global warming (Chapter 14) is enabling tropical diseases (such as malaria, dengue, and cholera) to gain footholds in temperate regions.

In our world of global mobility and dense human populations, novel infectious diseases (or new strains of old diseases) that emerge in one location are more likely to spread quickly to other locations. Recent examples include the H5N1 avian flu that emerged in 2004 and the H1N1 swine flu that spread across the globe in 2009–2010. Diseases like influenza, whose pathogens evolve rapidly, give rise to a variety of strains, making it more likely that one may turn exceedingly dangerous and cause a global pandemic (a widespread outbreak of a disease).

There are many ways to fight infectious disease. We can reduce the chance of infection through immunization, improved public sanitation, access to clean water, food security (p. 136), and public education campaigns that inform people of risks and how to avoid them. If infection occurs, the health effects and spread of infectious disease can be minimized by providing access to health care and medication.

## Toxicology is the study of chemical hazards

Although most indicators of human health are improving as the world's wealth increases, our modern society is exposing us to more and more synthetic chemicals. Some of these substances pose threats to human health, but figuring out which of them do—and how, and to what degree—is a complicated scientific endeavor. **Toxicology** is the science that examines how poisonous chemicals affect the health of humans and other organisms.

Toxicologists assess and compare substances to determine their **toxicity**, the degree of harm a chemical substance can inflict. A toxic substance, or poison, is called a **toxicant**, but any chemical substance may exert negative impacts if we expose ourselves to enough of it. Conversely, if the quantity is small enough, a toxicant may pose no health risk. These facts are often summarized in the catchphrase, "The dose makes the poison." In other words, a substance's toxicity depends not only on its chemical properties, but also on its quantity.

In recent decades, our ability to produce new chemicals has expanded, concentrations of chemical contaminants in the environment have increased, and public concern for health and the environment has grown. These trends have driven the rise of **environmental toxicology**, which deals specifically with toxic substances that come from or are discharged into the environment. Toxicologists generally focus on human health, using other organisms as models and test subjects. Environmental toxicologists study animals and plants to determine the ecological impacts of toxic substances and to find out whether other organisms can serve as indicators of health threats that could soon affect people.

## Many environmental health hazards exist indoors

Modern Americans spend roughly 90% of their lives indoors. Unfortunately, our homes and workplaces can be rife with physical, biological, chemical, and cultural hazards.

Cigarette smoke and radon (a radioactive gas that seeps up from the ground; p. 298) are leading indoor hazards and are the top two causes of lung cancer in developed nations. Homes and offices can have problems with toxic compounds produced by mold, which can flourish in wall spaces when moisture levels are high. Asbestos, used in the past as insulation in walls and other products, is dangerous when inhaled. Lead poisoning from water pipes or leaded paint can cause cognitive problems and behavioral abnormalities, damage to vital organs, and even death. Lead poisoning among U.S. children has greatly declined in recent years, however, as a result of education campaigns and the U.S. federal government mandating the phaseout of lead-based paints and leaded gasoline (p. 5) in the 1970s.

One recently recognized indoor hazard is *polybrominated diphenyl ethers* (PBDEs). These compounds are used as fire retardants in computers, televisions, plastics, and furniture, and they may evaporate at very slow rates throughout the lifetime of the product. Like bisphenol A, PBDEs appear to act as endocrine disruptors. Animal testing suggests that PBDEs may also cause cancer, affect thyroid hormones, and affect the development of the brain and nervous system. Concern about PBDEs rose after a study showed that their concentrations in the breast milk of Swedish mothers had increased exponentially from 1972 to 1997. The European Union banned PBDEs in 2003, and concentrations in the breast milk of European mothers have since fallen substantially. In the United States, however, there has so far been little movement on government regulation regarding PBDE exposure.

## Risks must be balanced against rewards

It is important to keep in mind that artificially produced chemicals have played a crucial role in giving us the high standard of living we enjoy today. These chemicals have helped create the industrial agriculture that produces our food, the medical advances that protect our health and prolong our lives, and many of the modern materials and conveniences we use every day. With most hazards, there is some tradeoff between risk and reward. In regard to bisphenol A, its usefulness for many purposes means that despite its health risks, we may as a society choose to continue using it in some products where we judge its benefits to outweigh its impacts on human health. Thus, for BPA and other toxic chemicals, it is appropriate to remember their benefits as we examine some of the unfortunate side effects these substances may elicit.

# Toxic Substances and Their Effects on Organisms

Our environment contains countless natural substances that may pose health risks. These include oil oozing naturally from the ground; radon gas seeping up from bedrock; and **toxins**, toxic chemicals manufactured in the tissues of living organisms—for example, chemicals that plants use to ward off herbivores or that insects use to defend themselves from predators. In addition, we are exposed to many synthetic (human-made) chemicals.

## Synthetic chemicals are all around us—and in us

Synthetic chemicals surround us in our daily lives, and each year in the United States we manufacture or import around 113 kg (250 lb) of chemical substances for every man, woman, and child. Many of these substances, particularly the pesticides we use to control insects and weeds, find their way into soil, air, and water—and into humans and other organisms (**FIGURE 10.3**).

As a result of all this exposure, every one of us carries traces of hundreds of industrial chemicals in our bodies. The U.S. government's latest National Health and Nutrition Examination Survey gathered data on 148 foreign compounds in Americans' bodies. Among these were several toxic persistent organic pollutants restricted by international treaty (p. 222). Depending on the pollutant, these were detected in 41–100% of the people tested. Our exposure to synthetic chemicals begins in the womb as substances our mothers ingest while pregnant are transferred to us. A 2009 study found 232 chemicals in the umbilical cords of 10 newborn babies it tested. Nine of the 10 umbilical cords contained bisphenol A, leading researchers to note that we are born "pre-polluted."

All this should not necessarily be cause for alarm. Not all synthetic chemicals pose health risks, and relatively few are known to be toxic. However, of the roughly 100,000 synthetic chemicals on the market today, very few have been thoroughly tested. For the vast majority, we simply do not know what effects, if any, they may have on us.

## *Silent Spring* began the public debate over synthetic chemicals

It was not until the 1960s that most people began to learn about the risks of exposure to pesticides. The key event was the publication of Rachel Carson's 1962 book *Silent Spring* (p. 104), which brought the insecticide dichlorodiphenyl-trichloroethane (DDT) to the public's attention. The book was written at a time when large amounts of pesticides virtually untested for health impacts were indiscriminately sprayed, on the assumption that the chemicals would do no harm to people.

Carson synthesized scientific studies, medical case histories, and other data to contend that DDT in particular, and artificial pesticides in general, were hazardous to people, wildlife, and ecosystems. The book became a best-seller and helped generate significant social change in views and actions toward the environment. The use of DDT was banned in the United States in 1973 and is now illegal in a number of nations.

Despite its damaging effects, DDT is still manufactured today because some developing countries with tropical climates use it to control disease vectors, such as mosquitoes that transmit malaria. In these countries, malaria represents a greater health threat than do the toxic effects of the pesticide. New research and technologies are showing promise for controlling mosquito populations without the use of DDT, however, giving hope that the chemical will soon no longer be needed for controlling disease vectors and could be phased out worldwide.

### WEIGHING THE ISSUES

**Ethics and Insecticides** Although many nations have banned the use of DDT, the compound is still manufactured in India and exported to developing nations that lack such bans. How do you feel about this? Is it unethical for a company to sell a substance that has been deemed toxic by so many nations? Or would it be unethical *not* to sell DDT to tropical nations if they desire it for improving public health, such as controlling mosquitoes that transmit malaria?

## Not all toxic substances are synthetic

Although many toxicologists focus on synthetic chemicals, toxic substances also exist naturally in the environment around us and in the foods we eat. Thus, it would be a mistake to assume that all artificial substances are unhealthy and that all natural substances are healthy. In fact, the plants and animals we eat contain many chemicals that can harm us. Many plants produce toxins to ward off animals that eat them. In domesticating crop plants, we have selected (p. 50) for strains with reduced toxin content, but we have not eliminated these dangers. Furthermore, when we consume animal meat, we ingest toxins the animals obtained from plants or animals they ate. Scientists are actively debating just how much risk natural toxicants pose, and it is clear that more research is needed to fully answer these questions.

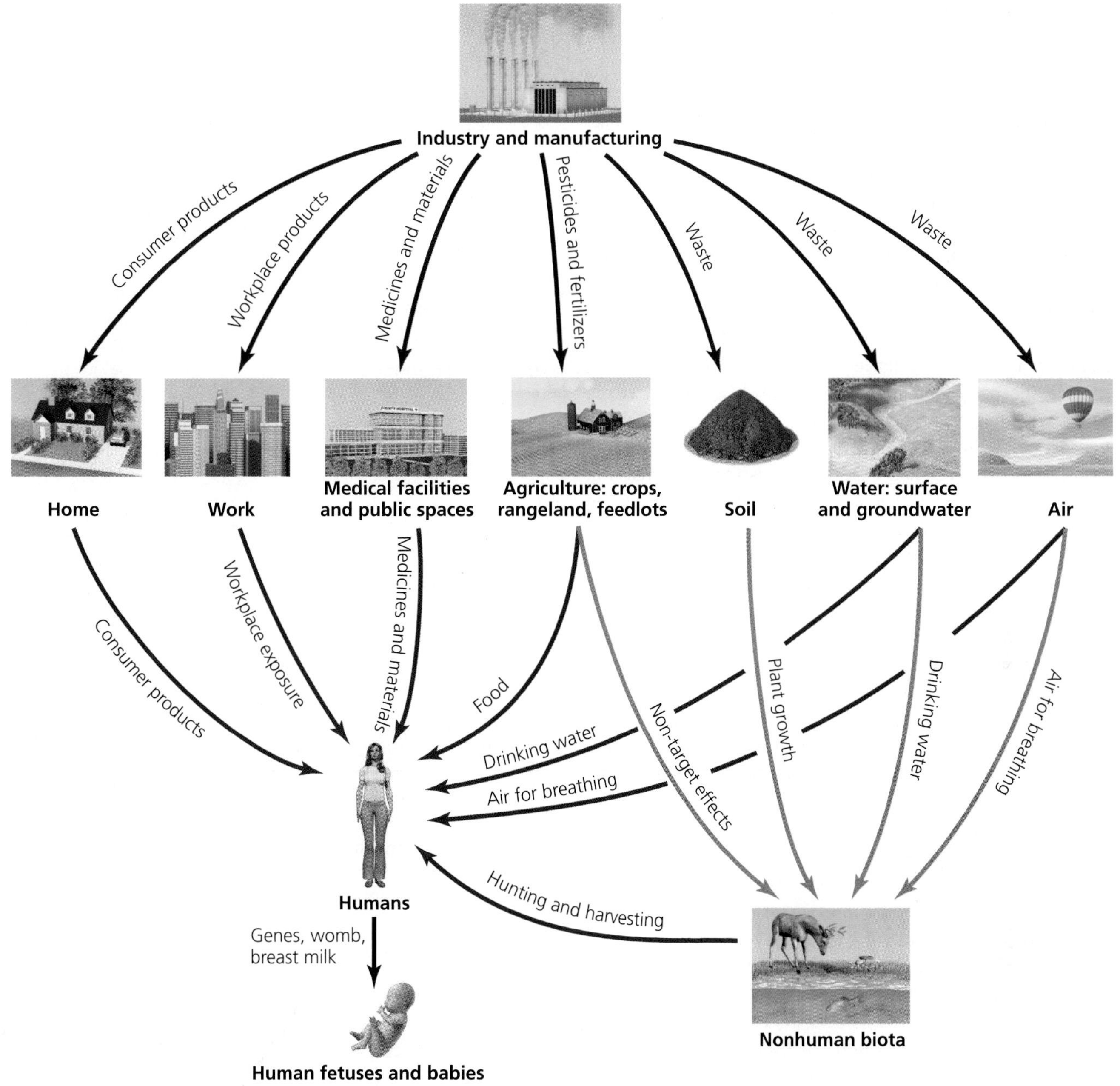


**FIGURE 10.3 Synthetic chemicals take many routes in traveling through the environment.** People take in only a tiny proportion of these compounds, and many compounds are harmless. However, people receive small amounts of toxicants from many sources, and developing fetuses and babies are particularly sensitive.

## Toxic substances come in different types

Toxic substances can be classified based on their particular impacts on health. The best-known toxicants are **carcinogens**, which are substances or types of radiation that cause cancer. In cancer, malignant cells grow uncontrollably, creating tumors, damaging the body, and often leading to death. Cancer frequently has a genetic component, but a wide variety of environmental factors are thought to raise the risk of cancer. In our society today, the greatest number of cancer cases is thought to result from carcinogens contained in cigarette smoke. Carcinogens can be difficult to identify because there may be a long lag time between exposure to the agent and the detectable onset of cancer—up to 15–30 years in the case of cigarette smoke—and because only a portion of the people exposed to a carcinogen eventually develop cancer.

**Mutagens** are substances that cause genetic mutations in the DNA of organisms (p. 48). Although most mutations have little or no effect, some can lead to severe problems, including cancer and other disorders. If mutations occur in an individual's sperm or egg cells, then the individual's offspring suffer the effects.

Chemicals that cause harm to the unborn are known as **teratogens**. Teratogens that affect development of human embryos in the womb can cause birth defects. One example is the drug thalidomide, developed in the 1950s as a sleeping pill and to prevent nausea during pregnancy. Tragically, the drug turned out to be a powerful teratogen. Its use caused birth defects in thousands of babies, and its use by pregnant women was banned in the 1960s.

Other toxicants, known as **neurotoxins**, assault the nervous system. Neurotoxins include venoms produced by animals, heavy metals such as lead and mercury, and some pesticides. A famous case of neurotoxin poisoning occurred in Japan, where a chemical factory dumped mercury waste into Minamata Bay between the 1930s and 1960s. Thousands of people there ate fish contaminated with the mercury and soon began suffering from slurred speech, loss of muscle control, sudden fits of laughter, and in some cases death.

The human immune system protects our bodies from disease. Some toxic substances weaken the immune system, reducing the body's ability to defend itself against bacteria, viruses, allergy-causing agents, and other attackers. **Allergens** overactivate the immune system, causing an immune response when one is not necessary. One hypothesis for the increase in asthma in recent years is that allergenic synthetic chemicals are more prevalent in our environment. Allergens are not universally considered toxicants, however, because they affect some people but not others and because one's response does not necessarily correlate with the degree of exposure.

**Pathway inhibitors** are toxicants that interrupt vital biochemical processes in organisms by blocking one or more steps in important biochemical pathways. Rat poisons, for example, cause internal hemorrhaging in rodents by interfering with the biochemical pathways that create blood-clotting proteins. Some herbicides, such as atrazine, kill plants by blocking steps in photosynthesis. Cyanide kills by interrupting chemical pathways that produce energy in mitochondria and depriving cells of life-sustaining energy.

Most recently, scientists have recognized **endocrine disruptors**, toxic substances that interfere with the endocrine system. The endocrine system consists of chemical messengers (hormones) that travel through the bloodstream at extremely low concentrations and perform many vital functions. Hormones stimulate growth, development, and sexual maturity, and they regulate brain function, appetite, sex drive, and many other aspects of our physiology and behavior. Some hormone-disrupting toxicants affect an animal's endocrine system by blocking the action of hormones or accelerating their breakdown. Others are so similar to certain hormones in their molecular structure and chemistry that they "mimic" the hormone by interacting with receptor molecules just as the actual hormone would (**FIGURE 10.4**).

Bisphenol A is one of many chemicals that appear to mimic the female sex hormone estrogen and bind to estrogen receptors. Indeed, emerging research is indicating that bisphenol A might not be the only estrogen-mimicking compound in plastics, calling into question the safety of all of the plastics that are ubiquitous in our lives.

Phthalates are another class of hormone-disrupting chemicals that are used widely in children's toys, perfumes and cosmetics, and other items. Health research on phthalates has linked them to birth defects, breast cancer, reduced sperm counts, and other reproductive effects. Like BPA, phthalates show how a substance can be a carcinogen, a mutagen, and an endocrine disruptor.

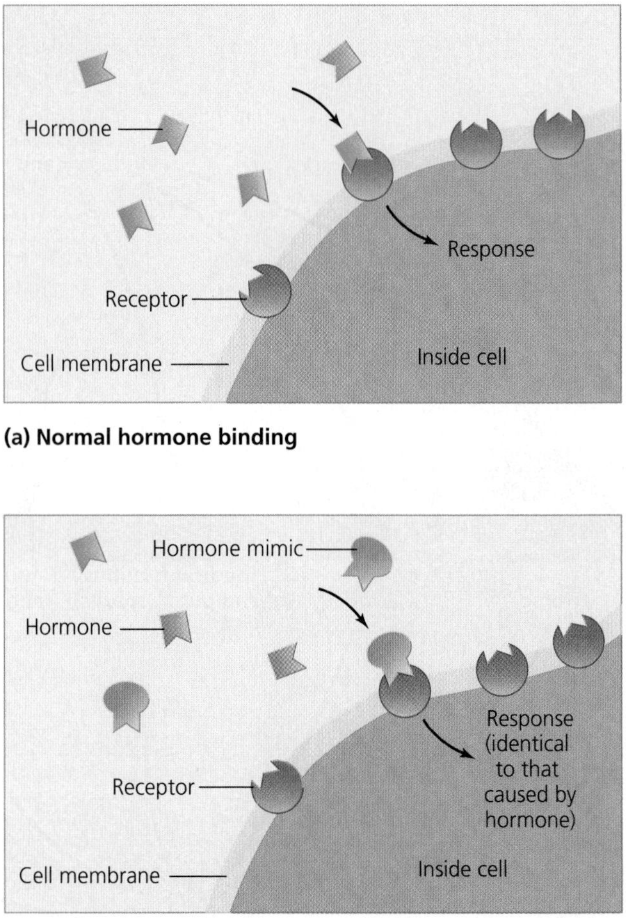


**(a) Normal hormone binding**

**(b) Hormone mimicry**

**FIGURE 10.4 Many endocrine-disrupting substances mimic the structure of hormone molecules.** Like a key similar enough to fit into another key's lock, the hormone mimic binds to a cellular receptor for the hormone, causing the cell to react as though it had encountered the hormone.

## Organisms have natural defenses against toxic substances

Although synthetic toxicants are new, organisms have long been exposed to natural toxicants. Mercury, cadmium, arsenic, and other harmful substances are found naturally in the environment. Additionally, some organisms produce biological toxins to deter predators or capture prey. Examples include venom in poisonous snakes and spiders, and the natural insecticide pyrethrin found in chrysanthemums. These exposures have provided selection pressure (pp. 48–50) for mechanisms that provide protection from toxins. Over time, organisms able to tolerate these harmful substances have gained an evolutionary advantage.

Skin, scales, and feathers are the first line of defense against toxic substances because they resist uptake from the surrounding environment. However, toxicants can circumvent these barriers and enter the body from vital activities such as eating, drinking, and breathing. Once inside the organism, they are distributed widely by the circulatory and lymph systems in animals, and by the vascular system in plants.

GO TO **PROCESS OF SCIENCE** ON MasteringEnvironmentalScience®

# Testing the Safety of Bisphenol A

**Dr. Patricia Hunt, Washington State University**

Of the many studies documenting health impacts of bisphenol A on lab animals, one of the first came about because a lab assistant reached for the wrong soap.

At a laboratory at Case Western Reserve University in Ohio in 1998, geneticist Patricia Hunt (now at Washington State University) was making a routine check of her female lab mice. As she extracted and examined developing eggs from the ovaries, she began to wonder what had gone wrong. About 40% of the eggs showed problems with their chromosomes, and 12% had irregular amounts of genetic material. This dangerous condition, called *aneuploidy*, can lead to miscarriages or birth defects in mice and people alike.

A bit of sleuthing revealed that a lab assistant had mistakenly washed the lab's plastic mouse cages and water bottles with an especially harsh soap. The soap damaged the cages so badly that parts of them seemed to have melted.

The cages were made from polycarbonate plastic, which contains BPA. Hunt knew at the time that BPA mimics estrogen and that some studies had linked the chemical to reproductive abnormalities in mice, such as low sperm counts and early sexual development. Other research indicated that BPA leaches out of plastic into water and food when the plastic is treated with heat, acidity, or harsh soap.

Hunt wondered whether the chemical might be adversely affecting the mice in her lab. Deciding to re-create the accidental cage-washing incident in a controlled experiment, Hunt instructed researchers in her lab to wash polycarbonate cages and water bottles using varying levels of the harsh soap. They then compared mice kept in damaged cages with plastic water bottles to mice kept in undamaged cages with glass water bottles.

The developing eggs of mice exposed to BPA through the deliberately damaged plastic showed significant problems during meiosis, the division of chromosomes during egg formation—just as they had in the original incident (**FIGURE 1**). In contrast, the eggs of mice in the control cages were normal.

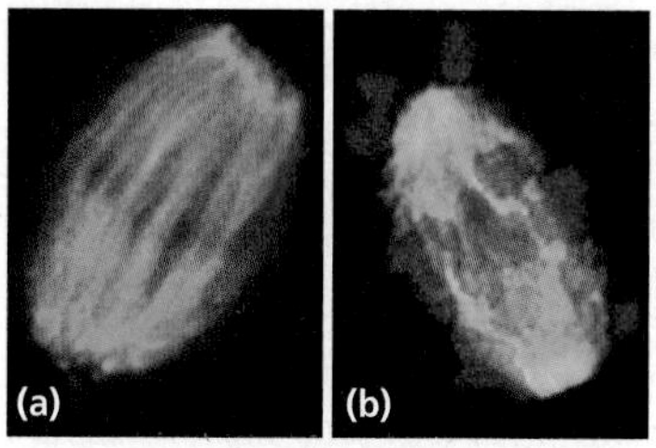


**FIGURE 1 In normal cell division (a), chromosomes (red) align properly.** Exposure to bisphenol A causes abnormal cell division **(b)**, whereby chromosomes scatter and are distributed improperly and unevenly between daughter cells.

In another round of tests, Hunt's team gave sets of female mice daily oral doses of BPA over 3, 5, and 7 days. They observed the same meiotic abnormalities in these mice, although at lower levels (**FIGURE 2**). The mice given BPA for 7 days were most severely affected.

Published in 2003 in the journal *Current Biology*, Hunt's findings set off a new wave of concern over the safety of BPA. The findings were disturbing because sex cells of mice and of people divide and function in similar ways. "We have observed meiotic defects in mice at exposure levels close to or even below those considered 'safe' for humans," the research paper stated. "Clearly, the possibility that BPA exposure increases the likelihood of genetically abnormal offspring is too serious to be dismissed without extensive further study."

Since that time, hundreds of other studies of BPA at low doses have documented harmful effects in lab animals, including reproductive disorders related to estrogen mimicry and other maladies ranging from thyroid problems to liver damage to elevated anxiety. Epidemiological studies are also beginning to shed light on the potentially far-reaching impacts of BPA on human health (see the opening Case Study, pp. 206–207). As a result, more and more scientists are urging regulators to adopt a precautionary position and restrict bisphenol A based on this diverse body of evidence. ■

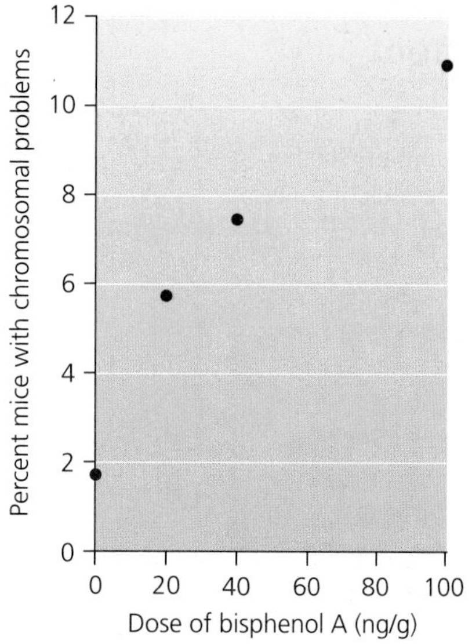


**FIGURE 2 In this dose-response experiment, the percentage of mice showing chromosomal problems during cell division rose with increasing dose of bisphenol A.** In the United States and Europe, regulators have set safe intake levels for people at doses of 50 ng/g of body weight per day. *Data from Hunt, P. A., et al., 2003. Bisphenol A exposure causes meiotic aneuploidy in the female mouse.* Current Biology *13: 546–553.*

**DATA Q** Using the figure, predict the percentage of mice in the study that would likely suffer chromosomal problems when exposed to a BPA dosage of 70 ng/g. What would this percentage be?

GO TO **INTERPRETING GRAPHS & DATA** ON MasteringEnvironmentalScience®

Organisms possess biochemical pathways that use enzymes to detoxify harmful chemicals once they enter the body. Some pathways break down toxic substances to render them inert. Other pathways make toxic substances water-soluble so they are easier to excrete through the urinary system. In humans, many of these pathways function in the liver. As a result, this organ is disproportionately affected by the intake of harmful substances, such as excessive alcohol.

Some toxic substances cannot be effectively detoxified or made water-soluble by detoxification enzymes. Instead, the body sequesters these chemicals in fatty tissues and cell membranes to keep them away from vital organs. Heavy metals, dioxins, and some insecticides (including DDT) are stored in body tissue in this manner.

Defense mechanisms for natural toxins have evolved over millions of years. For the synthetic chemicals that are so prevalent in today's environment, however, organisms have not had long-term exposure, so the impacts of these toxic substances can be severe and unpredictable.

## Individuals vary in their responses to hazards

Individuals may respond differently to identical exposures to hazards because they have different combinations of the genes that code for toxicant defenses. Additionally, sensitivity can vary with sex, age, and weight. Poorer health generally makes an individual more sensitive to biological and chemical hazards. And because of their smaller size and rapidly developing organ systems, younger organisms (for example, fetuses, infants, and young children) tend to be much more sensitive to toxicants than are adults. Regulatory agencies such as the U.S. Environmental Protection Agency (EPA) typically set human chemical exposure standards for adults and extrapolate downward for infants and children. However, many scientists contend that these linear extrapolations often do not offer adequate protection to fetuses, infants, and children. They argue that studies of the toxic effects of substances should examine large number of individuals, preferably of different ages, to gain a holistic view of a substance's toxicity.

## The type of exposure can affect the response

The risk posed by a hazard often varies according to whether a person experiences high exposure for short periods of time, known as **acute exposure**, or low exposure over long periods of time, known as **chronic exposure**. Incidences of acute exposure are easier to recognize because they often stem from discrete events, such as accidental ingestion, an oil spill, a chemical spill, or a nuclear accident. Toxicity tests in laboratories generally reflect effects of acute toxicity. However, chronic exposure is more common—yet more difficult to study, detect, and diagnose. Chronic exposure often affects organs gradually, as when smoking causes lung cancer, or when alcohol abuse leads to liver or kidney damage. Because of the long time periods involved, relationships between cause and effect may not be readily apparent.

# Toxic Substances and Their Effects on Ecosystems

When toxicants concentrate in environments and harm the health of many individuals, populations (p. 48) of the affected species may become smaller. This decline can then affect other species. For instance, if a predatory species experiences population decline due to toxicant exposure, the population size of its prey is likely to increase. Predators of the poisoned species, meanwhile, might decline as their food source became less abundant. Cascading impacts can cause changes in the composition of the biological community (p. 56) and threaten ecosystem functioning. Hence, there are many ways in which toxic substances can concentrate and persist in ecosystems and affect ecosystem services.

## Airborne substances can travel widely

Toxic substances are released around the world from agricultural, industrial, and domestic activities and may sometimes be redistributed by air currents (Chapter 13), exerting impacts on ecosystems far from their site of release.

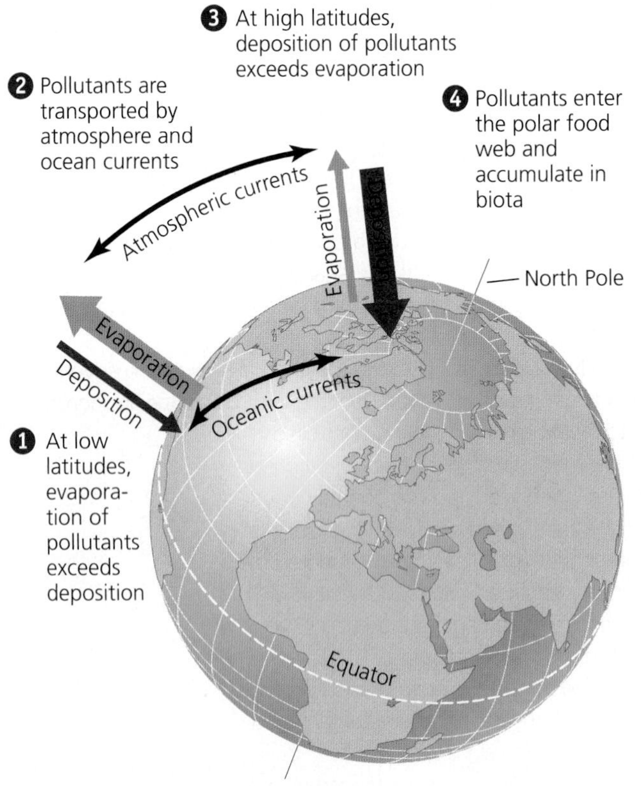

3 At high latitudes, deposition of pollutants exceeds evaporation
2 Pollutants are transported by atmosphere and ocean currents
4 Pollutants enter the polar food web and accumulate in biota
Atmospheric currents
Evaporation
North Pole
Evaporation
Deposition
Oceanic currents
1 At low latitudes, evaporation of pollutants exceeds deposition
Equator

**FIGURE 10.5 Air and water currents direct pollutants toward the poles.** In the process of "global distillation," pollutants that evaporate and rise high into the atmosphere at lower latitudes are carried toward the poles by atmospheric currents, while ocean currents carry pollutants deposited in the ocean toward the poles. This process exposes polar organisms to unusually concentrated levels of toxic substances.

Because so many substances are carried by the wind, synthetic chemicals have become ubiquitous worldwide, even in seemingly pristine areas. Air currents can carry pesticides to sites far away from agricultural fields in a process called *pesticide drift*. Frogs in the mountains of the Sierra Nevada, for example, have experienced population declines associated with pesticide drift from agriculture in California's nearby Central Valley.

Toxic substances can also be carried great distances through the air. Earth's polar regions are particularly contaminated because natural patterns of global atmospheric circulation (p. 282) tend to move airborne chemicals toward the poles (**FIGURE 10.5**). Thus, although we manufacture and apply synthetic substances mainly in temperate and tropical regions, contaminants have become strikingly concentrated in the tissues of Arctic polar bears, Antarctic penguins, and people living in Greenland.

## Toxic substances may concentrate in water

Runoff in watersheds (p. 24) concentrates contaminants from large expanses of land into relatively small volumes of water in creeks, rivers, lakes, and groundwater. Further, traces of toxic substances and pharmaceuticals excreted in human urine can enter waters from wastewater treatment plants (pp. 272–274), because the treatment process does not typically remove them. Many chemicals are soluble in water and enter organisms' tissues through drinking or absorption. For this reason, aquatic animals such as fish, frogs, and stream invertebrates are effective indicators of pollution. The contaminants that wash into streams and rivers also flow and seep into the water we drink. Once concentrated in water, toxicants can often move long distances through aquatic systems (pp. 250–251) and affect a diversity of organisms and ecosystems.

## Some toxicants persist in the environment

A toxic substance that is released into the environment may degrade quickly and become harmless, or it may remain unaltered and persist for many months, years, or decades. The rate at which a given substance degrades depends on its chemistry and on factors such as temperature, moisture, and sun exposure. The *Bt* toxin (p. 153) used in biocontrol and genetically modified crops has a very short persistence time, whereas chemicals such as DDT and PCBs persist for decades.

Persistent synthetic chemicals exist in our environment today because we have designed them to persist. The synthetic chemicals used in plastics, for instance, are used precisely because they resist breakdown. Sooner or later, however, most toxic substances degrade into simpler compounds called *breakdown products*. Often these are less harmful than the original substance, but sometimes they are just as toxic as the original chemical, or more so. For instance, DDT breaks down into DDE, a highly persistent and toxic compound in its own right.

## Toxic substances may accumulate and move up the food chain

Within an organism's body, some toxic substances are quickly excreted, and some are degraded into harmless breakdown products. Others persist intact in the body. Substances that are fat-soluble or oil-soluble (including organic compounds such as DDT and DDE) are absorbed and stored in fatty tissues. Such persistent toxicants accumulate in an organism's body in a process termed **bioaccumulation**, such that the organism's tissues have a greater concentration of the substance than exists in the surrounding environment.

Toxic substances that bioaccumulate in an organism's tissues may be transferred to other organisms as predators consume prey, resulting in a process called **biomagnification** (**FIGURE 10.6**). When one organism consumes another, the predator takes in any stored toxicants and stores them in its own body.

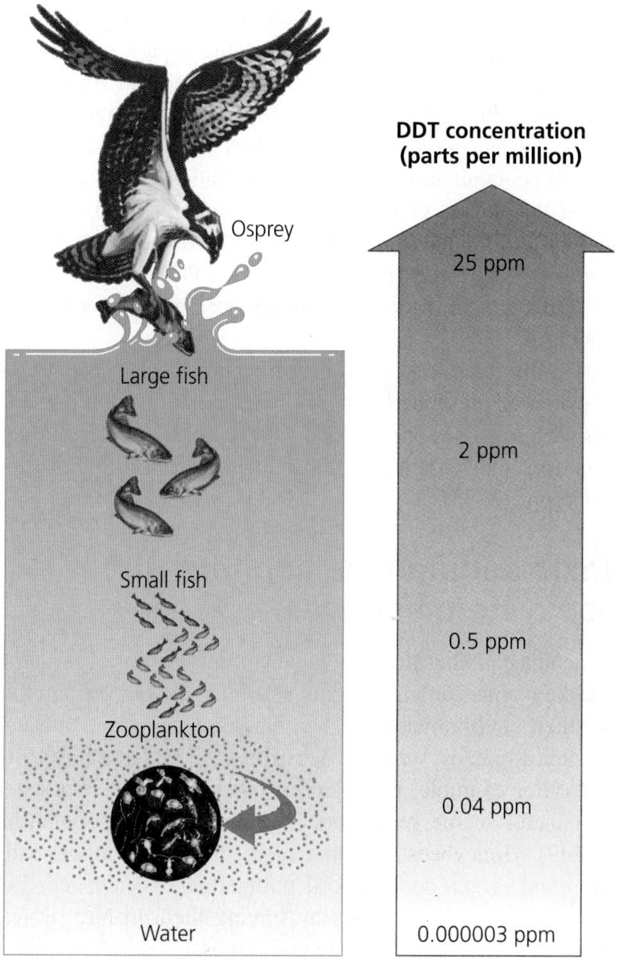

**FIGURE 10.6 In a classic case of biomagnification, DDT becomes highly concentrated in fish-eating birds such as ospreys.** Organisms at the lowest trophic level take in fat-soluble compounds such as DDT from water. As animals at higher trophic levels eat organisms lower on the food chain, each organism passes its load of toxicants up to its consumer, such that organisms on all trophic levels bioaccumulate the substance in their tissues.

Thus bioaccumulation takes place on all trophic levels. Moreover, each individual predator consumes many individuals from the trophic level beneath it, so with each step up the food chain, concentrations of toxicants become magnified.

The process of biomagnification occurred throughout North America with DDT. Top predators, such as birds of prey, ended up with high concentrations of the pesticide because concentrations became magnified as DDT moved from water to algae to plankton to small fish to larger fish and finally to fish-eating birds.

Biomagnification of DDT caused populations of many North American birds of prey to decline precipitously from the 1950s to the 1970s. The peregrine falcon was nearly wiped out in the eastern United States, and the bald eagle, the U.S. national bird, was virtually eliminated from the lower 48 states. Eventually scientists determined that DDT was causing these birds' eggshells to grow thinner, so that eggs were breaking in the nest and killing the embryos within. In a remarkable environmental success story, populations of all these birds have rebounded (pp. 178–179) since the United States banned DDT.

Impacts from biomagnification still persist, though. Unfortunately, DDT continues to impair wildlife in parts of the world where it is still used. Mercury bioaccumulates in some commercially important fish species, such as tuna. Polar bears of Svalbard Island in arctic Norway show extremely high levels of PCB contamination from biomagnification as a result of the global distillation process shown in Figure 10.5. Polar bear cubs suffer immune suppression, hormone disruption, and high mortality—and because the cubs receive PCBs in their mothers' milk, contamination persists and accumulates across generations.

In all these cases, biomagnification affects ecosystem composition and functioning. When populations of top predators such as eagles or polar bears are reduced, species interactions (pp. 68–70) change, and effects cascade through food webs (pp. 72–73).

## Toxic substances can threaten ecosystem services

Toxicants can alter the biological composition of ecosystems and the manner in which organisms interact with one another and their environments. In so doing, harmful compounds can threaten the ecosystem services (pp. 3, 35) that nature provides. For example, pesticide exposure has been implicated as a factor in the recent declines in honeybee populations (p. 149). Honeybees pollinate more than 100 economically important crops, and reduced pollination by wild bees has increased costs for farmers by forcing them to hire professional beekeepers to pollinate their crops.

Among the many services that healthy, functioning ecosystems provide is nutrient cycling. Decomposers and detritivores in the soil (p. 72) break down organic matter and replenish soils with nutrients that plants utilize. When soils are exposed to pesticides or antifungal agents, nutrient cycling rates can be altered. This can make nutrients less available to producers, affecting their growth and causing impacts that cascade throughout the ecosystem.

# Studying Effects of Hazards

Determining the effects of particular environmental hazards on organisms and ecosystems is a challenging job, and scientists rely on several different methods to do this, ranging from correlative surveys to manipulative experiments (pp. 9–10).

## Wildlife studies integrate work in the field and lab

Scientists study the impacts of environmental hazards on wild animals to help conserve animal populations and also to understand potential risks to people. Often, wildlife toxicologists work in the field with animals to take measurements, document patterns, and generate hypotheses before heading to the laboratory to run controlled manipulative experiments to test their hypotheses. For instance, in some of the early work tying environmental chemicals to endocrine disruption, biologist Louis Guillette and his collaborators discovered that alligators in lakes in Florida receiving agricultural runoff had higher rates of reproductive problems than alligators in less polluted lakes. Based on these field studies, he hypothesized that chemical contaminants were disrupting the endocrine systems of alligators during their development in the egg. Subsequent laboratory studies showed that contaminants found in alligator eggs, including the herbicide atrazine, could bind to receptors for estrogen or produce an enzyme that converts testosterone to estrogen.

Following Guillette's work, researcher Tyrone Hayes (**FIGURE 10.7**) found in laboratory experiments that male frogs raised in water containing very low doses of atrazine became

**FIGURE 10.7 Wildlife studies examine the effects of toxic substances in the environment.** Researcher Tyrone Hayes found that frogs show reproductive abnormalities that he attributes to endocrine disruption by pesticides.

feminized and hermaphroditic, developing both testes and ovaries. Field surveys then showed that leopard frogs across North America experienced hormonal problems in areas of heavy atrazine usage. These studies indicated that atrazine, which kills plants by blocking biochemical pathways in photosynthesis, can also act as an endocrine disruptor in animals.

## Human studies rely on case histories, epidemiology, and animal testing

In studies of human health, we gain much knowledge by studying sickened individuals directly. This process of observing and analyzing individual patients is known as a **case history** approach. Case histories have advanced our understanding of human illness, but they tell us little about probability and risk, such as how many extra deaths we might expect in a population due to a particular cause.

For such questions, which are common in environmental toxicology, we need **epidemiological studies**, large-scale comparisons among groups of people, usually contrasting a group known to have been exposed to some hazard against a group that has not. Epidemiologists track the fate of all people in the study for a long period of time (often years or decades) and measure the rate at which deaths, cancers, or other health problems occur in each group. The epidemiologists then analyze the data, looking for observable differences between the groups, and statistically test hypotheses accounting for differences. When a group exposed to a hazard shows a significantly greater degree of harm, it suggests that the hazard may be responsible. The epidemiological process is akin to a natural experiment (p. 10), in which an event creates groups of subjects that researchers can study (for example, people exposed to carcinogenic compounds in their drinking water versus those not similarly exposed).

Epidemiological studies measure a statistical association between a health hazard and an effect, but they do not confirm that the hazard "causes" the effect. To establish causation, manipulative experiments are needed. However, subjecting people to massive doses of toxic substances in a lab experiment would clearly be unethical. This is why researchers have traditionally used laboratory strains of rats, mice, and other mammals. Because of shared evolutionary history, substances that harm mice and rats are reasonably likely to harm us.

## Dose-response analysis is a mainstay of toxicology

The standard method of testing with lab animals in toxicology is **dose-response analysis**. Scientists quantify the toxicity of a substance by measuring the strength of its effects or the number of animals affected at different doses. The **dose** is the amount of substance the test animal receives, and the **response** is the type or magnitude of negative effects the animal exhibits as a result. The response is generally quantified by measuring the proportion of animals exhibiting negative impacts. The data are plotted on a graph, with dose on the *x*-axis and response on the *y*-axis (**FIGURE 10.8a**). The resulting curve is called a **dose-response curve**.

Once they have plotted a dose-response curve, toxicologists can calculate a convenient shorthand gauge of a substance's

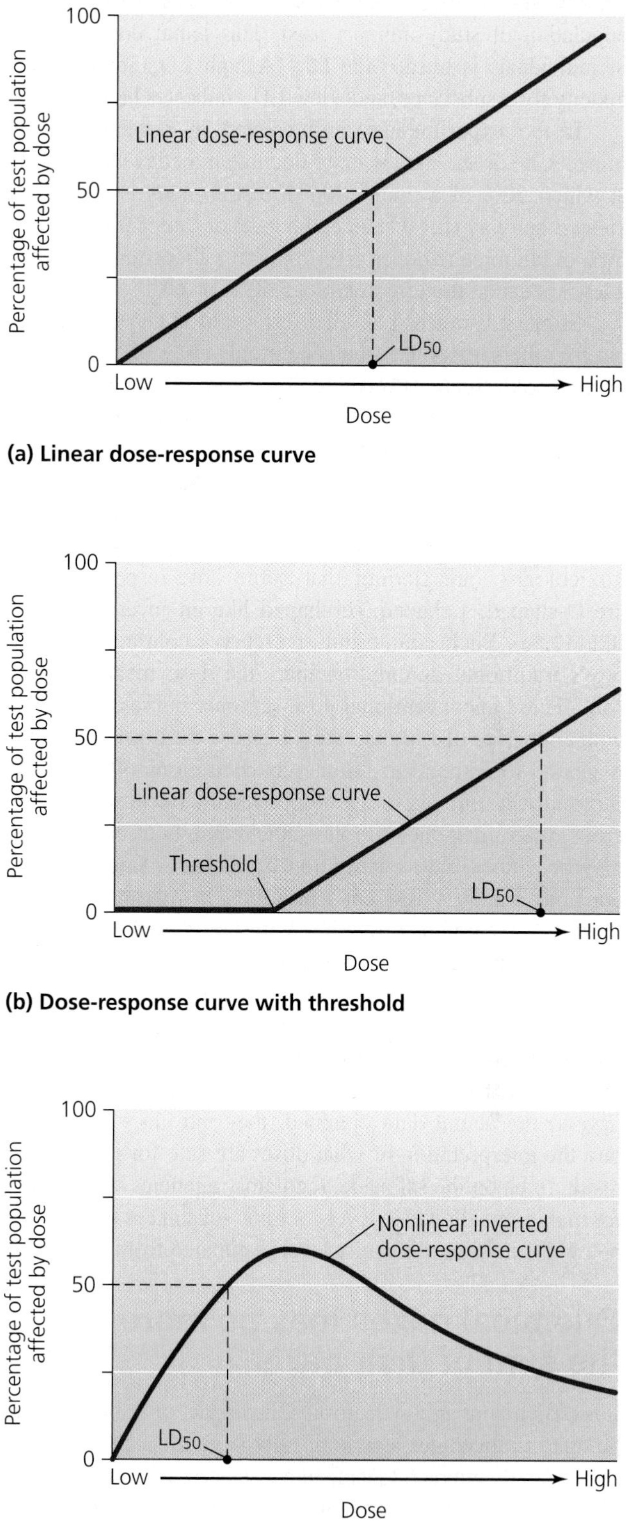


**FIGURE 10.8 Dose-response curves show that organisms' responses to toxicants may sometimes be complex.** In a classic linear dose-response curve **(a)**, the percentage of animals killed or otherwise affected by a substance rises with the dose. The point at which 50% of the animals are killed is labeled the *lethal-dose-50*, or $LD_{50}$. For some toxic substances, a threshold dose **(b)** exists, below which doses have no measurable effect. Some substances, in particular endocrine disruptors, show unconventional, nonlinear dose-response curves **(c)** that are U-shaped, J-shaped, or inverted.

toxicity: the amount of the substance it takes to kill half the population of study animals used. This lethal dose for 50% of individuals is termed the **$LD_{50}$**. A high $LD_{50}$ indicates low toxicity for a substance, and a low $LD_{50}$ indicates high toxicity.

If the experimenter is interested in nonlethal health impacts, he or she may want to document the level of toxicant at which 50% of a population of test animals is affected in some other way (for instance, the level of toxicant that causes 50% of lab mice to develop reproductive abnormalities). Such a level is called the effective-dose-50%, or **$ED_{50}$**.

Some substances can elicit effects at any concentration, but for others, responses may occur only above a certain dose, or threshold. Such a **threshold** dose (**FIGURE 10.8b**) might be expected if the body's organs can fully metabolize or excrete a toxicant at low doses but become overwhelmed at higher concentrations. It might also occur if cells can repair damage to their DNA only up to a certain point.

Sometimes a response may *decrease* as a dose increases. Toxicologists are finding that some dose-response curves are U-shaped, J-shaped, or shaped like an inverted U (**FIGURE 10.8c**). Such counterintuitive curves contradict toxicology's traditional assumption that "the dose makes the poison." These unconventional dose-response curves often occur with endocrine disruptors, likely because the hormone system is geared to respond to minute concentrations of substances (normally, hormones in the bloodstream). Because the endocrine system responds to minuscule amounts of chemicals, it may be vulnerable to disruption by contaminants that reach our bodies in very low concentrations. In research with bisphenol A, a number of studies with lab animals have found unconventional dose-response curves.

The shape of dose-response curves is important because estimating effects on humans often requires extrapolation—extending the dose-response curves beyond the doses tested with laboratory animals. Because extrapolations stretch beyond the actual data obtained, they introduce uncertainty into the interpretation of what doses are safe for people. As a result, to be on the safe side, regulatory agencies set standards for maximum allowable levels of toxic substances that are well below the minimum toxicity levels estimated from lab studies.

### Chemical mixes may be more than the sum of their parts

It is difficult enough to determine the impact of a single hazard, but the task becomes astronomically more difficult when multiple hazards interact. Chemical substances, when mixed, may act together in ways that cannot be predicted from the effects of each in isolation. Mixed toxicants may sum each other's effects, cancel out each other's effects, or multiply each other's effects. Interactive impacts that are greater than the simple sum of their constituent effects are called **synergistic effects**.

With Florida's alligators, lab experiments have indicated that the DDT breakdown product DDE can either promote or inhibit sex reversal, depending on the presence of other chemicals. Mice exposed to a mixture of nitrate, atrazine, and the insecticide aldicarb have been found to show immune, endocrine, and nervous system effects that were not evident from exposure to each of these chemicals alone.

Traditionally, environmental health has tackled impacts of single hazards one at a time. In toxicology, the complex experimental designs required to test interactions, and the sheer number of chemical combinations, have meant that single-substance tests have received priority. This approach is changing, but scientists in environmental health and toxicology will never be able to test all possible combinations.

### Endocrine disruption poses challenges for toxicology

Unconventional dose-response curves are presenting challenges for scientists studying toxic substances and for policymakers trying to set safety standards for them. Because so many novel synthetic chemicals exist in very low concentrations over wide areas, many scientists suspect that we may have underestimated the dangers of compounds that exert impacts at low concentrations.

Scientists first noted endocrine-disrupting effects decades ago, but the idea that synthetic chemicals might be altering the hormones of animals was not widely appreciated until the 1996 book *Our Stolen Future*, by Theo Colburn, Dianne Dumanoski, and J.P. Myers. Like *Silent Spring*, this book integrated scientific work from various fields and presented a unified picture that shocked many readers—and brought criticism from some scientists and from the chemical industry.

Today, thousands of studies have linked hundreds of substances to effects on reproduction, development, immune function, brain and nervous system function, and other hormone-driven processes. Evidence is strongest so far in nonhuman animals, but many studies suggest impacts on humans. Some researchers suggest that endocrine disruptors may account for rising rates of testicular cancer, undescended testicles, and genital birth defects in males. Others argue that the sharp rise in breast cancer rates (one in eight U.S. women today develops breast cancer) may be due to hormone disruption, because an excess of estrogen appears to feed tumor development in older women. Still other scientists attribute male reproductive problems to elevated BPA exposure. For example, studies found that workers in Chinese factories that manufactured BPA had elevated rates of erectile disfunction and reduced sperm counts when compared to workers in factories manufacturing other products.

Research on hormone disruption has brought about strident debate. This is partly because of the scientific uncertainty inherent in any emerging field of study, but also because of the economic value of the chemicals being tested. For instance, Tyrone Hayes's work has met with fierce criticism from scientists associated with atrazine's manufacturer, which would lose many millions of dollars if its widely used herbicide were to be banned or restricted.

## Risk Assessment and Risk Management

Policy decisions on whether to ban chemicals or restrict their use generally follow years of rigorous testing for toxicity. Likewise, strategies for combating disease and other health threats are based on extensive scientific research. However, policy and management decisions also incorporate economics and ethics—and all too often the decision-making process is heavily influenced by pressure from powerful corporate and political interests. The steps between the collection and interpretation of scientific data and the formulation of policy involve assessing and managing risk.

## We express risk in terms of probability

Exposure to an environmental health threat does not invariably produce a given consequence. Rather, it causes some probability of harm, a statistical chance that damage will result. To understand a health threat, a scientist must know more than just its identity and strength. He or she must also know the chance that one will encounter it, the frequency with which one may encounter it, the amount of substance or degree of threat to which one is exposed, and one's sensitivity to the threat. Such factors help determine the overall risk posed.

Risk can be measured in terms of *probability*, a quantitative description of the likelihood of a certain outcome. The probability that some harmful outcome (for instance, injury, death, environmental damage, or economic loss) will result from a given action, event, or substance expresses the risk posed by that phenomenon.

## Our perception of risk may not match reality

Every action we take and every decision we make involves some element of risk, some (generally small) probability that things will go wrong. We typically try to behave in ways that minimize risk, but our perceptions of risk do not always match statistical reality (**FIGURE 10.9**). People often worry unduly about small risks yet readily engage in activities that pose higher risks. For instance, most of us perceive flying in an airplane as a riskier activity than driving a car, but, statistically speaking, plane travel is much safer. Psychologists agree that this disconnect is because we feel more at risk when we are not controlling a situation and more safe when we are "at the wheel"—regardless of the actual risk involved.

This psychology may help account for people's anxiety over exposure to bisphenol A, nuclear power, toxic waste, and pesticide residues on foods—environmental hazards that are invisible or little understood and whose presence in our lives is largely outside our personal control. In contrast, people are more ready to accept and ignore the risks of smoking cigarettes, overeating, and not exercising—voluntary activities statistically shown to pose far greater risks to health.

## Risk assessment analyzes risk quantitatively

The quantitative measurement of risk and the comparison of risks involved in different activities or substances together are termed **risk assessment**. Risk assessment is a way to identify and outline problems. In environmental health, it helps ascertain which substances and activities pose health threats to people or wildlife and which are largely safe.

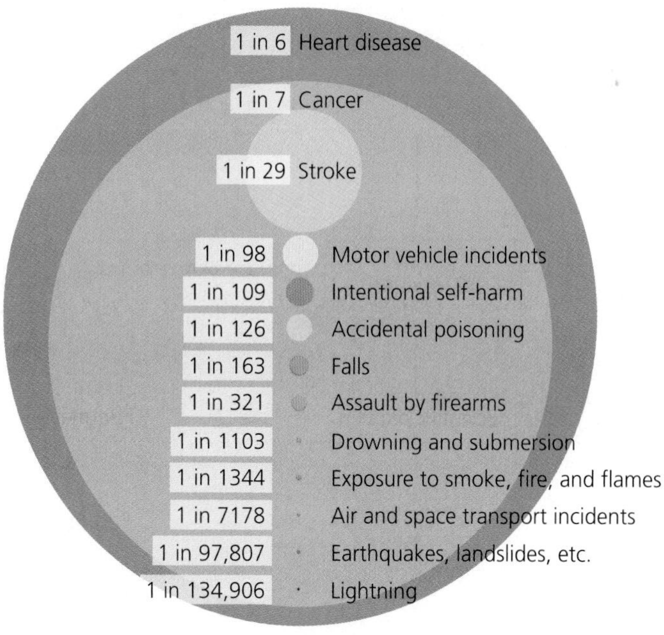


**FIGURE 10.9 Our perceptions of risk do not always match the reality of risk.** Shown are selected causes of death in the United States, along with a measure of the risk each poses. The larger the area of the circle, the greater the risk of dying from that cause. *Data are for 2008 from* Injury Facts, 2012, *National Safety Council, Itasca, IL.*

DATA Q People tend to view car travel as being safer than airplane travel, but a person is how many times more likely to die from a car accident than from an airplane crash?

GO TO **INTERPRETING GRAPHS & DATA** ON MasteringEnvironmentalScience®

Assessing risk for a chemical substance involves several steps. The first steps involve the scientific study of toxicity we examined above—determining whether a substance has toxic effects and, through dose-response analysis, measuring how effects vary with the degree of exposure. Subsequent steps involve assessing the individual's or population's likely extent of exposure to the substance, including frequency of contact, concentrations likely encountered, and length of encounter.

Some regulatory agencies are taking steps to help assess the health risks posed by endocrine-disrupting chemicals. For example, the U.S. National Institute of Environmental and Health Sciences devoted $30 million in 2011–2013 to advance research on the health impacts of bisphenol A.

## Risk management combines science and other social factors

Accurate risk assessment is a vital step toward effective **risk management**, which consists of decisions and strategies to minimize risk. In most nations, risk management is handled largely by federal agencies. In the United States, these agencies include the EPA, the CDC, and the Food and Drug Administration (FDA). In risk management, scientific assessments of risk are considered in light of economic, social, and political needs and values. Risk managers assess costs and benefits of addressing risk in various ways,

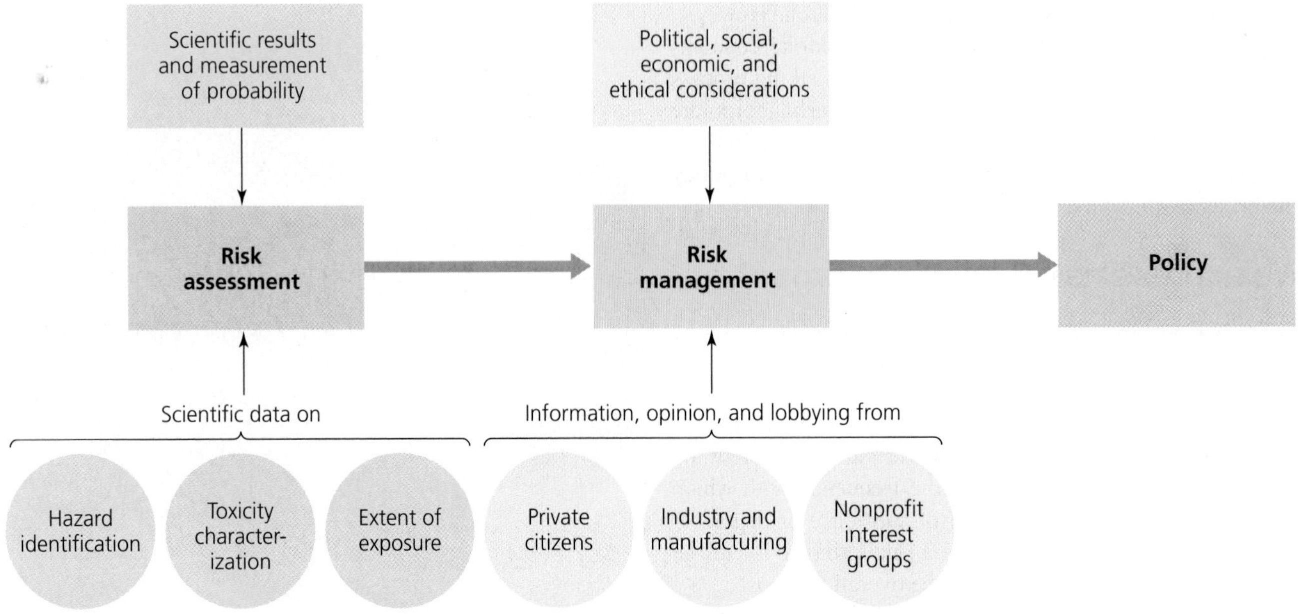


**FIGURE 10.10 The first step in addressing risks from an environmental hazard is risk assessment. Once science identifies and measures risks, then risk management can proceed.** In risk management, economic, political, social, and ethical issues are considered in light of the scientific data from risk assessment.

with regard to both scientific and nonscientific concerns, before making decisions on whether and how to reduce or eliminate risk (**FIGURE 10.10**).

In environmental health and toxicology, comparing costs and benefits (p. 94) can be difficult because the benefits are often economic, whereas the costs often pertain to health. Moreover, economic benefits are generally known, easily quantified, and of a discrete and stable amount, whereas health risks are hard-to-measure probabilities, often involving a small percentage of people likely to suffer greatly and a large majority likely to experience little effect. Because of the lack of equivalence in the way costs and benefits are measured, risk management frequently tends to stir up debate.

In the case of BPA, eliminating plastic linings in our food and drink cans could do more harm than good, because the linings help prevent metal corrosion and the contamination of food by pathogens. Alternative substances exist for most of BPA's uses, but replacing BPA with alternatives will entail economic costs to industry, and these costs get passed on to consumers in the prices of products. Such complex considerations can make risk management decisions difficult even if the science of risk assessment is fairly clear. This may help account for the observed hesitancy of U.S. regulatory agencies, so far, to issue stringent restrictions on the uses of BPA.

## Two approaches exist for testing the safety of new products

Because we cannot know a substance's toxicity until we measure and test it, and because there are so many untested chemicals and combinations, science will never eliminate the many uncertainties that accompany risk assessment. In such a world of uncertainty, there are two basic philosophical approaches to categorizing substances as safe or dangerous (**FIGURE 10.11**).

One approach is to assume that substances are harmless until shown to be harmful. We might nickname this the "innocent-until-proven-guilty" approach. Because thoroughly testing every existing substance (and combination of substances) for its effects is a hopelessly long, complicated, and expensive pursuit, the innocent-until-proven-guilty approach has the virtue of facilitating technological innovation and economic activity. However, it has the disadvantage of putting into wide use some substances that may later turn out to be dangerous.

The other approach is to assume that substances are harmful until shown to be harmless. This approach follows the **precautionary principle** (p. 155). This more cautious approach should enable us to identify troublesome toxicants before they are released into the environment, but it may also impede the pace of technological and economic advance.

These two approaches are actually two ends of a continuum of possible approaches. The two endpoints differ mainly in where they lay the burden of proof—specifically, whether product manufacturers are required to prove a product is safe or whether government, scientists, or citizens are required to prove a product is dangerous.

The choice of philosophical approach has direct implications for policy, and nations vary in how they blend the two approaches. European nations have recently embarked on a policy course that incorporates the precautionary principle, whereas the United States largely follows an innocent-until-proven-guilty approach.

In the United States, several federal agencies are assigned responsibility for tracking and regulating synthetic

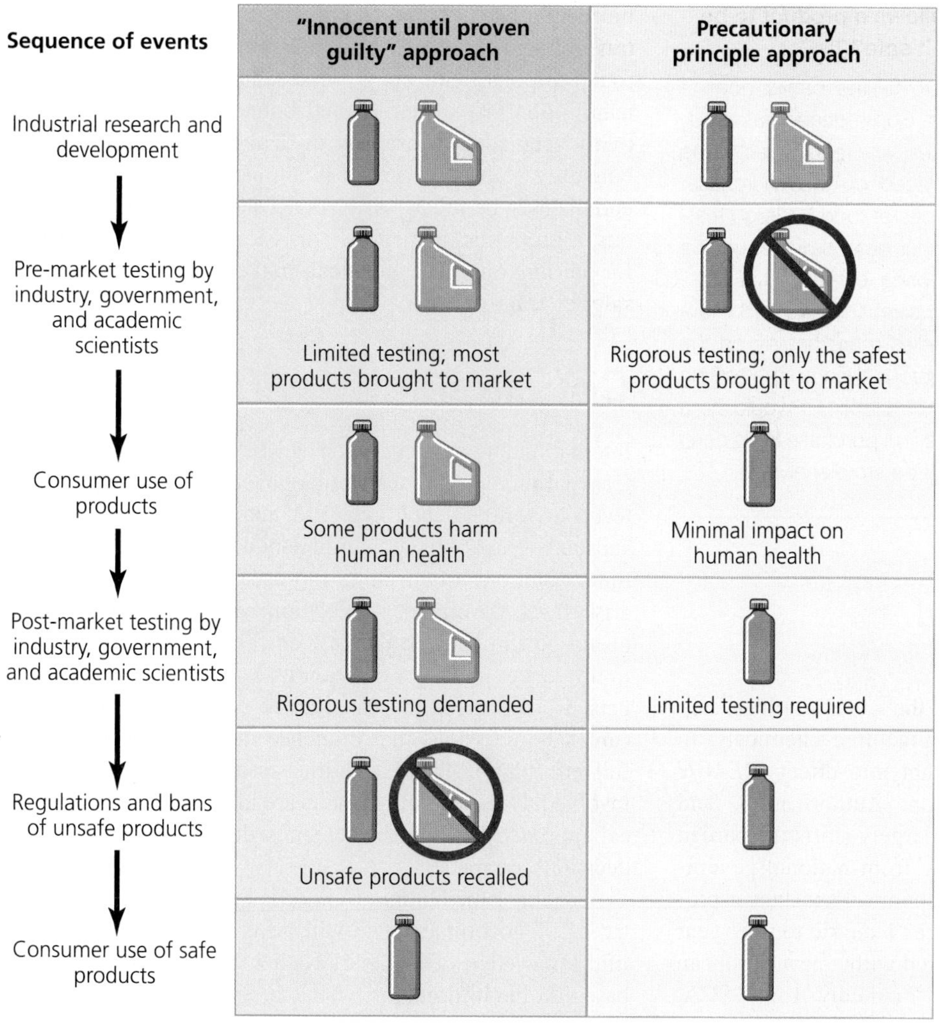

Sequence of events
"Innocent until proven guilty" approach
Precautionary principle approach
Industrial research and development
Pre-market testing by industry, government, and academic scientists
Limited testing; most products brought to market
Rigorous testing; only the safest products brought to market
Consumer use of products
Some products harm human health
Minimal impact on human health
Post-market testing by industry, government, and academic scientists
Rigorous testing demanded
Limited testing required
Regulations and bans of unsafe products
Unsafe products recalled
Consumer use of safe products


**FIGURE 10.11 Two main approaches can be taken to introduce new substances to the market.** In one approach, substances are "innocent until proven guilty"; they are brought to market relatively quickly after limited testing. Products reach consumers more quickly, but some fraction of them **(blue bottle in diagram)** may cause harm to some fraction of people. The other approach is to adopt the precautionary principle, bringing substances to market cautiously, only after extensive testing. Products that reach the market should be safe, but many perfectly safe products **(purple bottle diagram)** will be delayed in reaching consumers.

chemicals under various legislative acts. The FDA, under an act first passed in 1938, monitors foods and food additives, cosmetics, drugs, and medical devices. The EPA regulates pesticides under a 1947 act and its amendments. The Occupational Safety and Health Administration (OSHA) regulates workplace hazards under a 1970 act. Several other agencies regulate other substances. Synthetic chemicals not covered by other laws are regulated by the EPA under the 1976 **Toxic Substances Control Act (TSCA)**.

## WEIGHING THE ISSUES

**The Precautionary Principle**

Industry's critics say chemical manufacturers should bear the burden of proof for the safety of their products before they hit the market. Industry's supporters say that mandating more safety research will hamper the introduction of products that consumers want and increase the price of products. What do you think? Should government follow the precautionary principle and require proof of safety prior to a chemical's introduction into the market?

## EPA regulation is only partly effective

The Toxic Substances Control Act directs the EPA to monitor thousands of industrial chemicals manufactured in or imported into the United States, ranging from PCBs to lead to bisphenol A. The act gives the agency power to regulate these substances and ban them if they are found to pose excessive risk.

However, many public health advocates view TSCA as being far too weak. They note that the screening required of industry is minimal and that to mandate more extensive and meaningful testing, the EPA must show proof of the chemical's toxicity. In other words, the agency is trapped in a Catch-22: To push for studies looking for toxicity, it must have proof of toxicity already. The result is that most synthetic chemicals are not thoroughly tested before being brought to market. Of those that fall under TSCA, only 10% have been thoroughly tested for toxicity; only 2% have been screened for carcinogenicity, mutagenicity, or teratogenicity; fewer than 1% are regulated; and almost none have been tested for endocrine, nervous, or immune system damage, according to the U.S. National Academy of Sciences.

**FAQ** **If the government allows a product to be sold in stores, isn't it safe?**

Just because a product is available to the public doesn't mean it poses no risk to consumers. Some dangerous products, such as firearms and cigarettes, are recognized to be hazardous and have restrictions placed on their purchase. Medicines, cosmetics, and some types of food undergo testing for safety prior to release, but other potentially dangerous products, such as endocrine-disrupting chemicals in plastics, are not similarly tested. We've seen that the U.S. government's approach to regulation relies more heavily on the recall of products found to be dangerous than on the testing of products before they reach market. Hence, it would be a mistake to assume that the products you purchase have been thoroughly tested prior to their arrival on store shelves.

### Toxicants are regulated internationally

The European Union (EU) is taking the world's boldest step toward testing and regulating manufactured chemicals. In 2007, the EU's **REACH** program went into effect (*REACH* stands for *R*egistration, *E*valuation, *A*uthorization, and restriction of *CH*emicals). REACH largely shifts the burden of proof for testing chemical safety from national governments to industry and requires that chemical substances produced or imported in amounts of over 1 metric ton per year (some 30,000 chemicals) be registered with a new European Chemicals Agency. REACH differs markedly from TSCA, illustrating how the approach that Europe is pursuing is different from the approach being pursued in the United States.

The world's nations have also sought to address chemical pollution with international treaties. The *Stockholm Convention on Persistent Organic Pollutants* (*POPs*) came into force in 2004 and has been ratified by over 150 nations. POPs are toxic chemicals that persist in the environment, bioaccumulate and biomagnify up the food chain, and can travel long distances. The PCBs and other contaminants found in polar bears are a prime example. Because contaminants often cross international boundaries, an international treaty seemed the best way to deal fairly with such transboundary pollution. The Stockholm Convention aims first to end the use and release of 12 POPs shown to be most dangerous, a group nicknamed the "dirty dozen." It sets guidelines for phasing out these chemicals and encourages transition to safer alternatives.

## Conclusion

International agreements such as REACH and the Stockholm Convention represent a sign that governments may act to protect the world's people, wildlife, and ecosystems from toxic substances and other environmental hazards. At the same time, solutions often come more easily when they do not arise from government regulation alone. Consumer choice exercised through the market can often be an effective way to influence industry's decision making. Consumers of products, from plastics to pesticides to cosmetics to kids' toys, can make decisions that influence industry when they have full information from scientific research regarding the risks involved. Once scientific results are in, a society's philosophical approach to risk management will determine what policy decisions are made.

Whether the burden of proof is laid at the door of industry or of government, we will never attain complete scientific knowledge of any risk. Rather, we must make choices based on the information available. Synthetic chemicals have brought us innumerable modern conveniences, a larger food supply, and medical advances that save and extend human lives. Human society would be very different without them. Yet a better future, one that safeguards the well-being of both people and the environment, depends on knowing the risks that some hazards pose and on having in place the means to phase out harmful substances and replace them with safer ones.

# Testing Your Comprehension

1. What four major types of health hazards are examined by practitioners of environmental health?
2. In what way is disease the greatest hazard that people face? What kinds of interrelationships must environmental health experts study to learn how diseases affect human health?
3. Where does most exposure to lead, asbestos, radon, and PBDEs occur?
4. List and describe the seven general categories of toxic substances described in this chapter.
5. Explain the mechanisms within organisms that protect them from damage from toxic substances.
6. Describe and contrast the processes of bioaccumulation and biomagnification.
7. What are epidemiological studies, and how are they most often conducted?
8. Explain the dose-response curve. Why do endocrine-disrupting chemicals such as BPA pose challenges for toxicology?
9. What factors may affect an individual's response to a toxic substance?
10. How do scientists identify and assess risks from substances or activities?

# Seeking Solutions

1. Describe some environmental health hazards you may be living with indoors. How may you have been affected by indoor or outdoor hazards in the past? How could you best deal with these hazards in the future?
2. Do you feel that laboratory animals should be used in experiments in toxicology? Why or why not?
3. Describe differences in the policies of the United States and the European Union toward the study and management of the risks of synthetic chemicals. Which do you believe is more appropriate, the policies of the United States or those of the European Union? Why?
4. **THINK IT THROUGH** You are the parent of two young children, and you want to minimize the environmental health risks your kids are exposed to. Name five steps that you could take that would accomplish your goal.
5. **THINK IT THROUGH** After learning about the effects of bisphenol A, you choose to minimize your exposure to the chemical. You begin by examining your lifestyle and finding ways to use alternatives to BPA-containing products. Create a list of five ways you are exposed daily to BPA, and then list approaches that would avoid or minimize these exposures. Do these steps require more time? What are some costs of embracing these changes?

# Calculating Ecological Footprints

In 2007, the last year the EPA gathered and reported data on pesticide use (pp. 147–148), Americans used 1.13 billion pounds of pesticide active ingredients, and global use totaled 5.21 billion pounds. In that same year, the U.S. population was 302 million, and the world's population was 6.63 billion. In the table, calculate your share of pesticide use as a U.S. citizen in 2007 and the amount used by (or on behalf of) the average citizen of the world.

**Annual pesticide use**

| You | Pounds of active ingredients |
|---|---|
| Your class | |
| Your state | |
| United States | 1.13 billion |
| World (total) | 5.21 billion |
| World (per capita) | |

1. What is the ratio of your annual pesticide use to the world's per capita average?
2. In 2007, the average U.S. citizen had an ecological footprint of 8.0 hectares, and the average world citizen's footprint was 2.7 hectares (Chapter 1). Compare the ratio of pesticide usage with the ratio of the overall ecological footprints. How do these differ, and how would you account for the difference?
3. Does the per capita pesticide use for a U.S. citizen seem reasonable for you personally? Why or why not? Do you find this figure alarming, or of little concern? What else would you like to know to assess the risk associated with this level of pesticide use?


## MasteringEnvironmentalScience®

**STUDENTS**

Go to **MasteringEnvironmentalScience** for assignments, the etext, and the Study Area with practice tests, videos, current events, and activities.

**INSTRUCTORS**

Go to **MasteringEnvironmentalScience** for automatically graded activities, current events, videos, and reading questions that you can assign to your students, plus Instructor Resources.

# 11

# Geology, Minerals, and Mining

## Upon completing this chapter, you will be able to:

- ☐ Describe Earth's internal structure and explain how plate tectonics shapes its surface
- ☐ Identify the categories of rocks, and explain how the rock cycle shapes the landscape around us and the earth beneath our feet
- ☐ List the major types of geologic hazards and describe ways to reduce their impacts
- ☐ Outline types of mineral resources and how they contribute to our products and society
- ☐ Describe the major methods of mining
- ☐ Characterize the environmental and social impacts of mining
- ☐ Assess reclamation efforts and mining policy
- ☐ Evaluate ways to encourage the sustainable use of mineral resources



Photo: **Coltan mining in eastern Congo**

# CENTRAL CASE STUDY

## Mining for . . . Cell Phones?

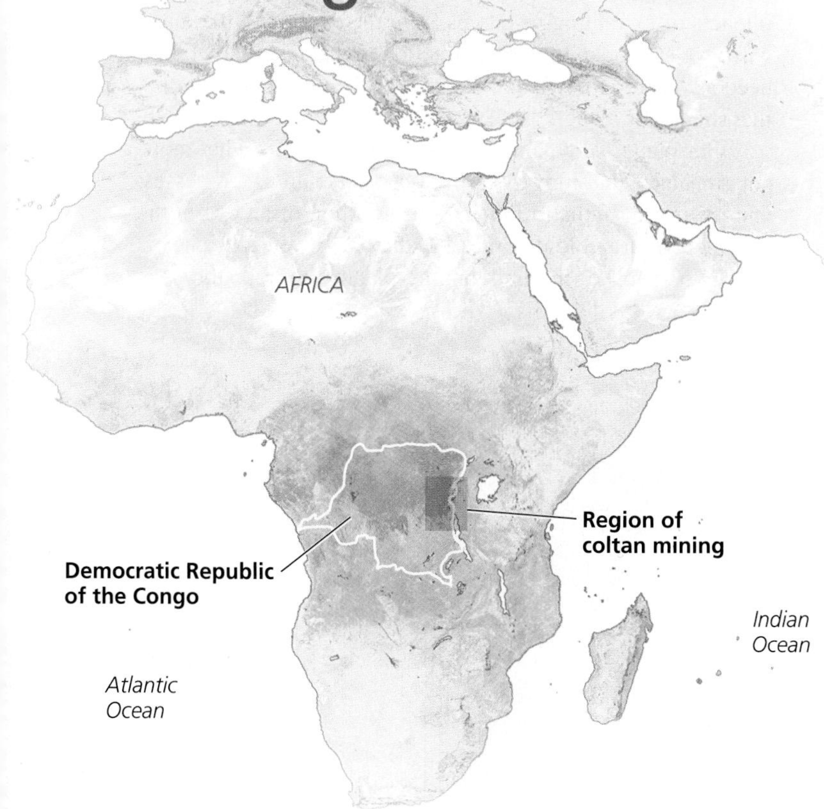


*"The conflict in the Democratic Republic of the Congo has become mainly about access, control, and trade of five key mineral resources: coltan, diamonds, copper, cobalt, and gold."*

**—Report to the United Nations Security Council, April 2001**

*"Coltan . . . is not helping the local people. In fact, it is the curse of the Congo."*

**—African journalist Kofi Akosah-Sarpong**

Pulling a cell phone from her pocket, a student on a college campus in the United States texts a friend. Inside her phone is a little-known metal called tantalum—just a tiny amount, but no cell phone could operate without it.

Half a world away, a miner in the heart of Africa toils all day in a jungle streambed, sifting sediment for nuggets of coltan ore, which contain tantalum. At nightfall, rebel soldiers take most of his ore, leaving him to sell what little remains to buy food for his family at the squalid mining camp where they live.

In bedeviling ways, tantalum links our glossy global high-tech economy with one of the most abused regions on Earth. The Democratic Republic of the Congo has long been embroiled in a sprawling conflict that has involved six nations and various rebel militias and has claimed over 5 million lives since 1998. It is the latest chapter in the sad history of a nation rich in natural resources—copper, cobalt, gold, diamonds, uranium, and timber—whose impoverished people keep losing control of those resources to others.

At the center of the recent conflict is tantalum (Ta), element number 73 on the periodic table (**APPENDIX D**). We rely on this metal for our cell phones, computer chips, DVD players, game consoles, and digital cameras. Tantalum powder is ideal for capacitors (the components that store energy and regulate current in miniature circuit boards) because it is highly heat resistant and readily conducts electricity.

Tantalum comes from a dull blackish mineral called tantalite, which often occurs with a mineral called columbite—so the ore is referred to as columbite–tantalite, or *coltan* for short. In eastern Congo, men dig craters in rainforest streambeds, panning for coltan much as early California miners panned for gold.

As information technology boomed in the late 1990s, global demand for tantalum rose, and market prices for the metal shot up. High prices led some Congolese men to mine coltan by choice, but many more were forced to work as miners. As the war began in 1998, local militias, supported by forces from neighboring Rwanda and Uganda, overran eastern Congo. Farmers were chased off their land, villages were burned, and civilians were raped, tortured, and killed. Soldiers from each army seized control of mining operations. They forced farmers, refugees, prisoners, and children to work, and skimmed profits from the coltan the people mined. Children and teachers abandoned school and worked in the mines, while prostitution spread AIDS and sexually transmitted disease through the mining camps. The turmoil also caused ecological havoc as people streamed into national parks, clearing rainforests and killing wildlife for food, including forest elephants, endangered gorillas, and okapi, a rare relative of the giraffe.

Most miners ended up with little, while rebels, soldiers, and bandits enriched themselves selling coltan to traders, who sold it to processing companies in the United States and Europe. These companies refine and sell tantalum powder to capacitor manufacturers, which in turn sell capacitors to Nokia, Motorola, Sony, Intel, Compaq, Dell, and other high-tech corporations.

In 2001, an expert panel commissioned by the United Nations (UN) Security Council concluded that coltan riches were fueling, financing, and prolonging the war. The panel urged a UN embargo on coltan and other minerals smuggled from Congo and exported by neighboring nations. A grassroots activist movement urged international action and advanced the slogan "No blood on my cell phone!"

Sony, Nokia, Ericsson, and other corporations rushed to assure consumers that they were not using tantalum from eastern Congo—and the region was in fact producing less than 10% of the world's supply.

Meanwhile, some observers felt an embargo could hurt the long-suffering Congolese people, rather than help them. The mining life may be miserable, they said, but it pays better than most jobs in a land where the average income is only 20 cents a day.

Today, the war is declared over, but militias and rebel groups continue to fight within Congo, bankrolled by the mineral riches of the region. While thousands of people continue to suffer from the conflict, recent success by Congolese troops and an African-led UN intervention brigade against a major rebel group in eastern Congo are helping to reduce conflict in the region.

Unfortunately, Congo is not the only source of "conflict minerals" in the world today. A thriving black market in coltan is emerging in remote portions of Brazil, Colombia, and Venezuela in the northern Amazon jungle. Armed gangs and narcotics smugglers in the region are accused of using women, children, and indigenous people to mine and smuggle coltan ore. The recent discovery of vast mineral reserves in Afghanistan (p. 243), coupled with that nation's political unrest, suggests that it too could become a significant source of conflict minerals in the near future.

Steps are now being taken to help support legitimate Congolese mines while preventing the exploitation that has defined mining in the Congo in the recent past. Industry groups, working with national governments and nongovernmental aid organizations, are creating a certification system for conflict-free coltan, and U.S. law now mandates that manufacturers of electronics report the origin of the tantalum in their products.

These efforts and others provide an opportunity to significantly reduce trade in conflict minerals while promoting trade of minerals sourced from legitimate mines in poor nations such as Congo. It is hoped that ongoing regulatory efforts to certify minerals will provide a framework that not only satisfies the world's demand for mineral resources, but also protects the people and ecosystems that provide them. ◻

# Geology: The Physical Basis for Environmental Science

Coltan provides just one example of how we extract raw materials from beneath our planet's surface and turn them into products we use every day. To understand the environmental impacts of extracting resources from the earth, and the many ways we can make mineral extraction less damaging, we first need a working knowledge of some of the physical processes that shape our planet.

Our planet is dynamic, and this dynamism is what motivates **geology**, the study of Earth's physical features, processes, and history. A human lifetime is just a blink of an eye in the long course of geologic time, and the Earth we experience is merely a snapshot in our changing planet's long history. We

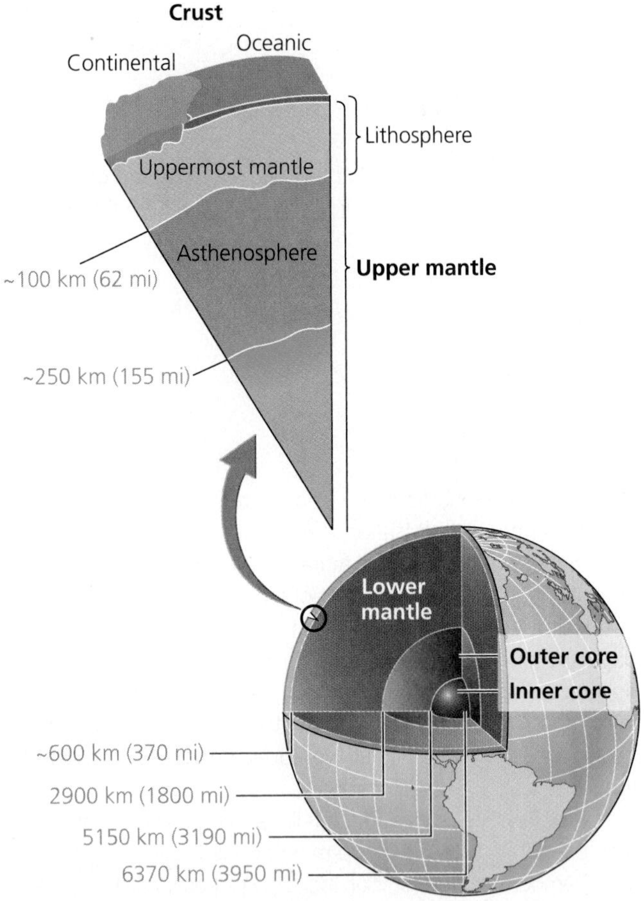


**FIGURE 11.1 Earth's three primary layers—core, mantle, and crust—are themselves layered.** The inner core of solid iron is surrounded by an outer core of molten iron, and the rocky mantle includes the molten asthenosphere near its upper edge. At Earth's surface, dense and thin oceanic crust abuts lighter, thicker continental crust. The lithosphere consists of the crust and the uppermost mantle above the asthenosphere.

DATA Q The driving distance from New York City to Denver, Colorado, is about 2850 km (1775 mi). If you were to tunnel into the Earth this same distance, where in Earth's interior would you find yourself?

GO TO **INTERPRETING GRAPHS & DATA** ON MasteringEnvironmentalScience®

can begin to grasp this long-term dynamism as we consider two processes of fundamental importance to geology—plate tectonics and the rock cycle.

## Earth consists of layers

Most geologic processes take place near Earth's surface, but our planet consists of multiple layers (**FIGURE 11.1**). At its center is a dense **core** consisting mostly of iron, solid in the inner core and molten in the outer core. Around the core is a thick layer of dense, elastic rock called the **mantle**. A portion of the upper mantle called the *asthenosphere* contains especially soft rock, melted in some areas. The harder rock above the asthenosphere is what we know as the **lithosphere**. The lithosphere includes the uppermost mantle and all of Earth's third major layer, the **crust**—the thin, brittle, low-density layer of rock that covers Earth's surface.

The intense heat in the inner Earth drives convection currents that flow in loops in the mantle, pushing its soft rock cyclically upward (as it warms) and downward (as it cools), like a gigantic conveyor belt system. As the soft rock moves, it drags large plates of lithosphere along its surface. This movement of lithospheric plates is known as **plate tectonics**, a process of extraordinary importance to our planet.

## Plate tectonics shapes Earth's geography

Our planet's surface consists of about 15 major tectonic plates, which fit together like pieces of a jigsaw puzzle (**FIGURE 11.2**). Imagine peeling an orange and then placing the pieces of peel back onto the fruit; the ragged pieces of peel are like the tectonic plates riding atop Earth's surface. However, the plates are thinner relative to the planet's size, more like the skin of an apple. These plates move at rates of roughly 2–15 cm (1–6 in.) per year. This slow movement has influenced Earth's climate and life's evolution throughout our planet's history as the continents combined, separated, and recombined in various configurations. By studying ancient rock formations throughout the world, geologists have determined that at least twice, all landmasses were joined together in a "supercontinent." Scientists have dubbed the one that occurred about 225 million years ago *Pangaea* (see Figure 11.2).

## There are three types of plate boundaries

The processes that occur at the boundaries between plates greatly influence Earth's surface. There are three types of plate boundaries: divergent, transform, and convergent.

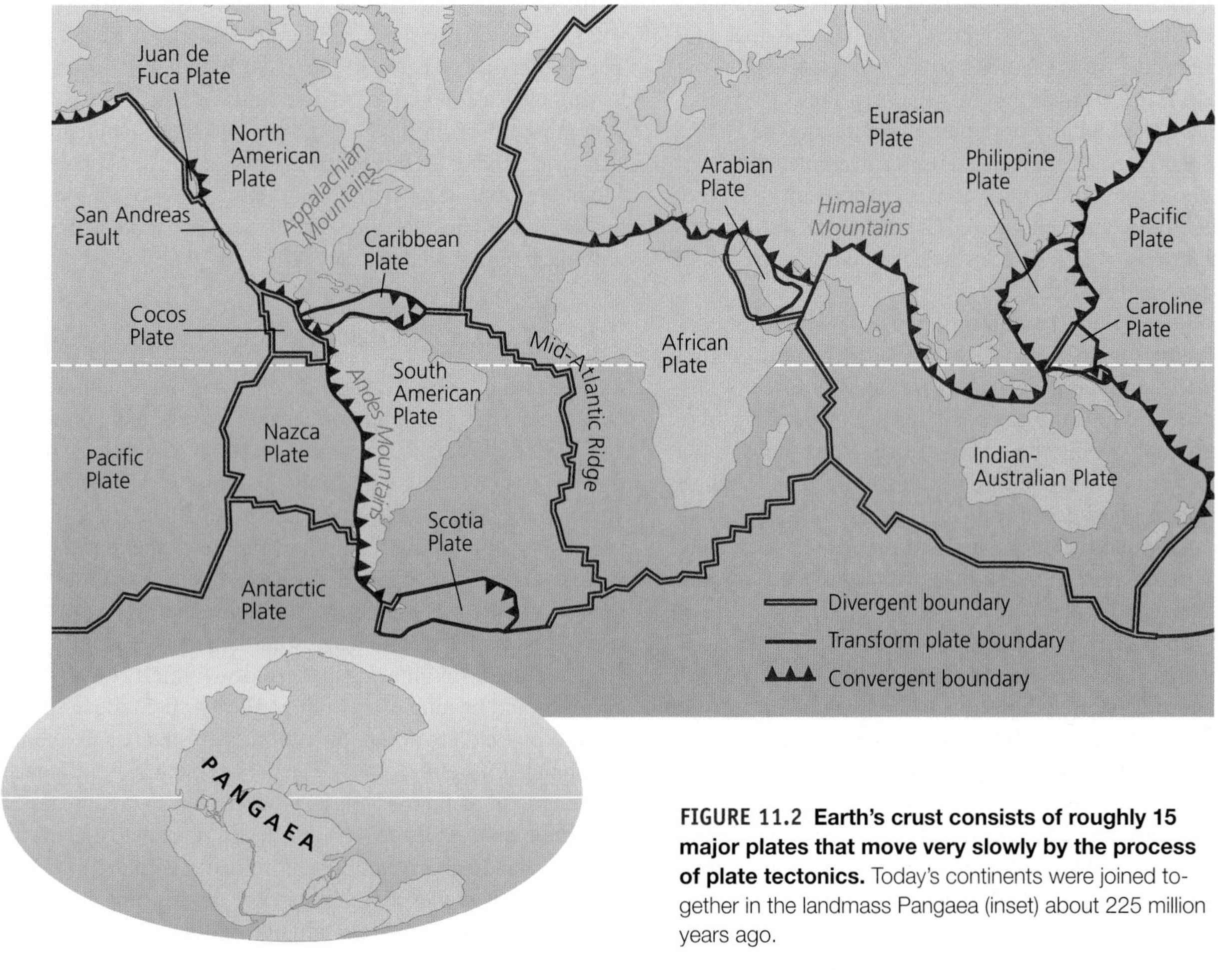


**FIGURE 11.2 Earth's crust consists of roughly 15 major plates that move very slowly by the process of plate tectonics.** Today's continents were joined together in the landmass Pangaea (inset) about 225 million years ago.

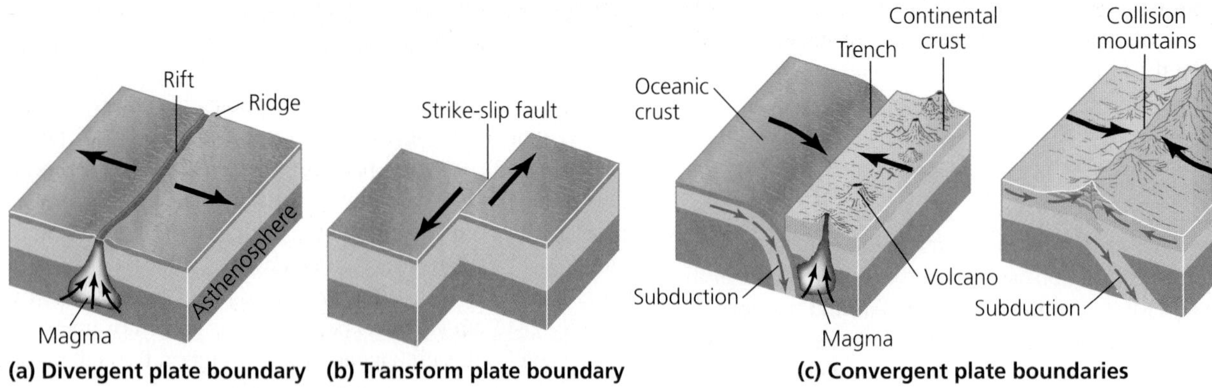


**FIGURE 11.3 There are three types of boundaries between tectonic plates, generating different geologic processes.** At a divergent plate boundary, such as a mid-ocean ridge on the seafloor **(a)**, the two plates move gradually away from the boundary in the manner of conveyor belts, and magma from beneath the crust may extrude as lava. At a transform plate boundary **(b)**, two plates slide alongside one another, creating friction that leads to earthquakes. Where plates collide at a convergent plate boundary **(c)**, one plate is subducted beneath another, leading either to volcanism or the to formation of mountain ranges.

At **divergent plate boundaries**, tectonic plates push apart from one another as **magma** (rock heated to a molten, liquid state) rises upward to the surface, creating new crust as it cools (**FIGURE 11.3a**). An example is the Mid-Atlantic Ridge, part of a 74,000-km (46,000-mi) system of divergent plate boundaries slicing across the floors of the world's oceans.

Where two plates meet, they may slip and grind alongside one another, forming a **transform plate boundary** (**FIGURE 11.3b**). This movement creates friction that generates earthquakes (pp. 230–231) along strike-slip faults. *Faults* are fractures in Earth's crust, and at strike-slip faults two landmasses move horizontally in opposite directions. The Pacific Plate and the North American Plate, for example, slide past one another along California's San Andreas Fault. Southern California is slowly inching its way northward along this fault, and the site of Los Angeles will eventually reach that of modern-day San Francisco.

**Convergent plate boundaries**, where two plates converge or come together, can give rise to different outcomes (**FIGURE 11.3c**). As plates of newly formed lithosphere push outward from divergent plate boundaries, this oceanic lithosphere gradually cools, becoming denser. Eventually (after millions of years), it becomes denser than the asthenosphere beneath it and dives downward into the asthenosphere in a process called **subduction**. As the lithospheric plate descends, it slides beneath a neighboring plate that is less dense, forming a convergent plate boundary. The subducted plate is heated and pressurized as it sinks, and water vapor escapes, helping to melt rock (by lowering its melting temperature). The molten rock rises, and this magma may erupt through the surface at volcanoes (pp. 231–232).

When one plate of oceanic lithosphere is subducted beneath another plate of oceanic lithosphere, the resulting volcanism may form arcs of islands, such as those in Japan or the Aleutian Islands of Alaska. Subduction zones may also create deep trenches, such as the Mariana Trench, Earth's deepest abyss. When oceanic lithosphere slides beneath continental lithosphere, volcanic mountain ranges that parallel coastlines may form (a process shown in the left-hand drawing of Figure 11.3c). The Cascades in the Pacific Northwest, for example, are fueled by magma from subduction.

When two plates of continental lithosphere meet, the continental crust on both sides resists subduction and instead crushes together, deforming layers of rock from both plates (shown in the right-hand drawing of Figure 11.3c). Portions of the accumulating masses of buckled crust are forced upward as they are pressed together, and mountain ranges form. The Himalayas, the world's highest mountains, result from the Indian-Australian Plate's collision with the Eurasian Plate beginning 40–50 million years ago, and these mountains are still rising today as these plates converge. The Appalachian Mountains of the eastern United States, once the world's highest mountains, resulted from a more ancient collision with the edge of what is today Africa.

## Tectonics produces Earth's landforms

In these ways, the processes of plate tectonics build mountains; shape the geography of oceans, islands, and continents; and give rise to earthquakes and volcanoes. The coltan mining areas of eastern Congo are situated along the western edge of Africa's Great Rift Valley system, a region where the African plate is slowly pulling itself apart. Some of the world's largest lakes have formed in the immense valley floors, far below towering volcanoes such as Mount Kilimanjaro.

The topography created by tectonic processes, in turn, shapes climate by altering patterns of rainfall, wind, ocean currents, and heating and cooling, all of which affect rates of weathering and erosion and the ability of plants and animals to inhabit different regions. Thus, plate tectonics influences the locations of biomes (p. 81). Moreover, tectonics has affected the history of life's evolution; for instance, the convergence of landmasses into supercontinents is thought to have contributed to widespread extinctions by limiting the extent of species-rich coastal areas and by creating an arid continental interior with extreme temperature swings.

Despite its importance in shaping Earth's surface, plate tectonics was completely unknown to humanity just half a century ago. Amazingly, our civilization was sending people to the moon by the time our geologists were explaining the movement of the land under our feet.

## The rock cycle alters rock

We tend to think of rock as pretty solid stuff. Yet in the long run, over geologic time, rocks and the minerals that comprise them are heated, melted, cooled, broken down, and reassembled in a very slow process called the **rock cycle** (**FIGURE 11.4**).

A **rock** is any solid aggregation of minerals. A **mineral**, in turn, is any naturally occurring solid element or inorganic compound with a crystal structure, a specific chemical composition, and distinct physical properties. The type of rock in a given region affects soil characteristics and thereby influences the region's plant community. Understanding the rock cycle enables us to better appreciate the formation and conservation of soils, mineral resources, fossil fuels, and other natural resources.

**Igneous rock** All rocks can melt. At high enough temperatures, rock will enter a molten, liquid state called **magma**. If magma is released through the lithosphere (as in a volcanic eruption), it may flow or spatter across Earth's surface as **lava**. Geologists call the rock that forms when magma or lava cools **igneous rock** (from the Latin *ignis,* meaning "fire") (**FIGURE 11.4a**).

**Sedimentary rock** All exposed rock weathers away with time. The relentless forces of wind, water, freezing, and thawing eat away at rocks, stripping off one tiny grain (or large chunk) after another. Through weathering (p. 140) and erosion (p. 141), particles of rock come to rest downhill, downstream, or downwind from their sources, forming **sediments**. Alternatively, some sediments form chemically from the precipitation of substances out of solution.

Over time, deep layers of sediment accumulate, causing the weight and pressure on the layers below them to increase. **Sedimentary rock** (**FIGURE 11.4b**) is formed as sediments are physically pressed together and as dissolved minerals seep through sediments and act as a kind of glue, binding sediment particles (a process called *lithification*).

Processes of physical compaction and chemical transformation in sedimentary layers also create the fossils of organisms (p. 53) and the fossil fuels we use for energy (p. 334). Because sedimentary layers, or strata, pile up in chronological order, geologists and paleontologists can assign relative dates to fossils they find in sedimentary rock and from that make inferences about Earth's history.

**Metamorphic rock** Geologic forces may bend, uplift, compress, or stretch rock. When rock is subjected to great heat or pressure, it may alter its form, becoming **metamorphic rock** (from the Greek for "changed form") (**FIGURE 11.4c**). The forces that metamorphose rock generally occur deep underground, at temperatures lower than the rock's melting point but high enough to change its appearance and physical properties.

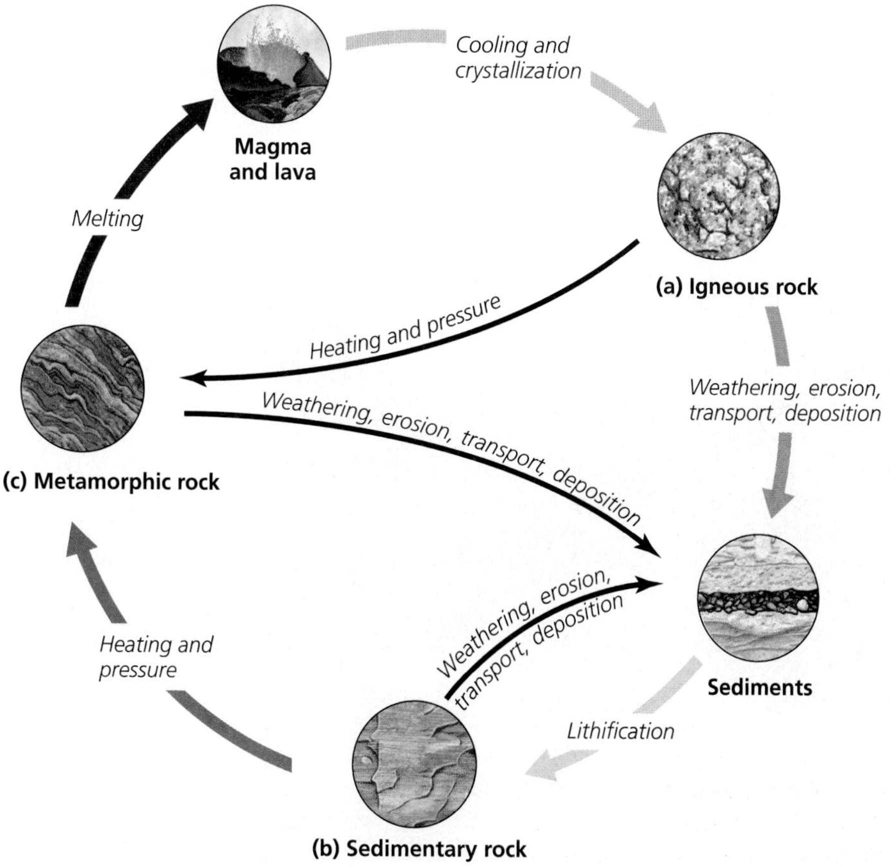


**FIGURE 11.4 The rock cycle.** Igneous rock **(a)** is formed when rock melts and the resulting magma or lava then cools. Sedimentary rock **(b)** is formed when rock is weathered and eroded and the resulting sediments are compressed to form new rock. Metamorphic rock **(c)** is formed when rock is subjected to intense heat and pressure underground. Through these processes, each type of rock can be converted into either of the other two types.

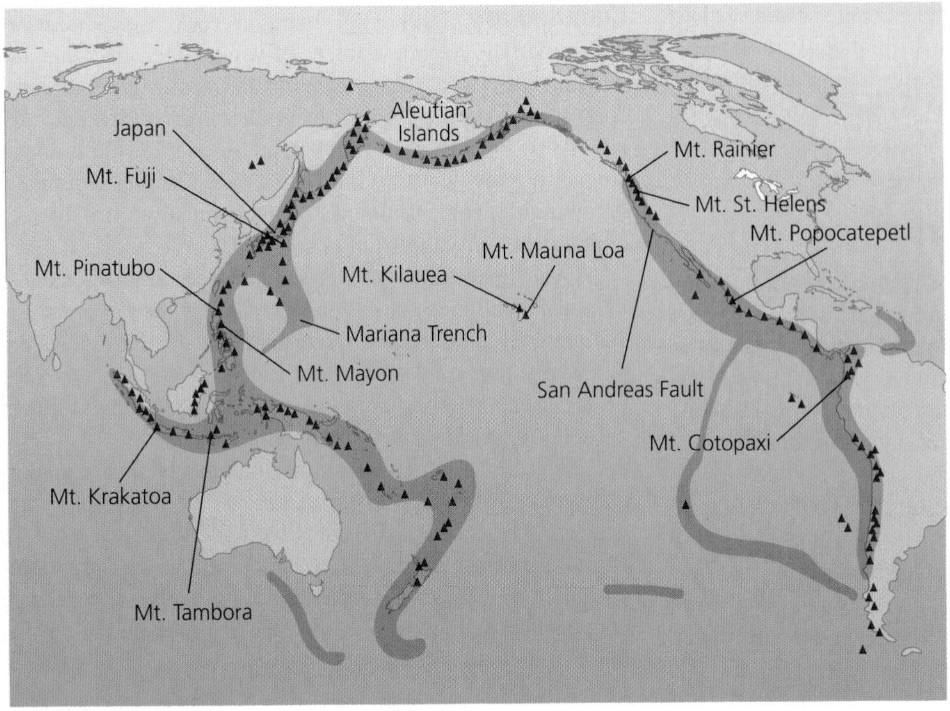


**FIGURE 11.5 Most of our planet's volcanoes and earthquakes occur along the circum-Pacific belt, or "ring of fire."** In this map, red symbols indicate major volcanoes, and gray-shaded areas indicate areas of greatest earthquake risk.

DATA **Q** What similarities do you note between the "ring of fire" around the edges of the Pacific Ocean and the boundaries of the tectonic plates shown in Figure 11.2? Which of type of plate boundary (see Figure 11.3) is most common along the length of the "ring of fire"?

GO TO **INTERPRETING GRAPHS & DATA** ON MasteringEnvironmentalScience®

# Geologic and Natural Hazards

Plate tectonics shapes our planet, but the consequences of tectonic movement can also pose hazards to us. Earthquakes and volcanic eruptions are examples of such geologic hazards. We can see how such hazards relate to tectonic processes by examining a map of the *circum-Pacific belt*, or "ring of fire" (**FIGURE 11.5**). Ninety percent of earthquakes and over half the world's volcanoes occur along this 40,000-km (25,000-mi) arc of subduction zones and fault systems.

## Earthquakes result from movement at plate boundaries and faults

Along tectonic plate boundaries, and in other places where faults occur, the earth may relieve built-up pressure in fits and starts. Each release of energy causes what we know as an **earthquake**. Most earthquakes are barely perceptible, but occasionally they are powerful enough to do tremendous damage to human life and property (**FIGURE 11.6**; **TABLE 11.1**). Damage is generally greatest where soils are loose or saturated with water; areas of cities built atop landfill are particularly susceptible.

**FIGURE 11.6 The 2010 earthquake in Haiti devastated the capital city of Port-au-Prince and killed an estimated 230,000 people.** One reason for the extensive loss of life was that many of Haiti's buildings were not constructed to withstand earthquakes.

| TABLE 11.1 Examples of Large or Recent Earthquakes | | | |
|---|---|---|---|
| **Year** | **Location** | **Fatalities** | **Magnitude[1]** |
| 1556 | Shaanxi Province, China | 830,000 | ~8 |
| 1755 | Lisbon, Portugal | 70,000[2] | 8.7 |
| 1906 | San Francisco, California | 3000 | 7.8 |
| 1923 | Kwanto, Japan | 143,000 | 7.9 |
| 1964 | Anchorage, Alaska | 128[2] | 9.2 |
| 1976 | Tangshan, China | 255,000+ | 7.5 |
| 1985 | Michoacan, Mexico | 9500 | 8.0 |
| 1989 | Loma Prieta, California | 63 | 6.9 |
| 1994 | Northridge, California | 60 | 6.7 |
| 1995 | Kobe, Japan | 5502 | 6.9 |
| 2004 | Northern Sumatra | 228,000[2] | 9.1 |
| 2005 | Kashmir, Pakistan | 86,000 | 7.6 |
| 2008 | Sichuan Province, China | 50,000+ | 7.9 |
| 2010 | Port-au-Prince, Haiti | 236,000 | 7.0 |
| 2010 | Maule, Chile | 500 | 8.8 |
| 2011 | Northern Japan | 19,000[2] | 9.0 |

[1]Measured by moment magnitude; each full unit is roughly 32 times as powerful as the preceding full unit.
[2]Includes deaths from the resulting tsunami.

To minimize damage from earthquakes, engineers have developed ways to protect buildings from shaking. They strengthen structural components while also designing points at which a structure can move and sway harmlessly with ground motion. Just as a flexible tree trunk bends in a storm while a brittle one breaks, buildings with built-in flexibility are more likely to withstand an earthquake's violent shaking. Such designs are an important part of new building codes in California, Japan, and other quake-prone regions, and many older structures are being retrofitted to meet these codes.

## Volcanoes arise from rifts, subduction zones, or hotspots

Where molten rock, hot gas, or ash erupts through Earth's surface, a **volcano** is formed, often creating a mountain over time as cooled lava accumulates. As we have seen, lava can extrude in rift valleys and along mid-ocean ridges, or above subduction zones as one tectonic plate dives beneath another. Lava may also be emitted at *hotspots*, localized areas where plugs of molten rock from the mantle erupt through the crust. As a tectonic plate moves across a hotspot, repeated eruptions from this source may create a linear series of volcanoes. The Hawaiian Islands provide an example of this process (**FIGURE 11.7a**).

At some volcanoes, lava flows slowly downhill, such as at Mount Kilauea in Hawaii (**FIGURE 11.7b**), which has been erupting continuously since 1983! Other times, a volcano

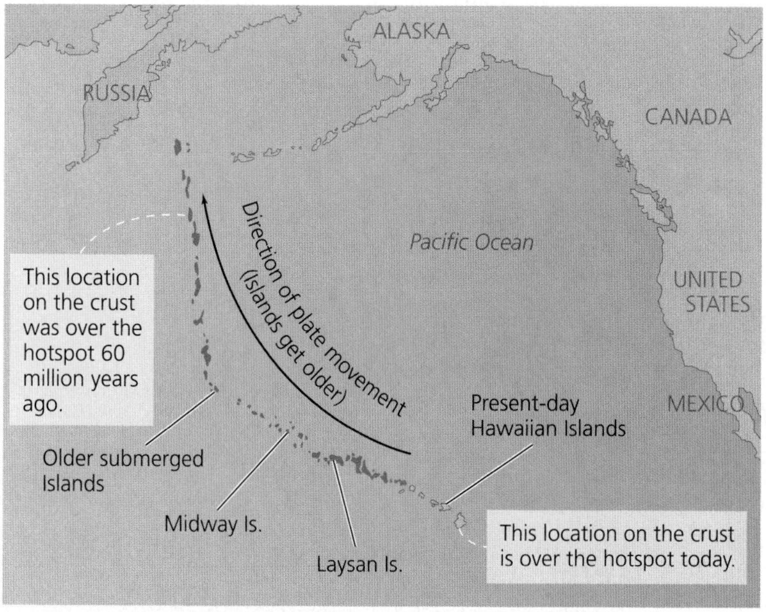


**(a) Current and former Hawaiian Islands, formed as crust moves over a volcanic hotspot**

**(b) Mt. Kilauea erupting**

**FIGURE 11.7 The Hawaiian Islands are the product of a hotspot on Earth's mantle.** The Hawaiian Islands **(a)** have been formed by repeated eruptions from a hotspot of magma in the mantle as the Pacific Plate passes over the hotspot. The Big Island of Hawaii is most recently formed, and it is still volcanically active. The other islands are older and have begun eroding away. To their northwest stretches a long series of former islands, now submerged. The active volcano Kilauea **(b)**, on the Big Island's southeast coast, is currently located above the hotspot.

**TABLE 11.2 Examples of Notable Volcanic Eruptions**

| Year | Location | Impacts | Magnitude[2] |
|---|---|---|---|
| 640,000 B.P.[1] | Yellowstone Caldera, Wyoming, United States | The most recent "mega-eruption" at site of Yellowstone National Park | 8 |
| 6870 B.P. | Mount Mazama, Oregon, United States | Created Crater Lake | 7 |
| A.D. 79 | Mount Vesuvius, Italy | Buried Pompeii and Herculaneum | 5 |
| 1815 | Mount Tambora, Indonesia | Created "year without a summer"; killed at least 70,000 people | 7 |
| 1883 | Krakatau, Indonesia | Killed over 36,000 people; heard 5000 km (3000 mi) away; affected weather for 5 years | 6 |
| 1980 | Mount St. Helens, Washington, United States | Blew top off mountain; sent ash 19 km (12 mi) into sky and into 11 U.S. states; 57 people killed | 5 |
| 1983–present | Kilauea, Hawaii, United States | Continuous lava flow | 1 |
| 1991 | Mount Pinatubo, Philippines | Sulfuric aerosols lowered world temperature 0.5°C (0.9°F) | 6 |
| 2010 | Eyjafjallajokull, Iceland | Ash cloud disrupted air travel throughout Europe | 1 |

[1]B.P. = years before the present.

[2]Measured by the Volcanic Explosivity Index, which ranges from 0 (least powerful) to 8 (most powerful).

may let loose large amounts of ash and cinder in a sudden explosion, as occurred during the 1980 eruption of Mount St. Helens (p. 283). Sometimes a volcano can unleash a *pyroclastic flow*—a fast-moving cloud of toxic gas, ash, and rock fragments that races down the slopes, enveloping everything in its path. Such a flow buried the inhabitants of the ancient Roman cities of Pompeii and Herculaneum in A.D. 79, when Mount Vesuvius erupted.

Volcanic eruptions also exert environmental impacts (TABLE 11.2). Ash blocks sunlight, and sulfur emissions lead to a sulfuric acid haze that blocks radiation and cools the atmosphere. Large eruptions—such as that of Mount Pinatubo in the Philippines in 1991—can depress temperatures throughout the world. When Indonesia's Mount Tambora erupted in 1815, it cooled the planet enough over the following year to cause worldwide crop failures and make 1816 "the year without a summer."

## Landslides are a form of mass wasting

At a smaller scale than volcanoes or earthquakes, a **landslide** occurs when large amounts of rock or soil collapse and flow downhill. Landslides are severe, often sudden, manifestations of the phenomenon of **mass wasting**, the downslope movement of soil and rock due to gravity. Mass wasting occurs naturally, but often it is brought about by human land use practices that expose or loosen soil, making slopes more prone to collapse. Heavy rains may saturate soils and trigger mudslides of soil, rock, and water (FIGURE 11.8).

Most often, mass wasting eats away at unstable hillsides, damaging property one structure at a time. Occasionally mass-wasting events can be colossal and deadly; mudslides that followed the torrential rainfall of Hurricane Mitch in Nicaragua and Honduras in 1998 killed over 11,000 people. Mudslides caused when volcanic eruptions melt snow and send huge volumes of destabilized mud racing downhill are called *lahars*, and they are particularly dangerous. A lahar following an eruption in 1985 buried the entire town of Armero, Colombia, killing 21,000 people.

## Tsunamis can follow earthquakes, volcanoes, or landslides

Earthquakes, volcanic eruptions, and large coastal landslides can all displace huge volumes of ocean water instantaneously and trigger a **tsunami**, an immense swell, or wave, of water that can travel thousands of miles across oceans. The world's attention was drawn to this hazard in 2004, when a massive tsunami, triggered by an earthquake off

FIGURE 11.8 **The town of Oso, Washington, was buried in 2014 by a massive landslide.** A rain-saturated hillside above the town suddenly gave way, sending tons of rock and soil streaming into the valley below. The landslide destroyed around 50 homes and claimed 43 lives. The barren hillside is visible in the background of the picture.

**FIGURE 11.9 Tsunami waves overtop a seawall following the Tohoku earthquake in Japan in 2011.** The tsunami caused a greater loss of life and property than the earthquake that generated it and led to a meltdown at the Fukushima Daiichi nuclear power plant.

Sumatra, devastated coastlines all around the Indian Ocean, from Indonesia to India to Africa. Roughly 230,000 people were killed, 1–2 million were displaced, and whole communities destroyed.

In 2011, a tsunami generated by an offshore earthquake devastated large portions of northeastern Japan (**FIGURE 11.9**). The tsunami and earthquake killed more than 19,000 people, caused over $300 billion in economic impacts, and contributed to the meltdown of the Fukushima Daiichi nuclear power plant (p. 360).

Since the 2004 tsunami, nations and international agencies have stepped up efforts to develop systems to give coastal residents advance warning of approaching tsunamis. In addition, we can lessen the impacts of tsunamis if we leave in place natural coastal vegetation such as mangrove forests (p. 256).

## We can worsen or lessen the impacts of natural hazards

Aside from geologic hazards, people face other types of natural hazards. Heavy rains can lead to *flooding* that ravages low-lying areas near rivers and streams (p. 262). *Coastal erosion* can eat away at beaches (p. 248). *Wildfires* can threaten life and property in fire-prone areas (p. 196). And tornadoes and hurricanes can cause extensive damage and loss of life.

Although we refer to such phenomena as "natural hazards," the magnitude of their impacts upon us often depends on choices we make. We tend to worsen the impacts of so-called natural hazards in various ways:

- As our population grows, more people live in areas susceptible to natural disasters.
- Many of us choose to live in areas that we deem attractive but that are also prone to hazards. For instance, coastlines are vulnerable to tsunamis and to erosion by storms, and mountainous areas are prone to volcanoes and mass wasting.
- We use and engineer landscapes around us in ways that can increase the frequency or severity of natural hazards. Damming and diking rivers to control floods can sometimes lead to catastrophic flooding (p. 262). Suppressing natural wildfires puts forests at risk of larger, highly damaging fires (p. 196). Clear-cutting on slopes (p. 195) and some mining practices (pp. 237, 240) can induce mass wasting, speed runoff, compact soil, and change drainage patterns.
- As we change Earth's climate by emitting greenhouse gases (Chapter 14), we alter patterns of precipitation, thereby increasing risks of drought, fire, flooding, and mudslides locally and regionally. Rising sea levels induced by global warming inundate low-lying coastal areas and increase coastal erosion and the potential for catastrophic flooding events.

Often we can lessen the impacts of hazards through the thoughtful use of technology, engineering, and policy, informed by a solid understanding of geology and ecology. Examples already noted include building earthquake-resistant structures; designing early warning systems for tornadoes, hurricanes, earthquakes, tsunamis, and volcanoes; and conserving coastal forests, reefs, and salt marshes to protect against tsunamis and coastal erosion (pp. 248–249). In addition, better forestry and mining practices can help prevent landslides. Zoning regulations, building codes, and insurance incentives that discourage development in areas prone to landslides, floods, fires, and storm surges can keep us out of harm's way and decrease taxpayer expense when cleaning up after natural disasters. Finally, addressing global climate change may help reduce the frequency of natural hazards in many regions.

## Earth's Mineral Resources

Both gradual geologic processes and catastrophic geologic hazards influence the distribution of rocks and minerals in the lithosphere and their availability to us. We mine and process a wide array of mineral resources. Indeed, without these resources—which we use to make building materials, wiring, clothing, appliances, fertilizers for crops, and so much more—civilization as we know it could not exist. Just consider a typical scene from a student lounge at a college or university (**FIGURE 11.10**), and note how many items are made with elements from the minerals we take from the earth.

### We obtain minerals by mining

We obtain the minerals we use in all these ways through the process of mining. The term *mining* in the broad sense describes the extraction of any resource that is nonrenewable on the timescale of our society. In this sense, we mine minerals, fossil fuels and even groundwater (when we extract groundwater faster than it is renewed by precipitation). When used specifically in relation to minerals, **mining** refers to the systematic removal of rock, soil, or other material for the purpose of extracting minerals of economic interest. Because most minerals of interest are widely spread and in low concentrations, miners and mining geologists first try to locate concentrated sources of minerals before mining begins.

**FAQ How do geologists "see" mineral deposits below the ground?**

Searching for reserves of underground minerals, also called *prospecting*, can be pursued in a number of ways. The earliest prospectors explored promising areas on foot, looking for exposed seams of mineral-containing rocks or for minerals carried into streams by runoff. Today, geologists direct vibrations into underground rock strata and capture them with sensors as they reflect back. This enables scientists to visualize the underlying rock layers and identify likely locations for reserves, just as they do for fossil fuel deposits (p. 337). Geologists also measure the magnetic fields in rock layers to look for metal ores, and they conduct chemical analyses of stream water to detect minerals of interest. If promising sites are located, cores can be drilled deep into the ground and inspected for the desired mineral.

### We use mined materials extensively

We often don't notice how many mined resources we use every day. Using data from the U.S. government, the Minerals Education Coalition estimated in 2013 that the average American consumed over 17,200 kg (38,000 lb) of new minerals and fuels every year. At current rates of use, a child born today will use over 1 million kg (2.2 million lb) during his or her lifetime (**FIGURE 11.11**).

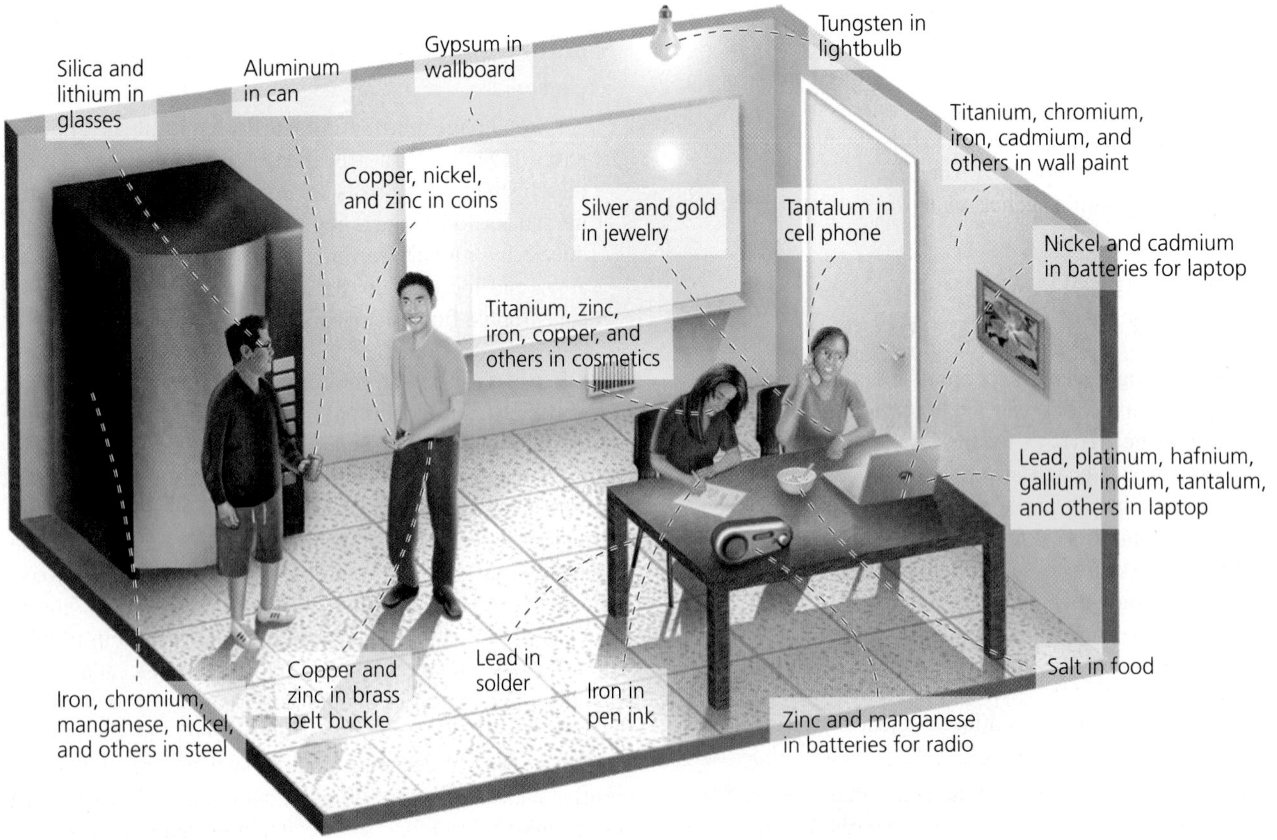


**FIGURE 11.10 Elements from minerals that we mine are everywhere in the products we use in our everyday lives.** This scene from a typical college student lounge points out just a few of the many elements from minerals that surround us.

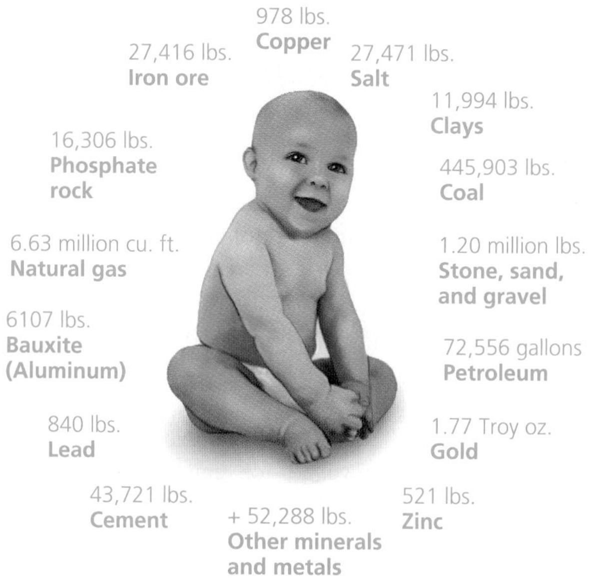


**FIGURE 11.11 At current rates of use, a baby born today is predicted to use over 1.3 million kg (3 million lb) of minerals over his or her lifetime.** *Data from Minerals Education Coalition, 2013.*

**(a) Coltan ore**

**(b) Capacitor containing tantalum**

**FIGURE 11.12 Tantalum is used to manufacture electronics.** Coltan ore **(a)** is mined from the ground and then processed to extract the pure metal tantalum. This metal is used in capacitors **(b)** and other electronic components.

More than half of our annual mineral and fuel use is from the coal, oil, and natural gas used to supply our intensive demands for energy. Much of the remaining mineral use is attributable to the sand, gravel, and stone used in constructing buildings, roads, bridges, and parking lots. Such high levels of consumption shows the potential for recycling and reuse (such as recycling stone and gravel from old highways into new construction) to make our modern, mineral-intensive lifestyle more sustainable.

## Metals are extracted from ores

Some minerals can be mined for metals. A **metal** is a type of chemical element that typically is lustrous, opaque, malleable, and can conduct heat and electricity. Most metals are not found in a pure state in Earth's crust, but instead are present within **ore**, a mineral or grouping of minerals from which we extract metals.

Copper, iron, lead, gold, and aluminum are among the many economically valuable metals we extract from mined ore. The tantalum used in the electronic components of computers, cell phones, video game consoles, and other devices is a metal that comes from the mineral tantalite (**FIGURE 11.12**). In nature, tantalite is often found with the mineral columbite within the ore called coltan.

## We process metals after mining ore

Extracting minerals from the ground is the first step in putting them to use. However, most minerals need to be processed in some way to become useful for our products. For example, after ores are mined, the rock is crushed and pulverized, and the desired metals are isolated by chemical or physical means. The material is then processed to purify the metals we desire. With coltan, processing facilities use acid solvents to separate tantalite from columbite. Other chemicals are then used to produce metallic tantalum powder. This powder can be consolidated by various melting techniques and can be shaped into wire, sheets, or other forms.

Sometimes we mix, melt, and fuse a metal with another metal or a nonmetal substance to form an *alloy*. For example, steel is an alloy of the metal iron that has been fused with a small quantity of carbon. The strength and malleability of this particular alloy make steel ideal for its many applications in buildings, vehicles, appliances, and more. To make steel, we first mine iron ore. Steelmakers then heat the ore and chemically extract the iron with carbon in a process known as **smelting** (heating ore beyond its melting point and combining it with other metals or chemicals). They then melt and reprocess the mixture, removing precise amounts of carbon and shaping the product into rods, sheets, or wires. During this melting process, certain other metals may be added to modify the strength, malleability, or other characteristics of the steel.

Processing minerals exerts environmental impacts. Most methods are water-intensive and energy-intensive. Moreover, extracting metals from ore emits air pollution, and smelting plants in particular can be hotspots of toxic air pollution. In addition, soil and water commonly become polluted by **tailings**, portions of ore left over after metals have been extracted. Tailings may leach heavy metals present in the ore waste as well as chemicals applied in the extraction process. For instance, we use cyanide to extract gold from ore, and we use sulfuric acid to extract copper. Mining operations often store toxic slurries of tailings in large reservoirs called **surface impoundments**. Impoundment walls are designed to prevent leaks, but accidents can occur if the structural integrity of the impoundment is compromised. In 2000, a breach of a coal tailings impoundment near Inez, Kentucky, released over 1 billion liters (250–300 million gal) of coal slurry, blackening 120 km (75 mi) of streams, killing aquatic wildlife, and affecting drinking water supplies for many communities.

## We also mine nonmetallic minerals and fuels

We mine and use many minerals that do not contain metals. **FIGURE 11.13** illustrates the nation of origin and uses for some economically important mineral resources, both metallic and nonmetallic. As you can see, many geologic resources in the products you use were mined in faraway nations.

Sand and gravel (the most commonly mined mineral resources) provide fill and construction materials. Phosphates provide us with fertilizer. We mine limestone, salt, potash, and other minerals for a number of diverse purposes.

Gemstones are treasured for their rarity and beauty. For instance, diamonds have long been prized—and like coltan, they have fueled resource wars. Besides the conflict in eastern Congo, the diamond trade has acted to fund, prolong, and intensify wars in Angola, Sierra Leone, Liberia, and elsewhere as armies exploit local people for mine labor and sell the diamonds for profit. This is the origin of the term "blood diamonds," just as coltan has been called a "conflict mineral."

We also mine substances for fuel. Uranium ore is a mineral from which we extract the metal uranium, which we use in nuclear power (pp. 356–358). One of the most common fuels we mine is coal. Coal (p. 338) is the modified remains of ancient swamp plants and is made up of the mineral carbon. Other fossil fuels—petroleum, natural gas, and alternative fossil fuels such as oil sands, oil shale, and methane hydrates—are also organic and are extracted from the earth (Chapter 15).

# Mining Methods and Their Impacts

Mining for minerals is an important industry that provides jobs for people and revenue for communities in many regions. Mining supplies the raw materials for countless products we use daily, so it is necessary for the lives we lead. However, mining also exerts a price in environmental and social impacts. Because minerals of interest often make up only a small portion of the rock in a given area, very large amounts of material are removed in order to obtain the desired minerals. This frequently means that mining disturbs huge swaths of land.

Depending on the nature of the mineral deposit, any of several mining methods may be employed to extract the resource from the ground. Mining companies select which method to use based largely on its economic efficiency.

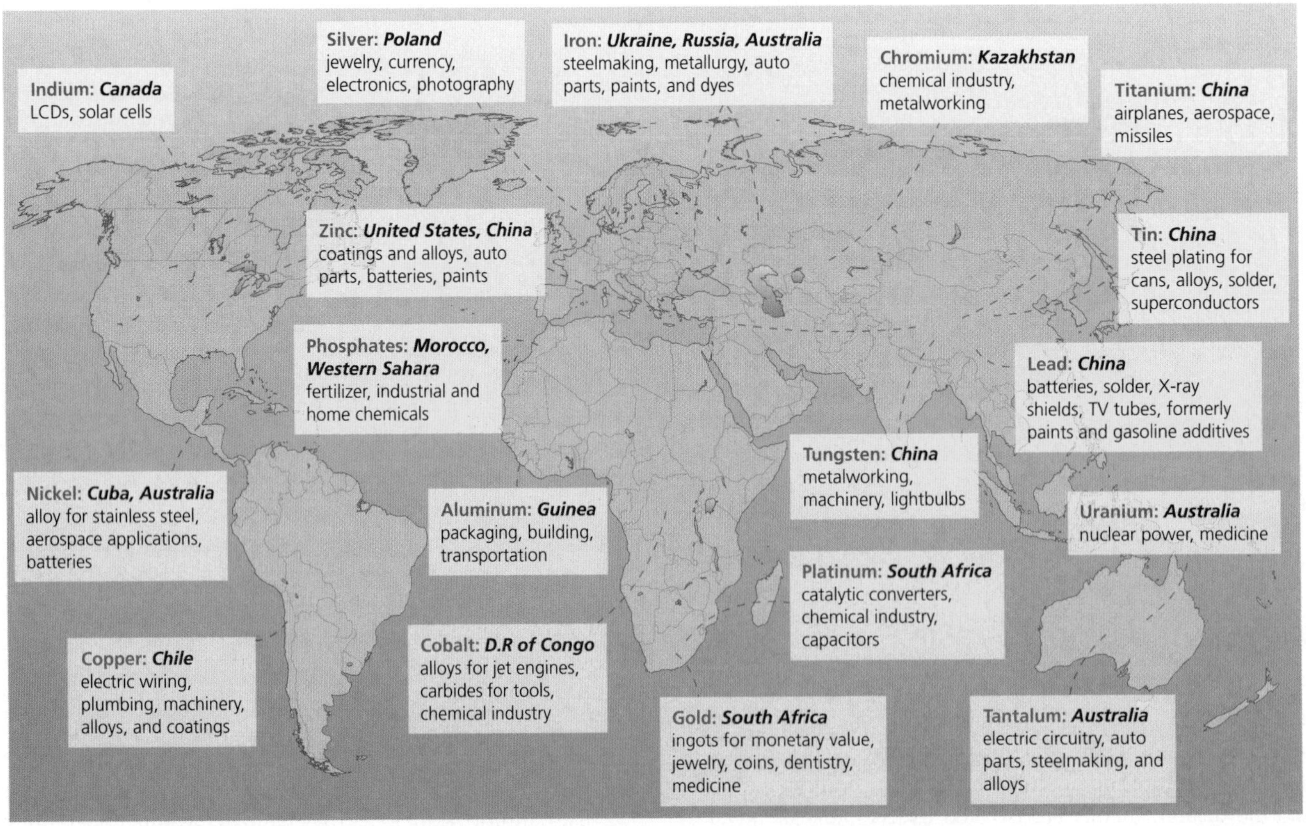


**FIGURE 11.13 The minerals we use come from all over the world.** Shown is a selection of economically important minerals (mostly metals, with several nonmetals), together with their major uses and their main nation of origin. Only a minority of minerals, uses, and origins is shown.

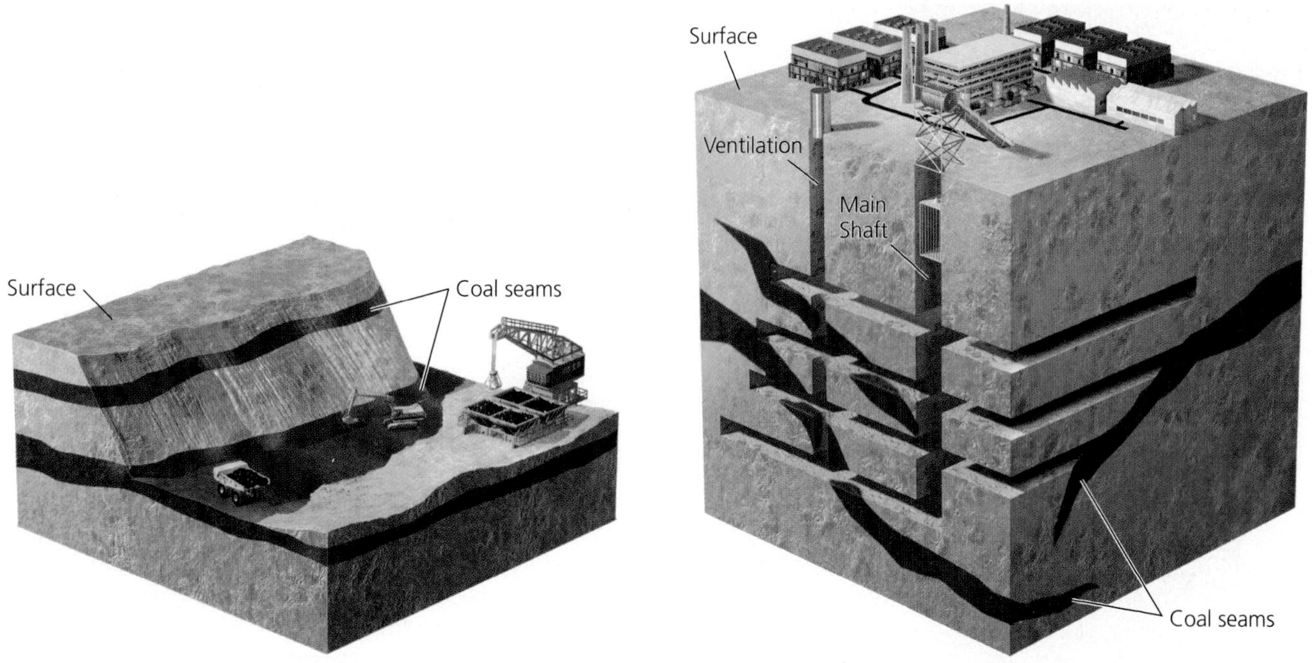


**(a) Strip mining** **(b) Subsurface mining**

**FIGURE 11.14 Coal mining illustrates two types of mining approaches.** In strip mining **(a)**, soil is removed from the surface in strips, exposing seams from which coal is mined. In subsurface mining **(b)**, miners work belowground in shafts and tunnels blasted through the rock. These passageways provide access to underground seams of coal or minerals.

## Strip mining removes surface layers of soil and rock

When a resource occurs in shallow horizontal deposits near the surface, the most effective mining method is often **strip mining**, whereby layers of surface soil and rock are removed from large areas to expose the resource. Heavy machinery removes the overlying soil and rock (termed *overburden*) from a strip of land, and the resource is extracted. This strip is then refilled with the overburden, and miners proceed to an adjacent strip of land and repeat the process. Strip mining is commonly used for coal (**FIGURE 11.14a**) and oil sands (p. 338), and sometimes for sand and gravel.

Strip mining can be economically efficient, but it obliterates natural communities over large areas, and the soil in refilled areas can easily erode away. Strip mining also pollutes waterways through the process of **acid mine drainage**, which occurs when sulfide minerals in newly exposed rock surfaces react with oxygen and rainwater to produce sulfuric acid (**FIGURE 11.15**). As the sulfuric acid runs off, it leaches metals from the rocks, and many of these metals are toxic to organisms. Acid drainage can affect fish and other aquatic organisms when it runs into streams and can pollute groundwater supplies people use for drinking water or irrigating crops. Although acid drainage is a natural phenomenon, mining greatly accelerates this process by exposing many new rock surfaces at once. New research has shown, however, that acidic mine drainage may be put to use to mitigate the effects of hydraulic fracturing (or "fracking") in areas where both occur (see **THE SCIENCE BEHIND THE STORY**, pp. 238–239).

## In subsurface mining, miners work underground

When a resource occurs in concentrated pockets or seams deep underground, and the earth allows for safe tunneling, then mining companies pursue **subsurface mining**. In this approach, shafts are excavated deep into the ground, and networks of tunnels are dug or blasted out to follow deposits of the mineral (**FIGURE 11.14b**). Miners remove the resource systematically and ship it to the surface.

**FIGURE 11.15 Acidic drainage flows from a coal mine in Scotland.** The orange color is due to iron from the drainage settling out on the soil surface and forming rust.

# Can Acid Mine Drainage Reduce Fracking's Environmental Impact?

**Dr. Avner Vengosh, Duke University**

In a strange twist of fate and chemistry, it appears that one environmental legacy of past mining for fossil fuels may help ameliorate the environmental impacts of mining today.

A study led by geologist Avner Vengosh of Duke University and published in 2014 in the scientific journal *Environmental Science & Technology* investigated whether the acid mine drainage leaking from old coal mines could be put to use to lessen the environmental impacts of the very modern process of using hydraulic fracturing, or "fracking," to produce natural gas (pp. 332–334, 344–345).

Hydraulic fracturing injects water laden with drilling chemicals into layers of shale rock deep underground, fracturing the rock and releasing natural gas trapped within. Along with natural gas, fracking wells pull up wastewater called "flowback fluid" made up of drilling chemicals mixed with water containing dissolved salts, toxic metals such as barium and strontium, and radioactive radium isotopes from deep underground. Previous work by the Duke team found that standard treatment of this wastewater was only partially effective in removing contaminants. Hence, when the treated wastewater was released into streams, its high salinity posed threats to freshwater organisms. The researchers also detected a buildup of radioactive radium in stream sediments near release sites, posing additional, long-term threats to wildlife. Clearly, a more effective method of treating wastewater was necessary to protect water quality near fracking sites.

To accomplish this goal, the team looked to something found in close proximity to many fracking wells—streams impacted by acid mine drainage. Acidic drainage from coal mines fouls over 20,000 km (12,400 mi) of streams in Pennsylvania alone (**FIGURE 1**). Vengosh's team assessed the chemical composition of acid mine drainage and hypothesized that it might chemically interact with dissolved substances in fracking wastewater and cause them to "drop out" of solution as a solid precipitate that could be collected and safely disposed of. To test this hypothesis, the team was provided with samples of fracking wastewater from three drilling sites and acid mine drainage from two sites in western Pennsylvania.

The researchers used two types of acid mine drainage: synthetic acid mine drainage (a solution mimicking natural acid

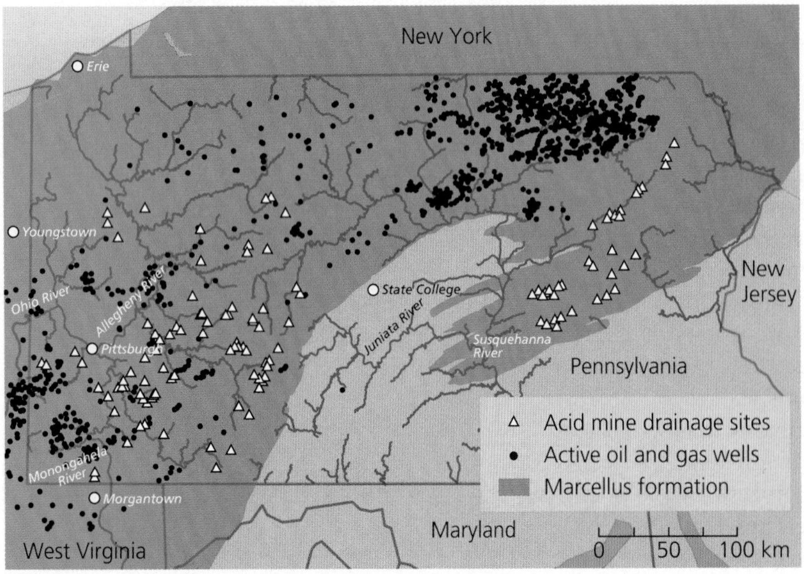


**FIGURE 1 In parts of Pennsylvania, active hydraulic fracturing operations are very near streams affected by acid mine drainage.**

*Source: Kondash, A. J., et al., 2014. Radium and barium removal through blending hydraulic fracturing fluids with acid mine drainage.* Environmental Science & Technology *48: 1334–1342.*

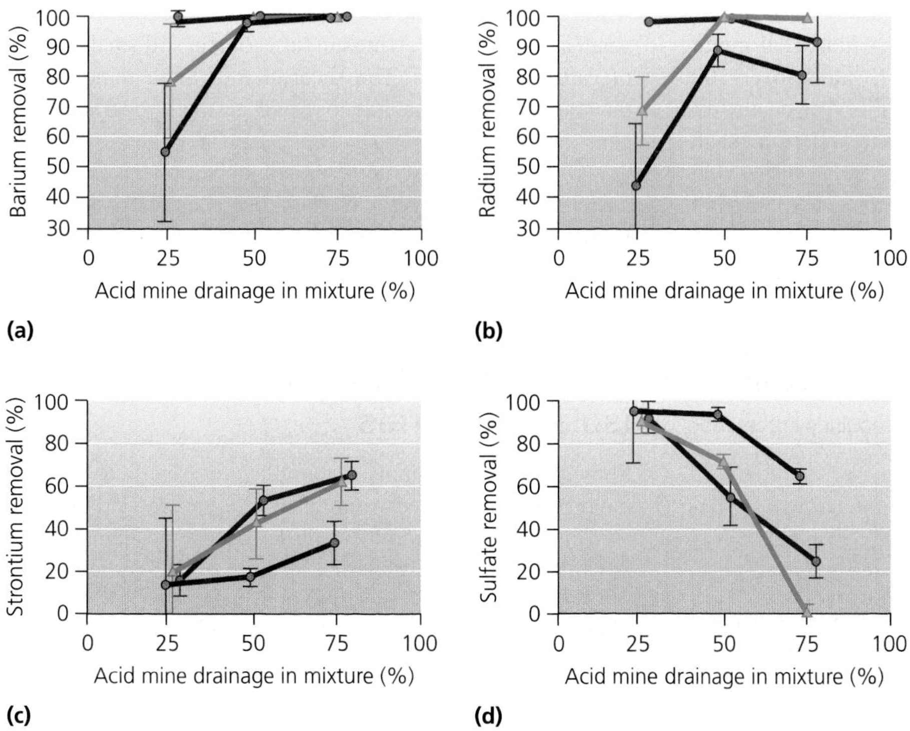


**FIGURE 2 Acid mine drainage was highly effective in removing (a) barium, (b) radioactive radium, (c) strontium, and (d) sulfate from wastewater produced by hydraulic fracturing.** Synthetic acid drainage **(orange lines)** was more effective at removing sulfate from wastewater than were field samples of lime-treated acid drainage from site #1 **(blue lines)** and site #2 **(green lines)**, but generally less effective at removing barium, strontium, and radium. Increasing concentrations of acid mine drainage removed higher percentages of barium, strontium, and radium, but sulfate was more effectively removed when acid mine drainage concentrations were lower. *Source: Kondash, A. J., et al., 2014. Radium and barium removal through blending hydraulic fracturing fluids with acid mine drainage.* Environmental Science & Technology *48: 1334–1342.*

mine drainage that researchers commonly use in laboratory tests) and acid mine drainage from the two field sites (these samples were modified on-site in Pennsylvania by the addition of the mineral lime, which raises the pH and lowers the iron content of the acidic drainage, reducing its impact on waterways). Working in the laboratory at Duke University, each of these samples of acidic drainage was combined with fracking wastewater in mixtures containing 25%, 50%, or 75% acid mine drainage. The samples were shaken for 48 hours to promote mixing, and their concentrations of select chemical constituents subsequently analyzed.

The effectiveness of treatment varied both by type of acid mine drainage (synthetic or field-collected) and by acid mine drainage concentration (25%, 50%, or 75%), but the team found that within 10 hours, very high percentages of the barium, strontium, sulfate (which gives acidic drainage its low pH (p. 28) and radioactive radium precipitated out of solution (**FIGURE 2**). The process was also effective in greatly reducing the salt content of the wastewater. After these reactions, the contaminants from wastewater were concentrated in precipitate, mostly in the form of strontium barite. Because they were not dissolved or suspended in solution, the contaminants were efficiently collected for disposal.

This study demonstrated that there is potential for treating fracking wastewater with mixtures of acid mine drainage, removing some of the contaminants as precipitates, and producing water that could be less damaging to the environment or reused in drilling applications. Although laboratory experiments were successful, the next step is to test the process at actual fracking sites where conditions are less controlled.

Conducting this process at a large scale in the field is not without cost, because acid mine drainage would need to be pumped to drilling sites (or vice versa). Also, the precipitate produced by the reaction is radioactive and highly concentrated with contaminants, so it would require disposal in a specialized hazardous waste landfill, increasing disposal costs. The process does save money, though, because it produces water clean enough to be reused for additional drilling, reducing drilling costs for companies. It also reduces demand on precious freshwater resources for fracking operations. The water savings from using recycled water are significant: A single fracking well can use 7–49 million liters (2–13 million gal) of water while in production.

While it would be far better for streams not to be polluted with mine drainage in the first place, this research shows that we may be able to "make lemonade out of lemons" by using a toxic legacy of past mining—acid mine drainage—to reduce the environmental impacts of present mining for fossil fuels.

We use subsurface mining to extract metals such as zinc, lead, nickel, tin, gold, copper, and uranium, as well as diamonds, phosphate, salt, and potash. In addition, a great deal of coal is mined using the subsurface technique. As with surface mining, the disturbance of rock in subsurface mining generates acid drainage that affects nearby waterways.

Subsurface mining is the most dangerous form of mining, and fatal accidents are not unusual. In China, coal-mining conditions are so dangerous that in 2013 alone over 1000 miners lost their lives. Besides risking injury or death from dynamite blasts, natural gas explosions, and collapsing shafts or tunnels, miners inhale toxic fumes and coal dust, which can lead to respiratory diseases, including fatal black lung disease.

**FAQ Why would anyone choose to work in a mine when it's such dangerous work?**

It seems that mining accidents appear regularly in the headlines. In late 2010, for example, 33 miners in Chile were rescued after being trapped 600 m (2000 ft) underground for 69 days in a gold and copper mine. Four days later, an explosion at an underground coal mine in China killed 37 miners. An explosion claimed the lives of 29 miners in a coal mine in West Virginia in 2010 and fire claimed 301 lives in coal mine in Turkey in 2014. Given these dangers, and the chronic health impacts of working in an underground mine, it's reasonable to wonder why anyone would accept such a job.

Many of the people who work in mines do so because they have few other options. Underground mining often occurs in economically depressed areas, such as Appalachia in the United States, where mining is one of the few jobs that pays well. And for most mining jobs, people can begin working right out of high school. So although the work is dangerous, many miners are willing to accept those risks to provide for themselves and their families because few other career opportunities are available.

## Open pit mining creates immense holes in the ground

When a mineral is spread widely and evenly throughout a rock formation, or when the earth is unsuitable for tunneling, the method of choice is **open pit mining**. This essentially involves digging a gigantic hole and removing the desired ore, along with waste rock that surrounds the ore. Some open pit mines are inconceivably enormous. The world's largest, the Bingham Canyon Mine near Salt Lake City, Utah, is 4 km (2.5 mi) across and 1.2 km (0.75 mi) deep (**FIGURE 11.16**). Conveyor systems and immense trucks with tires taller than a person carry out nearly half a million tons of ore and waste rock each day.

Open pit mines are terraced so that people and machinery can move about, and waste rock is left in massive heaps outside the pit. The pit is expanded until the resource runs out or becomes unprofitable to mine. Open pit mining is used to extract copper, iron, gold, diamonds, and coal, among other resources. We also use this technique to extract clay, gravel, sand, and stone (such as limestone, granite, marble, and slate), but we generally call these pits *quarries*.

Open pit mines are so large because huge volumes of waste rock need to be removed in order to extract relatively small amounts of ore, which in turn contain still smaller traces of valuable minerals. The sheer size of these mines means that the degree of habitat loss and aesthetic degradation is considerable.

Once mining is complete, abandoned pits generally fill up with groundwater, which soon becomes toxic as sulfides from the ore react and produce sulfuric acid. Acidic water from the pit can harm wildlife and can percolate into aquifers and spread through the region.

## Placer mining uses running water to isolate minerals

Some metals and gems accumulate in riverbed deposits, having been displaced from elsewhere and carried along by flowing water. To search for these metals and gems, miners sift through material in modern or ancient riverbed deposits, generally using running water to separate lightweight mud and gravel from heavier minerals of value (**FIGURE 11.17**). This technique is called **placer mining** (pronounced "plasser").

Placer mining is the method used by Congo's coltan miners, who wade through streambeds, sifting through large amounts of debris by hand with a pan or simple tools, searching for high-density tantalite that settles to the bottom while low-density material washes away. Today's African miners

**FIGURE 11.16 The Bingham Canyon open pit mine outside Salt Lake City, Utah, is the world's largest human-made hole in the ground.** This immense mine produces mostly copper.

**FIGURE 11.17 Miners in eastern Congo find coltan by placer mining.** Sediment is placed in plastic tubs, and water is run through them. A mixing motion allows the sediment to be poured off while the heavy coltan settles to the bottom.

practice small-scale placer mining similar to the method used by American miners who ventured to California in the Gold Rush of 1849, and later to Alaska in the Klondike Gold Rush of 1896–1899. Placer mining for gold is still practiced in areas of Alaska and Canada, although today miners use large dredges and heavy machinery.

Placer mining is environmentally destructive because most methods wash large amounts of debris into streams, making them uninhabitable for fish and other life for many miles downstream. This type of mining also disturbs stream banks, causing erosion and harming ecologically important plant communities.

## Mountaintop mining reshapes ridges and can fill valleys

When a resource occurs in underground seams near the tops of ridges or mountains, mining companies may practice **mountaintop removal mining**, in which several hundred vertical feet of mountaintop may be removed to allow recovery of entire seams of the resource (**FIGURE 11.18**). This method of mining is used primarily for coal in the Appalachian Mountains of the eastern United States. In mountaintop removal mining, a mountain's forests are clear-cut, the timber is sold, topsoil is removed, and then rock is repeatedly blasted away to expose the coal for extraction.

Afterwards, overburden is placed back onto the mountaintop, but this waste rock is unstable and typically takes up more volume than the original rock, so generally a great deal of waste rock is dumped into adjacent valleys (a practice called "valley filling"). So far, mountaintop removal has blasted away an area the size of Delaware and has buried nearly 3200 km (2000 mi) of streams.

Scientists are finding that dumping tons of debris into valleys degrades or destroys immense areas of habitat, clogs streams and rivers, and pollutes waterways with acid drainage. With slopes deforested and valleys filled with debris, erosion intensifies, mudslides become frequent, and flash floods ravage the lower valleys. Further, the Appalachian forests that are cleared in mountaintop mining are some of the richest forests for biodiversity in the nation.

People living in communities near the sites experience social and health impacts. Blasts from mines crack house foundations and wells, loose rock tumbles down into yards and homes, and floods tear through properties. Coal dust causes respiratory ailments, and contaminated water unleashes a variety of health problems. Studies have shown that people in mountaintop mining areas show elevated levels of birth defects, lung cancer, heart disease, kidney disease, pulmonary disorders, hypertension, and mortality.

Critics of mountaintop removal mining argue that valley filling violates the Clean Water Act (p. 105) because runoff flowing through waste rocks in valleys often contains high levels of salts and toxic metals that degrade water quality

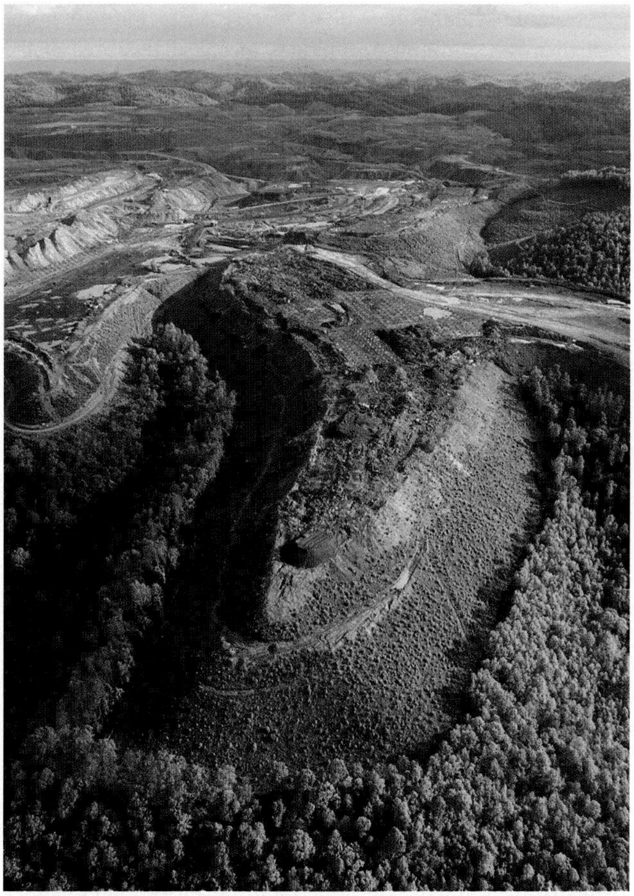

**FIGURE 11.18 Mountaintop mining removes entire mountaintops to obtain the coal underneath.** The rocks removed during the process are dumped into adjacent valleys, burying streams, promoting flooding, and contaminating drinking water supplies.

and impact aquatic organisms. While hundreds of permits for mountaintop mining were issued during the Clinton and G.W. Bush administrations, in 2010 the U.S. Environmental Protection Agency (EPA) announced new guidelines that prohibit valley fills unless strict measures of water quality can be attained. Critics of the policy argue that these new guidelines will essentially end the practice of mountaintop mining. In 2011, the EPA revoked the permit of an existing mountaintop mining operation in West Virginia, citing these new guidelines. A 2014 court ruling backed the EPA's decision, and the agency continued to reexamine other permits across Appalachia.

## Solution mining dissolves and extracts resources in place

When a deposit is especially deep underground and the resource can be dissolved in a liquid, miners may use a technique called *solution mining* or *in-situ recovery*. In this technique, a narrow borehole is drilled to reach the deposit, and water, acid, or another liquid is injected down the borehole to leach the resource from the surrounding rock and dissolve it in the liquid. The resulting solution is then sucked out, and the desired resource is removed from solution. Sodium chloride (table salt), lithium, boron, bromine, magnesium, potash, copper, and uranium can be mined in this way.

Solution mining generally exerts less environmental impact than other mining techniques, because less area at the surface is disturbed. The primary potential impacts involve accidental leakage of acids into groundwater surrounding the borehole, and the contamination of aquifers with acids, heavy metals, or uranium leached from the rock.

## Some mining occurs in the ocean

The oceans hold many minerals useful to our society. We extract some minerals from seawater, such as magnesium from salts held in solution. We extract other minerals from the ocean floor. For example, many minerals are concentrated in manganese nodules, small ball-shaped accretions that are scattered across parts of the ocean floor. More than 1.5 trillion tons of manganese nodules may exist in the Pacific Ocean alone, and their reserves of metal may exceed all terrestrial reserves. As land resources become scarcer and as undersea mining technology develops, mining companies may turn increasingly to the seas. The logistical difficulty of mining offshore resources, however, has kept their extraction limited so far.

## Restoration helps to reclaim mine sites

Because of the environmental impacts of mining, governments of the United States and other developed nations now require that mining companies restore, or reclaim, surface-mined sites following mining. The aim of such restoration, or **reclamation**, is to restore the site to a condition similar to its condition before mining.

To restore a site, companies are required to remove buildings and other structures used for mining, replace overburden, fill in shafts, and replant the area with vegetation (**FIGURE 11.19**). In the United States, the 1977 *Surface Mining Control and Reclamation Act* mandates restoration efforts, requiring companies to post bonds to cover reclamation costs before mining is approved. This ensures that if the company fails to restore the land, the government will have the money to do so. Many countries exercise less oversight, and in nations such as the Congo, there is essentially no regulation at all.

The mining industry has made great strides in reclaiming mined land, but even on sites that are restored, impacts from mining (such as soil and water damage from acid drainage) can be severe and long-lasting, because the soil is often acidic and can contain high levels of metals that are toxic to native plant life. It is therefore often difficult to regain the same biotic communities that were naturally present before mining. Research is continuing on new strains of plants that can tolerate conditions on reclaimed sites and pave the way for the return of native vegetation.

### WEIGHING THE ISSUES

**Restoring Mined Areas**

Mining has severe environmental impacts, and restoring a mined site to a condition similar to its state before mining is costly and difficult. How much do you think we should require mining companies to restore a site after a mine is shut down, and what criteria should we use to guide restoration? Should we require complete restoration? No restoration? What should our priorities be—to minimize water pollution, health impacts, biodiversity loss, or other factors? Should the amount of restoration we require depend on the mine's profitability? Explain your recommendations.

**FIGURE 11.19 More mine sites are now being restored.** Here, bison graze on land reclaimed from a tar sands mining operation in Alberta, Canada.

## An 1872 law still guides U.S. mining policy

The ways in which mining companies stake claims and use land in the United States is guided by a law that is well over a century old. The **General Mining Act of 1872** encourages people and companies to prospect for minerals on federally owned land by allowing any U.S. citizen, domestic company, or foreign company with permission to do business in the United States to stake a claim on any plot of public land open to mining. The person or company owning the claim gains the sole right to take minerals from the area. The claim-holder can also patent the claim (i.e., buy the land) for only about $5 per acre. Regardless of the profits they might make on minerals they extract, the law requires no payments of any kind to the public.

Supporters of the policy say that it is appropriate and desirable to continue encouraging the domestic mining industry, which must undertake substantial financial risk and investment to locate resources that are vital to our economy. Critics counter that the policy gives valuable public resources away to private interests nearly for free. Critics have tried to amend the law many times over the years, mostly without success. The latest effort, the Hardrock Mining and Reclamation Act of 2009, failed to get out of committee in both houses of Congress.

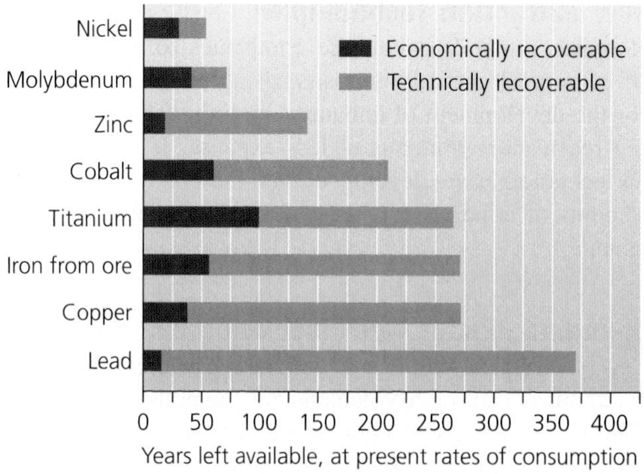


**FIGURE 11.20 Minerals are nonrenewable resources, so supplies of metals are limited.** Shown in red are the numbers of remaining years that certain metals are estimated to be economically recoverable at current prices. The entire lengths of the bars (red plus orange) show how long certain metals are estimated to be available using current technology on all known deposits, whether economically recoverable or not. *Data are for 2013, from U.S. Geological Survey, 2014.* Mineral commodity summaries 2014. *USGS, Washington, D.C.*

DATA Q For some metals, very little of the technically recoverable reserves are economically profitable to extract, but for others this percentage is higher. Which metal has the highest proportion of its technically recoverable reserves that are economically recoverable? What percentage is economically recoverable? Which metal has the smallest proportion of its technically recoverable reserves that are economically recoverable? What is this value?

GO TO **INTERPRETING GRAPHS & DATA** ON MasteringEnvironmentalScience®

# Toward Sustainable Mineral Use

Mining exerts plenty of environmental impacts, but we also have another concern to keep in mind: Minerals are nonrenewable resources (p. 94) in finite supply. As a result, it will benefit us to find ways to conserve the supplies we have left and to make them last.

## Minerals are nonrenewable resources in limited supply

Some minerals we use are abundant and will likely never run out, but others are rare enough that they could soon become unavailable. For instance, geologists in 2014 calculated that the world's known reserves of tantalum will last about 170 more years at today's rate of consumption. If demand for tantalum increases, it could run out faster. And if everyone in the world began consuming tantalum at the rate of U.S. citizens, then it would last for only 13 more years!

Most pressing may be dwindling supplies of indium. This obscure metal, which is used for LCD screens, might last only another 30 years. Because of these supply concerns and price volatility, industries now are working hard to develop ways of substituting other materials for indium. A lack of indium and gallium would threaten the production of high-efficiency cells for solar power. Platinum is dwindling too, and its unavailability would make it harder to develop fuel cells and catalytic converters for vehicles. However, platinum's high market price encourages recycling, which may keep it available, albeit as an expensive metal.

**FIGURE 11.20** shows estimated years remaining for selected minerals at today's consumption rates. Calculating how long a given mineral resource will be available to us is beset by a great deal of uncertainty. There are several major reasons why such estimates may increase or decrease over time.

**Discovery of new reserves** As we discover new deposits of a mineral, the known reserves—and thus the number of years this mineral is available to us—increase. As one recent example, in 2010 geologists associated with the U.S. military discovered that Afghanistan holds immense mineral riches. The newly discovered reserves of iron, copper, niobium, lithium, and many other metals are estimated to be worth over $1 trillion—enough to realign the entire Afghan economy around mining. (Note, however, that such riches are not guaranteed to make Afghanistan a wealthy nation; history teaches us that regions rich in nonrenewable resources, such as Congo and Appalachia, have often been unable to prosper from them.)

**New extraction technologies** Just as rising prices of scarce minerals encourage companies to expend more effort to reach difficult deposits, rising prices also may favor the development of enhanced mining technologies that can reach more minerals at less expense. If more powerful technologies are developed, then these may increase the amounts of minerals that are technically feasible for us to mine.

**Changing social and technological dynamics** New societal developments and new technologies in the marketplace can modify demand for minerals in unpredictable ways. Just as cell phones and computer chips boosted demand for tantalum, fiber-optic cables decreased demand for copper as they replaced copper wiring in communications applications.

**Changing consumption patterns** Changes in the rates and patterns of consumption also alter the speed with which we exploit mineral resources. For example, demand for minerals is depressed during periods of economic depression but then accelerates in times of economic growth.

**Recycling** Advances in recycling technologies and increased recycling rates can extend the lifetimes of mineral resources by allowing us to reuse them many times. Further progress in recycling will likely continue to do so.

## We can make our mineral use more sustainable

We can address both major challenges facing us regarding mineral resources—finite supply and environmental damage—by encouraging recycling of these resources. Municipal recycling programs handle used items that we as consumers place in recycling bins and divert from the waste stream (pp. 395–397). Currently, around 35% of metals in the U.S. municipal solid waste stream are diverted for recycling. Car battery recycling programs provide 80% of the lead we need for new products, and 33% of our copper comes from recycled copper sources such as pipes and wires.

In many cases, recycling can decrease energy use substantially. For instance, making steel by remelting recycled iron and steel scrap requires much less energy than producing steel from virgin iron ore. Similarly, over 40% of the aluminum in the United States today is recycled. This is beneficial because it takes over 20 times more energy to extract virgin aluminum from ore (bauxite) than it does to obtain it from recycled sources.

## We can recycle metals from e-waste

Electronic waste, or e-waste, from discarded computers, printers, cell phones, handheld devices, and other electronic products is rising fast—and that e-waste contains hazardous substances (pp. 400–401). Recycling old electronic devices helps keep them out of landfills and also helps us conserve valuable minerals such as tantalum.

In fact, each of the 1.7 billion cell phones sold each year contains about 200 chemical compounds and close to a dollar's worth of precious metals (**FIGURE 11.21**). Upgrades and improvements render over 130 million cell phones obsolete each year in the United States alone, and an estimated 500 million old cell phones are currently lying inactive in people's homes and offices.

When you turn in your old phone to a recycling and reuse center rather than discarding it, the phone may be refurbished and resold in a developing country. People in African nations in particular readily buy used cell phones, because they are inexpensive and because land-line phone service does not exist in many areas. Alternatively, the phone may be dismantled in a developing country and the various parts refurbished and reused or recycled for their metals. Either way, you are helping to extend the availability of resources through reuse and recycling and to decrease the waste of valuable minerals.

Today only about 10% percent of old cell phones are recycled, so we have room to improve. As we recycle our phones, computers, and other electronics, more tantalum and other metals may be recovered and reused. By recycling more, we can reduce demand for virgin ore and decrease pressure on people and ecosystems in central Africa and other regions where coltan is mined.

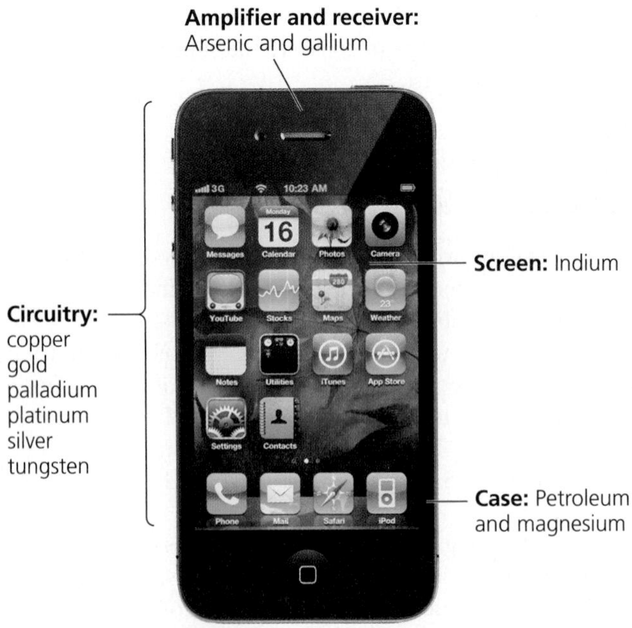


**FIGURE 11.21 Your cell phone contains a diversity of mined materials from around the world.** Figure 11.13 lists the nations that are major producers of these minerals.

## Conclusion

Physical processes of geology such as plate tectonics and the rock cycle are centrally important because they shape Earth's terrain and form the foundation for living systems. Geologic processes also generate phenomena that can threaten our lives and property, including earthquakes, volcanoes, landslides, and tsunamis. We depend on a diversity of minerals and metals from the Earth, and we mine these nonrenewable resources by various methods, according to how the minerals are distributed.

Economically efficient mining methods have greatly contributed to our material wealth, but they have also resulted in extensive environmental impacts, ranging from habitat loss to acid drainage. Restoration efforts and enhanced regulation can help to minimize the environmental and social impacts of mining. We can prolong our access to mineral resources and make our mineral use more sustainable by maximizing the recovery and recycling of key minerals.

# Testing Your Comprehension

1. Name the primary layers that make up our planet. Which portions does the lithosphere include?
2. Describe what occurs at a divergent plate boundary. What happens at a transform plate boundary? Compare and contrast the types of processes that can occur at a convergent plate boundary.
3. Name the three main types of rocks, and describe how each type may be converted to the others via the rock cycle.
4. Define and explain the processes that produce earthquakes, volcanoes, tsunamis, and mass wasting.
5. Define each of the following and contrast them with one another: (1) mineral, (2) metal, (3) ore, (4) alloy.
6. A mining geologist locates a horizontal seam of coal very near the surface of the land. What type of mining method will the mining company use to extract it? Describe one common environmental impact of this type of mining.
7. How does strip mining differ from subsurface mining? How do each of these approaches differ from open pit mining?
8. What is acid drainage, and why can it be toxic to aquatic organisms?
9. List five factors that influence how long global supplies of a given mineral will last, and explain how each affects the time span the mineral will be available to us.
10. Name three types of metal that we currently recycle, and identify the products or materials that are recycled to recover these metals.

# Seeking Solutions

1. For each of the following natural hazards, describe one thing that we can do to minimize or mitigate their impacts on our lives and property:
    - Earthquakes
    - Landslides
    - Flooding
2. List three impacts of mining on the natural environment, and describe how particular mining practices can lead to each of these impacts. How are these impacts being addressed? Can you think of additional solutions to prevent, reduce, or mitigate these impacts?
3. You have secured a grant from the U.S. Environmental Protection Agency (EPA) to work with a mining company to develop a more effective way of restoring a mine site that is about to be abandoned. Describe a few preliminary ideas for carrying out restoration more effectively than it is typically being done. Now describe a field experiment you would like to run to test one of your ideas.
4. **THINK IT THROUGH** The story of coltan in the Congo is just one example of how an abundance of exploitable resources can often worsen or prolong military conflicts in nations that are too poor or ineffectively governed to protect these resources. In such "resource wars," civilians often suffer the most as civil society breaks down. Suppose you are the head of an international aid agency that has earmarked $10 million to help address conflicts related to mining in the northern Amazon. You have access to the governments of Brazil, Colombia, and Venezuela; ambassadors of the world's nations in the United Nations; and representatives of international mining corporations. Based on what you know from this chapter, what steps would you consider taking to help improve the situation in the region?
5. **THINK IT THROUGH** As you finish your college degree, you learn that the mountains behind your childhood home in the hills of Kentucky are slated to be mined for coal using the mountaintop removal method. Your parents, who still live there, are worried for their health and safety and do not want to lose the beautiful forested creek and ravine behind their property. However, your brother is out of work and could use a mining job. What would you attempt to do in this situation?

# Calculating Ecological Footprints

As we saw in Figure 11.20, some metals are in such limited supply that, at today's prices, they could be available to us for only a few more decades. After that, prices will rise as they become scarcer. The number of years of total availability (at all prices) depends on a number of factors: On the one hand, metals will be available longer if new deposits are discovered, new mining technologies are developed, or recycling efforts are improved. On the other hand, if our consumption of metals increases, the number of years we have left to use them will decrease.

Currently the United States consumes metals at a much higher per-person rate than the world does as a whole. If one goal of humanity is to lift the rest of the world up to U.S. living standards, then this will sharply increase pressures on mineral supplies.

The chart shows currently known economically recoverable global reserves for several metals, together with the amount used per year (each figure in thousands of metric tons). For each metal, divide the reserves by the annual amount used to calculate the number of remaining years of supply at current prices. Enter the number of years in the fourth column.

The fifth column shows the amount that the world would use if everyone in the world consumed the metal at the rate that Americans do. Now calculate the number years of supply left at current prices for each metal if the world were to consume the metals at the U.S. rate, and enter these values in the sixth column.

| Metal | Known economic reserves | Amount used per year | Years of economic supply left | Amount used per year if everyone consumed at U.S. rate | Years of economic supply left if everyone consumed at U.S. rate |
|---|---|---|---|---|---|
| Titanium | 750,000 | 7550 | 99.3 | 32,975 | 22.7 |
| Copper | 690,000 | 17,900 | | 40,654 | |
| Nickel | 74,000 | 2490 | | 4562 | |
| Tin | 4700 | 230 | | 939 | |
| Tungsten | 3500 | 71 | | 314 | |
| Antimony | 1800 | 163 | | 542 | |
| Silver | 520 | 26 | | 152 | |
| Gold | 54 | 2.8 | | 3.6 | |

*Data are for 2013, from U.S. Geological Survey, 2014.* Mineral commodity summaries 2014. *USGS, Washington, D.C.*
*All numbers are in thousands of metric tons. World consumption data are assumed equal to world production data.*
*"Known economic reserves" include extractable amounts under current economic conditions. Additional reserves exist that could be mined at greater cost.*

1. Which of these eight metals will last the longest under current economic conditions and at current rates of global consumption? For which of these metals will economic reserves be depleted fastest?
2. If the average citizen of the world consumed metals at the rate that the average U.S. citizen does, which of these eight metals' economic reserves would last the longest? Which would be depleted fastest?
3. In this chart, our calculations of years of supply left do not factor in population growth. All else being equal, how do you think population growth will affect these numbers?
4. Describe two general ways that we could increase the years of supply left for these metals. What do you think it will take to accomplish this?

## MasteringEnvironmentalScience®

**STUDENTS**

Go to **MasteringEnvironmentalScience** for assignments, the etext, and the Study Area with practice tests, videos, current events, and activities.

**INSTRUCTORS**

Go to **MasteringEnvironmentalScience** for automatically graded activities, current events, videos, and reading questions that you can assign to your students, plus Instructor Resources.

# Fresh Water, Oceans, and Coasts

## Upon completing this chapter, you will be able to:

- ☐ Explain water's importance to people and ecosystems, and describe the distribution of fresh water on Earth
- ☐ Describe the freshwater, marine, and coastal portions of the interconnected aquatic system
- ☐ Discuss how we use water and alter aquatic systems
- ☐ Assess problems of water supply and propose solutions to address depletion of fresh water
- ☐ Describe the major classes of water pollution and propose solutions to address water pollution
- ☐ Describe legislation in the United States that addresses water quality
- ☐ Explain how we treat drinking water and wastewater
- ☐ Review the state of ocean fisheries and reasons for their decline
- ☐ Evaluate marine protected areas and reserves as innovative solutions for conserving biodivesity

**Photo: Louisiana's vanishing coastal wetlands support a diversity of wildlife, such as these Roseate Spoonbills.**

# CENTRAL CASE STUDY

# Starving the Louisiana Coast

*"The Louisiana and Mississippi coastal region is critical to the economic, cultural, and environmental integrity of the nation."*

**—Nancy Sutley, Chair of the White House Council on Environmental Quality**

*"What really screwed up the marsh is when they put the levees on the river. They should take the levees out and let the water run; that's what built the land."*

**—Frank "Blackie" Campo, Resident of Shell Beach, Louisiana**

The state of Louisiana is shrinking. Its coastal wetlands straddle the boundary between the land and the ocean, and these wetlands are disappearing beneath the waters of the Gulf of Mexico. Louisiana loses 65 $km^2$ (25 $mi^2$) of coastal wetlands each year—about the size of Manhattan Island in New York City. Comparisons of wetland area from the mid-1800s to the early 1990s show a drastic decrease in wetlands (**FIGURE 12.1a**). Since the 1930s alone, Louisiana has lost nearly 4900 $km^2$ (1900 $mi^2$) of coastal wetlands—an area roughly the size of Delaware.

Louisiana's coastal wetlands transition from communities of salt-tolerant grasses at the ocean's edge to freshwater bald cypress swamps further inland. These ecosystems support a diversity of animals, including eagles, pelicans, shrimp, oysters, black bears, alligators, and sea turtles. The state's coastal wetlands also protect New Orleans, Baton Rouge, and other communities from damaging storms. Vegetation in these wetlands acts as a windbreak against strong winds and as a buffer against storm surges coming inland from the Gulf.

Louisiana's millions of acres of coastal wetlands were created over the past 7000 years as the Mississippi River fanned out and deposited its sediments at its delta before emptying into the Gulf of Mexico. The Mississippi River accumulates large quantities of sediment from water flowing over land and into streams in the river's 3.2-million-$km^2$ (1.2-million-$mi^2$) watershed (**FIGURE 12.1b**). Much of this sediment originates from the Missouri River basin, which drains America's agricultural heartland.

The salt marshes in the river's delta naturally compact over time. This compaction lowers the level of the marsh bottom and submerges vegetation under increasingly deeper waters. When waters become too deep, the vegetation dies, and soils are then washed away by the Gulf of Mexico. The natural compaction is offset, however, by inputs of sediments from the river and from the deposition of organic matter from marsh grasses. These additions keep soil levels high, water depths relatively stable, and vegetation healthy.

So why are Louisiana's wetlands being swallowed by the sea? It's because people have modified the Mississippi River so extensively that much of its sediments no longer reach the wetlands that need them. The river's basin contains roughly 2000 dams, which slow river flow and allow sediments suspended in the water to settle in reservoirs. This not only prevents sediments from reaching the river's delta, but also slowly fills in each dam's reservoir, decreasing its volume and shortening its life span. Therefore, dams in Minnesota, Montana, and Pennsylvania and other parts of the Mississippi basin affect the Louisiana coastline hundreds of miles downriver.

The Mississippi River is also lined with thousands of miles of *levees* (long, raised mounds of earth). These structures prevent small-scale flooding, and the mouth of the Mississippi is lined with levees to provide a deep river channel for shipping into the Gulf of Mexico. These levees prevent the river from spilling into its delta and turn the lower Mississippi into a "barrel" that shoots sediments off the continental shelf into the deep waters of the Gulf (**FIGURE 12.1c**).

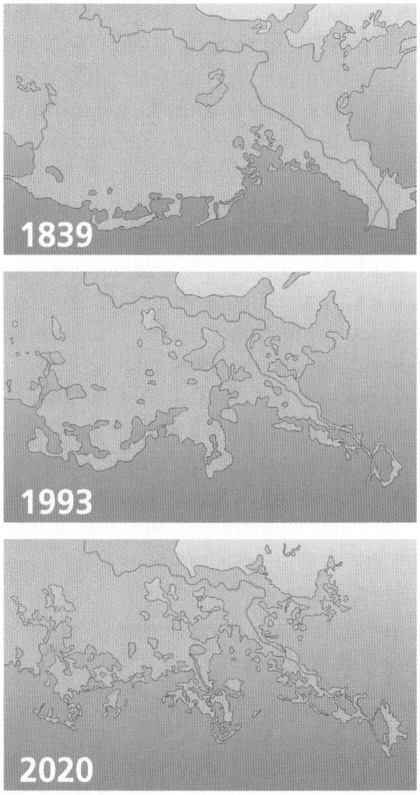


(a) Coastal wetland area in 1839, 1993, and 2020

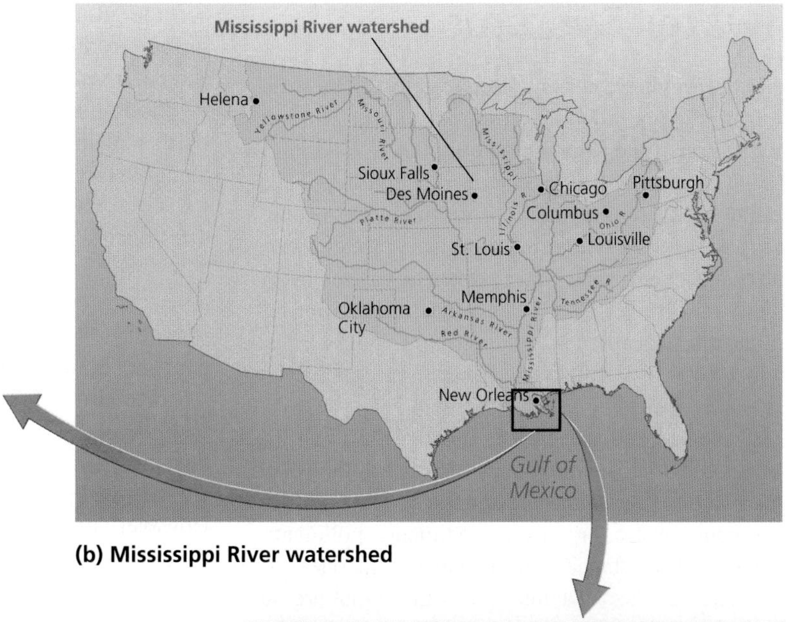


(b) Mississippi River watershed

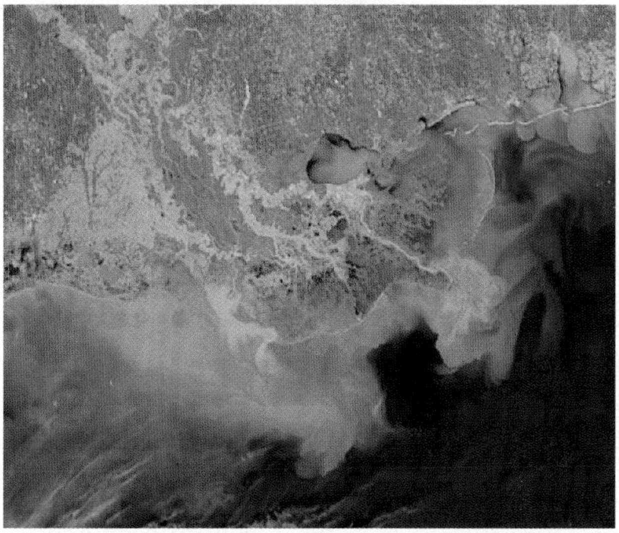

(c) Sediment plumes from Mississippi River entering Gulf

**FIGURE 12.1 Human modifications all along the Mississippi River affect coastal wetlands at the river's mouth.** Louisiana's coastal wetlands shrank substantially **(a)** from 1839 to 1993. They are predicted to shrink even more by 2020 because people have modified sediment deposition patterns in the river's delta by constructing dams and extensive levees along the river. The Mississippi River system **(b)** is the largest in the United States, draining over 40% of the land area of the lower 48 states. A satellite image of south Louisiana **(c)** shows the brown plumes of sediments being released into the deep waters of the Gulf of Mexico from the Mississippi River (plume on the right) and the Atchafalaya River (plume on the left). *(a) Adapted from Environmental Defense Fund.*

Although Louisiana's economy has benefited from oil and gas extraction, these activities have also promoted wetland losses. The extraction of large quantities of oil, natural gas, and saline groundwater associated with oil deposits causes the land to compact, lowering soil levels. Additionally, engineers have cut nearly 13,000 km (8000 mi) of canals through coastal wetlands to facilitate shipping and oil and gas exploration. These canals fragment the wetlands and increase erosion rates because they enable salty ocean water to penetrate inland and damage vegetation and wildlife in freshwater marshes.

Proposed solutions for coastal erosion center on restoring the system to its natural state by diverting large quantities of water from the Mississippi River into coastal wetlands instead of shooting it out into the Gulf in the river's main channel. Proponents of this approach point to the Atchafalaya River, which currently diverts one-third of the lower Mississippi River's volume and carries it to the Gulf. The Atchafalaya delta, fed by this water and sediment, is actually gaining coastal land area. Gulf restoration efforts are being bolstered by the 2012 Resources and Ecosystems Sustainability, Tourist Opportunities and Revived Economies of the Gulf Coast States Act (RESTORE Act), legislation passed in the aftermath of the *Deepwater Horizon* spill (pp. 345–346). The Act creates a comprehensive ecosystem restoration plan for the Gulf Coast, financed by 80% of the fines paid for Clean Water Act (p. 105) violations associated with the spill.

Given the conflicting demands we put on waterways for water withdrawal, shipping, and flood control, there are no easy solutions to the problems faced in the Mississippi River and southern Louisiana. But how we tackle problems like those in Louisiana's coastal wetlands will help determine the long-term sustainability of one of our most precious natural resources—the aquatic ecosystems that provide us life-sustaining water. ■

# Freshwater Systems

"Water, water, everywhere, nor any drop to drink." The well-known line from the poem *The Rime of the Ancient Mariner* is an apt description of the situation on our planet. Water may seem abundant, but water that we can drink is quite rare and limited (**FIGURE 12.2**). About 97.5% of Earth's water resides in the oceans and is too salty to drink or to use to water crops. Only 2.5% is considered **fresh water**, water that is relatively pure with few dissolved salts. Because most fresh water is tied up in glaciers, icecaps, and underground aquifers, just over 1 part in 10,000 of Earth's water is easily accessible for human use.

Water is renewed and recycled as it moves through the *water cycle* (pp. 38–39). The movement of water in the water cycle creates a web of interconnected aquatic systems (**FIGURE 12.3**) that exchange water, organisms, sediments, pollutants, and other dissolved substances. What happens in one system therefore affects other systems—even those that are far away. Precipitation falling from the sky either sinks into the ground or flows off the land to form rivers, which carry water to the oceans or large inland lakes. As they flow, rivers can interact with ponds, wetlands, and coastal aquatic ecosystems. Underground aquifers exchange water with rivers, ponds, and lakes through the sediments on the bottoms of these water bodies. Let's examine the components of this interconnected system, beginning with groundwater.

## Groundwater plays key roles in the water cycle

Liquid water occurs either as surface water or groundwater. **Surface water** is water located atop Earth's surface (such as a river or lake), and **groundwater** is water beneath the surface held within pores in soil or rock. Some of the precipitation reaching Earth's land surface infiltrates the surface to become groundwater. Groundwater flows slowly beneath the surface from areas of high pressure to areas of low pressure and can remain underground for long periods, in some cases for thousands of years. Groundwater makes up one-fifth of Earth's fresh water supply and plays a key role in meeting human water needs.

Groundwater is contained within **aquifers**: porous formations of rock, sand, or gravel that hold water (p. 38) (**FIGURE 12.4**). An aquifer's upper layer, or *zone of aeration*, contains pore spaces partly filled with water. In the lower layer, or *zone of saturation*, the spaces are completely filled with water. The boundary between these two zones is the **water table**.

The largest known aquifer is the Ogallala Aquifer, which underlies the Great Plains of the United States. Water from this massive aquifer has enabled American farmers to create the most productive grain-producing region in the world. However, unsustainable water withdrawals are threatening the long-term use of the aquifer for agriculture.

**FAQ Is groundwater found in huge underground caverns?**

As one of the "out of sight" elements of the water cycle, it's sometimes difficult for people to visualize how water exists underground. Many incorrectly assume that groundwater is always found in large, underground caves—essentially lakes beneath Earth's surface. The reality is far less glamorous. If you look at soil under a microscope, you'll see there are small pores between the particles of minerals and organic matter that compose the soil. Many types of rock, such as limestone and sandstone, have relatively large pores between the particles of minerals that compose the rock. So, when people extract groundwater with wells, we are simply sucking water out of the pores between soil particles or within rocks in the portion of the soil beneath the water table.

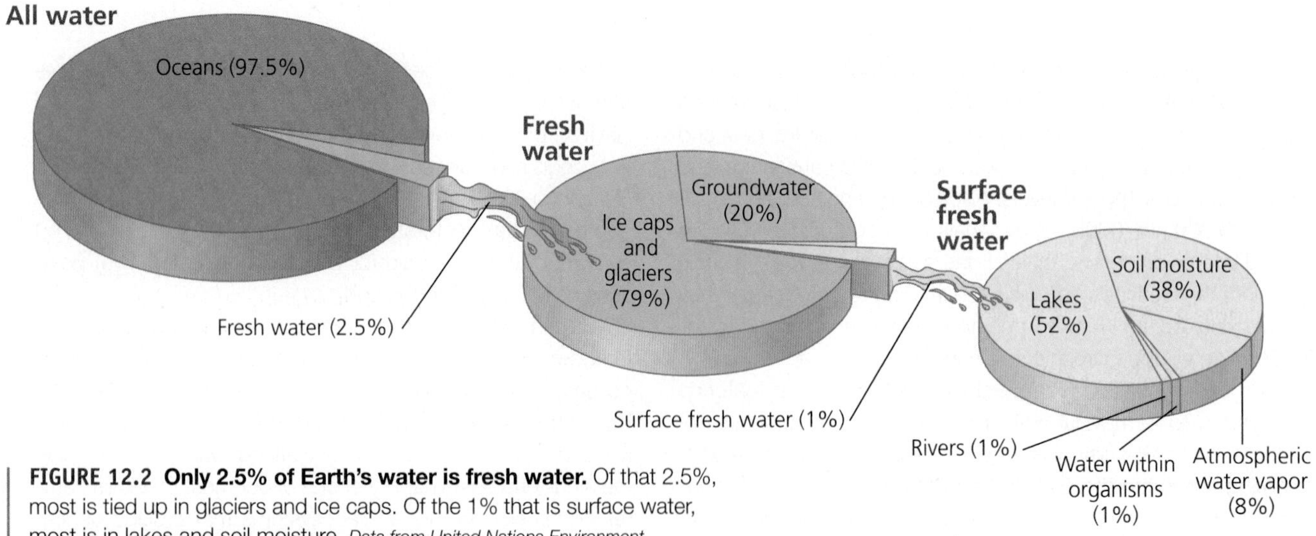


**FIGURE 12.2 Only 2.5% of Earth's water is fresh water.** Of that 2.5%, most is tied up in glaciers and ice caps. Of the 1% that is surface water, most is in lakes and soil moisture. *Data from United Nations Environment Programme (UNEP) and World Resources Institute.*

DATA Q What percentage of Earth's water is fresh water in lakes?
GO TO **INTERPRETING GRAPHS & DATA** ON MasteringEnvironmentalScience®

**FIGURE 12.3 Water flows through freshwater systems and marine and coastal aquatic systems that interact extensively with one another.** People affect the components of the system by constructing dams and levees, withdrawing water for human use, and introducing pollutants. Because the systems are closely connected, these impacts can cascade through the system and cause effects far from where they originated.

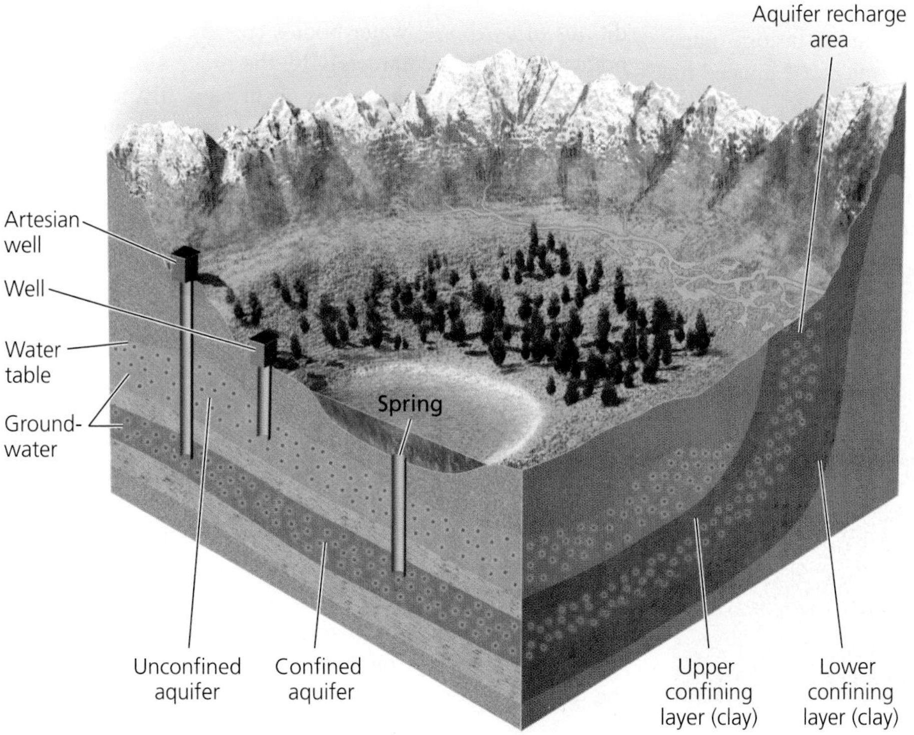


**FIGURE 12.4 Groundwater may occur in unconfined aquifers above impermeable layers or in confined aquifers under pressure between impermeable layers.** Water may rise naturally to the surface at springs and through the wells we dig. Artesian wells tap into confined aquifers to mine water under pressure.

## Surface water converges in river and stream ecosystems

Surface water accounts for just 1% of fresh water, but it is vital for our survival and for the planet's ecological systems. Groundwater and surface water interact, and water can flow from one type of system to the other. Surface water becomes groundwater by infiltration. Groundwater becomes surface water through springs (and human-drilled wells), often keeping streams flowing or wetlands moist when surface conditions are otherwise dry. Each day in the United States, 1.9 trillion L (492 billion gal) of groundwater are released into surface waters—nearly as much as the daily flow of the Mississippi River.

Water that falls from the sky as rain, emerges from springs, or melts from snow or a glacier and then flows over the land surface is called **runoff**. Runoff converges as it flows downhill and forms streams. These small watercourses may merge into rivers, whose water eventually reaches a lake or ocean. A smaller river flowing into a larger one is called a *tributary*. The area of land drained by a river and all its tributaries is the river's **watershed** (p. 22). If you could trace every drop of water in the Mississippi River back to the spot where it first fell as precipitation, you would have delineated the river's watershed, the area shown in Figure 12.1.

Landscapes determine where rivers flow, but rivers shape the landscapes through which they run. Over thousands or millions of years, a meandering river may shift from one course to another, back and forth over a large area, carving out a flat valley and picking up sediment that is later deposited in coastal wetlands (**FIGURE 12.5**). Areas nearest a river's course that are flooded periodically are said to be within the river's **floodplain**. Frequent deposition of silt (eroded soil) from flooding makes floodplain soils especially fertile. As a result, agriculture thrives in floodplains, and *riparian* (riverside) forests are productive and species-rich. A river's meandering is often driven by large-scale flooding events that scour new channels. However, extensive damming on the Mississippi and other rivers has reduced the rate of river meandering by 66–83% from its historic rate. Instead of coursing down the river, floodwaters are trapped in reservoirs by dams and contained in the river channel by levees.

**FIGURE 12.5 Rivers, such as the Wood River in Alaska shown here, shape the landscapes through which they flow.**

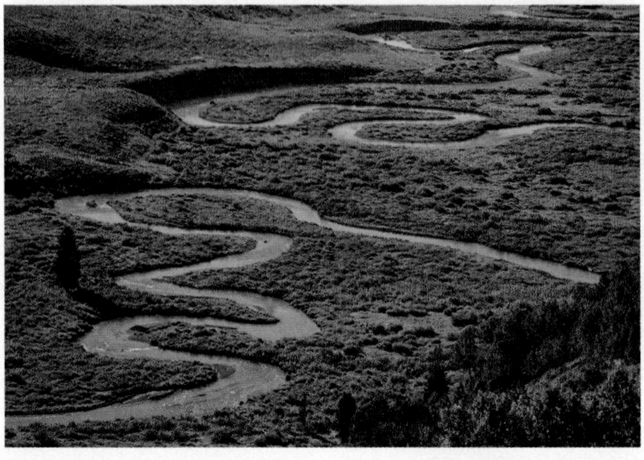

Rivers and streams host diverse biological communities. Algae and detritus (p. 71) support many types of invertebrates, from water beetles to crayfish. Fish and amphibians consume aquatic invertebrates and plants, and birds such as kingfishers, herons, and ospreys dine on fish and amphibians.

## Lakes and ponds are ecologically diverse systems

Lakes and ponds are bodies of standing surface water. The largest lakes, such as North America's Great Lakes, are sometimes known as inland seas. Although lakes and ponds can vary greatly in size, scientists have described several zones common to these waters (**FIGURE 12.6**).

Around the nutrient-rich edges of a water body, the water is shallow enough that aquatic plants grow from the mud and reach above the water's surface. This region, named the *littoral zone*, abounds in invertebrates—such as insect larvae, snails, and crayfish—that fish, birds, turtles, and amphibians feed on. The *benthic zone* extends along the bottom of the lake or pond, from shore to the deepest point. Many invertebrates live in the mud, feeding on detritus or on one another. In the open portion of a lake or pond, far from shore, sunlight penetrates shallow waters of the *limnetic zone*. Because light enables photosynthesis (pp. 31–32), the limnetic zone supports phytoplankton (algae, protists, and cyanobacteria; p. 22), which in turn support zooplankton (p. 24), both of which are eaten by fish. Below the limnetic zone lies the *profundal zone*, the volume of open water that sunlight does not reach. This zone lacks photosynthetic life and is lower in dissolved oxygen than upper waters.

Ponds and lakes change over time as streams and runoff bring them sediment and nutrients. *Oligotrophic* lakes and ponds, which are low in nutrients and high in oxygen, may slowly transition to the high-nutrient, low-oxygen conditions of *eutrophic* water bodies (p. 25). Eventually, water bodies may fill in completely by the process of aquatic succession (p. 78). These changes occur naturally, but eutrophication can also result from human-caused nutrient pollution (p. 22).

## Freshwater wetlands include marshes, swamps, bogs, and vernal pools

**Wetlands** are systems in which the soil is saturated with water and which generally feature shallow standing water with ample vegetation. There are many types of freshwater wetlands, and most are enormously rich and productive. In *freshwater marshes*, shallow water allows plants such as cattails and bulrushes to grow above the water surface. *Swamps* also consist of shallow water rich in vegetation, but they occur in forested areas (**FIGURE 12.7**). *Bogs* are ponds covered with thick floating mats of vegetation and can represent a stage in aquatic succession. *Vernal pools* are seasonal wetlands that form in early spring from rain and snowmelt and dry up once the weather becomes warmer.

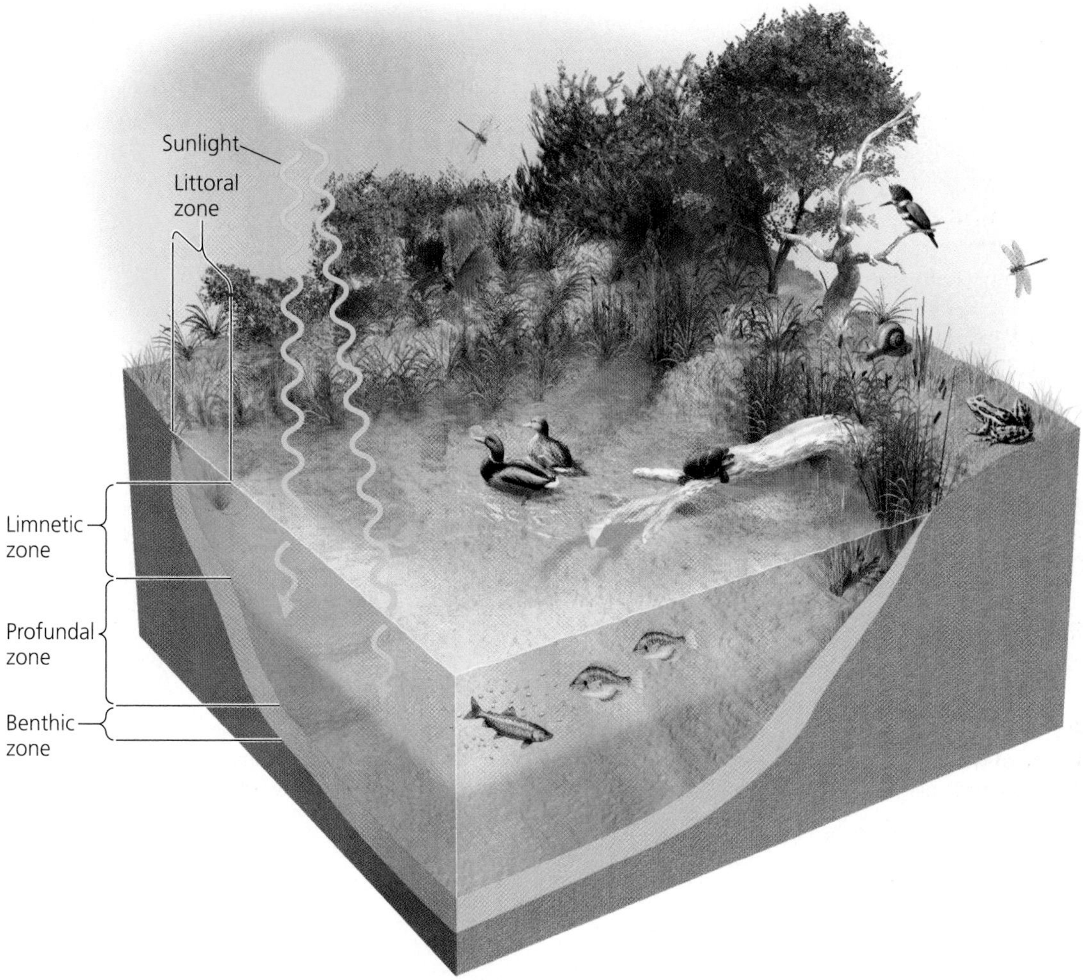


**FIGURE 12.6 In lakes and ponds, emergent plants grow along the shoreline in the littoral zone.** The limnetic zone is the layer of open, sunlit water, where photosynthesis takes place. Sunlight does not reach the deeper profundal zone. The benthic zone, at the bottom of the water body, often is muddy, rich in detritus and nutrients, and low in oxygen.

Wetlands are extremely valuable habitat for wildlife. Louisiana's coastal wetlands, for example, provide habitat for approximately 1.8 million migratory waterbirds each year.

**FIGURE 12.7 Freshwater wetlands such as this bald cypress swamp in Louisiana support biologically diverse and productive ecosystems.**

Wetlands also provide important ecosystem services by slowing runoff, reducing flooding, recharging aquifers, and filtering pollutants.

Despite the vital roles that wetlands play, people have drained and filled them extensively for agriculture. Many wetlands are lost when people divert and withdraw water, channelize rivers, and build dams. Southern Canada and the United States, for example, have lost well over half their wetlands since European colonization.

## The Oceans

The oceans are an important component of Earth's interconnected aquatic systems. The vast majority of rivers empty into oceans (a small number of rivers empty into inland seas), so the oceans receive most of the inputs of water, sediments, pollutants, and organisms carried by freshwater systems. The oceans influence virtually every environmental system and every human endeavor, so even if you live in a landlocked region far from the coast, the oceans affect you, and you affect the oceans.

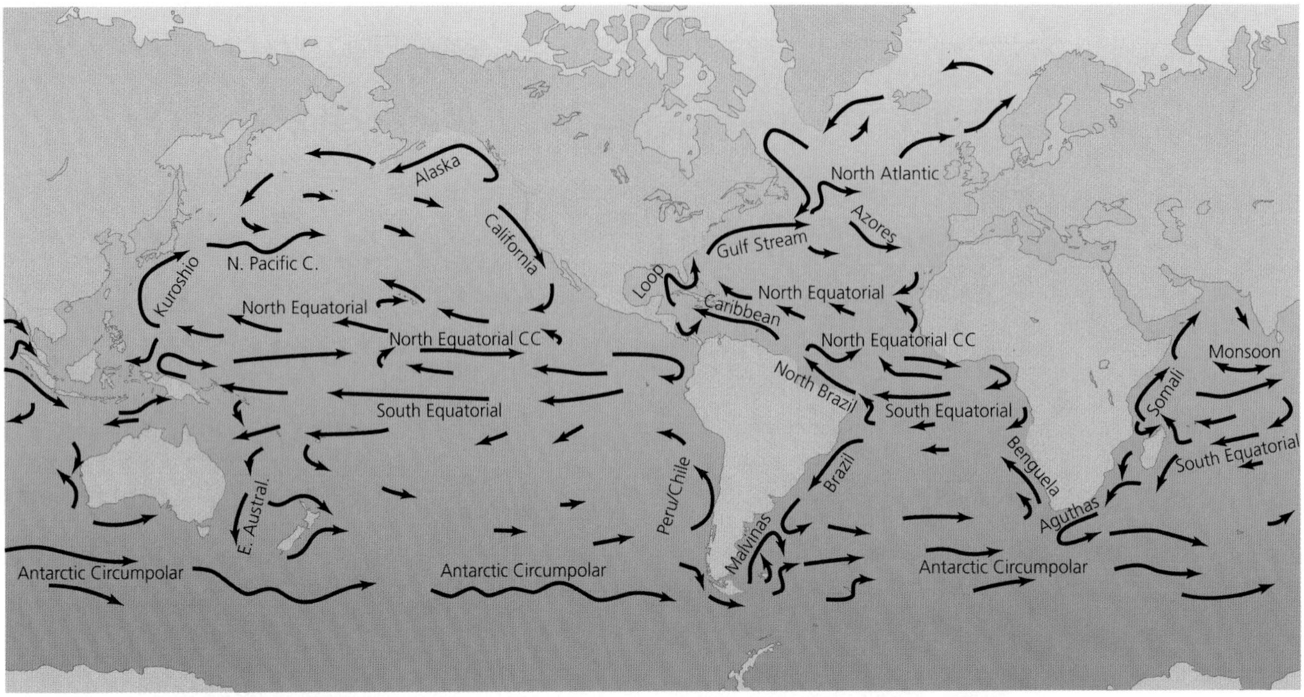


**FIGURE 12.8 The upper waters of the oceans flow in surface currents, long-lasting and predictable global patterns of water movement.** Warm-and cold-water currents interact with the planet's climate system, and people have used them for centuries to navigate the oceans. *Adapted from Rick Lumpkin (NOAA/AOML).*

DATA **Q** If you released a buoy into the Pacific Ocean from the southeastern coast of Japan that traveled on the surface ocean currents shown above, which would it likely reach first: the United States or Australia? On what currents would it be carried?

GO TO **INTERPRETING GRAPHS & DATA** ON MasteringEnvironmentalScience®

## The physical makeup of the ocean is complex

The world's five major oceans—Pacific, Atlantic, Indian, Arctic, and Antarctic—are all connected, comprising a single vast body of water that covers 71% of Earth's surface. Ocean water contains roughly 96.5% $H_2O$ by mass; most of the remainder consists of ions from dissolved salts. Ocean water is salty primarily because ocean basins are the final repositories for runoff that collects salts from weathered rocks and carries them to the ocean. Whereas the water in the ocean evaporates, the salts do not, and they accumulate in ocean basins (the salinity of ocean water generally ranges from 33,000 to 37,000 parts per million, whereas the salinity of freshwater runoff typically is less than 500 parts per million). If we were able to evaporate all the water from the oceans, a layer of dried salt 63 m (207 ft) thick would be left behind.

Surface waters of the oceans are warmer than subsurface waters because the sun heats them and because warmer water is less dense. Deep below the surface, water is dense and sluggish, unaffected by winds and storms, sunlight, and daily temperature fluctuations. Ocean water travels in **currents**, vast riverlike flows that move in the upper 400 m (1,300 ft) of water, horizontally and for great distances (**FIGURE 12.8**). Wind, solar heating and cooling, gravity, density differences, and the Coriolis effect (p. 282) drive the global system of ocean currents.

Surface winds and heating also create vertical currents in seawater. **Upwelling** is the rising of deep, cold, dense water toward the surface. Because this water is rich in nutrients from the bottom, upwellings often support high primary productivity (p. 32) and lucrative fisheries, such as those along the coasts of Peru and Chile. At **downwellings**, warm surface water rich in dissolved gases is displaced downward, providing an influx of oxygen for deep-water life and "burying" $CO_2$ in ocean sediments.

Many parts of the ocean floor are rugged and complex (**FIGURE 12.9**). Underwater volcanoes shoot forth enough magma to build islands above sea level, such as the Hawaiian Islands (see Figure 11.7, p. 231). Steep canyons lie just offshore of some continents. The deepest spot in the oceans—the Mariana Trench in the South Pacific—is deeper than Mount Everest is high, by over a mile. Our planet's longest mountain range is under water: the Mid-Atlantic Ridge (p. 227) runs the length of the Atlantic Ocean.

## Ocean currents affect Earth's climate

The horizontal and vertical movements of ocean water can have far-reaching effects on climate globally and regionally. The **thermohaline circulation** is a worldwide current system in which warmer, fresher water moves along the surface and colder, saltier water (which is denser) moves deep beneath the surface (**FIGURE 12.10**).

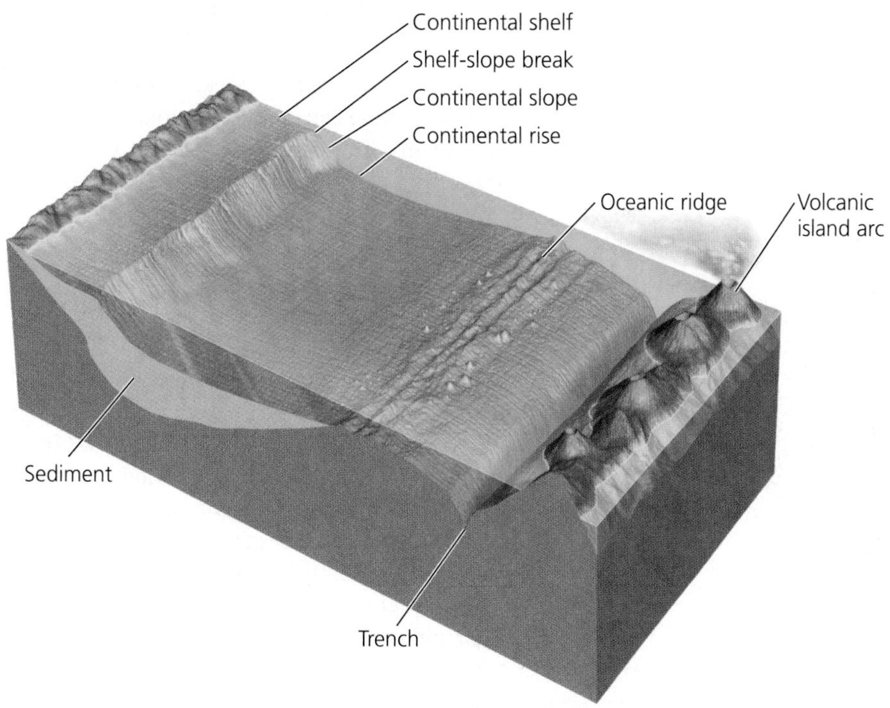


**FIGURE 12.9 A stylized bathymetric profile shows key geologic features of the submarine environment.** Shallow water exists around the edges of continents over the continental shelf, which drops off at the shelf-slope break. The steep continental slope gives way to the more gradual continental rise, all of which are underlain by sediments from the continents. Vast areas of seafloor are flat abyssal plain. Seafloor spreading occurs at oceanic ridges, and oceanic crust is subducted in trenches (p. 228). Volcanic activity along trenches may give rise to island chains such as the Aleutian Islands. Features on the left side of this diagram are more characteristic of the Atlantic Ocean, and features on the right side of the diagram are more characteristic of the Pacific Ocean. *Thurman, Harold V., and Alan P. Trujilo, 2002.* Essentials of Oceanography, *7th ed. Adapted and electronically reproduced by permission of Pearson Education, Inc., Upper Saddle River, New Jersey.*

One segment of this worldwide conveyor-belt system is the warm surface water in the Gulf Stream that flows across the Atlantic Ocean to Europe. As this water releases heat to the air, keeping Europe warmer than it would otherwise be, the water cools, becomes saltier through evaporation, becomes denser, and sinks. This creates a region of downwelling known as the *North Atlantic Deep Water (NADW).*

Scientists hypothesize that interrupting the thermohaline circulation could trigger rapid climate change. If global warming (Chapter 14) causes much of Greenland's ice sheet to melt, the resulting freshwater runoff into the North Atlantic would make surface waters less dense (because fresh water is less dense than saltwater). This could stop the NADW formation and shut down the northward flow of warm water, causing Europe to cool rapidly. Some data suggest that the thermohaline circulation in this region is already slowing, but other researchers maintain that Greenland will not produce enough runoff to cause a shutdown this century.

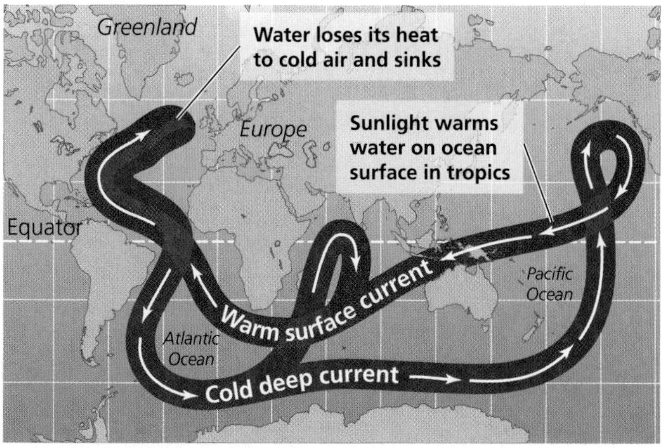


**FIGURE 12.10 As part of the oceans' thermohaline circulation, warm surface currents carry heat from equatorial waters northward toward Europe, where they warm the atmosphere.** The water then cools and sinks, forming the North Atlantic Deep Water (NADW).

Another interaction between ocean currents and the atmosphere that influences climate is the **El Niño–Southern Oscillation (ENSO)**, a systematic shift in atmospheric pressure, sea surface temperature, and ocean circulation in the tropical Pacific Ocean. Under normal conditions, prevailing winds blow from east to west along the equator, from a region of high pressure in the eastern Pacific to one of low pressure in the western Pacific, forming a large-scale convective loop in the atmosphere (**FIGURE 12.11a**, next page). The winds push surface waters westward, causing water to "pile up" in the western Pacific. As a result, water near Indonesia can be 50 cm (20 in.) higher and 8°C warmer than water near South America. The westward-moving surface waters allow cold water to rise up from the deep in a nutrient-rich upwelling along the coast of Peru and Ecuador.

**El Niño** conditions are triggered when air pressure decreases in the eastern Pacific and increases in the western Pacific, weakening the equatorial winds and allowing the warm water to flow eastward (**FIGURE 12.11b**, next page). This suppresses upwelling along the Pacific coast of the Americas, shutting down the delivery of nutrients that support marine life and fisheries. This phenomenon was called El Niño (Spanish for "little boy" or "Christ Child") by Peruvian fishermen because the arrival of warmer waters usually occurred shortly after Christmas. Coastal industries such as Peru's anchovy fisheries are devastated by El Niño events, and economic impacts from El Niño events can reach into many billions of dollars.

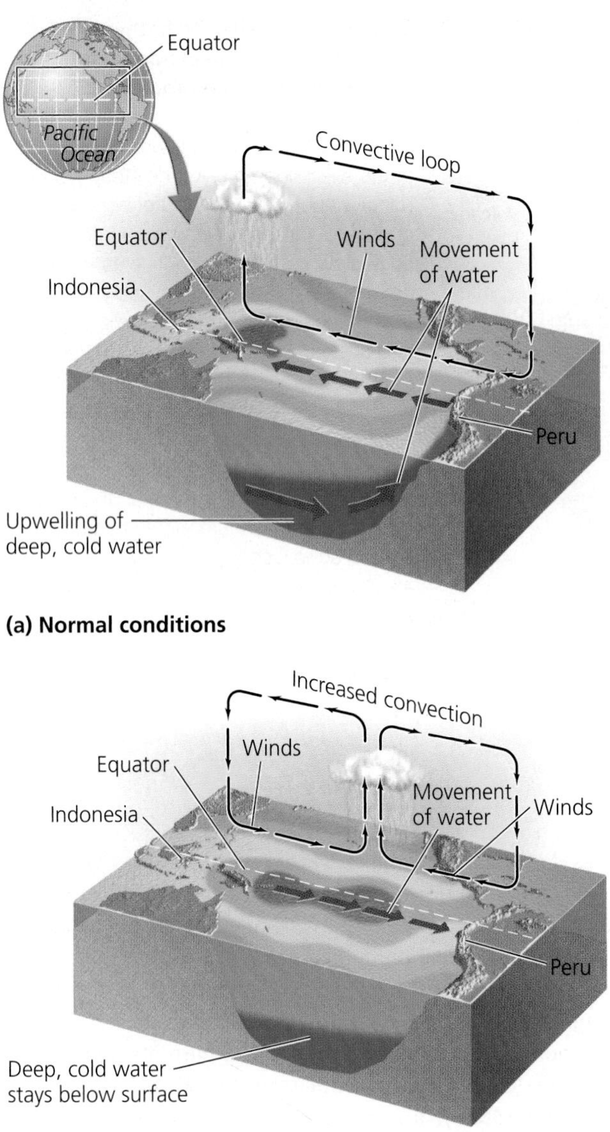


**FIGURE 12.11 El Niño conditions occur every 2 to 8 years, causing marked changes in weather patterns.** In these diagrams, red and orange colors denote warmer water, and blue and green colors denote colder water. Under normal conditions **(a)**, prevailing winds push warm surface waters toward the western Pacific. Under El Niño conditions **(b)**, winds weaken, and the warm water flows back across the Pacific toward South America, like water sloshing in a bathtub. *Adapted from National Oceanic and Atmospheric Administration, Tropical Atmospheric Ocean Project.*

El Niño events alter weather patterns around the world, creating rainstorms and floods in areas that are generally dry (such as southern California) and causing drought and fire in regions that are typically moist (such as Indonesia).

**La Niña** events are the opposite of El Niño events; in a La Niña event, cold waters rise to the surface and extend westward in the equatorial Pacific when winds blowing to the west strengthen, and weather patterns are affected in opposite ways.

ENSO cycles are periodic but irregular, occurring every 2–8 years. Scientists are trying to determine whether warming air and sea temperatures (Chapter 14) may be increasing the frequency and strength of these cycles.

# Marine and Coastal Systems

With their variation in topography, temperature, salinity, nutrients, and sunlight, marine and coastal environments feature a variety of ecosystems. These systems may not give us the water we need for drinking and growing crops, but they teem with biodiversity and provide many other necessary resources.

## Fresh water meets saltwater in estuaries

Water bodies where rivers flow into the ocean, mixing fresh water with saltwater, are called **estuaries**. Estuaries are biologically productive ecosystems that experience fluctuations in salinity with the daily and seasonal variations in tides and freshwater runoff. The shallow water of estuaries nurtures seagrass beds and other plant life and provides critical habitat for shorebirds and for many commercially important shellfish species.

Estuaries everywhere have been affected by coastal development, water pollution, habitat alteration, and overfishing. The Chesapeake Bay estuary (profiled in Chapter 2) is one such example. Estuaries and other coastal ecosystems have borne the brunt of human impact because two out of every three people choose to live within 160 km (100 mi) of the ocean.

## Salt marshes line temperate shorelines

Along many of the world's coasts at temperate latitudes, **salt marshes** occur where the tides wash over gently sloping sandy or silty substrates. Rising and falling tides flow into and out of channels called *tidal creeks* and at highest tide spill over onto elevated marsh flats, like those in coastal Louisiana (**FIGURE 12.12**). Marsh flats grow thick with salt-tolerant grasses, as well as rushes, shrubs, and other herbaceous plants. Salt marshes boast very high primary productivity and provide critical habitat for shorebirds, waterfowl, and many fish and shellfish species. Salt marshes also filter pollution and stabilize shorelines against storm surges.

## Mangrove forests line coasts in the tropics and subtropics

In tropical and subtropical latitudes, mangrove forests replace salt marshes along the coasts. **Mangroves** are salt-tolerant, and they have unique roots that curve upward like snorkels to attain oxygen or downward like stilts to support the tree in changing water levels (**FIGURE 12.13**). Fish, shellfish, crabs, snakes, and other organisms thrive among the root networks, and birds feed and nest in the dense foliage of these coastal forests. Mangroves protect shorelines from storm surges, filter pollutants, and capture eroded soils, protecting offshore coral reefs. They also provide materials that people use for food, medicine, tools, and construction. Half the world's mangrove forests have been destroyed as people have developed coastal areas, often for tourist resorts and shrimp farms (p. 151).

**FIGURE 12.12 Salt marshes occur in temperate intertidal zones where the substrate is muddy.** Tidal waters flow in channels called *tidal creeks* amid flat areas called *benches*, sometimes partially submerging the salt-adapted grasses.

**FIGURE 12.13 Mangrove forests line tropical and subtropical coastlines.** Mangrove trees, with their unique roots, are adapted for growing in saltwater and provide habitat for many fish, birds, crabs, and other animals.

## Intertidal zones undergo constant change

Where the ocean meets the land, **intertidal**, or **littoral**, ecosystems (**FIGURE 12.14**) spread between the uppermost reach of the high tide and the lowest limit of the low tide. **Tides** are the periodic rising and falling of the ocean's height at a given location, caused by the gravitational pull of the moon and sun. Intertidal organisms spend part of each day submerged in water, part of the day exposed to air and sun, and part of the day being lashed by waves.

Life abounds in the crevices of rocky shorelines, which provide shelter and pools of water (tide pools) during low tides. Sessile (stationary) animals such as anemones, mussels, and barnacles live attached to rocks, filter-feeding on plankton in the water that washes over them. Urchins, sea slugs, chitons, and limpets eat intertidal algae or scrape food from the rocks. Sea stars (starfish) prey upon the filter-feeders and

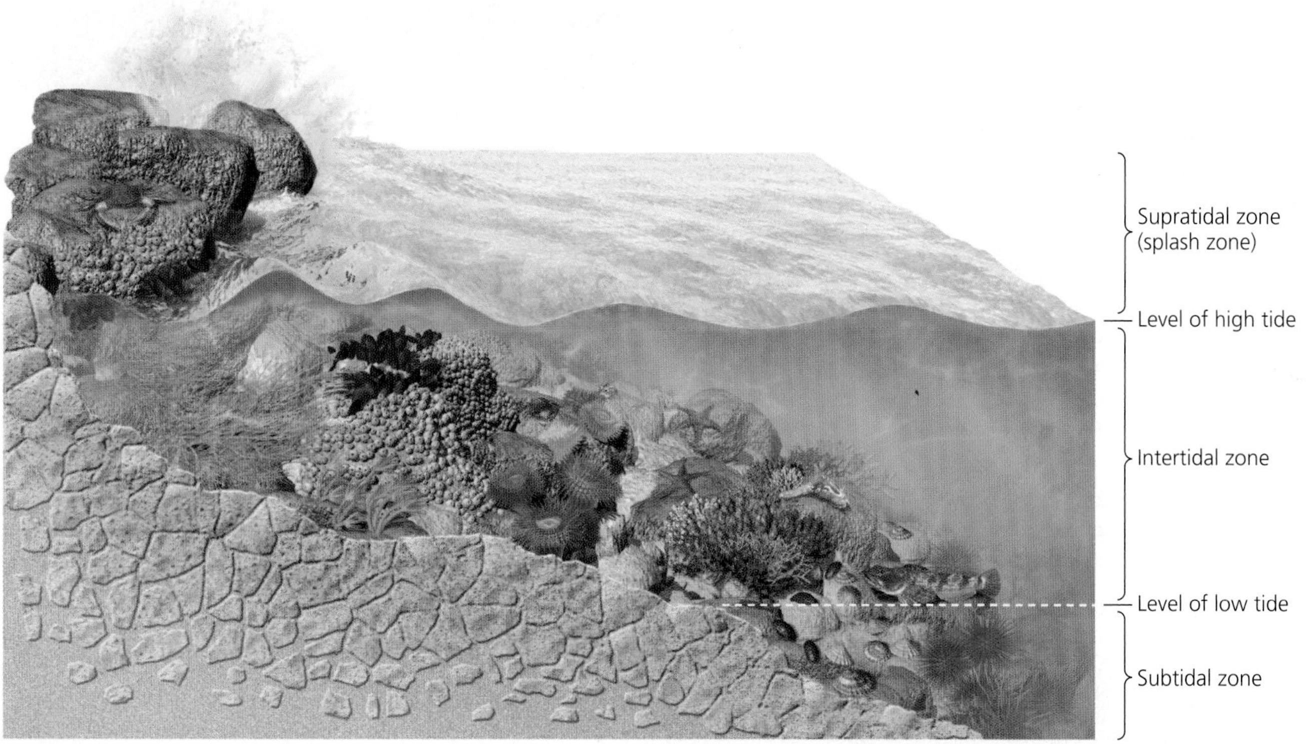


**FIGURE 12.14 The rocky intertidal zone stretches along rocky shorelines between the lowest and highest reaches of the tides.** The intertidal zone provides niches for a diversity of organisms, including sea stars (starfish), crabs, sea anemones, corals, chitons, mussels, nudibranchs (sea slugs), and sea urchins. Areas higher on the shoreline are exposed to the air more frequently and for longer periods, so organisms that tolerate exposure best specialize in the upper intertidal zone. The lower intertidal zone is exposed less frequently and for shorter periods, so organisms less tolerant of exposure thrive in this zone.

herbivores, while crabs scavenge detritus. The rocky intertidal zone is so diverse because environmental conditions such as temperature, salinity, and moisture change dramatically from the high to the low reaches.

## Kelp forests harbor many organisms

Along many temperate coasts, large brown algae, or **kelp**, grow from the floor of continental shelves, reaching up toward the sunlit surface. Some kelp reaches 60 m (200 ft) in height and can grow 45 cm (18 in.) per day. Dense stands of kelp form underwater "forests" (**FIGURE 12.15**). Kelp forests provide shelter and food for invertebrates and fish, which in turn provide food for larger predators. Kelp forests absorb wave energy and protect shorelines from erosion. People eat some types of kelp, and kelp provides compounds that serve as thickeners in cosmetics, paints, ice cream, and other consumer products.

## Coral reefs are treasure troves of biodiversity

Shallow subtropical and tropical waters are home to coral reefs. A coral reef is a mass of calcium carbonate composed of the skeletons of tiny invertebrate animals known as *corals*. Corals are related to jellyfish and capture passing food with stinging tentacles. They also derive nourishment from symbiotic algae known as *zooxanthellae*, which inhabit their bodies and produce food through photosynthesis (and provide the diversity of vibrant colors in reefs). Most corals are colonial, and the colorful surface of a coral reef consists of millions of densely packed individuals. As corals die, their skeletons remain part of the reef while new corals grow atop them, increasing the reef's size.

Like kelp forests, coral reefs protect shorelines by absorbing wave energy. They also host tremendous biodiversity (**FIGURE 12.16a**). This is because coral reefs provide complex physical structure (and thus many habitats) in shallow nearshore waters, which are regions of high primary productivity. Besides the staggering diversity of anemones, sponges, tubeworms, and other sessile invertebrates, innumerable mollusks, flatworms, and urchins patrol reefs, while thousands of fish species find food and shelter in the reef structure. The beauty and biodiversity of coral reefs also make them economically valuable destinations for tourists.

(a) Coral reef community

(b) Bleached coral

**FIGURE 12.16 Coral reefs provide food and shelter for a tremendous diversity (a) of fish and other creatures.** Today these reefs face multiple stresses from human impacts. Many corals have died as a result of coral bleaching **(b)**, in which corals lose their zooxanthellae. Bleaching is evident in the whitened portion of this coral.

**FIGURE 12.15 "Forests" of tall brown algae known as kelp grow from the floor of the continental shelf.** Numerous fish and other creatures eat kelp or find refuge among its fronds.

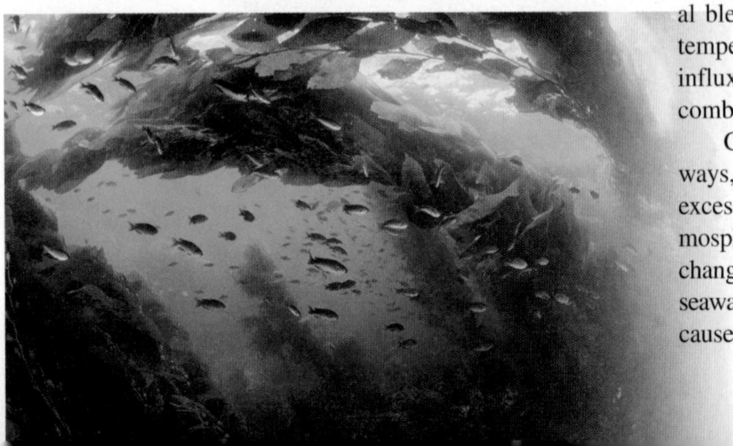

Coral reefs are experiencing alarming declines worldwide, however. An estimated three-fourths of coral reefs are currently being impacted by human activities. For example, nutrient pollution in coastal waters promotes the growth of algae, which smother reefs. Many reefs have undergone "coral bleaching," a process that occurs when zooxanthellae die or leave the coral, depriving it of nutrition (**FIGURE 12.16b**). Coral bleaching is thought to result from increased sea surface temperatures associated with global climate change, from the influx of pollutants, from unknown natural causes, or from combinations of these factors.

Coral reefs have been affected by climate change in other ways, too. The oceans have soaked up roughly a third of the excess carbon dioxide ($CO_2$) that people have added to the atmosphere so far. This has slowed the onset of global climate change; however, the excess $CO_2$ has lowered the pH (p. 28) of seawater, a phenomenon called **ocean acidification** (p. 318). This causes a series of chemical reactions that reduce the ocean's concentration of the carbonate ions that coral and other creatures need to build their shells. The increased acidity of ocean waters also dissolves the calcium carbonate in the shells of living coral. By dissolving shells at an increased rate and decreasing the rate at which new shells are formed, ocean acidification threatens the persistence of coral reefs. Ocean acidification increases along with atmospheric concentrations of $CO_2$. If atmospheric concentrations of $CO_2$ reach 500 parts per million (ppm) (up from 400 ppm today), very little ocean area will have carbonate ion concentrations sufficient to support coral reefs.

Acidification also threatens other aquatic creatures that form shells, such as oysters. Increased acidity of ocean water makes it more difficult for tiny, newly hatched oysters to form their first shell, which greatly elevates mortality rates in this sensitive life stage. This impacts populations of wild oysters and those in aquaculture operations (pp. 151–152), such as the oyster farms that generate $70 million a year in revenue for the Pacific Northwest in the United States (p. 318).

## Open-ocean ecosystems vary in their biodiversity

The uppermost 10 m (33 ft) of ocean water absorbs 80% of the solar energy that reaches its surface. For this reason, nearly all of the oceans' primary productivity occurs in the well-lit top layer, or *photic zone*. Generally, the warm, shallow waters of continental shelves are the most biologically productive and support the greatest species diversity. Habitats and ecosystems occurring between the ocean's surface and floor are classified as **pelagic**, whereas those that occur on the ocean floor are classified as **benthic**.

Biological diversity in pelagic regions of the open ocean is highly variable in its distribution. Primary production (p. 32) and animal life near the surface are concentrated in regions of nutrient-rich upwelling. Microscopic phytoplankton constitute the base of the marine food chain in the pelagic zone and are the prey of zooplankton (p. 24), which in turn become food for fish, jellyfish, whales, and other free-swimming animals (**FIGURE 12.17**). Predators at higher trophic levels include larger fish, sea turtles, and sharks.

**FIGURE 12.17 The uppermost reaches of ocean water contain billions upon billions of phytoplankton—tiny photosynthetic algae, protists, and bacteria that form the base of the marine food chain.** This part of the ocean is also home to zooplankton, small animals and protists that dine on phytoplankton and comprise the next trophic level.

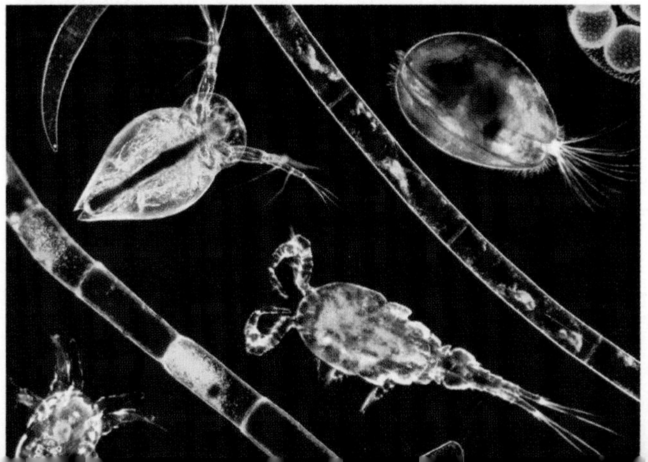

In the little-known deep-water ecosystems, animals have adapted to tolerate extreme water pressures and to live in the dark without food from autotrophs (p. 31). Many of these often bizarre-looking creatures scavenge carcasses or organic detritus that falls from above. Others are predators, and still others attain food from mutualistic (p. 70) bacteria. Ecosystems also form around hydrothermal vents, where heated water spurts from the seafloor, carrying minerals that precipitate to form rocky structures. Tubeworms, shrimp, and other creatures in these recently discovered deepwater systems use symbiotic bacteria to derive their energy from chemicals in the heated water rather than from sunlight.

## Aquatic systems are affected by human activities

Our tour of aquatic systems has shown the ecological and economic value of freshwater and marine ecosystems. We will now see how people affect these systems when we withdraw water for human use, build dams and levees, and introduce pollutants that alter water's chemical, biological, and physical properties.

# Effects of Human Activities on Waterways

Although water is a limited resource, it is also a renewable resource as long as we manage our use sustainably. However, people are withdrawing water at unsustainable levels and are depleting many sources of surface water and groundwater. Already, one-third of the world's people are affected by water shortages.

Additionally, people have intensively engineered freshwater waterways with dams, levees, and diversion canals to satisfy demands for water supplies, transportation, and flood control. As we have seen in our Central Case Study, dams and channelization in the Mississippi River basin have led to adverse impacts at the river's mouth. What we do in one part of the interconnected aquatic system affects many other parts, sometimes in significant ways.

## Fresh water and human populations are unevenly distributed across Earth

World regions differ in the size of their human populations and, as a result of climate and other factors, possess varying amounts of groundwater, surface water, and precipitation. Hence, not every human around the world has equal access to fresh water (**FIGURE 12.18**). Because of the mismatched distribution of water and population, human societies have frequently struggled to transport fresh water from its source to where people need it.

Fresh water is distributed unevenly in time as well as space. For example, India's monsoon storms can dump half of a region's annual rain in just a few hours. Rivers have seasonal differences in flow because of the timing of rains and snowmelt. For this reason, people build dams to store water

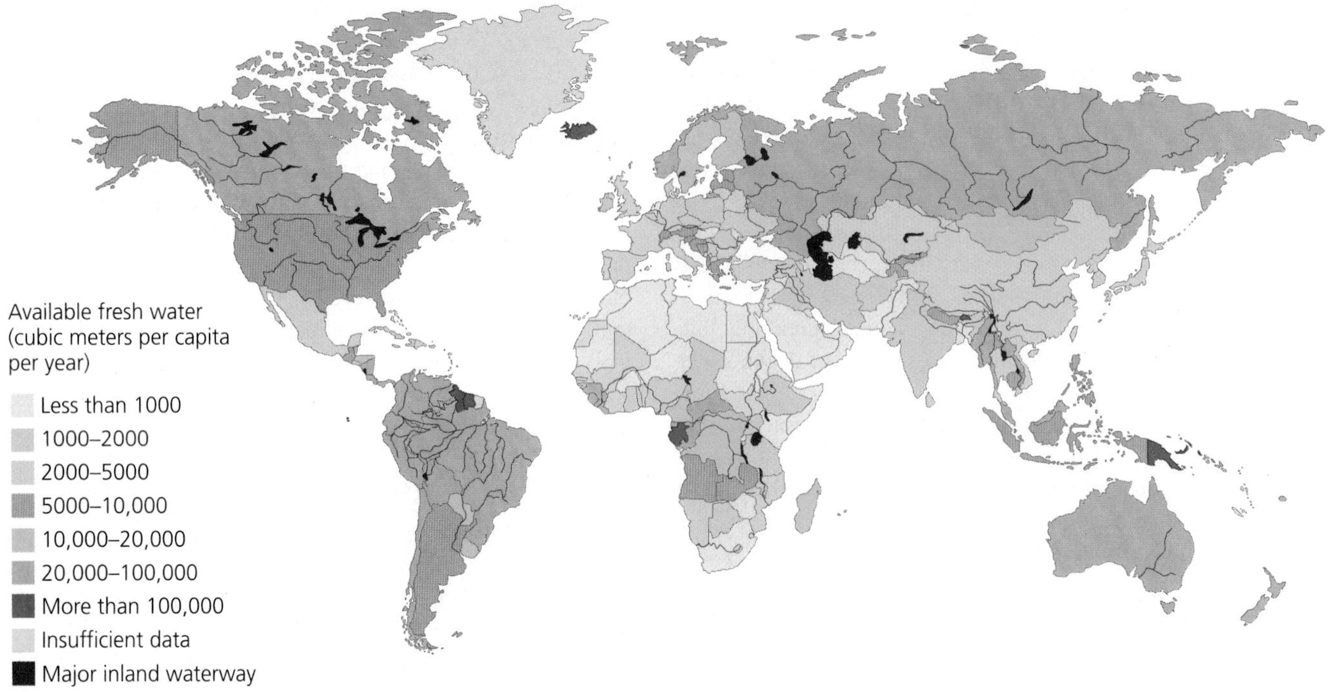


**FIGURE 12.18 Nations vary tremendously in the amount of fresh water per capita available to their citizens.** For example, Iceland, Papua New Guinea, Gabon, and Guyana (dark blue in this map) each have over 100 times more water per person than do many Middle Eastern and North African countries. *Data from Harrison, P., and F. Pearce, 2000.* AAAS atlas of population and the environment, *edited by the American Association for the Advancement of Science, © 2000 by the American Association for the Advancement of Science.*

DATA Q Compare the water availability per person in the developing regions of Africa, Asia, and Latin America and the Caribbean. Which region has the most water per person?

GO TO **INTERPRETING GRAPHS & DATA** ON MasteringEnvironmentalScience®

from wetter months that can be used in drier times of the year, when river flow is reduced.

As if the existing mismatches between water availability and human need were not enough, global climate change (Chapter 14) may worsen conditions in many regions by altering precipitation patterns, melting glaciers, causing early season runoff, and intensifying droughts and flooding.

## Water supplies households, industry, and especially agriculture

We all use water at home for drinking, cooking, and cleaning. Most mining, industrial, and manufacturing processes require water. Farmers and ranchers use water to irrigate crops and water livestock. Globally, we allot about 70% of our annual fresh water use to agriculture. Industry accounts for roughly 20%, and residential and municipal uses for only 10%.

The removal of water from an aquifer or from a body of surface water without returning it is called **consumptive use**. Our primary consumptive use of water is for agricultural irrigation (pp. 145–146). **Nonconsumptive use** of water, in contrast, does not remove, or only temporarily removes, water from an aquifer or surface water body. Using water to generate electricity at hydroelectric dams is an example of nonconsumptive use; water is taken in, passed through dam machinery to turn turbines, and released downstream.

We use so much water in agriculture because our rapid population growth requires us to feed and clothe more people each year. Overall, we withdraw 70% more water for irrigation today than we did 50 years ago and have doubled the amount of land under irrigation. Irrigation can more than double crop yields; as a result, the one-fifth of world farmland that we irrigate yields fully 40% of our produce, including 60% of the global grain crop.

## Excessive water withdrawals can drain rivers and lakes

In many places, we are withdrawing surface water at unsustainable rates and drastically reducing river flows. Because of excessive water withdrawals, many major rivers—the Colorado River in North America, the Yellow River in China, and the Nile in Africa—regularly run dry before reaching the sea. This reduction in flow not only threatens the future of the cities and farms that depend on these rivers, but also drastically alters the ecology of the rivers and their deltas, changing plant communities, wiping out populations of fish and invertebrates, and devastating fisheries.

Nowhere are the effects of surface water depletion so evident as at the Aral Sea, on the border of present-day Uzbekistan and Kazakhstan (**FIGURE 12.19a**). Once the fourth-largest lake on Earth, the Aral Sea lost over four-fifths of its volume

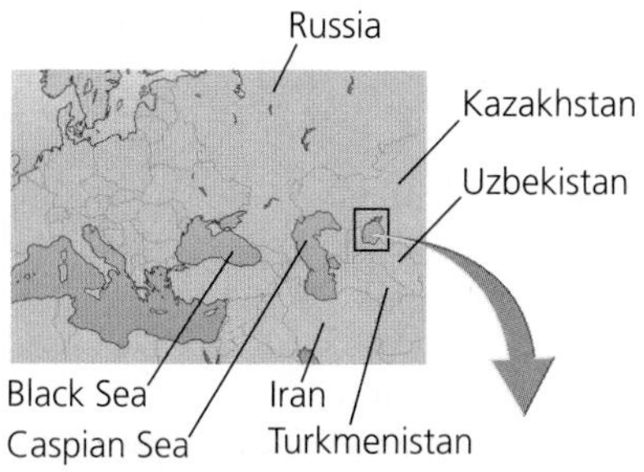


**FIGURE 12.19** **The Aral Sea in central Asia was once the world's fourth largest lake.** However, it has been shrinking **(a, b)** because so much water was withdrawn to irrigate cotton crops. Ships were stranded **(c)** along the former shoreline of the Aral Sea because the waters receded so far and so quickly. Today restoration efforts are beginning to reverse the decline in the northern portion of the sea and waters there are slowly rising.

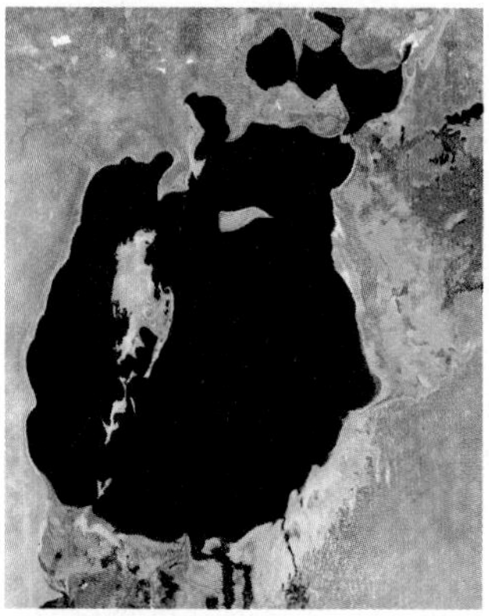

**(a) Satellite image of Aral Sea, 1987**

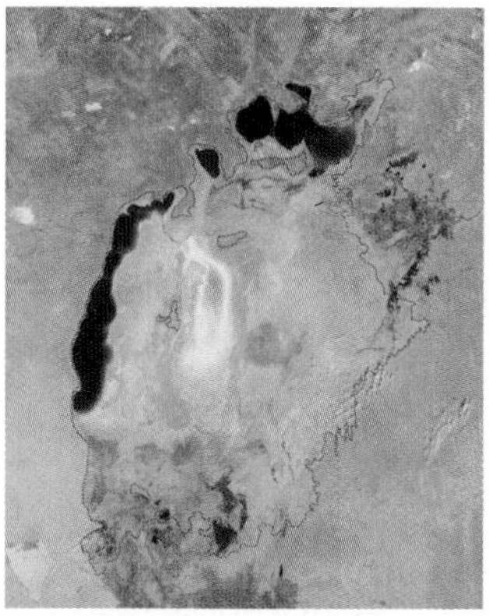

**(b) Satellite image of Aral Sea, 2009**

**(c) A ship stranded by the Aral Sea's fast-receding waters**

in just 45 years (**FIGURE 12.19b**) as water from the two rivers that feed it was withdrawn to provide irrigation for cotton farming in this arid region. The shrinking of the Aral Sea has caused the loss of 60,000 fishing jobs (**FIGURE 12.19c**). Scientists, engineers, and local people are struggling to save the northern portion of the Aral Sea, however, and their efforts may now have begun to reverse its decline.

Worldwide, roughly 15–35% of water withdrawals for irrigation are thought to be unsustainable (**FIGURE 12.20**). In areas where agriculture is demanding more fresh water than can be sustainably supplied, *water mining*—withdrawing water faster than it can be replenished—is taking place. In these areas, aquifers are being depleted or surface water is being piped in from other regions to meet the demand for irrigation.

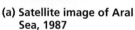

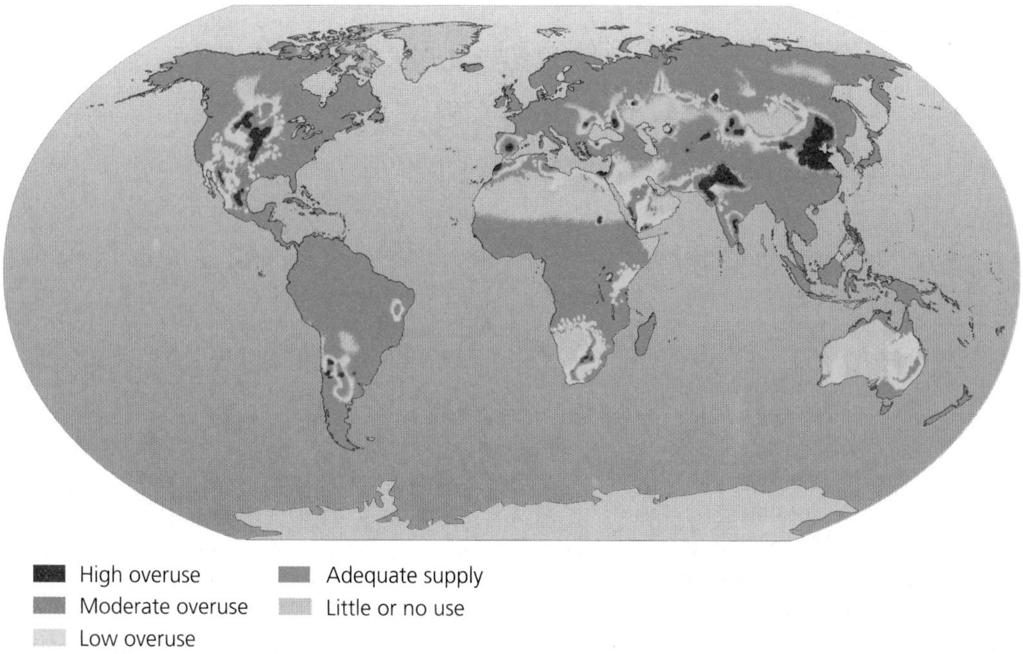


**FIGURE 12.20** **Irrigation for agriculture is the main contributor to unsustainable water use.** Mapped are regions where overall use of fresh water (for agriculture, industry, and domestic use) exceeds the available supply, requiring groundwater depletion or diversion of water from other regions. The map understates the problem, because it does not reflect seasonal shortages. *Data from UNESCO, 2006.* Water: A shared responsibility. *World Water Development Report 2. UNESCO and Berghahn Books.*

## Groundwater can also be depleted

Groundwater is more easily depleted than surface water because most aquifers recharge very slowly. If we compare an aquifer to a bank account, we are making more withdrawals than deposits, and the balance is shrinking. Today we are mining groundwater, extracting 160 $km^3$ (5.65 trillion $ft^3$) more water each year than returns to the ground. This is a problem because one-third of Earth's human population—including 99% of the rural population of the United States—relies on groundwater for its needs.

As aquifers are mined, water tables drop deeper underground. This deprives freshwater wetlands of groundwater inputs, causing them to dry up. Groundwater also becomes more difficult and expensive to extract, and eventually it may run out. In parts of Mexico, India, China, and multiple Asian and Middle Eastern nations, water tables are falling 1–3 m (3–10 ft) per year. In the United States, overpumping has drawn the Ogallala Aquifer down by more than 320 million liters (85 trillion gal), a volume equal to half the annual flow of the Mississippi River (**FIGURE 12.21**.)

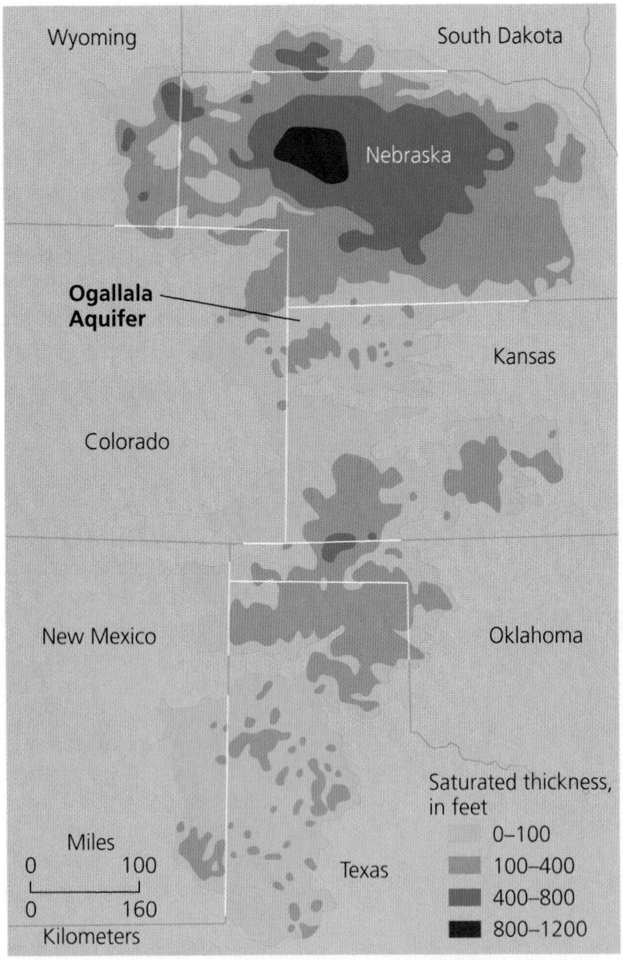


**FIGURE 12.21 The Ogallala Aquifer is the world's largest aquifer, and it held 3700 $km^3$ (881 $mi^3$) of water before pumping began.** This aquifer underlies 453,000 $km^2$ (175,000 $mi^2$) of the Great Plains beneath eight U.S. states. Overpumping for irrigation is currently reducing the volume and extent of this aquifer.

As aquifers lose water, they become less able to support overlying strata, and the land surface above may sink or collapse, creating **sinkholes**. Venice, Bangkok, Beijing, and Mexico City are slowly sinking; overpumping is causing the streets to buckle and undergound pipes to rupture. Once the ground sinks, soil and rock become compacted, losing the porosity that enabled them to hold water. Recharging a depleted aquifer thereafter becomes more difficult.

When groundwater is overpumped in coastal areas, saltwater can intrude into aquifers from the ocean, making water undrinkable. This has occurred in Florida, the Middle East, and other locations. In some cases, saline waters can penetrate many miles inland from the coast.

## Bottled water has ecological costs

These days, our groundwater is being withdrawn for a new purpose—to be packaged in plastic bottles and sold on supermarket shelves. The average American drinks over 30 gallons of bottled water a year, and sales top $15 billion annually in the United States and $65 billion worldwide.

Bottled water exerts substantial ecological impact. The energy costs of bottled water are estimated to be 1000–2000 times greater than the energy costs of tap water, mostly as a result of transportation costs. After their contents are consumed, at least three out of four bottles in the United States are thrown away, and not recycled. That's 30–40 billion containers per year (5 containers for every human being on Earth) and close to 1.5 million tons of plastic waste.

Many people who buy bottled water do so because they believe it is superior to tap water. However, in blind taste tests people think tap water tastes just as good, and chemical analyses show that bottled water is no safer or healthier than tap water.

Concerns about the environmental impacts of bottled water have led more than 90 colleges and universities in the United States and Canada to ban the sale of bottled water or restrict the use of plastic water bottles on campus. Similarly, New York City, San Francisco, Seattle, and other major U.S. cities prohibit using government funds to purchase bottled water. One reason is that drinking tap water instead of bottled water offers considerable cost savings.

## People build dikes and levees to control floods

**Flooding** is a normal, natural process that occurs when snowmelt or heavy rain swells the volume of water in a river so that water spills over the river's banks. This process naturally spreads nutrient-rich sediments over the floodplain, benefiting natural ecosystems and human agriculture.

As cities and towns grew on the floodplains of the Mississippi River and other waterways, however, people tired of the property damage caused by flooding. Communities and governments constructed *dikes* and *levees* (long raised mounds of earth) along riverbanks to hold water in main channels. These structures prevent flooding most of the time, but they can sometimes worsen flooding because they force water to stay in channels and accumulate, building up enormous energy and leading to occasional catastrophic flooding. Many of the levees along the Mississippi River were constructed after 1927, a year in which the river flooded catastrophically, causing the public to demand greater protection from flooding. Today, some of those same levees hold water in the river's channel and prevent its sediments from nourishing the coastal wetlands.

## We have erected thousands of dams

A **dam** is any obstruction placed in a river or stream to block its flow. Dams create **reservoirs**, artificial lakes that store water for human use. We build dams to prevent floods, provide drinking water, facilitate irrigation, and generate electricity.

Worldwide, we have erected more than 45,000 large dams (greater than 15 m, or 49 ft, high) across rivers in over 140 nations. We have built tens of thousands of smaller dams. Only a few major rivers in the world remain free-flowing. These undammed rivers run through the tundra and taiga of the Arctic and in remote areas of Latin America and Africa.

Dams produce a mix of benefits and costs, as illustrated in **FIGURE 12.22** and as mentioned in our Central Case Study. As an example of this complex mix, consider the world's largest dam: the Three Gorges Dam on China's Yangtze River. This dam was completed in 2008, and its reservoir stretches for 616 km (385 mi). This dam will help control the large floods that have historically occurred on the banks of the Yangtze. It will also facilitate shipping and will generate enough hydroelectric power to replace dozens of large coal or nuclear plants.

However, the Three Gorges Dam cost $39 billion to build and its reservoir displaced 1.24 million people and submerged productive farmlands, wildlife habitat, and 10,000-year-old archaeological sites. Tidal marshes at the Yangtze's mouth are eroding away, deprived of the sediments that now settle in the dam's reservoir as the river water slows. Many scientists also worry that the Yangtze's many pollutants will be trapped in the reservoir, making the water undrinkable.

People who feel that the costs of some dams outweigh their benefits are pushing for such dams to be dismantled. By removing dams and letting rivers flow free, they say, we can

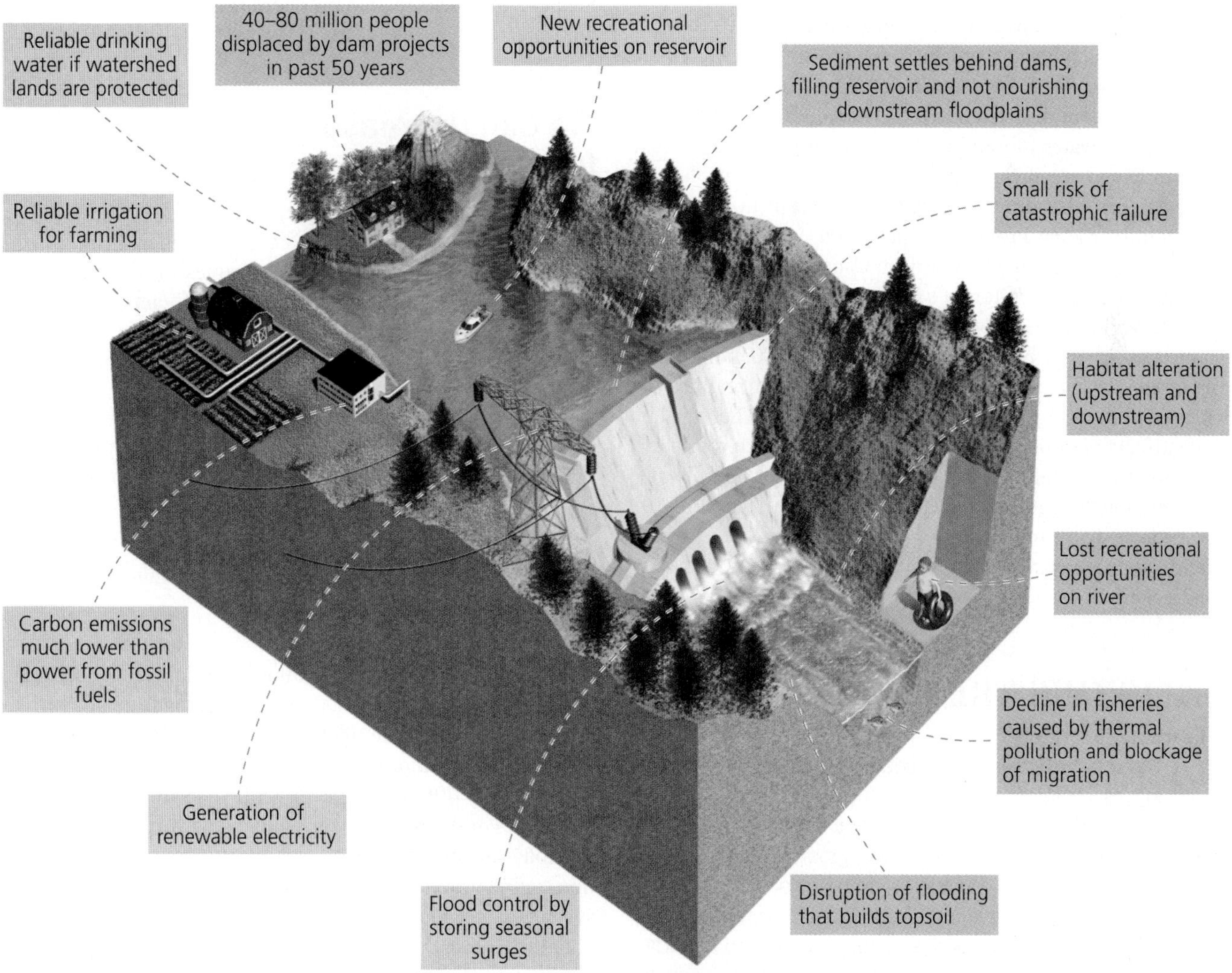


**FIGURE 12.22 Damming rivers has diverse consequences for people and the environment.** The generation of clean and renewable electricity is one of several major benefits **(green boxes)** of hydroelectric dams. Habitat alteration is one of several negative impacts **(red boxes)**.

restore ecosystems, reestablish economically valuable fisheries, and revive river recreation, such as fly-fishing and rafting. Many aging dams are in need of costly repairs or have outlived their economic usefulness, and these are candidates for removal. Some 400 dams have been removed in the United States in the past decade, and more will come down in the next 10 years, when the licenses of over 500 dams come up for renewal.

The results of dam removal can be stunning. For example, the 7.3-m (24-ft) high, 279-m (917-ft) Edwards Dam on Maine's Kennebec River was removed in 1999. Within a year, large numbers of 10 species of migratory fish ventured upstream and began using the 27-km (17-mi) stretch of river above the dam site.

### We divert surface water to suit our needs

People have long diverted water from rivers and lakes to farms, homes, and cities. Water in the Colorado River in the western United States is heavily diverted and utilized as the river flows toward the Pacific Ocean. Early in its course, some Colorado River water is piped through a mountain tunnel and down the Rockies' eastern slope to supply the city of Denver. More is removed for Las Vegas and other cities and for farmland as the water proceeds downriver. When the river reaches Parker Dam on the California–Arizona state line, large amounts are diverted into the Colorado River Aqueduct, which brings water to millions of people in the Los Angeles and San Diego areas via a long, open-air canal. From Parker Dam, Arizona also draws water, transporting it in canals of the Central Arizona Project. Further south, water is diverted into the Coachella and All-American Canals, destined for agriculture, mostly in California's Imperial Valley.

The world's largest diversion project is underway in China. Three sets of massive aqueducts, totaling 2500 km (1550 mi) in length, are being built to move trillions of gallons of water from the Yangtze River in southern China (where water is plentiful) to northern China's Yellow River, which routinely dries up at its mouth because the climate is drier and people withdraw much of its water. Many scientists say the $62 billion project won't transfer enough to make a difference, yet would cause extensive environmental impacts and displace hundreds of thousands of people.

#### WEIGHING THE ISSUES

**Reaching for Water**

The rapidly growing Las Vegas metropolitan area is exceeding its allotment of water from the Colorado River. To meet its growing demand, Nevada's largest city has proposed a $3.5 billion project that would divert groundwater from 450 km (280 mi) away. Do you think such diversions are ethically justified? If this project were to destroy rural communities and wetland ecosystems at the diversion site in eastern Nevada, would this be an acceptable cost given the economic activity generated in Las Vegas? How else might cities like Las Vegas meet their future water needs?

## Solutions to Depletion of Fresh Water

To address the depletion of fresh water, we can aim either to increase supply or to reduce demand. Increasing water supplies by constructing large dams was a common solution to water shortages in the past. However, we have already developed the vast majority of suitable sites for large dams, and most of those that remain are in remote locations. Building more dams therefore does not appear to be a viable solution in most locations for meeting people's demands for fresh water.

Another supply strategy is to generate fresh water by **desalination**, or *desalinization*, the removal of salt from seawater. Desalination can be accomplished by heating saltwater and condensing the water vapor that evaporates from it—essentially distilling fresh water. Over 20,000 desalination facilities are operating worldwide. However, desalination is expensive, requires large inputs of fossil fuel energy, kills aquatic life at water intakes, and generates concentrated salty waste. As a result, large-scale desalination is pursued mostly in wealthy oil-rich nations where water is extremely scarce. In Saudi Arabia, desalination produces half of the nation's drinking water.

### We can decrease our demand for water

Because supply-based strategies do not hold great promise for increasing water supplies, people are increasingly turning to demand-based solutions. Strategies for reducing fresh water demand include conservation and efficiency measures. Such strategies require changes in individual behaviors and can therefore be politically difficult, but they offer better economic returns and cause less ecological and social damage. Our existing shift from supply-based to demand-based solutions is already paying dividends. The United States, for example, decreased its water consumption by 5% from 1980 to 2005 thanks to conservation measures, even while its population grew 31%. Let's examine approaches that can conserve water in agriculture, households, industry, and municipalities.

**Agriculture** Farmers can improve efficiency by adopting more efficient irrigation methods. "Flood and furrow" irrigation, in which fields are liberally flooded with water, accounts for 90% of irrigation worldwide. However, crop plants end up using only 40% of the water applied, as much of the water evaporates or seeps into the ground far away from crops. Other methods are far more efficient. Low-pressure spray irrigation squirts water downward toward plants, and drip irrigation systems target individual plants and introduce water directly onto the soil (p. 146). Experts estimate that drip irrigation (in which as little as 10% of water is wasted) could cut water use in half while raising crop yields. This technology could produce as much as $3 billion in extra annual income for farmers of the developing world. Researchers are currently experimenting with various materials and approaches to develop reliable, inexpensive drip irrigation systems that could convey these benefits to poorer farmers.

Another way to reduce agricultural water use would be to eliminate government subsidies for the irrigation of crops that require a great deal of water (such as cotton and rice) in arid areas. Biotechnology may play a role by producing crop varieties that require less water through selective breeding (p. 50) and genetic modification (pp. 151–153).

**Households** In our residences, we can reduce water use by installing low-flow faucets, showerheads, washing machines, and toilets. High-efficiency toilets and showerheads provide the biggest savings because these are typically the two largest indoor uses of water. We can reduce outdoor water use by catching and storing rain runoff in a barrel. Replacing exotic vegetation with native plants adapted to the region's natural precipitation patterns can also reduce water demand. For example, more and more residents in the U.S. Southwest are practicing *xeriscaping*, a type of landscaping that uses plants adapted to arid conditions.

**Industry and municipalities** Manufacturers are shifting to processes that use less water and in doing so are reducing their costs along with their environmental impacts. Las Vegas is one of many cities that are recycling treated municipal wastewater for irrigation and industrial uses. Finding and patching leaks in pipes has saved some cities and companies large amounts of water—and money. Boston and its suburbs reduced water demand by 43% over 30 years by patching leaks, retrofitting homes with efficient plumbing, auditing industry, and promoting conservation. This program enabled Massachusetts to avoid an unpopular $500 million river diversion scheme.

## Nations often cooperate to resolve water disputes

Population growth, expansion of irrigated agriculture, and industrial development doubled our annual fresh water use between 1960 and 2000. Increased withdrawals of fresh water can lead to shortages, and resource scarcity can lead to conflict. Many security analysts predict that water's role in regional conflicts will increase as human population continues to grow and as climate change alters precipitation patterns. A total of 261 major rivers (whose watersheds cover 45% of the world's land area) cross national borders, and transboundary disagreements are common (**FIGURE 12.23**). Access to water is already a key element in the disagreements among Israel, the Palestinian people, and neighboring nations in the arid Middle East.

Yet on the positive side, many nations have cooperated to resolve water disputes. India has struck agreements to co-manage transboundary rivers with Pakistan, Bangladesh, Bhutan, and Nepal. In Europe, nations along the Rhine and Danube rivers have signed water-sharing treaties. Such progress gives reason to hope that "water wars" will be few and far between in coming decades.

# Water Pollution and Its Control

We have seen that people affect aquatic systems by withdrawing too much water and by erecting dams, diversions, and levees that alter the systems' natural processes. Introducing toxic substances and disease-causing organisms into surface waters and groundwater is yet another way that people adversely affect aquatic ecosystems—and threaten human health.

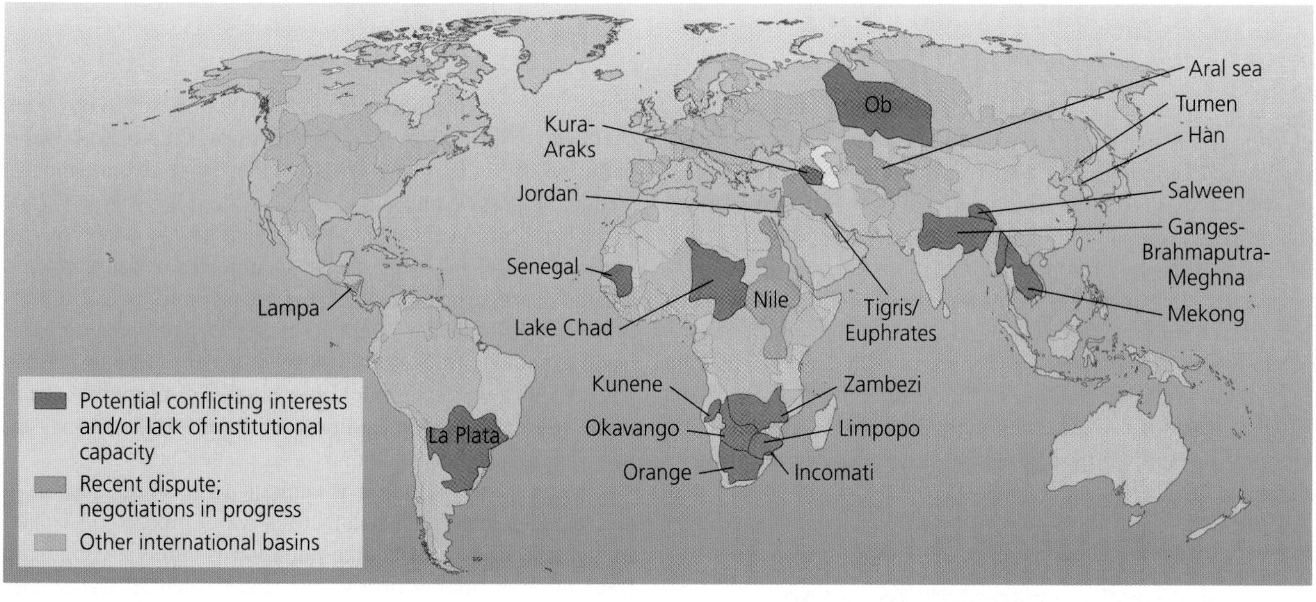


**FIGURE 12.23 Water basins that cross national boundaries (yellow) have the potential for conflict if water supplies become scarce.** Basins with higher potential for conflict **(red)** are found in regions with growing populations, but negotiations are underway on several international basins to prevent conflict **(orange)**.

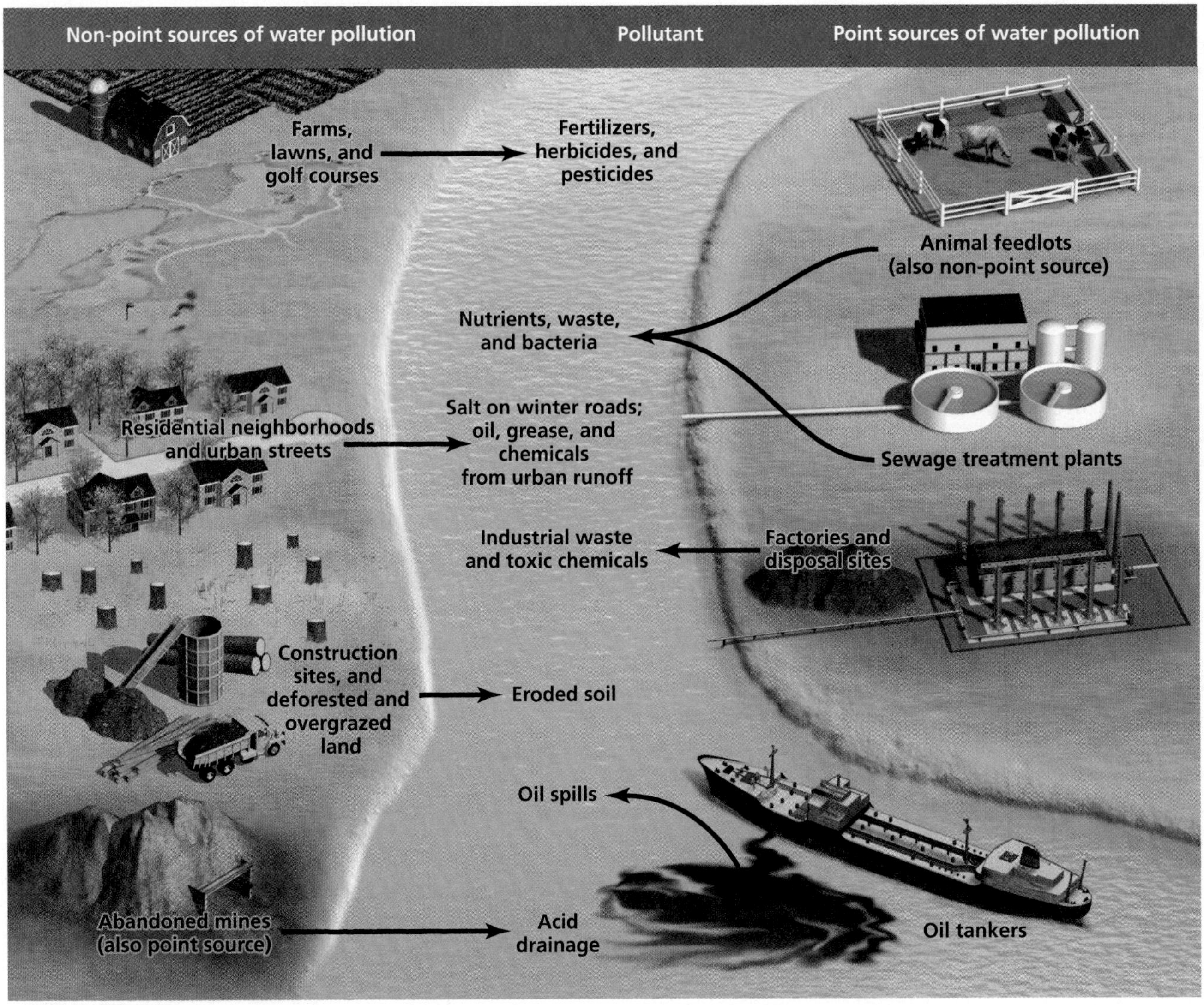


**FIGURE 12.24 Point-source pollution (on right) comes from discrete facilities or locations, usually from single outflow pipes.** Non-point-source pollution (such as runoff from streets, residential neighborhoods, lawns, and farms; **on left**) originates from numerous sources spread over large areas.

Developed nations have made admirable advances in cleaning up water pollution over the past few decades. Still, the World Commission on Water (an organization that focuses on water issues) recently concluded that over half the world's major rivers remain "seriously depleted and polluted." In 2013, the U.S. Environmental Protection Agency (EPA) reported that 55% of U.S. streams and rivers are in poor condition to support aquatic life. The largely invisible pollution of groundwater, meanwhile, has been termed a "covert crisis." Preventing pollution is easier and more effective than treating it later. Many of our current solutions to pollution therefore embrace preventive strategies rather than "end-of-pipe" treatment and cleanup.

## Water pollution comes from point sources and non-point sources

**Pollution** is the release into the environment of matter or energy that causes undesirable impacts on the health or well-being of people or other organisms. Pollution can be physical, chemical, or biological and can affect water, air, or soil. **Water pollution** comes in many forms and can cause diverse impacts on aquatic ecosystems and human health.

Most forms of water pollution are not conspicuous to the human eye, so scientists and technicians measure water's chemical properties (such as pH, plant nutrient concentrations, and dissolved oxygen levels), physical characteristics (such as temperature and turbidity—the density of suspended particles in a water sample), and biological properties (such as the presence of harmful microorganisms or the species diversity in aquatic ecosystems).

Some water pollution is emitted from **point sources**—discrete locations, such as a factory, sewer pipe, or oil tanker. In contrast, **non-point-source** pollution is cumulative, arising from multiple inputs over larger areas, such as farms, city streets, and residential neighborhoods (**FIGURE 12.24**). The U.S. Clean Water Act (p. 105) addressed point-source pollution with some success by targeting industrial discharges.

In the United States today, the greater threat to water quality comes from non-point-source pollution resulting from countless common activities, such as applying fertilizers and pesticides to lawns, applying salt to roads in winter, and changing automobile oil. To minimize non-point-source pollution of drinking water, governments limit development on watershed land surrounding reservoirs.

## Water pollution takes many forms

Water pollution comes in many forms that can impair waterways and threaten people and organisms that drink or live in affected waters. Let's survey the major classes of water pollutants affecting waters in the world today.

**Toxic chemicals** Our waterways and coastal ecosystems have become polluted with toxic organic substances of our own making, including pesticides, petroleum products, and other synthetic chemicals (pp. 210–212). Many of these can poison animals and plants, alter aquatic ecosystems, and cause an array of human health problems, including cancer. In addition, toxic metals such as arsenic, lead, and mercury damage human health and the environment, as do acids from acid precipitation (pp. 295–297) and from acid drainage from mining sites (p. 237). With its massive watershed encompassing rural, urban, and suburban areas, the Mississippi River is one waterway that carries toxic pollutants from sources in agriculture, industry, and homes and businesses.

Issuing and enforcing more stringent regulations of industry can help reduce releases of many toxic chemicals. We can also modify our industrial processes and our purchasing decisions to rely less on these substances.

**Pathogens and waterborne diseases** Disease-causing organisms (pathogenic viruses, protists, and bacteria) can enter drinking water supplies that become contaminated with human waste from inadequately treated sewage or animal waste from feedlots, chicken farms, or hog farms (pp. 150–151). Biological pollution by pathogens causes more human health problems than any other type of water pollution. In the United States, an estimated 20 million people fall ill each year from drinking water contaminated with pathogens. Worldwide, the United Nations estimates that 3800 children die every day from diseases associated with unsafe drinking water, such as cholera, dysentery, and typhoid fever. Although nearly 800 million people still lack reliable access to safe drinking water and one-third of all people do not have sanitation/sewer facilities, we are making progress worldwide in supplying these services.

We reduce the risks posed by waterborne pathogens by disinfecting drinking water (p. 272) and by treating wastewater (pp. 272–274). Other measures to lessen health risks include public education to encourage personal hygiene and government enforcement of regulations to ensure the cleanliness of food production, processing, and distribution.

**Nutrient pollution** The Chesapeake Bay's dead zone shows how nutrient pollution causes eutrophication and hypoxia (low dissolved oxygen concentrations) in surface waters (p. 22). When excess nitrogen and/or phosphorus enters a water body, it fertilizes algae and aquatic plants, boosting their growth. As algae die off, bacteria in sediments consume them. Because this decomposition requires oxygen, dissolved oxygen levels decline. These levels can drop too low to support fish and shellfish, leading to dramatic changes in aquatic ecosystems.

A "dead zone" of very low dissolved oxygen levels appears annually in the northern Gulf of Mexico, fueled by nutrients from Midwest farms carried by the Mississippi and Atchafalaya rivers (**FIGURE 12.25**). The low-oxygen conditions have adversely affected marine life and greatly reduced catches of shrimp and fish.

Excessive nutrient concentrations sometimes give rise to population explosions among several species of marine algae that produce powerful toxins. Blooms of these algae are known as **harmful algal blooms**. Some toxic algal species

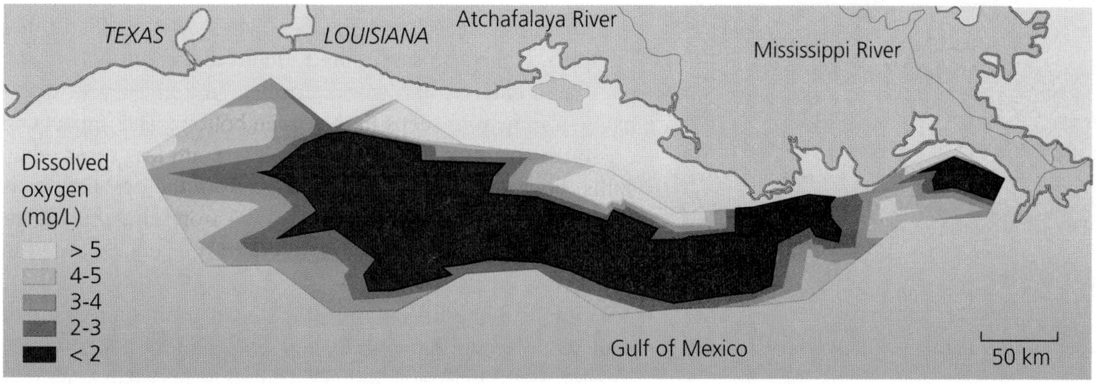


**FIGURE 12.25 Dissolved oxygen concentrations were mapped in the Gulf of Mexico off the Louisiana coast in 2013.** Areas in red indicate the lowest oxygen levels. Regions considered hypoxic (< 2 mg/L) are encircled with a black line. *Data from N. Rabalais, Louisiana Universities Marine Consortium (LUMCON), and R. E. Turner, Louisiana State University.*

produce a red pigment that discolors the water—hence the name **red tides**. Harmful algal blooms can cause illness and death in aquatic animals and people and adversely affect communities that rely on beach tourism and fishing.

Eutrophication is a natural process, but nutrient input from wastewater and fertilizer runoff from farms, golf courses, lawns, and sewage can dramatically increase the rate at which it occurs. We can reduce nutrient pollution by specially treating municipal wastewater to remove nutrients, reducing fertilizer applications, using phosphate-free detergents, and planting vegetation and protecting natural areas around streams and rivers to reduce nutrient inputs into waterways.

**Biodegradable wastes** Introducing large quantities of biodegradable materials into waters also decreases dissolved oxygen levels. When human wastes, animal manure, paper pulp from paper mills, or yard wastes (grass clippings and leaves) enter waterways, bacterial decomposition escalates as organic material is metabolized. As with elevated inputs of plant nutrients, biodegradable wastes lower dissolved oxygen levels in the water. **Wastewater** is water affected by human activities and is a source of biodegradable wastes. It includes water from toilets, showers, sinks, dishwashers, and washing machines; water used in manufacturing or industrial cleaning processes; and stormwater runoff. The widespread practice of treating wastewater to remove organic matter (pp. 272–274) has greatly reduced impacts from biodegradable wastes in rivers in developed nations. Oxygen depletion remains a major problem in some developing nations, however, where wastewater treatment is less common.

**Sediment** As we saw in the Central Case Study, eroded soils are carried to rivers by runoff and transported long distances by river currents (**FIGURE 12.26**). Clear-cutting, mining, clearing land for development, and cultivating farm fields all expose soil to wind and water erosion (pp. 141–142) and increase the amount of soil entering waterways. High sediment concentrations impair aquatic ecosystems by interfering with the respiration of fish and invertebrates and smothering benthic organisms. Sediment also clouds waters, blocking the sunlight needed by rooted aquatic plants.

We can reduce sediment pollution by adopting sustainable soil practices, avoiding large-scale disturbance of vegetation, and maintaining riparian vegetation to trap sediments at the water's edge.

**FIGURE 12.26 Sediments wash into the Pacific Ocean from a river in Panama.** Farming, construction, and other human activities can cause elevated levels of soil to enter waterways, affecting water quality and aquatic wildlife.

**Oil pollution** Oil pollution in freshwater and marine systems comes from spills of all sizes. Large spills occur infrequently, but their impacts can be staggering near the spill site. The danger that oil spills pose to fisheries, economies, and ecosystems became clear in 1989, when the oil tanker *Exxon Valdez* ran aground in Prince William Sound, spilling hundreds of thousands of barrels of oil and causing an ecological disaster along the Alaskan coast.

The dangers of oil pollution made headlines again in 2010, when British Petroleum's *Deepwater Horizon* offshore drilling platform exploded and sank into the Gulf of Mexico off the Louisiana coast (pp. 345–346). Oil gushed from the platform's underwater well, was spread widely by ocean currents, and washed up on coastal areas across the northern Gulf of Mexico. The economic and ecological impacts of the spill were visited on hundreds of miles of water, sediments, and shoreline along the coasts of Louisiana, Mississippi, Alabama, and Florida. Even organisms in the deep were affected by this spill. A 2012 study found that corals living 11 km (6.8 mi) from the oil well at a depth of 1370 m (4495 ft) were adversely affected by exposure to an underwater plume of oil emanating from the spill site.

The impacts could be long-lasting. A 2014 study found that when tuna embryos were exposed to concentrations of crude oil similar to those in Gulf waters following the spill, the fish demonstrated effects ranging from abnormal heartbeats to severely malformed hearts. Such problems could impair the fishes' ability to swim and feed, leading over time to population declines in this commercially important species.

Given the severity of the *Deepwater Horizon* and other oil spills, it may be surprising to learn that about half of the petroleum entering the world's oceans in a given year originates from natural seeps in the ocean bottom. The impacts on organisms from these petroleum releases are less severe than spills, however, because the seepage from the ocean floor is more diffuse. The spillage that results from an accident involving an oil tanker or oil rig, in contrast, is concentrated in a relatively small area.

Much of the oil entering the ocean from sources other than seeps accumulates in waters from innumerable, widely spread, small non-point sources. Shipping vessels and recreational boats can leak oil as they ply the waters of rivers, lakes, estuaries, and oceans. Motor oil from vehicles on roads and parking lots is washed into streams by rains and carried to

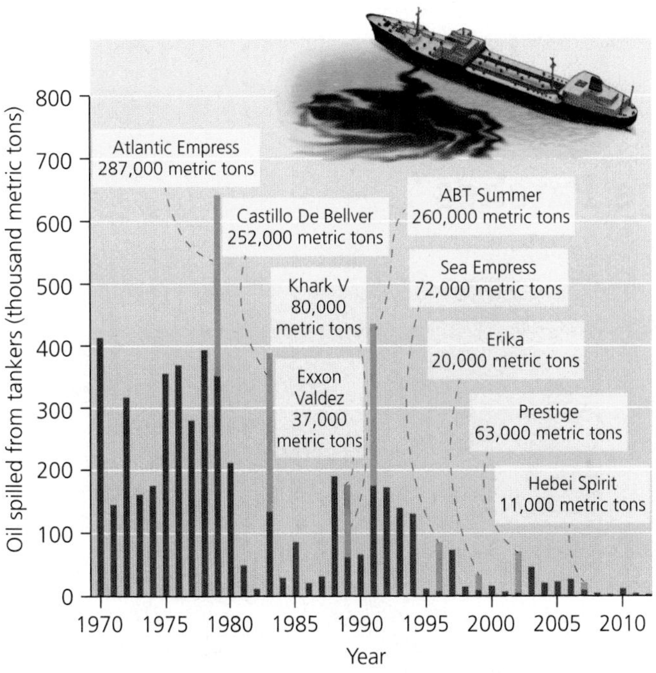


**FIGURE 12.27 Less oil is being spilled into ocean waters today in large tanker spills, thanks in part to regulations on the oil shipping industry and improved spill response techniques.** The bar chart shows cumulative quantities of oil spilled worldwide from nonmilitary spills over 7 metric tons, identifying larger spills by vessel name. *Data from International Tanker Owners Pollution Federation Ltd.*

rivers, coastal areas, and oceans. Spills from oil tankers account for 12% of oil pollution in an average year, and 3% comes from leakage that occurs during the extraction of oil by offshore oil rigs. Although the *Exxon Valdez* spill was catastrophic, the good news is that the amount of oil spilled from tankers worldwide has decreased over the past three decades, in part because of an increased emphasis on spill prevention and response (**FIGURE 12.27**).

To help deal with the problem of catastrophic oil spills, the U.S. Oil Pollution Act of 1990 created a $1 billion oil pollution prevention and cleanup fund. It also required that by 2015 all oil tankers in U.S. waters be equipped with double hulls as a precaution against puncture. Further, in the wake of the *Deepwater Horizon* spill, the U.S. government is considering tighter regulations on offshore drilling operations.

**Nets and plastic debris** Discarded fishing nets, fishing line, plastic bags and bottles, and other trash can harm aquatic organisms. Seabirds, fish, aquatic mammals, and sea turtles mistake floating plastic debris for food and can die as a result of ingesting material they cannot digest or expel. The chemicals that plastics contain also have toxic effects on organisms, and floating plastics can serve as "rafts" that facilitate the introduction of invasive species to new habitats.

In recent years, scientists have learned that plastic trash is accumulating in *gyres*, which are regions of the oceans where currents converge (see **THE SCIENCE BEHIND THE STORY**, pp. 270–271). One such area is the "Great Pacific Garbage Patch" in the northern Pacific, an area larger than Texas, in which tiny pieces of floating plastic outnumber organisms by a 6-to-1 margin.

Because plastics take 500–1000 years to degrade at sea and there is no viable way to collect the innumerable small bits of plastics that litter the oceans, preventing their entry into the oceans is key to remedying oceanic plastic pollution. Such efforts are aided by the Marine Debris Research, Prevention, and Reduction Act, passed by Congress in 2006, which utilizes a diversity of approaches to minimize oceanic plastic pollution.

**Thermal pollution** When we withdraw water from a river and use it to cool an industrial facility, we transfer heat from the facility back into the river where the water is returned. People also raise water temperatures by removing streamside vegetation that shades water from the sun. Elevated water temperatures can cause physiological stress in overheated plants and animals. Also, water's ability to hold dissolved oxygen decreases as temperature rises. Thus, when human activities raise water temperatures, some aquatic organisms experience hypoxic waters.

Too little heat can also cause problems. On the Mississippi River and its tributaries, as in many other dammed rivers, water at the bottoms of reservoirs is colder than water at the surface. When dam operators release water from the depths of a reservoir, downstream water temperatures drop suddenly and affect wildlife.

**FAQ** **Is the "Great Pacific Garbage Patch" a huge mass of floating debris?**

The vast majority of plastics that accumulate in the ocean are very small and are dispersed over relatively large areas, so they are difficult to detect with the naked eye. Hence, if you were expecting to see dense expanses of floating debris when traveling through the Great Pacific Garbage Patch and other hotspots of oceanic pollution, you would likely be disappointed. This might incorrectly lead you to believe the problem of oceanic debris is exaggerated.

Even though the plastic and other debris in these gyres is typically small (usually only several millimeters in length), it still poses threats to wildlife from ingestion and toxicity. In many ways, smaller pieces of plastic are more dangerous to marine life than larger pieces, because they resemble food items much more closely than do large items. So whereas a floating tire or fishing net certainly provides a striking visual impact, it's important to realize that the oceanic plastic pollution we cannot easily see is often of greatest concern.

THE SCIENCE BEHIND THE STORY

## Predicting the Oceans' "Garbage Patches"

In 1997, while sailing across a little-traveled portion of the Pacific Ocean as he returned from a recreational yacht race, Captain Charles Moore encountered a huge floating mass of debris he described as a "soupy" collection of items including tires, plastics, chemical drums, coat hangers, fishing nets, and other items. This was the first recorded visual confirmation of what is now called the "Great Pacific Garbage Patch" and brought firmly into the public eye the issue of plastic pollution in the oceans. Knowing where floating debris will accumulate in the oceans is important to marine biologists and oceanographers, but short of stumbling upon these areas in the ocean, as Captain Moore did, is there any way to know where to look for them? Thanks to the work of Nikolai Maximenko and his collaborators, the answer is yes.

Maximenko, a senior researcher of the International Pacific Research Center at the University of Hawaii at Manoa, studies the movements of currents in the oceans. In particular, his work examines *Ekman drift*, wind-driven currents in the water's upper layers first modeled by scientist Walfrid Ekman in the early 20th century. Although the large-scale ocean currents that are driven by pressure gradients and the Coriolis force (collectively called *geostrophic currents*) are fairly well described, the smaller-scale movements associated with Ekman currents are not. This lack of information hinders researchers' ability to predict movements of floating material in the oceans—such as plastics and other debris.

**Nikolai Maximenko, International Pacific Research Center at the University of Hawaii at Manoa**

Data gathered by the Global Drifter Program of the U.S. National Oceanic and Atmospheric Administration (NOAA) is helping scientists better understand these currents. Since 1979, the program has deployed more than 12,000 buoys in the world's oceans and tracked their movements to determine patterns in ocean currents. Each buoy (**FIGURE 1**) is composed of a 30- to 40-cm (12- to 16-in.) floating drifter, which contains sensors for detecting ocean temperature, salinity, and other properties, as well as a transmitter to send its information to satellites overhead. The drifter is tethered to a type of anchor, called a subsurface drogue, which hangs at a depth of 15 m (49 ft). The drogue keeps the drifter upright on the surface and provides an underwater "sail" for currents to push. Drifters typically operate for about 400 days before they stop transmitting, and the program aims to maintain an array of 1250 drifters at any given time in the world's oceans.

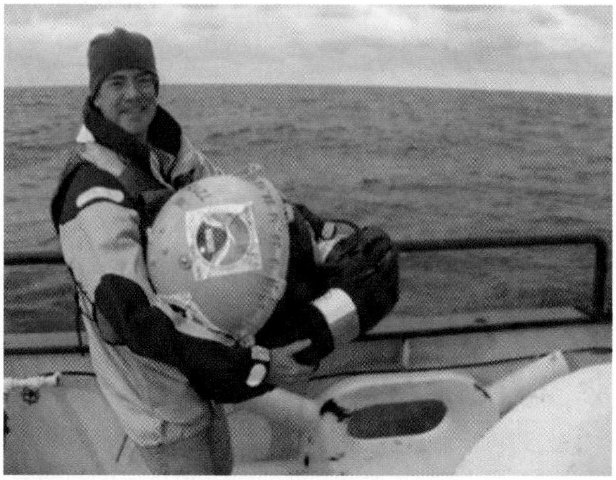

**FIGURE 1 A researcher prepares to deploy a buoy for the Global Drifter Program.** The cylindrical sea anchor (or drogue) extends vertically in the water column after deployment and is pushed along by shallow ocean currents.

In 2008, Maximenko partnered with Peter Niiler of the Scripps Institution of Oceanography to use data from the Global Drifter program to produce a more detailed map of surface ocean currents. The team combined data on drifter movements with satellite altimetry and wind currents, producing a highly detailed map of both geostrophic and Ekman currents. As the issue of oceanic plastic pollution gained attention, Maximenko

### Water pollutants can contaminate groundwater

Many pollutants that affect surface waters also affect groundwater. Groundwater pollution is difficult to detect. Moreover, pollutants may reside in groundwater longer than in surface waters. Because groundwater is not exposed to sunlight, contains fewer microbes and minerals, and holds less dissolved oxygen and organic matter than surface waters, pollutants decompose more slowly. For example, concentrations of the herbicide alachlor decline by half after 20 days in soil, but in groundwater this takes almost four years.

Some chemicals that are toxic at high concentrations, including aluminum, fluoride, nitrates, and sulfates, occur

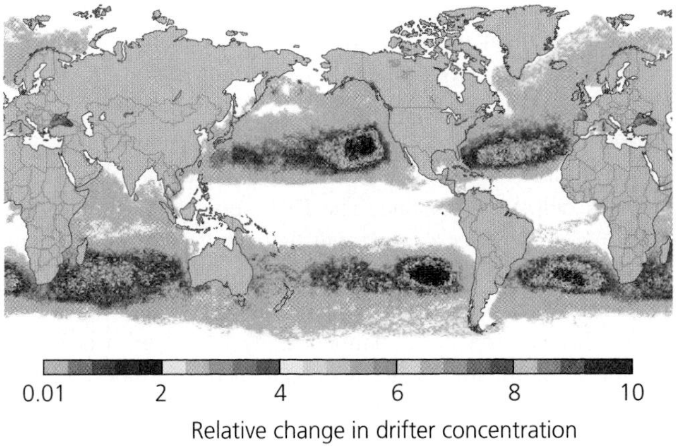


**FIGURE 2 Maximenko's computer simulation predicts where plastics and trash accumulate based on oceanic currents.** These currents include the North and South Atlantic Gyres, Indian Ocean Gyre, and the North and South Pacific Gyres. The North Pacific Gyre off the coast of California is home to the "Great Pacific Garbage Patch." Source: IPRC Climate, 2008.

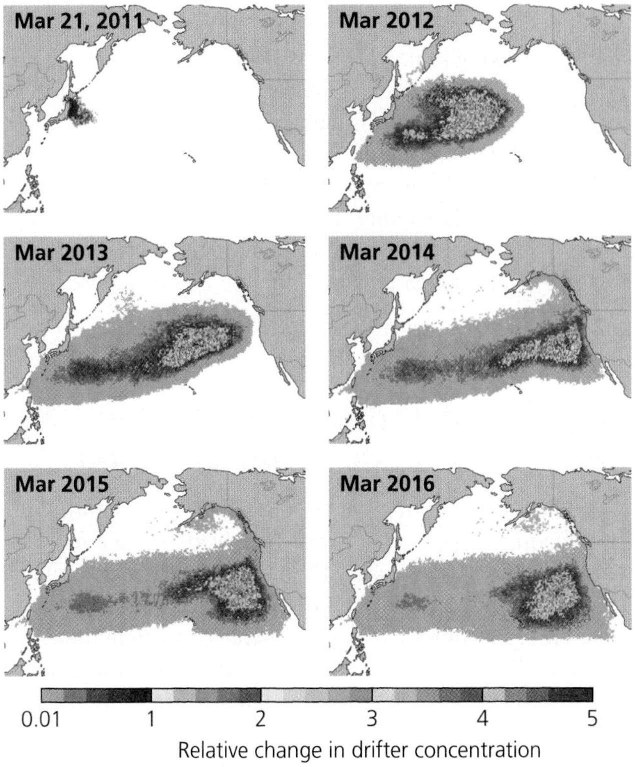


**FIGURE 3 Computer models predict the fate of heavier debris from the 2011 tsunami in northern Japan.** The bulk of the debris is expected to reach the shores of the western United States in 2014 and to remain off the coast of California in the North Pacific Gyre. *Source: Nikolai Maximenko, International Pacific Research Center.*

saw an opportunity to utilize his map to predict the areas of the ocean where floating debris is likely to accumulate. To do this, he partnered with his colleague, Jan Hafner, to create a computer model that predicts the movements of drifters over long time periods in the world's oceans. He then ran a simulation in which he uniformly distributed drifters in the oceans and saw where they concentrated over time.

The simulation's results, published in 2012 in *Marine Pollution Bulletin*, revealed that oceanic debris was likely to accumulate in portions of five subtropical gyres (**FIGURE 2**), including the area of the Great Pacific Garbage Patch. The simulation also revealed that 70% of the drifters remained at sea after 10 years, showing the lengthy life span of floating debris in the oceans. One of the model's predictions, which were first released in 2008, was verified when a 2010 study found high concentrations of plastic in the North Atlantic Gyre where Maximenko's model predicted it. The remaining "garbage patches" have all since been verified at their predicted locations.

Maximenko's model was also put to use to predict the fate of the debris washed into the Pacific by the tsunami that struck Japan in 2011 (p. 233). The original debris model (**FIGURE 3**) predicted how ocean currents would push debris across the Pacific. In 2012, the team updated their model to include the effects of winds pushing floating debris at faster rates than heavier debris. The model's predictions closely match observations of debris movements in the Pacific; the model is also aiding efforts to prepare for the arrival of debris, some of which harbors marine life that could act as invasive species (pp. 67–68) in the United States.

Thanks to the work of Maximenko and his colleagues, we now better understand the movements of plastics and other debris in the oceans. Armed with this knowledge, we will be better able to properly assess and, we hope, reduce the impacts of plastics on marine life. ☐

naturally in groundwater. However, groundwater pollution resulting from human activity is widespread. Industrial, agricultural, and urban wastes—from heavy metals to petroleum products to solvents to pesticides—can leach through soil and seep into aquifers. For example, nitrate from fertilizers has leached into aquifers across the United States and Canada. Nitrate in drinking water has been linked to cancers, miscarriages, and "blue-baby" syndrome, a condition in which the oxygen-carrying capacity of infants' blood is reduced.

Leakage of carcinogenic pollutants (such as chlorinated solvents and gasoline) from underground tanks of oil and industrial chemicals also poses a threat to groundwater. The EPA has embarked on a nationwide cleanup program to unearth and repair leaky tanks. After more than 15 years of work, by 2013 the EPA had confirmed leaks from 510,000 tanks and had completed cleanups on over 430,000 of them.

Leakage of radioactive compounds from underground tanks is also a source of groundwater pollution. In 2013, federal and state officials revealed that underground storage tanks at the Hanford Nuclear Reservation in Washington were leaking radioactive waste into the soil. The site, the most radioactively contaminated area in the United States, stores 60% of the United States' high-level radioactive waste in 177 underground tanks. It has been storing wastes since the 1940s, and billions of dollars have been spent on remediation efforts at the facility. Its cleanup, however, has experienced delays and cost overruns, and the radioactive material in aging underground tanks will not be completely removed until 2040.

## Legislation and regulation have helped to reduce water pollution

As numerous as our freshwater pollution problems may seem, it is important to remember that many were worse a few decades ago (pp. 104–106). Citizen activism and government response during the 1960s and 1970s in the United States resulted in legislation such as the Federal Water Pollution Control Act of 1972 (later amended and renamed the Clean Water Act in 1977). These acts set standards for industrial wastewater, set standards for contaminant levels in surface waters, funded construction of sewage treatment plants, and made it illegal to discharge pollution from a point source without a permit. Thanks to such legislation, point-source pollution in the United States was reduced, and rivers and lakes became noticeably cleaner.

The Great Lakes of Canada and the United States represent an encouraging success story in fighting water pollution. In the 1970s, these lakes were badly polluted with wastewater, fertilizers, and toxic chemicals. Lake Erie was even declared "dead." Today, efforts of the Canadian and U.S. governments have paid off. According to Environment Canada, releases of seven toxic chemicals into the lakes are down by more than 70%, municipal phosphorus has decreased by 80%, and chlorinated pollutants from paper mills are down by 82%. Levels of PCBs (industrial chemicals used in electrical equipment) and DDE (a chemical produced by the breakdown of the pesticide DDT) are down by 78% and 91%, respectively. Bird populations are rebounding, and Lake Erie is now home to the world's largest walleye fishery. The Great Lakes' troubles are by no means over, however—sediment pollution is still heavy, PCBs and mercury still settle from the air, and fish are not always safe to eat. Even so, the progress made thus far shows how conditions can improve when citizens push their governments to take action.

Such successes require effective enforcement of environmental regulations. A 2009 investigation by the *New York Times* revealed that violations of the Clean Water Act rose in the decade preceding the report and that underfunded and understaffed state and federal regulatory agencies acted on only a tiny percentage of these violations. As a result, 1 in 10 Americans has been exposed to unsafe drinking water. Soon after the report's release, EPA promised to strengthen enforcement to help remedy these issues.

## We treat our drinking water

The treatment of drinking water is a widespread practice in developed nations today. Before being sent to your tap, water from a reservoir or aquifer is treated with chemicals to remove particulate matter; passed through filters of sand, gravel, and charcoal; and/or disinfected with small amounts of an agent such as chlorine or ozone. The EPA sets standards for over 90 drinking water contaminants that local governments and private water suppliers are obligated to meet. The treatment of drinking water has greatly reduced mortality from the waterborne diseases that once claimed many lives, and it is one of the most significant technological advances of modern times.

## We treat our wastewater

Prior to the passage of the Clean Water Act, cities and towns regularly released untreated sewage into waterways. This led to oxygen depletion and pathogen contamination in many waters. Wastewater treatment is now a mainstream practice. In rural areas, **septic systems** are the most popular method of treating wastewater. In a septic system, wastewater runs from the house to an underground septic tank, inside which solids and oils separate from water. The clarified water proceeds downhill to a drain field of perforated pipes laid horizontally in gravel-filled trenches underground. Microbes decompose pollutants in the wastewater these pipes emit. Periodically, solid waste from the septic tank is pumped out and taken to a landfill (p. 392).

In more densely populated areas, municipal sewer systems carry wastewater from homes and businesses to centralized treatment locations, called *wastewater treatment plants* (**FIGURE 12.28**). At the plant, incoming wastewater is first passed through screens to capture large objects. It is then sent to **primary treatment**, where about 60% of the suspended solids in the wastewater settle out when the wastewater is allowed to sit in settling tanks. Wastewater then proceeds to **secondary treatment**, in which water is stirred and aerated so that aerobic bacteria consume most of the small particles of organic matter that remain in the wastewater. Roughly 90% of suspended solids may be removed after secondary treatment. In some cases, water is subject to additional treatments to remove pollutants of particular concern. Some wastewater treatments plants along the Mississippi River, for example, subject water to additional treatments that remove nitrogen and phosphorus inputs into the northern Gulf of Mexico. Some pollutants, such as pharmaceuticals excreted in people's urine, are not removed by treatment and are subsequently released into waterways, where they can impact aquatic organisms.

Finally, the clarified water is treated with chlorine, and sometimes with ultraviolet light, to kill bacteria. The treated water, called wastewater **effluent**, is then typically released into a river or the ocean. In some cases, effluent is "reclaimed" and used for watering lawns and golf courses, for irrigation, or for industrial purposes.

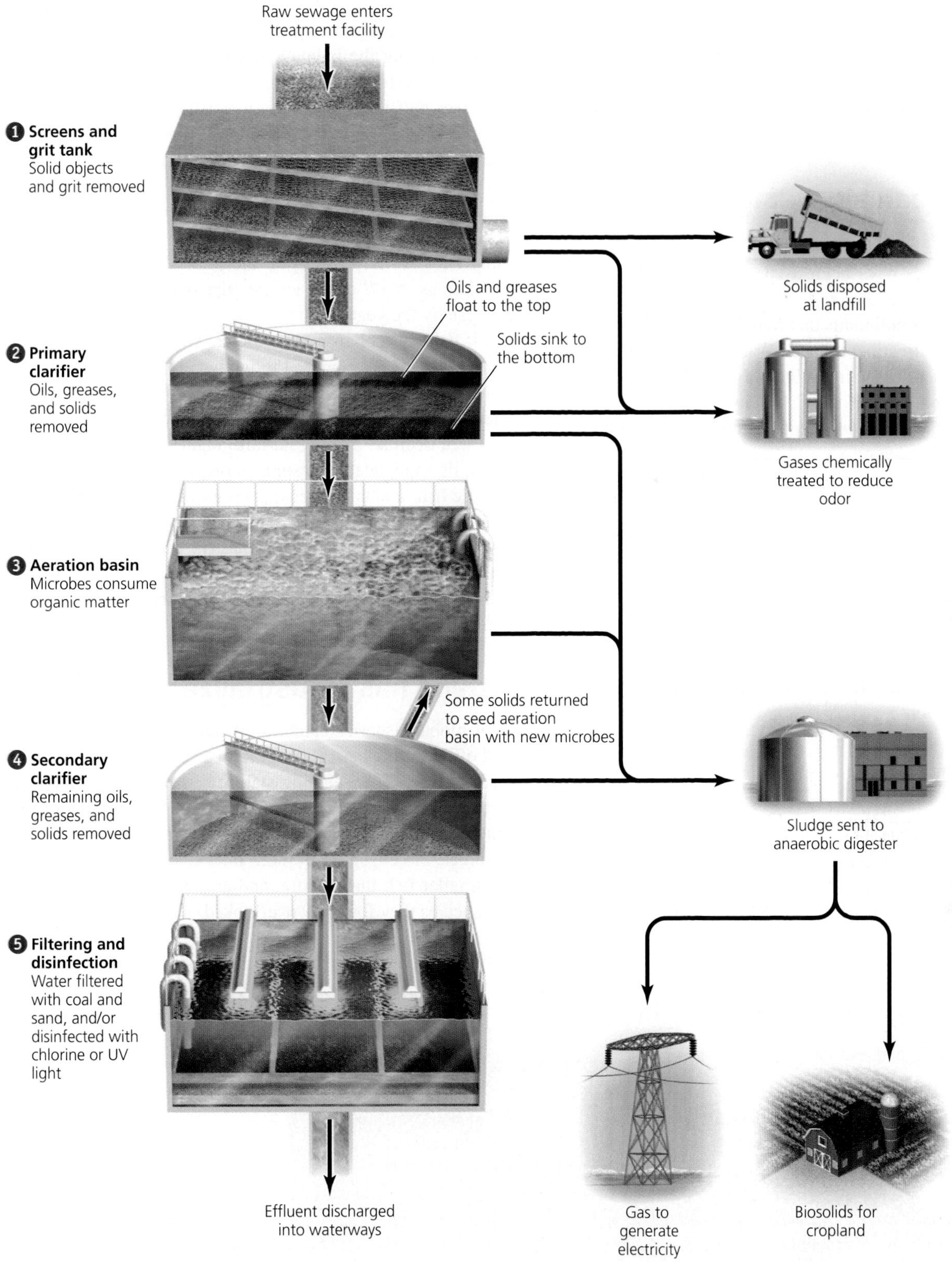


**FIGURE 12.28 Shown here is a generalized process from a modern, environmentally sensitive wastewater treatment facility.** Wastewater initially passes through screens to remove large debris and into grit tanks to let grit settle ❶ It then enters tanks called *primary clarifiers* ❷, in which solids settle to the bottom and oils and greases float to the top for removal. Clarified water then proceeds to aeration basins ❸ that oxygenate the water to encourage decomposition by aerobic bacteria. Water then passes into secondary clarifier tanks ❹ for removal of further solids and oils. Next, the water may be purified ❺ by chemical treatment with chlorine, passage through carbon filters, and/or exposure to ultraviolet light. The treated water (called *effluent*) may then be piped into natural water bodies, used for urban irrigation, flowed through a constructed wetland, or used to recharge groundwater. In addition, most treatment facilities use anaerobic bacteria to digest sludge removed from the wastewater. Biosolids from digesters may be sent to farm fields as fertilizer, and gas from digestion may be used to generate electric power.

The solids collected during the treatment process, called *sludge*, are sent to digesting vats, where microorganisms decompose much of the matter. The resultant "biosolids" are then dried and either landfilled, incinerated, or used as fertilizer on cropland.

### Wetlands can aid wastewater treatment

Artificial wetlands have been used to aid wastewater treatment. Most of these sites use microbes, algae, and aquatic plants to sequester pollutants that remain in the water released from wastewater treatment plants. Water cleansed in the wetland can then be released into waterways or allowed to percolate underground. One of the first constructed wetlands was established in the 1980s in Arcata, a town on northern California's scenic Redwood Coast. There are currently around 500 artificially constructed or restored wetlands performing this service in the United States.

The release of wastewater effluent even shows promise for preserving coastal wetlands along the Gulf Coast. At a study site in Louisiana, wastewater effluent was released into coastal wetlands, where its elevated nutrient concentrations facilitated growth in marsh grasses. The increased plant growth led to the deposition of more plant organic matter on marsh sediments, offsetting the water depth increases caused by soil compaction.

## Emptying the Oceans

We affect the oceans and their biological resources by engineering waterways with dams and levees and by introducing water pollutants directly into the ocean or into rivers that empty into the sea. In addition, we are putting pressure on oceans by overharvesting marine fish species and threatening the balance and functioning of marine and coastal ecosystems.

Over half the world's marine fish populations are fully exploited, meaning that we cannot harvest them more intensively without depleting them, according to the U.N. Food and Agriculture Organization (FAO). An additional 28% of marine fish populations are overexploited and already being driven toward extinction. If current trends continue, a comprehensive 2006 study in the journal *Science* predicted, populations of *all* ocean species that we fish for today will collapse by the year 2048.

If fisheries collapse as predicted, we will lose the ecosystem services they provide. Productivity will be reduced, ecosystems will become more sensitive to disturbance, and the filtering of water by vegetation and organisms (such as oysters) will decline, making harmful algal blooms, dead zones, fish kills, and beach closures more common. Aquaculture (raising fish in tanks or pens) is booming and is helping to relieve pressure on wild stocks, but fish farming comes with its own set of environmental dilemmas (p. 151). All this makes it vital, many scientists and fisheries managers say, that we turn immediately to more sustainable fishing practices.

### Fishing has industrialized

Total global fisheries catch, after decades of increases, leveled off after the late 1980s (**FIGURE 12.29**), despite increased fishing effort. This seeming stability in catch since then can be explained by several factors that conceal population declines: Fishing fleets are exploiting increasingly remote fishing areas, are engaging in more intensive fishing, are capturing smaller fish than before, and are targeting less desirable fish species they formerly overlooked.

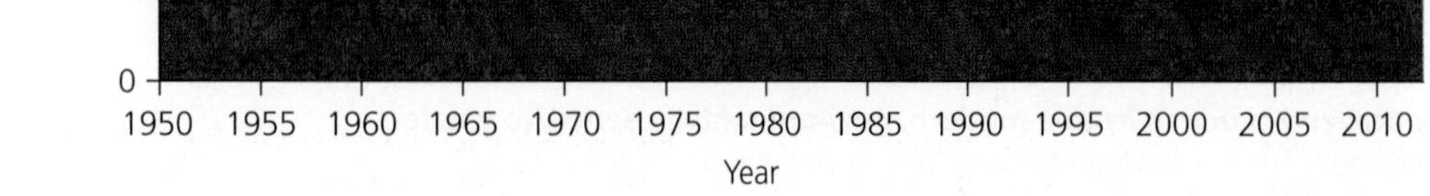

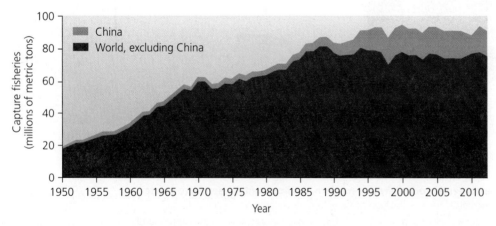

**FIGURE 12.29 After rising for decades, the total global fisheries catch has stalled for the past 20 years.** Many scientists fear that a global decline is imminent if conservation measures are not taken. The figure shows trends with and without China's data, because research suggests that China's data may be somewhat inflated. *Data from the Food and Agriculture Organization of the United Nations, 2014.* The state of world fisheries and aquaculture 2014.

DATA Q Roughly what percentage of the total global fishery harvest in 1950 was taken by China? What percentage of the 2012 harvest was taken by China?

GO TO **INTERPRETING GRAPHS & DATA** ON MasteringEnvironmentalScience®

Today's industrialized commercial fishing fleets employ huge vessels and powerful new technologies to capture fish in large numbers using several methods. Some ships set out long *driftnets* that span large expanses of water (**FIGURE 12.30a**). These chains of transparent nylon mesh nets are arrayed to drift with currents so as to capture passing fish; floats at the top and weights at the bottom keep the nets vertical. *Longline fishing* (**FIGURE 12.30b**) involves setting out extremely long lines (up to 80 km [50 mi] long) with up to several thousand baited hooks spaced along their lengths. *Trawling* entails dragging immense cone-shaped nets through the water, with weights at the bottom and floats at the top. Trawling in open water captures pelagic fish, whereas *bottom-trawling* (**FIGURE 12.30c**) involves dragging weighted nets across the floor of the continental shelf to catch benthic organisms.

Unfortunately, these fishing practices catch more than just the species they target. **Bycatch**, the accidental capture of nontarget animals, accounts for the deaths of millions of animals each year. Driftnetting captures dolphins, seals, and sea turtles, as well as countless nontarget fish. Longline fishing kills turtles, sharks, and seabirds. Bottom-trawling is often likened to clear-cutting (p. 195) and strip-mining (p. 237). It is especially destructive to structurally complex areas, such as reefs, that provide shelter and habitat for animals. The scale of bycatch can be stunning. A 2011 report from the National Oceanic and Atmospheric Administration (NOAA) reported that a staggering 17% of all commercially harvested fish were captured unintentionally.

## Modern fishing fleets deplete marine life rapidly

Throughout the world's oceans, today's industrialized fishing fleets are depleting marine populations. In a 2003 study, fisheries biologists Ransom Myers and Boris Worm analyzed fisheries data and concluded that the oceans today contain only one-tenth of the large-bodied fish and sharks they once did.

Many fisheries have collapsed in recent years. Groundfish (species that live in benthic habitats, such as Atlantic cod, haddock, halibut, and flounder) powered the economies of New England and Maritime Canada for 400 years. Yet fishing pressure became so intense that most stocks collapsed, bringing fishing economies down with them. The cod fishery of the northwestern Atlantic was decimated by over a century of overfishing, but after the Canadian and American governments banned cod harvests in the 1990s, the fishery is now on a slow path to recovery.

Red snapper stocks in the Gulf of Mexico off the Louisiana coast have been similarly depleted by overfishing. Snapper have also suffered when they are captured as bycatch on shrimping vessels. The species was identified as severely overfished in 1989, and current populations are at a mere 3% of their historical abundance. A comprehensive recovery plan has been developed for red snapper, but the road to recovery will be a slow one as stocks remain exceedingly low.

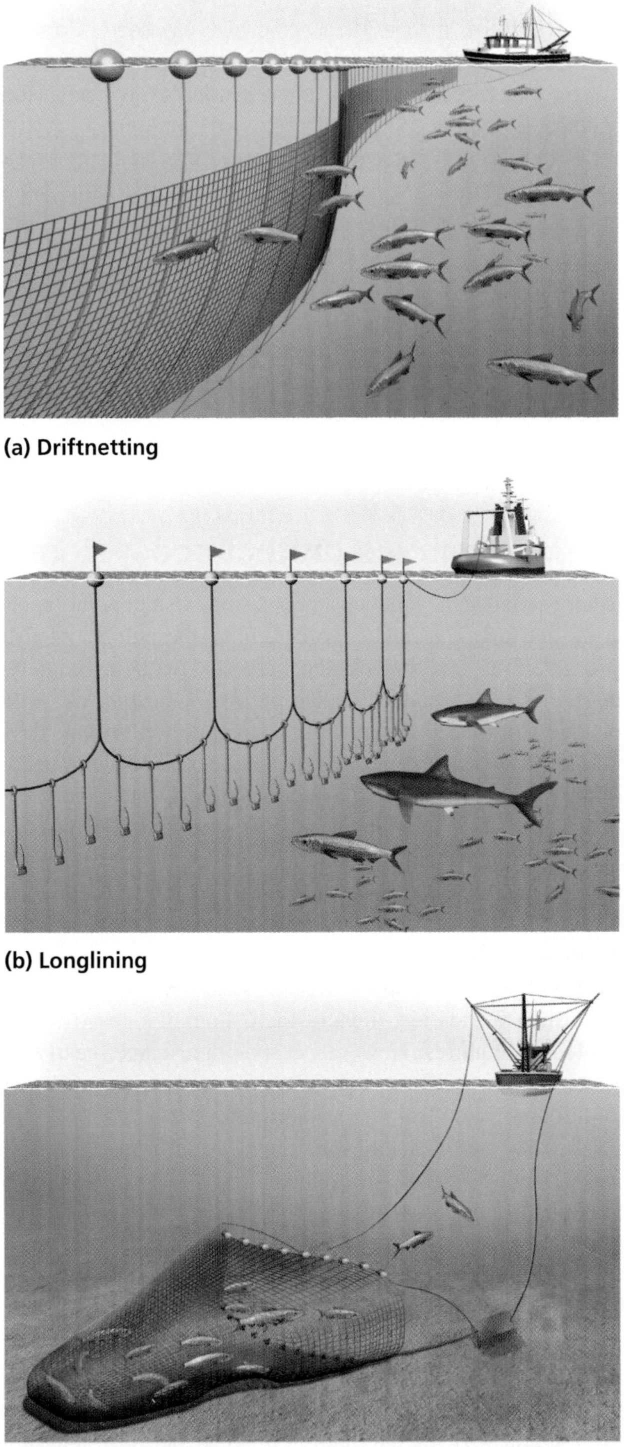

**(a) Driftnetting**

**(b) Longlining**

**(c) Bottom-trawling**

**FIGURE 12.30 Commercial fishing fleets use several methods of capture.** In driftnetting **(a)**, long transparent nylon nets are set out to drift through open water to capture schools of fish. In longlining **(b)**, lines with numerous baited hooks are set out in open water. In bottom-trawling **(c)**, weighted nets are dragged along the floor of the continental shelf. All methods result in large amounts of bycatch, the capture of nontarget animals. The illustrations above are schematic for clarity and do not portray the immense scale that these technologies can attain; for instance, industrial trawling nets can be large enough to engulf multiple Boeing 747 jumbo jets.

## FAQ I love seafood, so how can I make sustainable choices?

To most of us, marine fishing practices may seem a distant phenomenon over which we have no control, especially given that more than 80% of seafood sold in the United States is imported. But although we don't have control over the seafood presented to us, we have full control over which items we buy. Finding out how seafood items were caught is difficult, however, because this information is not readily available to consumers in most cases. Thus, several organizations, such as the Environmental Defense Fund, have devised concise guides and smartphone apps to help consumers differentiate fish and shellfish that are overfished or whose capture is ecologically damaging from those that are harvested more sustainably.

## Marine reserves protect ecosystems

Fisheries managers conduct surveys, study fish population biology, and monitor catches to determine the number of fish of a given species that can be harvested without reducing future catches—a concept called *maximum sustainable yield* (p. 193). Despite the use of this technique over several decades, a number of fish and shellfish stocks have plummeted. Thus, many scientists and managers feel it is time to shift the focus away from individual species and toward viewing marine resources as elements of larger ecological systems. One key aspect of such *ecosystem-based management* (p. 193) is to set aside areas of ocean where systems can function without human interference.

Hundreds of **marine protected areas (MPAs)** have been established worldwide, most of them along the coastlines of developed countries. However, despite their name, nearly all MPAs allow fishing or other extractive activities and so are not fully protected from impacts from people.

Because of the lack of true refuges from fishing pressure, many scientists want to establish areas where fishing is prohibited. Such "no-take" areas are called **marine reserves**. Designed to preserve ecosystems intact, marine reserves are also intended to improve fisheries. Scientists argue that marine reserves can act as production factories for fish for surrounding areas, because fish larvae produced inside reserves will disperse and stock other parts of the ocean. By serving both purposes, proponents maintain, marine reserves are a win-win proposition for environmentalists and fishers alike. However, many commercial and recreational fishers dislike the idea of no-take reserves and have opposed nearly every marine reserve that has been established.

But are MPAs successful in protecting marine organisms? A 2014 study in the journal *Nature* sought to answer this question by examining 87 MPAs around the world. The study found that fish biomass and fish species diversity was highest in MPAs that were large in size, remote in location, old (established at least 10 years prior to the study), and had strongly enforced no-take policies. This study provided a "blueprint" for creating successful MPAs and may lead us toward solutions to one of our most pressing environmental problems in today's oceans.

# Conclusion

Our planet's diverse aquatic systems comprise an interconnected web of ecosystems that exchange water. The introduction of pollutants and the engineering of waterways therefore cause impacts that cascade through the system. Our expanding population and increasing water use are straining water supplies and affecting surface waters and groundwater around the world. Water pollutants threaten human health and ecosystem stability, and overharvesting of marine fish populations threatens the oceans' biodiversity.

There is plenty of reason for optimism, however. Improvements in water use efficiency show promise for reducing demand for water, even with increasing human populations. Water quality in many freshwater bodies has improved in recent decades, thanks to legislative action from policymakers and the efforts of millions of concerned citizens. In the oceans, marine reserves give hope that we can restore ecosystems and fisheries at the same time.

# Testing Your Comprehension

1. Explain why the distribution of water on Earth makes it difficult for many people to access adequate fresh water.
2. Pick one of the aquatic systems profiled in this chapter, and provide three examples of ways it interacts with other aquatic systems.
3. Why are coral reefs biologically valuable? How are they being degraded by human impact? What is causing the disappearance of mangrove forests and salt marshes?
4. Describe three benefits and three costs of damming rivers. What are the costs and benefits of levees?
5. Why do the Colorado, Rio Grande, Nile, and Yellow rivers now slow to a trickle or run dry before reaching their deltas?
6. Name three major types of water pollutants, and provide an example of each. Explain which classes of pollutants you think are most important in your local area.
7. Define *groundwater*, and list some anthropogenic (human) sources of groundwater pollution. Why do many scientists consider groundwater pollution a greater problem than surface water pollution?
8. Describe and explain the major steps in the process of wastewater treatment. How can artificially constructed wetlands aid such treatment?

9. Name three industrial fishing practices, and explain how they create bycatch and harm marine life.

10. How does a marine reserve differ from a marine protected area? Why do many fishers oppose marine reserves? Explain why many scientists say no-take reserves will be good for fishers.

# Seeking Solutions

1. How can we lessen agricultural demand for water? Describe some ways we can reduce household water use.
2. Describe three ways in which your own actions contribute to water pollution. Now describe three ways in which you could diminish these impacts.
3. Describe the trends in global fish capture over the past 55 years and over the past 25 years. Describe and several factors that account for these trends.
4. **THINK IT THROUGH** Your state's governor has put you in charge of water policy for the state. The aquifer beneath your state has been overpumped, and many wells have run dry and last year crop production declined for the first time in 40 years. Meanwhile, the state's largest city is growing so fast that more water is needed for its burgeoning urban population. What policies would you consider to restore your state's water supply? Would you try to take steps to increase supply, decrease demand, or both? Explain why you would choose such policies.
5. **THINK IT THROUGH** You are mayor of a coastal town where some residents are employed as commercial fishers and others make a living serving ecotourists who come to snorkel and scuba dive at the nearby coral reef. In recent years, several fish stocks have crashed, and ecotourism is dropping off as fish disappear from the increasingly degraded reef. Scientists want to establish a marine reserve around portions of the reef, but most commercial fishers are opposed to this idea. What steps would you take to restore your community's economy and environment?

# Calculating Ecological Footprints

One of the single greatest personal uses of water is for showering. Old-style showerheads that were standard in homes and apartments built before 1992 dispense at least 5 gallons of water per minute, but low-flow showerheads produced after that year dispense just 2.5 gallons per minute. Given an average daily shower time of 8 minutes, calculate the amounts of water used and saved over the course of a year with old standard versus low-flow showerheads, and record your results in the table.

| | Annual water use with standard showerheads (gallons) | Annual water use with low-flow showerheads (gallons) | Annual water savings with low-flow showerheads (gallons) |
|---|---|---|---|
| You | | | |
| Your class | | | |
| Your state | | | |
| United States | | | |

1. In 2010, the EPA began promoting showerheads that produce still-lower flows of 2 gallons per minute (gpm). How much water would you save per year by using a 2-gpm showerhead instead of a 2.5-gpm showerhead?
2. How much water would you be able to save annually by shortening your average shower time from 8 minutes to 6 minutes? Assume you use a 2.5-gpm showerhead.
3. Compare your answers to questions 1 and 2. Do you save more water by showering 8 minutes with a 2-gpm showerhead or 6 minutes with a 2.5-gpm showerhead?
4. Can you think of any factors that are *not* being considered in this scenario of water savings? Explain.

## MasteringEnvironmentalScience®

**STUDENTS**

Go to **MasteringEnvironmentalScience** for assignments, the etext, and the Study Area with practice tests, videos, current events, and activities.

**INSTRUCTORS**

Go to **MasteringEnvironmentalScience** for automatically graded activities, current events, videos, and reading questions that you can assign to your students, plus Instructor Resources.

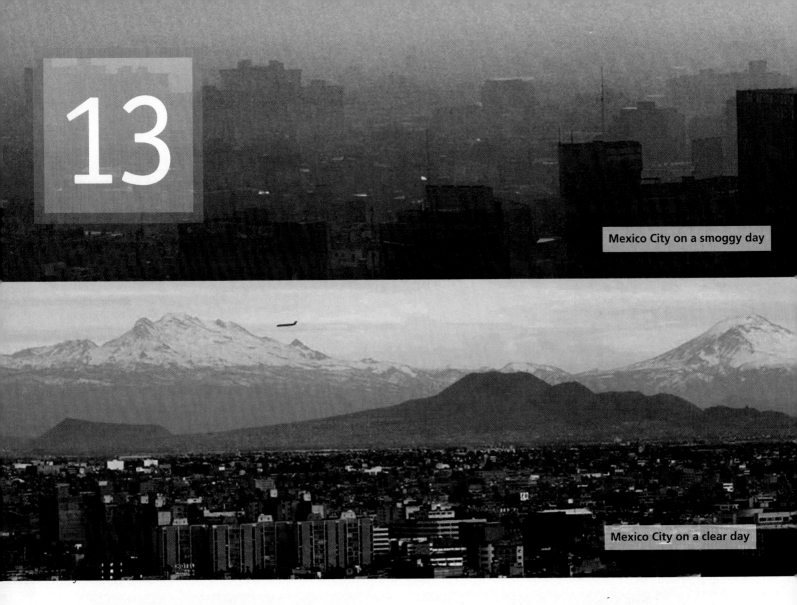

# 13

Mexico City on a smoggy day

Mexico City on a clear day

# Atmospheric Science, Air Quality, and Pollution Control

## Upon completing this chapter, you will be able to:

- ☐ Describe the composition, structure, and function of Earth's atmosphere
- ☐ Relate weather and climate to atmospheric conditions
- ☐ Identify major pollutants, outline the scope of outdoor air pollution, and assess strategies for pollution control
- ☐ Explain stratospheric ozone depletion and identify steps taken to address it
- ☐ Explain acid deposition, illustrate its consequences, and articulate how we are addressing it
- ☐ Characterize the scope of indoor air pollution and assess solutions



Photo: Mexico City on a clear day

## CENTRAL CASE STUDY

# Clearing the Air in L.A. and in Mexico City

*"I left L.A. in 1970, and one of the reasons I left was the horrible smog. And then they cleaned it up. That was one of the greatest things the government has ever done for me. You have beautiful days now. It's a much, much nicer place to live."*

**—Actor and comedian Steve Martin, speaking to *Los Angeles Magazine***

*"This city can be a model for others."*

**— Mexico City Mayor Miguel Ángel Mancera**

Los Angeles has long symbolized air pollution in Americans' minds. Smog blanketed the city in the 1970s, 1980s, and 1990s, fed by the exhaust from millions of automobiles clogging its freeways. But Los Angeles has improved its air quality, thanks to policy efforts and new technologies. Today L.A. still suffers the nation's worst smog, but its skies are clearer than in decades—and its air is cleaner than in many of its "sister cities" elsewhere in the world.

Devised to foster peace and understanding among nations, the Sister Cities program links hundreds of U.S. cities with foreign counterparts through shared government visits and cultural ties. One of L.A.'s sister cities is Mexico City, the capital of Mexico. One thing these two cities share is smog.

Not long ago Mexico City suffered the most polluted air in the world. On days of poor air quality throughout the 1990s, residents wore face masks on the streets, teachers kept students inside at recess, and outdoor sporting events were cancelled. Children drawing pictures would use brown crayons to color the sky. Each year thousands of deaths and tens of thousands of hospital visits were blamed on pollution. Mexican novelist Carlos Fuentes called his nation's capital "Makesicko City."

As in Los Angeles, traffic generates most of the pollution in Mexico City, where motorists in 6 million cars traverse miles of urban sprawl. And like L.A., Mexico City is surrounded by mountains, allowing thermal inversions to trap pollutants over the city. Moreover, at Mexico City's high altitude—2240 m (7350 ft) above sea level—solar radiation is intense, which worsens smog formed by the interaction of pollutants with sunlight. Mexico City environmental chemist Armando Retama likens his hometown to "a casserole dish with a lid on top."

Despite all the challenges facing the world's third-largest metropolis, Mexico City's 20 million people are fighting back. Several recent mayors have taken bold action to clean up the air, and today Mexico City is enjoying a renaissance. As the smog begins to clear, revealing beautiful views of the snow-capped peaks that ring the valley, Mexico City has become a model for other cities seeking to fight pollution.

Efforts began in the 1990s, when city leaders shut down an oil refinery and pushed factories and power plants to shift to cleaner-burning natural gas. Lead was removed from gasoline, the sulfur content of diesel fuel was reduced, and pollution control technologies such as catalytic converters were phased in for new vehicles.

Levels of many pollutants fell, but as the city grew, smog from automobile traffic persisted. In response, city officials stepped up vehicle emissions testing and upgraded taxis and city vehicles to cleaner models. To monitor air quality, 34 sampling stations were set up across the city, sending real-time data to city engineers.

In 2007 Mayor Marcelo Ebrard accelerated efforts as part of a 15-year sustainability plan he launched aiming to make Mexico City "the greenest city in the Americas." New lines were added to the subway system, and 800 exhaust-spewing minibuses were replaced with fuel-efficient buses that ran smoothly in designated lanes. Over 450,000 people now use these buses each day, cutting carbon dioxide emissions by an estimated 80,000 tons per year.

In 2010 Ebrard introduced a bicycle-sharing program to free short-distance commuters from dependence on cars. With 1000 bikes at rental stations throughout the city, people can rent a bike cheaply at one location and drop it off at another. In Ebrard's most popular initiative, every Sunday morning the city's main boulevard, the Paseo de la Reforma, is closed to car traffic, creating a safe and pleasant community space for pedestrians, bikers, joggers, and skateboarders.

Mexico City residents were so happy with the progress that in 2012, when Ebrard stepped down because of term limits, voters rewarded his political party by electing his ally Miguel Ángel Mancera in a landslide. Mancera announced plans to expand the bicycle program to 4000 bikes and to put electric buses and taxis on the roads. He also rolled out a car-sharing program that officials hope can remove 40,000 vehicles from circulation. Private entrepreneurs are now getting in on the act; recent university graduates have launched companies providing car-sharing and carpooling services, some using electric vehicles.

All these changes are paying off with cleaner air. In 1991, the city's air was deemed hazardous to breathe on all but 8 days of the year. Today, most pollutants have been slashed by over 75%, and the air meets health standards on one of every two days.

Plenty of hurdles remain. Rampant development is challenging efforts to plan for sustainable growth. New highway construction may induce more people to drive. The typical driver still spends three hours a day stuck in traffic that averages just 13 mph. Worst yet, smog still contributes to an estimated 4000 deaths each year. But Mexico City is following in the footsteps of Los Angeles, making steady progress toward cleaner air.

Indeed, L.A. and Mexico City typify cities of developed and developing nations today. Nations that are industrializing as they try to build wealth for their citizens are confronting the same air quality challenges that plagued the United States and other wealthy nations a generation and more ago. Los Angeles has 24 other sister cities, and many struggle with severe air pollution, including Athens, Greece; Jakarta, Indonesia; Mumbai, India; Guangzhou, China; Taipei, Taiwan; San Salvador, El Salvador; and Tehran, Iran. We will examine the solutions sought in Los Angeles, Mexico City, and elsewhere as we learn about Earth's atmosphere and how to reduce the pollutants we release into it. ■

# The Atmosphere

Every breath we take reaffirms our connection to the **atmosphere**, the layer of gases that envelops our planet. The atmosphere moderates our climate, provides oxygen, shields us from meteors and from hazardous solar radiation, and transports and recycles water and nutrients.

Earth's atmosphere consists of 78% nitrogen ($N_2$) and 21% oxygen ($O_2$), by volume of dry air. The remaining 1% is composed of argon (Ar) and minute concentrations of other gases (**FIGURE 13.1**). The atmosphere also contains water vapor ($H_2O$) in concentrations that vary with time and place from 0% to 4%. Today, human activity is altering the quantities of some atmospheric gases, such as carbon dioxide ($CO_2$), methane ($CH_4$), and ozone ($O_3$).

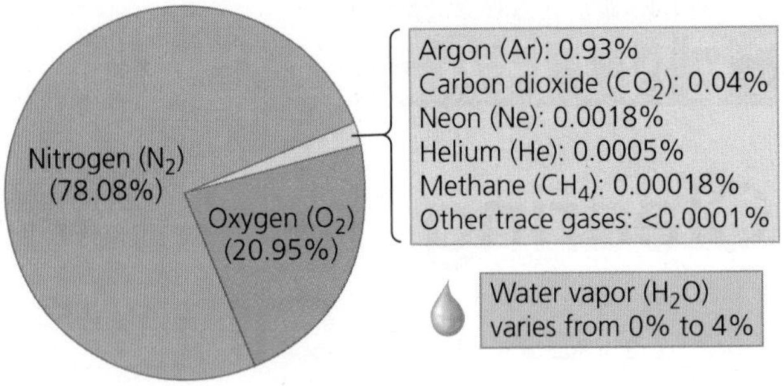


**FIGURE 13.1 Earth's atmosphere consists of nitrogen, oxygen, argon, and a mix of gases at dilute concentrations.**

## The atmosphere is layered

The atmosphere that stretches so high above us is actually just 1/100 of Earth's diameter, like the fuzzy skin of a peach. Air pressure and the density of molecules decrease exponentially as altitude increases, but temperature varies in a more complex way, allowing us to identify several layers (**FIGURE 13.2**).

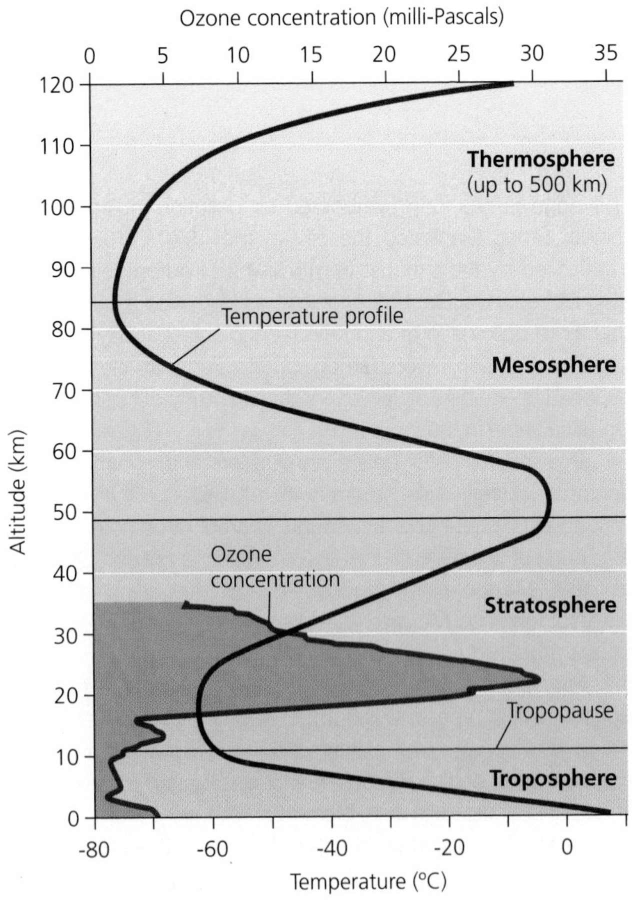


**FIGURE 13.2 The atmosphere is layered.** Temperature (red line) drops with altitude in the troposphere, rises with altitude in the stratosphere, drops in the mesosphere, and rises in the thermosphere. The tropopause separates the troposphere from the stratosphere. Ozone (blue area) is densest in the lower stratosphere. *Adapted from Jacobson, M. Z., 2002.* Atmospheric pollution: History, science, and regulation. *Cambridge: Cambridge University Press; and Parson, E. A., 2003.* Protecting the ozone layer: Science and strategy. *Oxford: Oxford University Press.*

Within the bottommost layer, the **troposphere**, air movement drives the planet's weather. The troposphere is thin (averaging 11 km [7 mi] high) relative to the atmosphere's other layers, but it contains three-quarters of the atmosphere's mass. The reason is that gravity pulls mass downward, making air denser near Earth's surface. Tropospheric air gets colder with altitude, dropping to roughly –52°C (–62°F) at the top of the troposphere. At this point, temperatures stabilize, marking a boundary called the *tropopause*. The tropopause acts like a cap, limiting mixing between the troposphere and the atmospheric layer above it, the stratosphere.

The **stratosphere** extends 11–50 km (7–31 mi) above sea level. Similar in composition to the troposphere, the stratosphere is drier and less dense. Its gases experience little vertical mixing, so once substances (including pollutants) enter it, they tend to remain for a long time. The stratosphere warms with altitude, because its ozone and oxygen absorb the sun's ultraviolet (UV) radiation (p. 31). Most of the atmosphere's ozone concentrates in a portion of the stratosphere roughly 17–30 km (10–19 mi) above sea level, a region we call the **ozone layer**. By absorbing and scattering incoming UV radiation, the ozone layer greatly reduces the amount of this radiation that reaches Earth's surface. UV light can damage living tissue and induce genetic mutations, and life has evolved to rely on the protective presence of the ozone layer.

Above the stratosphere lies the mesosphere, where temperatures decrease with altitude and where incoming meteors burn up. Above this the thermosphere extends to roughly 500 km (300 mi). Still higher, the atmosphere merges into space in a region called the exosphere.

## The sun influences weather and climate

An enormous amount of energy from the sun constantly bombards the upper atmosphere. Of this solar energy, about 70% is absorbed by the atmosphere and planetary surface, while the rest is reflected back into space (see Figure 14.1, p. 305).

Land and surface water absorb solar energy and then emit thermal infrared radiation, which warms the air and causes some water to evaporate. As a result, air near Earth's surface tends to be warmer and moister than air at higher altitudes. These differences set into motion a process of **convective circulation**. Warm air, being less dense, rises and creates vertical currents. As air rises into regions of lesser atmospheric pressure, it expands and cools, causing moisture to condense and fall as rain. Once the air cools, it descends and becomes denser, replacing warm air that is rising. The descending air picks up heat and moisture near ground level and prepares to rise again, continuing the process. Convective circulation influences both weather and climate.

Weather and climate each involve physical properties of the troposphere, such as temperature, pressure, humidity, cloudiness, and wind. **Weather** specifies atmospheric conditions in a location over minutes, hours, days, or weeks. In contrast, **climate** describes typical patterns of atmospheric conditions in a location over years, decades, centuries, or millennia. Mark Twain noted the distinction by remarking, "Climate is what we expect; weather is what we get." For example, Los Angeles has a climate characterized by warm dry summers and mild rainy winters, yet on some autumn days, dry Santa Ana winds blow in from the desert and bring extremely hot weather.

## Inversions affect air quality

Under most conditions, air in the troposphere becomes cooler as altitude increases. Because warm air rises, vertical mixing results (**FIGURE 13.3a**). Occasionally, however, a layer of cool air may form beneath a layer of warmer air. This departure from the normal temperature profile is known as a **temperature inversion**, or **thermal inversion** (**FIGURE 13.3b**). The band of air in which temperature rises with altitude is called an **inversion layer** (because the normal direction of temperature change is inverted). The cooler air at the bottom of the inversion layer is denser than the warmer air above, so it resists vertical mixing and remains stable. Thermal inversions can occur in different ways, sometimes involving cool air at ground level and sometimes involving an inversion layer

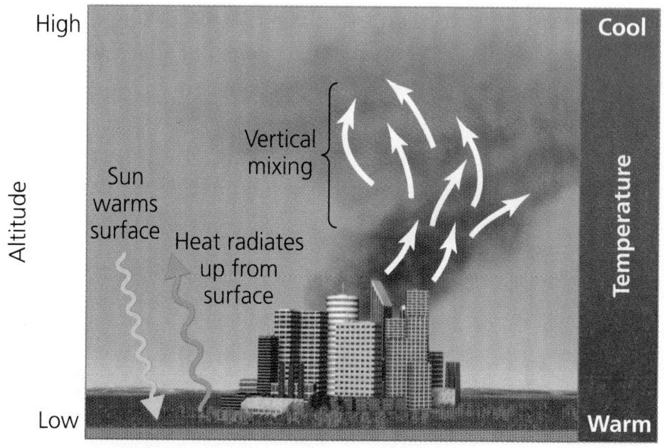


**(a) Normal conditions**

**(b) Thermal inversion**

**FIGURE 13.3 Thermal inversions trap air and pollutants.** Under normal conditions **(a)**, air becomes cooler with altitude; and air of different altitudes mixes, dispersing pollutants upward. In a thermal inversion **(b)**, dense cool air remains near the ground, and air warms with altitude within the inversion layer. Little mixing occurs, and pollutants are trapped.

higher above the ground. One common type of inversion (shown in Figure 13.3b) occurs in mountain valleys where slopes block morning sunlight, keeping ground-level air within the valley shaded and cool.

Vertical mixing allows pollutants in the air to be carried upward and diluted, but thermal inversions trap pollutants near the ground. As a result, cities such as Los Angeles and Mexico City suffer their worst pollution when inversions prevent pollutants from being dispersed. Both of these metropolitan areas are encircled by mountains that promote inversions, interrupt air flow, and trap pollutants. Los Angeles experiences inversions most often when a "marine layer" of air cooled by the ocean moves inland. In Mexico City in 1996, a persistent thermal inversion sparked a five-day crisis that killed at least 300 people and sent 400,000 to hospitals. Across the world today, inversions frequently concentrate pollution over major metropolitan areas in valleys ringed by mountains, from Tehran to Seoul to Río de Janeiro to São Paulo.

## Large-scale circulation systems produce global climate patterns

At large geographic scales, convective air currents contribute to broad climate patterns (**FIGURE 13.4**). Near the equator, solar radiation sets in motion a pair of convective cells known as *Hadley cells.* Here, where sunlight is most intense, surface air warms, rises, and expands. As it does so, it releases moisture, producing the heavy rainfall that gives rise to tropical rainforests. After releasing much of its moisture, this air diverges and moves in currents heading north and south. The air in these currents cools and descends at about 30 degrees latitude north and south. Because the descending air is now dry, the regions around 30 degrees latitude are quite arid, giving rise to deserts. Two further pairs of convective cells, *Ferrel cells* and *polar cells,* lift air and create precipitation around 60 degrees latitude north and south and cause air to descend at 30 degrees latitude and in the polar regions.

Together these three pairs of convective cells create wet climates near the equator, arid climates near 30° latitude, moist regions near 60° latitude, and dry conditions near the poles. These patterns, combined with temperature variation, help explain why biomes tend to be arrayed in latitudinal bands (see Figure 4.13, p. 81).

The Hadley, Ferrel, and polar cells interact with Earth's rotation to produce global wind patterns. As Earth rotates on its axis, regions of the planet's surface near the equator move west to east more quickly than locations near the poles. As a result, from the perspective of an Earth-bound observer, air currents of the convective cells that flow north or south appear to be deflected from a straight path. This deflection is called the *Coriolis effect,* and it results in the curving global wind patterns in Figure 13.4. For centuries, people made use of these patterns to facilitate ocean travel by wind-powered sailing ships.

Understanding how the atmosphere functions enables us to appreciate what drives Earth's climate and can help us predict our weather day by day. Such knowledge also helps us comprehend how our pollution of the atmosphere affects climate, ecological systems, economies, and human health.

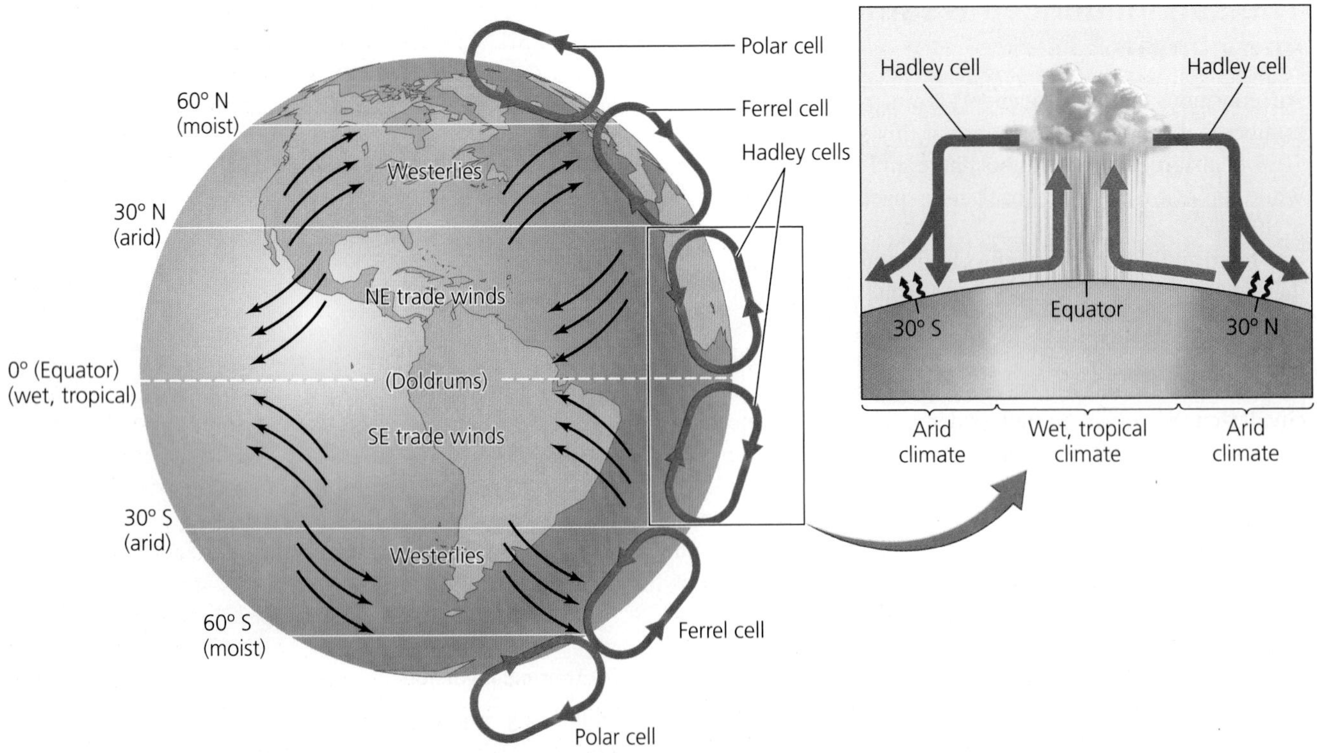


**FIGURE 13.4 Large-scale convective cells create global patterns of moisture and wind.** These cells give rise to a wet climate in tropical regions, arid climates around 30° latitude, moist climates around 60°, and dry climates near the poles. Surface air movement of these cells interacts with the Coriolis effect to create global wind currents.

(a) Natural fire in California (b) Mount Saint Helens eruption, 1980 (c) Dust storm blowing dust from Africa to the Americas

**FIGURE 13.5 Wildfire, volcanoes, and dust storms are three natural sources of air pollution.** In **(c)**, a satellite image shows trade winds blowing soil from Africa across the Atlantic Ocean to the Americas. This carries fungal and bacterial spores that harm Caribbean coral reefs, but it also brings nutrients to the Amazon rainforest.

# Outdoor Air Quality

Throughout history, we have made the atmosphere a dumping ground for our airborne wastes. Whether from simple wood fires or modern coal-burning power plants, we have generated **air pollutants**, gases and particulate material added to the atmosphere that can affect climate or harm people or other living things. At the same time, our efforts to control **air pollution**, the release of air pollutants, have brought some of our best successes in confronting environmental problems thus far.

In recent decades, public policy and improved technologies have helped us reduce most types of **outdoor air pollution** (often called *ambient air pollution*) in industrialized nations. However, outdoor air pollution remains a problem, particularly in industrializing nations and in urban areas. Globally, the World Health Organization (WHO) estimates that 3.3 million people die prematurely each year as a result of health problems caused by outdoor air pollution. Moreover, we face an enormous air pollution challenge today in our emission of greenhouse gases (p. 305), which contribute to global climate change. (We discuss this issue separately and in depth in Chapter 14.)

## Some pollution is from natural sources

When we think of outdoor air pollution, we tend to envision smokestacks belching smoke from industrial plants. However, natural processes produce a great deal of air pollution (**FIGURE 13.5**). Fires (p. 196) from burning vegetation generate soot and gases. Globally, over 60 million ha (150 million acres; an area the size of Texas) of forest and grassland burn in a typical year. Volcanic eruptions (pp. 231–232) release large quantities of particulate matter and sulfur dioxide that may spread over large regions. Ash from major eruptions can ground airplanes, destroy car engines, and pose respiratory health dangers. In 2012, Mexico City residents went on alert as Popocatepetl, a nearby volcano, let loose a series of eruptions, adding to the region's pollution challenges. Windblown dust is another natural pollution source. Dust storms occur naturally in arid regions and can sometimes send dust from one continent to another.

Some natural impacts are made worse by human activity and land use policies. Farming and grazing practices that strip vegetation from the soil promote wind erosion and lead to dust storms, like those that devastated America's Dust Bowl in the last century (p. 142). In the tropics, many farmers set fires to clear forest for farming and grazing (p. 141). In North American forests, the suppression of fire has allowed fuel to build up and eventually feed highly destructive fires (p. 196). Today scientists voice concern that global climate change (Chapter 14) is leading to drought in many regions, worsening dust storms and fires as a result.

## We create outdoor air pollution

Human activity introduces many sources of air pollution. As with water pollution, anthropogenic (human-caused) air pollution can emanate from point sources or non-point sources (p. 266). A point source describes a specific location from which large quantities of pollutants are discharged (such as a coal-fired power plant). Non-point sources are more diffuse, consisting of many small, widely spread sources (such as millions of automobiles).

Pollutants released directly from a source are termed **primary pollutants**. Ash from a volcano, sulfur dioxide from a power plant, or carbon monoxide from an engine are all primary pollutants. Often primary pollutants interact or react with constituents of the atmosphere and form other pollutants, called **secondary pollutants**. Examples include ozone formed from pollutants in urban smog, and the acids in acid rain, formed when certain primary pollutants react with water and oxygen.

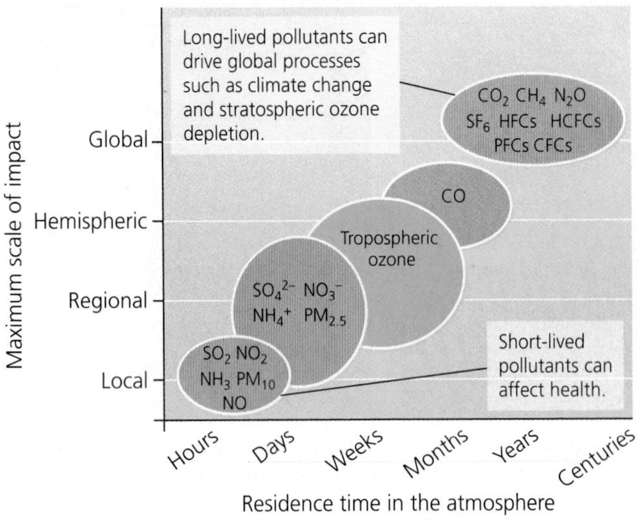


**FIGURE 13.6 Substances with short residence times affect air quality locally, whereas those with long residence times affect air quality globally.** *Data from United Nations Environment Programme, 2007.* Global environmental outlook (GEO-4), *Nairobi, Kenya.*

Because pollutants differ in how readily they react in air and in how quickly they settle to the ground, they differ in their **residence time**, the amount of time a pollutant spends in the atmosphere. Pollutants with brief residence times exert localized impacts over short time periods. Most particulate matter and most pollutants from automobile exhaust stay aloft only hours or days, which is why air quality in a city like Mexico City or Los Angeles changes from day to day. In contrast, pollutants with long residence times can exert impacts regionally or globally for long periods, even centuries (**FIGURE 13.6**). The pollutants that drive climate change and those that deplete Earth's ozone layer (two separate phenomena!—see **FAQ**, p. 295) are each able to cause long-lasting global impacts because they persist in the atmosphere for so long.

## The Clean Air Act addresses pollution

To address air pollution in the United States, Congress has passed a series of laws, most notably the Clean Air Act, first enacted in 1963 and amended multiple times since, particularly in 1970 and 1990. This body of legislation funds research into pollution control, sets standards for air quality, imposes limits on emissions from new sources, and enables citizens to sue parties violating the standards.

Under the Clean Air Act, the U.S. Environmental Protection Agency (EPA) sets nationwide standards for emissions and for concentrations of major pollutants in ambient air. It is largely up to the states to monitor emissions and air quality and to develop, implement, and enforce regulations within their borders. States submit implementation plans to the EPA for approval, and if a state's plans are not adequate, the EPA can take control of enforcement. If a region fails to clean up its air, the EPA can prevent it from receiving federal money for transportation projects.

## Agencies monitor emissions

State and local agencies monitor and report to the EPA emissions of six major pollutants—carbon monoxide, sulfur dioxide, nitrogen oxides, volatile organic compounds, particulate matter, and lead. Across the United States in 2013, human activity polluted the air with 92 million tons of these six pollutants.

**Carbon monoxide** **Carbon monoxide** (CO) is a colorless, odorless gas produced primarily by the incomplete combustion of fuel. Vehicles and engines account for most CO emissions in the United States. Other sources include industrial processes, waste combustion, and residential wood burning. Carbon monoxide is hazardous because it binds to hemoglobin in red blood cells, preventing the hemoglobin from binding with oxygen. Carbon monoxide poisoning induces nausea, headaches, fatigue, heart and nervous system damage, and potentially death.

**Sulfur dioxide** **Sulfur dioxide** ($SO_2$) is a colorless gas with a pungent odor. Most emissions result from the combustion of coal for electricity generation and for industry. During combustion, elemental sulfur (S), a contaminant in coal, reacts with oxygen ($O_2$) to form $SO_2$. Once in the atmosphere, $SO_2$ may react to form sulfur trioxide ($SO_3$) and sulfuric acid ($H_2SO_4$), which may settle back to Earth in acid deposition (p. 295).

**Nitrogen oxides** **Nitrogen oxides** ($NO_X$) are a family of compounds that include nitric oxide (NO) and nitrogen dioxide ($NO_2$). Most U.S. emissions of nitrogen oxides result when nitrogen and oxygen from the atmosphere react at high temperatures during combustion in vehicle engines. Fossil fuel combustion in industry and at electrical utilities accounts for most of the rest. $NO_X$ emissions contribute to smog, acid deposition, and stratospheric ozone depletion.

**Volatile organic compounds** **Volatile organic compounds** (VOCs) are carbon-containing chemicals emitted by vehicle engines and a wide variety of solvents, industrial processes, household chemicals, and consumer items. Examples range from benzene to acetone to formaldehyde. One group of anthropogenic VOCs consists of hydrocarbons (p. 29), whereas other VOCs are emitted naturally by plants. VOCs can react to produce a number of secondary pollutants, as occurs in urban smog.

**Particulate matter** **Particulate matter** is composed of solid or liquid particles small enough to be suspended in the atmosphere. Particulate matter includes primary pollutants such as dust and soot, as well as secondary pollutants such as sulfates and nitrates. Scientists classify particulate matter by the size of the particles. Smaller particles are more likely to get deep into the lungs and cause respiratory damage and heart problems. $PM_{10}$ pollutants consist of particles less than 10 microns in diameter (one-seventh the width of a human hair), whereas $PM_{2.5}$ pollutants consist of still-finer particles less than 2.5 microns in diameter. Most $PM_{10}$ pollution is from road dust, whereas most $PM_{2.5}$ pollution results from combustion.

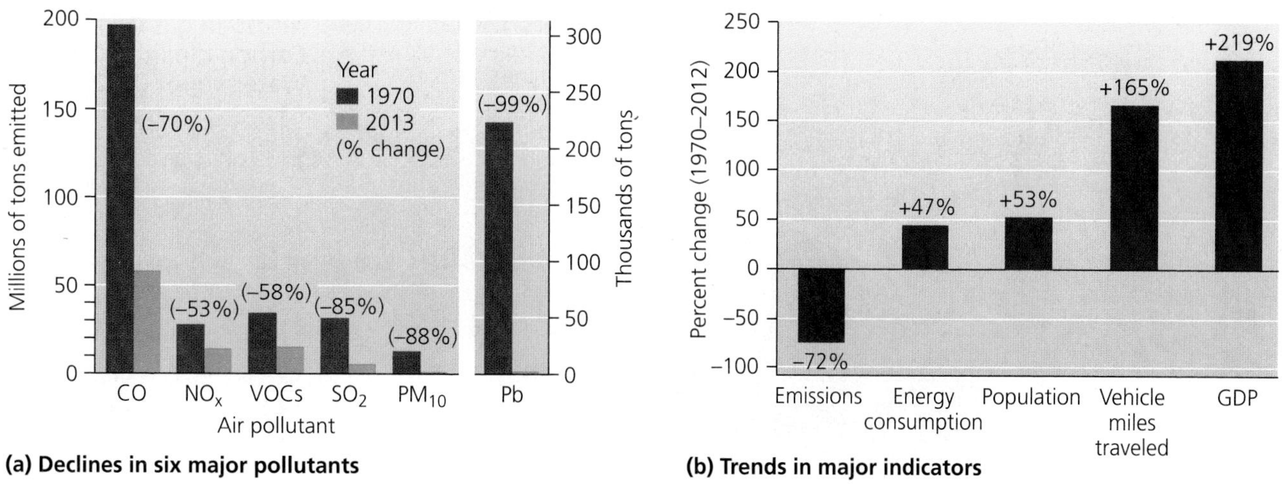


**(a) Declines in six major pollutants**

**(b) Trends in major indicators**

**FIGURE 13.7 U.S. emissions have declined sharply since 1970.** We have achieved reductions **(a)** in the six major pollutants tracked by the EPA, despite increases **(b)** in U.S. energy consumption, population, vehicle miles traveled, and gross domestic product. *Data from U.S. EPA.*

**DATA Q** By what percentage has population increased since 1970? By what percentage have emissions decreased? Using these two amounts, calculate the change in emissions per person.

GO TO **INTERPRETING GRAPHS & DATA** ON MasteringEnvironmentalScience®

**Lead** **Lead** (Pb) is a heavy metal that enters the atmosphere as a particulate pollutant. The lead-containing compounds tetraethyl lead and tetramethyl lead, when added to gasoline, improve engine performance. However, exhaust from the combustion of leaded gasoline emits airborne lead, which can be inhaled or can be deposited on land and water. When lead enters the food chain it accumulates in body tissues and can cause central nervous system malfunction and many other ailments (p. 209). Since the 1980s the United States and many other nations have phased out leaded gasoline (p. 5). Today most developing nations are following suit, although auto exhaust still creates significant lead pollution in many of them. In developed nations today, the main remaining source of atmospheric lead is industrial metal smelting.

## We have reduced pollutant emissions

Since the Clean Air Act of 1970, the United States has reduced emissions of each of the six monitored pollutants substantially (**FIGURE 13.7a**). These dramatic reductions in emissions have occurred despite significant increases in the nation's population, energy consumption, miles traveled by vehicle, and gross domestic product (**FIGURE 13.7b**). This success with pollution control is one of the top accomplishments of the United States in safeguarding human health and environmental quality. The EPA estimates that between 1970 and 1990, cleaner air saved the lives of 200,000 Americans. Similar advances have been made by most other wealthy nations in recent years.

This success in controlling pollution has come as a result of policy steps and technological developments, each motivated by grassroots social demand for cleaner air. In factories, power plants, and refineries, technologies such as baghouse filters, electrostatic precipitators, and **scrubbers** (**FIGURE 13.8**) chemically convert or physically remove pollutants before they are emitted from smokestacks. Cleaner-burning motor vehicle engines and automotive technologies such as catalytic converters have also played a large part. In a **catalytic converter**, engine exhaust reacts with metals that convert hydrocarbons, CO, and $NO_X$ into carbon dioxide, water vapor, and nitrogen

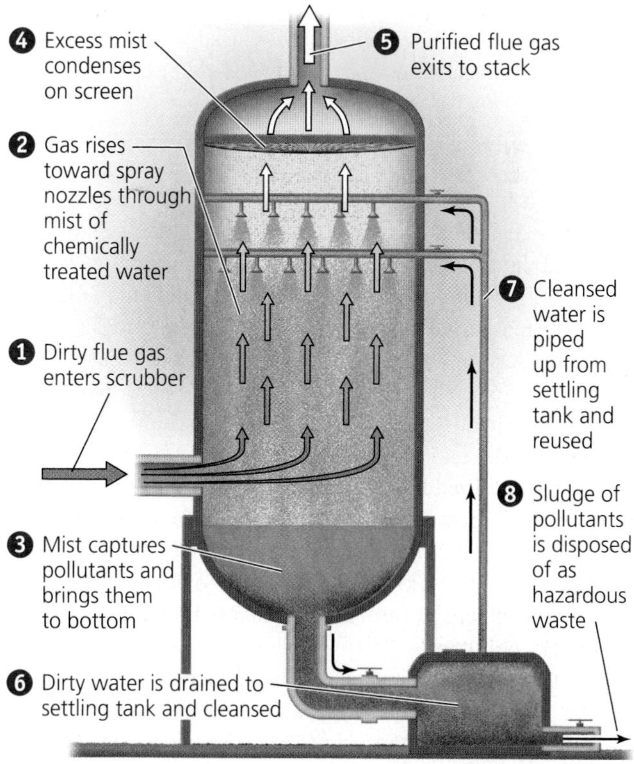


**FIGURE 13.8 Scrubbers typically remove at least 90% of particulate matter and gases such as sulfur dioxide.** Scrubbers and other pollution control devices come in many designs. In this spray-tower wet scrubber, polluted air rises through a chamber while nozzles spray a mist of water mixed with lime or other active chemicals to capture pollutants and wash them out of the air.

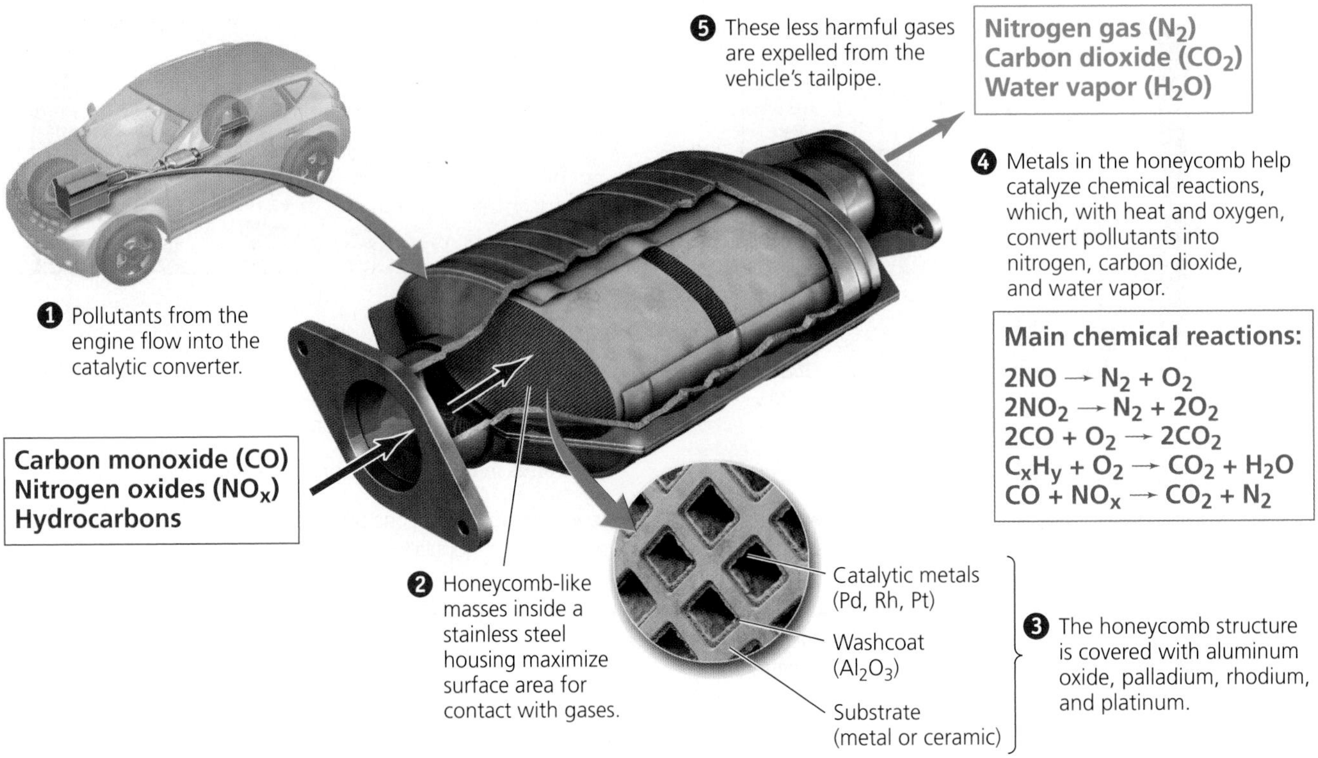
5 These less harmful gases are expelled from the vehicle's tailpipe.
Nitrogen gas ($N_2$)
Carbon dioxide ($CO_2$)
Water vapor ($H_2O$)
4 Metals in the honeycomb help catalyze chemical reactions, which, with heat and oxygen, convert pollutants into nitrogen, carbon dioxide, and water vapor.
Main chemical reactions:
$2NO \rightarrow N_2 + O_2$
$2NO_2 \rightarrow N_2 + 2O_2$
$2CO + O_2 \rightarrow 2CO_2$
$C_xH_y + O_2 \rightarrow CO_2 + H_2O$
$CO + NO_x \rightarrow CO_2 + N_2$
1 Pollutants from the engine flow into the catalytic converter.
Carbon monoxide (CO)
Nitrogen oxides ($NO_x$)
Hydrocarbons
2 Honeycomb-like masses inside a stainless steel housing maximize surface area for contact with gases.
Catalytic metals (Pd, Rh, Pt)
Washcoat ($Al_2O_3$)
Substrate (metal or ceramic)
3 The honeycomb structure is covered with aluminum oxide, palladium, rhodium, and platinum.

**FIGURE 13.9** **Catalytic converters improve air quality by filtering pollutants from vehicle exhaust.**

gas (**FIGURE 13.9**). The EPA's Acid Rain Program and its emissions trading system (pp. 296–297), along with clean coal technologies (p. 350), have reduced $SO_2$ and $NO_X$ emissions. Phaseouts of leaded gasoline have caused lead emissions to plummet; U.S. lead emissions fell by 93% in the 1980s alone.

## Air quality has improved

As a result of emissions reductions, air quality has improved markedly in industrialized nations. In the United States, the EPA and the states monitor outdoor air quality by measuring concentrations of six **criteria pollutants**, pollutants judged to pose substantial risk to human health. The six criteria pollutants include four of the six pollutants whose emissions are monitored—carbon monoxide, sulfur dioxide, particulate matter, and lead—as well as nitrogen dioxide (one type of nitrogen oxide) and tropospheric ozone.

**Nitrogen dioxide** ($NO_2$) is a foul-smelling, highly reactive, reddish brown gas that contributes to smog and acid deposition. Along with nitric oxide (NO), $NO_2$ belongs to the family of nitrogen oxides ($NO_X$). Nitric oxide reacts readily in the atmosphere to form $NO_2$, which is both a primary and secondary pollutant.

Although ozone in the stratosphere shields us from the dangers of UV radiation, ozone from human activity accumulates low in the troposphere. **Tropospheric ozone** ($O_3$; also called *ground-level ozone*) is a secondary pollutant, created by the reaction of nitrogen oxides and volatile carbon-containing chemicals in the presence of sunlight. A major component of photochemical smog (p. 291), this colorless gas poses health risks due to the instability of the $O_3$ molecule. This triplet of oxygen atoms will readily split into a molecule of oxygen gas ($O_2$) and a free oxygen atom. The oxygen atom may then participate in reactions that can injure living tissues and cause respiratory problems. Tropospheric ozone is the pollutant that most frequently exceeds its EPA standard.

Across the United States, over 4000 monitoring stations take hourly or daily air samples and share their data on pollutant concentrations with the EPA. The EPA compiles these data and calculates values on its **Air Quality Index** (AQI) for each site. Each of the six criteria pollutants receives an AQI value from 0 to 500 that reflects its current concentration. AQI values below 100 indicate satisfactory air conditions, and values above 100 indicate unhealthy conditions. The highest AQI value from the criteria pollutants on a particular day is reported as the overall AQI value for that day, and these values are made available online and reported in weather forecasts.

Thanks to the actions of scientists, policymakers, industrial leaders, and everyday citizens, outdoor air quality today is far better than it was a generation or two ago (**FIGURE 13.10**). However, plenty of room for improvement remains. Concerns over new pollutants are emerging, greenhouse gas emissions are altering the climate, and people in low-income communities often suffer from living near hotspots of pollution. In fact, many Americans live in areas where pollution continues to reach unhealthy levels. Residents of Los Angeles County, for instance, breathe air that violates health standards for five of the six criteria pollutants, while people in four adjacent southern California counties breathe air that violates four of the standards. As of 2012, 142 million Americans lived in counties that violated the national standard for at least one criteria pollutant. Still, L.A. and other pollution-choked metropolises are making perceptible headway toward cleaner air for their citizens (**FIGURE 13.11**).

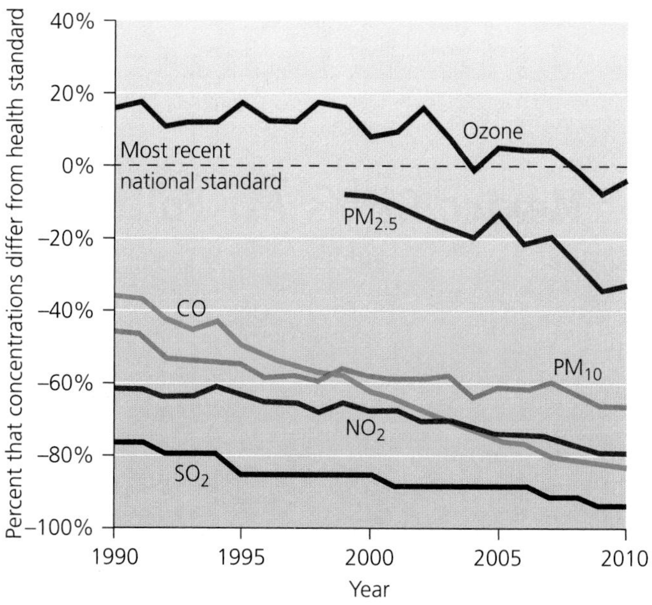


**FIGURE 13.10 Concentrations of criteria pollutants in ambient air across the United States have steadily fallen.** All except tropospheric ozone now average well below their health standards set by the EPA. Lead (not shown) has declined from 450% of its standard in 1990. However, local hotspots of pollution remain a problem, and U.S. standards are often more lax than standards set by European nations or the World Health Organization. *Data from U.S. EPA.*

## Air pollution is getting worse in industrializing nations

Although the United States and other industrialized nations have improved their air quality, outdoor air pollution is growing worse in many industrializing countries. In these societies, more and more citizens are driving automobiles while proliferating factories and power plants are emitting more pollutants. At the same time, many people continue to burn traditional sources of fuel, such as wood, charcoal, and coal, for cooking and home heating.

Mexico embodies these trends, and despite the progress in its capital, residents of Mexico City and many other Mexican cities and towns continue to suffer a variety of health impacts from polluted air (see **THE SCIENCE BEHIND THE STORY**, pp. 288–289).

People in the large sprawling cities of India, China, and other Asian nations suffer some of the world's worst air pollution. China has fueled its rapid industrial development with its abundant reserves of coal, the most-polluting fossil fuel. Power plants and factories often use outdated, inefficient technology because it is cheaper and quicker to build. Car ownership is skyrocketing; in the capital of Beijing alone, 1500 new cars hit the streets each day. As a result, in many Chinese cities, the haze is often too thick for people to see the sun.

In Beijing in the winter of 2013, smog became so severe that airplane flights were cancelled and people wore face

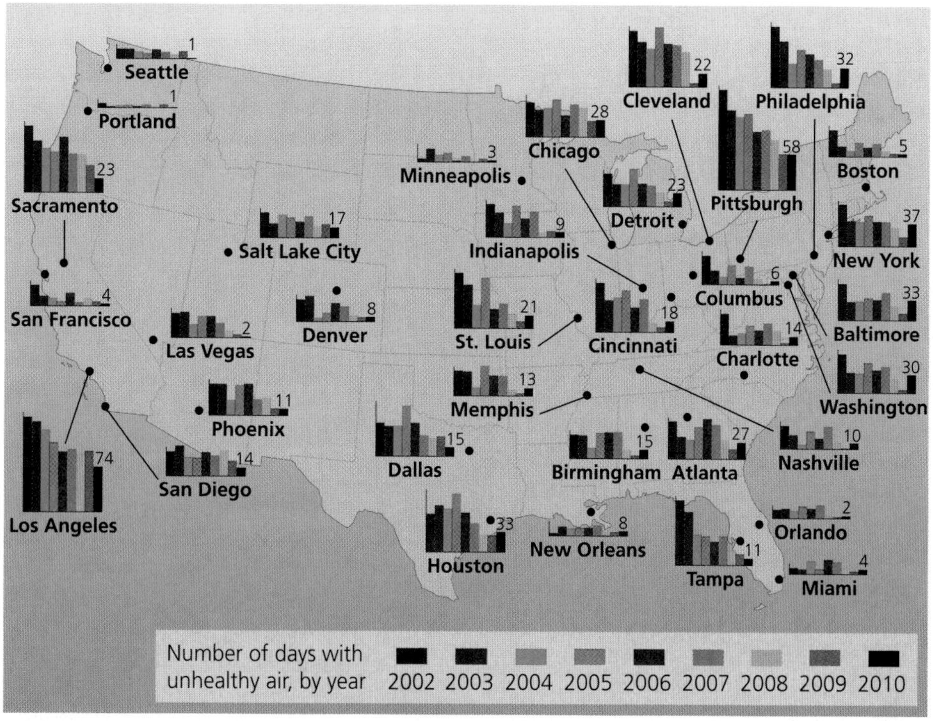


**FIGURE 13.11 In most U.S. cities, air has become cleaner.** This map shows numbers of days with unhealthy air from 2002 to 2010 for 35 metropolitan areas, according to the Air Quality Index (AQI). Each bar chart shows how many days in each year the city had AQI values above 100. The number of such unhealthy days in 2010 is printed for each city. *Data from U.S. EPA.*

DATA Q Locate where you live on this map. How does your nearest major city compare to others in its air quality? How has its air quality changed in recent years?

GO TO **INTERPRETING GRAPHS & DATA** ON MasteringEnvironmentalScience®

## Measuring the Health Impacts of Mexico City's Air Pollution

**Dr. Lilian Calderón-Garcidueñas**

"I know I'm inhaling poison," a 38-year-old candy vendor named Guadalupe told a reporter amid the fumes of a traffic-choked intersection in Mexico City. "But there is nothing I can do."

For as long as we have polluted our air, people have felt effects on their health. But quantifying those impacts poses a challenge for scientists. For researchers wanting to understand pollution's health impacts—and to design solutions for people like Guadalupe—what better place to go than Mexico City, long home to some of the world's worst air pollution?

A key first step is to determine what's in the air. One researcher who led the way is Mario Molina, a Nobel Prize–winning chemist who helped discover the cause of stratospheric ozone depletion (p. 293). Molina returned to his hometown of Mexico City to help address its pollution issues and in 2003 and 2006 organized air-sampling projects involving hundreds of scientists. Nearly 200 research papers later, these efforts have clarified many aspects of the city's pollution. One study used machines that could identify and record individual particles in real time. It found that metal-rich particulates from trash incinerators were peaking in the morning, whereas smoke from fires outside the city blew in during the afternoon. Other researchers discovered that VOCs control the amount of tropospheric ozone formed in smog (not nitrogen oxides, as was expected). City officials responded by choosing to target VOC emissions for reduction.

Few people understand Mexico City's air pollution better than Armando Retama, the city's director of atmospheric monitoring. But he may grasp its impacts best when he leaves town. "I can breathe better. I'm not all dry. My eyes aren't irritated. My skin doesn't crack," he says. "We have chronic symptoms that we aren't aware of."

Most health impacts of urban pollution affect the respiratory system. At high altitudes like Mexico City's, the "thin air" forces people to breathe deeply to obtain enough oxygen. This means they pull more air pollutants into their lungs than people at lower elevations. As a result, respiratory problems are commonplace. Many studies confirm that Mexico City residents show poorer lung function than people from less polluted areas and that respiratory problems and emergency room visits become more numerous when pollution is severe.

Most studies have looked at short-term exposure, but in 2007 a research team led by Isabelle Romieu of Mexico's National Institute of Public Health examined the effects of growing up amid polluted air. Her team measured lung function across 3 years in 3170 8-year-old children from 39 Mexico City schools and correlated this with their exposure to tropospheric ozone, nitrogen dioxide, and particulate matter. The children's ability to inhale and exhale deeply improved as they matured, but children from more polluted areas lagged behind those from cleaner areas, indicating smaller, weaker lungs.

Romieu and her colleagues also showed in a series of studies that the city's pollution worsens asthma in children, particularly those with certain genetic profiles. In 2008 her team analyzed data from 200 asthmatic and healthy children and found that children in areas with more traffic and pollutants coughed, wheezed, and used medication more often.

Another Mexican researcher, Lilian Calderón-Garcidueñas, has led several studies comparing chest X-ray films and medical records of Mexico City children with those of similar children from nonpolluted locations. Her team found hyperinflation and other problems with the lungs of Mexico City youth. Mexico City children also reported many respiratory problems, whereas rural children did not (**FIGURE 1**).

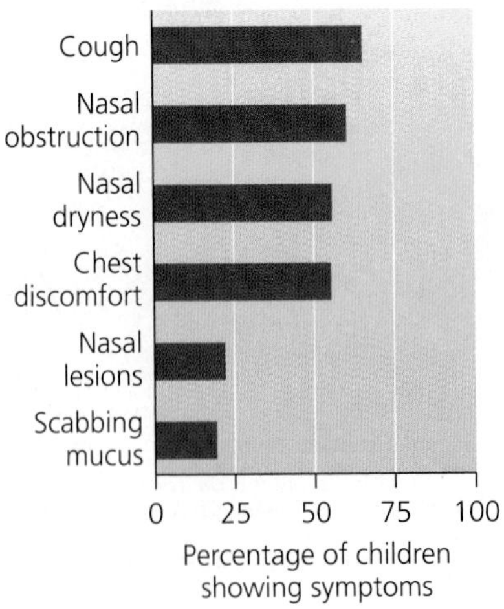


**FIGURE 1 Mexico City children show respiratory symptoms from air pollution.** These data are from 174 Mexico City children. Of 27 similar children from less polluted areas outside Mexico City, none of them showed any of these conditions. *Data from Calderón-Garcidueñas, L., et al., 2003. Respiratory damage in children exposed to urban pollution.* Pediatric Pulmonology *36: 148–161.*

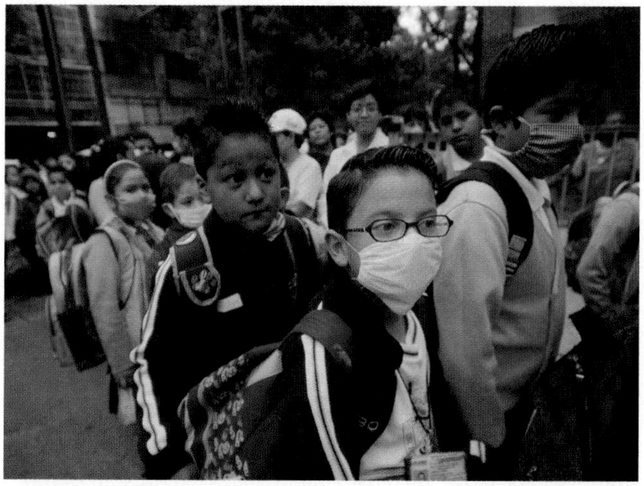

**FIGURE 2 Air pollution begins affecting the health of Mexico City schoolchildren at an early age.**

Air pollution harms the heart and the cardiovascular system, too. Multiple studies reveal that pollution can affect heart rate, blood pressure, blood clotting, blood vessels, and atherosclerosis. Epidemiological studies (p. 217) show that pollution correlates with emergency room admissions for heart attacks, chest pain, and heart failure, as well as death from heart-related causes.

Part of the reason smog hurts the cardiovascular system is that tiny particulates can work their way into the bloodstream, causing the heart to reduce blood flow or go out of rhythm. Even young people are at risk (**FIGURE 2**). One Mexican research team analyzed the hearts of 21 Mexico City residents who had died at an early age, and found that pollution exacts a toll before age 18. The heart mounts an inflammatory response against pollutant particles laden with dead bacteria in the blood, but because the pollution is persistent, the inflammation becomes chronic and stresses the heart.

Besides affecting the heart and the lungs, air pollution can affect the brain. Recent research shows it can damage children's brain tissue in ways similar to Alzheimer's disease. In one study, Calderón-Garciduenas and her colleagues used brain scans and found that 56% of Mexico City youth had lesions on the prefrontal cortex, whereas fewer than 8% did in a region with clean air. In another study, they compared 20 children aged 7–8 from Mexico City with 10 similar children from a Mexican city with clean air, measuring their cognitive skills and scanning their brains with magnetic resonance imagery (MRI). Results showed that the Mexico City children performed more poorly on most cognitive tests of reasoning, knowledge, and memory. The differences in cognition were consistent with differences in volume of white matter in key portions of the brain as revealed by the MRIs.

All these impacts can lead to higher rates of death. Studies by a U.S. and Mexican research team in Mexico City in the late 1990s confirmed this by comparing death certificate records against air pollution measurements. The team found that death rates rose on the day of and the day after severe pollution episodes, especially in response to particulate matter (**FIGURE 3a**). They also found that infant mortality was significantly higher in the days following strong pollution episodes (**FIGURE 3b**).

The extensive research showing a diversity of health impacts from air pollution has caught the attention of Mexico City's leaders. These scientific findings have helped motivate them to work hard cleaning up their city's air. ☐

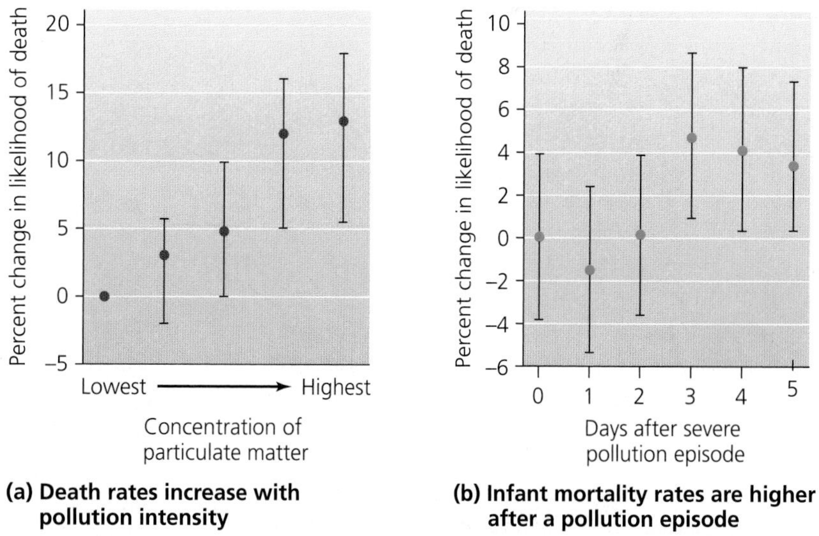


**FIGURE 3 Rates of (a) death and (b) infant mortality each increase in Mexico City with exposure to air pollution.** *Data from (a) Borja-Aburto, V., et al., 1997. Ozone, suspended particulates, and daily mortality in Mexico City.* Am. J. Epidemiology *145: 258–268; and (b) Loomis, D., et al., 1999. Air pollution and infant mortality in Mexico City.* Epidemiology *10: 118–123.*

**FIGURE 13.12 Air quality is poor in cities of today's developing nations.** At Tiananmen Square in Beijing, China, children wear face masks during the "airpocalypse" that gripped the city in 2013.

masks to breathe (**FIGURE 13.12**). Levels of particulate matter were literally off the charts; a pollution monitor atop the U.S. Embassy designed to measure the Air Quality Index detected record-breaking readings beyond the maximum value of 500, up to 755. During this "airpocalypse," a fire at a factory went unnoticed for three hours because the smog was so thick that no one could see the smoke! Thousands of people suffered ill health as the pollution soared 30 times past the WHO's safe limits. In winter 2014, the airpocalypse returned, and air quality readings exceeded 500 yet again.

Across China, the day-by-day health impacts of air pollution are enormous. A 2013 international research report blamed outdoor air pollution for 1.2 million premature deaths in China each year. Another 2013 study found that residents of polluted northern China die on average five years earlier than residents of southern China, where the air is cleaner. Moreover, prevailing westerly winds carry some of China's pollution across the Pacific Ocean to North America! A 2014 study quantified Chinese pollution reaching the U.S. West Coast and estimated that it adds at least one extra smoggy day per year to Los Angeles's total.

China's government is now striving to reduce pollution. It has closed down some heavily polluting factories and mines, phased out some subsidies for polluting industries, and installed pollution controls in power plants. It subsidizes efficient electric heaters for peoples' homes to replace dirty, inefficient coal stoves. It has mandated cleaner gasoline and diesel and has raised standards for fuel efficiency and emissions for cars above what the United States requires. In Beijing, mass transit is being expanded, many buses run on natural gas, and heavily polluting vehicles are kept out of the central city. China is also aggressively developing wind, solar, and nuclear power to substitute for coal-fired power.

## Smog poses health risks

Let's now examine one of our most prevalent types of air pollution: smog. *Smog* is an unhealthy mixture of air pollutants that often forms over urban areas as a result of fossil fuel combustion.

Since the onset of the industrial revolution, cities have suffered a type of smog known as **industrial smog** (**FIGURE 13.13a**).

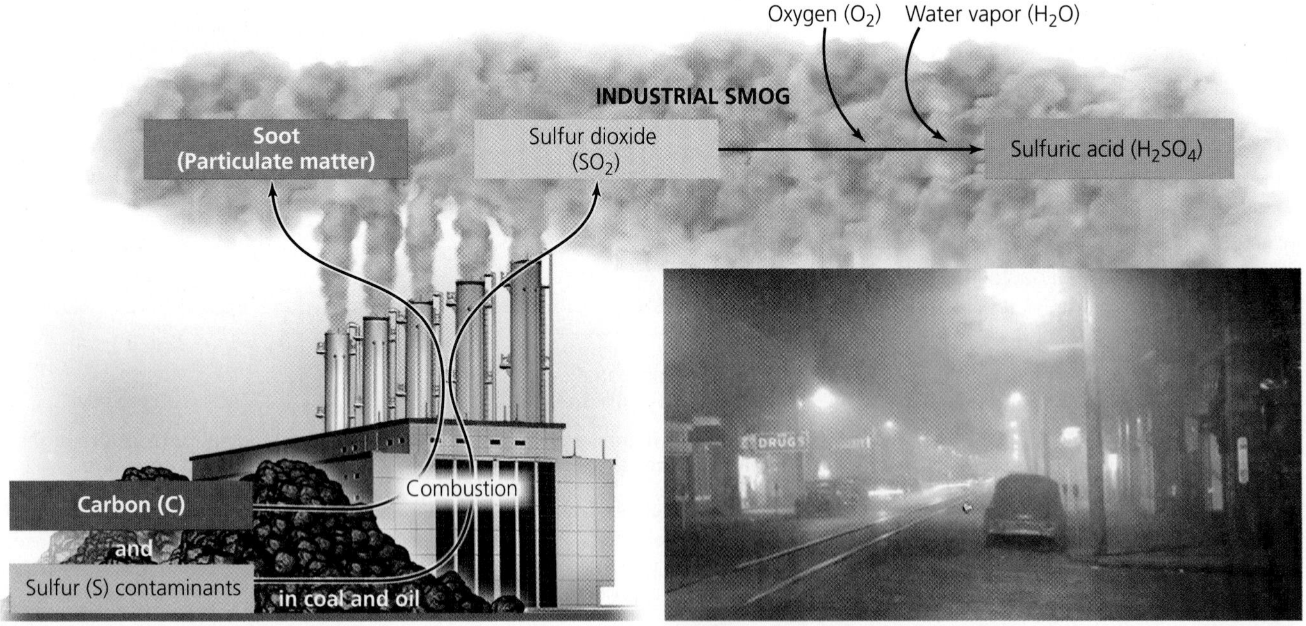


**(a) Formation of industrial smog**

**(b) Donora, Pennsylvania, at midday in its 1948 smog event**

**FIGURE 13.13 Industrial smog results from fossil fuel combustion.** When coal or oil is burned in a power plant or factory, soot (particulate matter of carbon) is released, and sulfur contaminants give rise to sulfur dioxide, which may react with atmospheric gases to produce further compounds **(a)**. Carbon monoxide and carbon dioxide are also emitted. Under certain weather conditions, industrial smog can blanket whole regions, as it did in Donora, Pennsylvania **(b)**, shown in the daytime during its deadly 1948 smog episode.

When coal or oil is burned, some portion is completely combusted, forming $CO_2$; some is partially combusted, producing CO; and some remains unburned and is released as soot (particles of carbon). Moreover, coal contains contaminants such as mercury and sulfur. Sulfur reacts with oxygen to form sulfur dioxide, which can undergo a series of reactions to form sulfuric acid and other chemicals. These substances, along with soot, are the main components of industrial smog.

America's most severe industrial smog event occurred in Pennsylvania, in a small town named Donora, in 1948 (**FIGURE 13.13b**). Donora is located in a mountain valley, and one day after air had cooled during the night, the morning sun did not reach the valley floor to warm and disperse the cold air. The resulting thermal inversion trapped smog from a steel and wire factory. Twenty-one people died, and over 6000 people—nearly half the town—became ill.

The world's worst industrial smog crisis occurred in London, England, in 1952, when a high-pressure system settled over the city for several days, trapping pollutants from factories and coal-burning stoves and creating foul conditions that killed 4000 people—and by some estimates up to 12,000. In the wake of "killer smog" episodes such as those in London and Donora, governments of developed nations began regulating industrial emissions and greatly reduced industrial smog. However, in industrializing regions such as China, India, and eastern Europe, coal burning and lax pollution control continue to result in industrial smog that poses significant health risks.

For most urban areas today, pollution results largely from automobile exhaust. In Mexico City, vehicles contribute 31% of volatile organic compounds, 50% of sulfur dioxide, and 82% of nitrogen oxides. Mexico City and Los Angeles also have sunny climates. As a result, they and cities like them suffer from a different type of smog. **Photochemical smog** forms when sunlight drives chemical reactions between primary pollutants and atmospheric compounds, producing a potent cocktail of over 100 different chemicals, tropospheric ozone often being the most abundant (**FIGURE 13.14a**). Because it also includes $NO_2$, photochemical smog generally appears as a brownish haze (**FIGURE 13.14b**).

Hot, sunny, windless days in urban areas provide perfect conditions for the formation of photochemical smog. Exhaust from morning traffic releases NO and VOCs into a city's air. Sunlight then promotes the production of ozone and other secondary pollutants, leading pollution typically to peak in midafternoon. Photochemical smog irritates people's eyes, noses, and throats, and over time can lead to asthma, lung damage, heart problems, decreased resistance to infection, and even cancer.

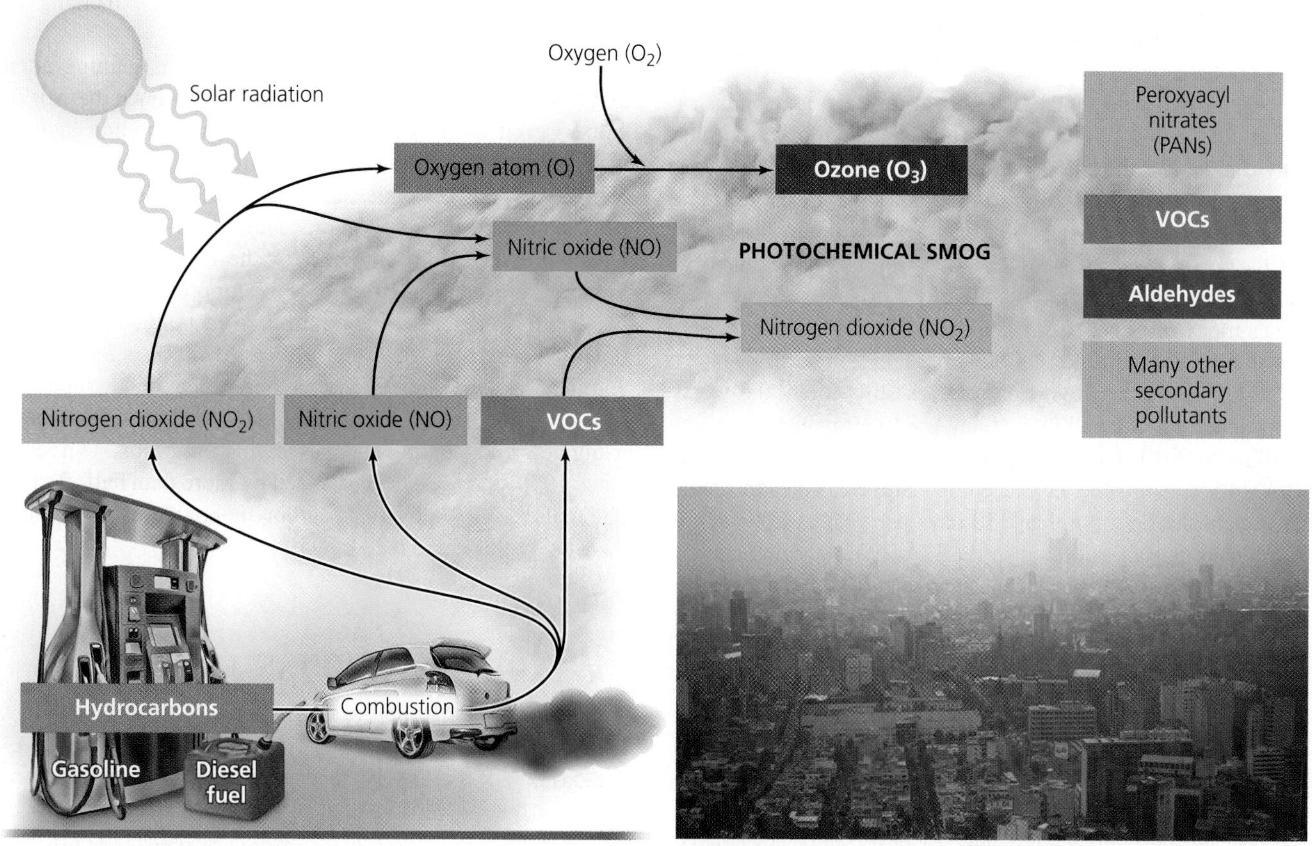


**(a) Formation of photochemical smog**

**(b) Photochemical smog over Mexico City**

**FIGURE 13.14 Photochemical smog results when pollutants from automobile exhaust react amid exposure to sunlight.** Nitrogen dioxide, nitric oxide, and VOCs initiate a series of chemical reactions **(a)** that produce a toxic brew of secondary pollutants including ozone, peroxyacyl nitrates (PANs), aldehydes, and others. Photochemical smog is common over Mexico City **(b)** and many other urban areas, especially those with hilly topography or frequent inversions.

## We can take steps to reduce smog

Smog afflicts countless American cities and suburbs, from Newark to Baltimore to Houston to Salt Lake City. Wherever you live, chances are you have experienced photochemical smog. Mayors, city councils, and state and federal regulators everywhere are trying to devise ways to clear their air.

Los Angeles's struggle with air pollution began in 1943, when the city's first major smog episode cut visibility to three blocks. With the city's image as a clean and beautiful coastal haven at risk, civic leaders confronted the problem head-on. In fact, Los Angeles's quest to understand and solve its smog problem spurred much of the original research into how photochemical smog forms and how automobiles might burn fuel more cleanly.

City and county officials began by passing ordinances restricting emissions from power plants, oil refineries, and the petrochemical industry, then continued with efforts to lower emissions from motor vehicles. Because air pollution spreads from place to place, responsibility for pollution control soon moved from the city and county levels to the state and federal levels.

California took the lead among U.S. states in adopting pollution control technology and in setting emissions standards for vehicles. In 1967 state leaders established the California Air Resources Board, the first state agency focused on regulating air quality. Today in California and 33 other states, drivers are required to have their vehicle exhaust inspected periodically. Inspection programs have cut vehicle emissions by 30% in these states.

California's demands also helped lead the auto industry to develop less-polluting cars. A study by the nonprofit group Environment California concluded that a new car today generates just 1% of the smog-forming emissions of a 1960s-era car. For this reason, the air is cleaner, even with more vehicles on the road. In Los Angeles, VOC pollution has declined by 98% since 1960, even though the city's drivers now burn 2.7 times more gasoline. L.A.'s peak smog levels have decreased substantially since 1980 (**FIGURE 13.15**).

Despite its progress, Los Angeles still suffers the worst tropospheric ozone pollution of any U.S. metropolitan area, according to 2013 rankings by the American Lung Association. L.A. residents breathe air exceeding California's health standard for ozone on more than 130 days per year. A 2008 study calculated that air pollution in the L.A. basin and the nearby San Joaquin Valley each year caused nearly 3900 premature deaths and cost society $28 billion (due to hospital admissions, lost workdays, etc.).

Many of L.A.'s sister cities are tackling photochemical smog. In Tehran, Iran, vehicle inspections are required, traffic into the city center is restricted, and drivers are paid to turn in old, polluting cars for newer, cleaner ones. The government reduced sulfur in diesel fuel and converted buses to run on (cleaner-burning) natural gas. To raise public awareness, 22 real-time pollution indicator boards were installed around the city, electronically displaying current pollutant levels. All these efforts helped reduce pollution, yet so many people were streaming into the city and buying cars that pollution soon grew worse again. In response, government officials lowered gasoline subsidies, rationed fuel, and began expanding the subway system.

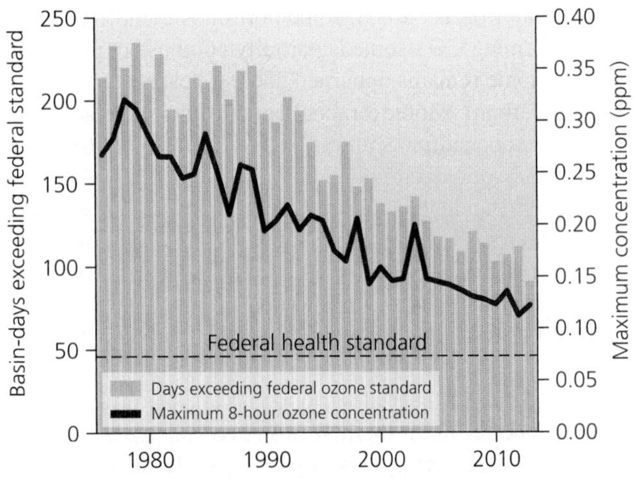


**FIGURE 13.15 In the Los Angeles region, tropospheric ozone in photochemical smog has decreased since the 1970s, thanks to public policy and improved automotive technology.** Ozone pollution still violates the federal health standard, however.
*Data from South Coast Air Quality Management District.*

**DATA Q** By roughly what percentage has Los Angeles reduced its ozone pollution since the late 1970s? Calculate changes in each data set shown. Do the two data sets show similar patterns?
GO TO **INTERPRETING GRAPHS & DATA** ON MasteringEnvironmentalScience®

Of all the world's cities, Mexico City is gaining attention today for its success in reducing smog—once the world's worst—even as its population, cars, and economic activity have grown. Regulations now require cars to have catalytic converters and get emissions tests. Some industrial facilities cleaned up their operations, and others were forced out. Under pressure from city leaders, the national oil company Pemex removed lead from gasoline, improved its refineries, imported cleaner gasoline, and removed pollutants from the liquefied petroleum gas that city residents use for cooking and heating. The subway system and a fleet of low-emission buses continue to be expanded today, while residents begin using new bike-sharing and car-sharing programs. As a result, since 1990 smog has been reduced by more than half. Particulate matter is down by 70%, carbon monoxide by 74%, sulfur dioxide by 86%, and lead by 95%.

### WEIGHING THE ISSUES

**Your Region's Air Quality**

What outdoor air pollution challenges exist in your region? Explore one of the EPA websites that lets you browse information on the air you breathe: www.airnow.gov, www.epa.gov/aircompare, or www.epa.gov/air/emissions/where.htm. How does your region's air quality compare to the rest of the nation? What factors influence the quality of your region's air? Propose three steps for reducing air pollution in your region. What benefits might your region enjoy after taking such steps?

## Should we regulate greenhouse gases as air pollutants?

Because most nations continue to release vast quantities of carbon dioxide and other greenhouse gases that warm the lower atmosphere and drive global climate change (Chapter 14), this is arguably today's biggest air pollution problem. Industry and utilities generate many of these emissions, but all of us contribute by living carbon-intensive lifestyles. Each year the average U.S. vehicle driver releases close to 6 metric tons of carbon dioxide, 275 kg (605 lb) of methane, and 19 kg (41 lb) of nitrous oxide, all of them greenhouse gases that drive climate change.

In 2007 the U.S. Supreme Court ruled that the EPA has legal authority under the Clean Air Act to regulate carbon dioxide and other greenhouse gases as air pollutants. President Barack Obama voiced his preference that Congress address greenhouse gas emissions through bipartisan legislation. When Congress failed to do so, Obama instructed the EPA to develop regulations for these emissions. In 2011, the EPA introduced moderate carbon emission standards for cars and light trucks, and in 2012 it announced that it would limit carbon emissions for new coal-fired power plants and cement factories (but not existing ones). The EPA decided to phase in regulations gradually, beginning with the largest emitters.

Several states joined the coal-mining and petrochemical industries in suing the EPA, but a court of appeals unanimously upheld the regulations in 2012. The automotive industry supported the EPA's regulations. U.S. automakers had already begun investing in fuel-efficient vehicles, and were glad to have one set of federal emissions standards, so as not to have to worry about meeting many differing state standards. The public also voiced strong support; 2.1 million Americans sent comments to the EPA in favor of its actions—a record number of public comments for any federal regulation.

In 2014, the EPA proposed a regulatory plan for existing power plants. Under the plan, states would be allowed to choose how to reduce their plants' emissions—by upgrading technology, switching from coal to natural gas, enhancing efficiency, promoting renewable energy, or by using carbon taxes or cap-and-trade programs. The plan aimed to cut carbon dioxide emissions by 30% (from 2005 levels) by the year 2030. A number of states and industries soon lined up to challenge the plan in court.

The EPA will no doubt continue to face formidable opposition from emitting industries and from policymakers who fear that regulations will hamper economic growth. Yet if we were able to reduce emissions of other pollutants sharply since 1970 while advancing our economy, we can hope to achieve similar results in reducing greenhouse gas emissions. Indeed, although U.S. carbon dioxide emissions rose by 51% from 1970 to 2007, they fell by 10% from 2007 to 2013 even as the economy grew. This decrease in emissions resulted partly from reduced energy use during recession years, but also from a shift from coal to cleaner-burning natural gas, and from improved fuel efficiency in automobiles and other technologies.

# Ozone Depletion

Although ozone in the troposphere is a pollutant in photochemical smog, ozone in the stratosphere (pp. 280–281) protects life on Earth by absorbing the sun's ultraviolet (UV) radiation, which can damage tissues and DNA. When scientists discovered that our planet's stratospheric ozone was being depleted, they realized this posed a major threat to human health and the environment. Years of research by hundreds of scientists revealed that certain airborne chemicals destroy ozone and that most of these **ozone-depleting substances** are human-made. Our subsequent campaign to halt degradation of the ozone layer stands as one of society's most successful efforts to address a major environmental problem.

## Synthetic chemicals deplete ozone

Researchers identifying ozone-depleting substances pinpointed primarily **halocarbons**—human-made compounds derived from simple hydrocarbons (p. 29) in which hydrogen atoms are replaced by halogen atoms, such as chlorine, bromine, or fluorine. In the 1970s, industry was producing over 1 million tons per year of one type of halocarbon, **chlorofluorocarbons (CFCs)**. CFCs were useful as refrigerants, as fire extinguishers, as propellants for aerosol spray cans, as cleaners for electronics, and for making polystyrene foam.

Unfortunately CFCs can linger in the stratosphere for a century or more. There, intense UV radiation from the sun breaks bonds in CFC molecules, releasing their chlorine atoms. In a two-step chemical reaction (**FIGURE 13.16**), each newly freed chlorine atom can split an ozone molecule and then ready itself to split more. During its long residence time, each free chlorine atom can catalyze the destruction of as many as 100,000 ozone molecules!

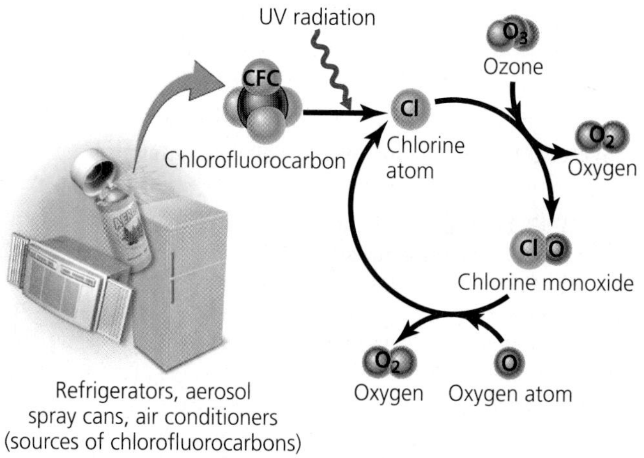


**FIGURE 13.16 CFCs destroy ozone in a multistep process, repeated many times.** A chlorine atom released from a CFC molecule in the presence of UV radiation reacts with an ozone molecule, forming one molecule of oxygen gas and one chlorine monoxide (ClO) molecule. The oxygen atom of the ClO molecule then binds with a stray oxygen atom to form oxygen gas, leaving the chlorine atom to begin the destructive cycle anew.

## The ozone hole appears each year

In 1985, researchers shocked the world by announcing that stratospheric ozone levels over Antarctica in the southern springtime had declined by nearly half during just the previous decade, leaving a thinned ozone concentration that was soon named the **ozone hole** (**FIGURE 13.17a, b**). During each Southern Hemisphere spring since then, ozone concentrations over this immense region have dipped to roughly half their historic levels.

Extensive scientific detective work has revealed why seasonal ozone depletion is so severe over Antarctica (and to a lesser extent, the Arctic). During the dark and frigid Antarctic winter, icy clouds form in the stratosphere containing condensed nitric acid, which splits chlorine atoms off from compounds such as CFCs. The freed chlorine atoms accumulate in the clouds, trapped over Antarctica by circular wind currents. In the Antarctic spring (starting in September), sunshine returns and dissipates the clouds. This releases the chlorine atoms, which begin destroying ozone. The solar radiation also catalyzes chemical reactions, speeding up ozone depletion as temperatures warm. The ozone hole lingers over Antarctica until December, when warmth weakens the circular currents, allowing ozone-depleted air to diffuse away and ozone-rich air from elsewhere to stream in. The ozone hole vanishes until the following spring.

By the time scientists had worked this out, plummeting ozone levels had become a serious international concern. Ozone depletion was occurring globally, year-round (not just in Antarctica) (**FIGURE 13.17c**). Scientists worried that intensified UV exposure at Earth's surface would promote skin cancer, damage crops, and kill off ocean phytoplankton, the base of the marine food chain.

## We addressed ozone depletion with the Montreal Protocol

Policymakers responded to the scientific concerns, and international efforts to restrict production of CFCs bore fruit in 1987 with the **Montreal Protocol**. In this treaty, the world's nations agreed to cut CFC production in half by 1998. Follow-up agreements deepened the cuts, advanced timetables for compliance, and added nearly 100 further ozone-depleting substances. Most substances covered by these agreements have now been phased out, and industry has been able to shift to alternative chemicals. As a result, we have evidently halted the advance of ozone depletion and stopped the Antarctic ozone hole from growing worse (**FIGURE 13.18**)—a success that all humanity can celebrate.

Earth's ozone layer is not expected to recover completely until about 2070. Much of the 5.5 million tons of CFCs emitted into the troposphere has not yet diffused into the stratosphere, so concentrations may not peak there until 2020. Because of this time lag and the long residence times of many halocarbons, we can expect many years to pass before our policies have the desired result. Such time lags are one reason scientists often

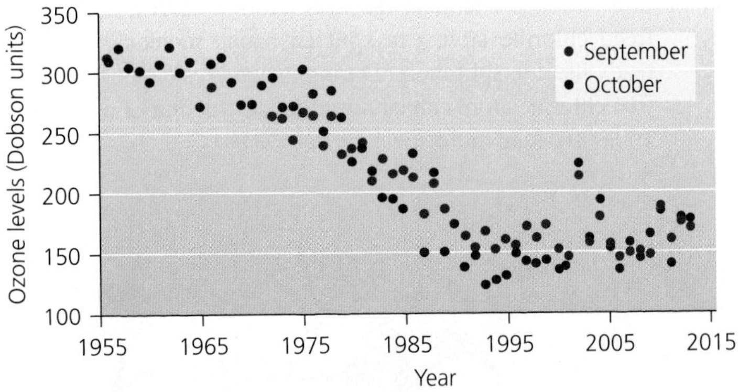


**(a) Monthly mean ozone levels at Halley, Antarctica**

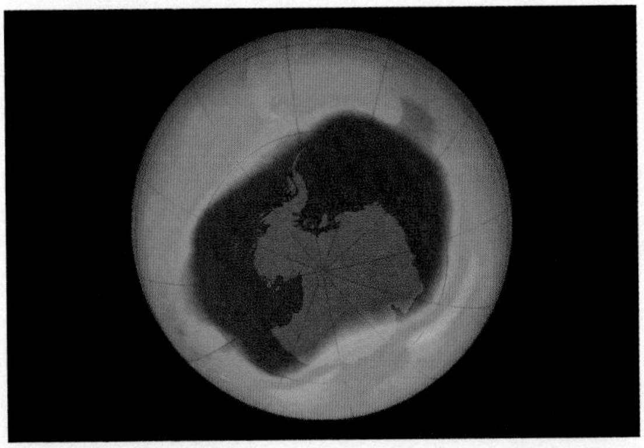

**(b) The ozone hole**

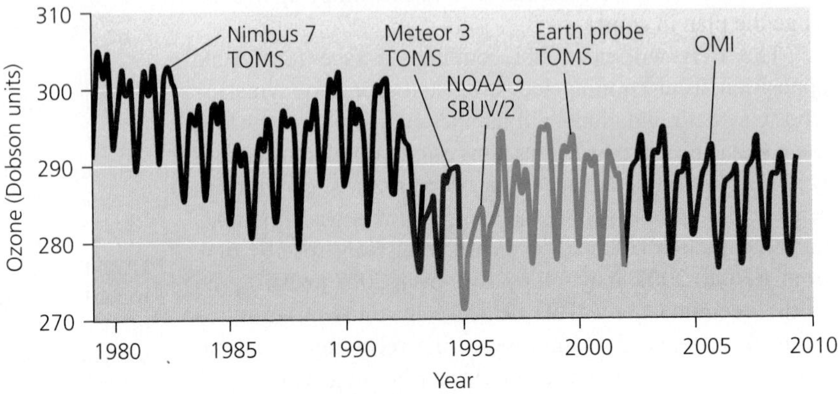


**(c) Global ozone readings across the world from 5 satellites**

**FIGURE 13.17 The "ozone hole" is a vast area of thinned ozone density in the stratosphere over the Antarctic region.** It has reappeared seasonally each September in recent decades. Data from Halley, Antarctica **(a)**, show a decrease in stratospheric ozone concentrations from the 1960s to 1990. Once ozone-depleting substances began to be phased out, ozone concentrations stopped declining. Colorized satellite imagery of Earth's Southern Hemisphere from September 24, 2006 **(b)**, shows the ozone hole **(purple/blue)** at its maximal recorded extent. Ozone decline and stabilization are also evident globally **(c)**, as seen in globally averaged data from five satellites. *Data from (a) British Antarctic Survey; and (c) NASA.*

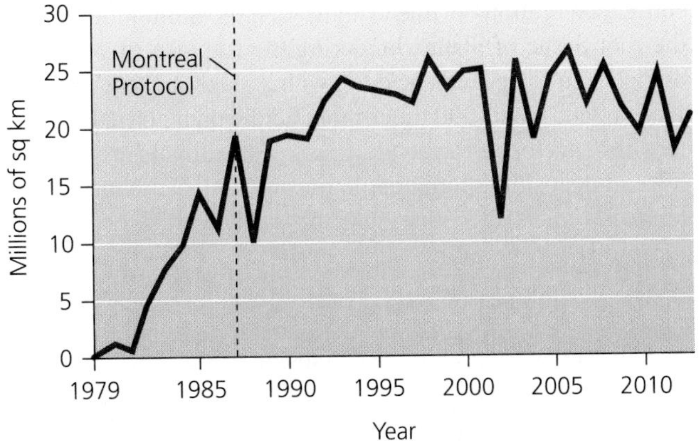


**FIGURE 13.18 The Antarctic ozone hole grew quickly after its appearance, but phase-outs of ozone-depleting substances since 1987 have halted its growth.** *Data from NASA are averages from 7 Sept. to 13 Oct. each year.*

## FAQ Is the ozone hole related to global warming?

This is a common misconception held by many people. In reality, stratospheric ozone depletion and global warming are completely different issues. Ozone depletion allows excess ultraviolet radiation from the sun to penetrate the atmosphere, but this does not significantly warm or cool the atmosphere. Conversely, global warming does not appreciably affect ozone loss. Although some recent research does suggest that the ozone hole may affect circulation and rainfall in the Southern Hemisphere, on the whole these two phenomena are unrelated. However, by coincidence many ozone-depleting substances banned by the Montreal Protocol also happen to be greenhouse gases that warm the atmosphere. Thus, although the Montreal Protocol was designed to combat ozone depletion, it is also helping us slow down climate change.

argue for proactive policy guided by the precautionary principle (p. 155), rather than reactive policy that risks responding too late.

Because of its success in addressing ozone depletion, the Montreal Protocol is widely viewed as a model for international cooperation on other global problems, from biodiversity loss (p. 179) to persistent organic pollutants (p. 222) to climate change (p. 325).

# Addressing Acid Deposition

Just as stratospheric ozone depletion crosses political boundaries, so does **acid deposition**, the deposition of acidic (p. 28) or acid-forming pollutants from the atmosphere onto Earth's surface. As with ozone depletion, we are having some success in addressing this challenge.

## Fossil fuel combustion spreads acidic pollutants far and wide

Acid deposition often takes place by precipitation (commonly referred to as **acid rain**) but also may occur by fog, gases, or the deposition of dry particles. Acid deposition is one type of **atmospheric deposition**, which refers broadly to the wet or dry deposition of a variety of pollutants, including mercury, nitrates, organochlorines, and others.

Acid deposition originates primarily with the emission of sulfur dioxide and nitrogen oxides, largely through fossil fuel combustion by automobiles, electric utilities, and industrial facilities. Once airborne, these pollutants react with water, oxygen, and oxidants to produce compounds of low pH (p. 28), primarily sulfuric acid and nitric acid. Suspended in the troposphere, these acids may travel days or weeks for hundreds of kilometers (**FIGURE 13.19**).

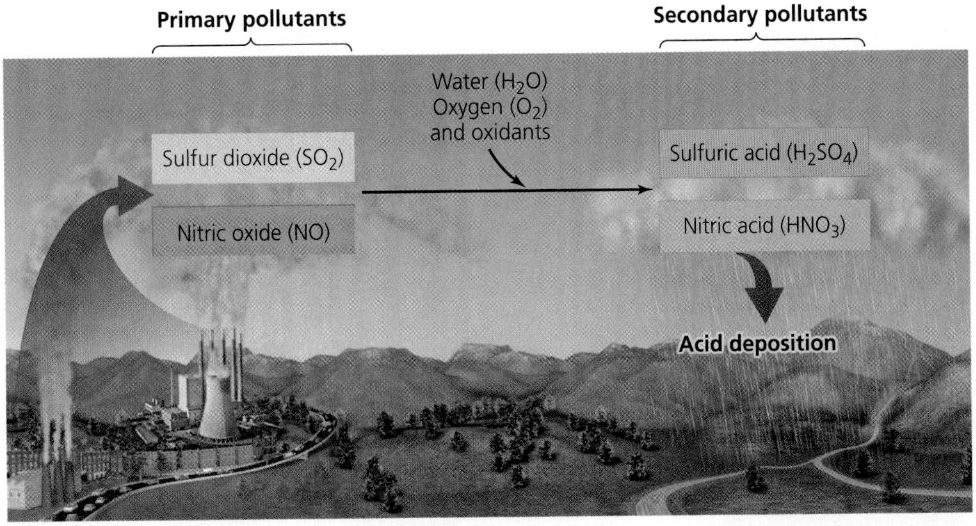


**FIGURE 13.19 Acid deposition has impacts far downwind from where pollutants are released.** Sulfur dioxide and nitric oxide emitted by industries, utilities, and vehicles react in the atmosphere to form sulfuric acid and nitric acid. These acidic compounds descend to Earth's surface in rain, snow, fog, and dry deposition.

**TABLE 13.1 Impacts of Acid Deposition**

| Acid deposition in northeastern U.S. forests has . . . |
|---|
| • accelerated leaching of base cations (ions such as $Ca^{2+}$, $Mg^{2+}$, $NA^{+}$, and $K^{+}$, which counteract acid deposition) from soil |
| • allowed sulfur and nitrogen to accumulate in soil, where excess N can encourage weeds |
| • increased dissolved inorganic aluminum in soil, hindering plant uptake of water and nutrients |
| • leached calcium from needles of red spruce, causing trees to die from wintertime freezing |
| • increased mortality of sugar maples due to leaching of base cations from soil and leaves |
| • acidified hundreds of lakes and diminished their capacity to neutralize further acids |
| • elevated aluminum levels in surface waters |
| • reduced species diversity and abundance of aquatic life, affecting entire food webs |

*Adapted from Driscoll, C. T., et al., 2001.* Acid rain revisited. *Hubbard Brook Research Foundation. © 2001 C. T. Driscoll. Used with permission.*

## Acid deposition has many impacts

Acid deposition has wide-ranging detrimental effects on ecosystems and on our infrastructure (TABLE 13.1). Acids leach nutrients such as calcium, magnesium, and potassium ions from the topsoil, altering soil chemistry and harming plants and soil organisms. This occurs because hydrogen ions from acid deposition take the place of calcium, magnesium, and potassium ions in soil compounds, and these valuable nutrients leach into the subsoil, where they become inaccessible to plant roots.

Acid deposition also "mobilizes" toxic metal ions such as aluminum, zinc, mercury, and copper by chemically converting them from insoluble forms to soluble forms. Elevated soil concentrations of metal ions such as aluminum damage the root tissue of plants, hindering their uptake of water and nutrients. In some areas, acid fog with a pH of 2.3 (equivalent to vinegar, and over 1000 times more acidic than normal rainwater) has enveloped forests, killing trees. Animals are affected by acid deposition, too; populations of snails and other invertebrates typically decline, and this reduces the food supply for birds.

When acidic water runs off from land, it affects streams, rivers, and lakes. Thousands of lakes in Canada, Europe, the United States, and elsewhere have lost their fish because acid deposition leaches aluminum ions out of soil and rock and into waterways. These ions damage the gills of fish and disrupt their salt balance, water balance, breathing, and circulation.

Besides altering ecosystems, acid deposition damages crops, erodes stone buildings, corrodes cars, and erases the writing from tombstones. Ancient cathedrals in Europe, temples in Asia, and monuments in Washington, D.C. are experiencing untold damage as their features dissolve away (**FIGURE 13.20**).

Because the pollutants leading to acid deposition can travel long distances, their effects may be felt far from their sources. Much of the pollution from power plants and factories in Pennsylvania, Ohio, and Illinois travels east with prevailing winds and falls out in states such as New York, Vermont, and New Hampshire. As a result, regions of greatest acidification tend to be downwind from heavily industrialized source areas of pollution.

## We are addressing acid deposition

The Acid Rain Program established under the Clean Air Act of 1990 has helped fight acid deposition in the United States. This program set up an emissions trading system (p. 111) for sulfur dioxide. Coal-fired power plants were allocated permits for emitting $SO_2$ and could buy, sell, or trade these

**(a) Before acid rain damage**

**(b) After acid rain damage**

**FIGURE 13.20 Acid deposition corrodes statues and buildings.** Shown is an Egyptian obelisk known as Cleopatra's Needle, in Central Park, New York City, **(a)** before and **(b)** after significant acid deposition.

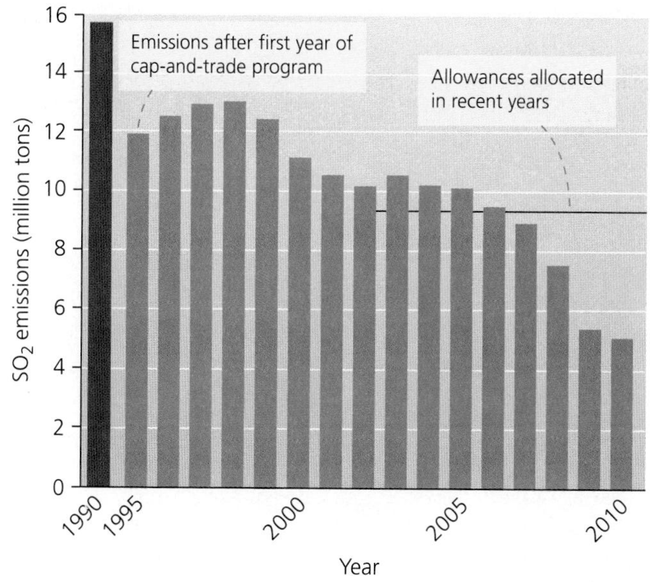


**FIGURE 13.21 Sulfur dioxide emissions fell 67% in the wake of an emissions trading system.** As of 2010, emissions from U.S. power plants participating in this EPA program mandated by the 1990 Clean Air Act had dropped well below the amount allocated in permits (black line). *Data from U.S. EPA.*

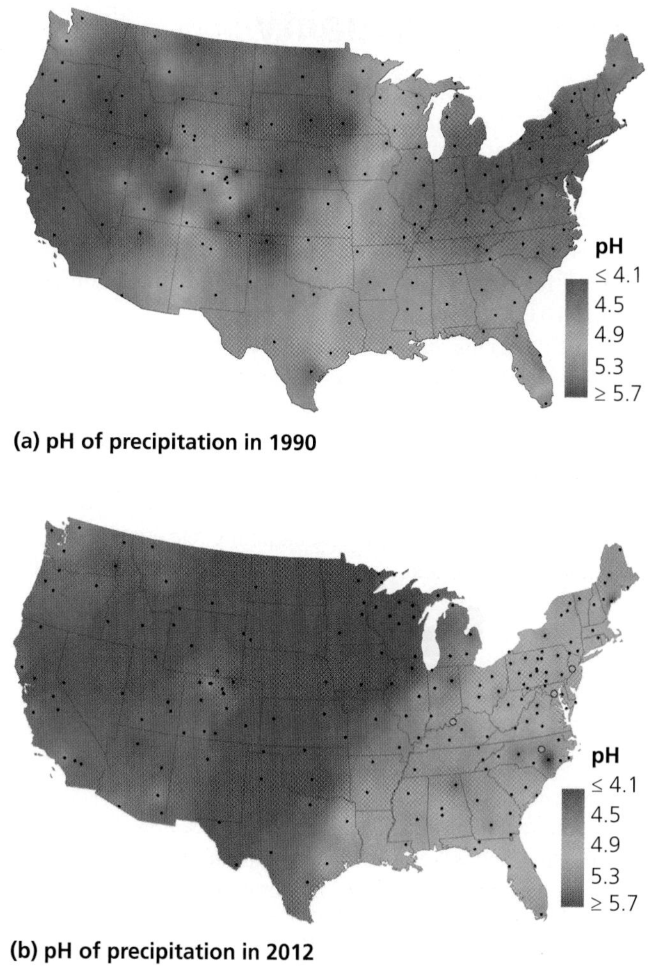


**FIGURE 13.22 Precipitation has become less acidic as air quality has improved under the Clean Air Act.** Average pH values for precipitation rose between **(a)** 1990 and **(b)** 2012. Precipitation remains most acidic in regions near and downwind from areas of heavy industry. *Data from the National Atmospheric Deposition Program.*

**DATA Q** In the area where you live, how did the pH of precipitation change between 1990 and 2012? Has precipitation become more acidic or less acidic?

GO TO **INTERPRETING GRAPHS & DATA** ON MasteringEnvironmentalScience®

allowances. Each year the overall amounts of allowed pollution were decreased. (See pp. 326–327 for further explanation of how such a system works.) The economic incentives created by this cap-and-trade program encouraged polluters to switch to low-sulfur coal, invest in technologies such as scrubbers (p. 285), and devise other ways to become cleaner and more efficient. During the course of the cap-and-trade program, $SO_2$ emissions across the United States fell by 67% (**FIGURE 13.21**)—further and faster than the Clean Air Act had required. As a result, average sulfate loads in precipitation across the eastern United States were 51% lower in 2008–2010 than in 1989–1991.

The Acid Rain Program also required power plants to reduce nitrogen oxide emissions, with the EPA allowing plants flexibility in how they did so. Emissions of $NO_X$ fell significantly as a result, and wet nitrogen deposition declined as well. Thanks to the declines in $SO_2$ and $NO_X$, air and water quality improved throughout the eastern United States (**FIGURE 13.22**).

Many have attributed this success to the Acid Rain Program, and the EPA has calculated that the program's economic benefits (in healthcare expenses avoided, for instance) outweighed its costs by 40 to 1. However, some experts maintain that pollution declined because cleaner fuels became less expensive and because simultaneous conventional regulation mandated emissions cuts. Indeed, during this time period European nations using command-and-control regulation instead of emissions trading reduced their $SO_2$ emissions by even more than the United States did.

In 2011, emissions trading ended once the EPA issued its Cross-State Air Pollution Rule, which aimed to limit pollution drifting from upwind states into downwind states. Although the Clean Air Act justifies tackling transboundary pollution in this way, industry and several states brought lawsuits claiming that the EPA was exceeding its authority, and a U.S. Court of Appeals vacated the rule in 2012. The EPA appealed to the U.S. Supreme Court, which in 2014 ruled that the Cross-State Air Pollution Rule should stand.

As with recovery of the ozone layer, there is a time lag before the positive consequences of emissions cuts kick in, so it will take time for acidified ecosystems to recover. Meanwhile, in industrializing nations, the problem is becoming worse. Today China emits the most sulfur dioxide of any nation, and it has the world's worst acid rain problem. Overall, data on acid deposition show advances in controlling outdoor air pollution, but they also show that more could be done. The same can be said for indoor air pollution, a source of human health threats that is less familiar to most of us, but statistically more dangerous.

# Indoor Air Quality

Indoor air generally contains higher concentrations of pollutants than does outdoor air. As a result, the health impacts from **indoor air pollution** in workplaces, schools, and homes outweigh those from outdoor air pollution. The World Health Organization (WHO) attributes nearly 3.5 million premature deaths each year to indoor air pollution (compared with 3.3 million for outdoor air pollution). Indoor air pollution takes nearly 10,000 lives each day.

If this seems surprising, consider that the average U.S. citizen spends at least 90% of his or her time indoors. Then consider the dizzying array of consumer products in our homes and offices. Many of these products are made of synthetic materials, and novel synthetic substances are not comprehensively tested for health effects before being brought to market (Chapter 10). Furniture, carpeting, cleaning fluids, insecticides, and the many varieties of plastics all exude volatile chemicals into the air.

Ironically, some attempts to be environmentally prudent during the "energy crises" of the 1970s (p. 353) worsened indoor air quality. To reduce heat loss, building managers sealed off ventilation, while designers constructed new buildings with limited ventilation and with windows that did not open. These steps saved energy, but they also trapped stable, unmixed air—and pollutants—inside. There is good news, however. In both developing and developed nations, we have feasible ways to address the primary causes of indoor air pollution.

## Risks differ in developing and developed nations

Indoor air pollution exerts most impact in the developing world, where poverty forces millions of people to burn wood, charcoal, animal dung, or crop waste inside their homes for cooking and heating, with little or no ventilation (**FIGURE 13.23**). As a result, people inhale soot, carbon monoxide, and other pollutants, which increase risks of premature death by pneumonia, bronchitis, and lung cancer, as well as allergies, sinus infections, cataracts, asthma, emphysema, and heart disease.

In developed nations, the primary indoor air health risks are cigarette smoke and radon. Smoking cigarettes irritates the eyes, nose, and throat; worsens asthma and other respiratory ailments; and greatly increases the risk of lung cancer. Inhaling secondhand smoke causes many of the same problems. Tobacco smoke is a brew of over 4000 chemical compounds, some of which are known or suspected to be toxic or carcinogenic. Although smoking has become less popular in developed nations in recent years, it is still estimated to cause 160,000 lung cancer deaths per year just in the United States.

Radon gas is the second-leading cause of lung cancer in the developed world, responsible for an estimated 21,000 deaths per year in the United States and for 15% of lung cancer cases worldwide. Radon (p. 209) is a colorless, odorless, radioactive gas resulting from the natural decay of uranium in soil, rock, or water. It seeps up from the ground and can penetrate buildings. The only way to determine whether radon is entering a building is to sample air with a test kit. The EPA estimates that 6% of U.S. homes exceed its safety standard for radon. Since the 1980s, millions of U.S. homes have been tested for radon, over a million have undergone mitigation, and new homes are being built with radon-resistant features.

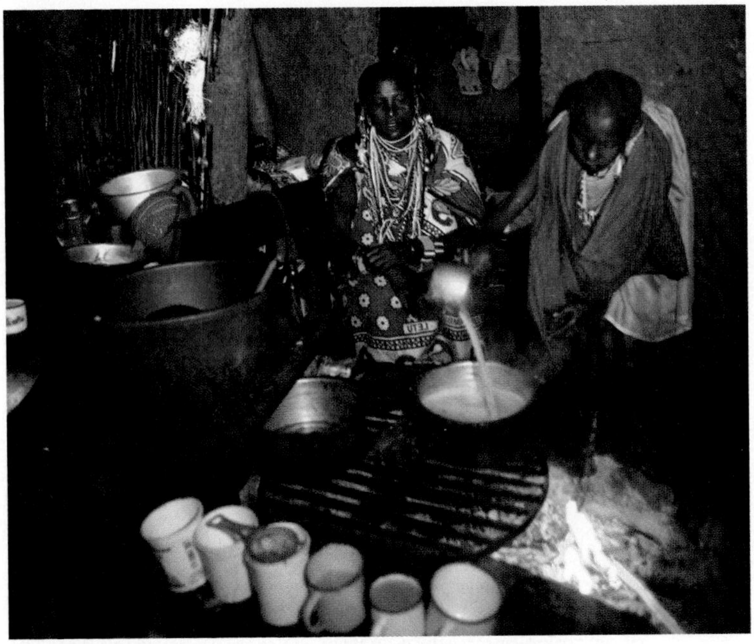

**FIGURE 13.23 In the developing world, many people build fires indoors for cooking and heating, as in this Maasai home in Kenya.** Indoor fires expose people to severe pollution from particulate matter and carbon monoxide.

## Many substances pollute indoor air

In our daily lives at home, we are exposed to many indoor air pollutants (**FIGURE 13.24**). The most diverse are volatile organic compounds (VOCs) (p. 284), airborne carbon-containing compounds released by plastics, oils, perfumes, paints, cleaning fluids, adhesives, and pesticides. VOCs evaporate from furnishings, building materials, carpets, laser printers, and fax machines. Some products, such as chemically treated furniture, release large amounts of VOCs when new and progressively less as they age. Others, such as photocopying machines, emit VOCs each time they are used. Formaldehyde—a VOC used in pressed wood, insulation, and other products—irritates mucous membranes, induces skin allergies, and causes other ailments. The "new car smell" that fills the interiors of new automobiles comes from a complex mix of dozens of VOCs as they outgas from the newly manufactured plastic, metal, and leather components of the car. Some scientific studies warn of health risks from this brew and recommend that you keep a new car well ventilated.

Another widespread source of indoor air pollution is tiny living organisms. Dust mites and animal dander can worsen asthma in children. The airborne spores of some fungi, molds, and mildews can cause allergies, asthma, and other respiratory ailments. Some airborne bacteria can cause infectious disease. Heating and cooling systems in buildings make ideal breeding grounds for microbes, providing moisture, dust, and foam insulation as substrates, along with air currents to carry the organisms aloft.

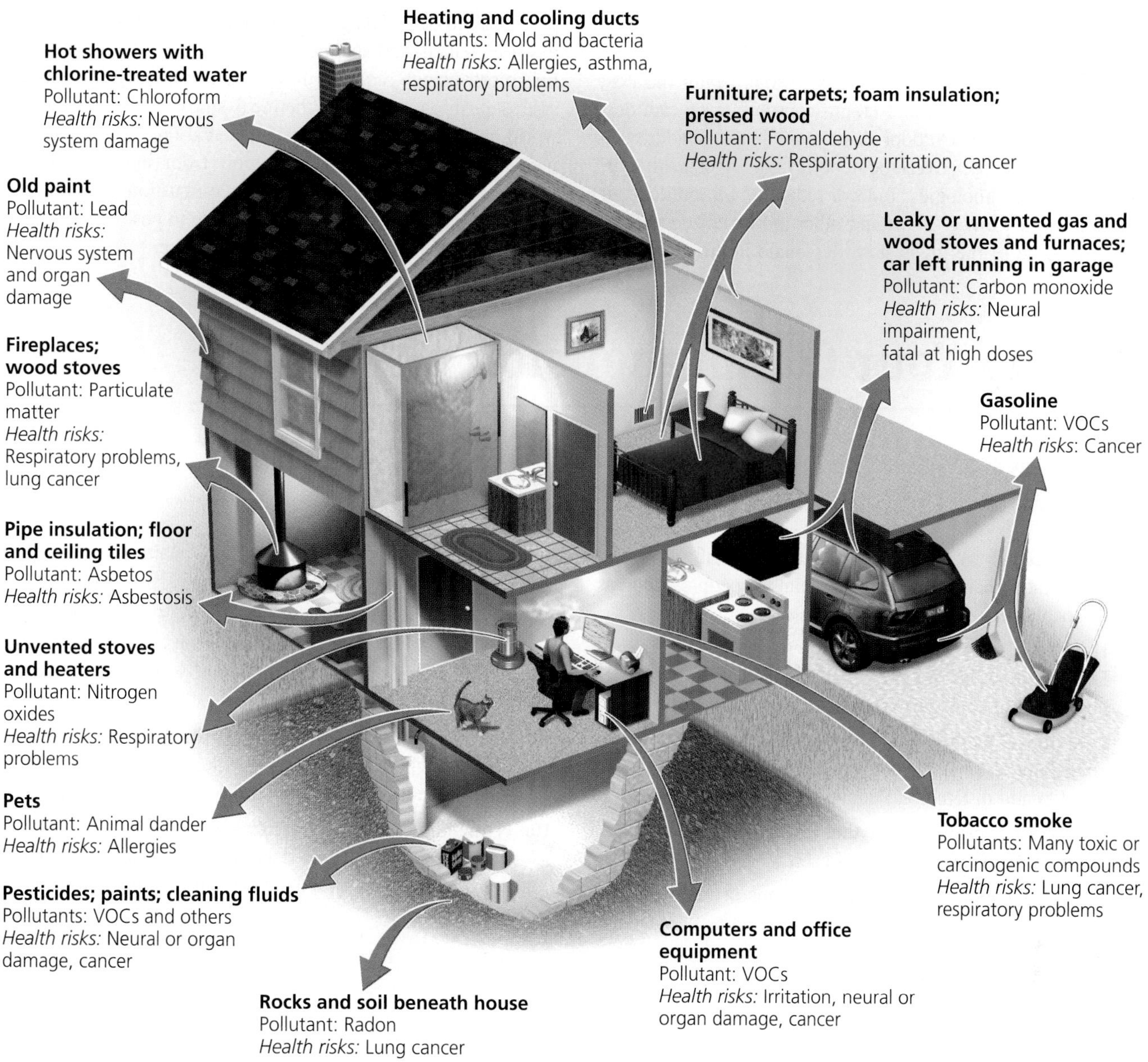


**FIGURE 13.24 The typical home contains many sources of indoor air pollution.** Shown are common sources, the major pollutants they emit, and some of the health risks they pose.

Microbes that induce allergic responses are thought to be a major cause of *building-related illness*, a sickness produced by indoor pollution. When the cause of such an illness is a mystery, and when symptoms are general and nonspecific, the illness is often called *sick-building syndrome.* The U.S. Occupational Safety and Health Administration (OSHA) estimates that 30–70 million Americans have suffered ailments related to the building in which they live.

## WEIGHING THE ISSUES

**How Safe Is Your Indoor Environment?**

Name some potential indoor air quality hazards in your home, work, or school environment. Are these spaces well ventilated? What could you do to improve the safety of the indoor spaces you use?

## We can enhance indoor air quality

Using low-toxicity materials, monitoring air quality, keeping rooms clean, and providing adequate ventilation are the keys to alleviating indoor air pollution in most situations. In the developed world, we can avoid cigarette smoke, limit our use of plastics and treated wood, and restrict our exposure to pesticides, cleaning fluids, and other toxic substances by keeping them in garages or outdoor sheds. The EPA recommends that we test our homes and offices for radon, mold, and carbon monoxide. Keeping rooms and air ducts clean will reduce potential irritants and allergens. Developing nations are making progress in reducing indoor air pollution through a variety of steps, especially by introducing cleaner-burning fuels and stoves. Researchers calculate that rates of premature death from indoor air pollution dropped nearly 40% from 1990 to 2010.

## Conclusion

Indoor air pollution poses potentially serious health hazards, but by keeping informed and taking appropriate precautions, we can minimize risk. Outdoor air pollution has been addressed more effectively by government policy, together with pollution control technologies. Indeed, reductions in outdoor air pollution in the United States and other industrialized nations represent some of the greatest strides made in environmental protection to date. The global depletion of stratospheric ozone has been halted, and acid deposition is being addressed. Room for improvement remains, however, particularly in reducing acid deposition and photochemical smog. In the developing world, air pollutant levels are higher and take a heavy toll on people's health. Reducing pollution from indoor fuelwood burning, automobile exhaust, coal combustion in outmoded facilities, and other sources will continue to pose challenges as the world's less wealthy nations industrialize.

# Testing Your Comprehension

1. About how thick is Earth's atmosphere? Name one characteristic of the troposphere and one characteristic of the stratosphere.
2. Where is the "ozone layer" located? Describe how and why stratospheric ozone is beneficial for people, whereas tropospheric ozone is harmful.
3. How does solar energy influence weather and climate? Describe how Hadley, Ferrel, and polar cells help to determine climate patterns and the location of biomes.
4. Describe a thermal inversion. Explain how inversions contribute to severe smog episodes such as the ones in London and in Donora, Pennsylvania.
5. How does a primary pollutant differ from a secondary pollutant? Give an example of each.
6. What has happened with the emissions of major pollutants in the United States in recent decades? What has happened with concentrations of "criteria pollutants" in U.S. ambient air?
7. How does photochemical smog differ from industrial smog? Give three examples of the health risks posed by the outdoor air pollutants in smog.
8. Explain how chlorofluorocarbons (CFCs) deplete stratospheric ozone. Why is this depletion considered a long-term international problem? What was done to address this problem?
9. Why are the effects of acid deposition often felt in areas far from where the primary pollutants are produced? List three impacts of acid deposition.
10. Name three common sources of indoor pollution and their associated health risks. For each pollution source, describe one way to reduce exposure to the source.

# Seeking Solutions

1. Describe several ways in which Los Angeles, Mexico City, and other metropolitan areas have tried to respond to photochemical smog. Now find an example of a city near you, not discussed in this chapter, that has reduced its air pollution. Describe how much pollution was reduced, and explain how this was accomplished.
2. Explain how and why emissions of major pollutants have been reduced by well over 50% in the United States since 1970, despite increases in population, energy use, and economic activity. Describe at least two ways you think air quality might be further improved.
3. International action through a treaty has helped to halt further stratospheric ozone depletion, but other transboundary pollution issues, including acid deposition, have not been addressed as effectively. What types of actions do you feel are appropriate for pollutants that cross political boundaries?
4. **THINK IT THROUGH** You are the head of your county health department, and the EPA informs you that your county has failed to meet the national ambient air quality standards for ozone, sulfur dioxide, and nitrogen dioxide. Your county is home to a city of 200,000 people and 10 sprawling suburbs. It has an aging coal-fired power plant, a number of factories with advanced pollution control technology, and no public transportation system. What steps would you urge the county and state governments to take to meet the air quality standards? Explain how you would prioritize these steps.
5. **THINK IT THROUGH** You have just taken a job at a medical clinic in your hometown. The nursing staff has asked you to develop a brochure for patients featuring tips on how to minimize health impacts from air pollution (both indoor and outdoor) in their daily lives. List the top five tips you will feature, and explain for each why you will include it in your brochure.

# Calculating Ecological Footprints

"While only some motorists contribute to traffic fatalities, all motorists contribute to air pollution fatalities." A researcher for the Earth Policy Institute wrote this to point out that air pollution kills far more people than vehicle accidents. According to EPA data, emissions of nitrogen oxides in the United States in 2013 totaled 12.6 million tons. Nitrogen oxides come from fuel combustion in motor vehicles, power plants, and other industrial, commercial, and residential sources, but fully 7.7 million tons of the 2013 total came from vehicles. The U.S. Census Bureau estimates the nation's population to have been 316.1 million in 2013 and projects that it will reach 346.4 million in 2025. Considering these data, calculate the missing values in the table below (1 ton = 2000 lb).

| | Total $NO_x$ emissions (lb) | $NO_x$ emissions from vehicles (lb) |
|---|---|---|
| You as a driver | | |
| Your class | | |
| Your state | | |
| United States | 25.2 billion | 15.4 billion |

*Data from U.S. EPA.*

1. By what percentage is the U.S. population projected to increase between 2013 and 2025? Do you think that $NO_x$ emissions will increase, decrease, or remain the same over that period of time? Why? (You may wish to refer to Figure 13.7.)
2. Assume you are an average American driver. Using the 2013 emissions totals, how many pounds of $NO_x$ emissions are you responsible for creating? How many pounds would you prevent if you were to reduce by half the vehicle miles you travel? What percentage of your total $NO_x$ emissions would that be?
3. How might a driver reduce his or her vehicle miles traveled by 50%? What other steps could you take to reduce the $NO_x$ emissions for which you are responsible?

## MasteringEnvironmentalScience®

**STUDENTS**

Go to **MasteringEnvironmentalScience** for assignments, the etext, and the Study Area with practice tests, videos, current events, and activities.

**INSTRUCTORS**

Go to **MasteringEnvironmentalScience** for automatically graded activities, current events, videos, and reading questions that you can assign to your students, plus Instructor Resources.

14

# Global Climate Change

## Upon completing this chapter, you will be able to:

- ☐ Describe Earth's climate system and explain the factors that influence global climate
- ☐ Identify greenhouse gases and characterize human influences on the atmosphere and on climate
- ☐ Summarize how researchers study climate
- ☐ Outline current and future trends and impacts of climate change in the United States and across the world
- ☐ Suggest and assess ways we may respond to climate change

Photo: **The Maldives' underwater cabinet meeting**

# CENTRAL CASE STUDY

# Rising Seas May Flood the Maldives

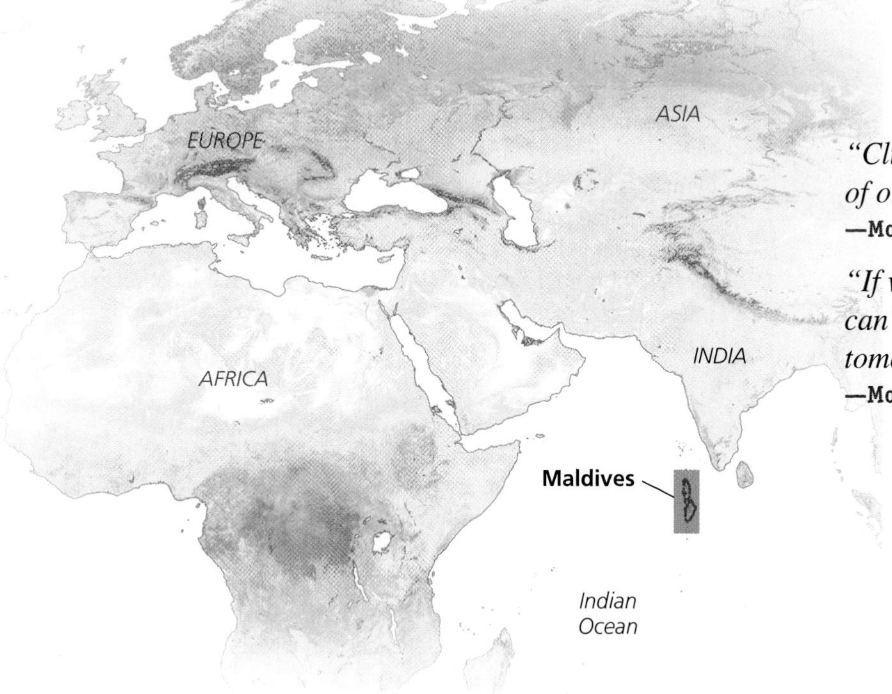


*"Climate change threatens the very existence of our country."*
**—Mohamed Waheed, recent president, Maldives**

*"If we can't save the Maldives today, we can't save London, New York, or Hong Kong tomorrow."*
**—Mohamed Nasheed, former president, Maldives**

With sun-drenched beaches, colorful coral reefs, and a spectacular tropical setting, the Maldives seems a paradise to the many tourists who visit. For its 370,000 residents, this island nation in the Indian Ocean is home. But residents and tourists alike now fear that the Maldives could soon be submerged by the rising seas brought by global climate change.

In this nation of 1200 islands, the highest point is just 2.4 m (8 ft) above sea level, and four-fifths of the land lies less than 1 m (39 in.) above sea level. The world's oceans rose 20 cm (8 in.) in the 20th century as warming temperatures expanded seawater and as melting polar ice discharged water into the ocean. Scientists predict that sea level will rise another 26–98 cm (10–39 in.) in this century.

Higher seas are beginning to flood large areas of the Maldives and cause salt water to contaminate drinking water supplies. Storms intensified by warmer water are eroding beaches and damaging the coral reefs that are vital to the nation's tourism and fishing industries. Residents have had to be evacuated from several low-lying islands, and political leaders have looked to buy land in mainland nations in case the Maldives' people one day need to abandon their homeland.

Small island nations like the Maldives are responsible for very few of the carbon emissions that drive global climate change, yet these nations are the ones bearing the earliest consequences. The Maldives' leaders have made sure the world knows this. In 2009, President Mohamed Nasheed donned scuba gear and dove into the blue waters of Girifushi Island lagoon, followed by his entire cabinet. These officials held the world's first underwater cabinet meeting. Sitting at a table beneath the waves, they signed a declaration reading:

> *SOS from the front line:* Climate change is happening and it threatens the rights and security of everyone on Earth. With less than one degree of global warming, the glaciers are melting, the ice sheets collapsing, and low-lying areas are in danger of being swamped. We must unite in a global effort to halt further temperature rises, by slashing carbon dioxide emissions to a safe level of 350 parts per million.

The underwater cabinet meeting was part of a campaign to draw global attention to the impacts of climate change. Nasheed then played a high-profile role at international climate talks in Copenhagen, where he pleaded with the United States, China, India, and other nations to unite in efforts to reduce emissions of polluting gases that warm the atmosphere. Back home in the Maldives, Nasheed announced a plan to make his nation carbon-neutral by 2020.

In 2012, Nasheed—the nation's first democratically elected president—was forced from power. His supporters called it a coup d'etat; his detractors said he had abused power. He was put on trial and narrowly lost the 2013 presidential election. Yet despite their nation's political turmoil, the people of the Maldives remain united in their concern over climate change. They are not alone in their predicament. Other island nations, from the Galápagos to Fiji to the Seychelles, also face a future of encroaching seawater. These island nations have organized to make their concern over climate change known to the world through AOSIS, the Alliance of Small Island States.

**(a) Malé, capital of the Maldives**

**(b) Miami, on the Florida coast**

FIGURE 14.1 **From the Maldives to Miami, rising sea levels threaten damage to vulnerable coastal areas.**

Coastal areas around the world will face challenges from rising sea levels (**FIGURE 14.1**). In the United States, the hurricane-prone shores of Florida, Louisiana, Texas, and the Carolinas are at risk, as are coastal cities such as San Francisco, Houston, Boston, and New York City. Superstorm Sandy in 2012 was a wake-up call. The lost lives and billions of dollars of damage this massive hurricane brought to New York, New Jersey, and other states made clear that the costs inflicted by rising seas can be enormous.

Indeed, scientists now know that the Atlantic seaboard and the Gulf coast of the United States are especially vulnerable to impacts from sea level rise (p. 316). From Maine to Miami, and from Cape Cod to Corpus Christi, millions of Americans who live in coastal communities will experience increasing expense, disruption, and property damage as beaches erode, neighborhoods flood, aquifers are contaminated, and storms strike with more force.

Impacts from rising seas are just a few of the many imminent consequences of global climate change. In one way or another, climate change will affect each and every one of us for the remainder of our lifetimes. Putting solutions into action stands as a central challenge for our society today and for the foreseeable future. ◻

# Our Dynamic Climate

Climate influences virtually everything around us, from the day's weather to major storms, from crop success to human health, and from national security to the ecosystems that support our economies. If you are a student in your teens or twenties, the accelerating change in our climate today may well be *the* major event of your lifetime and the phenomenon that most shapes your future.

Climate change is also the fastest-developing area of environmental science. New scientific studies that refine our understanding of climate are published every week, and policymakers and businesspeople make decisions and take actions just as quickly. By the time you read this chapter, some of its information will already be out of date. We urge you to explore further, with your instructor and on your own, the most recent information on climate change and the impacts it will have on your future.

## What is climate change?

**Climate** describes an area's long-term atmospheric conditions, including temperature, precipitation, wind, humidity, barometric pressure, solar radiation, and other characteristics. *Climate* differs from *weather* (p. 281) in that weather specifies conditions over hours or days, whereas climate summarizes conditions over years, decades, or centuries.

**Global climate change** describes an array of changes in aspects of Earth's climate, such as temperature, precipitation, and the frequency and intensity of storms. People often use the term *global warming* synonymously in casual conversation, but **global warming** refers specifically to an increase in Earth's average surface temperature. Global warming is only one aspect of global climate change, but warming does in turn drive other components of climate change.

Over the long term, our planet's climate varies naturally. However, today's climatic changes are unfolding at an exceedingly rapid rate, and they are creating conditions humanity has never experienced. Scientists agree that human activities, notably fossil fuel combustion and deforestation, are largely responsible. Some researchers point out that the term "climate change" is so mild-sounding as to be misleading, and that a more accurate term would be "climate disruption."

## Three factors influence climate

Three natural factors exert the most influence on Earth's climate. The first is the sun. Without the sun, Earth would be dark and frozen. The second is the atmosphere. Without this protective and buffering layer of gases, Earth would be as much as 33°C (59°F) colder on average, and temperature differences between night and day would be far greater than they are. The third is the oceans, which store and transport heat and moisture.

The sun supplies most of our planet's energy. Earth's atmosphere, clouds, land, ice, and water together absorb about 70% of incoming solar radiation and reflect the remaining 30% back into space (**FIGURE 14.2**).

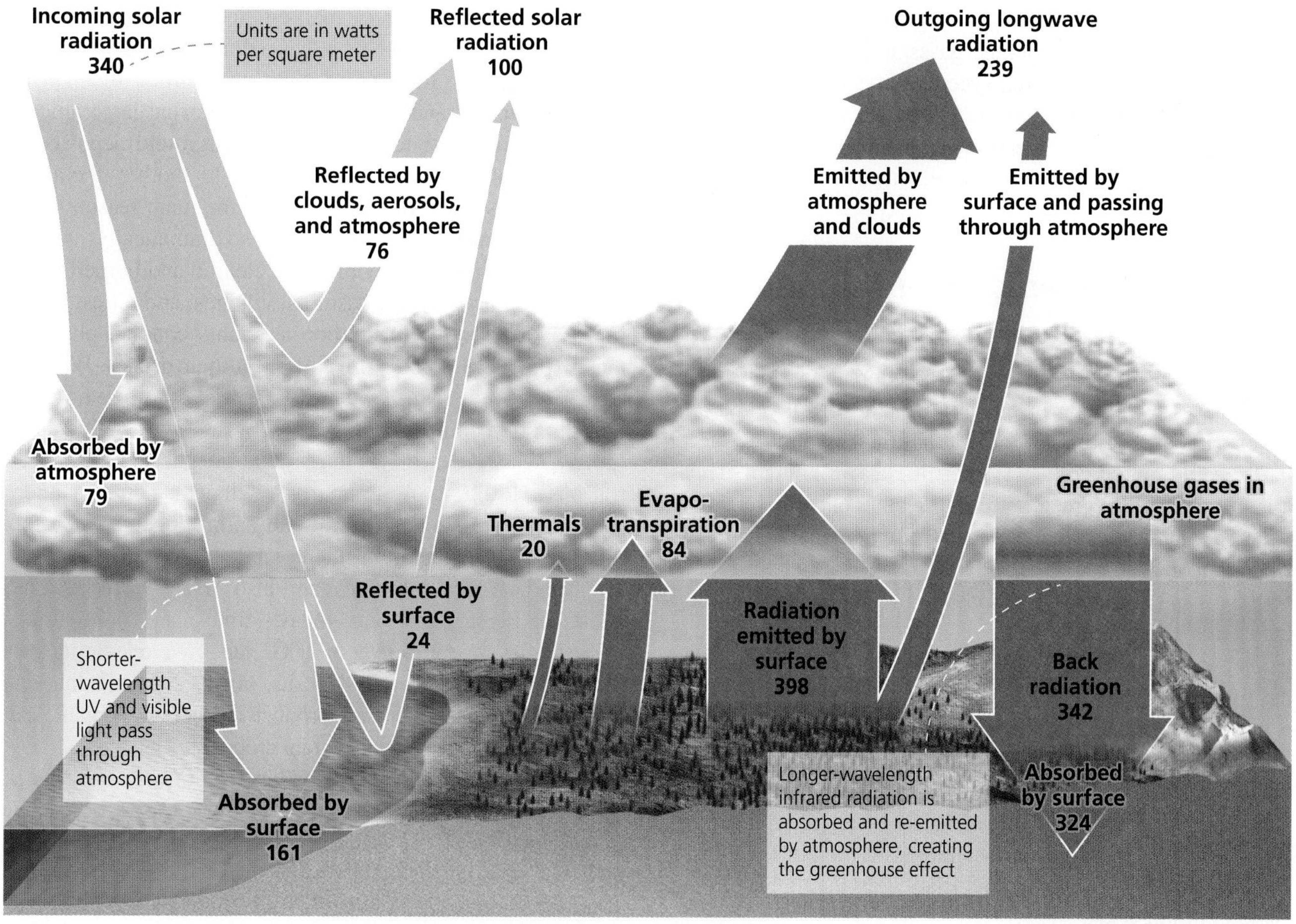


FIGURE 14.2 **Our planet receives about 340 watts of energy per square meter from the sun, and it naturally reflects and emits this same amount.** Earth absorbs 70% of the solar radiation it receives and reflects the rest back into space **(yellow arrows)**. The energy absorbed is re-emitted **(orange arrows)** as infrared radiation, which has longer wavelengths. Greenhouse gases in the atmosphere absorb a portion of this radiation and then re-emit it, sending some downward to warm the atmosphere and the surface by the greenhouse effect. *Data from Intergovernmental Panel on Climate Change (IPCC), 2013.* Fifth assessment report. The physical science basis: Contribution of Working Group I.

## Greenhouse gases warm the lower atmosphere

As Earth's surface absorbs solar radiation, the surface increases in temperature and emits infrared radiation (p. 31), radiation with wavelengths longer than those of visible light. Atmospheric gases having three or more atoms in their molecules tend to absorb infrared radiation. These include water vapor ($H_2O$), ozone ($O_3$), carbon dioxide ($CO_2$), nitrous oxide ($N_2O$), and methane ($CH_4$), as well as halocarbons, a diverse group of mostly human-made gases that includes chlorofluorocarbons (CFCs; p. 293). All these gases are known as **greenhouse gases.** After absorbing radiation emitted from the surface, greenhouse gases re-emit infrared radiation. Some of this re-emitted energy is lost to space, but most travels back downward, warming the lower atmosphere (specifically the troposphere; p. 281) and the surface, in a phenomenon known as the **greenhouse effect.**

Greenhouse gases differ in their ability to warm the troposphere and surface. *Global warming potential* refers to the relative ability of one molecule of a given greenhouse gas to contribute to warming. TABLE 14.1 shows global warming potentials for several gases. Values are expressed in relation to carbon dioxide, which is assigned a value of 1. For example, at a 20-year time horizon, a molecule of methane is 84 times more potent than a molecule of carbon dioxide. Yet because a typical methane molecule resides in the atmosphere for less time than a typical carbon dioxide molecule, methane's global warming potential is reduced at longer time horizons (it is 28 at a 100-year horizon).

**TABLE 14.1 Global Warming Potentials of Four Greenhouse Gases**

| GREENHOUSE GAS | RELATIVE HEAT-TRAPPING ABILITY (IN $CO_2$ EQUIVALENTS) | |
|---|---|---|
| | Over 20 years | Over 100 years |
| Carbon dioxide | 1 | 1 |
| Methane | 84 | 28 |
| Nitrous oxide | 264 | 265 |
| Hydrochlorofluorocarbon HFC-23 | 10,800 | 12,400 |

*Data are for 20-year and 100-year time horizons, from IPCC, 2013.* Climate change 2013: The physical science basis. Contribution of Working Group I to the Fifth assessment report.

Although carbon dioxide is less potent on a per-molecule basis than most other greenhouse gases, it is far more abundant in the atmosphere. Moreover, greenhouse gas emissions from human activity consist mostly of $CO_2$. For these reasons, carbon dioxide has caused nearly twice as much warming since the industrial revolution as methane, nitrous oxide, and halocarbons combined.

## Greenhouse gas concentrations are rising fast

The greenhouse effect is a natural phenomenon, and greenhouse gases have been in our atmosphere for all of Earth's history. It's a good thing, too. Without the greenhouse effect, Earth would be too cold to support life as we know it. Thus, it is not the natural greenhouse effect that concerns scientists today, but rather the *anthropogenic* (human-generated) intensification of the greenhouse effect. By adding novel greenhouse gases (certain halocarbons) to the atmosphere, and by increasing the concentrations of several natural greenhouse gases over the past 250 years (**FIGURE 14.3**), we are intensifying the greenhouse effect beyond what our species has ever experienced.

We have boosted Earth's atmospheric concentration of carbon dioxide from 280 parts per million (ppm) in the late 1700s to 399 ppm in 2014 (see Figure 14.3). Today the concentration of $CO_2$ in our atmosphere is far higher than it has been in over 800,000 years, and it is likely the highest in the last 20 million years.

Why have atmospheric carbon dioxide levels risen so rapidly? Most carbon is stored for long periods in the upper layers of the lithosphere (p. 227). The deposition, partial decay, and compression of organic matter (mostly plants and phytoplankton) in wetland or marine areas hundreds of millions of years ago led to the formation of coal, oil, and natural gas in buried sediments (p. 337). In the past two centuries, we have extracted these fossil fuels from the ground and burned them in our homes, factories, and automobiles, transferring large amounts of carbon from one reservoir (the underground deposits that stored the carbon for millions of years) to another (the atmosphere). This sudden flux of carbon is the main reason atmospheric $CO_2$ concentrations have risen so dramatically.

At the same time, people have cleared and burned forests to make room for crops, pastures, villages, and cities. Forests serve as a reservoir for carbon as plants conduct photosynthesis (pp. 31–32) and store carbon in their tissues. When we clear forests, we reduce the biosphere's ability to remove carbon dioxide from the atmosphere. In this way, deforestation (pp. 189–192) contributes to rising atmospheric $CO_2$ concentrations. **FIGURE 14.4** summarizes scientists' current understanding of the fluxes (natural and anthropogenic) of carbon dioxide between the atmosphere, land, and oceans.

Methane concentrations are also rising—2.6-fold since 1750 (see Figure 14.3)—and today's atmospheric concentration is the highest by far in over 800,000 years. We release methane by tapping into fossil fuel deposits, raising livestock that emit methane as a metabolic waste product, disposing of organic matter in landfills, and growing certain crops, such as rice.

Human activities have also elevated atmospheric concentrations of nitrous oxide. This greenhouse gas, a by-product of feedlots, chemical manufacturing plants, auto emissions, and synthetic nitrogen fertilizers, has risen by 20% since 1750 (see Figure 14.3).

Among other greenhouse gases, ozone concentrations in the troposphere (associated with photochemical smog, p. 291) have risen roughly 36% since 1750. The contribution of halocarbon gases to global warming has begun to slow because of the Montreal Protocol and subsequent controls on their production and use (p. 294). Water vapor is the most abundant greenhouse gas in our atmosphere and contributes most to the natural greenhouse effect. Its concentrations vary locally, but because its global concentration has not changed, it is not thought to have driven industrial-age climate change.

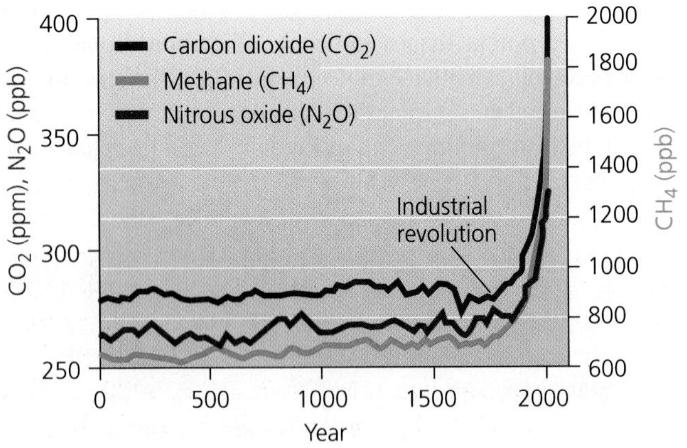


**FIGURE 14.3 Since the start of the industrial revolution around 1750, global concentrations of carbon dioxide, methane, and nitrous oxide in the atmosphere have increased markedly.** *Data from IPCC, 2013.* Fifth assessment report. The physical science basis: Contribution of Working Group I.

DATA Q By about what percentage has atmospheric carbon dioxide concentration increased since 1750?

GO TO **INTERPRETING GRAPHS & DATA** ON MasteringEnvironmentalScience®

## Other factors warm or cool the surface

Whereas greenhouse gases exert a warming effect on the atmosphere, **aerosols**, microscopic droplets and particles, can have either a warming or a cooling effect. Soot particles, or "black carbon aerosols," generally cause warming by absorbing solar energy, but most other tropospheric aerosols cool the atmosphere by reflecting the sun's rays. Sulfate aerosols produced by fossil fuel combustion may slow global warming, at least in the short term. When sulfur dioxide enters the atmosphere, it undergoes various reactions, some of which lead to acid deposition (pp. 295–297). These reactions can form a sulfur-rich aerosol haze in the upper atmosphere that blocks sunlight. For this reason, aerosols released by major volcanic eruptions can exert cooling effects on Earth's climate for up to several years. This occurred in 1991 with the eruption of Mount Pinatubo in the Philippines.

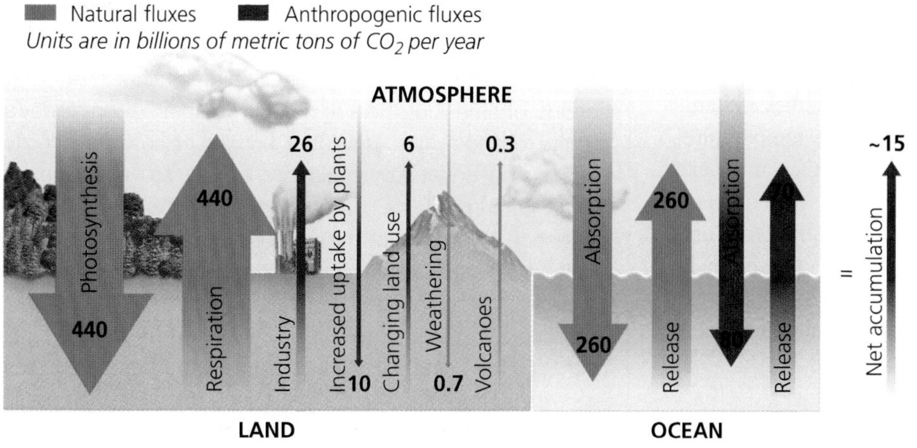


**FIGURE 14.4 Human activities are sending more carbon dioxide from Earth's surface to its atmosphere than is moving from the atmosphere to the surface.** Shown are all current fluxes of $CO_2$, with arrows sized according to mass. Green arrows are natural fluxes, and red arrows are anthropogenic fluxes. *Adapted from IPCC, 2007.* Fourth assessment report.

**DATA Q** For every metric ton of carbon dioxide we emit as a result of changes in land use (e.g. deforestation), how much do we emit from industry?

GO TO **INTERPRETING GRAPHS & DATA** ON MasteringEnvironmentalScience®

To measure the degree of impact that a given factor exerts on Earth's temperature, scientists calculate its **radiative forcing**, the amount of change in thermal energy that the factor causes. Positive forcing warms the surface, whereas negative forcing cools it. When scientists sum up the effects of all factors, they find that Earth is now experiencing radiative forcing of about 2.3 watts/m$^2$ (**FIGURE 14.5**). This means that our planet today is receiving and retaining 2.3 watts/m$^2$ more thermal energy than it is emitting into space. (By contrast, the pre-industrial Earth of 1750 was in balance, emitting as much radiation as it was receiving.) This extra amount is equivalent to the power converted into heat and light by 200 incandescent lightbulbs (or over 900 CFLs) across a football field. As Figure 14.2 shows, Earth is estimated naturally to receive and give off about 340 watts/m$^2$ of energy. Although 2.3 may seem like a small proportion of 340, heat from this imbalance accumulates, and over time this is enough to alter climate significantly.

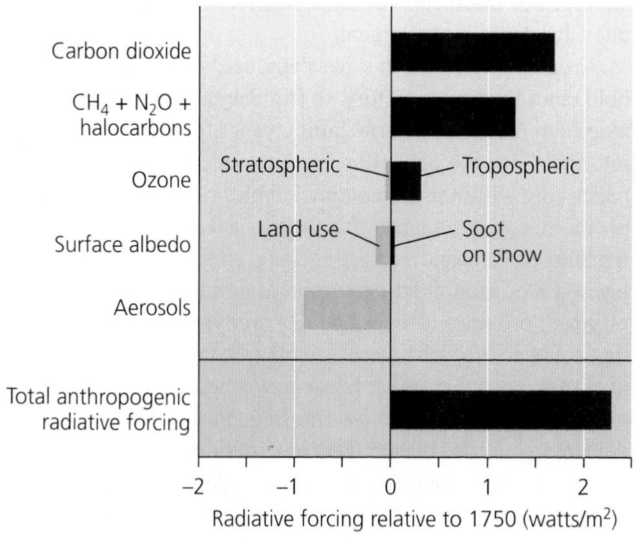


**FIGURE 14.5 Radiative forcing quantifies the influence that aerosols, greenhouse gases, and other factors exert over Earth's energy balance.** In this graph, radiative forcing is expressed as the warming or cooling effect that each factor has on temperature today relative to 1750, in watts/m$^2$. Red bars show positive forcing (warming), and blue bars show negative forcing (cooling). *Albedo* (p. 316) refers to the reflectivity of a surface. A number of more minor influences are not shown. *Data from IPCC, 2013.* Fifth assessment report. The physical science basis: Contribution of Working Group I.

## Climate varies naturally for several reasons

Atmospheric composition is one of several factors that influence climate. Other factors include variation in energy released by the sun, absorption of carbon dioxide by the oceans, ocean circulation patterns, and cyclic changes in Earth's rotation and orbit. However, scientific data indicate that none of these four natural factors can fully explain the rapid climate change that we are experiencing today.

**Solar output** The sun varies in the amount of radiation it emits, over both short and long timescales. However, scientists are concluding that the variation in solar energy reaching our planet in recent centuries has simply not been great enough to drive significant temperature change on Earth's surface. Estimates place the radiative forcing of natural changes in solar output at only about 0.05 watts/m$^2$—less than any of the anthropogenic causes shown in Figure 14.5.

**Ocean absorption** The oceans hold 50 times more carbon than the atmosphere. Oceans absorb carbon dioxide from the atmosphere when $CO_2$ dissolves directly in water and when marine phytoplankton use it for photosynthesis. However, the oceans are absorbing less $CO_2$ than we are adding to the atmosphere (see Figure 2.20, p. 40). Thus, oceanic carbon absorption is slowing global warming but is not preventing it. Moreover, as ocean water warms, it absorbs less $CO_2$ because gases are less soluble in warmer water—a positive feedback effect (pp. 23–24) that accelerates warming of the atmosphere.

**Ocean circulation** Ocean water exchanges heat with the atmosphere, and ocean currents move energy from place to place. For example, the oceans' thermohaline circulation system (pp. 254–255) moves warm tropical water northward toward Europe, providing Europe a far milder climate than it would otherwise have. Scientists are studying whether freshwater input from Greenland's melting ice sheet might shut down this warm-water flow—an occurrence that could plunge Europe into much colder conditions.

Multiyear climate variability results from the El Niño–Southern Oscillation (pp. 255–256), which involves systematic shifts in atmospheric pressure, sea surface temperature, and ocean circulation in the tropical Pacific Ocean. These shifts overlie longer-term variability from a phenomenon known as the Interdecadal Pacific Oscillation. El Niño and La Niña events alter weather patterns from region to region in diverse ways, often leading to rainstorms and floods in dry areas and drought and fire in moist areas. This leads to impacts on wildlife, agriculture, and fisheries.

**Milankovitch cycles** In the 1920s, Serbian mathematician Milutin Milankovitch described three types of periodic changes in Earth's rotation and orbit around the sun. Over thousands of years, our planet wobbles on its axis, varies in the tilt of its axis, and experiences change in the shape of its orbit, all in regular long-term cycles of different lengths. These variations, known as **Milankovitch cycles**, alter the way solar radiation is distributed over Earth's surface (**FIGURE 14.6**). By modifying patterns of atmospheric heating, these cycles trigger long-term climate variation. This includes periodic episodes of *glaciation* during which global surface temperatures drop and ice sheets advance from the poles toward the midlatitudes. These cycles are highly influential in the very long term, but science shows that they do not account for the very recent, rapid, extreme climate change we are experiencing today.

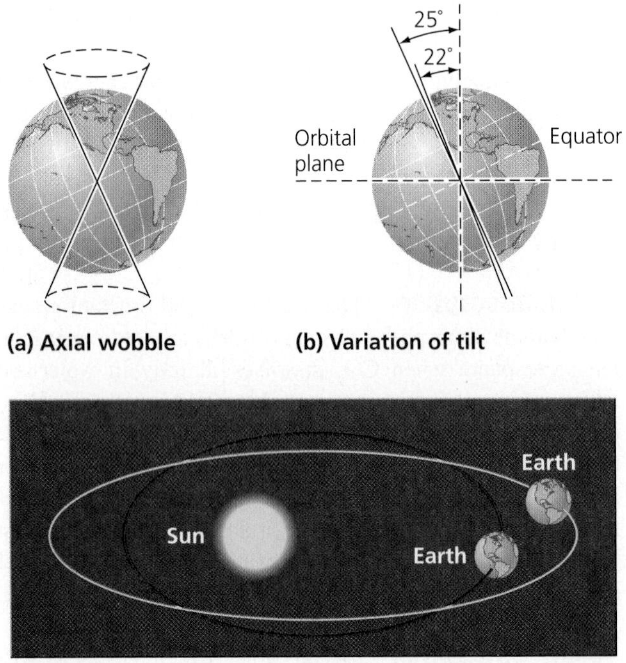


(a) Axial wobble (b) Variation of tilt

(c) Variation of orbit

**FIGURE 14.6 There are three types of Milankovitch cycles:** **(a)** an axial wobble that occurs on a 19,000- to 23,000-year cycle; **(b)** a 3-degree shift in the tilt of Earth's axis that occurs on a 41,000-year cycle; and **(c)** a variation in Earth's orbit from almost circular to more elliptical, which repeats every 100,000 years.

**FAQ The climate changes naturally, so why worry about climate change?**

Earth's climate does indeed change naturally across very long periods of time, but there is nothing "natural" about today's climate change. We know that human beings are directly causing the unnaturally rapid change we are now seeing. Moreover, humanity has never experienced the sheer amount of change predicted for this century. The quantity by which the world's temperature is forecast to rise is greater than the amount of cooling needed to bring on an ice age. Greenhouse gas concentrations are already higher than they've been in over 800,000 years, and they are rising. The human species has existed for only 200,000 years, and our civilization arose only in the last few thousand years during an exceptionally stable period in Earth's climate history. Unless we reduce our emissions, we will soon be challenged by climate conditions our species has never lived through.

# Studying Climate Change

To comprehend any phenomenon that changes, we must study its past, present, and future. Scientists monitor present-day climate, but they also devise clever means of inferring past change and sophisticated methods to predict future conditions.

## Proxy indicators tell us of the past

To understand past climate, scientists decipher clues from thousands or millions of years ago by taking advantage of the record-keeping capacity of the natural world. **Proxy indicators** are types of indirect evidence that serve as proxies, or substitutes, for direct measurement.

For example, Earth's ice caps, ice sheets, and glaciers hold clues to climate history. In frigid areas near the poles and atop high mountains, snow falling year after year compresses into ice. Over the ages, this ice accumulates to great depths, preserving within its layers tiny bubbles of the ancient atmosphere. Scientists can examine the trapped air bubbles by drilling into the ice and extracting long columns, or cores. The layered ice, accumulating season after season for thousands of years, provides a timescale. By studying the chemistry of the bubbles in each layer, scientists can determine atmospheric composition, greenhouse gas concentrations, temperature, snowfall, solar activity, and frequency of forest fires and volcanic eruptions during each time period.

Recently, researchers drilled and analyzed the deepest ice core ever (**FIGURE 14.7**). At a remote site in Antarctica, they drilled down 3270 m (10,728 ft) to bedrock and pulled out more than 800,000 years' worth of ice! This core chronicles Earth's history across eight glacial cycles. By analyzing air bubbles trapped in the ice, researchers discovered that over the past 800,000 years, atmospheric concentrations of carbon dioxide, methane, and nitrous oxide have never been as high as they are today. These data demonstrate that by emitting greenhouse gases since the industrial revolution, we have brought ourselves deep into uncharted territory.

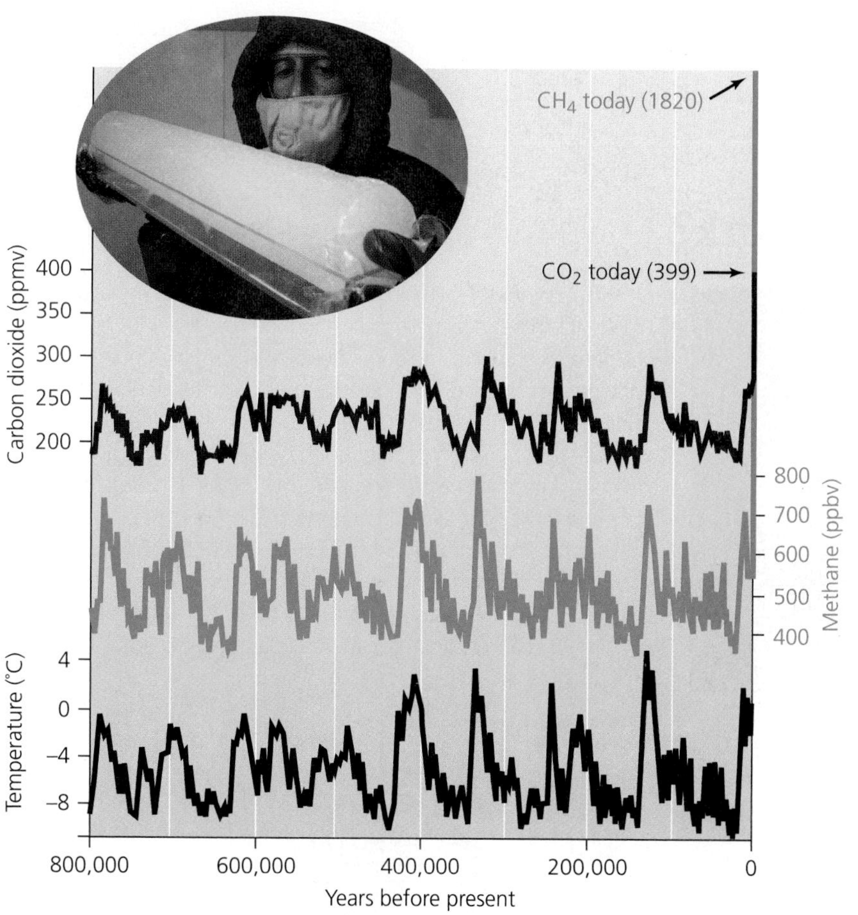


**FIGURE 14.7 Data from the EPICA ice core reveal changes across 800,000 years.** Shown are surface temperature **(black line)**, atmospheric methane concentration **(green line)**, and atmospheric carbon dioxide concentration **(red line)**. Concentrations of $CO_2$ and methane rise and fall in tight correlation with temperature. Today's current values are included at the top right of the graph, for comparison. *Adapted by permission of Macmillan Publishers Ltd: Brook, E. 2008. Paleoclimate: Windows on the greenhouse.* Nature *453: 291–292, Fig 1a.* www.nature.com.

The ice core results also confirm that temperature swings in the past were tightly correlated with greenhouse gas concentrations (compare the top two data sets in Figure 14.7 with the temperature data set at bottom). This bolsters the scientific consensus that our greenhouse gas emissions are causing Earth to warm today.

Researchers also drill cores into beds of sediment beneath bodies of water. Sediments often preserve pollen grains and other remnants from plants that grew in the past (as described in the study of Easter Island; pp. 6–7). Because climate influences the types of plants that grow in an area, knowing what plants were present can tell us a great deal about the climate at that place and time. Other types of proxy indicators include tree rings (which reveal year-by-year histories of precipitation and fire), pack-rat middens (rodent dens in which plant parts may be preserved for centuries in arid regions), and coral reefs (p. 258), which reveal aspects of ocean chemistry.

## Direct measurements tell us about the present

Today we measure temperature with thermometers, rainfall with rain gauges, wind speed with anemometers, and air pressure with barometers, using computer programs to integrate and analyze this information in real time. With these technologies and more, we document the fluctuations in weather day by day and hour by hour across the globe.

We also measure the chemistry of the atmosphere and the oceans. Direct measurements of carbon dioxide concentrations in the atmosphere reach back to 1958, when scientist Charles Keeling began analyzing hourly air samples from a monitoring station at Hawaii's Mauna Loa Observatory. These data show that atmospheric $CO_2$ concentrations have increased from 315 ppm in 1958 to 399 ppm in 2014.

## Models help us predict the future

To understand how climate systems function and to predict future climate change, scientists simulate climate processes with sophisticated computer programs. **Climate models** are programs that combine what is known about atmospheric circulation, ocean circulation, atmosphere–ocean interactions, and feedback cycles to simulate climate processes (see **THE SCIENCE BEHIND THE STORY**, pp. 310–311). This requires manipulating vast amounts of data with complex mathematical equations—a task not possible until the advent of modern computers.

Climate modelers provide starting information to the model, set up rules for the simulation, and then let it run. Researchers test the efficacy of a model by entering past climate data and running the model toward the present. If a model accurately reconstructs current climate, then we have reason to believe that it simulates climate mechanisms realistically and may accurately predict future climate.

Plenty of challenges remain for climate modelers, because Earth's climate system is so complex. Yet as scientific knowledge builds and computing power intensifies, climate models are improving in resolution and are making predictions region by region across the world.

## How Do Climate Models Work?

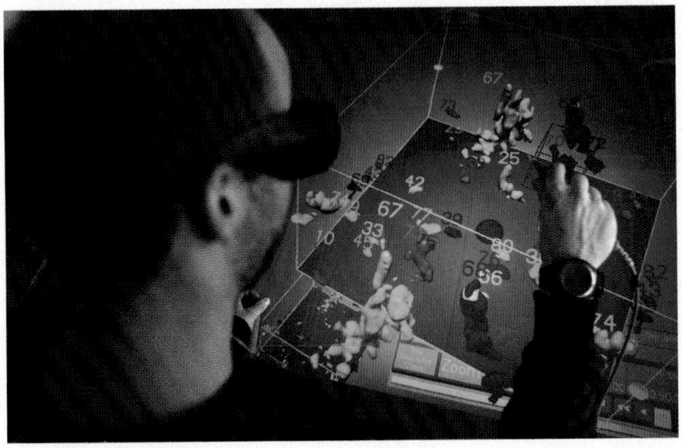

**A researcher manipulates a 3D cloud simulator to help in modeling climate.**

Models are indispensable for scientists studying climate today—and they are increasingly vital for society because they help predict what conditions will confront us in the future. Yet to most of us, a climate model is a mysterious black box. So how exactly do scientists create a climate model?

The colorful maps and data-rich graphs that scientists generate from a climate model are the end result, but the process begins when they put into the model a long series of mathematical equations. These equations describe how various components of Earth's systems function. Some equations are derived from physical laws such as those on the conservation of mass, energy, and momentum (p. 27). Others are derived from observational and experimental data on physics, chemistry, and biology. Converted into computing language, these equations are integrated with data on Earth's landforms, hydrology, vegetation, and atmosphere (**FIGURE 1**).

Earth's climate system is mind-bogglingly complex, but as computers grow more powerful, sophisticated models are including more and more factors that influence climate. Most models consist of submodels, each handling a different component—ocean water, sea ice, glaciers, forests, deserts, troposphere, stratosphere, and so on.

The processes within a model must be given equations to make them behave realistically in space and time. In the real climate system, time is continuous, and spatial effects reach down to the level of molecules interacting with one another. But the virtual reality of climate models cannot be so detailed—there is simply not enough computer power available. Instead, modelers approximate reality by dividing time into periods (called time steps) and by dividing Earth's surface into cells or boxes in a grid (called grid boxes) (**FIGURE 2**).

Each grid box contains land, ocean, or atmosphere, much like a digital photograph is made up of discrete pixels of certain colors. The grid boxes are arrayed in a three-dimensional layer

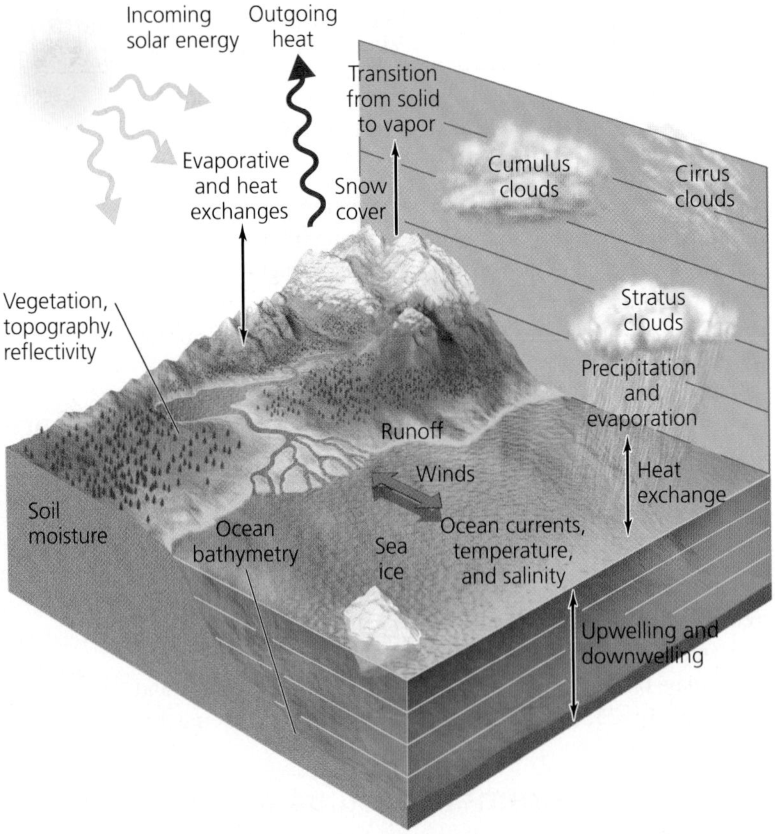


**FIGURE 1 Climate models incorporate a diversity of natural factors and processes.** Anthropogenic factors can then be added in.

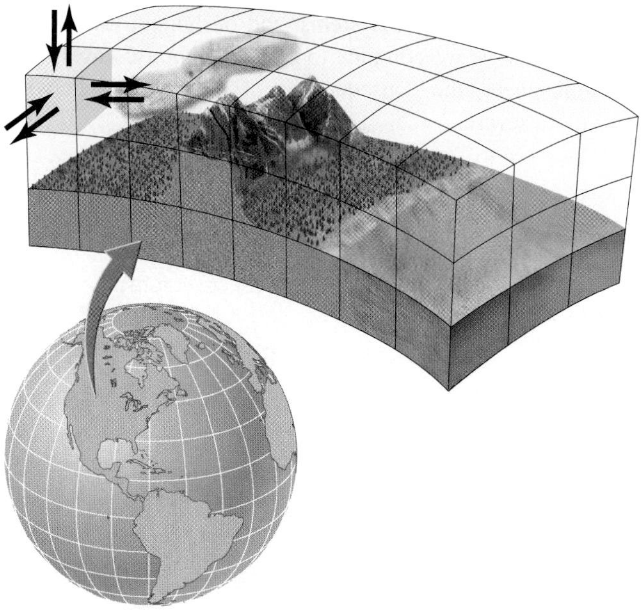

**FIGURE 2 Climate models divide Earth's surface into a layered grid.** Each grid box represents land, air, or water, and interacts with adjacent grid boxes via the flux of materials and energy. *Adapted from Bloom, Arnold J., 2010.* Global climate change: Convergence of disciplines. *Sinauer Associates, Sunderland, Mass.*

by latitude and longitude, or in equal-sized polygons. The finer the scale of the grid, the greater resolution the model will have, and the better it will be able to predict results region by region. However, more resolution requires more computing power, and climate models already strain the most powerful supercomputing networks. Today's best climate models feature dozens of grid boxes piled up from the bottom of the ocean to the top of the atmosphere, with each grid box measuring a few dozen miles wide, and time measured in periods of just minutes.

Once the grid is established, the processes that drive climate are assigned to each grid box, with their rates parceled out among the time steps. The model lets the grid boxes interact through time by means of the flux of materials and energy into and out of each grid box.

Once modelers have input all this information, learned from our study of Earth and the climate system, they let the model run through time and simulate climate, from the past into the future. If the computer simulation accurately reconstructs past and present climate, then that gives us confidence that it may predict future climate accurately as well.

A number of studies have compared model runs that include only natural processes, model runs that include only human-generated processes, and model runs that combine both. Repeatedly these studies find that the model runs that incorporate both human and natural processes are the ones that fit real-world climate observations the best (**FIGURE 3**). This supports the idea that human activities, as well as natural processes, are influencing our climate.

The major human influence on climate is our emission of greenhouse gases, and modelers need to select values to enter for future emissions if they want to predict future climate. Generally they will run their simulations multiple times, each time with a different emission rate according to a specified scenario. Differences between the results from such scenarios tell us what influence these different emission rates would have. You can see results from such a comparison in Figure 14.24 (p. 322).

Researchers are constantly testing and evaluating their models. They improve them by incorporating what is learned from new research and by taking advantage of what new computing technologies allow. As their work proceeds, we can expect increasingly precise and accurate predictions about future climate conditions. ☐

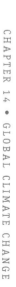

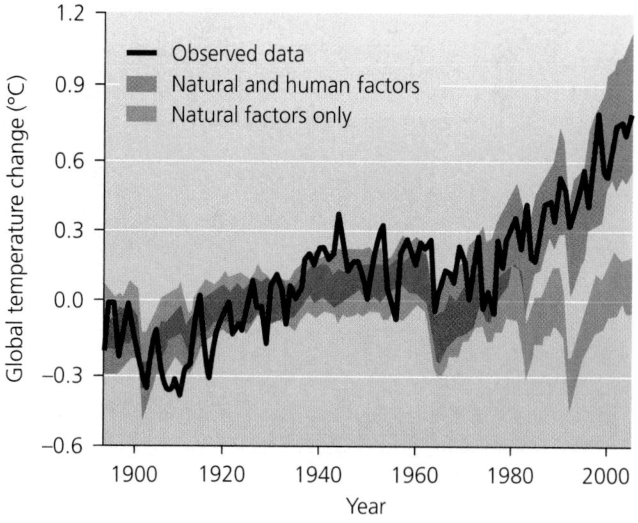


**FIGURE 3 Models that incorporate both natural and anthropogenic factors predict observed climate trends best.** *Adapted from Melillo, J. M., et al., eds., 2014.* Climate change impacts in the United States: The third national climate assessment. *U.S. Global Change Research Program.*

## Current and Future Trends and Impacts

Virtually everyone is noticing changes in the climate these days. Maldives fishermen note the seas encroaching on their home island. Texas ranchers and California farmers suffer multiyear droughts. Florida homeowners struggle to obtain insurance against hurricanes and storm surges. People from New York to Atlanta to Chicago to Los Angeles face one unprecedented weather event after another. Extreme weather events are indeed part of a real pattern backed by a tremendous volume of scientific evidence. Climate change has already had numerous impacts on the physical properties of our planet, on organisms and ecosystems, and on human well-being (**FIGURE 14.8**). If we continue to emit greenhouse gases into the atmosphere, the consequences of climate change will grow more severe.

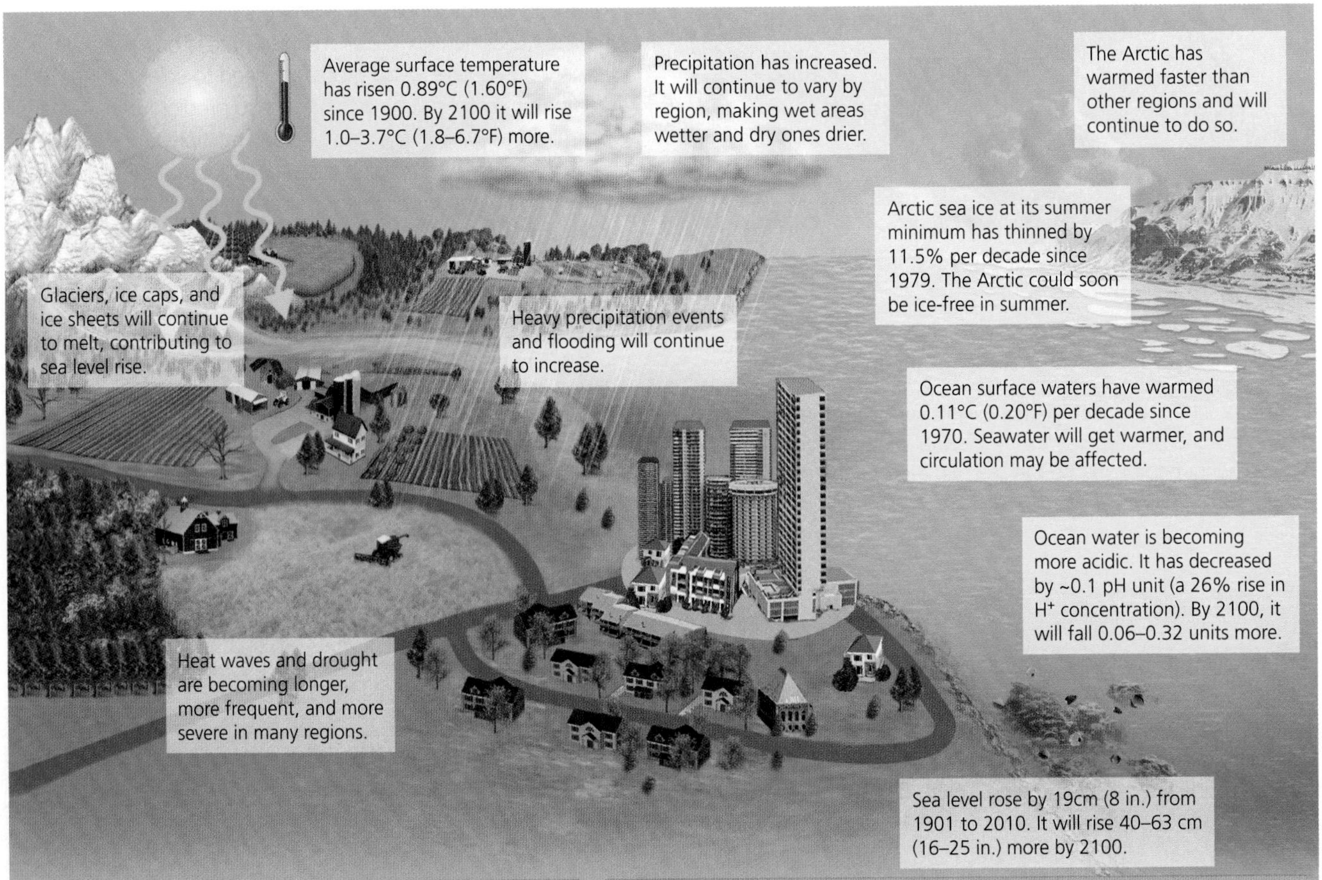

Average surface temperature has risen 0.89°C (1.60°F) since 1900. By 2100 it will rise 1.0–3.7°C (1.8–6.7°F) more.
Precipitation has increased. It will continue to vary by region, making wet areas wetter and dry ones drier.
The Arctic has warmed faster than other regions and will continue to do so.
Arctic sea ice at its summer minimum has thinned by 11.5% per decade since 1979. The Arctic could soon be ice-free in summer.
Glaciers, ice caps, and ice sheets will continue to melt, contributing to sea level rise.
Heavy precipitation events and flooding will continue to increase.
Ocean surface waters have warmed 0.11°C (0.20°F) per decade since 1970. Seawater will get warmer, and circulation may be affected.
Ocean water is becoming more acidic. It has decreased by ~0.1 pH unit (a 26% rise in H⁺ concentration). By 2100, it will fall 0.06–0.32 units more.
Heat waves and drought are becoming longer, more frequent, and more severe in many regions.
Sea level rose by 19cm (8 in.) from 1901 to 2010. It will rise 40–63 cm (16–25 in.) more by 2100.


**BIOLOGICAL TRENDS AND IMPACTS**

- Species ranges are shifting toward the poles and up in elevation.
- Seasonal timing (such as migration and breeding) is shifting.
- Some species are declining, and many could eventually face extinction.
- Ocean acidification is starting to affect marine life. Coral reefs could disappear, devastating marine ecosystems.
- Forests are being altered by drought, fire, and pest outbreaks.

**SOCIAL TRENDS AND IMPACTS**

- Ocean acidification could destroy major commercial fisheries.
- Droughts and flooding are leading to agricultural losses.
- Sea level rise will displace people and cause escalating expense.
- More intense storms are causing more property loss and loss of life.
- Farmers and foresters are struggling with altered growing seasons and disturbance regimes.
- Crop yields will likely fall in the dry tropics and subtropics.
- Impacts on biodiversity will cause losses of food, water, and ecosystem goods and services.
- Melting of mountain glaciers will reduce water supplies to millions.
- Economic costs will far outweigh benefits.
- Poorer nations and communities are suffering greater impacts.
- Increased warm-weather health hazards will outweigh decreased cold-weather hazards.

**FIGURE 14.8 Current trends and future impacts of climate change are extensive.** Shown are major physical, biological, and social trends and impacts (both observed and predicted) as reported by the IPCC. (Mean estimates are shown; the IPCC reports ranges and statistical probabilities as well.) *Data from IPCC, 2013. Fifth assessment report.*

## Scientific evidence for climate change is extensive

For years, scientists have studied climate change in enormous breadth, depth, and detail. As a result, researchers have gained a rigorous understanding of most aspects of climate change. To make this vast and growing research knowledge accessible to policymakers and the public, the **Intergovernmental Panel on Climate Change (IPCC)** has taken up the task of reviewing and summarizing it. This international body consists of many hundreds of scientists and governmental representatives. The IPCC was awarded the Nobel Peace Prize in 2007 for its work in informing the world of the trends and impacts of climate change.

In 2013 and 2014, the IPCC released its *Fifth Assessment Report*. By summarizing thousands of scientific studies, this report documents observed trends in surface temperature, precipitation patterns, snow and ice cover, sea levels, storm intensity, and other factors (see Figure 14.8). It also predicts future changes in these phenomena after considering a range of potential scenarios for future greenhouse gas emissions. The report addresses impacts of climate disruption on wildlife, ecosystems, and society. Finally, it discusses strategies we might pursue in response. To learn more, you and your instructor may wish to download the *Fifth Assessment Report* and explore its coverage of recent research findings. A series of three "working group" reports and a synthesis report are publicly accessible online at the IPCC's website.

## Temperatures continue to rise

Average surface temperatures on Earth have risen by nearly 0.9°C (1.6°F) in the past 100 years (**FIGURE 14.9**). Most of this increase has occurred since 1975—and just since 2000, we have experienced 13 of the 15 warmest years on record since global measurements began 150 years ago! The decade from 2001 to 2010 was the hottest ever recorded, and since the 1960s each decade has been warmer than the last. If you were born after 1985, you have never in your life lived through a month with average global temperatures lower than the 20th-century average.

In just the past two decades, temperatures in most areas of the United States have risen by more than 1 full degree Fahrenheit (**FIGURE 14.10**).

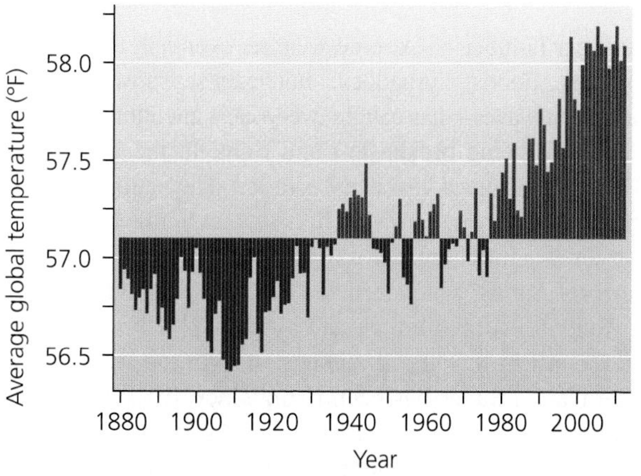


**(a) Global temperature measured since 1880**

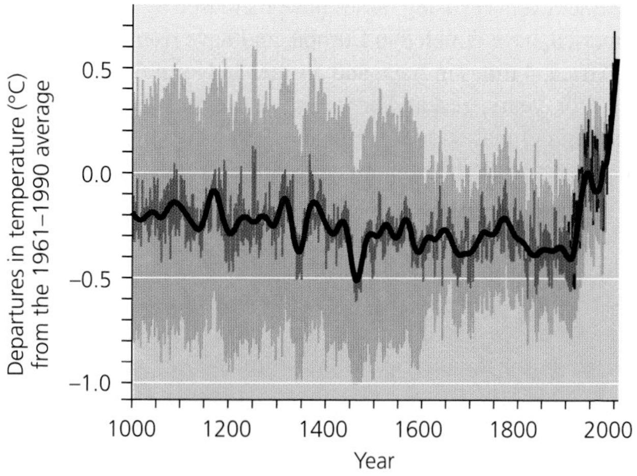


**(b) Northern Hemisphere temperature over the past 1000 years**

**FIGURE 14.9 Global temperatures have risen sharply in the past century.** Data from thermometers **(a)** show changes in Earth's average surface temperature since 1880. Since 1976, every single year has been warmer than average. In **(b)**, proxy indicators **(blue line)** and thermometer data **(red line)** together show average temperatures in the Northern Hemisphere over the past 1000 years. The gray-shaded zone represents the 95% confidence range. *Data from (a) Melillo, J.M., et al., eds., 2014.* Climate change impacts in the United States: The third national climate assessment. *U.S. Global Change Research Program; and (b) IPCC, 2001.* Third assessment report.

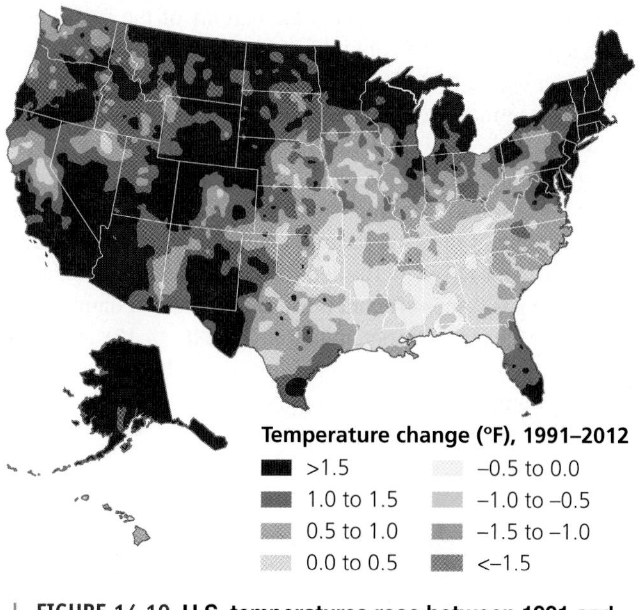


**FIGURE 14.10 U.S. temperatures rose between 1991 and 2011.** Although some areas of the Southeast cooled slightly, most of the nation warmed by more than 1 degree Fahrenheit (0.6°C). *Data from NOAA National Climatic Data Center as presented in Melillo, J.M., et al., eds., 2014.* Climate change impacts in the United States: The third national climate assessment. *U.S. Global Change Research Program.*

**DATA Q** By how much did the average temperature rise or fall where you live? How does this compare with changes in other parts of the country?

GO TO **INTERPRETING GRAPHS & DATA** ON MasteringEnvironmentalScience®

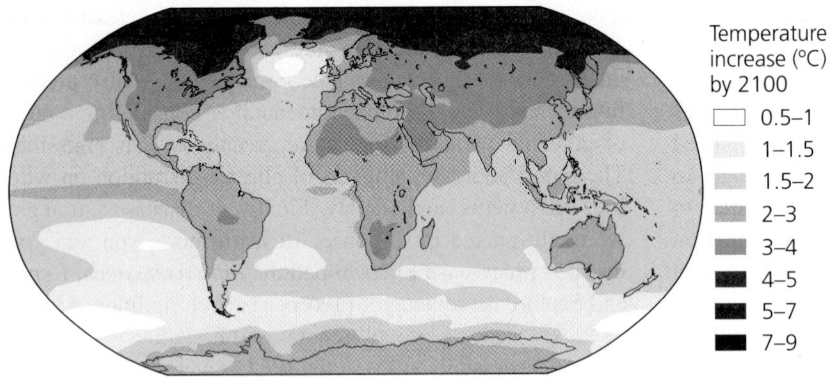


**FIGURE 14.11 Surface temperatures are projected to rise for the years 2081–2100, relative to 1986–2005.** Landmasses are expected to warm more than oceans, and the Arctic will warm the most. This map was generated using an intermediate emissions scenario with an average global temperature rise of 2.2°C (4.0°F). *Data from IPCC, 2013.* Fifth assessment report. The physical science basis: Contribution of Working Group I.

We can expect global surface temperatures to continue rising because we are still emitting greenhouse gases and because the greenhouse gases already in the atmosphere will continue warming the globe for decades to come. At the end of the 21st century, the IPCC predicts global temperatures will be 1.0–3.7°C (1.8–6.7°F) higher than today's, depending on how well we control our emissions. Unusually hot days and heat waves will become more frequent. Future changes in temperature are predicted to vary from region to region in ways that they already have (**FIGURE 14.11**). For example, polar regions will continue to experience the most intense warming.

## Precipitation is changing, too

A warmer atmosphere speeds evaporation and holds more water vapor, and precipitation has increased worldwide by 2% over the past century. Yet some regions of the world are receiving less rain and snow than usual while others receive more than usual. In the southwestern United States, droughts have become more frequent and severe, harming agriculture, worsening soil erosion, reducing water supplies, and triggering wildfire. In other regions, heavy rain events have increased, leading to flooding, such as the 2008 floods in Iowa and elsewhere in the U.S. Midwest and the 2011 floods along the Mississippi River that killed dozens of people, left thousands homeless, and inflicted billions of dollars in damage.

Future changes in precipitation (**FIGURE 14.12**) are predicted to intensify regional changes that have already occurred. Many wet areas will receive more rainfall, increasing the risk of flooding, while many dry areas become drier.

## Extreme weather is becoming "the new normal"

The sheer number of extreme weather events in recent years—droughts, floods, tornadoes, hurricanes, snowstorms, cold snaps, heat waves—has caught everyone's attention, and weather records are being broken left and right. In the United States, 2012 was the hottest year ever recorded. The nation experienced a freakish heat wave in March, a severe summer drought that devastated agriculture across three-fifths of the country, and Hurricane Sandy, which inflicted over $60 billion in damage.

Scientific data summarized by the U.S. Climate Extremes Index confirm that the frequency of extreme weather events in the United States has been rising since 1970. Scientists are not the only ones to notice this trend. The insurance industry is finely attuned to such patterns, because insurers pay out money each time a major storm, drought, or flood hits. A major German insurer, Munich Re, calculated that since 1980, extreme weather events causing losses have increased by 50% in South America, have doubled in Europe, and have risen by 2.5 times in Africa, 4 times in Asia, and 5 times in North America.

For years, researchers have conservatively stated that although climate trends influence the probability of what the weather may be like on any given day, no particular weather event can be directly attributed to climate change. In the aftermath of Hurricane Sandy, a metaphor spread across the Internet: When a baseball player takes steroids and starts hitting more home runs, you can't attribute any one particular home run to the steroids, but you *can* conclude that steroids are responsible for the increase in home runs. Our greenhouse gas

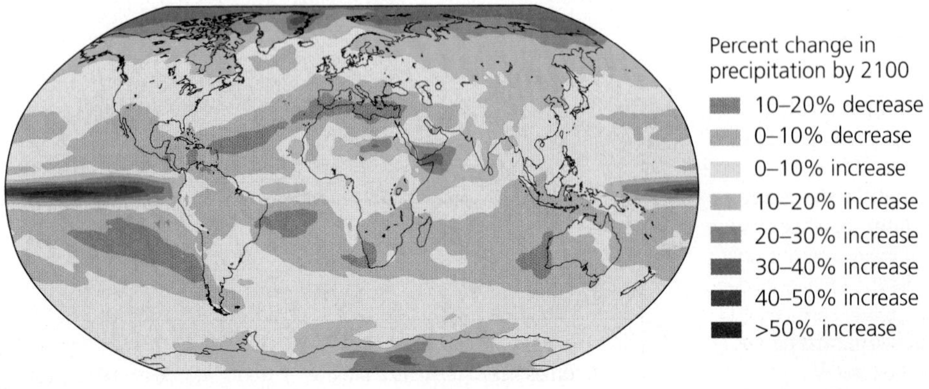


**FIGURE 14.12 Precipitation is projected to change for the years 2081–2100, relative to 1986–2005.** Browner shades indicate less precipitation; bluer shades indicate more. This map was generated using an intermediate emissions scenario with an average global temperature rise of 2.2°C (4.0°F). *Data from IPCC, 2013.* Fifth assessment report. The physical science basis: Contribution of Working Group I.

emissions are like steroids that are supercharging our climate and increasing the instance of extreme weather events.

In 2012, research by Jennifer Francis of Rutgers University and Stephen Vavrus of the University of Wisconsin revealed a mechanism that may explain how and why global warming leads to more extreme weather. Because warming has been greatest in the Arctic, this has weakened the intensity of the Northern Hemisphere's polar *jet stream*, a high-altitude air current that blows west-to-east and meanders north and south, influencing the weather across North America and Eurasia. As the jet stream slows down, its meandering loops become longer. These long lazy loops move west to east more slowly, and may get stuck in a north–south orientation for long periods of time. Meteorologists call this an *atmospheric blocking pattern* because it blocks the eastward movement of weather systems (**FIGURE 14.13**). When this happens, a rainy system that would normally move past a city in a day or two may instead be held in place for several days, causing flooding. Or dry conditions over a farming region might last two weeks instead of two days, resulting in drought. Hot spells last longer, and cold spells last longer, too.

Indeed, the record-breaking heat wave of March 2012 resulted after the jet stream became stuck in place (see Figure 14.13b). Atmospheric blocking patterns also were associated with the 2011 drought in Texas, the 2012 wildfires in Colorado, severe wintry weather in the eastern United States in 2014, both floods and heat waves in Europe, and other extreme weather events.

## Melting ice has far-reaching effects

As the world warms, mountaintop glaciers are disappearing (**FIGURE 14.14**). Between 1980 and 2012, the World Glacier Monitoring Service estimates that the world's major glaciers on average each lost mass equivalent to 16 m (52 ft) in vertical height of water. Many glaciers on tropical mountaintops have disappeared already. In Glacier National Park in Montana, only 25 of 150 glaciers present at the park's inception remain, and scientists estimate that by 2030 even these will be gone.

Mountains accumulate snow in winter and release meltwater gradually during summer. Over one-sixth of the world's people live in regions that depend on mountain meltwater. As warming temperatures diminish mountain glaciers, summertime water supplies will decline for millions of people, likely forcing whole communities to look elsewhere for water or to move.

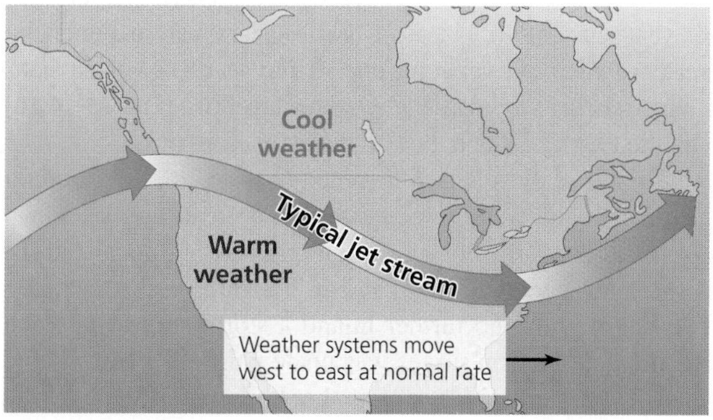


**(a) Normal jet stream**

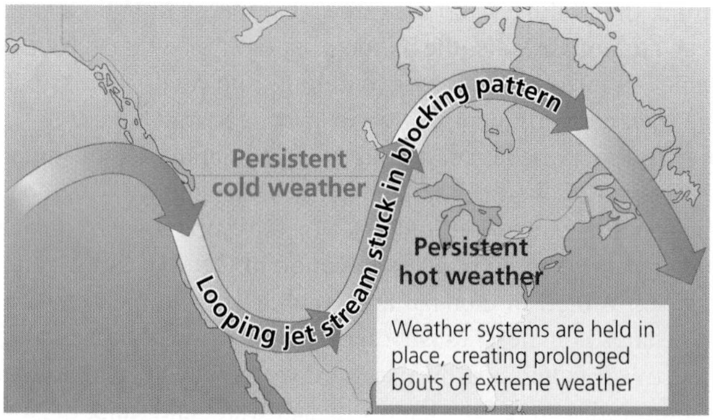


**(b) Jet stream in March 2012**

**FIGURE 14.13 Changes in the jet stream can cause extreme weather events.** When Arctic warming slows the jet stream, it departs from its normal configuration **(a)** and goes into a blocking pattern **(b)** that stalls weather systems in place. The blocking pattern shown here brought record-breaking heat to the eastern United States in March 2012.

**(a) Jackson Glacier in 1911**

**(b) Jackson Glacier in 2009**

**FIGURE 14.14 Glaciers are melting as global warming proceeds.** The Jackson Glacier in Glacier National Park, Montana, retreated substantially between **(a)** 1911 and **(b)** 2009. *Data from World Glacier Monitoring Service.*

Warming temperatures are also melting vast amounts of polar ice. In the Arctic, the immense ice sheet that covers Greenland is melting faster and faster. In Antarctica, coastal ice shelves the size of Rhode Island have disintegrated as a result of contact with warmer ocean water, and recent research suggests that the entire West Antarctic ice shelf may be on its way to unstoppable collapse, creating a 3-m (10-ft) rise in sea level.

Warming is accelerating in the Arctic because as snow and ice melt, darker, less reflective surfaces (such as bare ground and pools of meltwater) are exposed, and Earth's *albedo*, or capacity to reflect light, decreases. As a result, more of the sun's rays are absorbed at the surface, fewer rays reflect back into space, and the surface warms. In a process of positive feedback, this warming causes more ice and snow to melt, which in turn causes more absorption of radiation and more warming (see Figure 2.1b, p. 23).

As Arctic sea ice disappears, new shipping lanes open up for commerce, and governments and companies are rushing to exploit newly accessible underwater oil and mineral reserves. Russia, Canada, the United States, and other nations are jockeying for position, trying to lay claim to regions of the Arctic as the ice melts.

Warmer temperatures in the Arctic are now also causing *permafrost* (permanently frozen ground) to thaw. As ice crystals within permafrost melt, the thawing soil settles, destabilizing buildings, pipelines, and other infrastructure. When permafrost thaws, it also can release methane that has been stored for thousands of years. Because methane is a potent greenhouse gas, its release acts as a positive feedback mechanism that intensifies climate change.

## Rising sea levels may affect hundreds of millions of people

As glaciers and ice sheets melt, increased runoff causes sea levels to rise. Sea levels also are rising because ocean water is warming; water expands in volume as it warms. Worldwide, average sea levels have risen 21 cm (8.3 in.) in the past 130 years (**FIGURE 14.15**), reaching a rate of 3.2 mm/year from

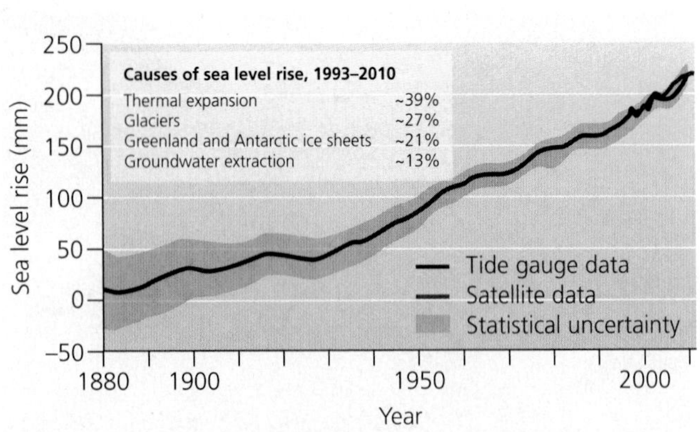


**FIGURE 14.15 Global average sea level has risen roughly 210 mm (8.3 in.) since 1880.** Sea levels rise because water expands as it warms, glaciers and ice sheets are melting, and groundwater we extract eventually reaches the ocean. *Data from IPCC, 2013.* Fifth assessment report. The physical science basis: Contribution of Working Group I.

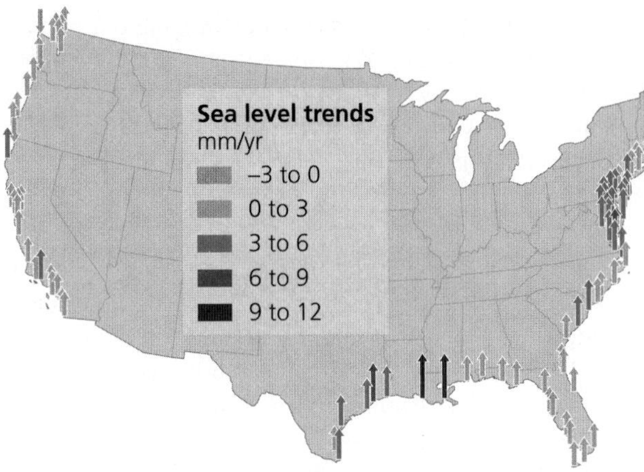


**FIGURE 14.16 Sea level is rising at varying rates along the U.S. coast.** Rates are highest where land is subsiding along the Gulf Coast and the central Atlantic Seaboard. *Data from National Oceanic and Atmospheric Administration.*

1993 to 2010. These numbers represent vertical rises in water level, and on most coastlines a vertical rise of a few inches means many feet of incursion inland.

Regions experience differing amounts of sea level change because land may be rising or subsiding naturally, depending on local geological conditions. The Maldives has fared better than many other island nations: It has seen sea levels rise about 3 mm per year since 1990, but most Pacific islands are experiencing greater rises in sea level, some up to 9 mm/year. The United States is experiencing varying degrees of sea level rise (**FIGURE 14.16**), with the East Coast and the Gulf Coast most at risk.

Higher sea levels lead to beach erosion, coastal flooding, intrusion of salt water into aquifers, and greater impacts from storm surges. A *storm surge* is a temporary and localized rise in sea level generated by a storm. The higher that sea level is to begin with, the further inland a storm surge can reach. The impact of storm surges was made painfully clear in 2013 when Typhoon Haiyan struck the Philippines. One of the strongest storms ever recorded, it took 7000 lives and caused over $1.5 billion in damage. In 1987, unusually high waves struck the Maldives and triggered a campaign to build a large seawall around Malé, the nation's capital. "The Great Wall of Malé" is intended to protect buildings and roads by dissipating the energy of incoming waves during storm surges.

In the United States, "Superstorm" Sandy demonstrated the impact that storm surges can have on highly developed metropolitan areas (**FIGURE 14.17**). This massive hurricane battered the eastern part of the nation in October 2012, causing $65 billion in damage and leaving 160 people dead and thousands homeless. In New Jersey, thousands of homes were destroyed, iconic boardwalks were washed away, and coastal communities were inundated with salt water and sand. In New York City, economic activity ground to a halt as tunnels and subway stations flooded and vehicles and buildings suffered damage. A fire broke out amid flooded homes in Queens and destroyed an entire neighborhood.

Like Typhoon Haiyan, Hurricane Sandy was not directly and solely caused by global warming, but in a statistical sense it was facilitated and strengthened by it. Warmer ocean water boosts the chances of large and powerful hurricanes. A warmer atmosphere retains more moisture that a hurricane can dump onto land. A blocking pattern in the jet stream contributed to Sandy's energy. And higher sea levels magnify the damage caused by storm surges.

Seven years before Sandy, the United States was hit by an even costlier storm. Hurricane Katrina slammed into New Orleans and the Gulf Coast in 2005, killing more than 1800 people and inflicting $80 billion in damage. Outside New Orleans today, marshes of the Mississippi River delta continue to disappear as rising seas eat away at coastal vegetation (pp. 248–249). More than 2.5 million ha (1 million acres) of Louisiana's coastal wetlands have vanished since 1940, weakening protection against future storm surges. Around the world, rising seas are eating away at the salt marshes, dunes, mangrove forests, and coral reefs that serve as barriers protecting our coasts.

The IPCC predicts that mean sea level will rise 26–82 cm (10–32 in.) higher by 2100, depending on emissions. It will continue rising after 2100, and some researchers are now using a 1-meter (39-in.) rise scenario to assess risk. More than half of the U.S. population lives in coastal counties, and 4 million Americans live within 1 vertical meter of the high tide line. Researchers recently estimated that a 1-m rise threatens 180 U.S. cities with losing an average of 9% of their land area. Miami, Tampa, New Orleans, and Virginia Beach are most at risk.

Whether sea levels this century rise a full meter or "only" 26 cm, hundreds of millions of people will be displaced or will need to invest in costly efforts to protect against storm surges. Densely populated regions on low-lying river deltas, such as Bangladesh, will be affected. So will storm-prone regions such as Florida; coastal cities such as Houston, Charleston, Boston, and New York City; and areas where land is subsiding, such as the U.S. Gulf Coast. Many islands may need to be evacuated. In the meantime, island nations such as the Maldives and coastal cities such as Tampa are vulnerable to shortages of fresh water as rising seas bring salt water into aquifers.

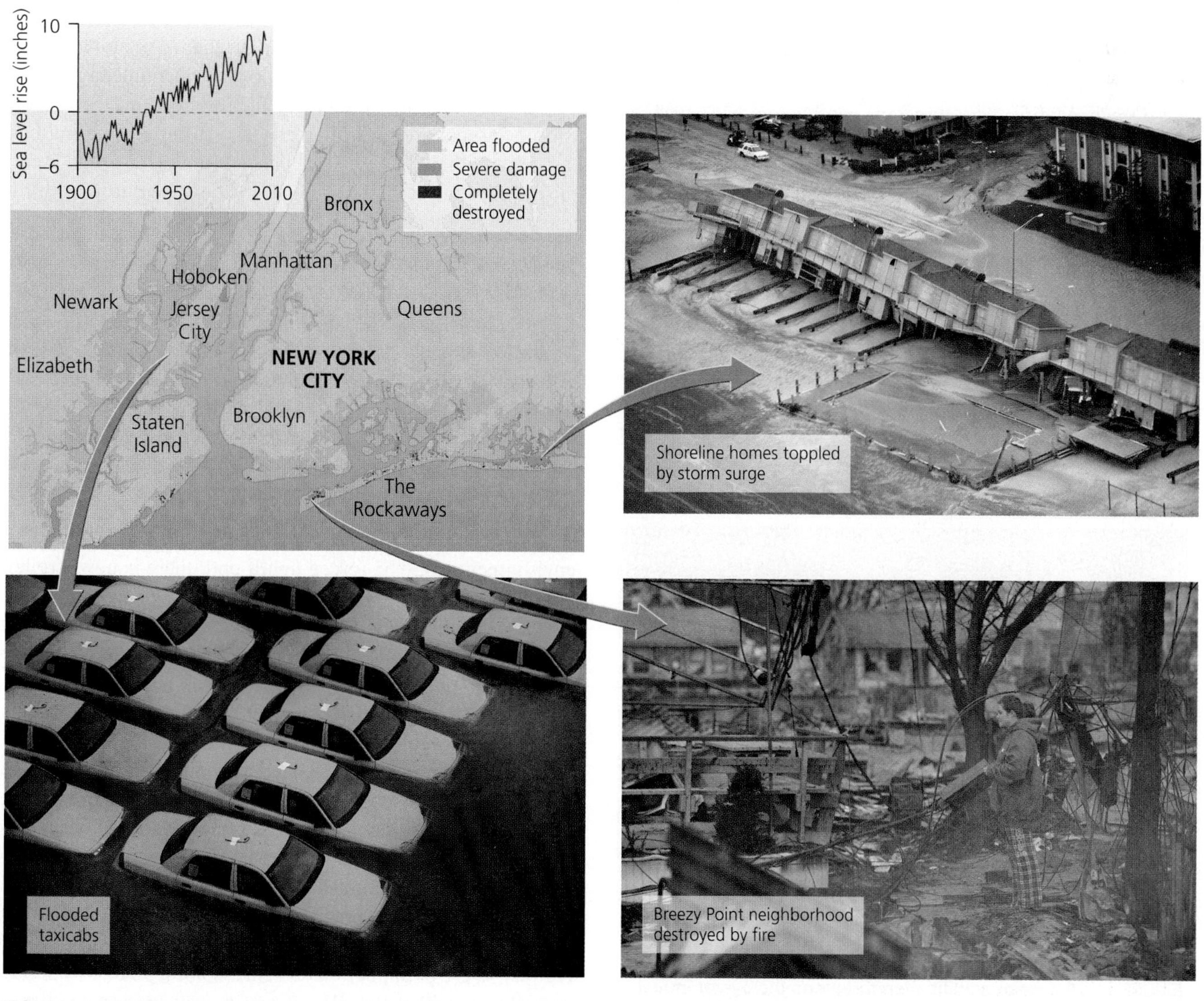


**FIGURE 14.17 Climate change contributes to the power and reach of devastating storms like Hurricane Sandy.** The map shows areas in New York City flooded by the storm. The graph shows sea level rise in New York City in the past century. *Map data from* The New York Times *as adapted from federal agencies; graph data from New York City Panel on Climate Change 2010.*

## WEIGHING THE ISSUES

**Environmental Refugees**

The Pacific island nation of Tuvalu has been losing 9 cm (3.5 in.) of elevation per decade to rising seas. Appeals from Tuvalu's 11,000 citizens were heard by New Zealand, which began accepting "environmental refugees" from Tuvalu in 2003. Do you think the rest of the world should grant such environmental refugees international status and assume some responsibility for taking care of them? Do you think a national culture can survive if its entire population is relocated? Think of the tens of thousands of refugees from Hurricane Katrina. How did their lives and culture fare in the wake of that tragedy?

## Acidifying oceans imperil marine life

As carbon dioxide concentrations in the atmosphere rise, the oceans absorb more $CO_2$. So far, the oceans have absorbed roughly one-quarter of the $CO_2$ we have added to the atmosphere. This is altering ocean chemistry, making seawater more acidic—a phenomenon referred to as **ocean acidification** (pp. 258–259).

Ocean acidification threatens marine animals such as corals, clams, oysters, mussels, and crabs, which pull carbonate ions out of seawater to build their exoskeletons of calcium carbonate. As seawater becomes more acidic, carbonate ions become less available, and calcium carbonate begins to dissolve, jeopardizing the existence of these animals.

Global ocean chemistry has already decreased by 0.1 pH unit, which corresponds to a 26% rise in hydrogen ion concentration. Initial impacts on corals and oysters are already apparent (**FIGURE 14.18**). By 2100, scientists predict that seawater will decline in pH by another 0.06–0.32 units—very possibly enough to destroy most of our planet's living coral reefs (p. 258). Such destruction could be catastrophic for marine biodiversity and fisheries, because so many organisms depend on coral reefs for food and shelter. Indeed, ocean acidification and the potential loss of marine life threaten to become one of the most far-reaching impacts of global climate change.

Coral reefs face two additional risks from climate change: Warmer waters contribute to deadly coral bleaching (p. 258), and stronger storms physically damage reefs. All these factors concern residents of places like the Maldives or south Florida. In each of these places, coral reefs provide habitat for fish consumed locally and exported for profit, offer snorkeling and scuba diving sites for tourism, and protect coastlines from erosion by reducing wave intensity.

**FIGURE 14.18 Ocean acidification threatens the oyster industry.** Since 2005, acidified seawater has killed billions of larval oysters in Washington and Oregon, jeopardizing the region's once-thriving industry. Scientists are helping producers find ways to respond, but many in the industry fear acidification will wreak havoc with many shellfish.

## Climate change affects organisms and ecosystems

As the developing crisis with marine life shows, changes in Earth's physical systems have consequences for living things. Organisms are adapted to their environments, so changes to those environments affect them. As global warming proceeds, it is modifying biological phenomena that rely on temperature. In the spring, plants are now leafing out earlier, insects are hatching earlier, birds are migrating earlier, and animals are breeding earlier. These shifts can create mismatches in seasonal timing. For example, European birds known as great tits had evolved to raise their young when caterpillars peak in abundance. Now caterpillars are peaking earlier, but the birds have been unable to adjust, and fewer young birds are surviving.

Biologists are also recording spatial shifts in the ranges of organisms, as plants and animals move toward the poles or upward in elevation (i.e., toward cooler areas) as temperatures warm (**FIGURE 14.19**). Some organisms will not be able to cope and could face extinction. Trees may not be able to shift their distributions fast enough. Rare species may be forced out of preserves and into developed areas where they cannot survive. Animals and plants adapted to mountainous environments may be forced uphill until there is nowhere left to go.

Effects on plant communities comprise an important component of climate change, because by drawing in carbon dioxide for photosynthesis, plants act as reservoirs for carbon. If an atmosphere richer in $CO_2$ enhances plant growth, then plants might in turn remove more $CO_2$ from the air. However, if climate change decreases plant growth (through drought, fire, or disease, for instance), then carbon flux to the atmosphere could increase. Large-scale outdoor experiments are revealing complex answers, showing that extra $CO_2$ can both augment and diminish plant growth.

In regions where precipitation and stream flow increase, erosion and flooding will pollute and alter aquatic systems. In regions where precipitation decreases, lakes, ponds, wetlands, and streams will shrink. The many impacts of climate change on ecological systems will diminish the ecosystem goods and services our societies depend on, from food to clean air to drinking water.

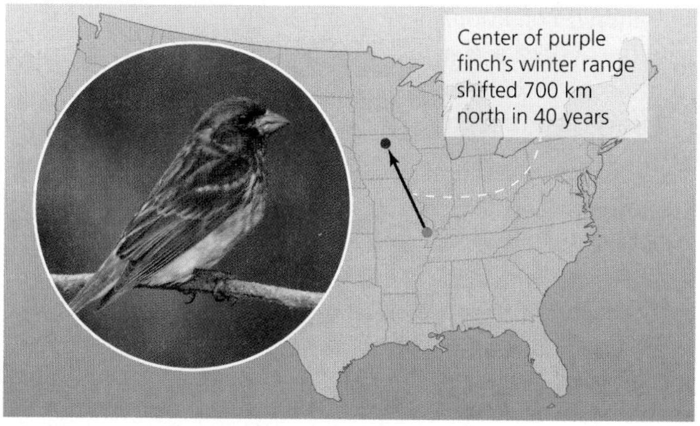
Center of purple
finch's winter range
shifted 700 km
north in 40 years

(a) Birds are moving north

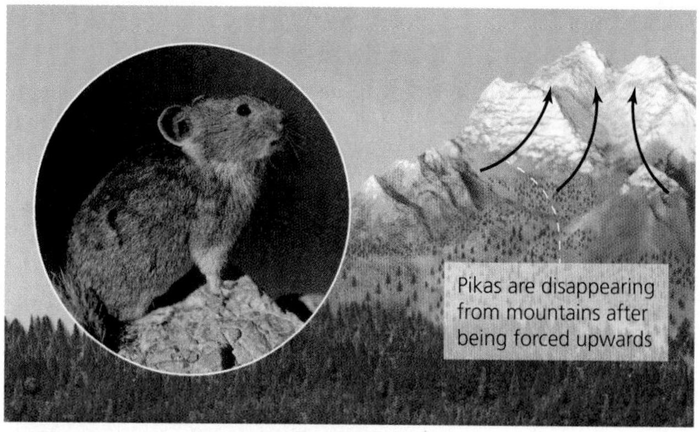
Pikas are disappearing
from mountains after
being forced upwards

(b) Pikas are being forced upslope

**FIGURE 14.19 Animal populations are shifting toward the poles and upward in elevation.** Fully 177 of 305 North American bird species have shifted their winter ranges significantly northward in the past 40 years, according to a 2009 analysis of Christmas Bird Count data by National Audubon Society researchers. The purple finch **(a)** has shown the greatest shift; its center of abundance moved 697 km (433 mi) north. Mountain-dwelling animals such as the pika **(b)**, a unique mammal of western North America, are being forced upslope (into more limited habitat) as temperatures warm. Many pika populations in the Great Basin have disappeared from mountains already.

## Climate change affects society

Drought, flooding, storm surges, and sea level rise are already taking a toll on the lives and livelihoods of millions of people. However, climate change will have still more consequences. These include impacts on agriculture, forestry, health, and economics.

**Agriculture** For some crops in the temperate zones, moderate warming may slightly increase production as growing seasons lengthen. Added carbon dioxide for photosynthesis may or may not boost yields, and research indicates that crops can become less nutritious when supplied with more $CO_2$. If rainfall continues to shift in space and time, intensified droughts and floods will likely cut into agricultural productivity (**FIGURE 14.20**). Considering all factors, the IPCC predicts that global crop yields will increase somewhat—but beyond a rise of 3°C (5.4°F), it expects crop yields to decline. It also expects crop production to fall in seasonally dry tropical and subtropical regions where drought and shorter growing seasons will worsen hunger in many developing nations.

**FIGURE 14.20 Drought induced by climate change decreases crop yields.** Withered corn fields like this one in Illinois were a common sight in 2012, when the U.S. government declared 1000 counties across 26 states to be disaster areas due to drought.

**Forestry** In the forests that provide our timber and paper products, enriched atmospheric $CO_2$ may spur greater growth, but drought, fire, and disease may eliminate these gains. Forest managers increasingly find themselves battling catastrophic fires, invasive species, and insect and disease outbreaks. Catastrophic fires are caused in part by decades of fire suppression (p. 196) but are also promoted by longer, warmer, drier fire seasons. Milder winters and hotter, drier summers are promoting outbreaks of bark beetles that are destroying millions of acres of trees (p. 197).

**Health** As climate change proceeds, we will face more heat waves—and heat stress can cause death. A 1995 heat wave in Chicago killed at least 485 people, and a 2003 heat wave in Europe killed 35,000 people. A warmer climate also exposes us to other health problems:

- Respiratory ailments from air pollution as hotter temperatures promote photochemical smog (p. 291)
- Expansion of tropical diseases, such as malaria and dengue fever, into temperate regions as disease vectors (such as mosquitoes) spread toward the poles
- Disease and sanitation problems when floods overcome sewage treatment systems
- Injuries and drowning as storms become more frequent or intense

**Economics** People will experience a variety of economic costs and benefits from the impacts of climate change, but on the whole researchers predict that costs will outweigh benefits. Climate change is also expected to widen the gap between rich and poor because poorer people lack the wealth and technology that help people adapt to change, and because poorer people rely more on resources (such as local food and water) that are sensitive to climate disruption.

From a variety of studies, the IPCC has estimated that climate change may cost 1–5% of GDP on average globally, with poor nations losing proportionally more than rich nations. Economists have proposed societal costs of anywhere from \$10 to \$350 per ton of carbon. The *Stern Review on the Economics of Climate Change*, commissioned by the British government, concluded that climate change could cost us 5–20% of GDP by the year 2200, but that investing just 1% of GDP starting now could enable us to avoid these future costs. A 2014 report titled *Risky Business* detailed hundreds of billions of dollars in likely damages from climate change. This report was issued by a politically diverse team of leading businesspeople and finance experts, chaired by former New York City Mayor Michael Bloomberg, former hedge fund manager Tom Steyer, and former Treasury Secretary Henry Paulson. Regardless of the precise numbers, economists are concluding that investing money now to fight climate change will save us a great deal more money in the future.

## Impacts vary by region

Each of us will experience the impacts of climate change differently, depending on where we live. Temperature changes have been greatest in the Arctic (**FIGURE 14.21**). Here, ice sheets are melting, sea ice is thinning, storms are increasing, and altered conditions are posing challenges for people and wildlife. As sea ice melts earlier, freezes later, and recedes from shore, it becomes harder for Inuit people and for polar bears alike to hunt the seals they each rely on for food. Permafrost is thawing, destabilizing buildings. As the strong Arctic warming melts ice caps and ice sheets, it contributes to sea level rise globally.

For the United States, impacts are assessed by the U.S. Global Change Research Program, which Congress created to coordinate federal climate research. Its 2014 *National Climate Assessment* summarizes current research, observed trends, and predicted future impacts of climate change on the United States. This report shows that average U.S. temperatures have increased by 0.7–1.1°C (1.3–1.9°F) since record keeping began in 1895, with the vast majority of this rise occurring just since 1970. Temperatures are predicted to rise by another 1.7–5.6°C (3–10°F) by the end of this century (**FIGURE 14.22**). Extreme weather events have become more frequent, and the costs they impose on farmers, city-dwellers, coastal communities, and taxpayers across the country are escalating.

Impacts vary regionally, and each U.S. region will face its own challenges (**FIGURE 14.23**). Winter and spring precipitation is projected to decrease in the South but increase in the North. Drought may strike in some regions and flooding in others. Sea level rise will affect the Atlantic and Gulf Coasts more than the West Coast. You can learn more about the predictions for your own region by consulting this publicly accessible report online.

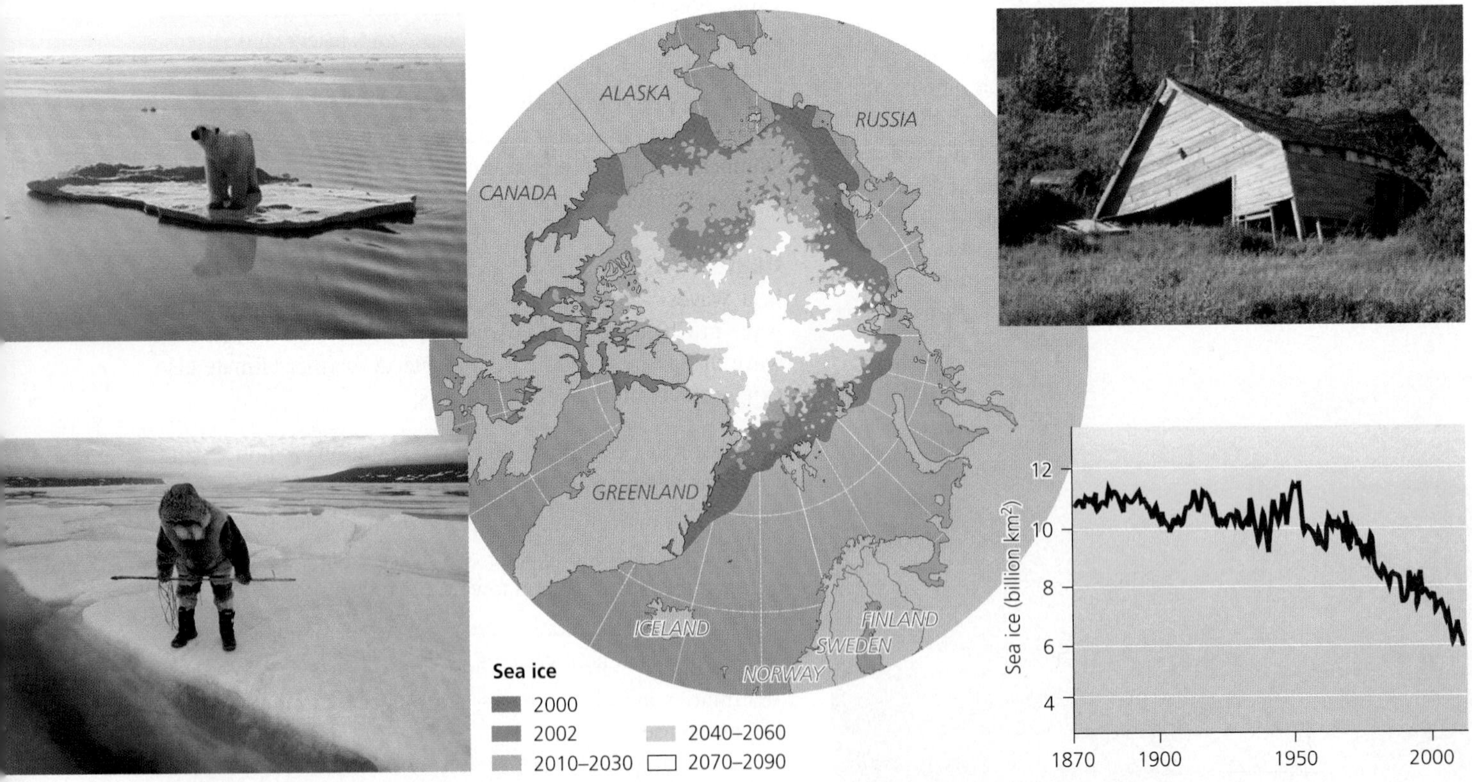


**FIGURE 14.21 The Arctic has borne the brunt of climate change.** As sea ice melts, it recedes from large areas. The map shows mean minimum summer extent of sea ice for the recent past, present, and future. The graph shows declines in sea ice averaged from six data sets. Inuit people find it difficult to hunt and travel in their traditional ways. Polar bears starve because they are less able to hunt seals. Structures are damaged as permafrost thaws beneath them. *Map data from National Center for Atmospheric Research and National Snow and Ice Data Center.*

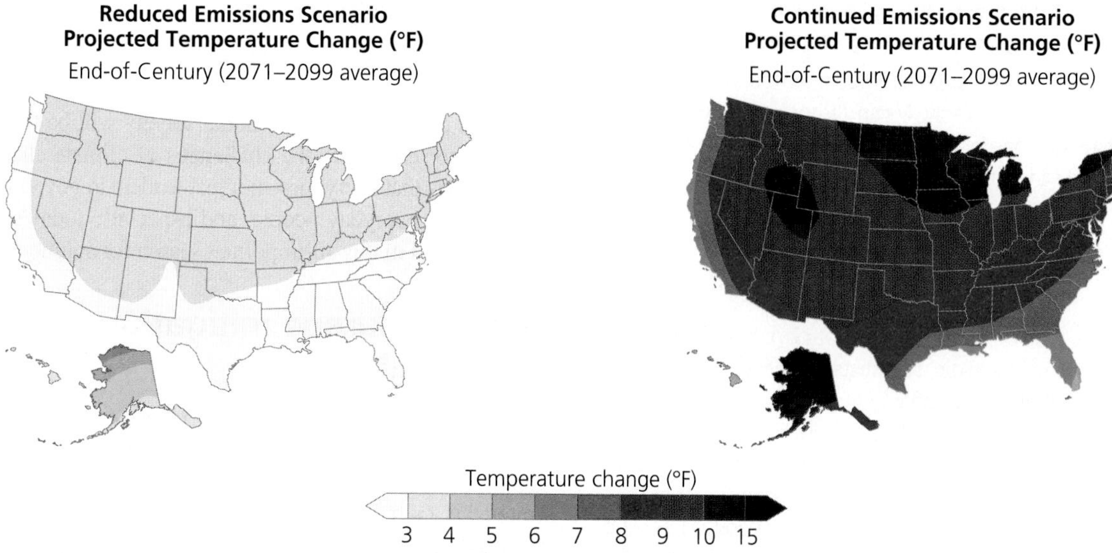


**FIGURE 14.22 Average temperatures across the United States are predicted to rise further by the end of this century.** Even in a scenario of sharply reduced emissions **(left)**, temperatures are predicted to rise by 3–4°F. In a scenario of business-as-usual emissions **(right)**, temperatures are predicted to rise by 7–11°F. *Data from Melillo, J. M., et al., eds., 2014.* Climate change impacts in the United States: The third national climate assessment. *U.S. Global Change Research Program.*

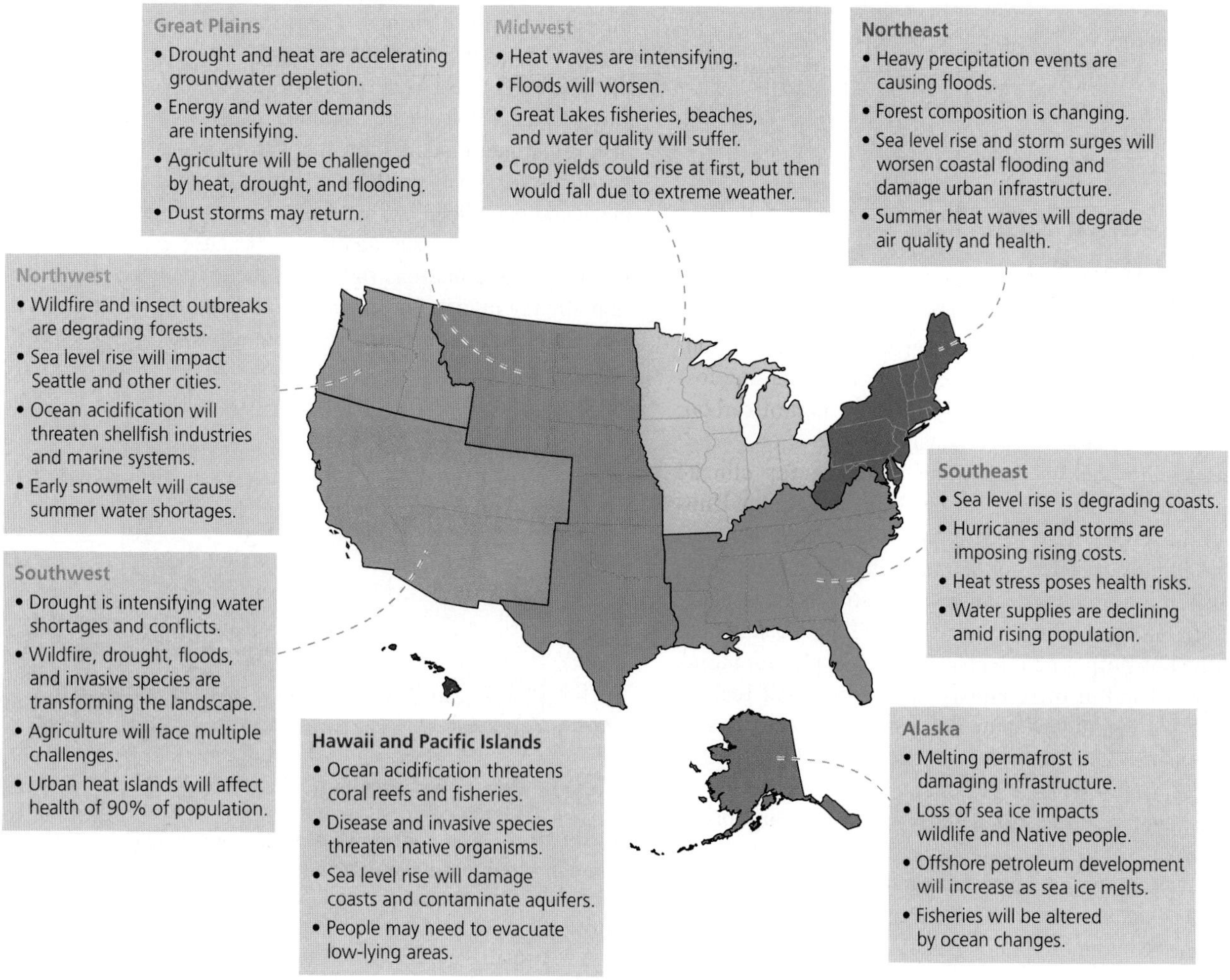

Great Plains
• Drought and heat are accelerating groundwater depletion.
• Energy and water demands are intensifying.
• Agriculture will be challenged by heat, drought, and flooding.
• Dust storms may return.
Midwest
• Heat waves are intensifying.
• Floods will worsen.
• Great Lakes fisheries, beaches, and water quality will suffer.
• Crop yields could rise at first, but then would fall due to extreme weather.
Northeast
• Heavy precipitation events are causing floods.
• Forest composition is changing.
• Sea level rise and storm surges will worsen coastal flooding and damage urban infrastructure.
• Summer heat waves will degrade air quality and health.
Northwest
• Wildfire and insect outbreaks are degrading forests.
• Sea level rise will impact Seattle and other cities.
• Ocean acidification will threaten shellfish industries and marine systems.
• Early snowmelt will cause summer water shortages.
Southeast
• Sea level rise is degrading coasts.
• Hurricanes and storms are imposing rising costs.
• Heat stress poses health risks.
• Water supplies are declining amid rising population.
Southwest
• Drought is intensifying water shortages and conflicts.
• Wildfire, drought, floods, and invasive species are transforming the landscape.
• Agriculture will face multiple challenges.
• Urban heat islands will affect health of 90% of population.
Hawaii and Pacific Islands
• Ocean acidification threatens coral reefs and fisheries.
• Disease and invasive species threaten native organisms.
• Sea level rise will damage coasts and contaminate aquifers.
• People may need to evacuate low-lying areas.
Alaska
• Melting permafrost is damaging infrastructure.
• Loss of sea ice impacts wildlife and Native people.
• Offshore petroleum development will increase as sea ice melts.
• Fisheries will be altered by ocean changes.


**FIGURE 14.23 Impacts of climate change will vary by region.** Shown are a few of the most important impacts scientists expect for each region by the end of the century. *Data from Melillo, J. M., et al., eds., 2014.* Climate change impacts in the United States: The third national climate assessment. *U.S. Global Change Research Program.*

## FAQ What's with the "global warming pause"? Has global warming stopped?

In recent years, the rate at which global temperatures are rising has slowed considerably. Mean temperatures from 1998 to 2012 rose less than one-sixth as much as in the preceding 14-year period, and at only one-third to one-half the rate they rose from 1951 to 2012. This so-called "pause" or "hiatus" in global warming has led many opponents of emissions reduction to declare that we no longer need to worry about climate change.

Everyone would like to believe that climate change is no longer occurring. However, scientists studying the situation are concluding that the pause is a temporary result of natural short-term variation, brought about by several coinciding factors: (1) a phase of the Interdecadal Pacific Oscillation (p. 308) in which the ocean absorbs more heat from the atmosphere than usual, (2) several volcanic eruptions that spread cooling aerosols through the atmosphere, and (3) decreasing irradiance from the sun during a downswing in the solar cycle. On top of this, various ozone-depleting substances that have been restricted by the Montreal Protocol (p. 294) happen to be greenhouse gases, and their reduction since 1987 has helped to slow climate change. Scientists predict that as natural cycles shift back into phases that intensify warming instead of dampening it, we can expect temperatures to swing upward more rapidly in coming years.

## Are we responsible for climate change?

Scientists agree that today's global warming is due to the well-documented recent increase in greenhouse gases in our atmosphere, and that this results primarily from burning fossil fuels for energy and secondarily from the loss of carbon-absorbing vegetation due to deforestation.

Yet despite the overwhelming evidence for climate change and its impacts, many people, especially in the United States, have long tried to deny that it is happening. Most of these "climate skeptics" or "climate change deniers" now admit that the climate is changing but still express doubt that we are the cause. Public debate over climate change has been fanned by corporate interests, spokespeople from think tanks, and a handful of scientists funded by fossil fuel industries, all of whom have aimed to cast doubt on the scientific consensus. Their views have been amplified by the U.S. news media, which seeks to present two sides to every issue, even when evidence does not equally support the arguments of each side.

However, as the evidence has mounted and as the economic and societal costs of climate change have grown more apparent, more and more policymakers, corporate executives, military leaders, national security experts, heads of business and industry, and everyday citizens have concluded that climate change is escalating and is causing impacts to which we must begin to respond.

# Responding to Climate Change

From this point onward, our global society will be focusing on how best to respond to the challenges of climate change. The good news is that everyone—not just leaders in government and business, but everyday people, and especially today's youth—can play a part in this all-important search for solutions.

## Shall we pursue mitigation or adaptation?

We can respond to climate change in two fundamental ways. One is to pursue actions that reduce greenhouse gas emissions, so as to lessen the severity of climate change. This strategy is called **mitigation** because the aim is to mitigate the problem; that is, to alleviate it or reduce its severity. Examples of mitigation include improving energy efficiency, switching to clean renewable energy sources, preserving forests, recovering landfill gas, and promoting farm practices that protect soil quality. The sooner we begin reducing emissions, the lower the level at which they will peak, and the less we will alter climate (**FIGURE 14.24**).

We can also pursue strategies to cushion ourselves from the impacts of climate change. This strategy is called **adaptation** because the goal is to adapt to change. Erecting a seawall like the Maldives' Great Wall of Malé is an example of adaptation. Other examples include restricting coastal development; adjusting farming practices to cope with drought; and modifying water management practices to deal with reduced river flows, glacial outburst floods, or salt contamination of groundwater.

We need to pursue adaptation because even if we were to halt all our emissions right now, the greenhouse gas pollution already in the atmosphere would continue driving global warming, with temperature rising an estimated 0.6°C (1.0°F)

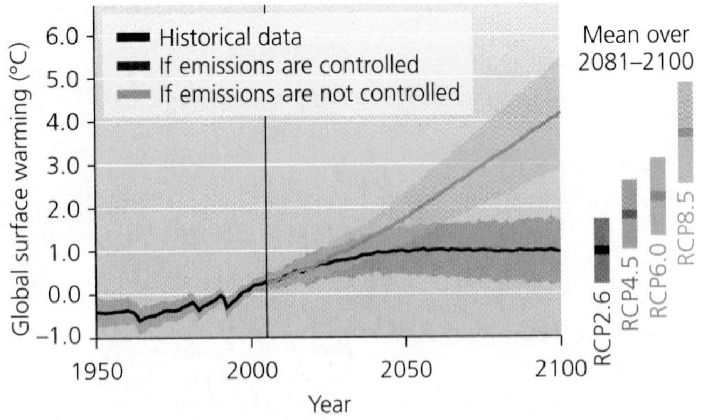


**FIGURE 14.24 The sooner we stabilize our emissions, the less climate change we will cause.** The red line shows temperature change we can expect if we strongly limit our carbon emissions. The orange line shows the change expected if we fail to control emissions effectively. Vertical bars show means and ranges of year-2100 temperatures for these two scenarios and two other intermediate ones studied by the IPCC. Predictions are based on a large number of climate models. *Data from IPCC, 2013.* Fifth assessment report. The physical science basis: Contribution of Working Group I.

more by the end of the century. Because we will face this change no matter what we do, it is wise to develop ways to minimize its impacts.

Yet we also need to pursue mitigation, because if we do nothing to diminish climate change, it will eventually overwhelm any efforts we might make to adapt. We will spend the remainder of our chapter examining approaches for the mitigation of climate change.

## We are developing solutions in electricity generation and transportation

The generation of electricity produces nearly 40% of U.S. carbon dioxide emissions, and transportation accounts for 35%.

**Electricity generation** From cooking to heating to lighting, much of what we do each day depends on electricity. Fossil fuel combustion generates two-thirds of U.S. electricity, and coal accounts for most of the resulting emissions. We can reduce electricity use by encouraging conservation and efficiency (pp. 354–356). Power producers can capture excess heat from electricity generation and put it to use (cogeneration; p. 355). Firms can manufacture and consumers can adopt energy-efficient appliances, lighting, windows, ducts, insulation, and heating and cooling systems. In addition, each of us can make lifestyle choices to lower electricity consumption.

We can also reduce greenhouse gas emissions by switching to cleaner energy sources. Natural gas generates the same amount of energy as coal, with half the emissions. Cleaner still are alternatives to fossil fuels, including nuclear power (Chapter 15), as well as bioenergy, hydropower, geothermal power, solar power, and wind power (Chapter 16).

While our society transitions to clean energy alternatives, we are also trying to capture emissions before they leak to the atmosphere. **Carbon capture** refers to technologies or approaches that remove carbon dioxide from emissions. The next step is **carbon storage** (also called **carbon sequestration**), in which the carbon is sequestered, or stored, under pressure in deep salt mines, depleted oil and gas deposits, or other underground reservoirs (see Figure 15.18, p. 351). However, we remain a long way from developing adequate technology and secure storage space to accomplish this without leakage—and it is questionable whether we will ever be able to store enough carbon to make a sizeable dent in our emissions.

**Transportation** The typical automobile is highly inefficient; only 14% of the energy from fuel we pump into our gas tanks actually moves our cars down the road (**FIGURE 14.25**). More aerodynamic designs, increased engine efficiency, and improved tire design can help make our vehicles more fuel-efficient (p. 355). Indeed, many nations use vehicles that are more fuel-efficient than those of the United States, and recent government mandates are now improving fuel efficiency in American vehicles (p. 356). New technology is also bringing us alternatives to the traditional combustion-engine automobile. These include electric vehicles, gasoline-electric hybrid vehicles (p. 355), hydrogen fuel cells (p. 385), and alternative fuels such as compressed natural gas and biodiesel (p. 383).

We also can make lifestyle choices that reduce our reliance on cars. Some people are choosing to live nearer to their workplaces. Others use mass transit such as buses, subway trains, and light rail. Still others bike or walk to work. Making automobile-based cities and suburbs more friendly to pedestrian and bicycle traffic and improving people's access to mass transit stand as central challenges for city and regional planners (pp. 412–415).

## We will need multiple strategies

Advances in agriculture, forestry, and waste management can help us mitigate climate change. In agriculture, sustainable management of cropland and rangeland enables soil to store more carbon. New techniques reduce methane emission from rice cultivation and from cattle and manure. In forest management, preserving forests, reforesting cleared areas, and pursuing sustainable forestry practices (p. 197) all help to absorb carbon from the air. Waste managers are cutting emissions by generating energy from waste in incinerators (p. 394); capturing methane seeping from landfills (p. 394); and encouraging recycling, composting, and reuse of materials and products (pp. 395–397).

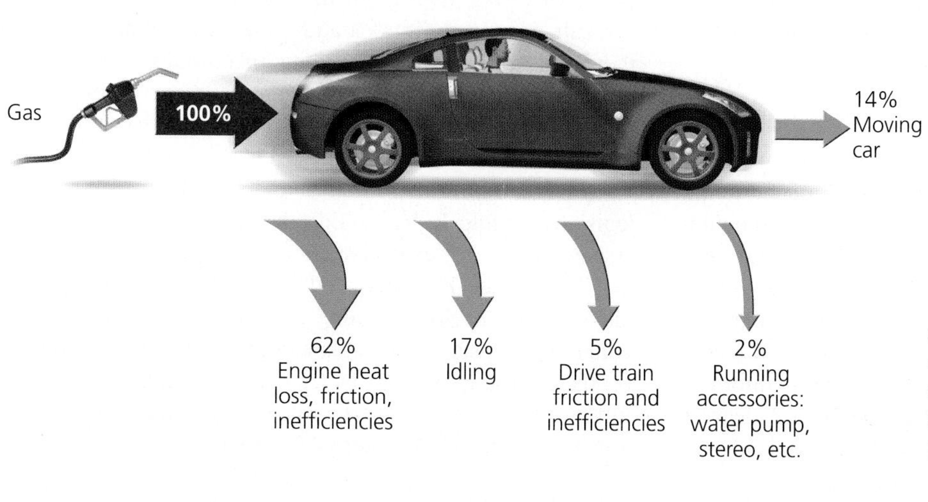


**FIGURE 14.25 Conventional automobiles are fuel-inefficient.** Only about 13–14% of the energy from a tank of gas actually moves the typical car down the road. Nearly 85% of useful energy is lost, primarily as heat. *Data from U.S. Department of Energy.*

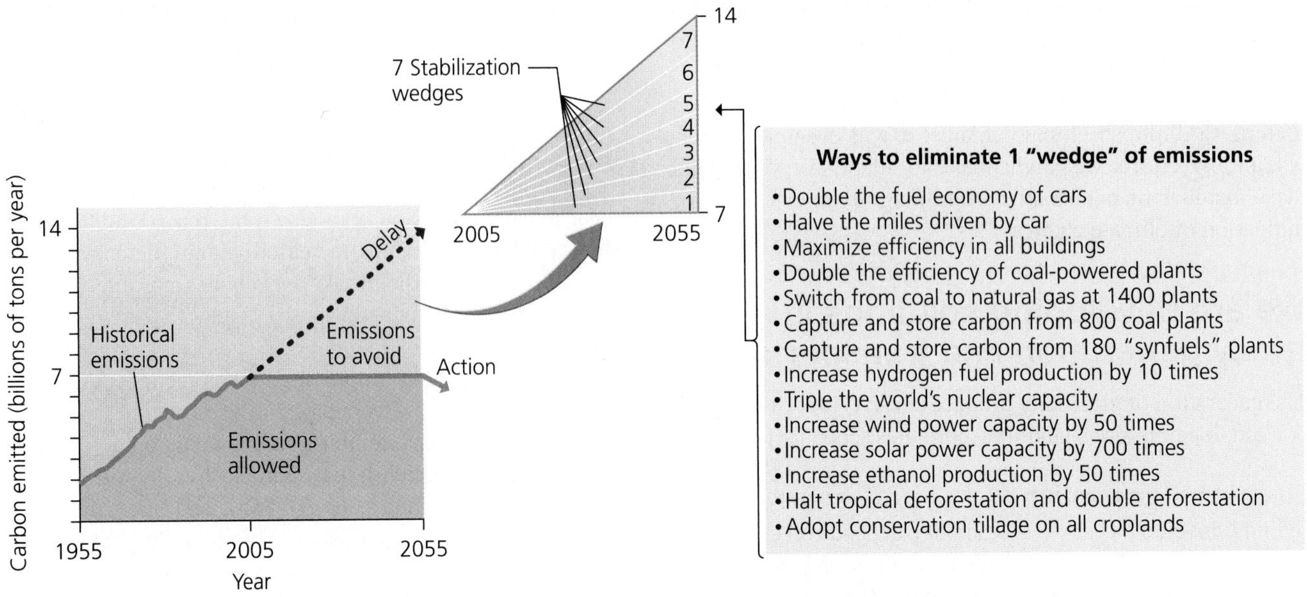


**FIGURE 14.26 We can stabilize emissions by breaking this large job into smaller steps.** Stephen Pacala and Robert Socolow began with a standard graph showing the doubling of $CO_2$ emissions that scientists expect to occur from 2005 to 2055. They added a flat line to represent the trend if emissions were held constant and then separated the graph into emissions allowed (below the line) and emissions to be avoided (the triangular area above the line). They then divided this "stabilization triangle" into seven equal-sized wedges. Each "stabilization wedge" represents 1 billion tons of $CO_2$ emissions in 2055 to be avoided. Finally, they identified a series of strategies, each of which could take care of one wedge. If we accomplish just 7 of these strategies, we could halt our growth in emissions for the next half-century. *Adapted from Pacala, S., and R. Socolow, 2004. Stabilization wedges: Solving the climate problem for the next 50 years with current technologies.* Science *305: 968–972. Reprinted by permission of AAAS and the author.*

We should not expect to find a single "magic bullet" for mitigating climate change. Reducing emissions will require steps by many people and institutions across many sectors of our economy. However, most reductions can be achieved using current technology, and we can begin implementing changes right away. For society to reduce emissions, environmental scientists Stephen Pacala and Robert Socolow advise that we follow some age-old wisdom: When the job is big, break it into smaller parts. Pacala and Socolow have identified 15 strategies (**FIGURE 14.26**) that could each eliminate 1 billion tons of carbon per year by 2050 if deployed at a large scale. Achieving just 7 of these 15 aims would stabilize our emissions. If we achieve more, then we reduce emissions.

## What role should government play?

Even if people agree on strategies and technologies to reduce emissions, they may disagree on what role government should play to encourage those strategies and technologies: Should it mandate change through laws and regulations? Should it impose no policies at all and hope that private enterprise will develop solutions on its own? Should it take the middle ground and design policies that give private entities financial incentives to reduce emissions? This debate has been vigorous in the United States and Canada, where many business leaders and politicians have opposed all government action to address climate change, fearing that emissions reductions will impose economic costs on industry and consumers.

In 2007, the U.S. Supreme Court ruled that carbon dioxide was a pollutant that the Environmental Protection Agency (EPA) could regulate under the Clean Air Act (p. 284). When Barack Obama became president, he instead urged that Congress craft laws to address emissions. In 2009, the House of Representatives passed legislation to create a **cap-and-trade** system (p. 111) in which industries and utilities would compete to reduce emissions for financial gain, and under which emissions were mandated to decrease 17% by 2020. However, legislation did not pass in the Senate. As a result, responsibility for addressing emissions passed to the EPA, which is now phasing in emissions regulations on industry and utilities, hoping to spur energy efficiency retrofits and renewable energy use at a pace that minimizes political opposition (p. 293).

In 2013 President Obama announced that because of legislative gridlock, he would take steps to address climate change using his executive authority. His "climate action plan" urged the EPA to speed its regulation of new power plants and to begin regulating existing power plants. It also aimed to jumpstart renewable energy development, modernize the electric grid, finance clean coal and carbon storage efforts, improve automotive fuel economy, protect and restore forests, and encourage energy efficiency.

## The Kyoto Protocol sought to limit emissions

Climate change is a global problem, so the world's policymakers have tried to tackle it with international treaties. In 1992, most nations signed the **U.N. Framework Convention on Climate Change**, which outlined a plan to reduce greenhouse gas emissions to 1990 levels by the year 2000 through a voluntary approach. Emissions kept rising, however, so nations forged a binding treaty to *require* emissions reductions. Drafted in 1997 in Kyoto, Japan, the **Kyoto Protocol** mandated signatory nations, by the period 2008–2012, to reduce emissions of six greenhouse gases to levels below those of 1990. The treaty took effect in 2005 after Russia became the 127th nation to ratify it.

The United States was the only developed nation not to ratify the Kyoto Protocol. U.S. leaders objected to how it required industrialized nations to reduce emissions but did not require the same of rapidly industrializing nations such as China and India. Proponents of the Kyoto Protocol countered that the differential requirements were justified because industrialized nations created the current problem and therefore should take the lead in resolving it.

As of 2011 (the most recent year with full international data), nations that signed the Kyoto Protocol had decreased their emissions by 9.3% from 1990 levels (**FIGURE 14.27**). However, much of this reduction was due to economic contraction in Russia and former Soviet-Bloc nations following the breakup of the Soviet Union. When these nations are factored out, the remaining signatories showed a 3.2% *increase* in emissions. Nations not parties to the accord, including China, India, and the United States, increased their emissions still more.

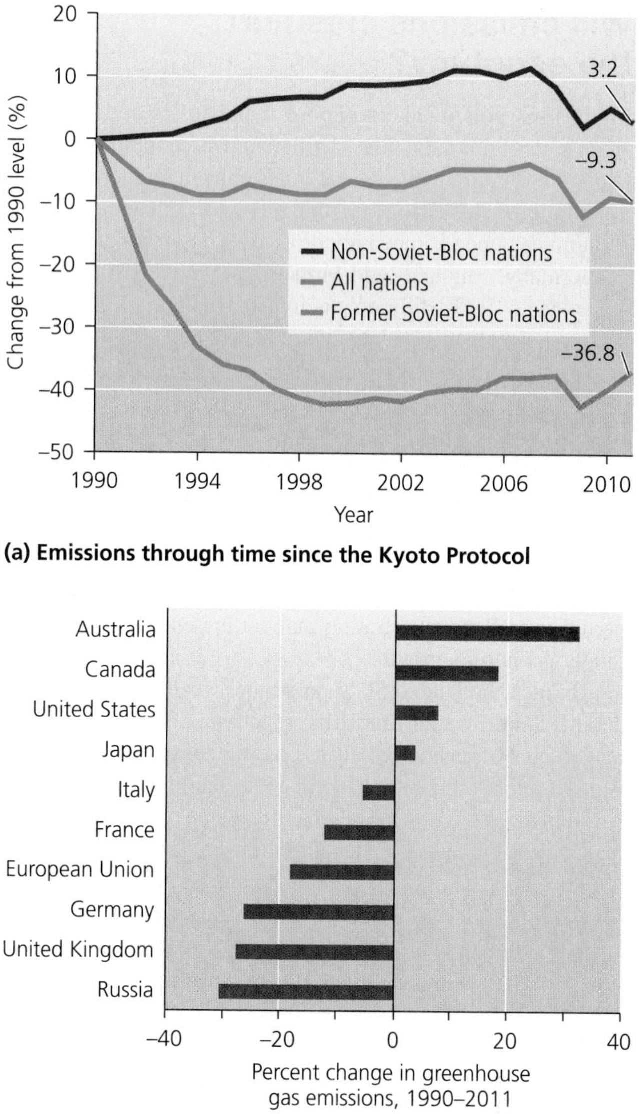


**(a) Emissions through time since the Kyoto Protocol**

**(b) Changes in emissions since the Kyoto Protocol**

**FIGURE 14.27 The Kyoto Protocol produced mixed results.** Nations ratifying it decreased their emissions of six greenhouse gases by 9.3% by 2011 **(a)**, but this was largely because of unrelated economic contraction in the former Soviet-Bloc countries. A selection of major nations **(b)** shows varied outcomes in reducing emissions. The United States did not ratify the Protocol, Australia joined it late, and Canada left early. Values do not include influences of land use and forest cover. *Data from U.N. Framework Convention on Climate Change, 2014.*

DATA Q In part (b), compare the nations whose emissions have increased with those whose emissions have decreased. What difference(s) do you note between these two groups that might explain why their emissions differ?

GO TO **INTERPRETING GRAPHS & DATA** ON MasteringEnvironmentalScience®

## International climate negotiations seek a way forward

In recent years, representatives of the world's nations have met at a series of conferences, trying to design a treaty to succeed the Kyoto Protocol. Delegates from European nations and small island nations have generally taken the lead, while China, India, and the United States have been reluctant to make commitments.

A contentious 2009 conference in Copenhagen, Denmark, ended without specific targets or solid commitments. The process got back on track in Cancún, Mexico, in 2010, where developed nations promised to pay developing nations to assist mitigation and adaptation efforts. Nations also broadly agreed on a plan, nicknamed *REDD* (p. 192), to help tropical nations reduce forest loss. Developed nations agreed to transfer clean energy technology to developing nations, and China and India agreed in principle to emission targets and international monitoring. However, most of these plans and promises have not yet come to pass.

In Durban, South Africa, in 2011, negotiators failed to design a new treaty. Instead, nations agreed to a "road map" toward a legally binding international deal in 2015, which would come into force only after 2020. This plan was reaffirmed at the 2012 conference in Doha, Qatar, and the 2013 conference in Warsaw, Poland. In Doha, negotiators also extended the Kyoto Protocol until 2020. However, a number of nations backed out of this Kyoto extension, and the treaty now applies to only about 15% of the world's emissions.

Most scientists reacted with disappointment and alarm at having to wait until 2020 for a meaningful agreement, warning that this is creating a "lost decade" during which climate change is intensifying.

## Will emissions cuts hurt the economy?

Many U.S. policymakers oppose mandates to reduce emissions out of fear that they will hamper economic growth. China and India have resisted emissions cuts under the same assumption. This is understandable, given that our current economies depend so heavily on fossil fuels. Yet nations such as Germany, England, and France have reduced their emissions since 1990 while enhancing their economies and providing their citizens high standards of living. Wealthy nations from Denmark to New Zealand to Hong Kong to Switzerland to Sweden emit fewer than half the greenhouse gases per person as the United States does.

Indeed, the United States reduced its carbon dioxide emissions 11% from 2007 to 2013. This occurred as a result of efficiency measures, a shift from coal to natural gas, and a recession that caused fossil fuel use to decline temporarily. Despite the recession, the U.S. economy grew during this period overall, demonstrating that cutting emissions need not hinder economic growth.

In fact, the United States and other wealthy industrialized nations are the nations most likely to *gain* economically from major energy transitions, because they are best positioned to invent, develop, fund, and market new technologies to power the world in a post-fossil-fuel era. Germany, Japan, and China have realized this and are now leading the world in production, deployment, and sales of solar energy technology (**FIGURE 14.28**). If the United States does not act more assertively to develop energy technologies for the future, then the future could belong to nations like China, Germany, and Japan.

**FIGURE 14.28 China is racing to become the world's leader in renewable energy technology.** Here, workers at a Chinese factory produce photovoltaic solar panels.

## States and cities are advancing climate change policy

In the absence of legislative action at the federal level to address climate change, state and local governments across the United States are advancing policies to limit emissions. Mayors from over 1000 cities have signed the U.S. Mayors Climate Protection Agreement, committing their cities to pursue policies to "meet or beat" Kyoto Protocol guidelines. Former New York City Mayor Michael Bloomberg launched a panel on climate change in 2008, and this helped prepare the city to tackle challenges soon posed by Superstorm Sandy.

A number of U.S. states have enacted targets or mandates for renewable energy production, seeking to boost alternatives to fossil fuels. The boldest state-level action so far has come in California. In 2006 that state's legislature worked with Governor Arnold Schwarzenegger to pass the Global Warming Solutions Act, which seeks to cut California's greenhouse gas emissions 25% by the year 2020. This law established a cap-and-trade program for carbon emissions and followed earlier efforts in California to mandate higher fuel efficiency for automobiles.

Action is also being taken by nine northeastern states in the Regional Greenhouse Gas Initiative. In this effort, Connecticut, Delaware, Maine, Maryland, Massachusetts, New Hampshire, New York, Rhode Island, and Vermont run a cap-and-trade program for power plant emissions. From 2005 to 2012, these states cut their $CO_2$ emissions from power plants by 40%, even as their economies grew. It is estimated that investment of the auction proceeds will save more than $2 billion in energy costs and eliminate 8 million tons of $CO_2$ emissions.

## Market mechanisms can be used to address climate change

Emissions trading programs (p. 111) seek to harness the economic efficiency of market capitalism to control pollution by allowing business, industry, or utilities flexibility in how they do so. Supporters of emissions trading argue that this approach provides the fairest, least expensive, and most effective method of reducing emissions. Polluters choose how to cut their emissions and are given financial incentives for reducing emissions below the legally required amount (**FIGURE 14.29**). Cap-and-trade programs are intended to be self-sustaining. The price of permits fluctuates freely in the market, creating the same kinds of financial incentives as any other commodity that is bought and sold. As an example of how a cap-and-trade program works, consider the Regional Greenhouse Gas Initiative:

1. Each state decided what polluting sources it would require to participate.
2. Each state set a cap on the total $CO_2$ emissions it would allow, equal to its 2009 level.
3. Each state distributed to each polluter one permit for each ton it emits, up to the amount of the cap.
4. Each state is lowering its cap progressively—10% by 2018.
5. Sources with too few permits to cover their pollution must reduce emissions, buy permits from other sources, or pay for carbon offsets (p. 328). Sources with excess permits may sell them.
6. Any source emitting more than its permitted amount faces penalties.

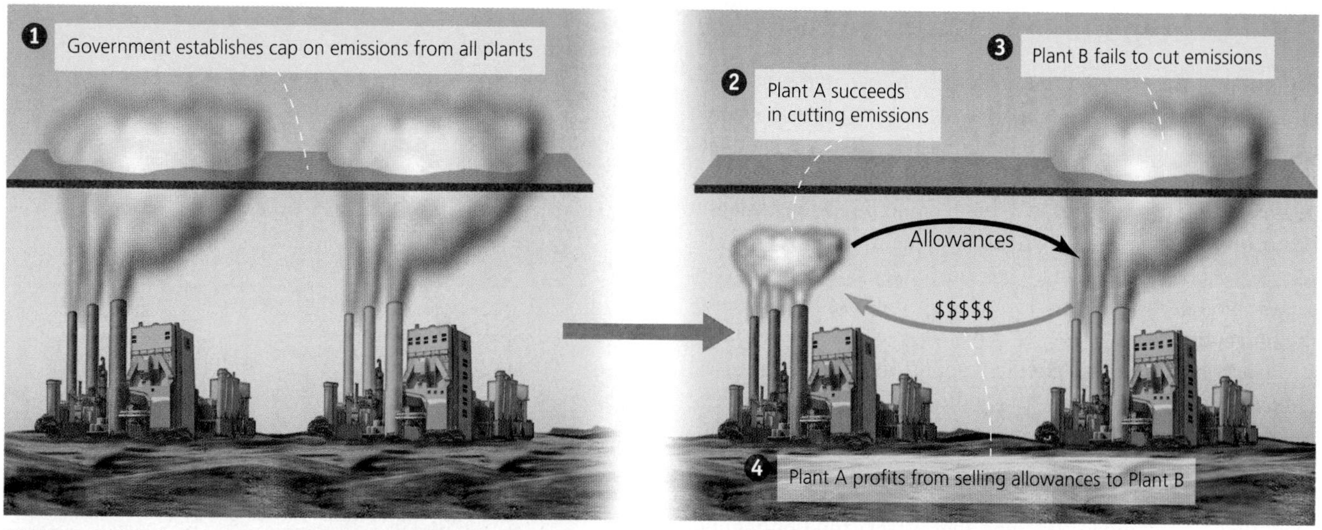


**FIGURE 14.29 A cap-and-trade emissions trading system harnesses the efficiency of market capitalism to achieve the goal of reducing emissions.** In such a system, (1) government first sets an overall cap on emissions. As polluting facilities respond, some will have better success reducing emissions than others. In this figure, (2) Plant A succeeds in cutting its emissions well below the cap, whereas (3) Plant B fails to cut its emissions at all. As a result, (4) Plant B pays money to Plant A to purchase allowances that Plant A is no longer using. Plant A profits from this sale, and the government cap is met, reducing pollution overall. Over time, the cap can be progressively lowered to achieve further emissions cuts.

The world's largest cap-and-trade program is the European Union Emission Trading Scheme. This market began in 2005, but investors soon realized that national governments had allocated too many permits to their industries. The overallocation gave companies little incentive to reduce emissions, so permits lost their value, and prices in the market took a nosedive. Europeans partly addressed these problems by making emitters pay for permits and setting emissions caps across the entire European Union while expanding the program. In the long run, permits will retain value and an emissions trading market will work only if government policies are in place to limit emissions.

## Carbon taxes are another option

As the world's carbon trading markets show mixed results early in their growth, some economists, scientists, and policymakers are saying that cap-and-trade systems are not effective enough, don't work quickly enough, or leave too much to chance. Many of these critics would prefer that governments enact a **carbon tax** instead. In this approach, governments charge polluters a fee for each unit of greenhouse gases they emit. This gives polluters a financial incentive to reduce emissions. Carbon taxes have been introduced in over 20 nations. In the United States, Boulder, Colorado, taxes electricity consumption and Montgomery County, Maryland, taxes power plants.

The downside of a carbon tax is that polluters pass the cost along to consumers by charging higher prices for the products or services they sell. Proponents of carbon taxes have responded by proposing an approach called **fee-and-dividend**. In this approach, funds from the carbon tax, or "fee," are transferred as a tax refund, or "dividend," given to taxpayers. This way, if polluters pass their costs along to consumers, those consumers will be reimbursed. In theory, the system should provide polluters a financial incentive to reduce emissions while imposing no financial burden on taxpayers and no drag on the economy. The fee-and-dividend approach is a type of **revenue-neutral carbon tax**, because there is no net transfer of revenue from taxpayers to the government. For this reason, the approach is gaining broad political appeal.

The Canadian province of British Columbia introduced a revenue-neutral carbon tax in 2008. In the five years that followed, the province reduced its emissions by 10% as fuel consumption declined 17% and the province's economy grew. The tax began at $10 per ton of $CO_2$-equivalent, was gradually raised to $30, and has been gathering over a billion dollars a year. The tax revenues replace revenues from personal income taxes and corporate taxes, which are lowered by the same amount. Thus, rather than receiving a dividend check, citizens see a decline in their income taxes. The end result so far has been that emissions are lower, the economy remains just as strong, and taxpayers are happy to have lower income taxes.

### WEIGHING THE ISSUES

**Cap-and-Trade or a Carbon Tax?**

What advantages and disadvantages do you see in using a cap-and-trade system to reduce greenhouse gas emissions? What pros and cons do you see in using carbon taxes to achieve this goal? What do you think of the idea of a revenue-neutral carbon tax? If you were a U.S. senator, what type of policy would you support in order to address emissions in the United States, and why?

## Offsets help achieve carbon-neutrality

Emissions trading programs generally allow participants to buy **carbon offsets**, voluntary payments intended to enable another entity to help reduce the emissions that one is unable to reduce. The payment thus offsets one's own emissions. For example, a coal-burning power plant could pay a reforestation project to plant trees that will soak up as much carbon dioxide as the coal plant emits. Or a university could fund clean renewable energy projects to make up for fossil fuel energy the university uses.

Carbon offsets are popular among utilities, businesses, universities, governments, and individuals trying to achieve **carbon-neutrality**, a condition in which no net carbon is emitted. In principle, carbon offsets are a powerful idea, but rigorous oversight is needed to make sure that offset funds achieve what they are intended for—and that offsets fund only emissions cuts that would not occur otherwise.

Businesses and corporations seeking to make their practices more sustainable also can find ways to reduce their carbon footprints directly. An excellent example is Pearson Education, the publisher of your textbook! In 2009 Pearson achieved carbon-neutrality after a concerted two-year effort. Pearson reduced its energy consumption and carbon footprint directly by 12% by upgrading buildings for energy efficiency, designing more efficient computer servers, reducing the number of vehicles in its fleets, increasing the proportion of hybrid vehicles, and cutting back on employee business travel while enhancing the use of video conferencing. Pearson eliminated a further 47% of its emissions by purchasing clean renewable energy instead of fossil fuel energy and by installing large solar panel arrays at two of its sites in New Jersey and a wind turbine at a Minnesota site. To offset the remaining 41% of its emissions, the company is funding a number of programs to preserve forest and replant trees in various areas of the world, from England to Costa Rica.

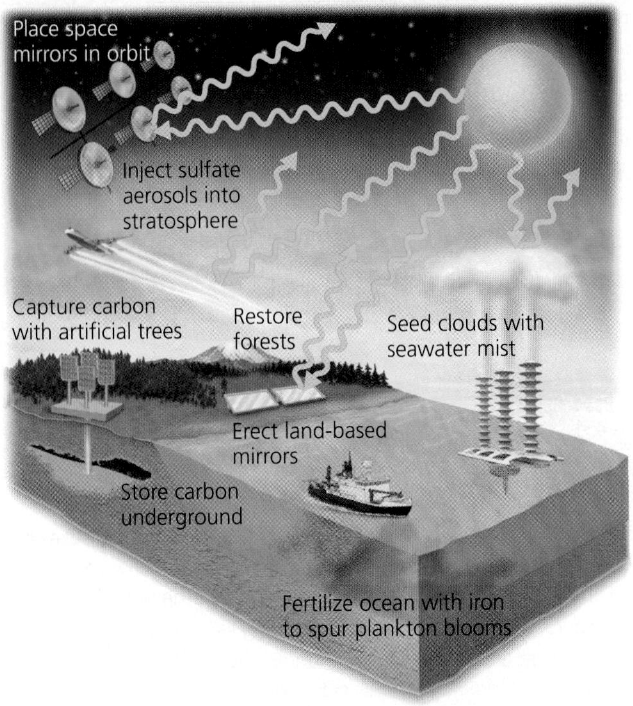


**FIGURE 14.30 Geoengineering proposals seek to use technology to remove carbon dioxide from the air or reflect sunlight away from Earth.** However, most geoengineering ideas would take years to develop, may not work well, or might cause undesirable side effects. Thus, they are not a substitute for reducing emissions.

## Should we engineer the climate?

What if all our efforts to reduce emissions are not adequate to rein in climate change? As climate change becomes more severe, some scientists and engineers are reluctantly considering drastic steps to alter Earth's climate in a last-ditch attempt to reverse global warming—an approach called **geoengineering** (**FIGURE 14.30**).

One geoengineering approach would be to suck carbon dioxide out of the air. We might enhance photosynthesis in natural systems by planting trees or by fertilizing ocean phytoplankton with nutrients such as iron. A more high-tech method would be to design "artificial trees," structures that chemically filter $CO_2$ from the air. A different approach would be to block sunlight before it reaches Earth, thereby cooling the planet. We might deflect sunlight by injecting sulfates or other fine dust particles into the stratosphere, by seeding clouds with seawater, or by deploying fleets of reflecting mirrors on land, at sea, or in orbit in space.

Scientists were long reluctant even to discuss the notion of geoengineering. The potential methods are technically daunting, would take years or decades to develop, and might pose unforeseen risks. Blocking sunlight does not reduce greenhouse gases, so ocean acidification would continue. And any method would work only as long as society has money and ability to maintain it. Moreover, many experts are wary of promulgating hope for easy technological fixes, lest politicians lose incentive to try to reduce emissions. However, as climate change intensifies, scientists are beginning to assess the risks and benefits of geoengineering, so that we can be ready to take action if climate change becomes severe enough to justify it.

## You can address climate change

Government policies, corporate actions, international treaties, technological innovations—and perhaps even geoengineering—will all play roles in addressing climate change. But in the end the most influential factor may be the collective decisions of millions of regular people. Just as we each have an ecological footprint (p. 4), we each have a **carbon footprint** that expresses the amount of carbon we are responsible for emitting. To reduce emissions, each of us can take steps in our everyday lives—from choosing energy-efficient appliances, to eating less meat, to deciding where to live and how to get to work.

College students are vital to driving the personal and societal changes needed to reduce carbon footprints and address climate change. Today a groundswell of interest is sweeping across campuses, and many students are pressing their administrations to seek carbon-neutrality or to divest from fossil fuel investments and promote renewable energy (pp. 18, 422). College students have played a role in many grassroots events and organizations, including 350.org's International Day of Climate Action on October 24, 2009. This event—kicked off by the Maldives' underwater cabinet meeting—featured 5200 events in 181 nations and was deemed "the most widespread day of political action in the planet's history."

Global climate change may be the biggest challenge we face, but halting it would be our biggest victory. With concerted action, there is still time to avert the most severe impacts. Through outreach, education, innovation, and lifestyle choices, we have the power to turn the tables on climate change and help bring about a bright future for humanity and our planet.

## Conclusion

Many factors influence Earth's climate, and human activities have come to play a major role. Climate change is well underway, and further greenhouse gas emissions will intensify global warming and cause increasingly severe and diverse impacts. Sea level rise and other consequences of global climate change will affect locations worldwide from the Maldives to Bangladesh to Alaska to New York to Florida. As scientists and policymakers come to better understand anthropogenic climate change and its environmental, economic, and social consequences, more and more of them are urging immediate action. Reducing greenhouse gas emissions and taking other steps to mitigate and adapt to climate change represent the foremost challenges for our society in the coming years.

# Testing Your Comprehension

1. What is the fate of solar radiation after it reaches Earth? How do greenhouse gases warm the lower atmosphere?
2. Why is carbon dioxide considered the main greenhouse gas? Why are carbon dioxide concentrations increasing in the atmosphere?
3. What evidence do scientists use to study the ancient atmosphere? Describe what a proxy indicator is, and give two examples.
4. Has simulating climate change with computer programs been effective in helping us predict climate? Briefly describe how these programs work.
5. List three major trends in climate that scientists have documented so far. Now list three future trends that they predict, along with their potential consequences.
6. Describe how rising sea levels, caused by global warming, can create problems for people. How is climate change affecting marine ecosystems?
7. How might a warmer climate affect agriculture? How is it affecting distributions of plants and animals? How might it affect human health?
8. What are the largest two sources of greenhouse gas emissions in the United States? How can we reduce these emissions?
9. What roles have international treaties played in addressing climate change? Give two specific examples.
10. List two economic market-based approaches for reducing greenhouse gas emissions. Discuss advantages and disadvantages of each approach.

# Seeking Solutions

1. Some people argue that we need "more proof" or "better science" before we commit to changes in our energy economy. How much certainty do you think we need before we take action regarding climate change? How much certainty do you need in your own life before you make a major decision? Should nations and elected officials follow a different standard? Do you feel that the precautionary principle (pp. 155, 220) is an appropriate standard in the case of global climate change? Why or why not?
2. Suppose that you have decided to make your own lifestyle carbon-neutral. You plan to begin by making a 25% reduction in the emissions for which you are responsible. What are the first three actions you would take to achieve your goal?
3. How might your campus reduce its greenhouse gas emissions? Develop three concrete proposals for ways to reduce emissions on your campus that you feel would be effective and feasible. How would you present these proposals to campus administrators to gain their support?
4. **THINK IT THROUGH** You have been appointed as the U.S. representative to an international conference to negotiate terms of an emissions reduction treaty to replace the Kyoto Protocol in 2020. The U.S. government has instructed you to take a leading role in designing the new treaty and to engage constructively with other nations' representatives while protecting America's economic and political interests. What type of agreement will you try to shape? Describe at least three components that you would propose or agree to, and at least one that you would oppose.

5. **THINK IT THROUGH** You have just been elected governor of a medium-sized U.S. state. Citizens want you to take bold action to reduce greenhouse gas emissions—but they do not want prices of gasoline or electricity to rise. Industries in your state are wary of emissions reductions being required of them but are willing to explore ideas with you. The state legislature will support your efforts as long as you remain popular with voters. The state to your west has just passed ambitious legislation mandating steep emissions cuts. The state to your east has joined a regional emissions trading consortium. The state to your north has just established a revenue-neutral carbon tax. What actions will you take in your first year as governor, and why? What effects would you expect each action to have?

# Calculating Ecological Footprints

Global climate change is something to which we all contribute, because fossil fuel combustion plays such a large role in supporting the lifestyles we lead. Conversely, as individuals, each one of us can help to address climate change through personal decisions and actions in how we live our lives. Several online calculators enable you to calculate your own personal carbon footprint, the amount of carbon emissions for which you are responsible. Go to http://www.nature.org/greenliving/carboncalculator/, take the quiz, and enter the relevant data in the table.

| | Carbon footprint (tons per person per year) |
|---|---|
| World average | |
| U.S. average | |
| Your footprint | |
| Your footprint with three changes (see question 3) | |

1. How does your personal carbon footprint compare to that of the average U.S. resident? How does it compare to that of the average person in the world? Why do you think your footprint differs in the ways it does?
2. As you took the quiz and noted the impacts of various choices and activities, which one surprised you most?
3. Think of three changes you could make in your lifestyle that would lower your carbon footprint. Now take the footprint quiz again, incorporating these three changes. Enter your resulting footprint in the table. By how much did you reduce your yearly emissions?
4. What do you think would be an admirable yet realistic goal for you to set as a target value for your own footprint? Would you choose to purchase carbon offsets to help reduce your impact? Why or why not?

# Nonrenewable Energy Sources, Their Impacts, and Energy Conservation

## Upon completing this chapter, you will be able to:

- Identify the energy sources that we use
- Understand the value of the EROI concept
- Describe the formation of coal, natural gas, and crude oil, and evaluate how we extract, process, and use these fossil fuels
- Assess concerns over the future decline of conventional oil supplies
- Outline ways in which we are extending our reach for fossil fuels and exploring unconventional new fossil fuel sources
- Examine and assess environmental, political, social, and economic impacts of fossil fuel use, and explore potential solutions
- Specify strategies for enhancing energy efficiency and conserving energy
- Describe nuclear energy and how we harness it
- Assess the benefits and drawbacks of nuclear power, and outline the societal debate over this energy source

**Photo: Rachel Farnelli rides a swing in her backyard overlooking a natural gas well in Dimock, Pennsylvania.**

# CENTRAL CASE STUDY

# Hydrofracking the Marcellus Shale

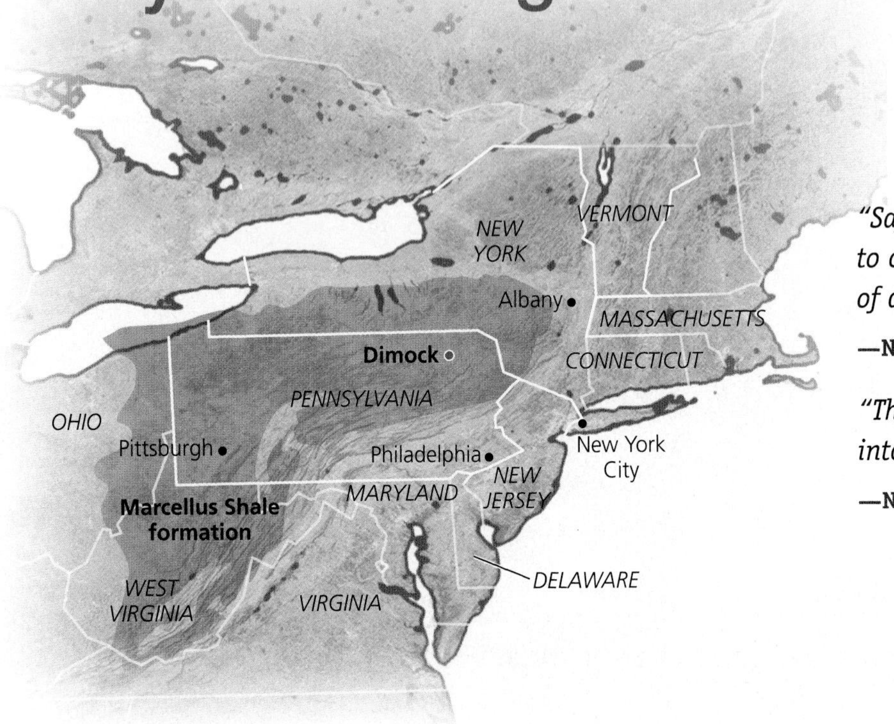


*"Safe drilling for natural gas has the potential to create thousands of new jobs and millions of dollars in economic investment."*

**—New York Senate majority leader Dean G. Skelos**

*"There's no safe way to put toxic chemicals into the ground and control them."*

**—New York schoolteacher Elizabeth Bouiss**

When the men from Cabot Oil and Gas Corporation came to the small town of Dimock in rural Pennsylvania, many of Dimock's 1500 residents were happy to sign on to the contracts the company offered. In exchange for the right to drill for natural gas on their land, Cabot would pay them royalties on sales of the gas extracted from their property. For some in this small, rural community, the gas money seemed like a ticket to economic security.

Soon the drilling sites around Dimock were producing the most natural gas from anywhere in the Marcellus Shale, the vast gas-bearing rock formation that underlies portions of Pennsylvania, New York, West Virginia, and Ohio. Money and jobs from the gas boom kept Dimock economically afloat, while other towns reeled from recession and cut funding for schools and basic services.

Yet despite the economic gains, some Dimock residents were having second thoughts about drilling. Their once-peaceful community was now experiencing noise, lights, air pollution, heavy truck traffic, and toxic wastewater spills. Then people's drinking water began to turn brown, gray, or cloudy with sediment, and strange chemical smells began wafting from their wells. On New Years Day, 2009, Norma Fiorentino's well exploded. Methane had built up in her well water, and a spark from a motorized pump set off a potentially lethal blast.

Residents blamed the drilling technique that Cabot Oil and Gas was using: hydraulic fracturing. Citizens who could no longer drink their own well water appealed for help. Retired schoolteacher Victoria Switzer approached Cabot, local political leaders, and the Pennsylvania Department of Environmental Protection (DEP) but was turned away by them all. Then she went to the news media, and the story got out. Documentary filmmaker Josh Fox came to town and filmed residents setting their methane-contaminated tap water on fire. His 2010 film *Gasland* won numerous awards, and Dimock became Ground Zero in the burgeoning national debate over hydraulic fracturing.

In the United States, virtually all the easily accessible oil and natural gas has already been extracted. To extract more, we've needed to develop ever-more powerful technology to reach petroleum deposits deeper underground, deeper under water, and at lower concentrations. In formations such as the Marcellus Shale, natural gas is locked up in tiny bubbles dispersed throughout the shale rock. The technique of hydraulic fracturing is now making this **shale gas** accessible.

**Hydraulic fracturing** (also called **hydrofracking** or simply **fracking**) involves drilling deep into the earth and then angling the drill horizontally once it meets a shale formation. An electric charge sets off targeted explosions that perforate the drilling pipe and create fractures in the shale. Drillers then pump a slurry of water, sand, and chemicals down the pipe under great pressure. The sand lodges in the fractures and holds them open, while some of the liquids return to the surface. Natural gas trapped in the shale migrates into the fractures and rises to the surface through the drilling pipe (**FIGURE 15.1**).

By boosting production of natural gas, hydrofracking employs people and keeps the price of gas low. Increased use of natural gas has reduced the United States' reliance on coal for electricity. Because natural gas is cleaner-burning than coal, burning it in place of coal releases fewer of the greenhouse gas emissions that lead to climate change.

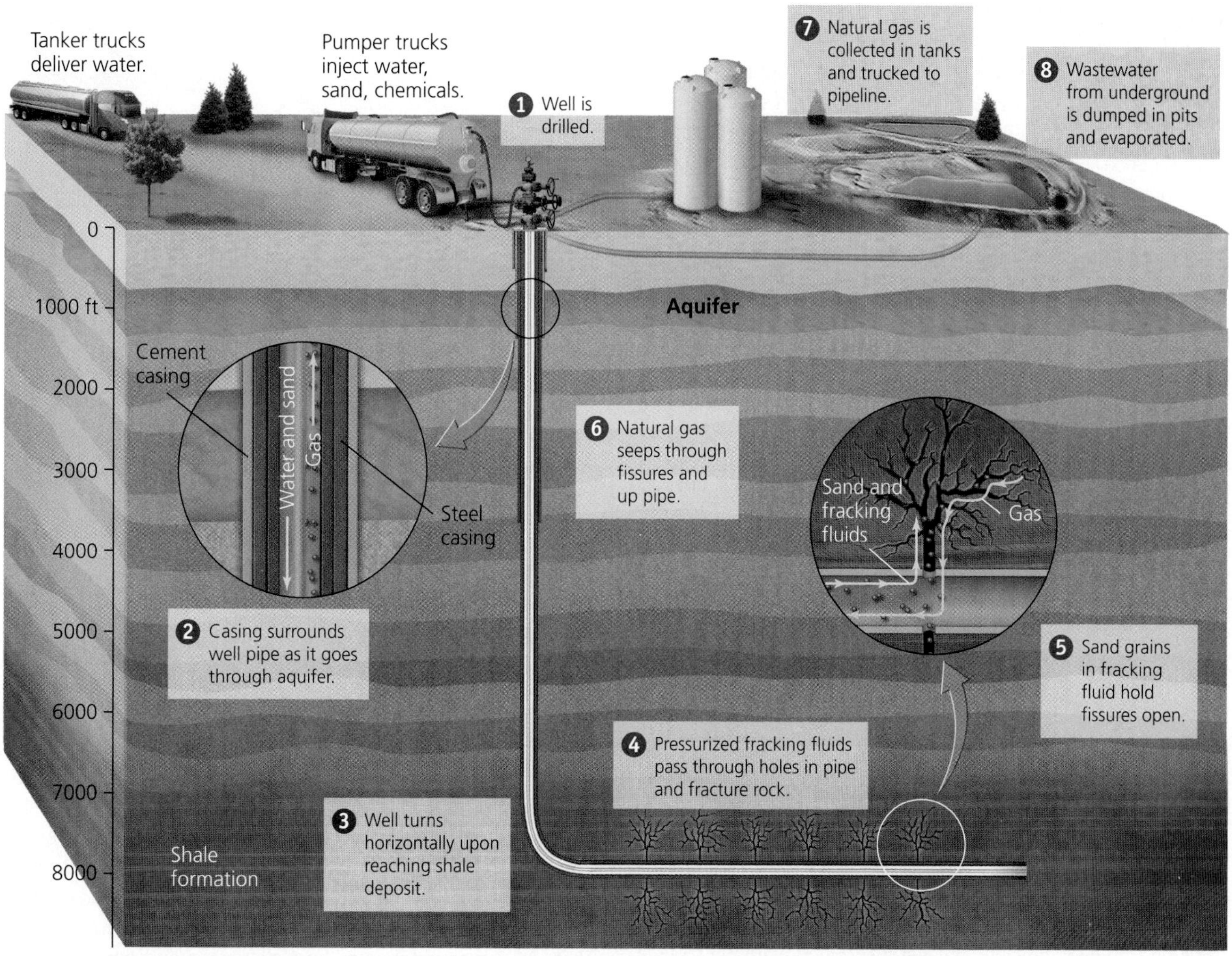
Tanker trucks
deliver water.
Pumper trucks
inject water,
sand, chemicals.
1 Well is
drilled.
7 Natural gas is
collected in tanks
and trucked to
pipeline.
8 Wastewater
from underground
is dumped in pits
and evaporated.
0
1000 ft
2000
3000
4000
5000
6000
7000
8000
Aquifer
Cement
casing
Water and sand
Gas
Steel
casing
2 Casing surrounds
well pipe as it goes
through aquifer.
6 Natural gas
seeps through
fissures and
up pipe.
Sand and
fracking
fluids
Gas
5 Sand grains
in fracking
fluid hold
fissures open.
4 Pressurized fracking fluids
pass through holes in pipe
and fracture rock.
3 Well turns
horizontally upon
reaching shale
deposit.
Shale
formation

**FIGURE 15.1 Hydraulic fracturing is used to extract natural gas trapped in shale deposits deep underground.** A well is drilled ❶ with protective casing ❷ and is turned horizontally ❸ upon reaching a shale deposit. Pressurized fluids fracture the rock ❹, and sand lodges in the cracks ❺, holding them open for natural gas ❻ to seep into them and rise through the pipe to the surface ❼. Polluted fluids that rise to the surface are piped to wastewater pits ❽.

For all these reasons—and also because of the powerful political influence of oil and gas corporations—policymakers have encouraged fracking. They have freed it from many regulatory constraints that would normally apply. Fracking has been exempted from seven major federal environmental laws that protect public health, including the National Environmental Policy Act and the Safe Drinking Water Act.

As a result, gas companies do not need to report the chemical additives they use in fracking, nor do they need to test for chemical compounds drawn up in fracking wastewater. Much of this wastewater is radioactive because drillers add radioisotopes (p. 28) as tracers to the fracking fluids they inject, and also because naturally occurring radioisotopes are brought up from deep underground. Yet regulators, policymakers, scientists, and homeowners do not have access to data to make fully informed judgments about potential health or environmental effects of fracking.

In such a climate of uncertainty, people in places like Dimock are left to wonder, worry, and argue. Victoria Switzer and several dozen other families in town eventually brought lawsuits against Cabot, but some of their neighbors are angry with them. Residents whose water has not been affected—and those who feel that the benefits of receiving gas payments outweigh the health risks from pollution—blame townspeople who have spoken out about water quality for bringing media attention to Dimock and driving down property values.

After recognizing that for some Dimock families the water was undrinkable, Pennsylvania's DEP eventually fined Cabot and required it to pay for clean drinking water to be hauled in from outside town. However, Pennsylvania Governor Tom Corbett, an avid supporter of fracking, soon ordered an end to the water shipments. As publicity built and activists complained that Pennsylvania was not safeguarding its citizens, the U.S. Environmental Protection Agency (EPA) sent in federal researchers to run tests of Dimock's water. Results showed elevated levels of several chemicals that could threaten health in 5 out of 64 wells tested, but the EPA stated that Cabot was addressing these impacts. Some residents and scientists suspected that industry and politicians influenced the EPA's conclusions. The EPA denied such influence and embarked on a nationwide study of the health and environmental impacts of hydraulic fracturing, due out in late 2014.

State governments have approached hydraulic fracturing in various ways. Pennsylvania and New York both sit atop the Marcellus Shale, which holds some of the richest gas deposits in America. Pennsylvania's political leaders welcomed the gas industry with open arms, exempting it from regulations. New York's leaders, in contrast, placed a moratorium on fracking until they could conduct further studies and reach political agreements. Ohio, Texas, Louisiana, Wyoming, North Dakota, and other states have encouraged fracking, whereas Vermont has banned it. In 2014, Coloradans debated going to the polls to give communities local control over whether to allow fracking. In California, drought has intensified debate over fracking, which uses large amounts of water.

Today many policymakers, scientists, and engineers are seeking ways to minimize the impacts of hydrofracking, so that it can continue to boost the production of shale gas. Their hope is that natural gas can bring economic and national security benefits and that this fossil fuel can serve as a "bridge" from the more-polluting fossil fuels of yesterday and today to the cleaner renewable energy sources we will need to develop for tomorrow. ☐

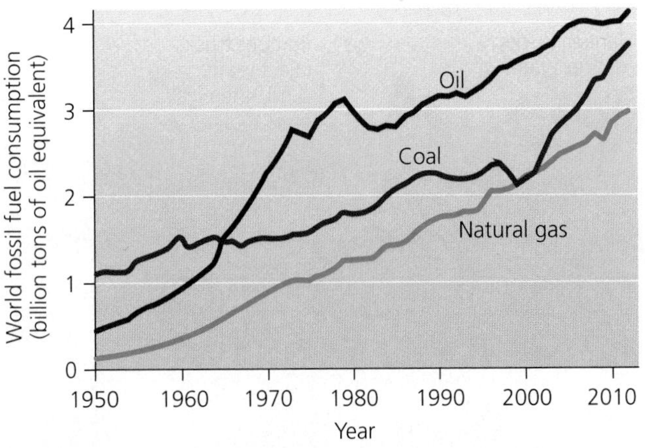


**FIGURE 15.2 Annual global consumption of fossil fuels has risen greatly over the past half-century.** Oil remains our leading energy source. *Data from U.S. Energy Information Administration.*

DATA Q By roughly what percentage has the annual consumption of oil increased since the year you were born?

GO TO **INTERPRETING GRAPHS & DATA** ON MasteringEnvironmentalScience®

## Sources of Energy

Humanity has devised many ways to harness the various forms of energy available on our planet (TABLE 15.1). We have taken most of our energy from nonrenewable sources—fossil fuels and nuclear power—but we also have available to us a diversity of renewable sources. We use Earth's energy sources to heat and light our homes; power our machinery; fuel our vehicles; produce plastics, pharmaceuticals, and synthetic fibers; and provide the comforts and conveniences we've grown accustomed to in our industrial age.

### We rely mostly on fossil fuels

Of all the energy sources shown in Table 15.1, we rely the most on **fossil fuels**, highly combustible substances formed underground over millions of years from the buried remains of ancient organisms. We use three main fossil fuels, in the form of a solid (coal), liquid (oil), and gas (natural gas). Fossil fuels provide most of the energy that we buy, sell, and consume because their high energy content makes them efficient to ship, store, and burn. A single gallon of oil contains as much energy as a person would expend in nearly 600 hours of human labor.

We use fossil fuels for transportation, heating, and cooking, and also to generate **electricity**, a secondary form of energy that we can transfer over long distances and apply to many uses. Global consumption of coal, oil, and natural gas is now at its highest level ever (**FIGURE 15.2**). Across the world today, over 80% of our energy and two-thirds of our electricity come from these three main fossil fuels (**FIGURE 15.3**).

Given our accelerating consumption, we risk using up these nonrenewable fuel resources. Some energy sources, such as sunlight, geothermal energy, and tidal energy, are perpetually renewable because natural processes readily replenish them (pp. 2–3). In contrast, energy sources such as coal, oil, and natural gas are nonrenewable. These fuels take so long to form

**TABLE 15.1 Energy Sources We Use**

| ENERGY SOURCE | DESCRIPTION | TYPE OF ENERGY |
|---|---|---|
| Crude oil | Fossil fuel extracted from ground (liquid) | Nonrenewable |
| Natural gas | Fossil fuel extracted from ground (gas) | Nonrenewable |
| Coal | Fossil fuel extracted from ground (solid) | Nonrenewable |
| Nuclear energy | Energy from atomic nuclei of uranium | Nonrenewable |
| Biomass energy | Energy stored in plant matter from photosynthesis | Renewable |
| Hydropower | Energy from running water | Renewable |
| Solar energy | Energy from sunlight directly | Renewable |
| Wind energy | Energy from wind | Renewable |
| Geothermal energy | Earth's internal heat rising from core | Renewable |
| Tidal and wave energy | Energy from tides and ocean waves | Renewable |

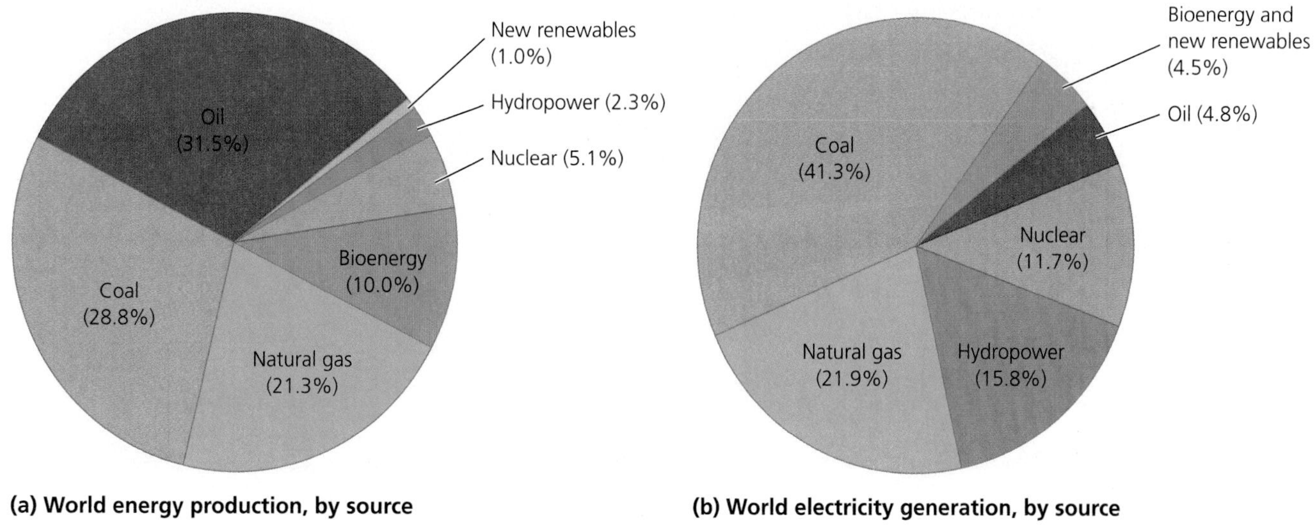


**FIGURE 15.3 Fossil fuels dominate the global energy supply.** Together, oil, coal, and natural gas account for **(a)** 82% of the world's energy production and **(b)** two-thirds of global electricity. *Data from International Energy Agency, 2013.* Key world energy statistics 2013. *Paris: IEA.*

that, once depleted, they cannot be replenished within any time span useful to our civilization. It takes a thousand years for the biosphere to generate the amount of organic matter that must be buried to produce a single day's worth of fossil fuels for our society. To replenish the fossil fuels we have used so far would take many millions of years.

At current rates of consumption, we will use up Earth's easily accessible conventional fossil fuels in just decades. This is why fossil fuel companies have developed approaches such as hydraulic fracturing to exploit less-accessible deposits. This is also why measures to conserve energy and improve energy efficiency are so vital.

## Energy is unevenly distributed

Some regions of the globe have substantial reserves of oil, coal, or natural gas, whereas others have very few. Half the world's proven reserves of crude oil lie in the Middle East. The Middle East is also rich in natural gas, as is Russia. The United States possesses the most coal of any nation (**TABLE 15.2**).

Consumption rates across the world are also unequal. Per person, the most industrialized nations use up to 100 times more energy than do the least industrialized nations. The United States has only 4.4% of the world's population, but it consumes nearly 18% of the world's energy.

Societies differ in how they use energy, as well. Industrialized nations apportion roughly one-third to transportation, one-third to industry, and one-third to all other uses. In contrast, industrializing nations devote more energy to subsistence activities such as growing and preparing food and heating homes. Because industrialized nations rely more on mechanized equipment and technology, they use more fossil fuels. In the United States, oil, coal, and natural gas supply 82% of energy demand (**FIGURE 15.4**, next page).

**TABLE 15.2 Nations with the Largest Proven Reserves of Fossil Fuels**

| OIL (% world reserves) | | NATURAL GAS (% world reserves) | | COAL (% world reserves) | |
|---|---|---|---|---|---|
| Venezuela* | 17.7 | Iran | 18.2 | United States | 26.6 |
| Saudi Arabia | 15.8 | Russia | 16.8 | Russia | 17.6 |
| Canada* | 10.3 | Qatar | 13.3 | China | 12.8 |
| Iran | 9.3 | Turkmenistan | 9.4 | Australia | 8.6 |
| Iraq | 8.9 | United States | 5.0 | India | 6.8 |

*Most reserves in Venezuela and Canada consist of oil sands, which are included in these figures. *Data from BP p.l.c., 2014.* Statistical review of world energy 2014.

## It takes energy to make energy

We do not simply get energy for free. To harness, extract, process, and deliver energy, we need to invest substantial inputs of energy. For instance, fracturing shale layers deep underground requires the use of powerful machinery, large quantities of water, and specialized chemicals, as well as vehicles, storage tanks, pipelines, waste ponds, processing facilities, equipment for workers, and more—all requiring the consumption of energy. Thus, when evaluating an energy source, it is important to subtract costs in energy invested from the benefits in energy received. **Net energy** expresses the difference between energy returned and energy invested:

$$\text{Net energy} = \text{Energy returned} - \text{Energy invested}$$

When we assess energy sources, it is also useful to use a ratio known as **EROI—energy returned on investment**. EROI ratios are calculated as follows:

$$\text{EROI} = \text{Energy returned} / \text{Energy invested}$$

Higher EROI ratios mean that we receive more energy from each unit of energy that we invest. Fossil fuels are widely used because their EROI ratios have historically been high. However, EROI ratios can change over time. Ratios rise as technologies to extract and process fuels become more efficient, and they fall as resources are depleted and become harder to extract.

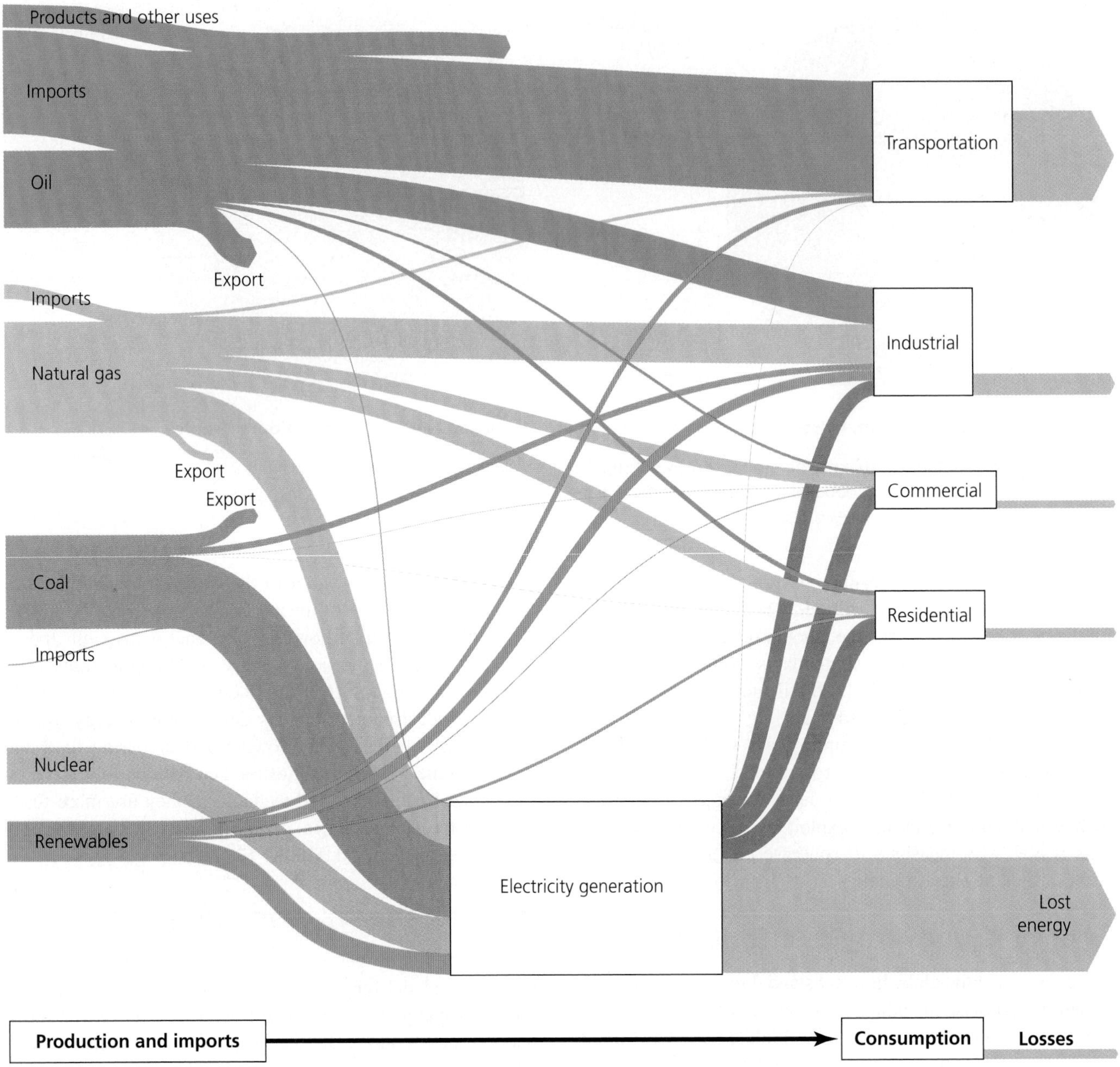


**FIGURE 15.4 Total energy flow of the United States.** Amounts are represented by the thickness of each bar. Domestic production and imports are shown on the left, and destinations of the energy are shown on the right. Portions of each energy source are used directly in the residential, commercial, industrial, and transportation sectors. Other portions are used to generate electricity, which in turn powers these sectors. The large amounts of energy lost as waste heat are shown on the right. *Data are for 2013, from U.S. Energy Information Administration and Lawrence Livermore National Laboratory.*

EROI ratios for producing conventional oil and natural gas in the United States declined from roughly 30:1 in the 1950s to about 11:1 today (**FIGURE 15.5**). This means we used to gain 30 units of energy for every unit of energy expended, but now we gain only 11. EROI ratios for oil and gas declined because we extracted the easiest deposits first and now must work harder to extract the remaining amounts.

## Where will we turn for energy?

Since the onset of the industrial revolution, coal, oil, and natural gas have powered the astonishing advances of our civilization. These extraordinarily rich sources of energy have helped to bring us a standard of living our ancestors could scarcely have imagined. Yet because fossil fuel deposits are finite and nonrenewable, easily accessible supplies of the three main fossil fuels have dwindled, EROI ratios have fallen, and fuels have become more expensive.

In response, today we are devoting enormous amounts of money, energy, and technology to reach deeper and farther for fossil fuels. We are using potent extraction methods such as hydraulic fracturing to free gas and oil tightly bound in rock layers. We are using powerful new machinery and techniques to squeeze more fuel from sites that were already extracted.

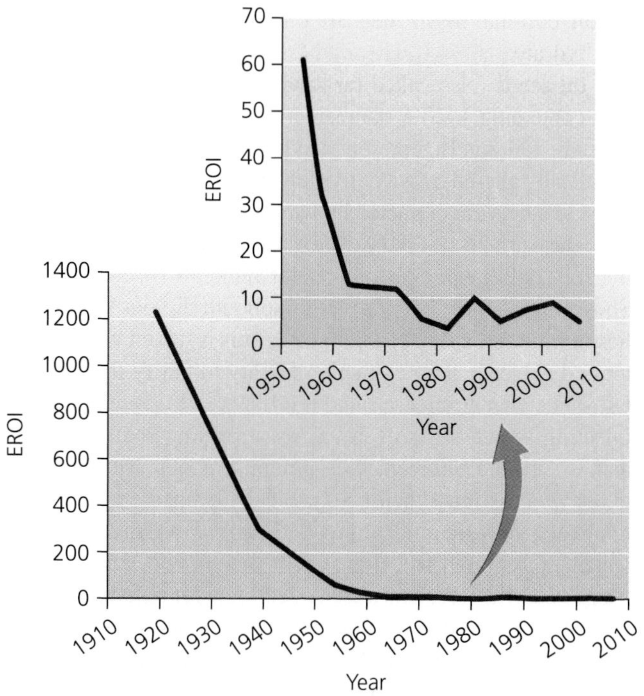


**FIGURE 15.5 EROI values for discovering oil and gas in the United States have declined over the past century.**
*Data from Guilford, M., et al., 2011. A new long term assessment of energy return on investment (EROI) for U.S. oil and gas discovery and production. Pp. 115–136 in* Sustainability, Special Issue, 2011, eds. C. Hall and D. Hansen, New studies in EROI (Energy return on investment).

**DATA Q** Why is it useful to zoom in on a portion of the graph as shown here? Describe what you can learn from the inset graph that you might not appreciate from the main graph.

GO TO **INTERPRETING GRAPHS & DATA** ON MasteringEnvironmentalScience®

We are drilling deeper underground, further offshore, and into the Arctic seabed. And we are pursuing new types of fossil fuels.

There is, however, a different way we can respond to the depletion of conventional fossil fuel resources: We can hasten the development of renewable energy sources to replace them (Chapter 16). By transitioning to clean and renewable alternatives, we can gain energy that is sustainable in the long term while reducing pollution, health impacts, and the emission of greenhouse gases that drive climate change.

# Fossil Fuels: Their Formation, Extraction, and Use

To grapple effectively with the energy issues that face us, it is important to understand how fossil fuels are formed, how we locate deposits, how we extract these resources, and how our society puts them to use.

## Fossil fuels are formed from ancient organic matter

Fossil fuels form only after organic material is broken down over millions of years in an *anaerobic* environment, one with little or no oxygen. Such environments include the bottoms of lakes, swamps, and shallow seas. The fossil fuels we burn today in our vehicles, homes, industries, and power plants were formed from the tissues of organisms that lived 100–500 million years ago. When organisms were buried quickly in anaerobic sediments after death, chemical energy in their tissues became concentrated as the tissues decomposed and their hydrocarbon compounds (p. 29) were altered and compressed (**FIGURE 15.6**).

Because fossil fuels form only under certain conditions, they occur in isolated deposits. For example, the Marcellus Shale holds rich reserves of natural gas that other nearby rock formations do not. Geologists searching for fossil fuels drill cores and conduct ground, air, and seismic surveys to map underground rock formations and predict where fossil fuel deposits might occur.

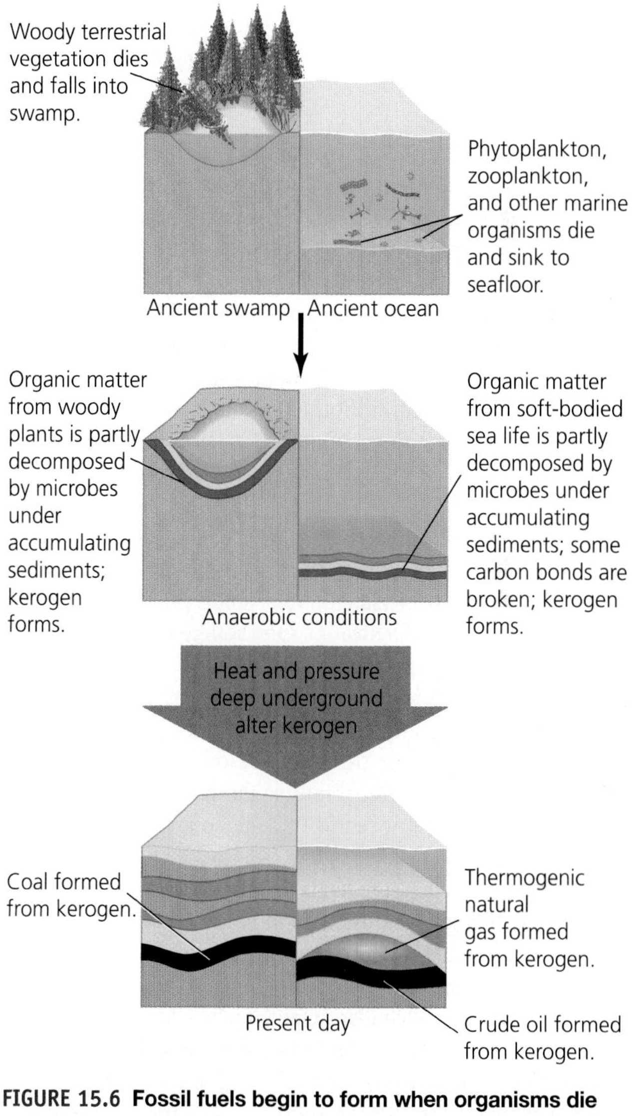


**FIGURE 15.6 Fossil fuels begin to form when organisms die and end up in oxygen-poor conditions.** This can occur when trees fall into lakes and are buried by sediment, or when phytoplankton and zooplankton drift to the seafloor and are buried **(top)**. Organic matter that undergoes slow anaerobic decomposition deep under sediments forms kerogen **(middle)**. Coal results when plant matter is compacted so tightly that there is little decomposition **(bottom left)**. The action of geothermal heating on kerogen may create crude oil and natural gas **(bottom right)**, which come to reside in porous rock layers beneath dense, impervious layers.

**Coal** The world's most abundant fossil fuel is **coal**, a hard blackish substance formed from organic matter (generally woody plant material) compressed under very high pressure, creating dense, solid carbon structures. Coal typically results when water is squeezed out of such material as pressure and heat increase over time, and when little decomposition takes place. The proliferation 300–400 million years ago of swamps where organic material was buried created coal deposits in many regions of the world.

To extract coal from deposits near the surface, we use *strip mining*, in which heavy machinery scrapes away huge amounts of earth. For deposits deep underground, we use *subsurface mining*, digging vertical shafts and blasting out networks of horizontal tunnels to follow seams, or layers, of coal. (Strip mining and subsurface mining are illustrated in Figure 11.14, p. 237.) We are also now mining coal on immense scales in the Appalachian Mountains, blasting away entire mountaintops in a process called *mountaintop removal mining* (pp. 241, 344).

**Oil and natural gas** The sludgelike liquid we know as **oil**, or **crude oil**, contains a mix of various hydrocarbon molecules (p. 29). **Natural gas** is a gas consisting of methane ($CH_4$) and lesser, variable, amounts of other volatile hydrocarbons. Oil is also known as **petroleum**, although this term is commonly used to refer to oil and natural gas collectively.

Both oil and natural gas are formed from organic material (especially dead plankton) that drifted down through coastal marine waters millions of years ago and was buried in sediments on the ocean floor. This organic material was transformed by time, heat, and pressure into today's natural gas and crude oil.

Two processes give rise to natural gas. *Biogenic* gas is created at shallow depths by the anaerobic decomposition of organic matter by bacteria. You may smell this gas when stepping into the muck of a swamp. One source of biogenic natural gas is the decay process in landfills, and landfill operators are now capturing this gas to sell as fuel (p. 394). *Thermogenic* gas results from compression and heat deep underground. It may form directly or from coal or oil altered by heating. Most gas extracted commercially is thermogenic and is found above deposits of oil or seams of coal, so it is often extracted along with those fuels.

Underground pressure tends to drive oil and natural gas upward through cracks and fissures in porous rock until they become trapped under a dense, impermeable rock layer. Oil and gas companies employ geologists to study rock formations to identify promising locations. Once such a location is identified, a company typically conducts *exploratory drilling*, drilling small holes to great depths. If enough oil or gas is encountered, extraction may begin. Because oil and gas are under pressure while in the ground, they rise to the surface when a deposit is tapped. Once pressure is relieved and some portion has risen to the surface, the remainder will need to be pumped out.

**Unconventional fossil fuels** Besides the three conventional fossil fuels—coal, oil, and natural gas—other types of fossil fuels exist, often called "unconventional" because we are not (yet) using them as widely. Three examples of unconventional fossil fuels are oil sands, oil shale, and methane hydrate.

**Oil sands** (also called **tar sands**) consist of moist sand and clay containing 1–20% bitumen, a thick and heavy form of petroleum. Oil sands represent crude oil deposits degraded and chemically altered by water erosion and bacterial decomposition.

Oil sands are extracted by two main methods. For deposits near the surface (**FIGURE 15.7a**), a process akin to strip mining for coal or open-pit mining for minerals (p. 240) is used. Shovel-trucks peel back layers of soil and dig out vast quantities of bitumen-soaked sand or clay. This is mixed with hot water and solvents at an extraction facility to purify the bitumen. Oil sands deeper underground (**FIGURE 15.7b**) are extracted by injecting steam and solvents down a drilling shaft to liquefy and isolate the bitumen, then pumping it out. Bitumen from either process must then be chemically refined and processed to create synthetic crude oil (called *syncrude*). Three barrels of water are required to extract each barrel of oil, and the toxic wastewater that results is discharged into vast reservoirs.

The second type of unconventional fossil fuel, **oil shale**, is sedimentary rock (p. 229) filled with organic matter that can be processed to produce a liquid form of petroleum called **shale oil**. Oil shale is formed by the same processes that form crude oil but occurs when the organic matter was not buried deeply enough or subjected to enough heat and pressure to form oil.

Oil shale is extracted using strip mines or subsurface mines. It can be burned directly like coal, or it can be processed in several ways. One way is to bake it in the presence of hydrogen and in the absence of air to extract liquid petroleum (a process called *pyrolysis*). The world's known deposits of oil shale may contain 3 trillion barrels of petroleum (more than all the world's conventional crude oil), but oil shale is costly to extract, and its EROI is low, ranging from 4:1 down to just 1.1:1.

The third unconventional fossil fuel, **methane hydrate** (also called *methane clathrate* or *methane ice*), is an ice-like solid consisting of molecules of methane embedded in a crystal lattice of water molecules. Methane hydrate occurs in sediments in the Arctic and on the seafloor because it is stable at temperature and pressure conditions found there.

Scientists estimate there are enormous amounts of this substance on Earth, holding perhaps twice as much carbon as all known deposits of oil, coal, and natural gas combined. Japan recently extracted methane hydrate from the seafloor by sending down a pipe and lowering pressure within it so that the methane turned to gas and rose to the surface. However, we do not yet know whether extraction is safe and reliable. If extraction were to destabilize a methane hydrate deposit on the seafloor and lead to a sudden release of gas, this could cause a landslide and tsunami and emit large amounts of methane, a potent greenhouse gas, into the atmosphere.

## Economics determines how much will be extracted

As we develop more powerful technologies, the proportion of a fossil fuel that is physically accessible to us—the "technically recoverable" portion—tends to increase. Yet while technology

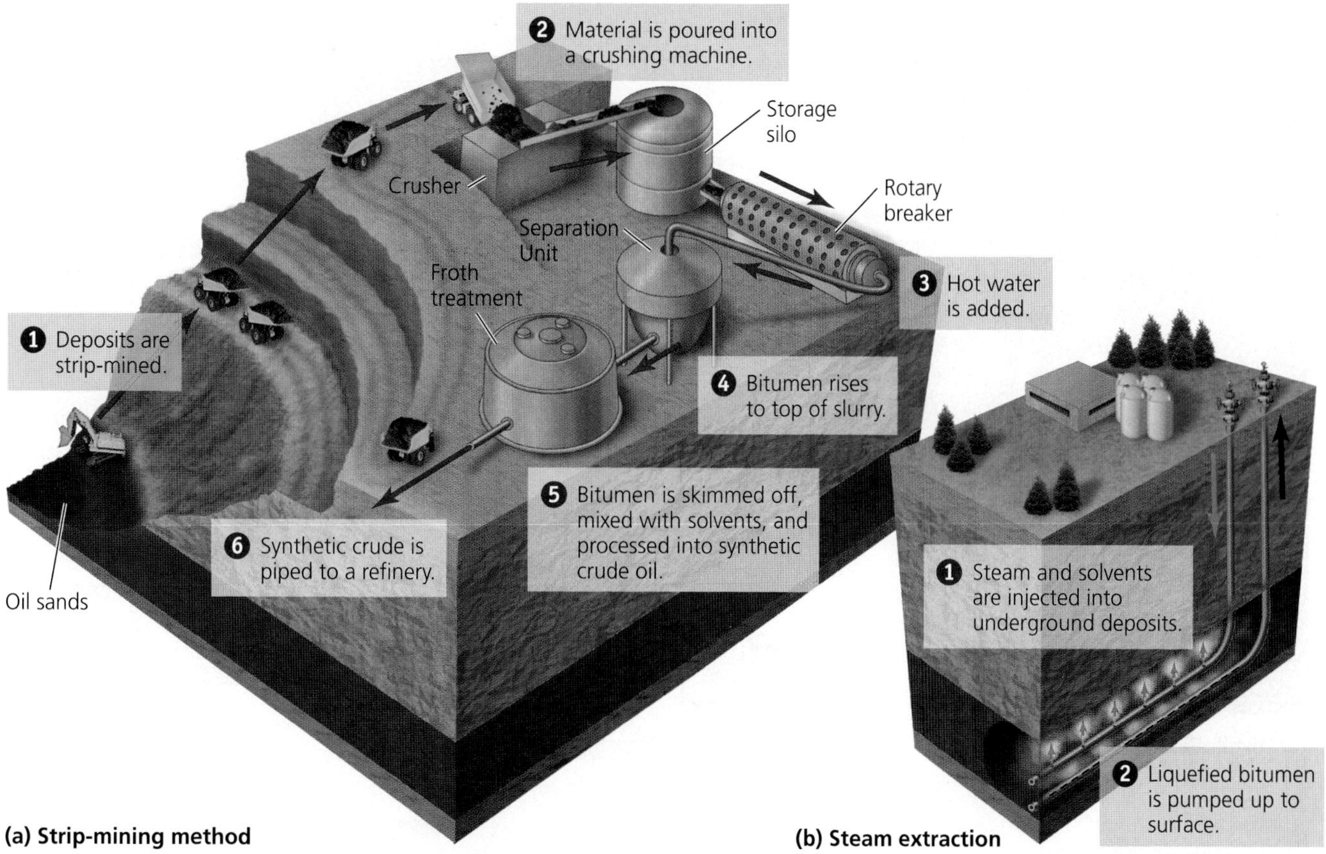


**FIGURE 15.7 Oil sands are extracted by two processes.** Near-surface deposits **(a)** are strip-mined. The deposits are dug out ❶ with gigantic shovels and poured into a crushing machine ❷. The material is mixed with hot water ❸, and piped to a facility where the bitumen floats in a froth atop water in a tank ❹ while sand and clay settle out. The bitumen froth is skimmed off, mixed with chemical solvents, and processed into synthetic crude oil ❺; it is then piped ❻ to a refinery. Deeper deposits of oil sands **(b)** are extracted by injecting pressurized stem into a well, ❶ which liquefies the bitumen and allows it to be pumped ❷ to the surface.

determines how much fuel *can* be extracted, economics determines how much *will* be extracted. This is because extraction becomes increasingly expensive as a resource is removed, so companies rarely find it profitable to extract the entire amount. Instead, a company will consider its costs and balance these against the income it expects from sale of the fuel. Because market prices of fuel fluctuate, the portion of fuel from a given deposit that is "economically recoverable" fluctuates as well. As market prices rise, economically recoverable amounts approach technically recoverable amounts.

The amount of a fossil fuel that is technologically and economically feasible to remove under current conditions is termed its **proven recoverable reserve**. Proven recoverable reserves increase as extraction technology improves or as market prices of the fuel rise. Proven recoverable reserves decrease as fuel deposits are depleted or as market prices fall (making extraction unprofitable).

Some examples of proven recoverable reserves are shown in Table 15.2 (p. 335). The amounts of a fossil fuel "produced" (extracted and processed) from a nation's reserves depends on many factors. TABLE 15.3 shows amounts produced and amounts consumed (as a percentage of global production and consumption) by leading nations.

**TABLE 15.3 Top Producers and Consumers of Fossil Fuels**

| PRODUCTION (% world production) | | CONSUMPTION (% world consumption) | |
|---|---|---|---|
| **COAL** | | | |
| China | 47.4 | China | 50.3 |
| United States | 12.9 | United States | 11.9 |
| Australia | 6.9 | India | 8.5 |
| Indonesia | 6.7 | Japan | 3.4 |
| India | 5.9 | Russia | 2.4 |
| **OIL** | | | |
| Saudi Arabia | 13.1 | United States | 19.9 |
| Russia | 12.9 | China | 12.1 |
| United States | 10.8 | Japan | 5.0 |
| China | 5.0 | India | 4.2 |
| Canada | 4.7 | Russia | 3.7 |
| **NATURAL GAS** | | | |
| United States | 20.6 | United States | 22.2 |
| Russia | 17.9 | Russia | 12.3 |
| Iran | 4.9 | Iran | 4.8 |
| Qatar | 4.7 | China | 4.8 |
| Canada | 4.6 | Japan | 3.5 |

*Data from BP p.l.c., 2014.* Statistical review of world energy 2014.

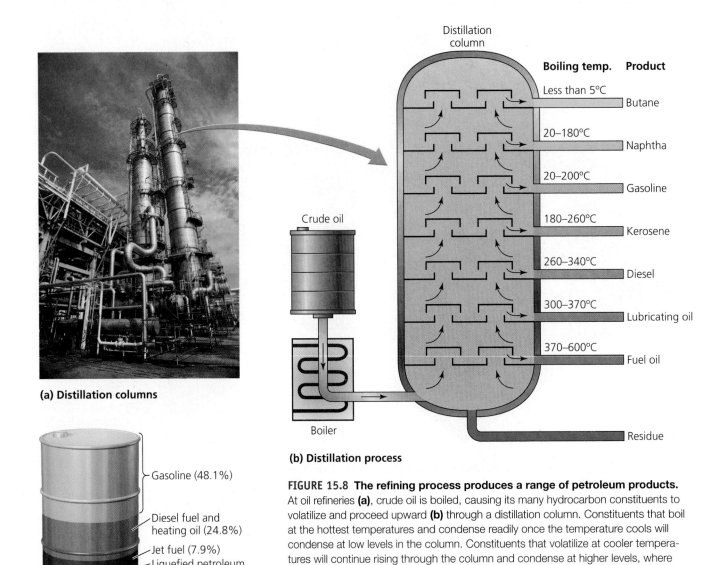

(a) Distillation columns

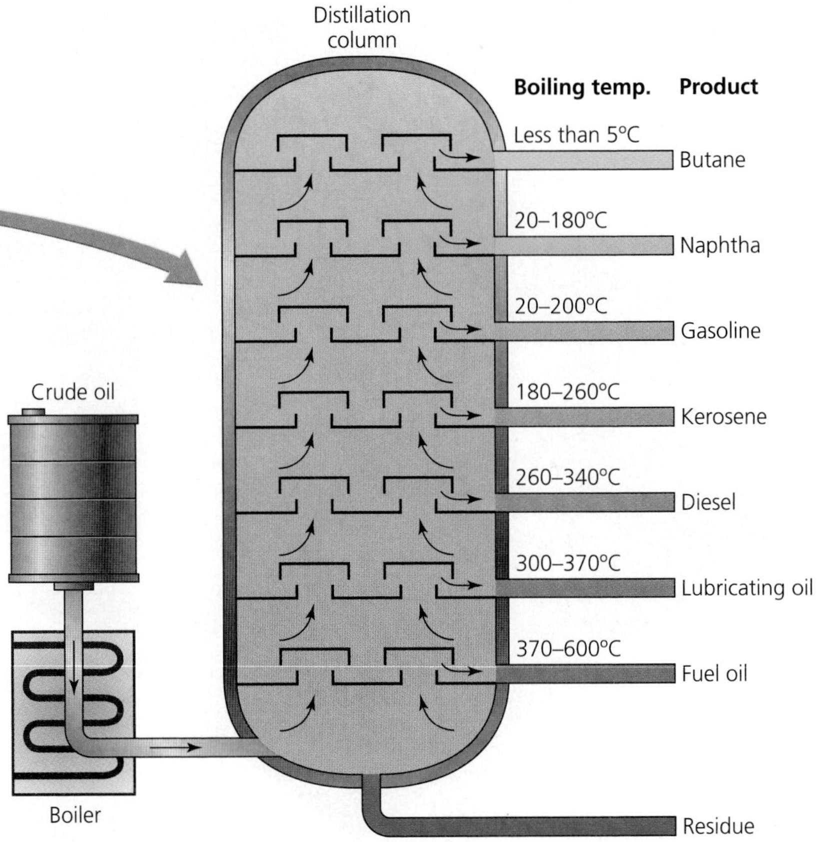


(b) Distillation process

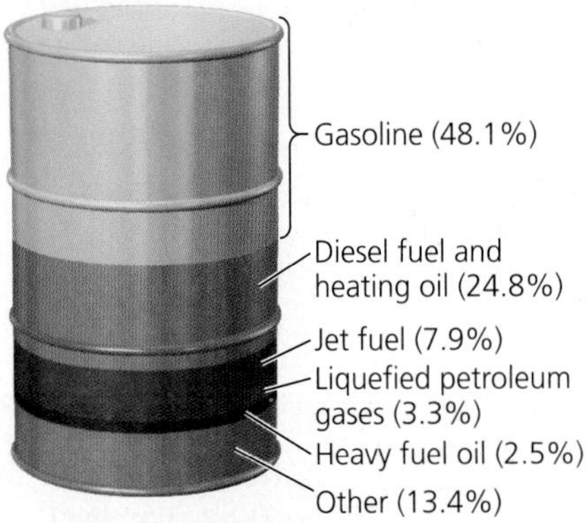


(c) Typical composition of refined oil

**FIGURE 15.8 The refining process produces a range of petroleum products.** At oil refineries **(a)**, crude oil is boiled, causing its many hydrocarbon constituents to volatilize and proceed upward **(b)** through a distillation column. Constituents that boil at the hottest temperatures and condense readily once the temperature cools will condense at low levels in the column. Constituents that volatilize at cooler temperatures will continue rising through the column and condense at higher levels, where temperatures are cooler. In this way, heavy oils (generally those with hydrocarbon molecules with long carbon chains) are separated from lighter oils (generally those with short-chain hydrocarbon molecules). Shown in **(c)** are percentages of each major category of product typically generated from a barrel of crude oil. *Data (c) from U.S. Energy Information Administration.*

## Refining produces a diversity of fuels

Once we extract oil or gas, it must be processed and refined (**FIGURE 15.8**). Crude oil is a mix of hundreds of types of hydrocarbon molecules characterized by carbon chains of different lengths (p. 29). Chain length affects a substance's chemical properties, which has consequences for human use, such as whether a given fuel burns cleanly in a car engine. At a refinery, hydrocarbon molecules are separated by size and are chemically transformed to create specialized fuels for heating, cooking, and transportation and to create lubricating oils, asphalts, and the precursors of plastics and other petrochemical products.

## Fossil fuels have many uses

Each major type of fossil fuel has its own mix of uses.

**Coal** People have burned coal to cook food, heat homes, and fire pottery for thousands of years. Coal-fired steam engines helped drive the industrial revolution by powering factories, trains, and ships, and coal fueled the furnaces of the steel industry. Today we burn coal largely to generate electricity. In coal-fired power plants, coal combustion converts water to steam, which turns a turbine to create electricity (**FIGURE 15.9**). Coal provides 40% of the electrical generating capacity of the United States, and it powers China's surging economy.

**Natural gas** We use natural gas to generate electricity in power plants, to heat and cook in our homes, and for much else. Converted to a liquid at low temperatures (*liquefied natural gas*, or *LNG*), it can be shipped long distances in refrigerated tankers. Versatile and clean-burning, natural gas emits just half as much carbon dioxide per unit of energy produced as coal and two-thirds as much as oil. For this reason, many energy experts view natural gas as a climate-friendly "bridge fuel" that can help us transition from today's polluting fossil-fuel economy toward a clean renewable energy economy. However, many

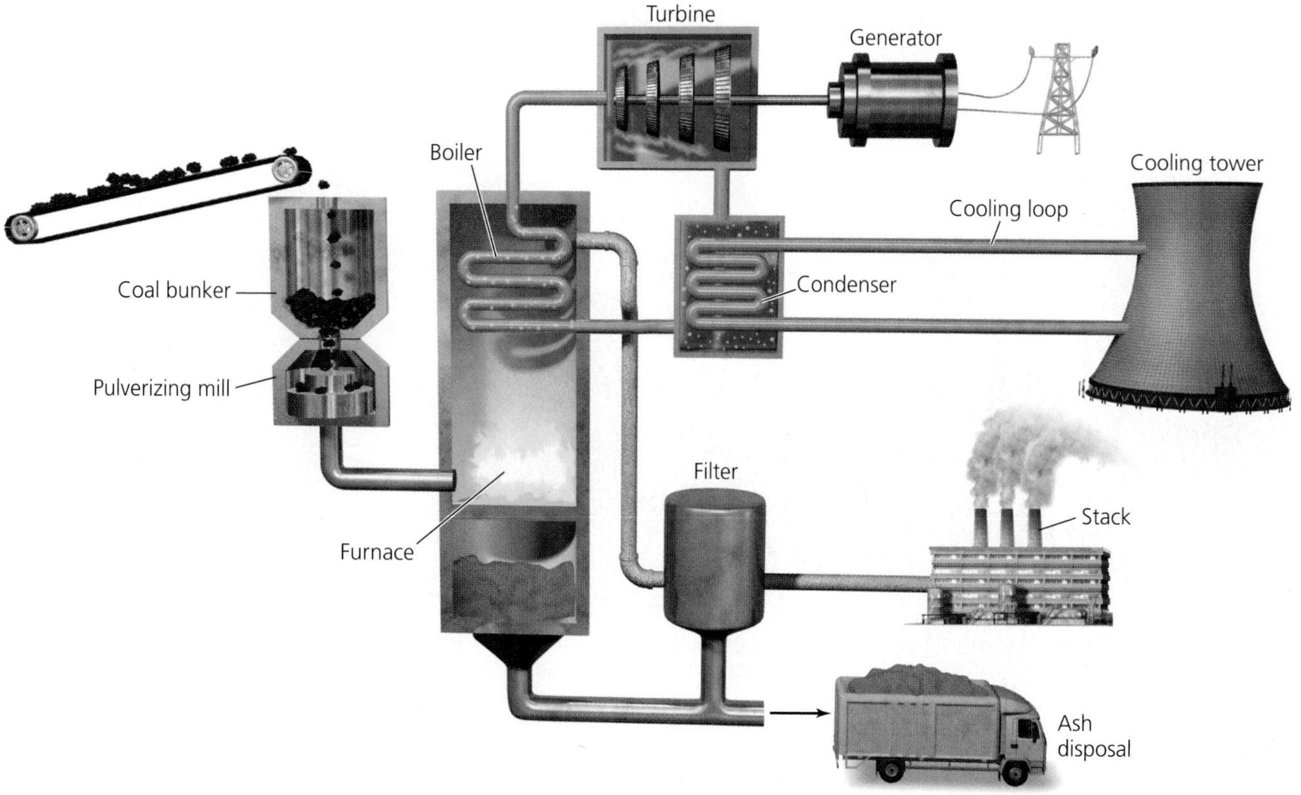


**FIGURE 15.9 At a coal-fired power plant, coal is pulverized and blown into a high-temperature furnace.** Heat from the combustion boils water, and the resulting steam turns a turbine, generating electricity by passing magnets past copper coils. The steam is then cooled and condensed in a cooling loop and returned to the furnace. "Clean coal" technologies (p. 350) help filter pollutants from the combustion process, and toxic ash residue is taken to hazardous waste disposal sites.

other people worry that investing in natural gas (whose extraction causes water pollution and can leak methane to the atmosphere) will simply delay the transition to renewables and instead deepen our reliance on fossil fuels.

**Oil** Our global society consumes nearly 750 L (200 gal) of oil each year for every man, woman, and child. Most is used as fuel for vehicles, including gasoline for cars, diesel for trucks, and jet fuel for airplanes. Fewer homes burn oil for heating these days, but industry and manufacturing continue to use a great deal.

Refining techniques and chemical manufacturing have greatly expanded our uses of petroleum to include a wide array of products and applications, from plastics to lubricants to fabrics to pharmaceuticals. Today, petroleum-based products are all around us in our everyday lives (**FIGURE 15.10**, next page). The fact that petroleum is used to create so many items and materials we have come to rely on makes it vital that we conserve our remaining oil reserves.

## We are depleting fossil fuel reserves

Because fossil fuels are nonrenewable, the total amount available on Earth declines as we use them. Many scientists and oil industry analysts calculate that we have extracted nearly half the world's conventional oil reserves. So far we have used about 1.1 trillion barrels of oil, and most estimates hold that somewhat more than 1 trillion barrels of proven recoverable reserves remain. Adding proven reserves of oil sands brings the total remaining to nearly 1.7 trillion barrels.

To estimate how long this remaining oil will last, analysts calculate the **reserves-to-production ratio**, or **R/P ratio**, by dividing the amount of remaining reserves by the annual rate of production (extraction and processing). At the current levels of production (31 billion barrels globally per year), 1.7 trillion barrels would last about 53 more years. Applying the R/P ratio to natural gas, we find that the world's proven reserves of this resource would last 55 more years. For coal, the latest R/P ratio estimate is 113 years.

The actual number of years remaining for these fuels may turn out to be less than these figures suggest if our demand and production continue to increase. Alternatively, the actual number of years may end up being *more* than the figures suggest if we reduce demand by enhancing efficiency. The actual number of years may also turn out to be greater because proven recoverable reserves increase as technology becomes more powerful and as market prices rise.

For instance, hydraulic fracturing for natural gas in the Marcellus Shale and elsewhere in the United States has expanded the nation's proven reserves of natural gas considerably in just the past several years. Likewise, hydraulic fracturing of the Bakken Formation, layers of shale and dolomite that

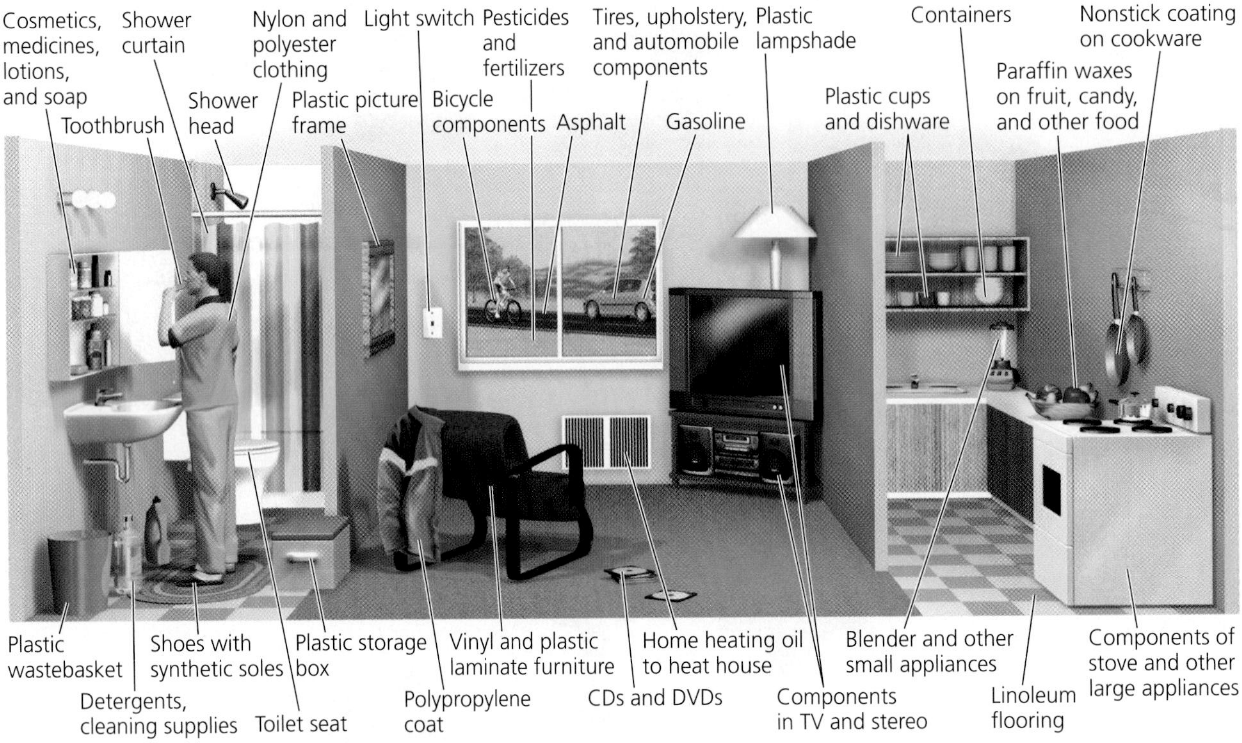


**FIGURE 15.10 Petroleum products are everywhere in our daily lives.** Besides the fuels we use for transportation and heating, petroleum is used to make many of the fabrics we wear, the materials we consume, and the plastics in countless items we use every day.

underlie parts of North Dakota, Montana, and Canada, is allowing us to extract oil trapped tightly in this rock. By accessing this *tight oil*, conventional oil held tightly in or near shale (which differs from shale oil, a petroleum liquid from specially processed oil shale), the United States is boosting its proven recoverable reserves of oil. In fact, the current oil boom in North Dakota, along with recent increases in drilling deep offshore, led the International Energy Agency to predict that the United States could become the world's biggest oil producer by the 2020s.

Eventually, however, production of any nonrenewable resource will come to a peak and then decline. In general, production tends to decline once reserves are depleted halfway. If demand for the resource holds steady or rises while production declines, a shortage will result. With oil, this scenario has come to be nicknamed **peak oil**. Because we have already used roughly half of Earth's conventional oil reserves, many experts calculate that a peak oil crisis could begin soon.

## Peak oil will pose challenges

To understand concerns about peak oil, we must turn back the clock to 1956. In that year, Shell Oil geologist M. King Hubbert calculated that U.S. oil production would peak around 1970. His prediction was ridiculed at the time, but it proved to be accurate; U.S. production peaked in that very year (**FIGURE 15.11a**). This peak in production came to be known as **Hubbert's peak**.

In 1974, Hubbert analyzed data on technology, economics, and geology, and predicted that worldwide oil production would peak in 1995. In fact it grew past 1995, but many scientists using newer data today predict that global production will soon begin to decline (**FIGURE 15.11b**). Predicting an exact date for peak oil is difficult, however. Many companies and governments do not reveal their data on oil reserves, and estimates differ as to how much oil we can continue extracting from existing deposits. Indeed, the U.S. Geological Survey recently estimated 2 trillion barrels of conventional oil remaining in the world, rather than 1 trillion, and some estimates predict greater amounts. A 2007 report by the U.S. General Accounting Office found that of 21 studies estimating the timing of an oil production peak, most predicted dates before 2040.

Whenever it occurs, a divergence of supply and demand could have momentous consequences that profoundly affect our lives. Writer James Howard Kunstler has sketched a frightening scenario of our post-peak world during what he calls "the long emergency": Lacking cheap oil with which to transport goods long distances, today's globalized economy would collapse into isolated local economies. Large cities would need to run urban farms to feed their residents, and with less mechanized farming and fewer petroleum-based fertilizers and pesticides, we might feed only a fraction of the world's 7 billion people. The American suburbs would be hit particularly hard because of their dependence on the automobile.

More-optimistic observers argue that as oil supplies dwindle, rising prices will create powerful incentives for businesses, governments, and individuals to conserve energy and to develop alternative energy sources—and that this will save us from major disruptions.

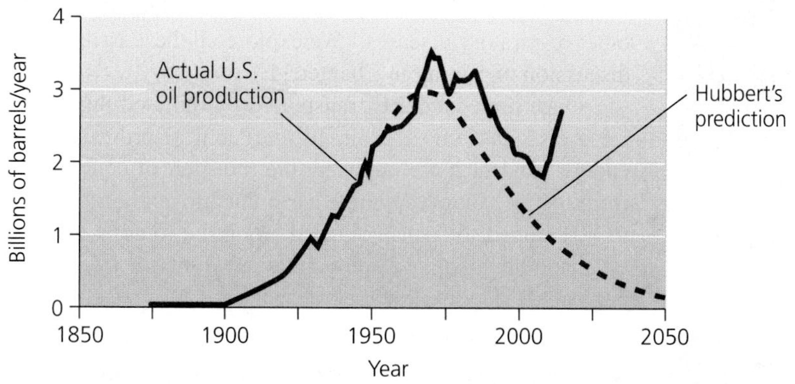


**(a) Hubbert's prediction of peak in U.S. oil production, with actual data**

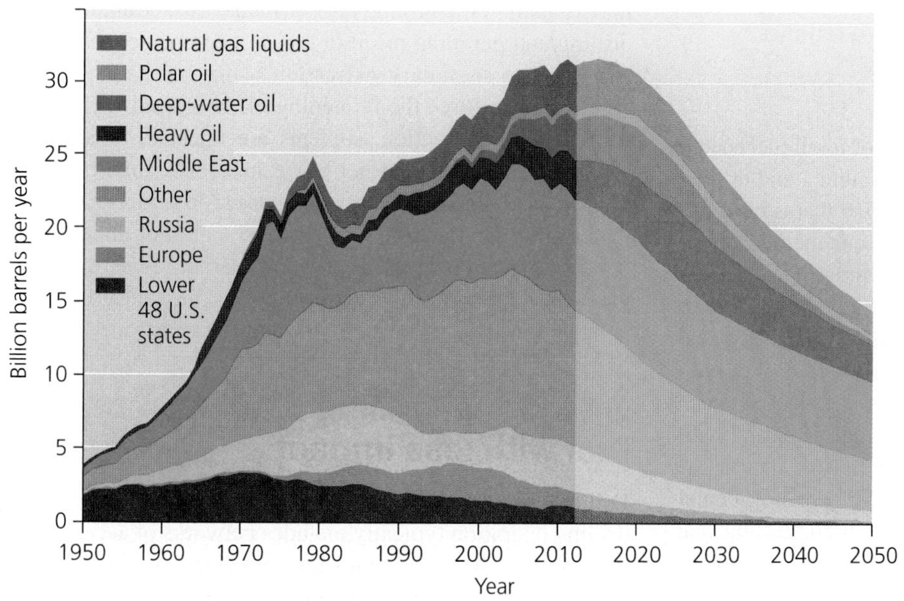


**(b) Modern prediction of peak in global oil production**

**FIGURE 15.11 Peak oil describes a peak in production.** U.S. oil production peaked in 1970 **(a)**, just as geologist M. King Hubbert had predicted. Since then, success in Alaska, in the Gulf of Mexico, and with secondary extraction has enhanced production during the decline. Today U.S. oil production is spiking upward as deep offshore drilling and hydraulic fracturing make new deposits accessible. Many analysts believe global production of conventional oil will soon peak; shown in **(b)** is one recent projection, from a 2011 analysis by scientists at the Association for the Study of Peak Oil. *Data from (a) Hubbert, M. K., 1956.* Nuclear energy and the fossil fuels. *Shell Development Co. Publ. No. 95, Houston, TX; and U.S. Energy Information Administration; and (b) Campbell, C.J., and Association for the Study of Peak Oil. By permission of Dr. Colin Campbell.*



**FAQ** **Why should I worry about "peak oil" if there is still a lot of oil remaining?**

If demand for oil continues to rise as supply falls following a global peak in oil production, the divergence of demand and supply could drive up oil prices dramatically, triggering major economic ripple effects. High oil prices would create financial incentive to develop alternative energy sources and better conservation measures, but in a depressed economy we might find it hard to devote adequate resources to develop new renewable sources quickly.

Alternatively, if we discover and exploit enough new deposits to continue expanding oil production, we might postpone our day of reckoning for decades. If this occurs, however, we will find ourselves wrestling with another concern: trying to avoid runaway climate change driven by greenhouse gas emissions from the combustion of all that additional oil!

# Reaching Further for Fossil Fuels . . . and Coping with the Impacts

To stave off the day when production of oil, gas, and coal begin to decline, we are investing more and more funding, technology, and energy into locating and extracting new fossil fuel deposits. We are extending our reach for fossil fuels in several ways:

- Mountaintop mining for coal
- Secondary extraction from existing wells
- Directional drilling
- Hydraulic fracturing for oil and gas
- Offshore drilling in deeper waters
- Moving into ice-free waters of the Arctic
- Exploiting new "unconventional" fossil fuel sources

**FIGURE 15.12 In mountaintop removal mining for coal, entire mountain peaks are leveled, and fill is dumped into adjacent valleys,** as shown in this aerial view over many square miles in West Virginia.

All these pursuits are expanding the amount of fossil fuel energy available to us. However, as we extend our mining and drilling efforts into less-accessible places to obtain fuel that is harder to extract, we also reduce the EROI ratios of our fuels, drive up fuel prices, intensify pollution, and worsen climate change.

Our society's love affair with fossil fuels has helped to ease constraints on travel, increase our life expectancy, and boost our material standard of living dramatically. Yet continued reliance on fossil fuels poses growing risks to human health, environmental quality, and social, political, and economic stability. Indeed, the risks from climate change alone are great enough that the International Energy Agency's chief economist recently joined some scientists in suggesting that we'd be better off leaving most oil, gas, and coal in the ground!

## Mountaintop mining extends our reach for coal

Coal mining has long been an economic mainstay of the Appalachian region, but mountaintop removal mining has brought coal extraction—and its impacts—to a whole new level (**FIGURE 15.12** and p. 241). The massive scale of mountaintop removal mining makes it economically efficient. However, the technique can cause staggering volumes of rock and soil to slide downslope, degrading or destroying entire hillsides, polluting or burying streams, and disrupting life for people living nearby.

Mountaintop mining magnifies many of the impacts of traditional strip mining for coal, which unleashes soil erosion and destroys large areas of habitat. These mining methods also send chemical runoff into waterways in the form of acid drainage (p. 237), whereby sulfide minerals in newly exposed rock surfaces react with oxygen and rainwater to produce sulfuric acid. In most developed nations, mining companies are required to restore affected areas after mining, but reclamation rarely is able to recreate the ecological communities that preceded mining (p. 242). Subsurface coal mining, for its part, has long posed health risks to miners. They are at risk of accidents and they breathe coal dust and hazardous gases in confined spaces, which can lead to black lung disease and other respiratory diseases. (We explore all these issues in our discussion of mining in Chapter 11.)

Once mined, coal is transported by rail, and this can release coal dust into the air. In the Pacific Northwest, clean energy advocates are opposing the transport of coal by train from the interior West where it is mined to coastal terminals, to be shipped to Asia. Pollution along the route is one concern; another is that we are facilitating China's heavy reliance on coal, a driver of global climate change.

## Secondary extraction produces more fuel

At a typical oil or gas well, as much as two-thirds of a deposit may remain in the ground after **primary extraction**, the initial drilling and pumping of oil or gas. So, companies may return and conduct **secondary extraction** using new technology or approaches to force the remaining oil or gas out by pressure. In secondary extraction, solvents are injected, underground rocks are flushed with water or steam, or hydraulic fracturing may be used. Because secondary extraction is more expensive than primary extraction, many deposits did not undergo secondary extraction in the past when market prices of oil and gas were too low to make it profitable. Once oil prices rose, companies reopened sites for secondary extraction.

## Directional drilling reaches more fuel with less impact

Drilling for oil or gas requires extensive infrastructure. A drilling operation typically includes networks of access roads, housing for workers, transport pipelines, waste piles for removed soil, and ponds to collect toxic sludge. All this tends to pollute soil, air, and water; fragment habitat; and disturb wildlife and people. Today's **directional drilling** technology helps to lessen some of these impacts by allowing drillers to bore down vertically and then curve to drill horizontally. This enables them to follow horizontal layered deposits such as the Marcellus Shale or the Bakken Formation. By allowing access to a large underground area (up to several thousand meters in radius) around each drill pad, fewer drill pads are needed, and the surface footprint of drilling is smaller.

## Hydraulic fracturing expands our access to oil and gas

For oil and natural gas trapped tightly in shale or other rock, oil and gas companies now use hydraulic fracturing (see Figure 15.1). Chemically treated water under high pressure is pumped into layers of rock to crack them, and sand or small glass beads hold the cracks open as the water is withdrawn. Gas or oil then travels upward through the system of fractures. By unlocking formerly inaccessible deposits of shale gas and tight oil, hydrofracking has ignited a boom in extraction in the United States.

The resulting abundance of natural gas has enabled many power plants to switch from coal to gas, and this is a main reason that U.S. carbon dioxide emissions have fallen since 2007. Whether the gas boom can continue to help reduce the emissions that cause climate change will depend on whether natural gas truly turns out to be a "bridge fuel" helping us transition to renewables, or whether it locks us into a deeper reliance on fossil fuels.

At the local level, fracking has engendered debate wherever it has occurred (**FIGURE 15.13**). Above the Marcellus Shale, it is affecting the landscapes, economies, politics, and everyday lives of people in Pennsylvania, Ohio, New York, and neighboring states. The choices people here face between financial gain and protecting their health, drinking water, and environment have been dramatized in popular films such as *Promised Land* and *Gasland*. In North Dakota, fracking for oil has supercharged the economy and drawn young men from around the nation for good-paying jobs, but it is also intensifying competition for water in an arid region. Like all energy booms before it, today's fracking rush brings jobs and money to small towns, but it can also spark social upheaval and leave communities with a legacy of pollution.

One pollution risk from hydraulic fracturing is that fracking fluids (whose chemicals are not required to be publicly disclosed) may leak out of drilling shafts and into aquifers that people rely on for drinking water. Another concern is that methane may contaminate groundwater if it travels up fractures or leaks through a shaft. Hydrofracking also gives rise to air pollution as methane and volatile toxic components of fracking fluids seep up from drilling sites. Some of the unhealthiest air in the United States was recently found in a remote region of Wyoming near fracking operations. Like the people of Dimock, Pennsylvania, residents of areas near fracking sites across North America have experienced polluted air and fouled drinking water.

Hydrofracking also consumes immense volumes of fresh water. Injected water often returns to the surface laced with salts, radioactive elements such as radium, and toxic chemicals such as benzene from deep underground. This wastewater may be sent to sewage treatment plants that are not designed to handle all the contaminants and that do not test for radioactivity. In Pennsylvania, shale gas extraction has been sending millions of gallons of drilling waste to treatment plants, which then release their water into rivers that supply drinking water for people in Pittsburgh, Harrisburg, and other cities. In response to public outcry and government pressure, the oil and gas industry is beginning to reduce its water use by reusing its wastewater in multiple injections.

**FIGURE 15.13 Hydraulic fracturing is expanding U.S. production of oil and natural gas but is also sparking debates in communities where it takes place.** This drill rig is fracking for shale gas on private land among homes in rural Pennsylvania.

**FIGURE 15.14 The explosion at British Petroleum's *Deepwater Horizon* drilling platform in 2010 unleashed the world's largest accidental oil spill.** Here vessels try to put out the blaze.

## We are drilling farther offshore

Roughly 35% of the oil and 10% of the natural gas extracted in the United States today come from offshore sites, primarily in the Gulf of Mexico and also off southern California. Geologists estimate that most U.S. gas and oil remaining occurs offshore and that deepwater sites in the Gulf may hold 59 billion barrels of oil. As oil and gas are depleted at shallow-water sites and as drilling technology improves, the industry is moving into deeper and deeper water.

Deep offshore drilling is boosting oil and gas production, but it poses risks. In the *Deepwater Horizon* oil spill of 2010 (p. 268), faulty equipment allowed natural gas accompanying an oil deposit to shoot up a well shaft and ignite on a British Petroleum platform off the Louisiana coast, killing 11 workers (**FIGURE 15.14**). The platform sank, emergency shut-off systems failed, and oil began gushing out of a broken pipe on the ocean floor at a rate of 30 gallons per second. British Petroleum's engineers tried one solution after another, but the flow of oil and gas continued out of control for three months, spilling roughly 4.9 million barrels (206 million gallons) of oil. BP's Macondo well, where the accident took place, lay beneath 1500 m (5000 ft) of water. The deepest wells in the Gulf of Mexico are now twice that depth.

As oil from the Macondo well spread through the Gulf and washed ashore, the region suffered a wide array of impacts

(a) Brown pelican coated in oil

(b) Beach cleanup

**FIGURE 15.15 Impacts of the *Deepwater Horizon* spill were many.** This brown pelican, coated in oil **(a)**, was one of countless animals killed. For months, volunteers and workers labored **(b)** to clean oil from the Gulf's beaches.

(**FIGURE 15.15**). Unknown numbers of fish, birds, shrimp, corals, and other marine animals were killed, affecting coastal and ocean ecosystems in complex ways. Plants in coastal marshes died, causing erosion that put New Orleans and other cities at greater risk from storm surges and flooding. Beach tourism suffered, as did Gulf Coast fisheries, which supply much of the nation's seafood. Thousands of fishermen and shrimpers were put out of work. Throughout this process, scientists studied the spill's many impacts on the region (see **THE SCIENCE BEHIND THE STORY**, pp. 348–349).

The *Deepwater Horizon* spill was the largest accidental oil spill in world history, far eclipsing the spill that resulted when the *Exxon Valdez* tanker ran aground in 1989 and damaged ecosystems and economies in Alaska's Prince William Sound. The *Exxon Valdez* event had led U.S. policymakers to tighten regulation and improve spill response capacity. But in 2008, responding to rising gasoline prices and a desire to reduce dependence on foreign oil, Congress lifted a long-standing moratorium on offshore drilling along much of the nation's coast. The Obama administration then opened vast areas for drilling, including most waters from Delaware to Florida, more of the Gulf of Mexico, and most waters off Alaska's North Slope. Once the *Deepwater Horizon* spill occurred, however, public reaction forced Obama's administration to backtrack. Later, the administration sought a middle path, opening access to areas holding 75% of technically recoverable offshore oil and gas reserves while banning drilling offshore from states that did not want it. Drilling leases were expanded off Alaska and in the Gulf of Mexico, but not along the East and West Coasts.

Globally, pollution from large oil spills has declined in recent decades (p. 269), thanks to government regulations (such as requirements for double-hulled ships) and improved spill response efforts. Most water pollution from oil today results from innumerable small non-point sources to which we all contribute (p. 266). Oil from automobiles, homes, gas stations, and businesses runs off roadways and enters rivers and wastewater facilities, being discharged eventually into the ocean.

## Melting ice is opening up the Arctic

Today all eyes are on the Arctic. As global climate change melts the sea ice that covers the Arctic Ocean (pp. 316, 320), new shipping lanes are opening and nations and companies are jockeying for position, hoping to stake claim to oil and gas deposits that lie beneath the seafloor. Offshore drilling in Arctic waters, however, poses severe pollution risks. Frigid temperatures, ice floes, winds, waves, and brutal storms make conditions challenging and accidents likely. If a spill were to occur, icebergs, pack ice, storms, cold, and wintertime darkness would hamper response efforts, while frigid water temperatures would slow the natural breakdown of oil.

The risks from expansion into Arctic waters became clear in 2012–2013, when Royal Dutch Shell's *Kulluk* drilling rig ran aground while being towed south from Alaska in a winter storm. Damage to the rig raised fresh questions as to whether Arctic Ocean drilling can be conducted safely.

## We are exploiting new fossil fuel sources such as oil sands

Three sources of "unconventional" fossil fuels—oil sands, oil shale, and methane hydrate—are abundant, and together could theoretically supply our civilization for centuries. However, they are difficult and expensive to extract and process, and their net energy values and EROI ratios are very low (p. 383). Extracting oil sands and oil shale consumes large volumes of water, devastates landscapes, and pollutes waterways. Burning these fossil fuels would likely emit more greenhouse gases than our use of coal, oil, and natural gas currently does, worsening air pollution and climate change.

Oil sands are becoming a major fuel source—and a focus of debate over our energy future. Much of the world's oil sands underlie a vast, thinly populated region of boreal forest in northern Alberta in Canada. These tar-like deposits produce a low-quality fuel that requires a great deal of energy to extract and process. Most scientific estimates for the EROI ratio of Alberta's oil sands range from around 3:1 to 5:1.

Yet rising oil prices have made mining oil sands profitable, and today companies produce nearly 2 million barrels of oil per day. Each truckload of oil sands leaving a mine carries oil worth close to $20,000.

To extract these resources, companies clear vast areas of forest and dig enormous open pits miles wide and hundreds of feet deep (**FIGURE 15.16a**). The immense volumes of water used become polluted and are piped to gigantic reservoirs, where the toxic oily water kills waterfowl. The Syncrude company's wastewater reservoir near Fort McMurray, Alberta, is so massive that it is held back by the world's second-largest dam.

To sell its oil, Canada looked south to the United States. TransCanada Corporation built the Keystone Pipeline to pipe diluted bitumen 3500 km (2200 mi) to Illinois and Oklahoma. TransCanada then proposed the Keystone XL extension, a two-part project consisting of (1) a southern leg to get oil from Oklahoma to Texas refineries for export overseas and (2) a northern leg cutting across the Great Plains to shave off distance and add capacity to the line (**FIGURE 15.16b**). The Keystone XL pipeline proposal met opposition from people living along the route who were concerned about health, environmental protection, and property rights. It also faced nationwide opposition from advocates of action to address global climate change.

Pipeline proponents argued that the Keystone XL project would create jobs and guarantee a dependable oil supply for decades. They stressed that buying oil from Canada—a stable, friendly, democratic neighbor—could help reduce U.S. reliance on oil-producing nations with authoritarian governments and poor human rights records, such as Saudi Arabia and Venezuela.

Opponents of the pipeline extension expressed dismay at forest destruction in Alberta and anxiety about transporting oil over the Ogallala Aquifer (p. 262), where spills might contaminate drinking water for millions of people and irrigation water for America's breadbasket. They also sought to prevent extraction of a vast new source of fossil fuels whose combustion would emit huge amounts of greenhouse gases. By buying a source of oil that is energy-intensive to extract and that burns 14–20% less cleanly than conventional oil, they held, the United States would prolong fossil fuel dependence and worsen climate change.

**(a) Massive oil sands mine in Alberta**

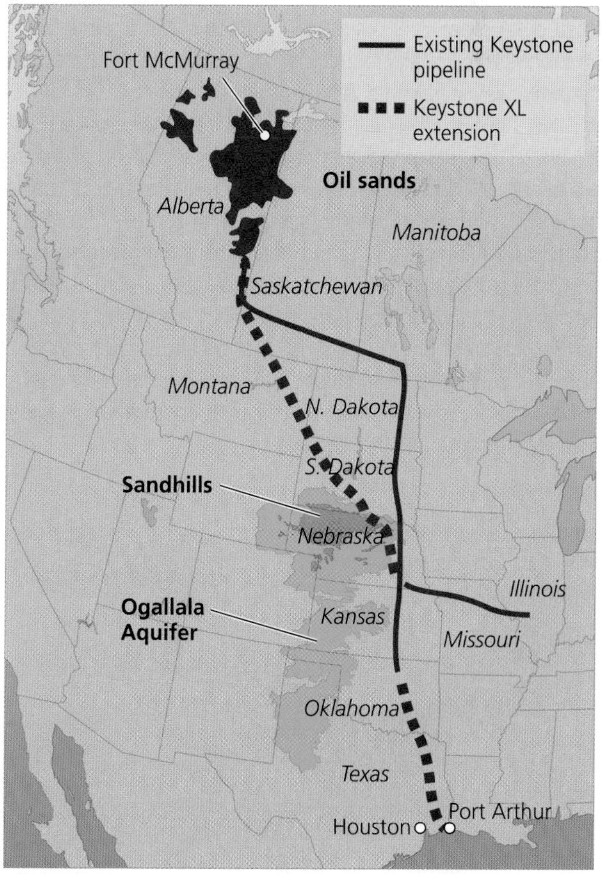


**(b) Keystone pipeline and Keystone XL extension**

**(c) Protest at the White House**

**FIGURE 15.16 Canadian oil sands and the Keystone XL pipeline extension have been a focus of debate.** In Alberta **(a)**, oil sands are mined from enormous pits. The Keystone pipeline brings the oil into the United States **(b)**, where its proposed extensions set off a complex debate. Tens of thousands of activists protested in front of the U.S. White House **(c)** in a series of increasingly large rallies.

# THE SCIENCE BEHIND THE STORY

## Discovering Impacts of the Gulf Oil Spill

President Barack Obama echoed the perceptions of many Americans when he called the *Deepwater Horizon* oil spill "the worst environmental disaster America has ever faced." But what has scientific research told us about the actual impacts of the Gulf oil spill?

**A scientist rescues an oiled Kemp's ridley sea turtle.**

We don't yet have all the answers, because the deep waters affected by the spill have been difficult for scientists to study. A great deal will remain unknown. Yet the intense and focused scientific response to the spill demonstrates the dynamic way in which science can assist society.

As the spill took place, government agencies called on scientists to help determine how much oil was leaking. Researchers eventually determined the maximum rate at 62,000 barrels per day. Using underwater imaging, aerial surveys, and shipboard water samples, researchers tracked the movement of oil up through the water column and across the Gulf. These data helped predict when and where oil might reach shore, thereby helping to direct prevention and cleanup efforts. Meanwhile, as engineers struggled to seal off the well using remotely operated submersibles, researchers helped government agencies assess the fate of the oil (**FIGURE 1**).

University of Georgia biochemist Mandy Joye, who had studied natural seeps in the Gulf for years, documented that the leaking wellhead was creating a plume of oil the size of Manhattan. She also found evidence of low oxygen concentrations, or hypoxia (pp. 22, 26, 267), because some bacteria consume oil and gas, depleting oxygen from the water and making it uninhabitable for fish and other creatures.

Joye and other researchers feared that the thinly dispersed oil might devastate plankton (the base of the marine food chain) and the tiny larvae of shrimp, fish, and oysters (the pillars of the fishing industry). Scientists taking water samples documented sharp drops in plankton during the spill, but it will take years to learn whether the impact on larvae will diminish populations of adult fish and shellfish. Studies on the condition of living fish in the region show gill damage, tail rot, lesions, and reproductive problems at much higher levels than is typical.

What was happening to life on the seafloor was a mystery, because only a handful of submersible vehicles in the world are able to travel to the crushing pressures of the deep sea. Luckily, a team of researchers led by Charles Fisher of Penn State University was scheduled to embark on a regular survey of deepwater coral across the Gulf of Mexico in late 2010. Using the three-person submersible *Alvin* and the robotic vehicles *Jason* and *Sentry*, the team found healthy coral communities at sites far away from the Macondo well but found dying corals and brittlestars covered in a brown material at a site 11 km from the Macondo well.

Eager to determine whether this community was contaminated by the BP oil spill, the research team added chemist Helen White of Haverford College and returned a month later,

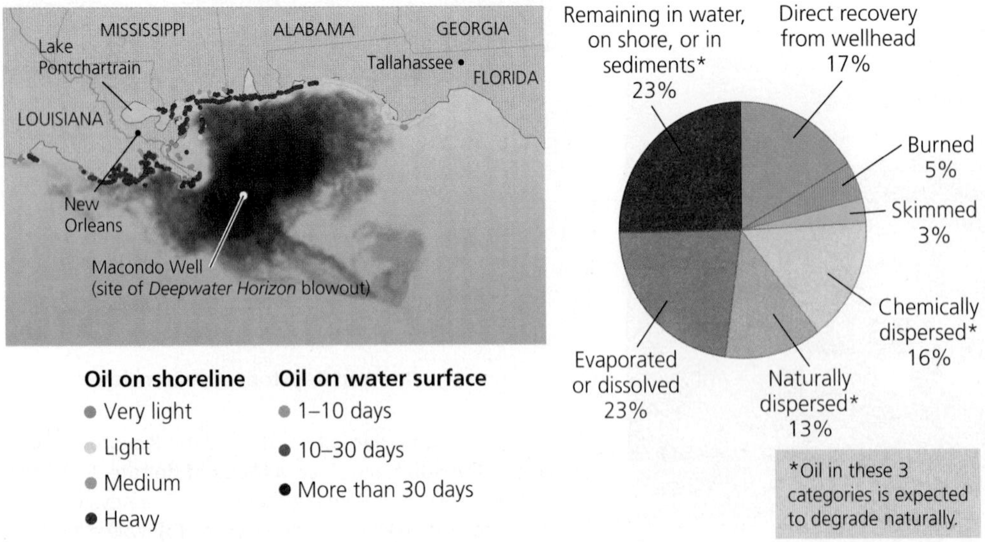


**(a) Extent of the spill** **(b) Fate of the oil**

**FIGURE 1 Scientists helped track oil from the *Deepwater Horizon* spill.** The map **(a)** shows areas polluted by oil. The pie chart **(b)** gives a breakdown of the oil's fate.
*Source (a): National Geographic and NOAA; (b) NOAA.*

thanks to a National Science Foundation program that funds rapid response research. On this trip, chemical analysis of the brown material showed it to match oil from the BP spill, rather than from any other known source.

Other questions revolve around the chemical dispersant that BP used to break up the oil, called Corexit 9500. Work by biologist Philippe Bodin following the *Amoco Cadiz* oil spill in France in 1978 had found that Corexit 9500 appeared more toxic to marine life than the oil itself. BP threw an unprecedented amount of this chemical at the *Deepwater Horizon* spill, injecting a great deal directly into the path of the oil at the wellhead. This caused the oil to dissociate into trillions of tiny droplets that dispersed across large regions. Many scientists worried that this caused the oil to affect far more plankton, larvae, and fish.

Impacts of the oil on birds, sea turtles, and marine mammals were easier to assess. Officially confirmed deaths numbered 6104 birds, 605 turtles, and 97 mammals—and hundreds more animals were cleaned and saved by wildlife rescue teams—but a much larger, unknown, number succumbed to the oil. What impact this mortality may have on populations in coming years is unclear. (After the *Exxon Valdez* spill in Alaska in 1989, populations of some species rebounded, but others have never come back.) Researchers are following the movements of marine animals in the Gulf with radio transmitters to try to learn what effects the oil may have had.

As images of oil-coated marshes saturated the media, researchers worried that death of marsh grass would leave the shoreline vulnerable to severe erosion by waves. Louisiana has already lost many coastal wetlands to subsidence, dredging, sea level rise, and silt capture by dams on the Mississippi River (p. 248). Fortunately, researchers found that oil did not penetrate to the roots of most plants and that oiled grasses were sending up new growth. Indeed, Louisiana State University researcher Eugene Turner said that loss of marshland from the oil "pales in comparison" with marshland lost each year due to other factors.

The ecological damage caused by the spill had measurable consequences for people. The region's mighty fisheries were shut down, forcing thousands of fishermen out of work. The government tested fish and shellfish for contamination and reopened fishing once they were found to be safe, but consumers balked at buying Gulf seafood. Beach tourism remained low all summer as visitors avoided the region. Together, losses in fishing and tourism totaled billions of dollars.

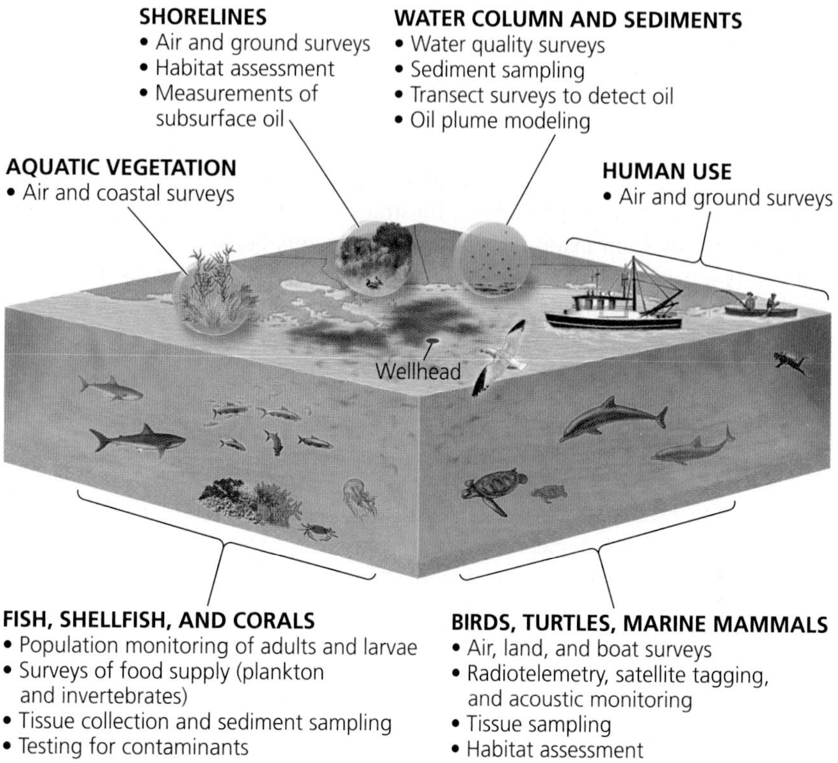


**FIGURE 2 Thousands of researchers continue to help assess damage to natural resources from the *Deepwater Horizon* oil spill.** They are surveying habitats, collecting samples and testing them in the lab, tracking wildlife, monitoring populations, and more.

Scientists expect some consequences of the Gulf spill to be long-lasting. Oil from the similar *Ixtoc* blowout off Mexico's coast in 1979 continues to lie in sediments near dead coral reefs, and fishermen there say it took 15–20 years for catches to return to normal. After the *Amoco Cadiz* spill, it took seven years for oysters and other marine species to recover. In Alaska, oil from the 1989 *Exxon Valdez* spill remains embedded in beach sand today.

However, many researchers are hopeful about the Gulf of Mexico's recovery from the *Deepwater Horizon* spill. The Gulf's warm waters and sunny climate speed the natural breakdown of oil. In hot sunlight, volatile components of oil evaporate from the surface and degrade in the water, so that fewer toxic compounds such as benzene, naphthalene, and toluene affect marine life. In addition, bacteria that consume hydrocarbons thrive in the Gulf because some oil has always seeped naturally from the seafloor and because leakage from platforms, tankers, and pipelines is common. These microbes give the region a natural self-cleaning capacity.

Researchers continue to conduct a wide range of scientific studies (**FIGURE 2**). A consortium of federal and state agencies is coordinating research and restoration efforts in the largest ever Natural Resource Damage Assessment, a process mandated under the Oil Pollution Act of 1990. Answers to questions will come in gradually as long-term impacts become clear. ☐

The northern leg of the Keystone XL extension required a permit from the U.S. State Department. Starting in 2008, this spurred an escalating political drama that came to include lawsuits, conflict-of-interest charges, high-stakes quarrels between President Obama and Congress, and street protests at the White House (**FIGURE 15.16c**).

Meanwhile, oil from Canada's oil sands (which is more corrosive than conventional oil) leaked out of other pipelines, creating spills along the Kalamazoo River in Michigan, in a residential neighborhood of Mayflower, Arkansas, and elsewhere. At the same time, a series of explosive derailments of trains carrying North Dakota crude oil illustrated the dangers of carrying oil by train instead of by pipeline. As this book went to press, a decision to approve or deny the Keystone XL pipeline had not yet been made. We will leave it to you and your instructor to flesh out the rest of this story!

## Emissions pollute air and drive climate change

As our society extends its reach for fossil fuel energy in so many ways, scientists, engineers, and policymakers are seeking solutions for air pollution and climate change. When we burn fossil fuels, we alter fluxes in Earth's carbon cycle (pp. 39–40). We essentially remove carbon from a long-term reservoir underground and release it into the air (**FIGURE 15.17**). This occurs as carbon from the hydrocarbon molecules of fossil fuels unites with oxygen from the atmosphere during combustion, producing carbon dioxide ($CO_2$). Carbon dioxide is a greenhouse gas (p. 305), and $CO_2$ released from fossil fuel combustion warms our planet (Chapter 14). Because climate change is beginning to have diverse and severe impacts, carbon dioxide pollution may become the biggest impact of fossil fuel use. Methane is also a potent greenhouse gas that drives climate warming.

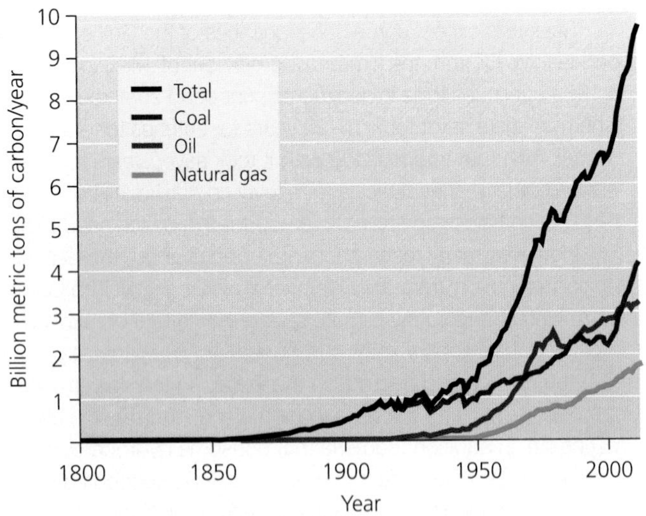


**FIGURE 15.17 Emissions from fossil fuel combustion have risen dramatically as nations have industrialized and as population and consumption have grown.** Here, global emissions of carbon from carbon dioxide are subdivided by source (oil, coal, or natural gas). *Data from Carbon Dioxide Information Analysis Center, Oak Ridge National Laboratory, U.S. Department of Energy, Oak Ridge, TN.*

DATA Q By what percentage have carbon emissions increased since the year your mother or father was born?

GO TO **INTERPRETING GRAPHS & DATA** ON MasteringEnvironmentalScience®

Fossil fuel emissions affect our health as well. Combusting coal high in mercury content emits mercury that bioaccumulates in organisms' tissues, poisoning animals as it moves up food chains (pp. 215–216) and posing health risks to people. Gasoline combustion in automobiles releases cancer-causing pollutants such as benzene and toluene. Workers at drilling operations, refineries, and in other jobs that entail frequent exposure to oil pollutants such as hydrogen sulfide, lead, and arsenic can develop serious health problems, including cancer.

The combustion of oil in vehicles and coal in power plants releases sulfur dioxide and nitrogen oxides, which contribute to smog (pp. 290–291) and acid deposition (pp. 295–297). Air pollution from fossil fuel combustion is intensifying in developing nations that are industrializing rapidly—but it has been reduced in developed nations as a result of laws and regulations to protect public health (Chapter 13). In these nations, public policy has encouraged industry to develop and install technologies that reduce pollution, such as catalytic converters that cleanse vehicle exhaust (see Figure 13.9, p. 286).

## Clean coal technologies aim to reduce air pollution from coal

Burning coal emits a variety of pollutants, including sulfur, mercury, arsenic, and other trace metals. The particular mix depends on chemical impurities in the coal, and coal deposits vary in the impurities they contain. **Clean coal technologies** refer to techniques, equipment, and approaches that aim to remove chemical contaminants during the generation of electricity from coal at power plants. Among these technologies are scrubbers, devices that chemically convert or physically remove pollutants (see Figure 13.8, p. 285). Another approach is to dry coal that has high water content, making it burn cleaner. We can also gain more power from coal with less pollution through a process called *gasification*, in which coal is converted into a cleaner synthesis gas, or *syngas*, by reacting it with oxygen and steam at a high temperature. Syngas from coal can be used to turn a gas turbine or to heat water to turn a steam turbine.

The U.S. government and the coal industry have each invested billions of dollars in clean coal technologies for new power plants. These efforts have helped to reduce air pollution from sulfates, nitrogen oxides, mercury, and particulate matter (pp. 285–287). If the many older plants that still pollute our air were retrofitted with these technologies, pollution could be reduced even more. However, the coal industry spends a great deal of money fighting regulations and mandates on its practices. As a result, many new power plants are built with little in the way of pollution control technologies. Moreover, these technologies will never result in energy production that is completely clean. Many energy analysts argue that coal is an inherently dirty way of generating power and should be replaced outright with cleaner energy sources.

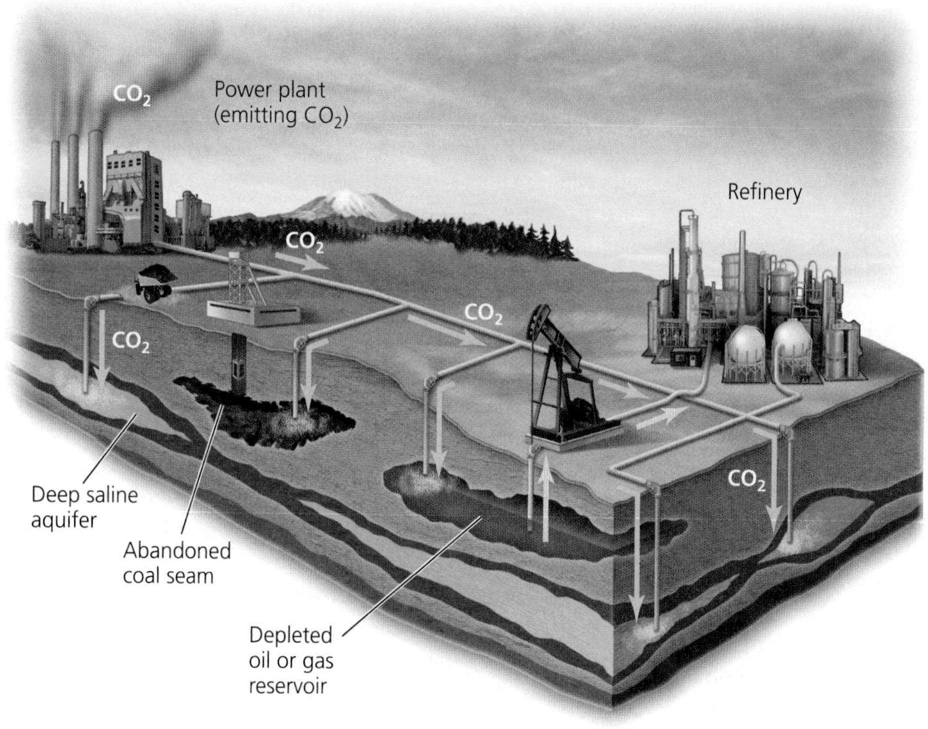


**FIGURE 15.18 Carbon capture and storage schemes propose to inject liquefied carbon dioxide emissions underground.** The $CO_2$ may be injected into depleted fossil fuel deposits, deep saline aquifers, or oil or gas deposits undergoing secondary extraction.

## Can we capture and store carbon?

Even if clean coal technologies were able to remove every last contaminant from power plant emissions, coal combustion would still pump huge amounts of carbon dioxide into the air, intensifying the greenhouse effect and climate change. This is why many current efforts focus on **carbon capture** followed by **carbon storage**, or **carbon sequestration** (p. 323). This approach involves capturing $CO_2$ emissions, converting the gas to a liquid, and then sequestering (storing) it in the ocean or underground in a geologically stable rock formation (**FIGURE 15.18**).

Carbon capture and storage (abbreviated as CCS) is being attempted at a variety of facilities. The world's first coal-fired power plant to approach zero emissions opened in 2008 in Germany. This plant removes its sulfate pollutants and captures its carbon dioxide, then compresses the $CO_2$ into liquid form, trucks it away, and injects it 900 m (3000 ft) underground into a depleted natural gas field. In North Dakota, the Great Plains Synfuels Plant gasifies its coal and sends half the $CO_2$ through a pipeline to Canada, where an oil company injects it into an oilfield to help it pump out the remaining oil.

Currently the U.S. Department of Energy is teaming up with five energy companies to build a prototype of a near-zero-emissions coal-fired power plant. The $1.65-billion *FutureGen 2.0* project aims to design, construct, and operate a power plant that burns coal, produces electricity, captures 90% of its carbon dioxide emissions, and sequesters the $CO_2$ underground. The project, located in Meredosia, Illinois, plans to pump $CO_2$ more than 1200 m (three-quarters of a mile) underground beneath layers of impermeable rock. If this showcase project succeeds, it could be a model for a new generation of power plants.

At present carbon capture and storage is too unproven to be the central focus of a clean energy strategy. We do not know how to ensure that carbon dioxide will stay underground once injected there. Injection might in some cases contaminate groundwater supplies or trigger earthquakes. Injecting carbon dioxide into the ocean would further acidify its waters (pp. 258, 318). Moreover, CCS is energy-intensive and decreases the EROI of coal, adding to its cost and the amount we consume. Finally, many renewable energy advocates fear that CCS takes the burden off emitters and prolongs our dependence on fossil fuels rather than facilitating a shift to renewables.

### WEIGHING THE ISSUES

**Clean Coal and Carbon Capture**

Do you think we should be spending billions of dollars to try to find ways to burn coal cleanly and to sequester carbon emissions from fossil fuels? Or is our money better spent on developing new clean and renewable energy sources that don't yet have enough infrastructure to produce power at the scale that coal can? What pros and cons do you see in each approach?

## We all pay external costs

The costs of addressing the many health and environmental impacts of fossil fuel extraction and use are generally not internalized in the market prices of fossil fuels. Instead, we all pay these external costs (pp. 94–95, 102) through medical expenses, costs of environmental cleanup, and impacts on our quality of life. The prices we pay at the gas pump or on our monthly utility bill have been kept inexpensive as a result of government subsidies to extraction companies. The profitable fossil fuel industries receive far more financial support from taxpayers than do the young and struggling renewable energy sources (see Figure 5.14, p. 110, and Figure 16.5, p. 370). In this way, we all pay extra for fossil fuel energy through our taxes, generally without realizing it.

## Fossil fuel extraction has mixed consequences for local people

Wherever fossil fuels are extracted, people living nearby must weigh the environmental, health, and social drawbacks of extractive development against the financial benefits they may gain. Communities where fossil fuel extraction takes place generally experience a flush of high-paying jobs and economic activity, and for many people these benefits far outweigh other concerns. Perceptions may change with time, however. Economic booms often prove temporary, whereas residents may be left with a polluted environment for generations to come.

Today debate is occurring in Pennsylvania, New York, and other states above the Marcellus Shale. Many working-class residents of small New York towns see jobs and economic activity across the border in Pennsylvania and wish their state were encouraging fracking for shale gas as well. Other New Yorkers fear drinking water contamination and feel the short-term economic benefits are not worth the long-term health and environmental impacts (**FIGURE 15.19**).

People are wrestling with similar dilemmas in North Dakota and parts of the West in response to new oil and gas drilling. In Appalachia, the debate has gone on for years over mountaintop removal mining. And along the Gulf of Mexico, the oil and gas industry employs 100,000 people and helps fund local economies—yet far more people are employed in tourism, fishing, and service industries, all of which were hurt by the *Deepwater Horizon* spill.

The newest flashpoint for debate is along the proposed route of the Keystone XL pipeline. TransCanada negotiated with thousands of landowners, offering money for the right to install the pipeline across their land. Those who declined TransCanada's offer found their land rights taken away by **eminent domain**—the policy by which courts set aside private property rights to make way for projects judged to be for the public good. Under U.S. law, private companies (even foreign ones) can usurp land rights of American citizens. The landowner is paid an amount determined by a court to be fair and cannot appeal the decision.

In Alaska, the oil industry gains support for drilling by paying the Alaskan government a portion of its revenues. Since the 1970s, the state of Alaska has received over $65 billion in oil revenues. One-quarter of these revenues are placed in the Permanent Fund, which pays yearly dividends to all citizens. Since 1982, each Alaska resident has received annual payouts ranging from $331 to $2069.

In most parts of the world where fossil fuels are extracted, however, local residents suffer pollution without compensation. When multinational corporations pay governments of developing nations for access to oil or gas, the money generally does not trickle down to the people who live where the extraction takes place. Moreover, oil-rich developing nations such as Ecuador, Venezuela, and Nigeria tend to have few environmental regulations, and governments may not enforce regulations if doing so would jeopardize the large sums of money associated with oil development.

In Ecuador, local people brought suit against Chevron for environmental and health impacts from years of oil extraction in the nation's rainforests. An Ecuadorian court in 2011 found the oil company guilty and ordered it to pay $9.5 billion for cleanup—the largest-ever such judgment. Chevron refused, and the court battle proceeded to the United States, where a judge threw out the ruling. The ongoing legal battle has now moved to other nations.

In Nigeria, the Shell Oil Company extracted $30 billion of oil from land of the native Ogoni people. Oil spills, noise, and gas flares caused chronic illness among them, but oil profits went to Shell and to the military dictatorships of Nigeria, while the Ogoni remained in poverty with no running water or electricity. Ogoni activist and leader Ken Saro-Wiwa worked for fair compensation to the Ogoni. After 30 years of persecution by the Nigerian government, he was arrested in 1994, given a trial universally regarded as a sham, and put to death by military tribunal.

**FIGURE 15.19 Pollution from shale gas drilling creates external costs.** This Pennsylvania homeowner can set fire to her tap water because it is contaminated with methane.

## Dependence on foreign energy affects the economies of nations

Putting all your eggs in one basket is always a risky strategy. Because virtually all of our modern technologies and services depend somehow on fossil fuels, we are vulnerable to supplies becoming costly or unavailable. Nations with few fossil fuel reserves of their own are especially vulnerable (**FIGURE 15.20**). In the wake of its 1970 oil production peak, the United States began relying more on foreign supplies.

Such reliance means that seller nations can control energy prices, forcing buyer nations to pay more as supplies dwindle. This became clear in 1973, when the *Organization of Petroleum Exporting Countries (OPEC)* resolved to stop selling oil to the United States. The predominantly Arab nations of OPEC opposed U.S. support for Israel in the Arab–Israeli Yom Kippur War and sought to raise prices by restricting supply.

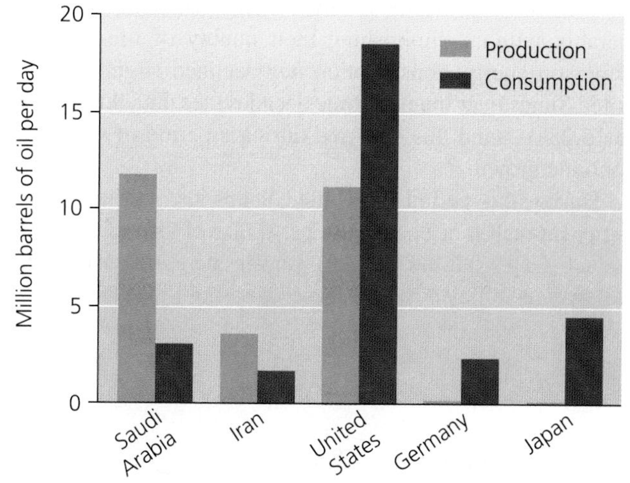


**FIGURE 15.20 Japan, Germany, and the United States are among nations that consume more oil than they produce.** Iran and Saudi Arabia produce more oil than they consume and are able to export oil to high-consumption countries. *Data from U.S. Energy Information Administration.*

DATA Q For every barrel of oil produced in the United States, how many barrels are consumed in the United States?

GO TO **INTERPRETING GRAPHS & DATA** ON MasteringEnvironmentalScience®

The embargo created panic in the West and caused oil prices to skyrocket (**FIGURE 15.21**), spurring inflation. Fear of oil shortages drove Americans to wait in long lines at gas pumps. A similar supply shock occurred in 1979 in response to the Iranian revolution.

With the majority of world oil reserves located in the politically volatile Middle East, crises in this region are a constant concern for policymakers. The Middle East has a long history of events that have affected oil supplies and prices, from the democratic street uprisings of the "Arab Spring" that began in 2011, stretching back through the two U.S.-led wars in Iraq to the Iran–Iraq war of the 1980s to the 1973 OPEC embargo. For many U.S. leaders, this troubled history enhances the allure of using hydraulic fracturing to boost domestic oil and gas extraction. By producing more of its own energy, the United States becomes less prone to get caught in foreign entanglements. Indeed, the United States now imports just one-third of its oil and has diversified its sources of imported petroleum considerably (**FIGURE 15.22**).

The U.S. shale gas boom could have further geopolitical consequences. Many European nations rely heavily on Russia for natural gas. In the wake of Russia's 2014 annexation of Crimea, some European nations that opposed this incursion into Ukraine nonetheless hesitated to respond. Increased U.S. gas exports to Europe—and the prospect of European nations fracking for shale gas themselves—could reduce Europe's reliance on Russian gas.

While diversifying sources of foreign energy in response to the 1973 OPEC embargo, U.S. leaders also enacted conservation measures, funded research on renewable energy sources, and established an emergency stockpile (which today stores one month of oil) deep underground in salt caverns in Louisiana, called the Strategic Petroleum Reserve. They also called for secondary extraction and the development of more domestic sources.

Since then, the desire to reduce reliance on foreign oil by boosting domestic production has driven the expansion of offshore drilling into deeper water. It has repeatedly driven a proposal to open the Arctic National Wildlife Refuge on Alaska's North Slope to oil extraction, despite arguments that drilling there would spoil America's last true wilderness while adding little to the nation's oil supply. Today it is driving the push to drill for oil in Arctic waters, despite the

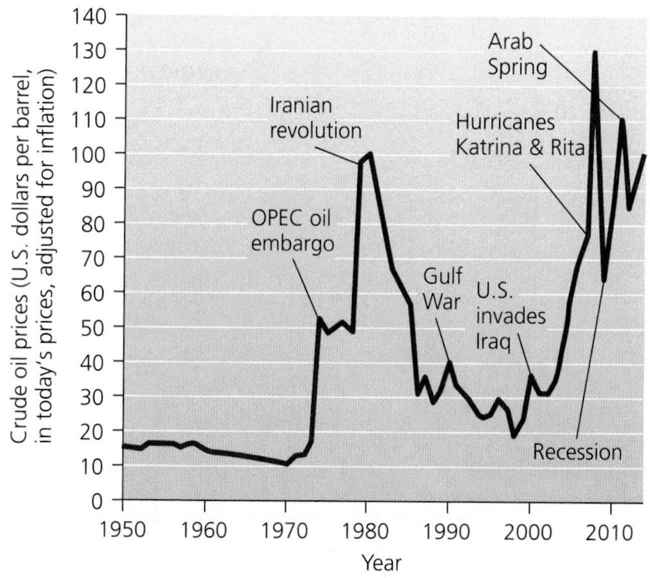


**FIGURE 15.21 World oil prices have gyrated over the decades.** Often this has resulted from political and economic events in oil-producing countries, particularly in the Middle East. *Data from U.S. Energy Information Administration and BP p.l.c., 2014,* Statistical review of world energy 2014.

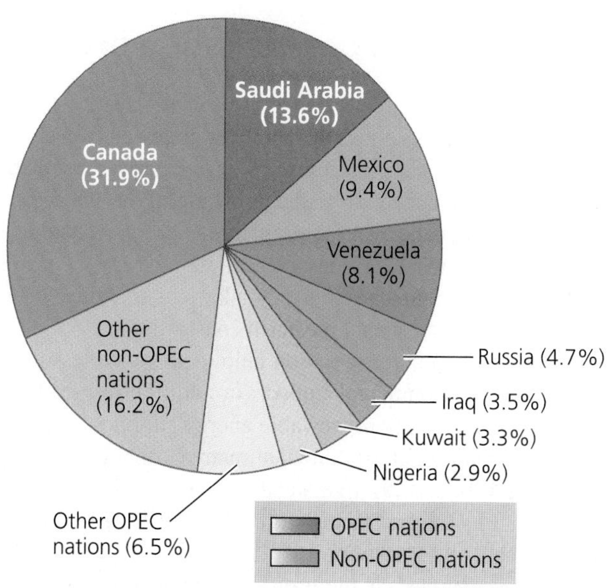


**FIGURE 15.22 The United States now receives most of its imported oil from non-OPEC nations and from non-Middle-Eastern nations.** *Data from U.S. Energy Information Administration, 2014.*

risks. As domestic oil and gas production rises with enhanced drilling, the United States becomes freer to make geopolitical decisions without being hamstrung by dependence on foreign energy. At the same time, however, accelerating climate change is posing new national security concerns. As we reach further for fossil fuels, our society will continue to debate the complex mix of social, political, economic, and environmental costs and benefits.

### WEIGHING THE ISSUES

**Drill, Baby, Drill?**

Do you think the United States should encourage hydraulic fracturing by subsidizing it and loosening regulations? Do you think the United States should open more of its offshore waters to oil and gas extraction? In each case, what benefits and costs do you foresee? Would the benefits likely exceed the costs? How strongly should government regulate oil and gas extraction once drilling begins? Give reasons for your answers.

# Energy Efficiency and Conservation

Fossil fuels are limited in supply, and their use has health, environmental, political, and socioeconomic consequences. For these reasons, many people have concluded that fossil fuels are not a sustainable long-term solution to our energy needs. They see a need to shift to clean and renewable sources of energy that exert less impact on climate and human health. As our society transitions to renewable energy, it will benefit us to extend the availability of fossil fuels. We can do this by conserving energy through lifestyle changes that reduce energy use and through technological advances that improve efficiency.

## Efficiency and conservation bring benefits

**Energy efficiency** describes the ability to obtain a given amount of output while using less energy input. **Energy conservation** describes the practice of reducing wasteful or unnecessary energy use. In general, efficiency results from technological improvements, whereas conservation stems from behavioral choices. Because greater efficiency allows us to reduce energy use, efficiency is a primary means of conservation.

Efficiency and conservation help us to waste less and to reduce our environmental impact. In addition, by extending the lifetimes of our nonrenewable energy supplies, efficiency and conservation help to alleviate many of the difficult individual choices and divisive societal debates related to fossil fuels, from hydraulic fracturing to Arctic drilling to oil sands development and transport.

Americans use far more energy per person than people in most other nations (**FIGURE 15.23a**). Citizens of many European nations enjoy standards of living similar to those of U.S. citizens yet use less energy per capita. This indicates that Americans could reduce their energy consumption considerably without diminishing their quality of life. Indeed, per-person energy consumption *has* declined slightly in the United States over the past three decades (see data line in Figure 15.23a)—and this occurred during a period of sustained economic growth.

During this period the United States also cut in half its **energy intensity**, or energy use per dollar of Gross Domestic Product (GDP) (**FIGURE 15.23b**). Lower energy intensity indicates greater efficiency, and these data show that the United

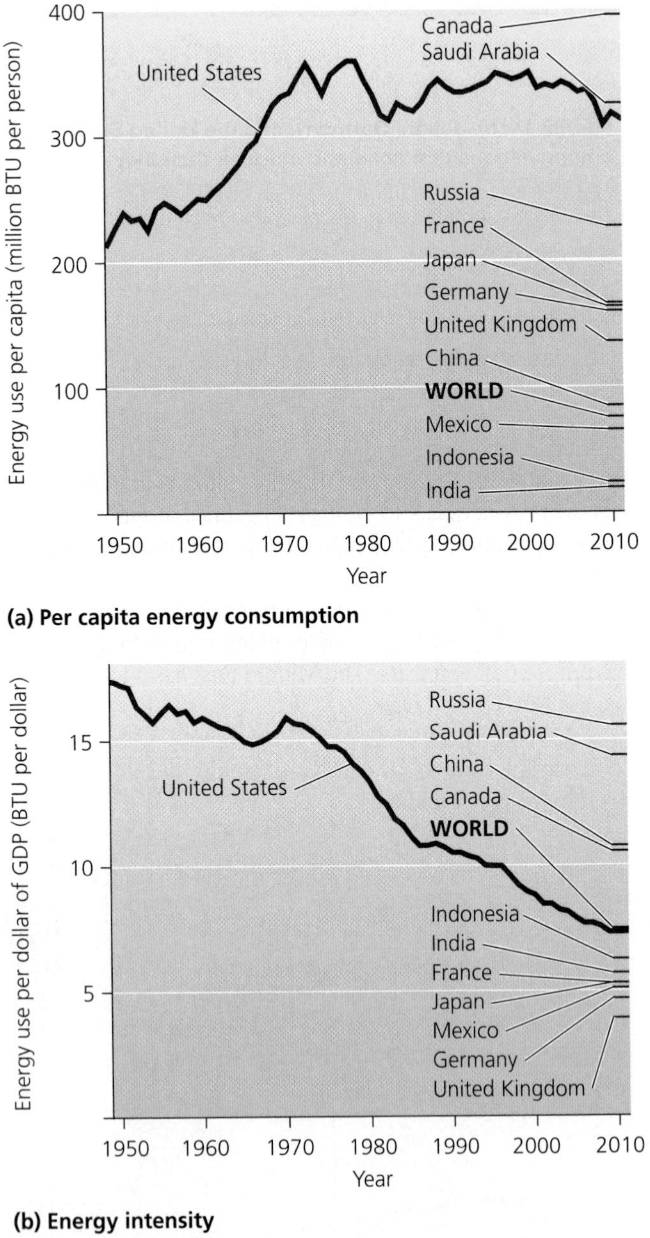


**FIGURE 15.23 The United States trails other developed nations in energy efficiency but has made much progress.** U.S. per-person energy use **(a)** has fallen slightly since 1979 but remains greater than that of most other nations. U.S. energy intensity **(b)** has fallen steeply and now approaches that of other developed nations. Energy intensity is energy use per inflation-adjusted 2005 dollars of GDP, using purchasing power parities, which control for differences among nations in purchasing power. *Data from U.S. Energy Information Administration.*

States now gets twice as much economic bang for its energy buck. Thus, although the United States continues to burn through more energy per dollar of GDP than most other industrialized nations, Americans have achieved tremendous gains in efficiency already and should be able to make further progress.

## Personal choice and efficient technologies are two ways to conserve

As individuals, we can make conscious choices to reduce our energy consumption by driving less, dialing down thermostats, turning off lights when rooms are not in use, and cutting back on the use of energy-intensive machines and appliances. For any given individual or business, reducing energy consumption saves money while helping to conserve resources.

As a society, we can conserve energy by developing technologies and strategies to make devices and processes more efficient. Currently, more than two-thirds of the fossil fuel energy we use is simply lost, as waste heat, in automobiles and power plants (see Figure 15.4).

One way we can improve the efficiency of power plants is through **cogeneration**. In cogeneration, excess heat produced during electricity generation is captured and used to heat nearby workplaces and homes and to produce other kinds of power. Cogeneration can almost double the efficiency of a power plant. The same is true of coal gasification and combined cycle generation, in which coal is treated to create hot gases that turn a gas turbine, while exhaust from this turbine heats water to drive a steam turbine.

In homes, offices, and public buildings, a significant amount of heat is needlessly lost in winter and gained in summer because of poor design and inadequate insulation (**FIGURE 15.24**). Improvements in design can reduce the energy required to heat and cool buildings. Such improvements may involve passive solar design (pp. 370–371), better insulation, a building's location, the vegetation around it, and the color of its roof (lighter colors keep buildings cooler by reflecting the sun's rays).

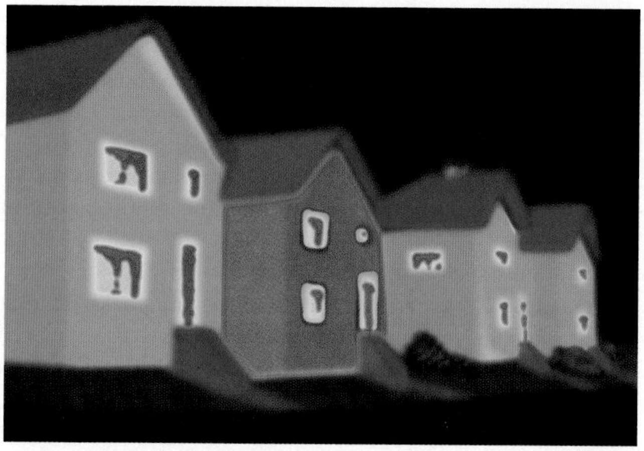

**FIGURE 15.24 A thermogram reveals heat loss from buildings by recording energy in the infrared portion of the electromagnetic spectrum (p. 31).** In this image, one house is uninsulated; its red color signifies warm temperatures where heat is escaping. Green shades signify cool temperatures, where heat is being conserved. Also note that in all houses, more heat is escaping from windows than from walls.

Many consumer products, from lightbulbs to appliances, have been reengineered through the years to enhance efficiency. Energy-efficient lighting, for example, can reduce energy use by 80%. Compact fluorescent bulbs are more efficient than incandescent light bulbs, which is why the U.S. and many other nations are phasing out incandescent bulbs.

Federal standards for energy-efficient appliances have already reduced per-person home electricity use below what it was in the 1970s. The U.S. EPA's Energy Star program labels refrigerators, dishwashers, and other appliances for their efficiency, enabling consumers to take energy use into account when shopping for them. Studies show that savings on utility bills readily offset the higher prices of energy-efficient appliances. The EPA estimates that if all U.S. households purchased energy-efficient appliances, the nation's annual energy expenditure would fall by $200 billion.

Automotive technology represents perhaps our best opportunity to conserve large amounts of fossil fuels fairly easily. We can accomplish this with vehicles such as electric cars, electric/gasoline hybrids, plug-in hybrids, or vehicles that use hydrogen fuel cells (p. 385). Among electric/gasoline hybrids, current U.S. models obtain fuel-economy ratings of up to 50 miles per gallon (mpg)—twice as much as the average American car. Aside from such alternative vehicles, we already possess the means to enhance fuel efficiency for gasoline-powered vehicles by using lightweight materials, continuously variable transmissions, and more efficient engines.

## Automobile fuel efficiency is a key to conservation

One of the ways in which U.S. leaders responded to the OPEC embargo of 1973 was to mandate an increase in the mile-per-gallon (mpg) fuel efficiency of automobiles and a reduction in the national speed limit to 55 miles per hour. Over the next three decades, however, as market prices for oil fell, many of the conservation initiatives of this time were abandoned. Without high market prices and an immediate threat of shortages, people lacked economic motivation to conserve. Government funding for research into alternative energy sources dwindled, speed limits rose, and U.S. policymakers repeatedly failed to raise the *corporate average fuel efficiency (CAFE) standards*, which set benchmarks for auto manufacturers to meet. The average fuel efficiency of new vehicles fell from 22.0 mpg in 1987 to 19.3 mpg in 2004, as sales of sport-utility vehicles increased relative to sales of cars.

In 2007, Congress mandated that automakers raise average fuel efficiency to 35 mpg by the year 2020. Since 2007, fuel economy has climbed substantially (**FIGURE 15.25**). Yet despite this advance, American automobiles continue to lag behind the vehicles of most other developed nations. When automakers requested a government bailout during the recent recession, President Obama forced them to agree to boost average fuel economies to 54.5 mpg by 2025. If this improvement comes to pass, it will enable a huge reduction in oil use. New technologies will add over $2000 to the average price of

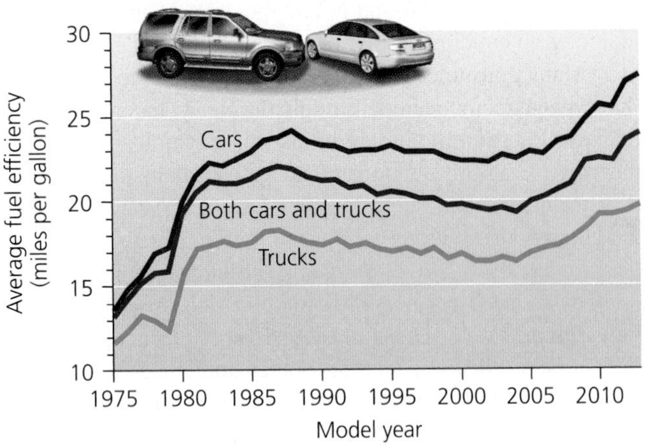


**FIGURE 15.25 Automotive fuel efficiencies have responded to public policy.** Fuel efficiency for vehicles in the United States rose dramatically in the late 1970s as a result of legislative mandates but then stagnated once no further laws were enacted to improve fuel economy. Recent legislation is now improving it again. *Data from U.S. Environmental Protection Agency, 2013.* Light-duty automotive technology, carbon dioxide emissions, and fuel economy trends: 1975 through 2013.

a car, but drivers will save perhaps $6000 in fuel costs over the car's lifetime.

U.S. policymakers could do more to encourage oil conservation. The United States has kept taxes on gasoline extremely low, relative to most other nations. Americans pay two to three times *less* per gallon of gas than drivers in many European countries. In fact, gasoline in the United States is sold more cheaply than bottled water! As a result, U.S. gasoline prices do not account for the substantial external costs (p. 94) that oil production and consumption impose on society. Some experts estimate that if all costs to society were taken into account, the price of gasoline would exceed $13/gallon. Instead, our artificially low gas prices diminish our economic incentive to conserve.

## WEIGHING THE ISSUES

**More Miles, Less Gas**

If you drive an automobile, what gas mileage does it get? How does it compare to the vehicle averages in Figure 15.25? If your vehicle's fuel efficiency were 10 mpg greater, and if you drove the same amount, how many gallons of gasoline would you no longer need to purchase each year? How much money would you save?

Do you think U.S. leaders should mandate further increases in the CAFE standards? Should the government raise taxes on gasoline sales as an incentive for consumers to conserve energy? What effects on economics, on health, and on environmental quality might each of these steps have?

### The rebound effect cuts into efficiency gains

Energy efficiency is a vital pursuit, but it may not always save as much energy as we expect. This is because gains in efficiency from better technology may be partly offset if people engage in more energy-consuming behavior as a result. For instance, a person who buys a fuel-efficient car may choose to drive more because he or she feels it's okay to do so now that less gas is being used per mile. This phenomenon is called the **rebound effect**, and studies indicate that it is widespread. In some instances, the rebound effect may completely erase efficiency gains, and attempts at energy efficiency may end up actually causing greater energy consumption!

Nonetheless, efficiency will play a necessary role in the conservation efforts we make toward reducing energy use. It is often said that reducing energy use is equivalent to finding a new oil reserve. Some estimates hold that energy conservation and efficiency in the United States could save 6 million barrels of oil a day—nearly the amount gained from all offshore drilling, and considerably more than would be gained from Canada's oil sands. In fact, conserving energy is *better* than finding a new reserve because it alleviates health and environmental impacts while at the same time extending our future access to fossil fuels. Yet regardless of how much we conserve, we will still need energy. Among the alternatives to fossil fuels for our energy economy is nuclear power.

## Nuclear Power

Nuclear power is free of the air pollution produced by fossil fuel combustion and thereby offers us a powerful means of combating climate change. Yet nuclear power's promise has been clouded by nuclear weaponry, the thorny dilemma of radioactive waste disposal, and the long shadow of accidents at Chernobyl and Fukushima. As a result, public safety concerns and the costs of addressing them have constrained nuclear power's spread.

### Fission releases nuclear energy in reactors to generate electricity

**Nuclear energy** is the energy that holds together protons and neutrons (p. 27) in the nucleus of an atom. We harness this energy by converting it to thermal energy inside **nuclear reactors**, facilities contained within nuclear power plants. This thermal energy is then used to generate electricity by turning turbines with steam (**FIGURE 15.26**).

The reaction that drives the release of nuclear energy inside nuclear reactors is **nuclear fission**, the splitting apart of atomic nuclei (**FIGURE 15.27**). In fission, the nuclei of large, heavy atoms, such as uranium or plutonium, are bombarded with neutrons. Ordinarily neutrons move too quickly to split nuclei when they collide with them, but if neutrons are slowed down, they can break apart nuclei. Each split nucleus emits energy in the form of heat, light, and radiation, and it also releases neutrons. These neutrons (two to three in the case of uranium-235) can bombard other nearby uranium-235 ($^{235}U$) atoms, resulting in a self-sustaining chain reaction.

If not controlled, this chain reaction becomes a runaway process of positive feedback (pp. 23–24)—the process that creates the explosive power of a nuclear bomb. Inside a power plant, however, fission is controlled so that, on average, only one of the two or three neutrons emitted with each fission

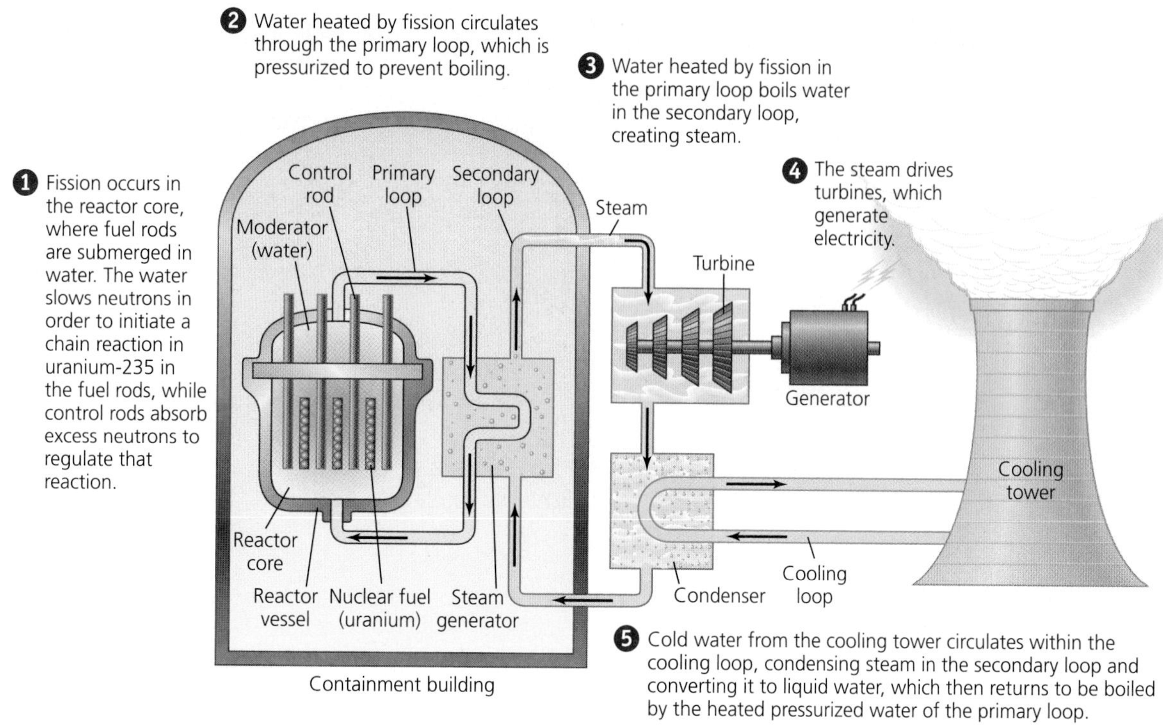

2 Water heated by fission circulates through the primary loop, which is pressurized to prevent boiling.
3 Water heated by fission in the primary loop boils water in the secondary loop, creating steam.
1 Fission occurs in the reactor core, where fuel rods are submerged in water. The water slows neutrons in order to initiate a chain reaction in uranium-235 in the fuel rods, while control rods absorb excess neutrons to regulate that reaction.
Control rod
Primary loop
Secondary loop
Steam
4 The steam drives turbines, which generate electricity.
Moderator (water)
Turbine
Generator
Cooling tower
Reactor core
Cooling loop
Reactor vessel
Nuclear fuel (uranium)
Steam generator
Condenser
5 Cold water from the cooling tower circulates within the cooling loop, condensing steam in the secondary loop and converting it to liquid water, which then returns to be boiled by the heated pressurized water of the primary loop.
Containment building


**FIGURE 15.26 Nuclear reactors produce electricity.** In a pressurized light water reactor (the most common type of nuclear reactor), uranium fuel rods are placed in water, which slows neutrons so that fission can occur ❶. Control rods are moved into and out of the reactor core, absorbing excess neutrons to regulate the chain reaction. Water heated by fission circulates through the primary loop ❷ and warms water in the secondary loop, which turns to steam ❸. Steam drives turbines, which generate electricity ❹. The steam is then cooled in the cooling tower by water from an adjacent river or lake and returns to the containment building ❺, to be heated again by heat from the primary loop.

event goes on to induce another fission event. In this way, the chain reaction maintains a constant output of energy.

For fission to begin, the neutrons bombarding uranium are slowed down with a substance called a *moderator*, most often water or graphite. To soak up the excess neutrons produced when uranium nuclei divide, *control rods* made of a metallic alloy that absorbs neutrons are placed among the water-bathed fuel rods of uranium. Engineers move the control rods in and out of the water to maintain fission at the desired rate. All this takes place within the reactor core and is the first step in the electricity-generating process of a nuclear power plant.

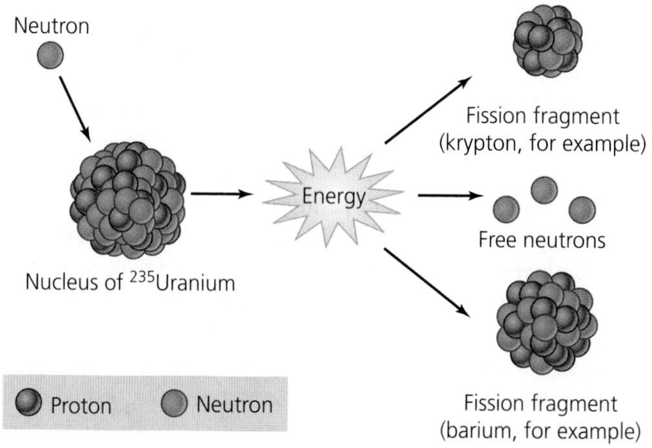


**FIGURE 15.27 Nuclear fission drives modern nuclear power.** In nuclear fission, the nucleus of an atom of uranium-235 is bombarded with a neutron. The collision splits the uranium atom into smaller atoms and releases two or three neutrons, along with heat, light, and radiation. The neutrons can continue to split uranium atoms and set in motion a runaway chain reaction, so engineers at nuclear plants must absorb excess neutrons with control rods to regulate the rate of the reaction.

## Nuclear energy comes from processed and enriched uranium

We use the element uranium for nuclear power because its atoms are radioactive, emitting subatomic particles and high-energy radiation as they decay into a series of daughter isotopes (p. 28). We obtain uranium by mining. Uranium-containing minerals are uncommon and in finite supply, so nuclear power is generally considered a nonrenewable energy source.

Over 99% of the uranium in nature occurs as the isotope uranium-238. Uranium-235 (with three fewer neutrons) makes up less than 1% of the total. Because $^{238}U$ does not emit enough neutrons to maintain a chain reaction, we use $^{235}U$ for commercial nuclear power. Therefore, mined uranium must be processed to enrich the concentration of $^{235}U$ to at least 3%. This enriched uranium is formed into pellets of uranium dioxide, which are used in fuel rods.

**TABLE 15.4 Top Producers of Nuclear Power**

| NATION | NUCLEAR POWER CAPACITY (GIGAWATTS) | NUMBER OF REACTORS | PERCENTAGE ELECTRICITY FROM NUCLEAR POWER |
|---|---|---|---|
| United States | 99.1 | 100 | 19.0 |
| France | 63.1 | 58 | 74.8 |
| Japan | 44.2 | 50 | 2.1 |
| Russia | 23.6 | 33 | 17.8 |
| South Korea | 20.7 | 23 | 30.4 |
| China | 14.9 | 19 | 2.0 |
| Canada | 13.5 | 19 | 15.3 |
| Ukraine | 13.1 | 15 | 46.2 |
| Germany | 12.1 | 9 | 16.1 |
| Sweden | 9.5 | 10 | 38.1 |
| United Kingdom | 9.2 | 16 | 18.1 |

*Data from International Atomic Energy Agency.*

After several years in a reactor, the decayed uranium fuel no longer generates adequate energy, so it must be replaced with new fuel. In some countries, the spent fuel is reprocessed to recover the remaining energy. However, this is costly relative to the market price of uranium, so most spent fuel is disposed of as radioactive waste.

The United States generates over a quarter of the world's nuclear power production yet receives only 19% of its electricity from nuclear power. A number of other nations rely more heavily on nuclear power (**TABLE 15.4**). Today 436 nuclear power plants operate in 31 nations.

## Nuclear power delivers clean energy

Using fission, nuclear power plants generate electricity without creating the air pollution that fossil fuels do. Of course, the construction of plants and equipment has a large carbon footprint, but the actual power-generating process is essentially emission-free. All told, scientists estimate that using nuclear power in place of fossil fuels helps the world avoid emissions of 2.5 billion metric tons of carbon dioxide per year, about 7% of global $CO_2$ emissions.

Nuclear power offers a mix of advantages and disadvantages compared with fossil fuels—coal in particular (**TABLE 15.5**). For residents living downwind from power plants, scientists calculate that nuclear power poses far fewer chronic health risks from pollutants (such as nitrogen oxides and sulfur dioxide). And because uranium generates far more power than coal by weight or volume, less of it needs to be mined, so uranium mining causes less damage to landscapes and generates less waste than coal mining. A drawback of nuclear power is that arranging for safe disposal of radioactive waste is challenging. Another is that if an accident occurs at a power plant, the consequences can potentially be catastrophic.

### WEIGHING THE ISSUES

**Choose Your Risk**

Examine Table 15.5. Given the choice of living next to a nuclear power plant or living next to a coal-fired power plant, which would you choose? What would concern you most about each option?

## Nuclear power poses small risks of large accidents

Although nuclear power delivers clean energy, the possibility of catastrophic accidents has spawned a great deal of public anxiety. Three events have been influential in shaping public opinion about nuclear power: Three Mile Island, Chernobyl, and Fukushima.

**Three Mile Island** In Pennsylvania in 1979, a combination of mechanical failure and human error at the **Three Mile Island** plant caused coolant water to drain from the reactor vessel, temperatures to rise inside the reactor core, and metal surrounding the fuel rods to melt, releasing radiation. This process is termed a **meltdown**, and at Three Mile Island it proceeded through half of one reactor core. Area residents stood ready to be evacuated as the nation held its breath, but fortunately most radiation remained inside the containment building.

Once this accident was brought under control, the damaged reactor was shut down, and multi-billion-dollar cleanup efforts stretched on for years. Three Mile Island is best regarded as a near miss; the emergency could have been far worse had the meltdown proceeded through the entire stock of uranium fuel or had the containment building not contained the radiation.

**TABLE 15.5 Risks and Impacts of Coal-Fired versus Nuclear Power Plants**

| TYPE OF IMPACT | COAL | NUCLEAR |
|---|---|---|
| **Land and ecosystem disturbance from mining** | Extensive, on surface or underground | Less extensive |
| **Greenhouse gas emissions** | Considerable emissions | None from plant operation; much less than coal over the entire life cycle |
| **Other air pollutants** | Sulfur dioxide, nitrogen oxides, particulate matter, and other pollutants | No pollutant emissions |
| **Radioactive emissions** | No appreciable emissions | Possibility of dangerous emissions if severe accident occurs |
| **Occupational health among workers** | More known health problems and fatalities | Fewer known health problems and fatalities |
| **Health impacts on nearby residents** | Air pollution impairs health | No appreciable known health impacts under normal operation |
| **Effects of accident or sabotage** | No widespread effects | Potentially catastrophic widespread effects |
| **Solid waste** | More generated | Less generated |
| **Radioactive waste** | None | Radioactive waste generated |
| **Fuel supplies remaining** | Should last several hundred more years | Uncertain; supplies could last longer or shorter than coal supplies |

*For each type of impact, the more severe impact is highlighted in yellow.*

**Chernobyl** In 1986 the **Chernobyl** plant in Ukraine (part of the Soviet Union at the time) suffered the most severe nuclear accident yet (**FIGURE 15.28**). Engineers had turned off safety systems to conduct tests, and human error, along with unsafe reactor design, led to explosions that destroyed the reactor and sent clouds of radioactive debris billowing into the atmosphere. Winds carried radioactive fallout across much of the Northern Hemisphere, particularly Ukraine, Belarus, and parts of Russia and Europe. For 10 days radiation escaped while emergency crews risked their lives putting out fires. The Soviet government evacuated more than 100,000 residents of the area.

Afterwards, workers erected a gigantic concrete sarcophagus around the demolished reactor, scrubbed buildings and roads, and removed irradiated materials. However, the landscape for at least 30 km (19 mi) around the plant remains contaminated, the demolished reactor is still full of dangerous fuel and debris, and radioactivity leaks from the hastily built, deteriorating sarcophagus. Today an international team is trying to build a larger sarcophagus around the original one.

The accident killed 31 people directly and sickened thousands more. Exact numbers are uncertain because of inadequate data and the difficulty of determining long-term radiation effects. Health authorities estimate that most of the 6000-plus cases of thyroid cancer diagnosed in people who were children at the time resulted from radioactive iodine. An international consensus effort 20 years after the event estimated that radiation raised cancer rates among exposed people by as much as a few percentage points, resulting in up to several thousand fatal cancer cases.

**FIGURE 15.28 The world's worst nuclear accident unfolded in 1986 at Chernobyl.** The destroyed reactor (shown here after the explosion) was later encased in a massive concrete sarcophagus to contain further radiation leakage.

(a) The tsunami barrels toward the Fukushima reactors

Cesium-137 radiation: Intensity of fallout relative to the plant site

| | | |
|---|---|---|
| ■ 0.9–1 | ■ 0.4–0.6 | ▨ 0.01–0.1 |
| ■ 0.7–0.9 | ■ 0.3–0.4 | ▨ 0.005–0.01 |
| ■ 0.6–0.7 | ■ 0.1–0.3 | ▨ 0.001–0.005 |

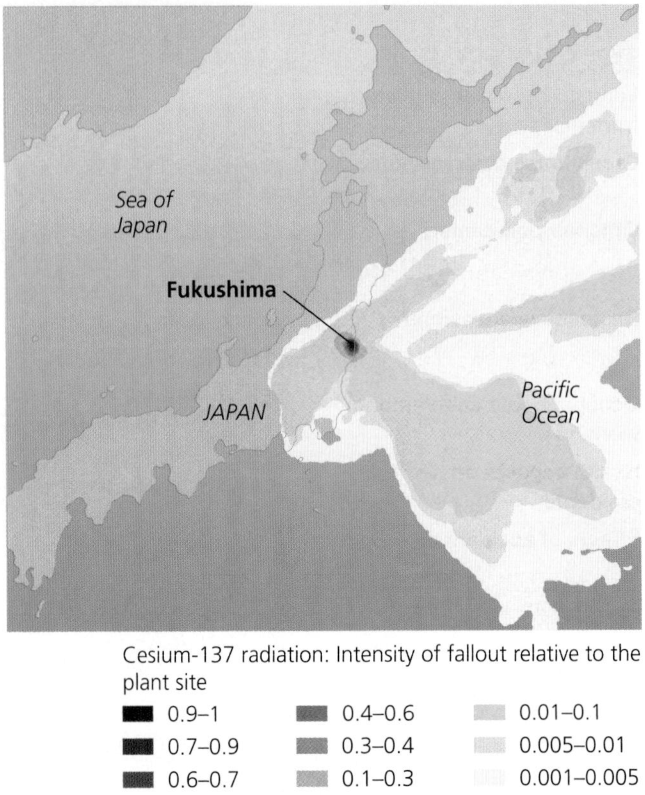

(c) Most radiation drifted eastward over the ocean

(b) A Japanese child is screened for radiation

**FIGURE 15.29 The Fukushima Daiichi crisis was unleashed after an earthquake generated a massive tsunami.** The tsunami tore through a seawall **(a)** and inundated the plant's nuclear reactors. Children evacuated from the region **(b)** were screened for radiation exposure. About 80% of the radiation that escaped from the plant drifted over the ocean, as shown in this map **(c)** of cesium-137 isotopes in the 9 days following the accident.

*Data in (c) from Yasunari, T. J., et al. 2011. Cesium-137 deposition and contamination of Japanese soils due to the Fukushima nuclear accident. Proc. Natl. Acad. Sci. 108: 19530–19534.*

**Fukushima Daiichi** On March 11, 2011, a magnitude 9.0 earthquake struck eastern Japan and sent an immense tsunami roaring onshore (pp. 232–233). Nearly 19,000 people were killed, and many thousands of buildings were destroyed. This natural disaster affected the operation of several of Japan's nuclear plants, most notably **Fukushima Daiichi**. Here, the earthquake shut down power, and the tsunami flooded the plant's emergency power generators (**FIGURE 15.29a**). Without electricity, workers could not use moderators and control rods, and the fuel began to overheat as fission proceeded, uncontrolled.

Amid the chaos across the region, help was slow to arrive, so workers flooded the reactors with seawater in a desperate effort to prevent meltdowns. Several explosions and fires occurred. Three reactors experienced full meltdowns, and three others were seriously damaged. Parts of the plant remained inaccessible for months because of radioactive water. It will likely require decades to fully clean up the site.

Radioactivity was released during and after these events at levels about one-tenth of those from Chernobyl. Thousands of area residents were evacuated and screened for radiation

(**FIGURE 15.29b**), and restrictions were placed on food and water from the region. Much of the radiation spread by air or water into the Pacific Ocean, and trace amounts were detected around the world (**FIGURE 15.29c**). In the years following the event, minor releases of radioactivity continued. Long-term health effects on the region's people remain uncertain.

In the aftermath of the disaster, the Japanese government idled all 50 of the nation's nuclear reactors and embarked on safety inspections. Efforts to restart them were met with public debate and street protests. Across the world, many nations reassessed their nuclear programs. Germany reacted most strongly, shutting down half of its nuclear power plants and deciding to phase out the rest by 2022.

The calamity at Fukushima could likely have been avoided had the emergency generators not been located in the basement where a tsunami could flood them. And the design of most modern reactors is safer than Chernobyl's. Yet natural disasters and human error will always pose risks—and as plants age, they require more maintenance and become less safe. Moreover, radioactive material could be stolen from plants and used in terrorist attacks. This possibility is especially worrisome in the cash-strapped nations of the former Soviet Union, where hundreds of former nuclear sites have gone without adequate security for years. In a cooperative international agreement, the U.S. government has been buying up some of this material and diverting it to peaceful use in power generation. As a result, in recent years up to 10% of America's electricity has been generated from fuel recycled from Russian warheads that used to be atop missiles pointed at American cities!

## Waste disposal remains a challenge

Even if nuclear power could be made completely safe, we would still be left with the conundrum of what to do with spent fuel rods and other radioactive waste. This waste will continue emitting radiation for as long as our civilization exists—uranium-235 has a half-life (p. 28; the time it takes for half the atoms to decay and give off radiation) of 700 million years. Currently, such waste is held in temporary storage at nuclear power plants. Spent fuel rods are sunken in pools of cooling water or encased in thick casks of steel, lead, and concrete to minimize radiation leakage.

In total, U.S. power plants are storing nearly 70,000 metric tons of high-level radioactive waste (such as spent fuel)—enough to fill a football field to the depth of 7 m (21 ft)—as well as much more low-level radioactive waste (such as contaminated clothing and equipment). This waste is held at more than 120 sites spread across 39 states (**FIGURE 15.30**). A 2005 National Academy of Sciences report judged that most of these sites were vulnerable to terrorist attacks. Over half of U.S. citizens live within 125 km (75 mi) of temporarily stored waste.

Because storing waste at many dispersed sites creates a large number of potential hazards, nuclear waste managers would prefer to send all waste to a single, secure, central repository. In the United States, government scientists selected Yucca Mountain, a remote site in the desert of southern Nevada, 160 km (100 mi) from Las Vegas (**FIGURE 15.31**). Choice of this site followed extensive study, and $13 billion was spent on its development. Most Nevadans opposed the choice, however, and concerns were raised about seismic activity. In 2010 President Barack Obama's administration ended support for the project, although lawsuits are challenging this decision.

Until the United States establishes a central repository for radioactive waste, this waste will remain spread among many locations. However, one concern with a centralized repository is that waste would need to be transported there from all 120-plus current storage areas and from current and future nuclear plants and military installations. Because this would involve many thousands of shipments by rail and truck across hundreds of public highways through almost every state of the union, some people worry that the risk of an accident is unacceptably high.

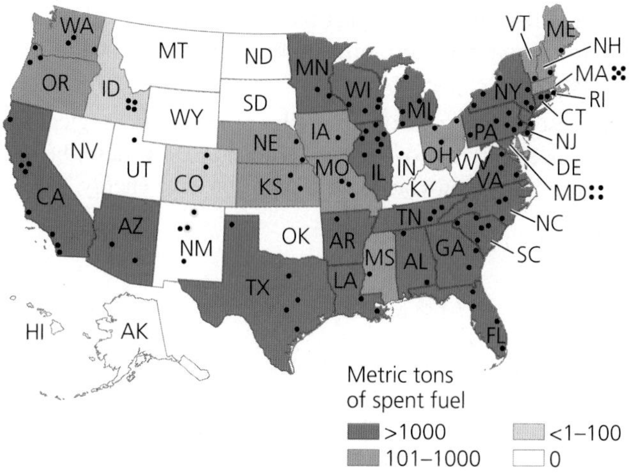


**FIGURE 15.30 High-level radioactive waste from civilian reactors is currently stored at over 120 sites in 39 states across the United States.** In this map, dots indicate storage sites, and colors indicate the amount of waste stored in each state. *Data from Nuclear Energy Institute.*

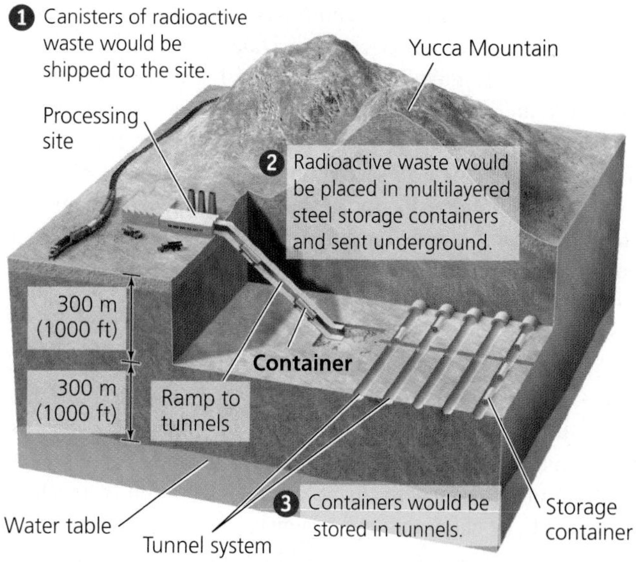


**FIGURE 15.31 Yucca Mountain has not been approved.** This site, in a remote part of Nevada, was being developed as the central repository for all commercial nuclear waste in the United States until support was withdrawn in 2010.

### WEIGHING THE ISSUES

**How to Store Waste?**

Which do you think is a better option—to transport nuclear waste cross-country to a single repository, or to store it permanently at numerous power plants and military bases scattered across the nation? Would your opinion be affected if you lived near the repository site? Near a power plant? On a highway route along which waste is transported?

## Nuclear power's growth has slowed

Dogged by concerns over waste disposal, safety, and cost overruns, nuclear power's growth has slowed. Public anxiety in the wake of Chernobyl made utilities less willing to invest in new plants, and reaction to Fukushima stalled a resurgence in the industry. Building, maintaining, operating, and ensuring the safety of nuclear facilities is enormously expensive, and almost every nuclear plant has overrun its budget. In addition, plants have aged more quickly than expected because of problems that were underestimated, such as corrosion in coolant pipes. The plants that have been shut down—well over 100 around the world to date—have served on average less than half their expected lifetimes. Moreover, shutting down, or decommissioning, a plant is sometimes more expensive than the original construction. As a result of these financial issues, electricity from nuclear power remains more expensive than electricity from fossil fuels, even after generous subsidies.

Nonetheless, nuclear power remains one of the few currently viable alternatives to fossil fuels with which we can generate large amounts of electricity in short order. This is why more and more environmental advocates propose expanding nuclear capacity using a new generation of reactors designed to be safer and less expensive. Yet with slow growth expected for nuclear power, fossil fuels in limited supply, and climate change worsening, where will we turn for clean and sustainable energy? Increasingly, people are turning to renewable energy sources (Chapter 16), those that cannot be depleted by our use.

# Conclusion

Over the past two centuries, fossil fuels have helped us build the complex industrialized societies we enjoy today. Yet our production of conventional fossil fuels is about to decline. We can respond to this challenge by expanding our search for new sources of fossil fuels—and paying ever-higher economic, health, and environmental costs. Or, we can encourage conservation and efficiency while developing alternative energy sources that are clean and renewable. The path we choose will have far-reaching consequences for Earth's climate and for the stability and progress of our civilization.

Nuclear power provides a climate-friendly alternative to fossil fuels, but high costs and public fears over safety in the wake of accidents have stalled its growth. Today's debate over hydraulic fracturing for shale gas is a microcosm of the debate over our energy future. A key question is whether natural gas will be a bridge fuel to renewables or a heavy anchor that keeps us in the fossil fuel age. As renewable energy sources become increasingly economical, it becomes easier to envision freeing ourselves from a reliance on fossil fuels and charting a bright future for humanity and the planet with renewable energy.

# Testing Your Comprehension

1. Why are fossil fuels our most prevalent source of energy today? How are fossil fuels formed? Why are they considered nonrenewable?
2. Describe how *net energy* differs from *energy returned on investment (EROI)*. Why are these concepts important when evaluating energy sources?
3. Define the term *proven recoverable reserve*. Explain why the proven recoverable reserve of a given fuel could increase over time. Now explain why it could decrease.
4. Describe how coal is used to generate electricity. Now, describe how we create petroleum products. Provide examples of several of these products.
5. Why do many experts think we are about to pass the global production peak for conventional oil? What are two ways in which we could respond to *peak oil*?
6. Describe three ways in which we are now extending our reach for fossil fuels. List several impacts (positive or negative) that these actions might have.
7. Give an example of a clean coal technology. Now describe how carbon capture and storage is intended to work.
8. In what ways did the events at Three Mile Island, Chernobyl, and Fukushima Daiichi differ? What consequences resulted from each incident? Now list several concerns about the disposal of radioactive waste. What has been done so far about its disposal?
9. Based on data in this chapter, explain two reasons why it is reasonable to expect that Americans should be able to use energy more efficiently in the future as the U.S. economy expands.
10. Describe two specific examples of how technological advances can improve energy efficiency. Now describe one specific action you could take to conserve energy.

# Seeking Solutions

1. Summarize the main arguments for and against hydraulic fracturing for natural gas in the Marcellus Shale. What problems is this extraction helping to address? What problems is it creating? If a gas company offered you money to drill for gas in your backyard, how would you respond?

2. Compare the health and environmental effects of coal extraction and use with those of oil and gas extraction and use. What steps could governments, industries, and individuals take to alleviate some of these impacts?
3. Contrast the experiences of the Ogoni people of Nigeria with those of the citizens of Alaska. How have they been similar and different? Do you think businesses or governments should take steps to ensure that local people benefit from oil drilling operations? How could they do so?
4. **THINK IT THROUGH** You are elected governor of the state of Florida as the federal government is debating opening new waters to offshore drilling for oil and natural gas. Drilling in Florida waters would create jobs for Florida citizens and revenue for the state in the form of royalty payments from oil and gas companies. However, there is always the risk of a catastrophic oil spill, with its ecological, social, and economic impacts. Would you support or oppose offshore drilling off the Florida coast? Why? What, if any, regulations would you insist be imposed on such development? What questions would you ask of scientists before making your decision? What factors would you consider in making your decision?
5. **THINK IT THROUGH** You are the mayor of a rural Pennsylvania town above the Marcellus Shale. Some of your town's residents are eager to have jobs they believe that hydraulic fracturing for shale gas will bring. Others are fearful that leaks of methane and fracking fluids from drilling shafts could contaminate the water supply. Some of your town's landowners are looking forward to receiving payments from the gas company for use of their land, whereas others dread the prospect of noise and pollution. If the company receives too much local opposition in your town it says it will drill elsewhere instead. What information would you seek from the gas company, from your state regulators, and from scientists and engineers before deciding whether support for hydraulic fracturing is in the best interest of your town? How would you make your decision? How might you try to address the diverse preferences of your town's residents?

# Calculating Ecological Footprints

Scientists at the Global Footprint Network calculate the energy component of our ecological footprint by estimating the amount of ecologically productive land and sea required to absorb the carbon released from fossil fuel combustion. This translates into 4.9 ha of the average American's 7.2-ha ecological footprint. Another way to think about our footprint, however, is to estimate how much land would be needed to grow biomass with an energy content equal to that of the fossil fuel we burn.

Assume that you are an average American who burns about 6.3 metric tons of oil-equivalent in fossil fuels each year and that average terrestrial net primary productivity (p. 32) can be expressed as 0.0037 metric tons/ha/year. Calculate how many hectares of land it would take to supply our fuel use by present-day photosynthetic production.

| | Hectares of land for fuel production |
|---|---|
| You | 1703 |
| Your class | |
| Your state | |
| United States | |

1. Compare the energy component of your ecological footprint calculated in this way with the 4.9 ha calculated using the method of the Global Footprint Network. Explain why results from the two methods may differ.
2. Earth's total land area is approximately 15 billion hectares. Compare this to the hectares of land for fuel production from the table.
3. In the absence of stored energy from fossil fuels, how large a human population could Earth support at the level of consumption of the average American, if all of Earth's area were devoted to fuel production? Do you consider this realistic? Provide two reasons why or why not.

## MasteringEnvironmentalScience®

**STUDENTS**

Go to **MasteringEnvironmentalScience** for assignments, the etext, and the Study Area with practice tests, videos, current events, and activities.

**INSTRUCTORS**

Go to **MasteringEnvironmentalScience** for automatically graded activities, current events, videos, and reading questions that you can assign to your students, plus Instructor Resources.

# 16 Renewable Energy Alternatives

## Upon completing this chapter, you will be able to:

- ☐ Discuss reasons for seeking alternatives to fossil fuels
- ☐ Outline the major sources of renewable energy and assess their potential for growth
- ☐ Describe solar energy and the ways it is harnessed, and evaluate its advantages and disadvantages
- ☐ Explain wind power and how we harness it, and evaluate its benefits and drawbacks
- ☐ Describe geothermal energy and the ways we make use of it, and assess its advantages and disadvantages
- ☐ List potential ocean energy sources and how we can harness them
- ☐ Clarify the scale, methods, and impacts of hydroelectric power
- ☐ Describe the major sources and techniques involved in bioenergy, along with their benefits and shortcomings
- ☐ Explain hydrogen fuel cells and weigh options for energy storage and transportation



**Photo:** Homes in Freiburg, Germany, which produce excess solar power and sell it to the grid.

CENTRAL CASE STUDY

# Germany Goes Solar

*"[Renewable energy] will provide millions of new jobs. It will halt global warming. It will create a more fair and just world. It will clean our environment and make our lives healthier."*

**—Hermann Scheer, energy expert and member of the German parliament, 2009**

*"The nation that leads the clean energy economy will be the nation that leads the global economy."*

**—U.S. President Barack Obama, 2010**

When we think of solar energy, most of us envision a warm sunny place such as Arizona or southern California. Yet the country that produces the most solar power is Germany, a northern European nation that receives less sun than Alaska! Germany is the world's top user of photovoltaic (PV) solar technology, which produces electricity from sunshine. In recent years Germany has installed nearly half the world's total of this technology, and it now obtains more of its electricity from solar power than any other nation.

How is this happening in such a cool and cloudy country? A bold federal policy is using economic incentives to promote solar power and other forms of renewable energy. Germany has a **feed-in tariff** system whereby utilities are required to buy power from anyone who can generate power from renewable energy sources and feed it into the electric grid. Under this system, utilities must pay guaranteed premium prices for this power under long-term contract. As a result, German homeowners and businesses have rushed to install PV panels and are selling their excess solar power to utilities at a profit.

The feed-in tariffs apply to all forms of renewable energy. As a result, Germany became the world leader in wind power in the 1990s, until being overtaken by the United States and China. It ranks second in the world in using power from biomass, and third in solar water heating. Germany ranks third globally in electrical power capacity from renewable sources overall, trailing only China and the United States, which have far more people and businesses. In renewable energy generated per person, Germany is number one in the world.

Germany's push for renewable energy dates back to 1990. In the wake of the disaster at the Chernobyl nuclear power plant (p. 359), Germany decided to phase out its nuclear power plants. However, by shutting these down, the nation would lose virtually all its clean energy and would become utterly dependent on oil, gas, and coal imported from Russia and the Middle East.

Enter Hermann Scheer, a German parliament member and an expert on renewable energy. While everyone else assumed that solar, wind, and geothermal energy were costly, risky, and underdeveloped, Scheer saw them as a great economic opportunity—and as the only long-term answer. In 1990, Scheer helped push through a landmark law establishing feed-in tariffs. Ten years later, the law was revised and strengthened: The Renewable Energy Sources Act of 2000 aimed to promote renewable energy production and use, enhance the security of the energy supply, reduce carbon emissions, and lessen the many external costs (pp. 351–352) of fossil fuel use.

Under the law, each renewable source is assigned its own payment rate according to market considerations. These rates have been gradually reduced to encourage increasingly efficient means of producing power. In 2010, the German government slashed PV solar tariff rates to reduce the cost of the subsidies to taxpayers and because PV market prices had already fallen by half. In response, sales of PV modules skyrocketed as Germans rushed to lock in the old rates. In 2010, 2011, and 2012, Germans installed over 7 gigawatts of PV solar capacity each year—more than the total cumulative capacity of the United States at the time. Solar installations slowed in 2013, and the government plans to end subsidies once the nation reaches 52 gigawatts of cumulative capacity, roughly 45% beyond its total as of 2014.

By reducing the subsidies, the government aims to create a strong and sustainable solar industry that can outcompete foreign companies for international business. Indeed, German industries have become leaders in "green tech," designing and selling renewable energy technologies around the world. Germany is second in PV production behind China, leads the world in biodiesel production, and has built several cellulosic ethanol (p. 383) facilities. Renewable energy industries today employ nearly 400,000 German citizens.

By 2020, Germany plans to obtain 35% of its electricity from renewable sources and aims to increase this percentage to 50% by 2030, 65% by 2040, and 80% by 2050. To achieve this, the government has been allotting more public money to renewable energy than any other nation—over $25 billion annually in recent years. Following the Fukushima nuclear disaster in Japan in 2011 (p. 360), the German government responded to anti-nuclear demonstrations by shutting down 7 of its 15 nuclear power plants and promising to phase out the rest by 2022. The resulting dramatic reduction in electricity supply from nuclear power is causing rates to rise and is forcing an increase in coal combustion for power—thereby putting added pressure on Germany to develop renewable energy sources quickly.

By replacing some of its fossil fuel use with renewable energy, Germany has reduced its emissions of carbon dioxide by 140 million tons per year—equal to taking 24 million cars off the road. Since 1990, $CO_2$ emissions from German energy sources have fallen by 24%, and emissions of seven other major pollutants ($CH_4$, $N_2O$, $SO_2$, $NO_x$, CO, VOCs, and dust) have been reduced by 12–95%. At least half of these reductions are attributed to renewable energy paid for under the feed-in tariff system.

Germany's success is serving as a model for other countries. As of 2014, more than 90 nations had implemented some sort of feed-in tariff. Spain and Italy ignited their wind and solar development as a result. In North America, Vermont and Ontario established feed-in tariff systems similar to Germany's, while California, Hawaii, Oregon, Washington, and New York conduct more-limited programs. In 2010, Gainesville, Florida, became the first U.S. city to establish feed-in tariffs, and solar power is growing quickly there as a result. Moreover, utilities in 43 U.S. states now offer *net metering*, in which utilities credit customers who produce renewable power and feed it into the grid. As more nations, states, and cities encourage renewable energy, we may soon experience a historic transition in the way we meet our energy demands. ■

# Renewable Energy Sources

Germany's bold federal policy is just one facet of a global shift toward renewable energy sources. Across the world, nations are searching for ways to move away from fossil fuels while ensuring a reliable, affordable, and sustainable supply of energy.

## We have alternatives to fossil fuels

Fossil fuels drove the industrial revolution and helped to create the unprecedented material prosperity we enjoy today. Our global economy is still powered by coal, oil, and natural gas, which together provide over four-fifths of the world's energy and two-thirds of our electricity (see Figure 15.3, p. 335). However, easily extractable supplies of these nonrenewable energy sources are in decline, while the economic and social costs, security risks, and health and environmental impacts of fossil fuel dependence continue to intensify (Chapter 15). For these reasons, most energy experts accept that we will need to shift to energy sources that are less easily depleted and gentler on our health and environment.

We have developed a range of alternatives to fossil fuels (see Table 15.1, p. 334). These alternatives include a diverse array of renewable energy sources, as well as nuclear power (which is nonrenewable because it relies on uranium, a mineral in finite supply).

Compared with fossil fuels, renewable sources are used relatively less for transportation and relatively more for generating electricity. In the United States, renewable sources account for 9.4% of total energy use (**FIGURE 16.1a**) and 12.9% of electricity generated (**FIGURE 16.1b**).

Of these renewable sources, hydroelectric power and energy from biomass are well established and already

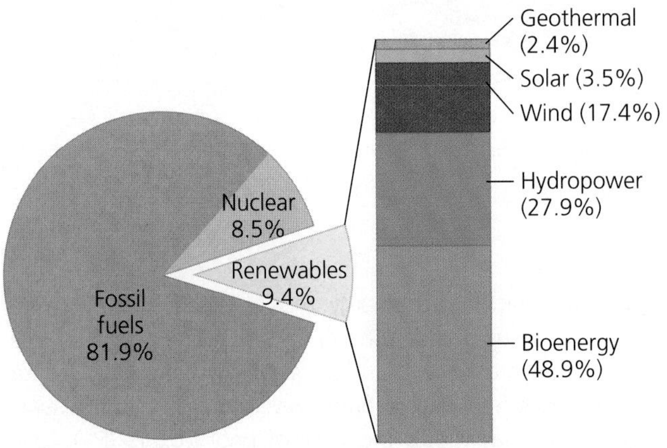


**(a) U.S. consumption of energy, by source**

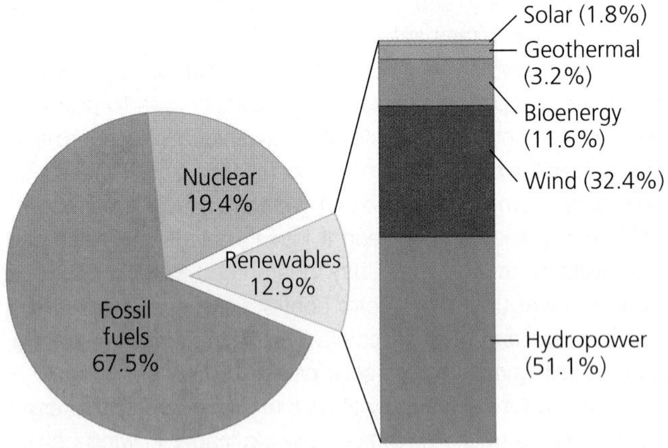


**(b) U.S. electricity generation, by energy source**

**FIGURE 16.1 Nine percent of the energy consumed in the United States comes from renewable sources.** Of this amount **(a)**, most derives from bioenergy and hydropower. Of electricity generated in the United States **(b)**, 13% comes from renewable energy sources, predominantly hydropower and wind. *Data are for 2013, from Energy Information Administration.*

**DATA Q** What percentage of overall U.S. energy consumption does bioenergy contribute?

GO TO **INTERPRETING GRAPHS & DATA** ON MasteringEnvironmentalScience®

play substantial roles in our energy and electricity budgets. Although hydropower and bioenergy are renewable, supplies of water and biomass can be locally depleted if overharvested.

Perpetually renewable sources include energy from the sun, wind, Earth's geothermal heat, and ocean water. These energy sources are often called "new renewables" because (1) they are just beginning to be used on a wide scale in our modern industrial society, (2) they are harnessed using technologies still in a rapid phase of development, and (3) they will likely play much larger roles in the future.

## Renewable sources are growing fast

Although solar, wind, and geothermal energy sources comprise a small proportion of our energy budget, over the past four decades they have grown far faster than other sources. Because these "new renewable" sources started from such low levels of use, however, it will take them some time to catch up to fossil fuels. One hurdle constraining their growth is that so far we lack infrastructure to transfer huge volumes of power from them inexpensively on a continent-wide scale. However, rapid growth in renewable energy seems likely to continue as technology improves and prices fall, as population and consumption rise, as conventional fossil fuel supplies decline, and as people demand cleaner environments.

## Renewable energy offers advantages

Replacing fossil fuels with clean, renewable energy sources helps to reduce air pollution (Chapter 13) and the greenhouse gas emissions that drive global climate change (Chapter 14) (**FIGURE 16.2**). Renewable alternatives help diversify an economy's energy mix, reducing both price volatility and reliance on imported fuels. Some renewable sources can generate income and property tax for rural communities, and some help people in developing regions of the world to produce their own energy. And unlike fossil fuels, most renewable sources are inexhaustible on time scales relevant to our society.

Shifting to renewable energy also creates employment opportunities. The design, installation, maintenance, and management required to develop technologies and to rebuild and operate our society's energy infrastructure are becoming major sources of employment for young people today, through **green-collar jobs** (**FIGURE 16.3**). Nearly 6.5 million people work in renewable energy jobs around the world already.

Germany and a number of other industrialized nations are showing that wealthy and advanced economies can replace fossil fuels gradually with alternative sources while continuing to raise living standards for their citizens. Sweden provides an example. Since 1970, it has decreased its fossil fuel use from 81% to 38% of its national energy budget. Today bioenergy, hydropower, and nuclear power together provide Sweden with over 60% of its energy and virtually all of its electricity.

As energy technologies evolve and novel sources are developed, we can better assess the benefits and drawbacks of each. Understanding their particular contributions and impacts should enable us to make smart energy choices (see **THE SCIENCE BEHIND THE STORY**, pp. 368–369).

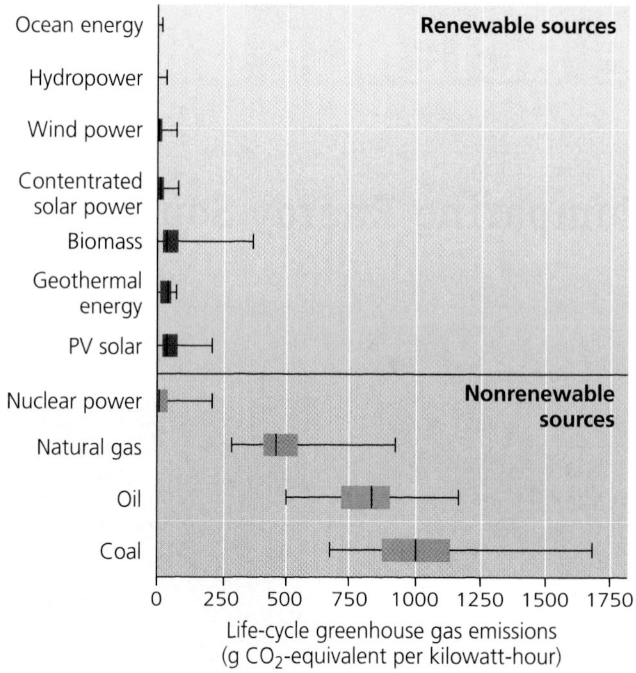


**FIGURE 16.2 Renewable energy sources release far fewer greenhouse gas emissions than fossil fuels.** Shown are medians and ranges of estimates from scientific studies of each source when used to generate electricity. *Data from Intergovernmental Panel on Climate Change, 2012.* Renewable energy sources and climate change mitigation. *Special report. Cambridge Univ. Press, New York.*

**FIGURE 16.3 Renewable energy creates new green-collar jobs.** This worker is installing solar panels at Arizona State University. As of 2013, about 6.5 million people worldwide were employed, directly or indirectly, in renewable energy jobs. *Data from REN21, 2014.* Renewables 2014: Global status report. *REN21, UNEP, Paris.*

THE SCIENCE BEHIND THE STORY

## Comparing Energy Sources

In a world facing climate change, air pollution, and risks to energy security, the choices we make as a society about energy are far-reaching. With the stakes so high, which energy sources should we pursue?

Many researchers have compared energy sources. Let's examine one recent study that sought to analyze all the biggest issues in a comprehensive way. In 2009, Mark Jacobson of Stanford University dug deep into the published literature on alternative energy to calculate impacts of various energy sources across their entire life cycles (including their materials, construction, operation, and so on). Reviewing over 100 studies, his life-cycle analysis (p. 400) was published in the journal *Energy and Environmental Science*.

**Mark Jacobson of Stanford University**

It is difficult to compare energy sources because some are used primarily to generate electricity, some are used mostly for heating and cooling, and others provide fuel for transportation. To make an apples-to-apples comparison, Jacobson calculated impacts for each source as if it alone were being used to power all the vehicles in the United States. For electricity-producing sources such as wind or PV solar, he calculated how much was needed to power electric vehicles for the entire nation, and he compared the impacts to those from powering the same number of combustion-engine vehicles with liquid fuels.

In all, Jacobson evaluated and ranked nine electric power sources and two liquid fuel sources, in conjunction with three vehicle types: battery-electric vehicles (p. 355), hydrogen-fuel-cell vehicles (p. 384), and E85 flex-fuel vehicles (p. 382).

He first examined carbon dioxide emissions. His analysis showed that wind power was lowest in $CO_2$ emissions but that nearly all renewable options eliminated the vast majority of emissions from vehicles, thus reducing total U.S. emissions by close to 30%. The exceptions were corn-based ethanol and cellulosic ethanol (p. 383), both of which, he found, would be likely to *increase* emissions.

Next, Jacobson analyzed air pollution from seven major pollutants. Using data on U.S. deaths attributed to fossil fuel pollution each year, he calculated the number of deaths expected to result from using each energy source to fuel vehicles. The number of deaths was close to zero for all renewable sources except for biofuels. Deaths from ethanol emissions were equivalent in number to those currently attributed to gasoline. Nuclear power was equivalent to most renewables in having a low death rate, but Jacobson also calculated impacts from potential nuclear proliferation and weapons use by a government or by terrorists. When these factors were included, nuclear power emerged as the least safe option.

Jacobson next compared each energy source's footprint in land area, water consumption, effects on wildlife, thermal pollution, chemical pollution of water, radioactive waste, operating reliability, and energy supply disruption. With regard to supply disruption, he concluded that although many renewables are intermittent in their supply, they are least prone to major disruptions because they are less centralized than coal-fired plants, nuclear plants, or oil and ethanol refineries. Moreover, he argued that intermittency could be overcome by integrating sources across large geographic areas and by combining them strategically through the day, taking advantage of peaks in power from solar and wind and balancing the troughs with geothermal, pumped-storage hydropower, batteries, and hydrogen fuel (**FIGURE 1**).

When Jacobson at last went to assess overall results and rank the energy sources by a weighted average of all the above issues, he found that wind power was most desirable, followed by concentrated solar power, geothermal power, tidal power, and PV solar (**FIGURE 2**). Worst were corn ethanol and cellulosic ethanol.

Is it realistic to think we could power all U.S. vehicles with wind power? Jacobson says yes. To do so would require between 73,000 and 144,000 5-megawatt wind turbines. This sounds like a lot, yet it is fewer than the 300,000 airplanes that the United States built during World War II, and fewer than the

### Policy and investment can accelerate our transition

Although prices are falling, most renewable energy remains more expensive than fossil fuel energy (**FIGURE 16.4**, p. 370). However, renewables can overcome technological and economic barriers if they are lent political support. Feed-in tariffs like Germany's hasten the spread of renewable energy by creating financial incentives for businesses and individuals. Governments can also set goals or mandate that certain percentages of power come from renewable sources. As of 2014, 144 nations and 29 U.S. states had set targets for renewable energy use. Governments also invest in research and technology development, lend money to businesses as they start up, and offer tax credits and tax rebates to companies and individuals who produce or buy renewable energy.

When a government boosts an industry with such policies, the private sector often responds with investment of its own as investors recognize an enhanced chance of success and profit.

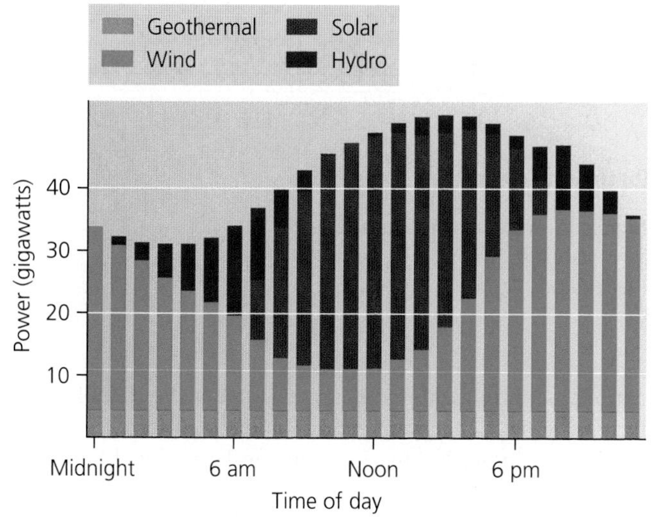


**FIGURE 1 Intermittent wind and solar power can be balanced with stable and flexible geothermal and hydropower.** This model for California shows how such a combination might reliably supply our daily energy needs. *Data from Jacobson, M., and M. Delucchi, 2009. A path to sustainable energy by 2030.* Scientific American *Nov. 2009 pp. 58–65.*

These researchers judged that our main potential barrier was not the scale of development but rather the availability of a handful of rare-earth metals in limited supply (e.g., silver, platinum, lithium, indium, tellurium, and neodymium).

Critics of Jacobson's work have pointed to such articles in popular magazines, and to public speaking events, to question Jacobson's objectivity and paint him as an advocate. Critics who objected to his inclusion of impacts from a nuclear weapons disaster in his comparison of energy sources pointed out that he has spoken out against nuclear power in public forums.

Any scientist who steps out of the lab, field, or office and advocates publicly for particular solutions is liable to be criticized by those who disagree with his or her positions. Whether such criticisms are fair—and whether they are in society's best interest—are complicated questions. On one hand, remaining objective and free of advocacy gives science its strength and credibility. On the other hand, shouldn't we want our advocates to be the people who have studied the issues themselves and are the best informed about them? ☐

number of (mostly smaller) turbines already erected worldwide. One reason so few turbines are needed is that battery-electric vehicles make use of power much more efficiently than gasoline-powered vehicles, so that far less energy is required.

As his research paper was being published, Jacobson also co-authored an article in the popular magazine *Scientific American* with Mark Delucchi of the University of California–Davis, analyzing how the world could meet all its energy needs from renewable energy. The pair concluded that we could do this with a combination of:

- 490,000 tidal turbines
- 5350 geothermal plants
- 900 hydroelectric plants
- 3.8 million wind turbines
- 720,000 wave converters
- 1.7 billion rooftop PV systems
- 49,000 concentrated solar power plants
- 40,000 PV power plants

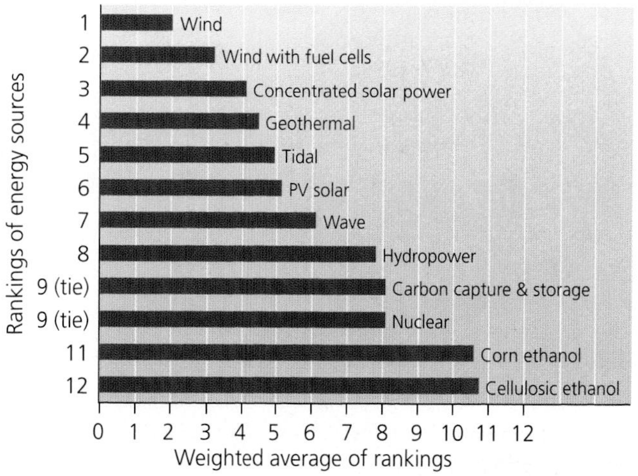


**FIGURE 2 Jacobson found wind power to be the most desirable alternative energy source overall, and ethanol to be least desirable.** *Data from Jacobson, M., 2009. Review of solutions to global warming, air pollution, and energy security.* Energy & Environmental Science *2009 (2): 148–173.*

Global investment (public plus private) in renewable energy rose to $254 billion in 2013—six times the amount just eight years earlier. Yet the economics of renewable energy have been erratic. Technologies evolve quickly, and policies vary from place to place and can change unpredictably. As a result, renewable markets have been volatile, with steep price fluctuations, bursts of rapid growth, and bankruptcies of promising companies.

Critics of public subsidies for renewable energy complain that funneling taxpayer money to particular energy sources is inefficient and skews the market. Instead, they propose, we should let energy sources compete freely. Proponents of renewable energy point out that governments have long subsidized fossil fuels and nuclear power far more than they now subsidize renewables. As a result, there has never been a level playing field, or a truly free market.

In one recent study, researchers dug into data on the U.S. government's many energy subsidies and tax breaks (p. 110) over the past century. Their report revealed that in total, oil and

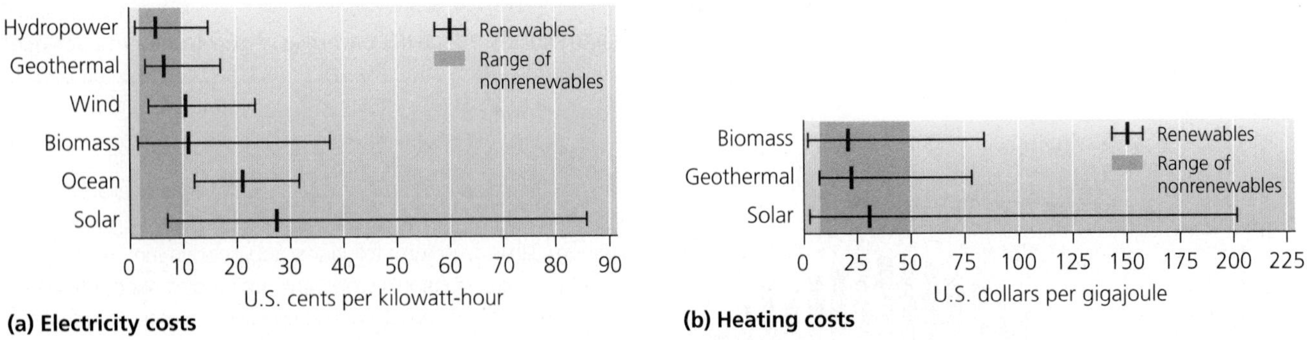


**FIGURE 16.4 Most renewable energy sources have market prices higher than those for nonrenewable sources, but some are competitive.** Shown are medians and ranges of market prices for each source used for **(a)** electricity and **(b)** heating. *Data from Intergovernmental Panel on Climate Change, 2012.* Renewable energy sources and climate change mitigation. *Special report. Cambridge Univ. Press, New York.*

gas have received 75 times more subsidies—and nuclear power has received 31 times more—than new renewable sources (**FIGURE 16.5a**). Moreover, on a per-year basis (controlling for the differing amounts of time each source has been used), oil and gas have received 13 times more subsidies—and nuclear power over 9 times more—than solar, wind, and geothermal power (**FIGURE 16.5b**).

Many feel that the subsidies showered on nonrenewable energy sources have helped to enhance America's economy, national security, and international influence by establishing thriving global energy industries dominated by U.S. firms. Yet by this logic, if America wants to be a global leader to rival nations such as China and Germany in the transition to renewable energy, then political and financial support will likely need to be redirected toward these new energy sources.

Let's now begin our tour of renewable energy sources. We'll start with the new renewables (solar, wind, geothermal, and ocean energy), proceed through the well-established hydropower and bioenergy sources, and then briefly examine hydrogen fuel cells.

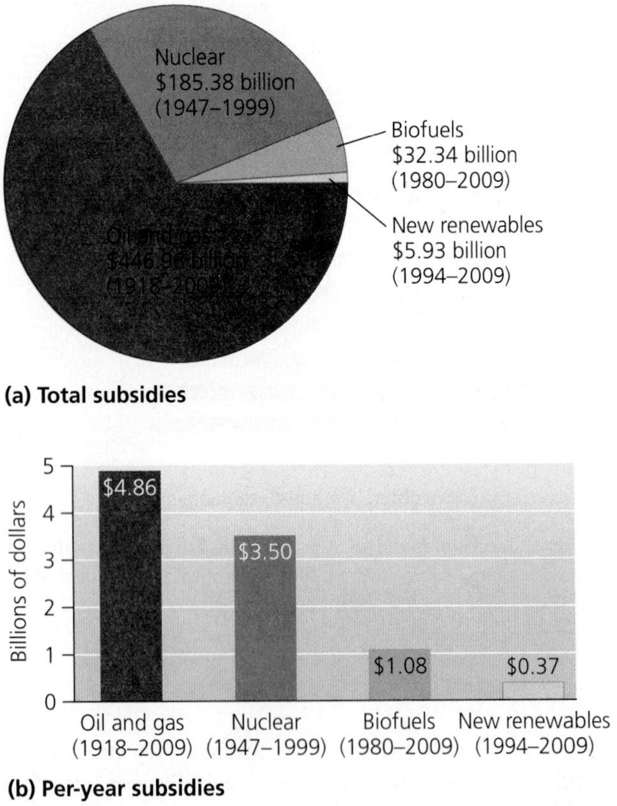


**FIGURE 16.5 Fossil fuels and nuclear power have received far more in U.S. government subsidies (mostly tax breaks) than have renewable energy sources.** This is true both for **(a)** total amounts over the past century and for **(b)** average amounts per year. *Data are in 2010 dollars, from Pfund, N., and B. Healey, 2011.* What would Jefferson do? The historical role of federal subsidies in shaping America's energy future. *DBL Investors.*

# Solar Energy

The sun releases astounding amounts of energy by converting hydrogen to helium through nuclear fusion. The tiny proportion of this energy that reaches Earth is enough to drive most processes in the biosphere, helping to make life possible on our planet. Each day, Earth receives enough **solar energy**, or energy from the sun, to power human consumption for a quarter of a century. On average, each square meter of Earth's surface receives about 1 kilowatt of solar energy—17 times the energy of a lightbulb. As a result, a typical home has enough roof area to meet all its power needs with rooftop panels that harness solar energy.

## We collect solar energy using passive or active methods

The simplest way to harness solar energy is through the use of **passive solar energy collection**. In this approach, buildings are designed to maximize absorption of sunlight in winter yet to keep the interior cool in summer (**FIGURE 16.6**). South-facing windows maximize the capture of winter sunlight. Overhangs shade windows in summer, when the sun is high in the sky and cooling is desired. Planting vegetation around a building buffers it from temperature swings. Passive solar techniques also use materials that absorb heat, store it, and release it later. Such *thermal mass* (of straw, brick, concrete, or other materials) can make up floors, roofs, and walls, or

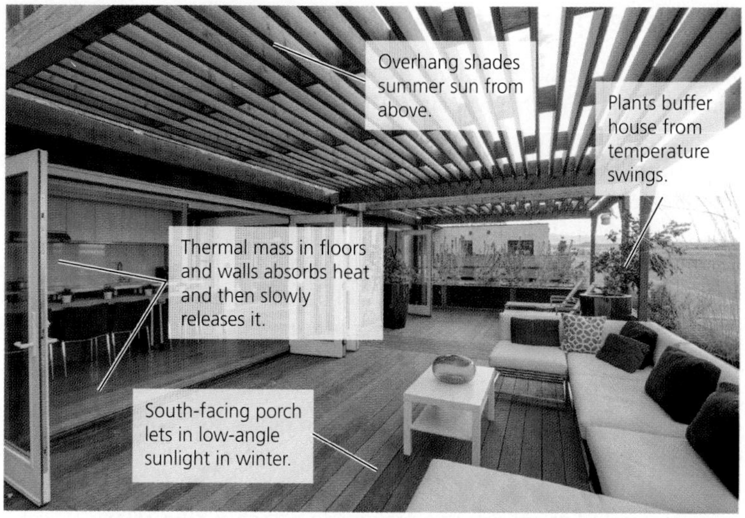
Overhang shades summer sun from above.
Plants buffer house from temperature swings.
Thermal mass in floors and walls absorbs heat and then slowly releases it.
South-facing porch lets in low-angle sunlight in winter.

**FIGURE 16.6 Passive solar design elements can be seen in this house built by college students.** Stanford University students and staff designed and constructed this solar-powered house as part of the sixth biannual Solar Decathlon in 2013. In this event, college and university teams compete to build the best houses fully powered by solar energy.

can be used in portable blocks. All these approaches conserve energy and reduce costs.

**Active solar energy collection** makes use of devices to focus, move, or store solar energy. For instance, *flat plate solar collectors* (**FIGURE 16.7**) heat water and air for homes and businesses. These generally consist of dark, heat-absorbing metal plates mounted on rooftops in flat glass-covered boxes. Water, air, or antifreeze runs through tubes that pass through the collectors, transferring heat from them to the building or its water tank. Heated water can be stored for later use and passed through pipes designed to release the heat into the building.

Over 200 million households worldwide heat water with solar collectors. Active solar heating is used most widely in China and in Europe, where Germans motivated by feed-in tariffs installed 200,000 new systems in 2008 alone.

## Concentrating sunlight focuses energy

We can harness energy from sunlight to produce electricity at **concentrated solar power (CSP)** plants, which generate electricity at large centralized facilities and transmit power to homes and businesses via the electric grid. CSP facilities gather sunlight from a large area and focus it on a small area. This is the principle behind *solar cookers*, simple portable ovens that use reflectors to cook food and that are proving useful in the developing world. CSP facilities apply this principle on a huge scale, in developments that sprawl across wide stretches of land in sunny desert regions such as southern California.

In one method, curved mirrors focus sunlight onto synthetic oil in pipes (**FIGURE 16.8**, next page). The superheated oil is piped to a facility where it heats water, creating steam that drives turbines to generate electricity. In another method, hundreds of mirrors focus sunlight onto a central receiver atop a tall "power tower." From here, air or fluids carry heat through pipes to a steam-driven generator.

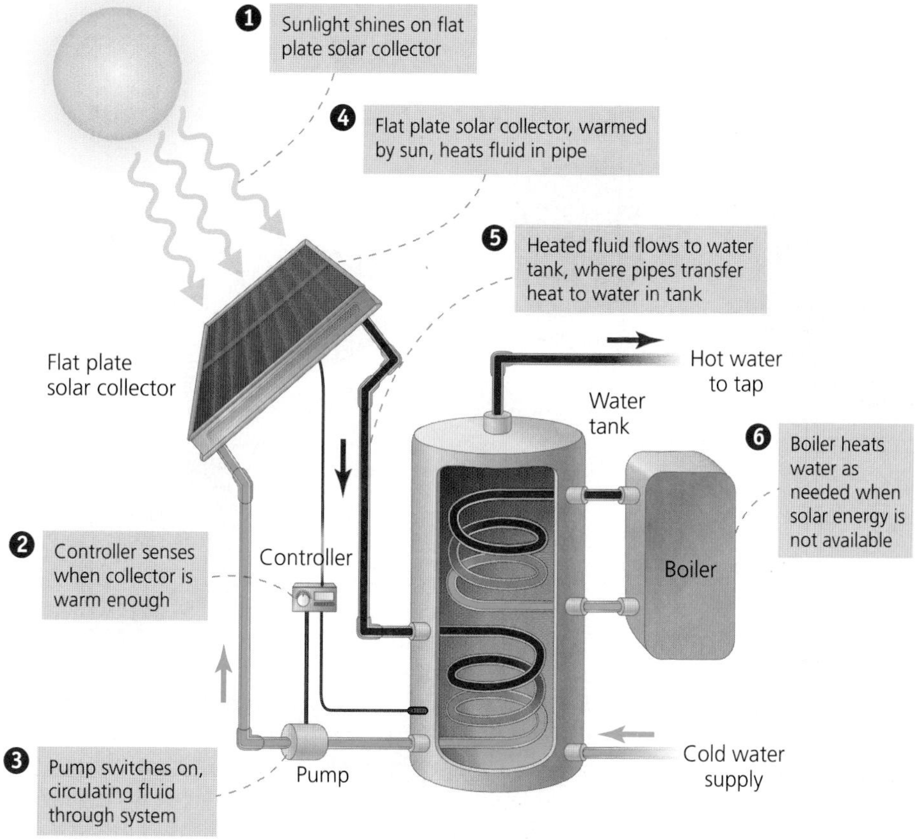
1 Sunlight shines on flat plate solar collector
4 Flat plate solar collector, warmed by sun, heats fluid in pipe
5 Heated fluid flows to water tank, where pipes transfer heat to water in tank
Flat plate solar collector
Hot water to tap
Water tank
6 Boiler heats water as needed when solar energy is not available
Boiler
2 Controller senses when collector is warm enough
Controller
3 Pump switches on, circulating fluid through system
Pump
Cold water supply

**FIGURE 16.7 Solar collectors provide heating.** Systems for heating water vary in design, but typically ❶ sunlight is gathered on a flat plate solar collector, until a controller ❷ switches on a pump ❸ to circulate fluid through pipes to the collector. The sunlit collector heats the fluid ❹, which flows through pipes ❺ to a water tank. The hot fluid in the pipes transfers heat to the water in the tank, and this heated water is available for the taps of the home or business. Generally an external boiler ❻ kicks in to heat water when solar energy is not available.

**FIGURE 16.8 Utilities concentrate solar power to generate electricity at large scales.** This solar plant in California is one of nine that spreads across more than 650 ha (1600 acres) of the Mojave Desert, providing power for over 230,000 homes. Such facilities require vast areas of land as well as precious water, yet researchers estimate that impacts are still less than those exerted by fossil fuels.

Concentrated solar power holds promise for producing tremendous amounts of energy, but large CSP developments pose considerable environmental impacts. All land beneath the mirrors is cleared and graded, destroying sensitive habitat for desert species, while maintenance requires enormous amounts of water, which is scarce and precious in desert regions.

## PV cells generate electricity

We can also generate electricity from sunlight using photovoltaic (PV) systems. **Photovoltaic (PV) cells** convert sunlight to electrical energy when light strikes one of a pair of plates made primarily of silicon, a semiconductor that conducts electricity (**FIGURE 16.9**). The light causes one plate to release electrons, which are attracted by electrostatic forces to the opposing plate. Connecting the two plates with wires enables the electrons to flow back to the original plate, creating an electrical current (direct current, DC), which can be converted into alternating current (AC) and used for residential and commercial electrical power. Small PV cells may already power your watch or your calculator. Arrays of PV panels can be seen on the roofs of the German houses in the photo that opens this chapter (p. 364).

Researchers are experimenting with variations on PV technology, and manufacturers are developing **thin-film solar cells**, photovoltaic materials compressed into ultra-thin sheets. Although less efficient at converting sunlight to electricity, they are cheaper to produce. Thin-film technologies can be incorporated into roofing shingles and potentially many other surfaces, even highways!

Photovoltaic cells of all types can be connected to batteries that store the accumulated charge until needed. Or, producers of PV electricity can sell power to their local utility if they are connected to the regional electric grid. In parts of 46 U.S. states, homeowners can sell power to their utility in a process called **net metering**, in which the value of the power the consumer provides is subtracted from the consumer's monthly utility bill. Feed-in tariff systems like Germany's go a step further by paying producers more than the market price of the power, offering producers the hope of turning a profit.

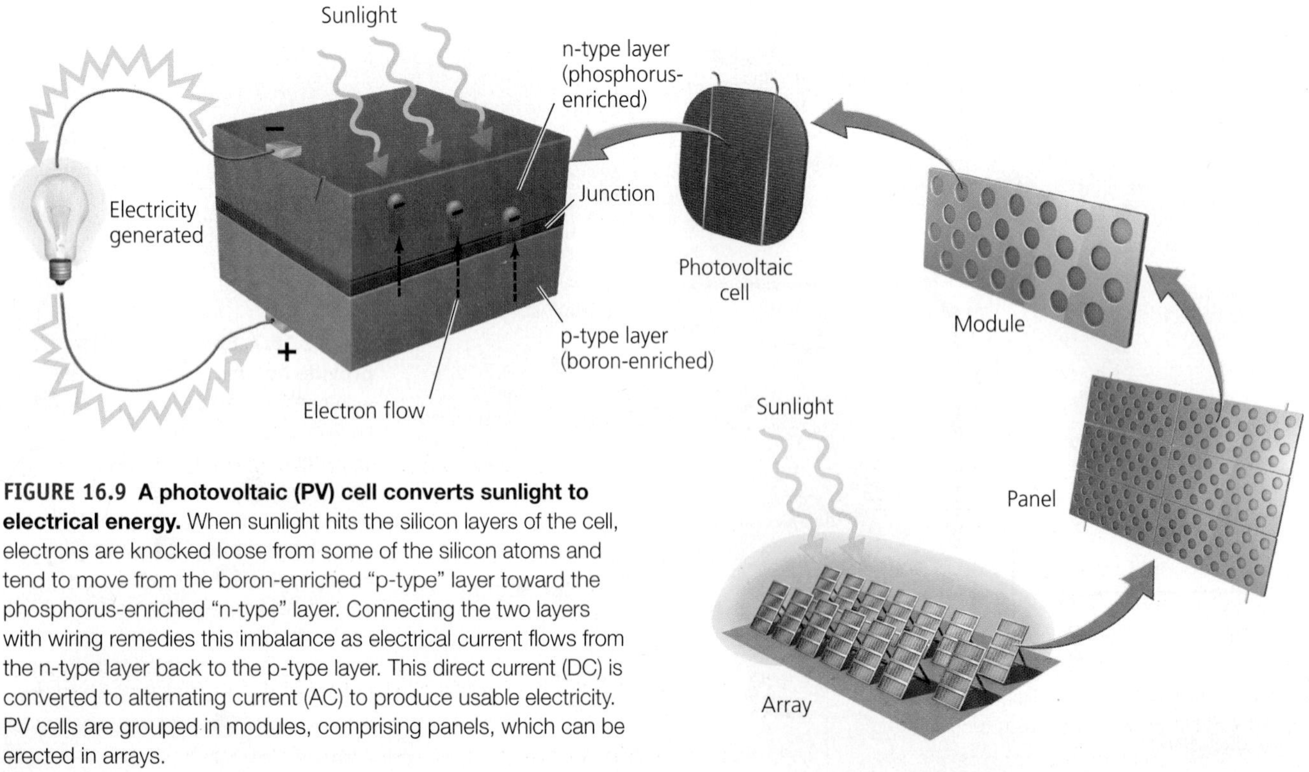


**FIGURE 16.9 A photovoltaic (PV) cell converts sunlight to electrical energy.** When sunlight hits the silicon layers of the cell, electrons are knocked loose from some of the silicon atoms and tend to move from the boron-enriched "p-type" layer toward the phosphorus-enriched "n-type" layer. Connecting the two layers with wiring remedies this imbalance as electrical current flows from the n-type layer back to the p-type layer. This direct current (DC) is converted to alternating current (AC) to produce usable electricity. PV cells are grouped in modules, comprising panels, which can be erected in arrays.

## Solar energy offers many benefits

The fact that the sun will continue burning for another 4–5 billion years makes it inexhaustible as an energy source for human civilization. Moreover, the amount of solar energy reaching Earth should be enough to power our society once we deploy technology adequate to harness it. These advantages of solar energy are clear, but the technologies themselves also provide benefits. PV cells and other solar technologies use no fuel, are quiet and safe, contain no moving parts, require little maintenance, and do not require a turbine or generator to create electricity. An average unit can produce energy for 20–30 years.

Solar systems allow for local, decentralized control over power. Homes, businesses, rural communities, and isolated areas in developing nations can use solar power to produce electricity without being near a power plant or connected to a grid. In developed nations, most PV systems are connected to the regional electric grid, and homeowners can sell excess solar energy to their local utility through feed-in tariffs or net metering.

Developing, manufacturing, and deploying solar technology also creates new green-collar jobs; PV-related jobs now employ nearly 2.3 million people worldwide. Finally, a major advantage of solar energy over fossil fuels is that it does not emit greenhouse gases and other air pollutants (see Figure 16.2). The manufacture of equipment currently requires fossil fuels, but once up and running, a solar system produces no emissions.

## Location, timing, and cost can be drawbacks

Solar energy currently has three main disadvantages. One is that not all regions are equally sunny (**FIGURE 16.10**). People in Seattle or Anchorage will find it more challenging to harness solar energy than people in Tucson or San Diego. However, observe in Figure 16.10 that Germany receives even less sun than Alaska, and yet it is the world leader in solar power!

A second drawback is that solar energy is an intermittent resource. Daily or seasonal variation in sunlight can limit stand-alone solar systems if storage capacity in batteries or fuel cells is not adequate or if backup power is not available from a municipal electric grid. However, pumped-storage hydropower (p. 380) and other sources can help compensate for periods of low solar power production (see Figure 1 in *The Science Behind the Story*, p. 369).

The third disadvantage of current solar technology is the up-front cost of the equipment. Because of the investment cost, solar power remains the most expensive way to produce electricity (see Figure 16.4). However, declines in price and improvements in efficiency are encouraging, even in the absence of significant funding from government and industry. Solar systems are becoming less expensive and now can sometimes pay for themselves in less than 10–20 years. After that time, they provide energy virtually for free as long as the equipment lasts.

## Solar energy is expanding

Solar energy contributes just 0.33%—33 parts in 10,000—of the U.S. energy supply, and just 0.2% of U.S. electricity generation. Even in Germany, which gets more of its energy from solar than any other nation, the percentage is only 5%. However, solar energy use has grown by 30% each year worldwide in the past four decades. Solar is proving especially attractive in developing countries rich in sun but poor in infrastructure, where hundreds of millions of people still live without electricity.

PV technology is the fastest-growing power source today, having recently doubled every two years. Germany leads the world in installation of PV technology, with over one-third of all PV cells in the world. Germany's investment began in 1998 when Hermann Scheer spearheaded a "100,000 Rooftops" program to install PV panels atop 100,000 German roofs. The popular program easily surpassed this goal, and today over 1.2 million German rooftops have PV systems.

China leads the world in production of PV cells, followed by Germany and Japan, while U.S. firms account for less than 3% of the industry. The Chinese government's support of its solar industry led to so much production that supply has outstripped global demand in recent years. Highly subsidized

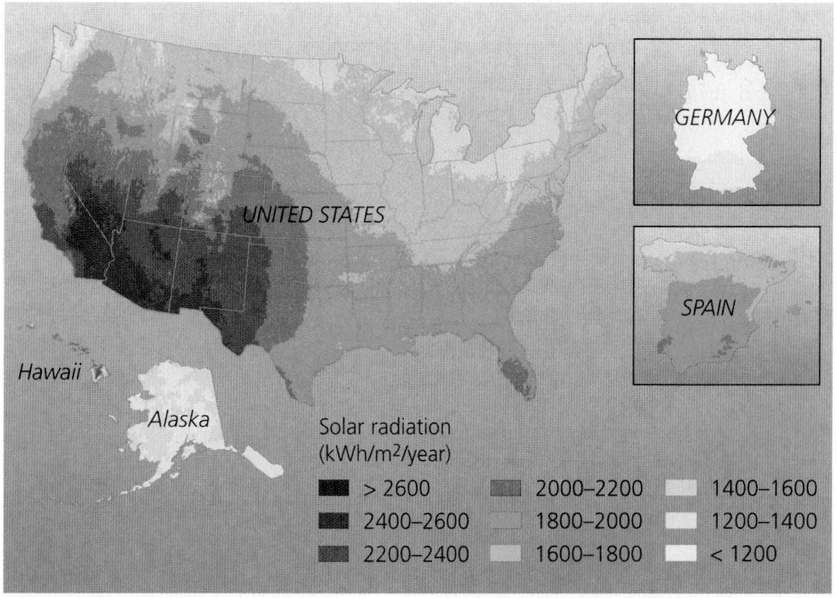


**FIGURE 16.10 Solar radiation varies from place to place.** Harnessing solar energy is more profitable in sunny regions such as the southwestern United States than in cloudier regions such as Alaska and the Pacific Northwest. However, compare solar power leaders Germany and Spain with the United States. Spain is similar to Kansas in the amount of sunlight it receives, and Germany is cloudier than Alaska! This suggests that solar power can be used with success just about anywhere. *Data from National Renewable Energy Laboratory, U.S. Department of Energy.*

DATA Q Roughly how much more solar radiation does southern Arizona receive than Germany?

GO TO **INTERPRETING GRAPHS & DATA** ON MasteringEnvironmentalScience®

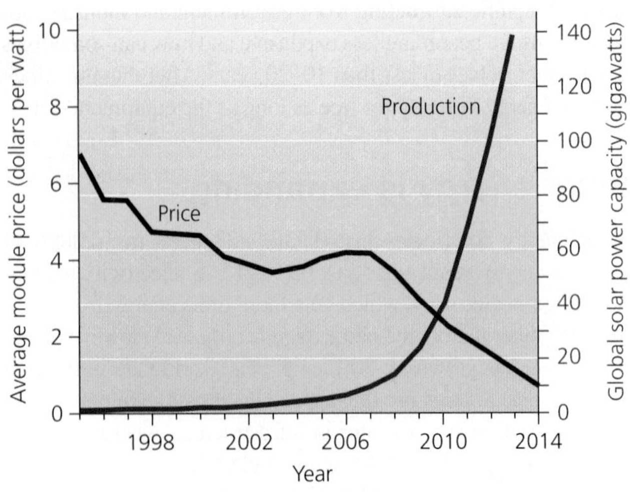


**FIGURE 16.11 Global production of PV cells has grown exponentially, and prices have fallen rapidly.** *Data from REN21, 2014.* Renewables 2014: Global status report. *REN21, UNEP, Paris; U.S. Department of Energy, EERE, 2011.* 2010 Solar technologies market report; and pvXchange, 2014.

Chinese firms have been selling solar products abroad at low prices (often at a loss), driving American and European solar manufacturers out of business. In response, the United States and European nations have slapped tariffs on Chinese imports, while both sides have filed complaints with the World Trade Organization.

As production of PV cells rises, prices fall (**FIGURE 16.11**). At the same time, efficiencies are increasing, making each unit more powerful. In the 1950s, solar technologies had efficiencies of around 6%, but today's PV cells boast up to 20% efficiency commercially and 40% in lab research, suggesting that future solar cells could be more efficient than any energy technology we have today.

# Wind Power

As the sun heats the atmosphere, it causes air to move, producing wind. We can harness **wind power** by using **wind turbines**, mechanical assemblies that convert wind's kinetic energy (p. 30), or energy of motion, into electrical energy.

## Wind turbines convert kinetic energy to electrical energy

Wind blowing into a turbine turns the blades of the rotor, which rotate machinery inside a compartment called a *nacelle*, which sits atop a tower (**FIGURE 16.12**). Inside the nacelle are a gearbox, a generator, and equipment to monitor and control the turbine's activity. Today's towers average 80 m (260 ft) in height, and the largest are taller than a football field is long. Higher is generally better, to minimize turbulence (and potential damage) while maximizing wind speed. Engineers design turbines to rotate in response to changes in wind direction, so that the motor faces into the wind at all times. They also design them to begin turning at specified wind speeds to harness energy efficiently. Turbines are often erected in groups called **wind farms**. The world's largest wind farms contain hundreds of turbines.

## Wind power is growing fast

Like solar energy, wind provides just a small proportion of the world's power needs, but wind power is growing fast—doubling every three years (**FIGURE 16.13**). Five nations account for nearly three-quarters of the world's wind power output (**FIGURE 16.14**), but dozens of nations now produce wind power. Germany had long produced the most, but the United States overtook it in 2008, and China surpassed the United States two years later.

Denmark leads the world in obtaining the highest percentage of its energy from wind power. In this small European nation, wind supplies nearly 30% of Danish electricity needs. Texas generates the most wind power of all U.S. states, while Iowa and South Dakota each obtain nearly 25% of their electricity from wind.

Wind power's growth in the United States has been haphazard because Congress has not committed to a long-term tax credit for wind development, but instead has passed a series of short-term renewals, leaving the industry uncertain about how to invest. However, experts agree that wind power's growth will continue, because only a small portion of this resource is currently being tapped and because wind power at favorable locations already generates electricity at prices nearly as low as fossil fuels (see Figure 16.4).

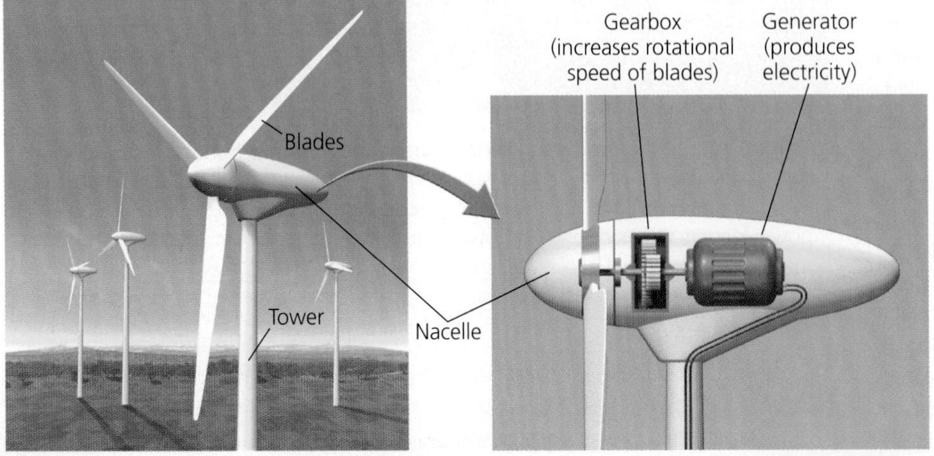


**FIGURE 16.12 A wind turbine converts wind's energy of motion into electrical energy.** Wind spins a turbine's blades, turning a shaft that extends into the nacelle. Inside the nacelle, a gearbox converts the rotational speed of the blades into much higher speeds, providing adequate motion for the generator to produce electricity.

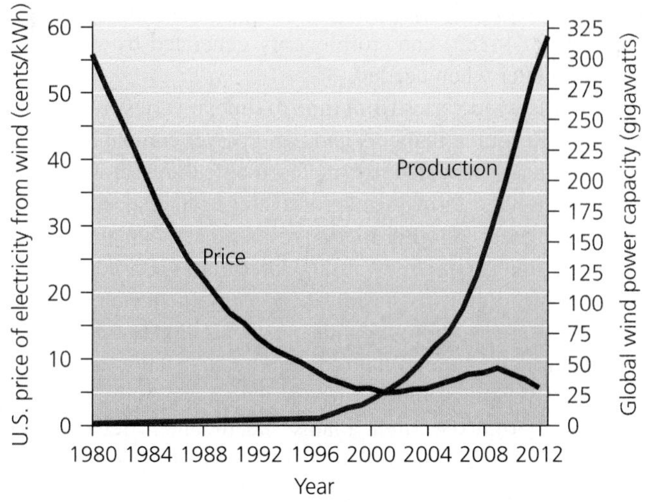


**FIGURE 16.13 Global production of wind power has been doubling every three years in recent years, and prices have fallen.** *Data from Global Wind Energy Council; and U.S. Department of Energy, 2013.* Revolution now: The future arrives for four clean energy technologies.

**FIGURE 16.15 More and more wind farms are being developed offshore.** Offshore winds tend to be stronger yet less turbulent.

## Offshore sites hold promise

Wind speeds on average are 20% greater over water than over land, and air is less turbulent over water. For these reasons, offshore wind turbines are becoming popular (**FIGURE 16.15**). Costs to erect and maintain turbines in water are higher, but the stronger, less turbulent winds make offshore wind potentially more profitable.

Denmark erected the first offshore wind farm in 1991, and soon more came into operation across northern Europe. Germany raised its feed-in tariff rate for offshore wind by 67% in 2009, and within just five years 14 new wind farms were operating. By 2014, nearly 2100 wind turbines were powering 69 wind farms in the waters of 11 European nations.

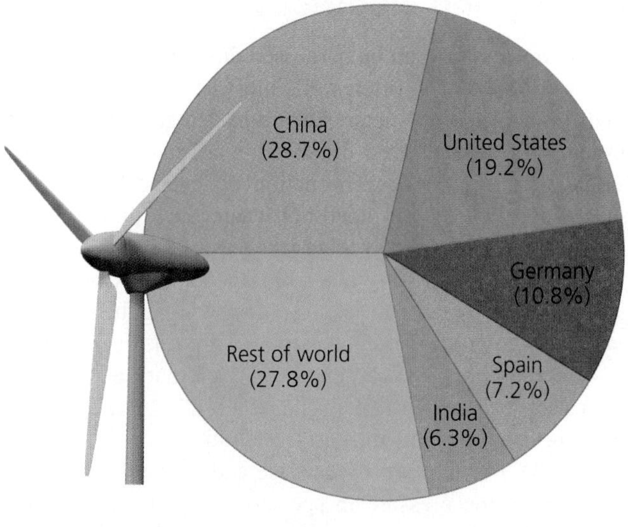


**FIGURE 16.14 Most of the world's wind power capacity has been developed in China, the United States, and Germany.** *Data from Global Wind Energy Council.*

In the United States, no offshore wind farms have yet been constructed, but development of the first was approved in 2010 after nine years of debate. The Cape Wind offshore wind farm, if constructed, will feature 130 turbines rising from Nantucket Sound 8 km (5 mi) off the coast of Cape Cod in Massachusetts. As of 2014, ten offshore wind developments were in the planning stages off the Northeast coast, the Texas coast, and in Lake Erie.

## Wind power has many benefits

Like solar power, wind power produces no emissions once the equipment is manufactured and installed. As a replacement for fossil fuel combustion in the average U.S. power plant, running a 1-megawatt wind turbine for 1 year prevents the release of more than 1500 tons of carbon dioxide, 6.5 tons of sulfur dioxide, 3.2 tons of nitrogen oxides, and 60 lb of mercury, according to the U.S. Environmental Protection Agency. The amount of carbon pollution that all U.S. wind turbines together prevent from entering the atmosphere is equal to the emissions from nearly 10 million cars or from combusting the cargo of a 750-car freight train of coal each and every day.

Under optimal conditions, wind power appears efficient in its energy returned on investment (EROI; p. 335). Most studies find that wind turbines produce roughly 20 times more energy than they consume—an EROI value better than that of most energy sources.

Wind turbine technology can be used on many scales, from a single tower for local use to large farms that supply whole regions. Small-scale turbine development can help make local areas more self-sufficient, just as solar energy can. Farmers and ranchers can also make money leasing their land for wind development. A single large turbine can bring in $2000 to $4500 in annual royalties while occupying just a quarter-acre of land. This also increases property tax income for the rural community.

Lastly, wind power creates job opportunities. Over 85,000 Americans and 800,000 people globally are employed in the wind industry. More than 100 colleges and universities now offer programs and degrees that train people in the skills needed for jobs in wind power and other renewable energy fields.

## Wind power has limitations

Wind is an intermittent resource; we have no control over when it will occur. This limitation is alleviated, though, if wind is one of several sources contributing to a utility's power generation. Pumped-storage hydropower (p. 380) can help to compensate during windless times, and batteries or hydrogen fuel (pp. 384–385) can store energy generated by wind and release it later when needed.

Just as wind varies from time to time, it varies from place to place. Resource planners and wind power companies study wind patterns revealed by meteorological research before planning a wind farm. A map of average wind speeds across the United States (**FIGURE 16.16a**) reveals that mountain regions are best, along with areas of the Great Plains. Based on such data, the wind industry has located much of its generating capacity in states with high wind speeds (**FIGURE 16.16b**) and is expanding in the Great Plains and mountain states.

However, most of North America's people live near the coasts, far from the Great Plains and mountain regions that have the best wind resources. Thus, continent-wide transmission networks would need to be enhanced to send wind power's electricity to these population centers. When wind farms *are* proposed near population centers, local residents often oppose them. Turbines are generally located in exposed, conspicuous sites, and some people object to them for aesthetic reasons.

Turbines also pose a hazard to birds and bats, which are killed when they fly into the blades. Large open-country raptors such as golden eagles are known to be at risk, and turbines located on ridges along migratory flyways are likely most damaging. More research on wildlife impacts is urgently needed.

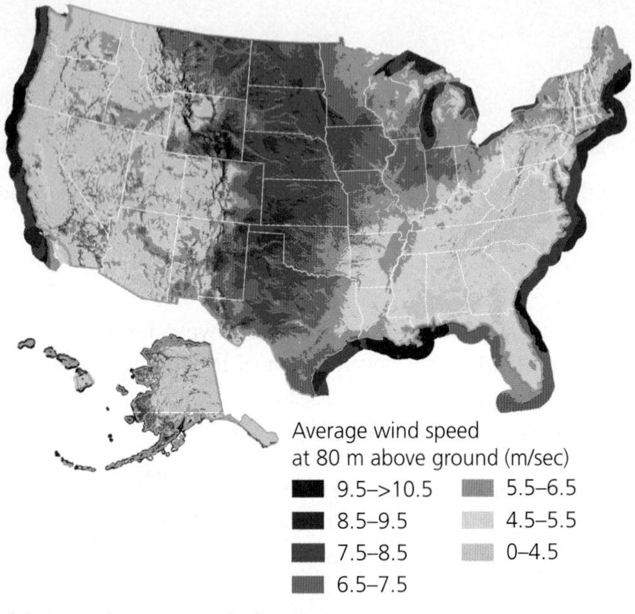


**(a) Annual average wind power**

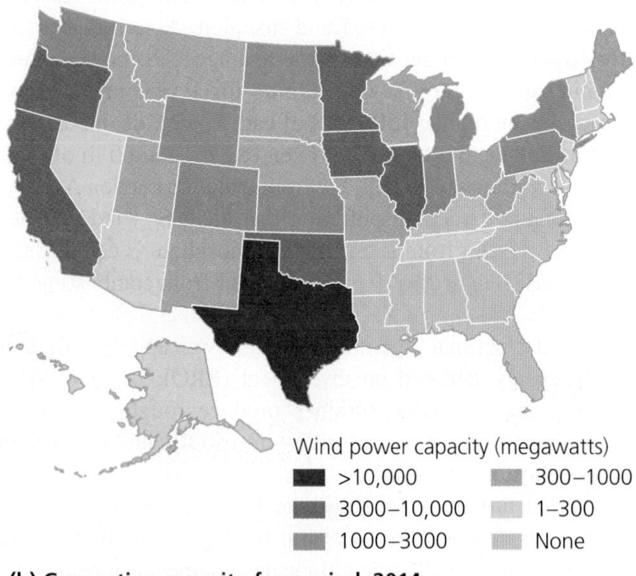


**(b) Generating capacity from wind, 2014**

**FIGURE 16.16 Wind speed varies from place to place.** Maps of average wind speeds **(a)** help guide the placement of wind farms. Another map **(b)** shows wind-power-generating capacity developed in each U.S. state through the end of 2013. *Sources: (a) U.S. National Renewable Energy Laboratory; (b) American Wind Energy Association, 2014.* 2nd quarter 2014 market report.

DATA **Q** Compare parts (a) and (b). Which states or regions have high wind speeds but are not yet heavily developed with commercial wind power?

GO TO **INTERPRETING GRAPHS & DATA** ON MasteringEnvironmentalScience®

# Geothermal Energy

**Geothermal energy** is thermal energy that arises from beneath Earth's surface. The radioactive decay of elements (p. 28) amid high pressures deep in the interior of our planet generates heat that rises to the surface in molten rock and through cracks and fissures. In regions where this energy heats groundwater, spurts of water and steam rise from below and may erupt through the surface as geysers or as submarine hydrothermal vents (p. 259).

## We harness geothermal energy for heating and electricity

Geothermal energy can be harnessed directly from geysers at the surface, but most often wells must be drilled down hundreds or thousands of meters toward heated groundwater. Hot groundwater can be used directly for heating buildings and for industrial processes. The nation of Iceland heats nearly 90% of its homes with piped hot water. Such direct heating is efficient and inexpensive, but it is feasible only where geothermal energy is readily available. Iceland has a wealth of geothermal resources because it is located along the spreading boundary of two tectonic plates (pp. 227–228).

Geothermal power plants harness naturally heated water and steam to generate electricity (**FIGURE 16.17**). A power plant brings groundwater at temperatures of 150–370°C (300–700°F) or more to the surface and converts it to steam by lowering the pressure in specialized compartments. The steam turns turbines to generate electricity. The world's largest geothermal plants, The Geysers in California, provide electricity for 725,000 homes.

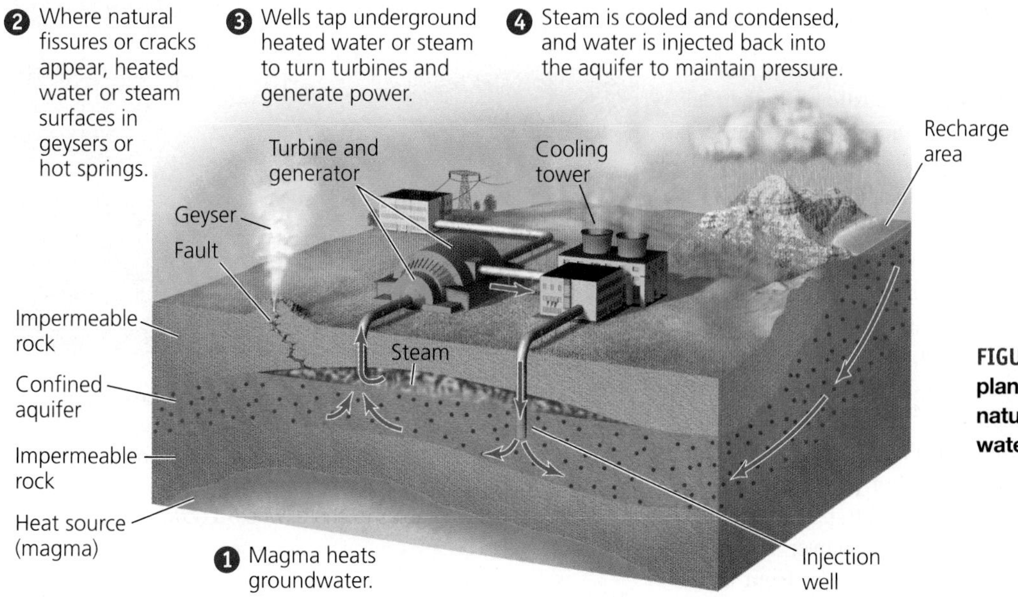
2 Where natural fissures or cracks appear, heated water or steam surfaces in geysers or hot springs.
3 Wells tap underground heated water or steam to turn turbines and generate power.
4 Steam is cooled and condensed, and water is injected back into the aquifer to maintain pressure.
Turbine and generator
Cooling tower
Recharge area
Geyser
Fault
Impermeable rock
Steam
Confined aquifer
Impermeable rock
Heat source (magma)
1 Magma heats groundwater.
Injection well

**FIGURE 16.17 Geothermal power plants generate electricity using naturally heated underground water.**

## Heat pumps make use of temperature differences

Heated groundwater is available only in certain areas, but we can take advantage of the mild temperature differences that exist naturally between the soil and the air just about anywhere. Soil varies in temperature from season to season less than air does, and **ground-source heat pumps** (GSHPs) make use of this fact. These pumps provide heating in the winter by transferring heat from the ground into buildings, and they provide cooling in the summer by transferring heat from buildings into the ground. This heat transfer is accomplished with a network of underground plastic pipes that circulate water and antifreeze (**FIGURE 16.18**). Because heat is simply moved from place to place rather than being produced using outside energy inputs, heat pumps can be highly efficient.

More than 600,000 U.S. homes use GSHPs. Compared to conventional electric heating and cooling systems, GSHPs heat spaces 50–70% more efficiently, cool them 20–40% more efficiently, can reduce electricity use by 25–60%, and can reduce emissions by up to 70%.

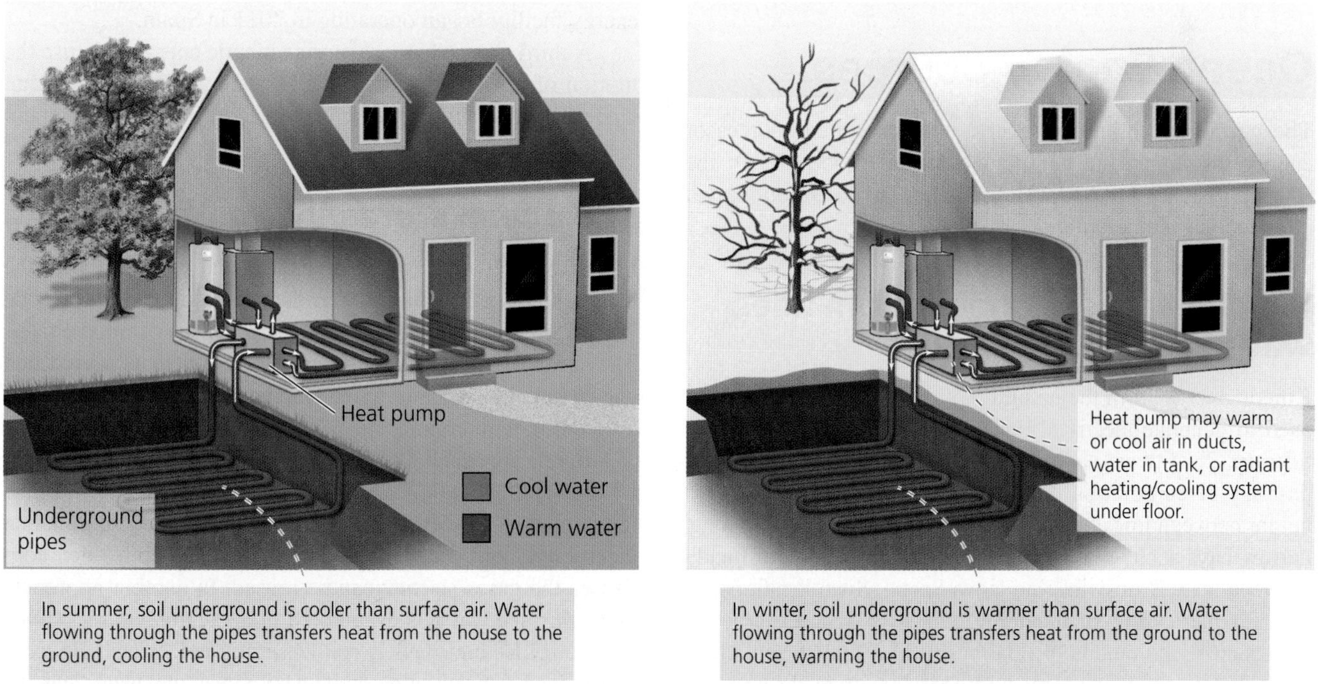
Heat pump
Cool water
Warm water
Underground pipes
In summer, soil underground is cooler than surface air. Water flowing through the pipes transfers heat from the house to the ground, cooling the house.
Heat pump may warm or cool air in ducts, water in tank, or radiant heating/cooling system under floor.
In winter, soil underground is warmer than surface air. Water flowing through the pipes transfers heat from the ground to the house, warming the house.

**FIGURE 16.18 Ground-source heat pumps provide an efficient way to heat and cool air and water in a home.** A network of pipes filled with water and antifreeze extends underground. Soil is cooler than air in the summer **(left)**, and warmer than air in the winter **(right)**, so by running fluid between the house and the ground, these systems adjust temperatures inside.

## Geothermal power has pros and cons

Like wind and solar, geothermal heating and power greatly reduce emissions relative to fossil fuel combustion. Geothermal energy is renewable in that its use does not affect the amount of thermal energy produced underground. However, not every geothermal power plant will be able to operate indefinitely. If a plant uses heated water more quickly than groundwater is recharged, it will eventually run out of water. This was occurring at The Geysers in California, so operators began injecting municipal wastewater into the ground to replenish the supply! Moreover, patterns of geothermal activity in Earth's crust shift naturally over time, so an area that produces hot water may not always do so. In addition, some hot groundwater is laced with salts and minerals that corrode equipment and pollute the air. These factors may shorten the lifetime of plants, increase maintenance costs, and add to pollution.

The greatest limitation of geothermal power is that it is restricted to regions where we can tap energy from naturally heated groundwater. Places such as Iceland, northern California, and Yellowstone National Park are rich in naturally heated groundwater, but most areas of the world are not. Engineers are now trying to overcome this limitation by developing **enhanced geothermal systems** (EGS), in which we drill deeply into dry rock, fracture the rock, and pump in cold water. The water becomes heated deep underground and is then drawn back up and used to generate power. In theory we could use EGS widely in many locations. Germany, for instance, has little heated groundwater, but feed-in-tariffs have enabled an EGS facility to operate profitably there. However, EGS appears to trigger occasional minor earthquakes. Unless we can develop ways to use EGS safely and reliably, our use of geothermal power will remain localized.

# Ocean Energy Sources

The oceans are home to several underexploited sources of energy resulting from continuous natural processes. Of the four approaches being developed, three involve motion, and one involves temperature.

## We can harness energy from tides, waves, and currents

Kinetic energy from the natural motion of ocean water can be used to generate electrical power. The rise and fall of ocean tides (p. 257) twice each day moves large volumes of water past any given point on the world's coastlines. Differences in height between low and high tides are especially great in long, narrow bays such as Alaska's Cook Inlet or the Bay of Fundy between New Brunswick and Nova Scotia. Such locations are best for harnessing **tidal energy**, which is done by erecting dams across the outlets of tidal basins. As tidal currents pass through the dam, water turns turbines to generate electricity.

The world's largest tidal generating station is South Korea's Sihwa Lake facility, which opened in 2011 and is just larger than the La Rance facility in France, which has operated for nearly 50 years. The first U.S. tidal station began operating in 2012 in Maine, and one is now being installed in New York City's East River. Tidal stations release few or no pollutant emissions, but they can affect the ecology of estuaries and tidal basins.

**FIGURE 16.19 Various technologies harness energy from ocean waves.** This wave energy converter stretches across the water off the coast of Scotland. As waves flex segments of the machinery, hydraulic equipment in the joints generates electricity and sends it by undersea cables to shore, where it powers 500 homes.

Scientists and engineers are also working to harness the motion of ocean waves and convert this kinetic energy into electricity. Many designs for machinery to harness **wave energy** have been invented, but few have been adequately tested. Some designs for offshore facilities involve floating devices that move up and down with the waves (**FIGURE 16.19**). Some designs for onshore facilities funnel waves from large areas into narrow channels and elevated reservoirs, from which water then flows out, generating electricity. Other coastal designs use rising and falling waves to push air into and out of chambers, turning turbines. The first commercial wave energy facility began operating in 2011 in Spain.

A third way to harness marine kinetic energy is to use the motion of ocean currents (p. 254), such as the Gulf Stream. Underwater turbines have been erected in European waters to test this approach.

## The ocean stores thermal energy

Each day the tropical oceans absorb solar radiation with the heat content of 250 billion barrels of oil—enough to provide 20,000 times the electricity used daily in the United States. The ocean's sun-warmed surface is warmer than its deep water, and **ocean thermal energy conversion** (OTEC) relies on this gradient in temperature.

In one approach, warm surface water is piped into a facility to evaporate chemicals, such as ammonia, that boil at low temperatures. The evaporated gases spin turbines to generate electricity. Cold water piped up from ocean depths then condenses the gases so they can be reused. In another approach, warm surface water is evaporated in a vacuum, and its steam turns turbines and then is condensed by cold water. Because ocean water loses salts as it evaporates, the water can be recovered, condensed, and sold as desalinized fresh water for drinking or agriculture. OTEC research has been conducted in Hawaii and elsewhere, but costs remain high, and so far no facility operates commercially.

## Hydroelectric Power

In **hydroelectric power**, or **hydropower**, we use the kinetic energy of flowing river water to turn turbines and generate electricity. Next to biomass, we draw more renewable energy from the motion of fresh water than from any other resource. In total, hydropower provides nearly one-sixth of the electricity generated throughout the world. We examined hydroelectric power and its environmental impacts in our discussion of fresh water resources (pp. 263–264). Now we will take a closer look at hydropower as an energy source.

### Hydropower uses three approaches

Most hydroelectric power today comes from impounding water in reservoirs behind concrete dams that block the flow of river water and then letting that water pass through the dam. Because water is stored behind dams, this is called the **storage** technique. As reservoir water passes through a dam, it turns the blades of turbines, which cause a generator to generate electricity (**FIGURE 16.20**). Electricity generated in the powerhouse is transmitted to the grid by transmission lines, while the water flows into the riverbed below

**(a) Ice Harbor Dam, Snake River, Washington**

**(b) Turbine generator inside McNary Dam, Columbia River**

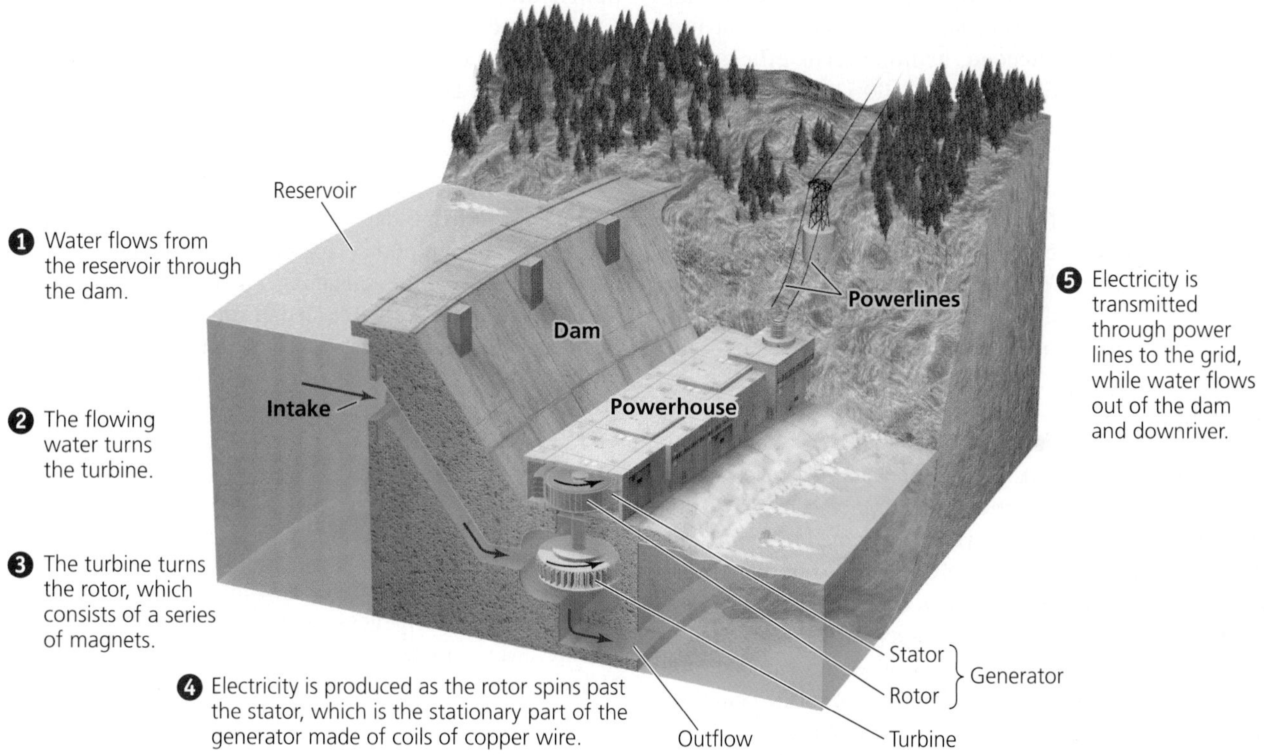
Reservoir
❶ Water flows from the reservoir through the dam.
Dam
Intake
Powerlines
Powerhouse
❷ The flowing water turns the turbine.
❸ The turbine turns the rotor, which consists of a series of magnets.
❹ Electricity is produced as the rotor spins past the stator, which is the stationary part of the generator made of coils of copper wire.
❺ Electricity is transmitted through power lines to the grid, while water flows out of the dam and downriver.
Stator
Rotor
Generator
Outflow
Turbine

**(c) Hydroelectric power**

**FIGURE 16.20 We generate hydroelectric power with large dams.** Inside these dams **(a)**, flowing water is used to turn massive turbines **(b)**. Water is funneled from the reservoir through the dam **(c)** and into the riverbed below to generate electricity by this process.

the dam and continues downriver. By storing water in reservoirs, dam operators can ensure a steady and predictable supply of electricity, even during periods of naturally low river flow.

An alternative approach is the **run-of-river** technique, which generates electricity without greatly disrupting a river's flow. One method is to divert a portion of a river's flow through a pipe or channel, passing it through a powerhouse and returning it to the river. Run-of-river systems are useful in areas remote from established electrical grids and in regions without the economic resources to build and maintain large dams. This approach cannot guarantee reliable water flow in all seasons, but it minimizes many impacts of the storage technique.

To better control the timing of flow, pumped-storage hydropower can be used. In **pumped storage**, water is pumped from a lower reservoir to a higher reservoir at times when demand for power is weak and prices are low. When demand is strong and prices are high, water is sent downhill through a turbine, generating electricity. Although energy must be input to pump the water, pumped storage can be profitable. When paired with intermittent sources such as solar and wind power, it can help balance a region's power supply by compensating for dips in power.

## Hydropower is clean and renewable, yet has impacts

Hydroelectric power has three clear advantages over fossil fuels. First, it is renewable; as long as precipitation falls from the sky and fills rivers and reservoirs, we can use water to turn turbines. Second, hydropower is efficient. It is thought to have an EROI ratio of 100:1 or more, higher than any other modern-day energy source. Third, no carbon compounds are burned in the production of hydropower, so no carbon dioxide or other pollutants are emitted into the atmosphere. Fossil fuels *are* used in constructing and maintaining dams—and large reservoirs may release the greenhouse gas methane as a result of anaerobic decay in deep water. But overall, hydropower accounts for only a small fraction of the greenhouse gas emissions typical of fossil fuel combustion.

Damming rivers (p. 263), however, destroys habitat for wildlife as riparian areas above dam sites are submerged and those below often are starved of water. Because water discharge is regulated to optimize electricity generation, the natural flooding cycles of rivers are disrupted. Suppressing flooding prevents river floodplains from receiving fresh nutrient-laden sediments. Instead, sediments become trapped behind dams, where they begin filling the reservoir. Dams also cause thermal pollution (p. 269) by modifying water temperatures, and this, along with habitat alteration, has diminished or eliminated many native fish populations in dammed waterways. In addition, dams generally block the passage of fish and other aquatic creatures, fragmenting the river and reducing biodiversity in each stretch. These ecological impacts generally translate into negative social and economic impacts on local communities.

**TABLE 16.1 Top Producers of Hydropower**

| Nation | Hydropower produced (Percentage of world total) | Percentage of nation's electricity generation from hydropower |
|---|---|---|
| China | 19.6 | 14.8 |
| Brazil | 12.0 | 80.6 |
| Canada | 10.5 | 59.0 |
| United States | 9.7 | 7.9 |
| Russia | 4.7 | 15.9 |
| India | 3.7 | 12.4 |
| Norway | 3.4 | 95.3 |
| Japan | 2.6 | 8.7 |
| Venezuela | 2.3 | 68.6 |
| Sweden | 1.9 | 44.3 |
| Rest of world | 29.6 | 13.6 |

*Data from International Energy Agency, 2013.* Key world energy statistics. *IEA, Paris, France.*

## Hydroelectric power is widely used, but it may not expand much more

Hydropower accounts for 16% of the world's electricity production (see Figure 15.3b, p. 335). For nations with large amounts of river water and the economic resources to build dams, hydroelectric power has been a keystone of their development and wealth. Canada, Brazil, Norway, Austria, Venezuela, and other nations obtain large amounts of their energy from hydropower (TABLE 16.1).

Today the world is witnessing some gargantuan hydroelectric projects. China's recently completed Three Gorges Dam (p. 263) is the world's largest. However, hydropower may not expand much more, because most of the world's large rivers are already dammed. In addition, as people have grown aware of the ecological impacts of dams, residents of some regions are resisting dam construction. In the United States, 98% of rivers appropriate for dam construction already are dammed. Many of the remaining 2% are protected under the Wild and Scenic Rivers Act, and efforts are underway to dismantle some dams and restore river habitats (pp. 263–264).

# Bioenergy

**Bioenergy**—also known as **biomass energy**—is energy obtained from biomass. **Biomass** (p. 72) consists of organic material derived from living or recently living organisms, and it contains chemical energy that originated with sunlight and photosynthesis. We harness bioenergy by burning biomass for heating, using biomass to generate electricity, and processing biomass to create liquid fuels for transportation (TABLE 16.2).

## Fuelwood is used widely in the developing world

Over 1 billion people use wood from trees as their principal energy source. In rural regions of developing nations, people

**TABLE 16.2 Sources and Uses of Bioenergy**

| | |
|---|---|
| **Biomass for direct combustion for heating** | |
| • | Wood cut from trees (fuelwood) |
| • | Charcoal |
| • | Manure from farm animals |
| **Biomass for generating electricity (biopower)** | |
| • | Crop residues (such as cornstalks) burned at power plants |
| • | Forestry residues (wood waste from logging) burned at power plants |
| • | Processing wastes (from sawmills, pulp mills, paper mills, etc.) burned at power plants |
| • | "Landfill gas" burned at power plants |
| • | Livestock waste from feedlots for gas from anaerobic digesters |
| **Biofuels for powering vehicles** | |
| • | Corn grown for ethanol |
| • | Bagasse (sugarcane residue) grown for ethanol |
| • | Soybeans, rapeseed, and other crops grown for biodiesel |
| • | Used cooking oil for biodiesel |
| • | Algae grown for biofuels |
| • | Plant matter treated with enzymes to produce cellulosic ethanol |

**FIGURE 16.21 Well over a billion people in developing countries rely on wood from trees for heating and cooking.** In theory, biomass is renewable, but in practice it may not be if forests are overharvested.

(generally women) gather fuelwood to burn in their homes for heating, cooking, and lighting (**FIGURE 16.21**). Although fossil fuels and electricity are replacing traditional energy sources as developing nations industrialize, fuelwood, charcoal, and livestock manure constitute three-quarters of all renewable energy used worldwide.

These traditional biomass sources are renewable only if they are not overharvested. Harvesting fuelwood at unsustainably rapid rates can lead to deforestation, soil erosion, and desertification (pp. 189, 141, 142). Burning fuelwood for cooking and heating also poses health hazards from indoor air pollution (p. 298).

## We use biomass to generate electricity

Biomass can be burned in power plants to produce heat and generate electricity. This production of **biopower** can be achieved using a variety of sources and techniques.

**Waste products** The waste products of various industries and processes may be combusted for biopower. These include woody debris from logging, liquid waste from pulp mills, organic waste from landfills or feedlots, and residue from crops (such as cornstalks and corn husks).

**Crops** We are beginning to grow certain plants as crops specifically to generate biopower. These include fast-growing grasses such as bamboo, fescue, and switchgrass, as well as trees, such as specially bred willows and poplars.

**Combustion strategies** At small scales, farmers, ranchers, or villages can operate modular biopower systems that use livestock manure to generate electricity. Small household biodigesters provide portable and decentralized energy production for remote rural areas.

At large scales, power plants built to run on biomass operate like those fired by fossil fuels; combustion heats water, creating steam to turn turbines and generators, thereby generating electricity. These facilities are often located where they can take advantage of forestry waste.

In some coal-fired power plants, wood chips, wood pellets, or other biomass is introduced with coal into a high-efficiency boiler in a process called *co-firing*. We can substitute biomass for up to 15% of the coal with only minor equipment modification and no appreciable loss of efficiency. Co-firing is a relatively easy and inexpensive way for utilities to expand their use of renewable energy.

We also harness biopower by *gasification*, in which biomass is vaporized at high temperatures in the absence of oxygen, creating hydrogen, carbon monoxide, carbon dioxide, and methane. This mixture generates electricity when used to turn a gas turbine to propel a generator in a power plant. Another method of heating biomass in the absence of oxygen results in *pyrolysis* (p. 338), producing a liquid fuel that can be burned to generate electricity.

**Benefits and drawbacks** By enhancing energy efficiency and putting waste products to use, biopower helps move our utilities and industries in a sustainable direction. Gaining power from biomass also helps alleviate climate change by reducing carbon dioxide emissions. Capturing landfill gas reduces emissions of methane, a potent greenhouse gas. By replacing coal in co-firing and direct combustion, biopower reduces emissions of sulfur dioxide as well, because plant matter, unlike coal, contains no appreciable sulfur. One disadvantage is that when we burn plant matter for power, we deprive the soil of nutrients and organic matter it would have gained from the plant matter's decomposition, thus requiring practices to restore soil fertility.

(a) Corn grown for ethanol

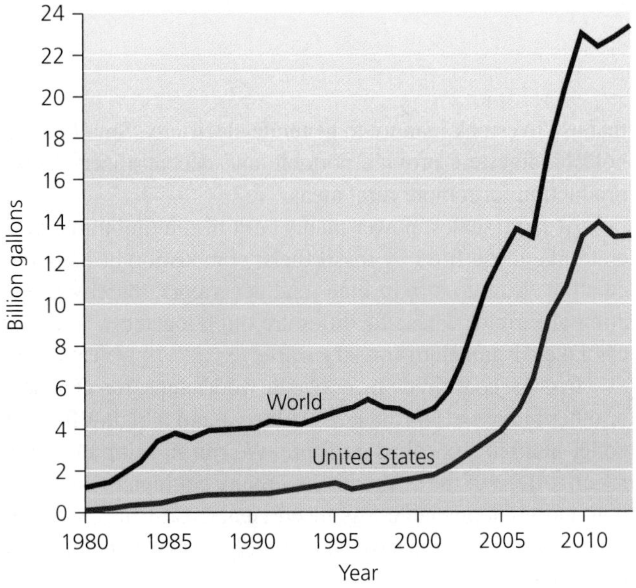


**(b) Ethanol production, 1980–2013**

**FIGURE 16.22 Ethanol has been booming.** About 40% of the U.S. corn crop **(a)** is used to produce ethanol to add to gasoline (28% after by-products are put to use). Brazil produces most of the rest of the world's ethanol, from bagasse (sugarcane residue). Ethanol production **(b)** has grown rapidly in recent years. *Data from Renewable Fuels Association.*

DATA Q Roughly what percentage of the world's ethanol is produced by the United States?

GO TO **INTERPRETING GRAPHS & DATA** ON MasteringEnvironmentalScience®

## Biofuels can power vehicles

Some biomass sources can be converted into **biofuels**, liquid fuels used primarily to power automobiles. The two primary biofuels developed so far are ethanol (for gasoline engines) and biodiesel (for diesel engines).

**Ethanol** **Ethanol** is the alcohol in beer, wine, and liquor. It is produced as a biofuel by fermenting biomass, generally from carbohydrate-rich crops, in a process similar to brewing beer. In fermentation, carbohydrates are converted to sugars and then to ethanol. Spurred by the 1990 Clean Air Act amendments and generous government subsidies, ethanol is widely added to gasoline in the United States to reduce automotive emissions. In 2013 in the United States, over 50 billion L (13.3 billion gal) of ethanol were produced—42 gallons for every American—mostly from corn (**FIGURE 16.22**). This amount will continue to grow, as Congress in 2007 mandated production and use of 136 billion L (36 billion gal) per year of ethanol by 2022.

Any vehicle with a gasoline engine runs well on gasoline blended with up to 10% ethanol, but automakers are also producing *flexible-fuel* vehicles that run on E-85, a mix of 85% ethanol and 15% gasoline. Over 15 million such cars have been built in the United States since 1998. In Brazil, almost all new cars are flexible-fuel vehicles, and ethanol from crushed sugarcane residue (called *bagasse*) accounts for half of all fuel that Brazil's drivers use.

The enthusiasm for corn-based ethanol shown by U.S policymakers is not widely shared by scientists. Growing corn to produce ethanol intensifies pesticide use, fertilizer use, fresh water depletion, and other impacts of industrial agriculture (p. 138). Fully 40% of the U.S. corn crop today is used to make ethanol. (Some by-products of ethanol production are used in livestock feed; with this accounted for, 28% of the U.S. corn crop goes solely toward ethanol.) Corn ethanol crops take up millions of acres of land. To produce all the automotive fuel used in the United States with ethanol from U.S. corn, the nation would need to expand its already immense corn acreage by more than four times (**FIGURE 16.23**). Even at our current level of production, ethanol already competes with food production and drives up food prices.

Growing corn for ethanol also requires substantial inputs of fossil fuel energy (for running farm equipment, making petroleum-based pesticides and fertilizers, transporting corn to processing plants, and heating water in refineries to distill

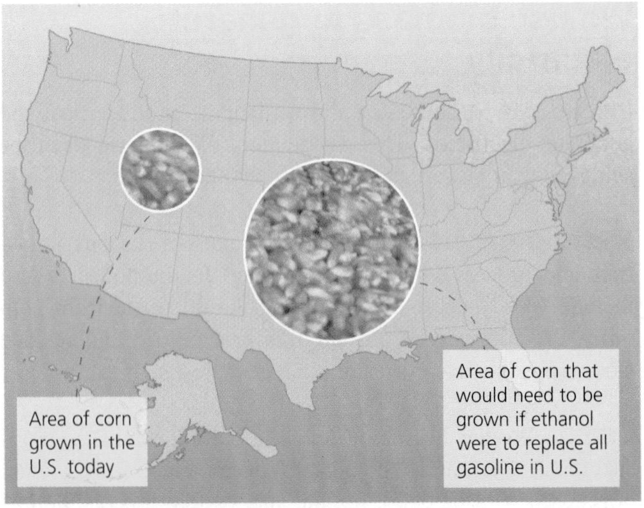


**FIGURE 16.23 Growing corn to produce ethanol uses up a great deal of land.** To produce enough ethanol to replace all gasoline used by U.S. drivers, the United States would need to more than quadruple its area planted in corn.

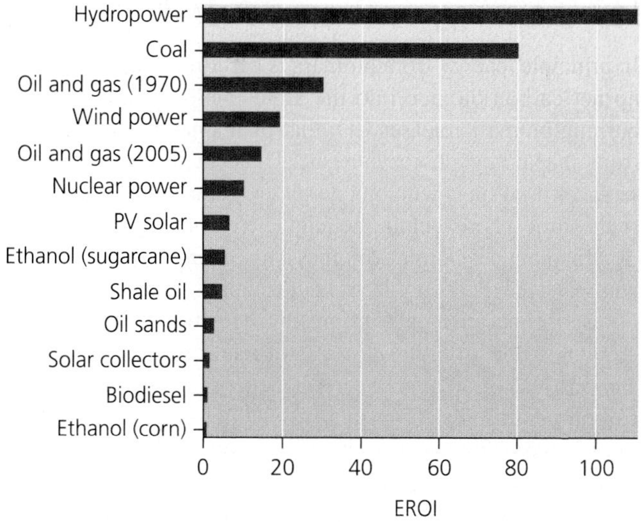


**FIGURE 16.24 EROI values vary from over 100 : 1 for hydropower down through 1.3 : 1 for biodiesel and corn ethanol.** *Adapted from Murphy, D., and C. Hall, 2010. Year in review—EROI or energy return on (energy) invested.* Annals New York Acad. Sci. *1185: 102–118.*

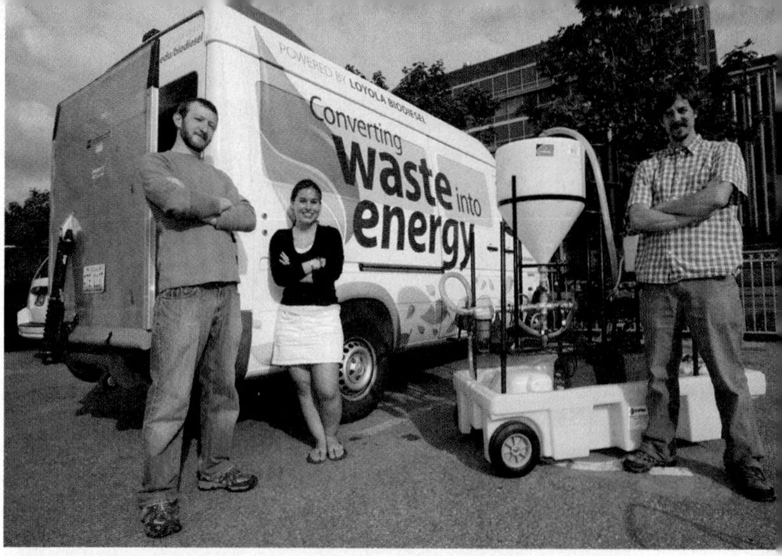


**FIGURE 16.25 At Loyola University Chicago, students and staff produce biodiesel from waste vegetable oil from the dining halls and use it to fuel this van.** They transport this mini-diesel reactor to local high schools to teach students about alternative fuels.

ethanol). In fact, corn ethanol yields only a modest amount of energy relative to the energy that needs to be input. The EROI (energy returned on investment) ratio (p. 335) for corn-based ethanol is variable and controversial, but recent estimates place it around 1.3:1 (**FIGURE 16.24**). This means that to gain 1.3 units of energy from ethanol, we need to expend 1 unit of energy.

**FAQ** **If we substitute ethanol for gasoline, won't that solve most of our problems with oil dependence?**

Increasing the proportion of ethanol in gasoline does indeed help to conserve oil and reduce reliance on foreign imports. However, obtaining the amount of corn ethanol needed to replace gasoline entirely would require that impractically large amounts of land be converted to corn production. Moreover, so much corn would likely be diverted from food to fuel that food prices would rise sharply. This is why researchers are studying other plants as more efficient sources of ethanol and are trying to produce cellulosic ethanol from crop and forestry wastes.

**Biodiesel** Drivers of diesel-fueled vehicles can use **biodiesel**, a fuel produced from vegetable oil, used cooking grease, or animal fat. The oil or fat is mixed with small amounts of ethanol or methanol in the presence of a chemical catalyst. In Europe, where most biodiesel is used, rapeseed oil is the oil of choice, whereas U.S. biodiesel producers use mostly soybean oil. Vehicles with diesel engines can run on 100% biodiesel, or biodiesel can be mixed with conventional diesel; a 20% biodiesel mix (called B20) is common.

Replacing diesel with biodiesel cuts down on emissions. Biodiesel's fuel economy is nearly as good, it costs just slightly more, and it is nontoxic and biodegradable. Increasing numbers of people are fueling their cars with biodiesel from waste oils. Indeed, some college students are creating biodiesel from food waste from dining halls and restaurants (**FIGURE 16.25**). Some buses and recycling trucks now run on biodiesel, and many state and federal fleets use biodiesel blends.

Using waste oil for biofuel is sustainable, but most biodiesel today, like most ethanol, comes from crops grown specifically for the purpose—and this has environmental impacts. Growing soybeans in Brazil and oil palms in Southeast Asia hastens the loss of tropical rainforest (pp. 191–192). Growing soybeans in the United States and rapeseed in Europe takes up large areas of land as well.

**Novel biofuels** Because the major crops grown for biodiesel and ethanol exert heavy impacts on the land, farmers and scientists are experimenting with other crops—from wheat, sorghum, cassava, and sugar beets to less known plants such as hemp, jatropha, and the grass miscanthus.

One promising next-generation biofuel crop is algae. Several species of these photosynthetic organisms produce large amounts of lipids that can be converted to biodiesel. Alternatively, carbohydrates in algae can be fermented to create ethanol. In fact, we can use algae to produce a variety of fuels, even jet fuel. Algae can be grown outdoors in open circulating ponds or in labs in closed tanks or transparent tubes (**FIGURE 16.26**). Algae grow faster and produce more oil than terrestrial biofuel crops, and they can grow in seawater, saline water, or nutrient-rich wastewater from sewage treatment plants. Because carbon dioxide speeds their growth, algae can even make use of smokestack emissions piped in from power plants—thereby providing a source of carbon capture (pp. 323, 351). Producing biofuels from algae is expensive, but private investment may bring costs down.

Relying on any monocultural crop for energy may not be sustainable. With this in mind, researchers are refining techniques to produce **cellulosic ethanol** by using enzymes to produce ethanol from cellulose, the substance that gives structure to all plant material. This would be a substantial advance because ethanol as currently made from corn or sugarcane uses starch, which is nutritionally valuable to us. Cellulose,

**FIGURE 16.26 Algae are a leading candidate for next-generation biofuels.** Here algae are seen growing at a demonstration facility.

in contrast, is of no food value to people yet is abundant in all plants. If we can produce cellulosic ethanol in commercially feasible ways, then ethanol could be made from low-value crop waste (such as corn stalks and husks), rather than from high-value crops.

Moreover, certain plants hold promise as crops specifically for cellulosic ethanol. Switchgrass (**FIGURE 16.27**) might potentially be a sustainable choice for the United States because it is a native grass of the North American prairies.

### WEIGHING THE ISSUES

**Biofuels** Do you think producing and using ethanol from corn is a good idea? Do the benefits outweigh the drawbacks? Explain why or why not. Should we invest billions of dollars into developing next-generation biofuels such as algae and cellulosic ethanol? Can you suggest ways of using biofuels that would minimize environmental impacts?

**FIGURE 16.27 Switchgrass is being studied as a crop to produce cellulosic ethanol, and also provides fuel for biopower.**

### Is bioenergy carbon-neutral?

In principle, energy from biomass is carbon-neutral, releasing no net carbon dioxide into the atmosphere. This is because burning biomass releases $CO_2$ that plants had recently pulled from the air by photosynthesis. Thus, in theory, when we replace fossil fuels with bioenergy, we reduce net carbon flux to the atmosphere, helping to alleviate climate change.

In practice, burning biomass for energy is not carbon-neutral if forests are destroyed to plant bioenergy crops. Forests sequester more carbon (in vegetation and soil) than croplands do, so cutting forests to plant crops will increase carbon flux to the atmosphere. Bioenergy also is not strictly carbon-neutral if we consume fossil fuel energy to produce the biomass (for instance, by using tractors, fertilizers, and pesticides to grow biofuel crops). We have a diversity of options in the realm of bioenergy, however, and researchers are working hard to develop ways of using bioenergy that are truly renewable and sustainable.

## Hydrogen and Fuel Cells

Each renewable energy source we have discussed can be used to generate electricity more cleanly than can fossil fuels. However, electricity cannot be stored easily in large quantities for use when and where it is needed. This is why most vehicles rely on gasoline for power. The development of fuel cells and of fuel consisting of hydrogen—the simplest and most abundant element in the universe—holds promise as a way to store sizeable quantities of energy conveniently, cleanly, and efficiently. Like electricity and like batteries, hydrogen is an energy carrier, not a primary energy source. It holds energy that can be converted for use at later times and in different places.

### Some yearn for a "hydrogen economy"

Some energy experts envision that hydrogen fuel, along with electricity, could serve as the basis for a clean, safe, and efficient energy system. In such a system, electricity generated from intermittent renewable sources, such as wind or solar energy, could be used to produce hydrogen. Fuel cells (**FIGURE 16.28**) could then use hydrogen to produce electricity to power vehicles, computers, cell phones, home heating, and more. NASA's space programs have used fuel-cell technology since the 1960s.

Basing an energy system on hydrogen could alleviate dependence on foreign fuels and help fight climate change. For these reasons, governments have funded research into hydrogen and fuel-cell technology, and auto companies have developed vehicles that run on hydrogen. Today, Germany is one of several nations with hydrogen-fueled city buses, and it is planning a network of hydrogen filling stations for hydrogen cars that are being designed.

**FIGURE 16.28 Hydrogen fuel drives electricity generation in a fuel cell, creating water as a waste product.** Atoms of hydrogen are split ❶ into protons and electrons. The protons, or hydrogen ions ❷, pass through a proton exchange membrane. The electrons, meanwhile, move from a negative electrode to a positive one via an external circuit ❸, creating a current and generating electricity. The protons and electrons then combine with oxygen ❹ to form water molecules.

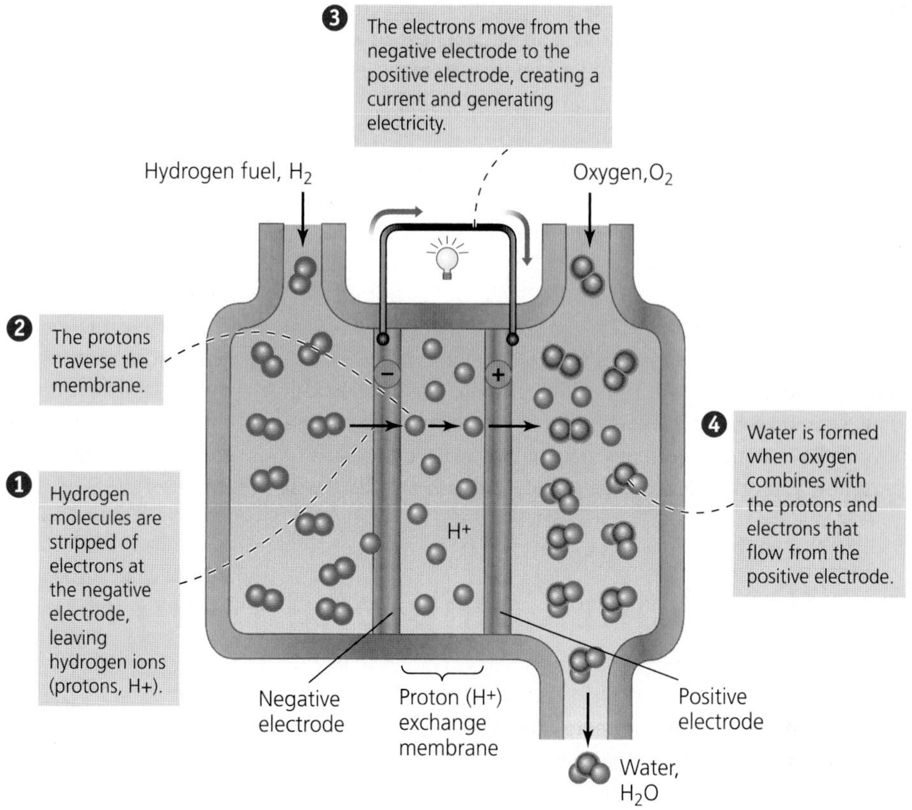

❸ The electrons move from the negative electrode to the positive electrode, creating a current and generating electricity.
Hydrogen fuel, $H_2$
Oxygen, $O_2$
❷ The protons traverse the membrane.
❶ Hydrogen molecules are stripped of electrons at the negative electrode, leaving hydrogen ions (protons, H+).
❹ Water is formed when oxygen combines with the protons and electrons that flow from the positive electrode.
H+
Negative electrode
Proton (H+) exchange membrane
Positive electrode
Water, $H_2O$

## Hydrogen fuel may be produced from water or from other matter

Hydrogen gas ($H_2$) tends not to exist freely on Earth. Instead, hydrogen atoms bind to other molecules, becoming incorporated in everything from water to organic molecules. To obtain hydrogen gas for fuel, we must force these substances to release their hydrogen atoms, and this requires an input of energy. In **electrolysis**, electricity is input to split hydrogen atoms from the oxygen atoms of water molecules:

$$2H_2O \rightarrow 2H_2 + O_2$$

Electrolysis produces pure hydrogen, and it does so without emitting the carbon- or nitrogen-based pollutants of fossil fuel combustion. However, whether this strategy for producing hydrogen will cause pollution over its life cycle depends on the source of electricity used for electrolysis. If coal is burned to generate the electricity, then the process will not reduce emissions. The "cleanliness" of a hydrogen economy would, therefore, depend largely on the source of electricity used in electrolysis.

The environmental impact of hydrogen production also depends on the source material for the hydrogen. Besides water, hydrogen can be obtained from biomass and from fossil fuels. This generally requires less energy input but results in pollution. For instance, extracting hydrogen from methane ($CH_4$) in natural gas produces one molecule of the greenhouse gas carbon dioxide for every four molecules of hydrogen gas:

$$CH_4 + 2H_2O \rightarrow 4H_2 + CO_2$$

Thus, whether a hydrogen-based energy system is cleaner than a fossil fuel system will depend on how the hydrogen is extracted.

Once isolated, hydrogen gas can be used as a fuel to produce electricity within fuel cells. The chemical reaction involved in a fuel cell is simply the reverse of that for electrolysis:

$$2H_2 + O_2 \rightarrow 2H_2O$$

Figure 16.28 shows how this occurs within one common type of fuel cell.

## Hydrogen and fuel cells have costs and benefits

One drawback of hydrogen at this point is a lack of infrastructure. To convert a nation such as Germany or the United States to hydrogen would require massive and costly development of facilities to transport, store, and provide the fuel. Another concern is that some research suggests that leakage of hydrogen could potentially deplete stratospheric ozone (pp. 293–295) and lengthen the atmospheric lifetime of the greenhouse gas methane.

Hydrogen's benefits include the fact that we will never run out of it, because it is the most abundant element in the universe. Hydrogen can be clean and nontoxic to use, and—depending on its source and the source of electricity for its extraction—it may produce few greenhouse gases and other pollutants. Water and heat are the only waste products from a hydrogen fuel cell, along with negligible traces of other compounds. Fuel cells are also energy-efficient; depending on the type, 35–70% of the energy released in the reaction can be used—or up to 90% if the system is designed to capture heat as well as electricity. Unlike batteries (which also produce electricity through chemical reactions), fuel cells generate electricity whenever hydrogen fuel is supplied, without ever needing recharging. For all these reasons, hydrogen fuel cells could soon be used to power cars, much as they already power buses on the streets of some German cities.

## Conclusion

Rising concern over air pollution, climate change, health impacts, and security risks resulting from our reliance on fossil fuels have convinced many people that we need to shift to renewable energy sources that pollute far less and that will not run out. Solar energy, wind power, geothermal energy, and ocean energy sources hold promise to sustain our civilization far into the future without greatly degrading our environment. Hydropower is clean and renewable, but it cannot grow much further and has ecological impacts. Bioenergy sources are diverse and can be carbon-neutral but are not all strictly renewable. By using electricity from renewable sources to produce hydrogen fuel, we could use fuel cells to produce electricity when and where it is needed, helping to create a nonpolluting transportation sector.

Renewable energy sources have been held back by limited funding for research and development and by competition with established nonrenewable fuels whose market prices do not cover external costs. Despite these obstacles, recent progress offers hope that we can shift from fossil fuels to renewable energy.

# Testing Your Comprehension

1. What proportion of our energy today comes from renewable sources? What is the most prevalent form of renewable energy we use? What form of renewable energy is most used to generate electricity?
2. What factors are causing renewable energy use to expand? Which two renewable sources are experiencing the most rapid growth?
3. Describe several passive solar approaches. Now explain how photovoltaic (PV) cells function and are used.
4. List several advantages of solar power. What are some disadvantages?
5. How do wind turbines generate electricity? Describe several environmental and economic benefits of wind power. What are some drawbacks?
6. Define *geothermal energy*, and explain three main ways in which it is obtained and used. Describe one sense in which it is renewable and one sense in which it is not.
7. List and describe four approaches for obtaining energy from ocean water.
8. Compare and contrast the three major approaches to generating hydroelectric power. List one benefit and one negative impact of hydropower.
9. List five sources of bioenergy. What is the world's most used source of bioenergy? Describe two potential benefits and two potential drawbacks of bioenergy.
10. How is hydrogen fuel produced? Is this a clean process? What factors determine the amount of pollutants hydrogen production will emit?

# Seeking Solutions

1. Explain how Germany accelerated its development of renewable energy by establishing a system of feed-in tariffs. Do you think the United States should adopt a similar system? Why or why not?
2. For each source of energy discussed in this chapter, what factors stand in the way of an expedient transition from fossil fuel use? What could be done in each case to ease a shift to these alternative sources?
3. What steps would we need to take to develop renewable energy sources enough to largely replace fossil fuels? Will market forces alone suffice to bring about this transition, or will we also need government? Do you think such a shift will be good for our economy? Why or why not?
4. **THINK IT THROUGH** You are an investor seeking to invest in alternative energy. You are considering buying stock in companies that (1) build corn ethanol refineries, (2) are developing algae farms for biofuels, (3) construct turbines for hydroelectric dams, (4) produce PV solar panels, (5) install wind turbines, and (6) plan to build a wave energy facility. For each company, what questions would you research before deciding how to invest your money? How do you expect you might apportion your investments, and why?

5. **THINK IT THROUGH** You are the president of a nation the size of Germany, and your legislature is asking you to propose a national energy policy. Your country is located along a tropical coastline. Your geologists do not yet know whether there are fossil fuel deposits or geothermal resources under your land. However, your country gets a lot of sunlight and a fair amount of wind, and broad, shallow shelf regions line its coasts. Your nation's population is moderately wealthy but is growing fast, and importing fossil fuels from other nations is becoming expensive.

   What approaches would you propose in your energy policy? Name some specific steps you would urge your legislature to fund. Are there trade relationships you would seek to establish with other countries? What questions would you fund scientists to research?

# Calculating Ecological Footprints

Energy sources vary tremendously in their energy returned on investment (EROI) ratios. Examine the EROI data for each of the energy sources as provided in Figure 16.24, and enter the data in the table below.

| Energy source | Energy Returned on Investment (EROI) |
|---|---|
| Coal | 80 |
| Oil and gas (2005) | |
| Nuclear power | |
| Hydroelectric power | |
| Ethanol from corn | |
| Biodiesel | |
| PV solar | |
| Wind power | |

1. How many units of energy would you generate by investing 1 unit of energy into producing hydroelectric power? To generate that same amount of energy, about how many units of energy would you need to invest into producing nuclear power? Roughly how many units of energy would you need to invest into producing corn ethanol if you wanted to generate that same amount of energy?
2. Based on EROI values, is it more efficient to get energy from oil and gas or from PV solar? Which source would you guess has a larger ecological footprint, based on EROI values?
3. Let's say you wanted to generate 130 units of energy from biodiesel. How many units of energy would you need to invest?
4. Based on EROI ratios alone, which energy sources would you advocate that we further develop? Which would you urge that we avoid? What other issues, besides EROI, are worth considering when comparing energy sources?


## MasteringEnvironmentalScience®

**STUDENTS**

Go to **MasteringEnvironmentalScience** for assignments, the etext, and the Study Area with practice tests, videos, current events, and activities.

**INSTRUCTORS**

Go to **MasteringEnvironmentalScience** for automatically graded activities, current events, videos, and reading questions that you can assign to your students, plus Instructor Resources.

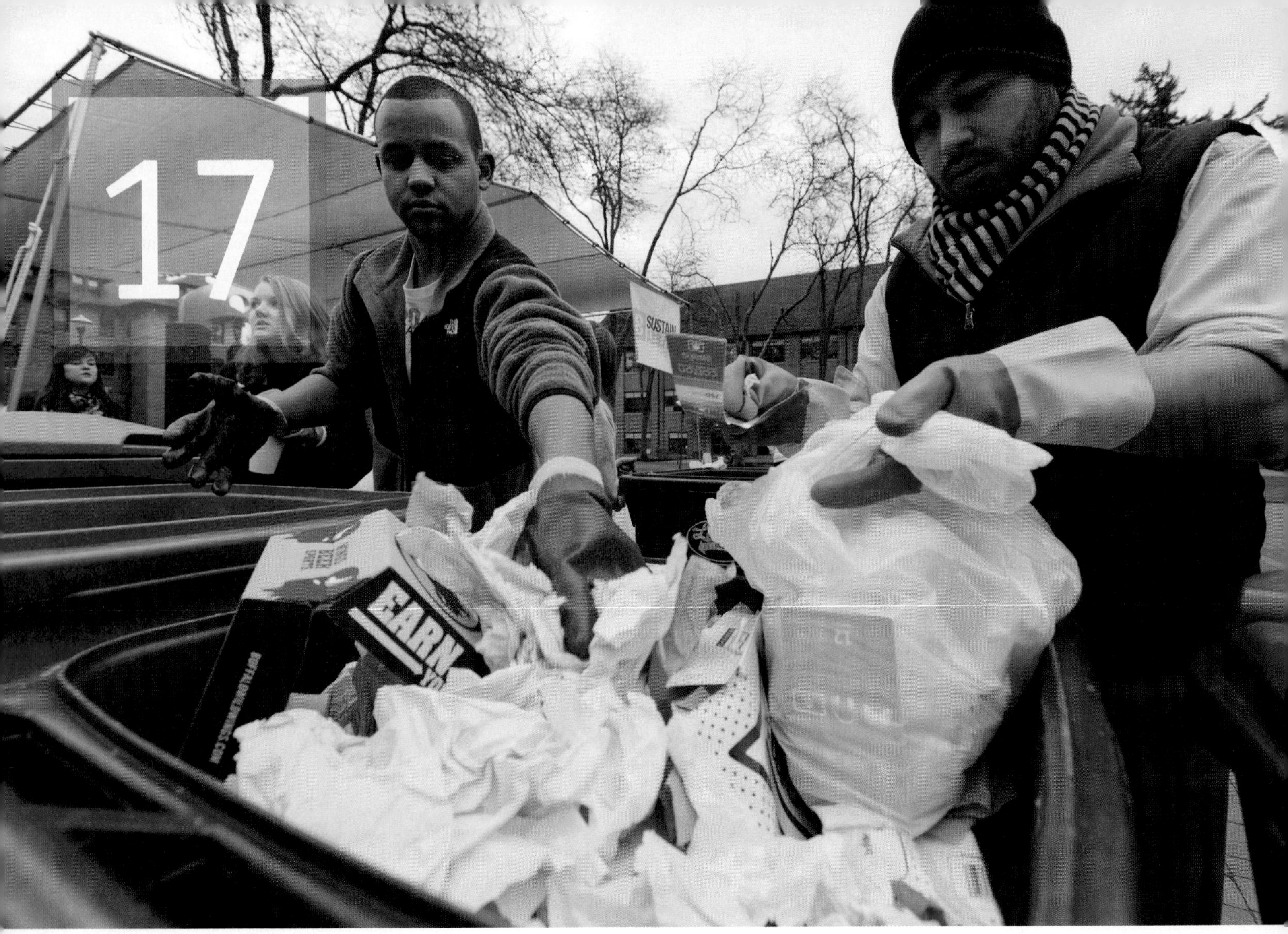


# 17

# Managing Our Waste

## Upon completing this chapter, you will be able to:

- Summarize and compare the types of waste we generate
- List the major approaches to managing waste
- Delineate the scale of the waste dilemma
- Describe conventional waste disposal methods: landfills and incineration
- Evaluate approaches for reducing waste: source reduction, reuse, composting, and recycling
- Discuss management of industrial solid waste and principles of industrial ecology
- Assess issues in managing hazardous waste

**Photo: Students at Pacific Lutheran University compete in the Recyclemania tournament.**

# CENTRAL CASE STUDY

## A Mania for Recycling on Campus

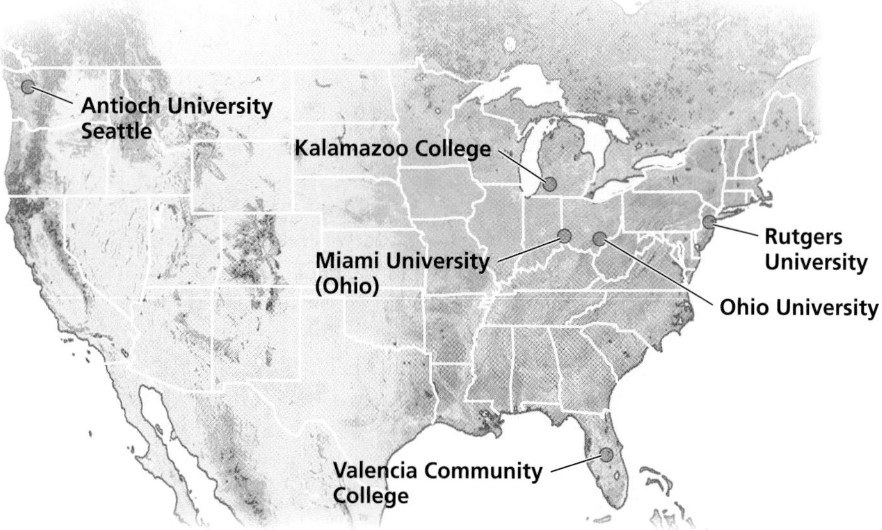


*"An extraterrestrial observer might conclude that conversion of raw materials to wastes is the real purpose of human economic activity."*

**—Gary Gardner and Payal Sampat, Worldwatch Institute**

*"Recycling is one of the best environmental success stories of the late 20th century."*

**—U.S. Environmental Protection Agency**

At that time of year when NCAA basketball fever sweeps America's campuses, there's another kind of March Madness now taking hold: a mania for recycling.

It began in 2001, when waste managers at two Ohio campuses got the idea to use their schools' long-standing athletics rivalry to jump-start their recycling programs. Ed Newman of Ohio University, in Athens, and Stacy Edmonds Wheeler of Miami University, across the state in Oxford, challenged one another to see whose campus could recycle more in a 10-week competition. Come April, Miami University had taken the prize, recycling 41.2 pounds per student. Recyclemania was born.

Students at other colleges and universities heard about the event and wanted to get in on the action, and year by year more schools joined. Today Recyclemania pits several hundred institutions against one another, involving several million students and staff across North America. The event has grown to have a board of directors and major corporate sponsors.

Student leaders rouse their campuses to compete in two divisions and 11 different categories over 8 weeks each spring. Each week, recycling bins are weighed and campuses report their data, which is compiled online at the Recyclemania website as the competition proceeds. The all-around winner gets a funky trophy made of recycled materials (a figure nicknamed "Recycle Dude," whose body is a rusty propane tank)—and more important, global bragging rights for a year.

The spring of 2014 saw a typical battle, as 461 colleges and universities slugged it out from February to April. In the end, the battlefield was littered with stories of the victors and the vanquished.

Antioch University Seattle took top honors as Grand Champion, recycling an amazing 93 percent of its waste, topping runners-up University of Missouri–Kansas City and Richland College. Antioch also nearly won the competition for least waste per capita but was edged out in the final week by Valencia Community College, whose students limited their waste to just 2.87 pounds per person. Kalamazoo College won the competition for most recyclables per capita, with 48.6 pounds per student. The Gorilla category measures total weight of items recycled, and Rutgers University won this prize, having recycled a staggering 1,353,940 pounds.

Campuses also compete to see which can collect the most of certain types of items per person. In 2014, Westfield State University collected the most paper, Bard College saved the most food waste, Kalamazoo College gathered the most bottles and cans, and the U.S. Military Academy recycled the most corrugated cardboard. Southwestern College recycled the most electronic waste per capita and Salt Lake Community College the most film plastic. Lastly, Temple University, Brigham Young University, Minnesota State University–Mankato, and University of California, Berkeley were champions in the competition to see who can best reduce waste at a home basketball game.

By encouraging all this recycling, Recyclemania cuts down on pollution from the mining of new resources and the manufacture of new goods. Each year students in the event help to prevent the release of over 120,000 tons of carbon dioxide—equal to the emissions output of over 25,000 cars or the electricity use of over 15,000 homes. By focusing the attention of administrators on waste issues, Recyclemania facilitates the expansion of campus waste reduction programs. Most important, it gets a new generation of young people revved up about the benefits of recycling.

Recyclemania is the biggest of a growing number of campus competitions in the name of sustainability. The Campus Conservation Nationals are gaining quickly on Recyclemania's predominance. In this 3-week springtime event, schools compete against one another for savings in water use and energy use, and buildings within campuses compete with one another as well. In the 2014 Campus Conservation Nationals, 265,000 students in 1330 buildings at 109 colleges and universities took part, saving 476,000 gallons of water (equal to

3,000 hours in the shower) and 2.2 million kilowatt-hours of electricity (like taking 201 homes off the grid for a year).

Recyclemania has led this trend because recycling is the most widespread activity among campus sustainability efforts (pp. 18, 422). These efforts include water conservation, energy efficiency, green buildings, transportation options, and sustainable food in dining halls and campus gardens. Students restore native plants and habitats, promote renewable energy, and advocate for carbon neutrality on campus. A growing movement, campus sustainability is thriving because for students, faculty, staff, and administrators, it's satisfying to do the right thing and pitch in to help make your campus more sustainable. . . . And it's even more fun when you can compete and show that you can do it better than your rival school across the state! ☐

# Approaches to Waste Management

As the world's population rises, and as we produce and consume more material goods, we generate more waste. **Waste** refers to any unwanted material or substance that results from a human activity or process. Waste can degrade water quality, soil quality, air quality, and human health. Waste also indicates inefficiency—so reducing waste can save money and resources. For these reasons, waste management has become a vital pursuit.

For management purposes, we divide waste into several categories. **Municipal solid waste** is nonliquid waste that comes from homes, institutions, and small businesses. **Industrial solid waste** includes waste from production of consumer goods, mining, agriculture, and petroleum extraction and refining. **Hazardous waste** refers to solid or liquid waste that is toxic, chemically reactive, flammable, or corrosive. Another type of waste is wastewater, water we use in our households, businesses, industries, or public facilities and drain or flush down our pipes, as well as the polluted runoff from our streets and storm drains (pp. 268, 272–274).

## We have several ways to manage waste

There are three main components of **waste management**:

1. Minimizing the amount of waste we generate
2. Recovering discarded materials and finding ways to recycle them
3. Disposing of waste effectively and safely

We have several ways to reduce the amount of material in the **waste stream**, the flow of waste as it moves from its sources toward disposal destinations (**FIGURE 17.1**). Minimizing waste at its source—called **source reduction**—is the preferred approach. The next-best strategy is **recovery**, which consists of recovering, or removing, waste from the waste stream. Recovery includes recycling and composting. **Recycling** is the process of collecting used goods and sending them to facilities that extract and reprocess raw materials that can then be used to manufacture new goods. **Composting** is the practice of recovering organic waste (such as food and yard waste) by converting it to mulch or humus (p. 140) through natural biological processes of decomposition.

Once we lessen the waste stream through source reduction and recovery, there will still be waste left to dispose of. Disposal methods include burying waste in landfills and burning waste in incinerators. We will first examine how waste managers use these three major approaches (reduction, recovery, and disposal) to manage municipal solid waste, and then we will address industrial solid waste and hazardous waste.

# Municipal Solid Waste

Municipal solid waste—waste we generate in our homes, public facilities, small businesses, and colleges and universities—is what we commonly refer to as "trash" or "garbage." In the United States, paper, food scraps, yard trimmings, and plastics are the principal components of municipal solid waste

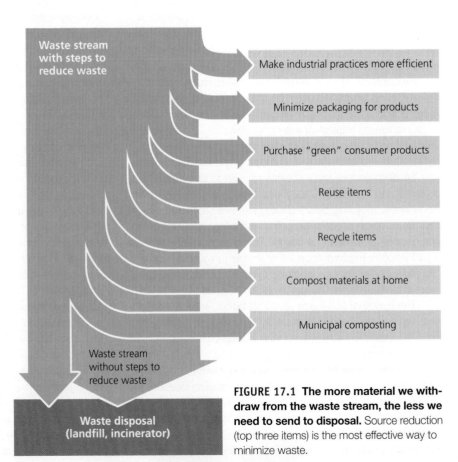

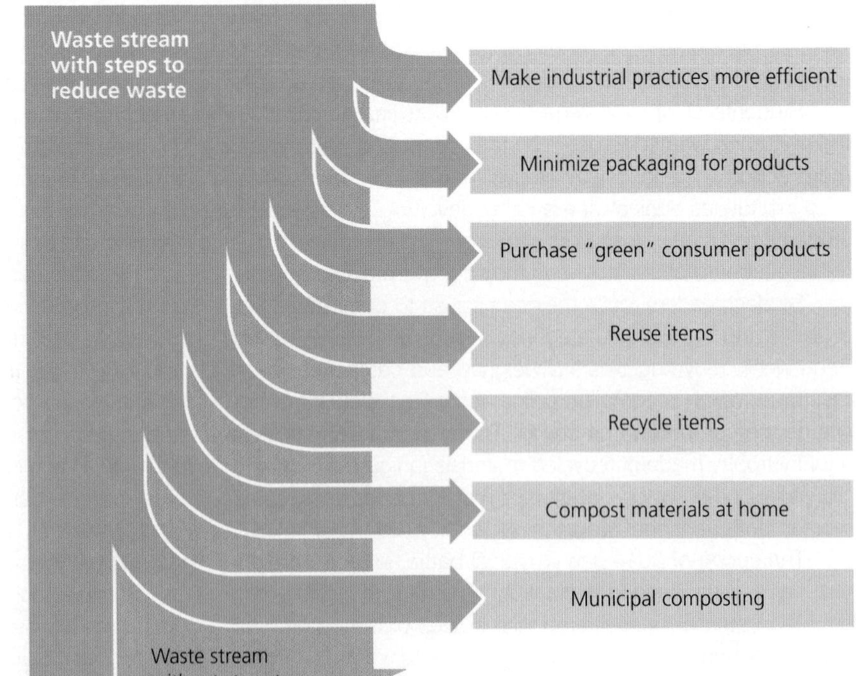


**FIGURE 17.1 The more material we withdraw from the waste stream, the less we need to send to disposal.** Source reduction (top three items) is the most effective way to minimize waste.

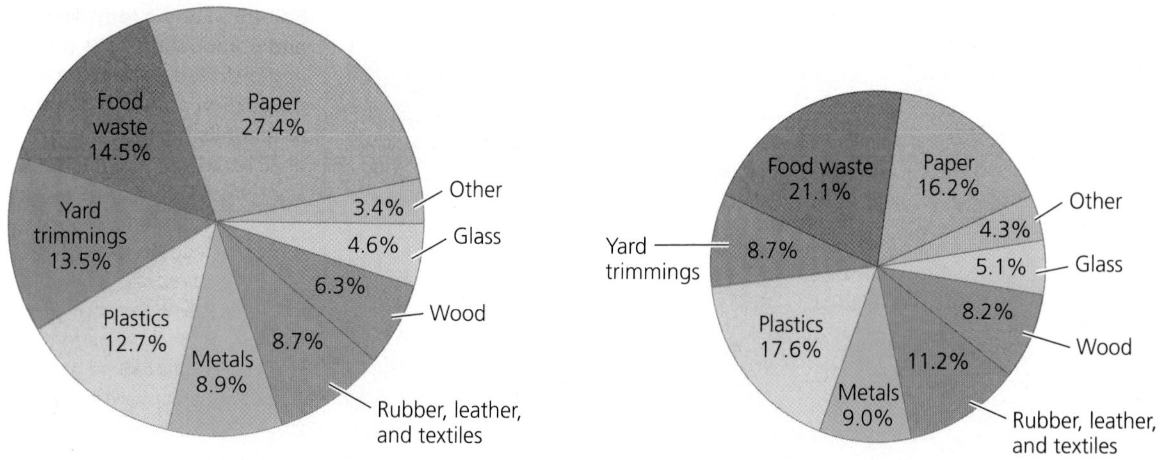


**(a) Before recycling and composting**

**(b) After recycling and composting**

**FIGURE 17.2 Components of the municipal solid waste stream in the United States.** Paper products comprise the greatest portion by weight **(a)**, but after recycling and composting removes many items **(b)**, the waste stream becomes one-third smaller. Food scraps are now the largest contributor, because so much paper is recycled and yard waste is composted. *Data from U.S. Environmental Protection Agency, 2014.* Municipal solid waste generation, recycling, and disposal in the United States: Tables and figures for 2012. *EPA, Washington, D.C.*

(**FIGURE 17.2a**). Paper is recycled at a high rate and yard trimmings are composted at a high rate, so after recycling and composting reduce the waste stream, food scraps and plastics are left as the largest components of U.S. municipal solid waste (**FIGURE 17.2b**).

Most municipal solid waste comes from packaging and nondurable goods (products meant to be discarded after a short period of use). In addition, consumers throw away old durable goods and outdated equipment as they purchase new products. Plastics, which came into wide consumer use only after 1970, have accounted for the greatest relative increase in the waste stream during the last several decades.

## Consumption leads to waste

As we acquire more goods, we generate more waste. In 2012, U.S. citizens produced 251 million tons of municipal solid waste (before recovery)—close to 1 ton per person. The average American generates 2.0 kg (4.4 lb) of trash per day. These figures come from the U.S. Environmental Protection Agency (EPA), whereas a rigorous biennial survey that produces the "State of Garbage in America" report uses different methods and comes up with larger estimates—389 million tons, or 3.2 kg (7.1 lb) per person per day. Regardless of how this is calculated, the data show that Americans generate considerably more waste than people in most other industrialized nations. The relative wastefulness of the U.S. lifestyle, with its excess packaging and reliance on nondurable goods, has caused critics to label the United States "the throwaway society."

However, Americans are beginning to turn this around. Thanks to source reduction and reuse (especially by businesses looking to cut costs), total waste generation decreased slightly between 2005 and 2012. Americans now generate less waste per capita than they have at any time since the late 1980s.

In developing nations, people consume fewer resources and goods, and as a result, generate less waste. However, consumption is intensifying in developing nations as they become more affluent, and these nations are generating more and more waste. As a result, trash is piling up and littering the landscapes of countries from Mexico to Kenya to Indonesia. Like U.S. consumers in the "throwaway society," wealthy people in developing nations often discard items that can still be used. In fact, at many dumps and landfills in the developing world, poor people support themselves by selling items that they scavenge (**FIGURE 17.3**).

In many industrialized nations, per capita generation rates have begun to decline in recent years. Wealthier nations also can afford to invest more in waste collection and disposal, so they are often better able to manage their waste and minimize

**FIGURE 17.3 Affluent consumers discard so much usable material that some people in developing nations support themselves by scavenging items from dumps.** Tens of thousands of people used to scavenge each day from this dump outside Manila in the Philippines, selling material to junk dealers for 100–200 pesos (U.S. $2–$4) per day. The dump was closed in 2000 after an avalanche of trash killed hundreds of people.

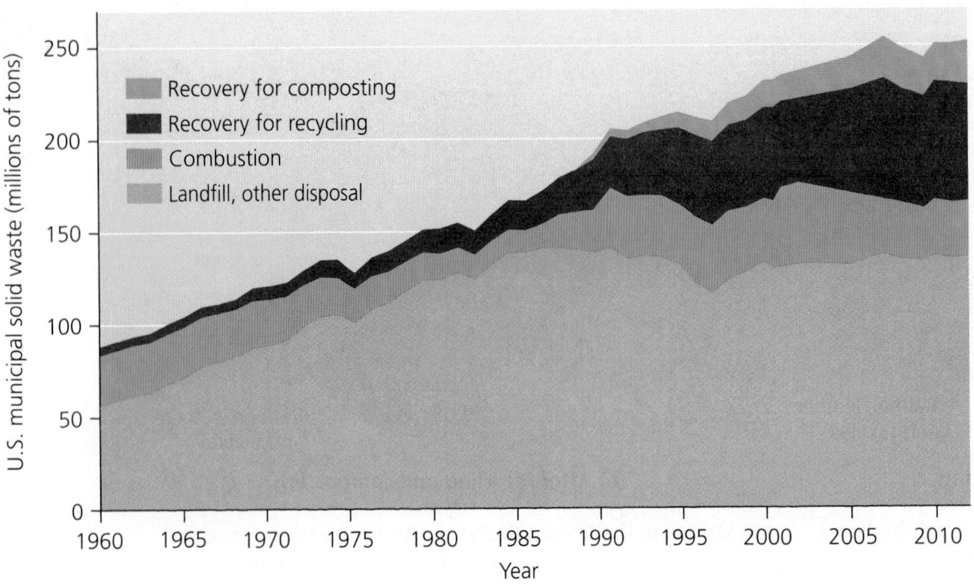


**FIGURE 17.4 As recycling and composting have grown in the United States, the proportion of waste going to landfills has declined.** As of 2012, 54% of U.S. municipal solid waste went to landfills and 12% to incinerators, whereas 35% was recovered for composting and recycling. *Data from U.S. Environmental Protection Agency, 2014.* Municipal solid waste generation, recycling, and disposal in the United States: Tables and figures for 2012. *EPA, Washington, D.C.*

DATA Q Has the amount of solid waste that is combusted (incinerated) increased or decreased since 1960?

GO TO **INTERPRETING GRAPHS & DATA** ON MasteringEnvironmentalScience®

impacts on health and the environment. Moreover, enhanced recycling and composting efforts—fed by a conservation ethic growing among a new generation on today's campuses—have begun to reduce the waste stream (**FIGURE 17.4**). We will examine reduction, reuse, recycling, and composting shortly, but let's first assess how we dispose of waste.

## Landfills provide our main method of disposal

In modern **sanitary landfills**, waste is buried in the ground or piled up in mounds engineered to prevent waste from contaminating the environment and threatening public health (**FIGURE 17.5**). Most municipal landfills in the United States are regulated locally or by the states, but they must meet national standards set by the EPA under the **Resource Conservation and Recovery Act** (p. 105), a major federal law enacted in 1976 and amended in 1984.

In a sanitary landfill, waste is partially decomposed by bacteria and compresses under its own weight to take up less space. Soil is layered along with the waste to speed decomposition, reduce odor, and lessen infestation by pests. Some infiltration of rainwater into the landfill is good, because it encourages biodegradation by bacteria—yet too much is not good, because the excess water can flow out, allowing contaminants to escape.

To protect against environmental contamination, U.S. regulations require that landfills be located away from wetlands and earthquake-prone faults and be at least 6 m (20 ft) above the water table. The bottoms and sides of sanitary landfills must be lined with heavy-duty plastic and 60–120 cm (2–4 ft) of impermeable clay to help prevent contaminants from seeping into aquifers. Sanitary landfills also have systems of pipes, ponds, and treatment facilities to collect and treat **leachate**, liquid that results when substances from the trash dissolve in

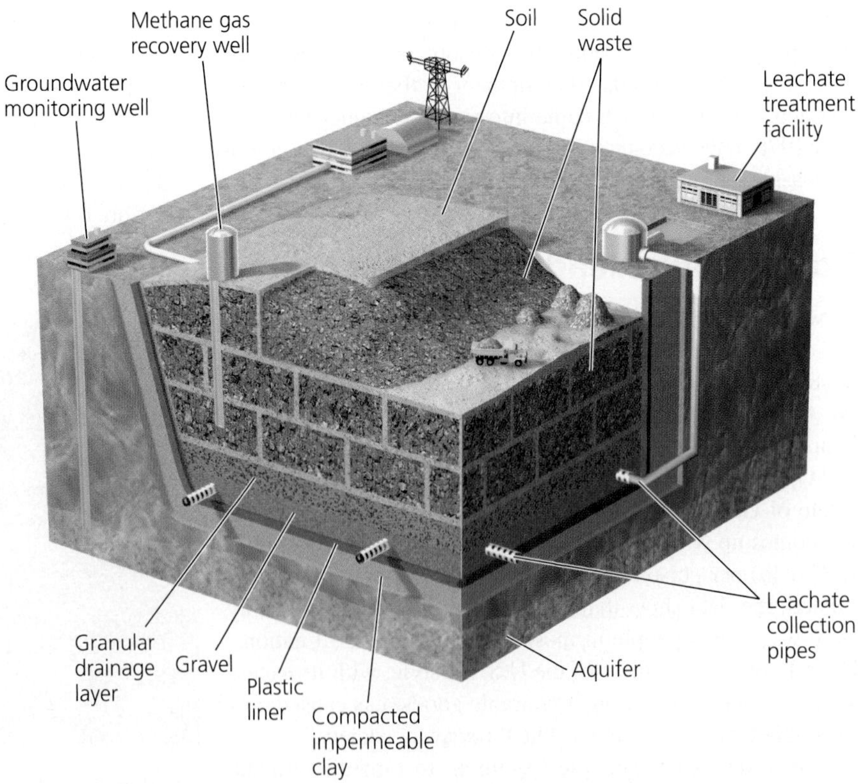


**FIGURE 17.5 Sanitary landfills are engineered to prevent waste from contaminating soil and groundwater.** Waste is laid in a large depression lined with plastic and impervious clay designed to prevent liquids from leaching out. Pipes draw out these liquids from the bottom of the landfill. Waste is layered with soil, filling the depression, and then is built into a mound until the landfill is capped. Landfill gas produced by anaerobic bacteria may be recovered, and waste managers monitor groundwater for contamination.

water as rainwater percolates downward. Once a landfill is closed, it is capped with an engineered cover consisting of layers of plastic, gravel, and soil, and managers are required to maintain leachate collection systems for 30 years thereafter.

Despite improvements in liner technology and landfill siting, however, liners can be punctured and leachate collection systems eventually cease to be maintained. Moreover, landfills are kept dry to reduce leachate, but dryness slows waste decomposition. In fact, the low-oxygen conditions of most landfills turns trash into a sort of time capsule. Researchers examining landfills find many of their contents perfectly preserved, even after years or decades.

THEN: Fresh Kills Landfill in operation
NOW: Fresh Kills Landfill site today


**FIGURE 17.6 Old landfills, once capped, can serve other purposes.** Visitors to Fresh Kills Park will enjoy this panoramic view of the Manhattan skyline from atop what used to be an immense mound of trash.

**FAQ How much does garbage decompose in a landfill?**

You might assume that a banana peel you throw in the trash will soon decay away to nothing in a landfill. However, it just might survive longer than you do! This is because surprisingly little decomposition occurs in landfills. Researcher William Rathje, a recently retired archaeologist known as "the Indiana Jones of Solid Waste," made a career out of burrowing into landfills and examining their contents to learn about what we consume and what we throw away. His research teams would routinely come across whole hot dogs, intact pastries that were decades old, and grass clippings that were still green. Newspapers 40 years old were often still legible, and the researchers used them to date layers of trash.

A second challenge is finding suitable locations for landfills, because most communities do not want them nearby. As a result of this *not-in-my-backyard (NIMBY)* reaction, landfills are rarely sited in neighborhoods that are home to wealthy and educated people with the political clout to keep them out. Instead, they end up disproportionately sited in poor and minority communities, as environmental justice advocates (pp. 14–15) have frequently pointed out.

In 1988, the United States had nearly 8000 landfills, yet today, it has fewer than 2000. Waste managers have consolidated the waste stream into fewer landfills of larger size. Some now-closed landfills are being converted into public parks or other uses (**FIGURE 17.6**). The world's largest landfill conversion project is at New York City's former Fresh Kills Landfill. This site, on Staten Island, was the primary repository of New York City's garbage for half a century, and its mounds rose higher than the nearby Statue of Liberty! Today New York is transforming the site into a world-class public park—a verdant landscape of ball fields, playgrounds, jogging trails, rolling hills, and wetlands teeming with wildlife. Almost three times bigger than Central Park, the mounds offer panoramic views of the Manhattan skyline.

## Incinerating trash reduces pressure on landfills

Just as sanitary landfills are an improvement over open dumping, incineration in specially constructed facilities is better than open-air burning of trash. **Incineration**, or combustion, is a controlled process in which garbage is burned at very high temperatures. At incineration facilities, waste is generally sorted and metals removed. Metal-free waste is chopped into small pieces and then is burned in a furnace. Incinerating waste reduces its weight by up to 75% and its volume by up to 90%.

The ash remaining after trash is incinerated contains toxic components and must be disposed of in hazardous waste landfills (p. 403). Moreover, when trash is burned, hazardous chemicals—including dioxins, heavy metals, and polychlorinated biphenyls (PCBs) (Chapter 10)—can be created and released into the atmosphere. Such emissions caused a backlash against incineration from citizens concerned about health hazards.

Most developed nations now regulate incinerator emissions, and some have banned incineration outright. Engineers have also developed technologies to mitigate emissions. Scrubbers (see Figure 13.8, p. 285) chemically treat the gases produced in combustion to neutralize acids and remove hazardous components. Particulate matter, called *fly ash*, contains some of the worst dioxin and heavy metal pollutants in incinerator emissions. To physically remove these tiny particles, facilities may use a huge system of filters known as a *baghouse*. In addition, burning garbage at especially high temperatures can destroy certain pollutants, such as PCBs. Even all these measures, however, do not fully eliminate toxic emissions.

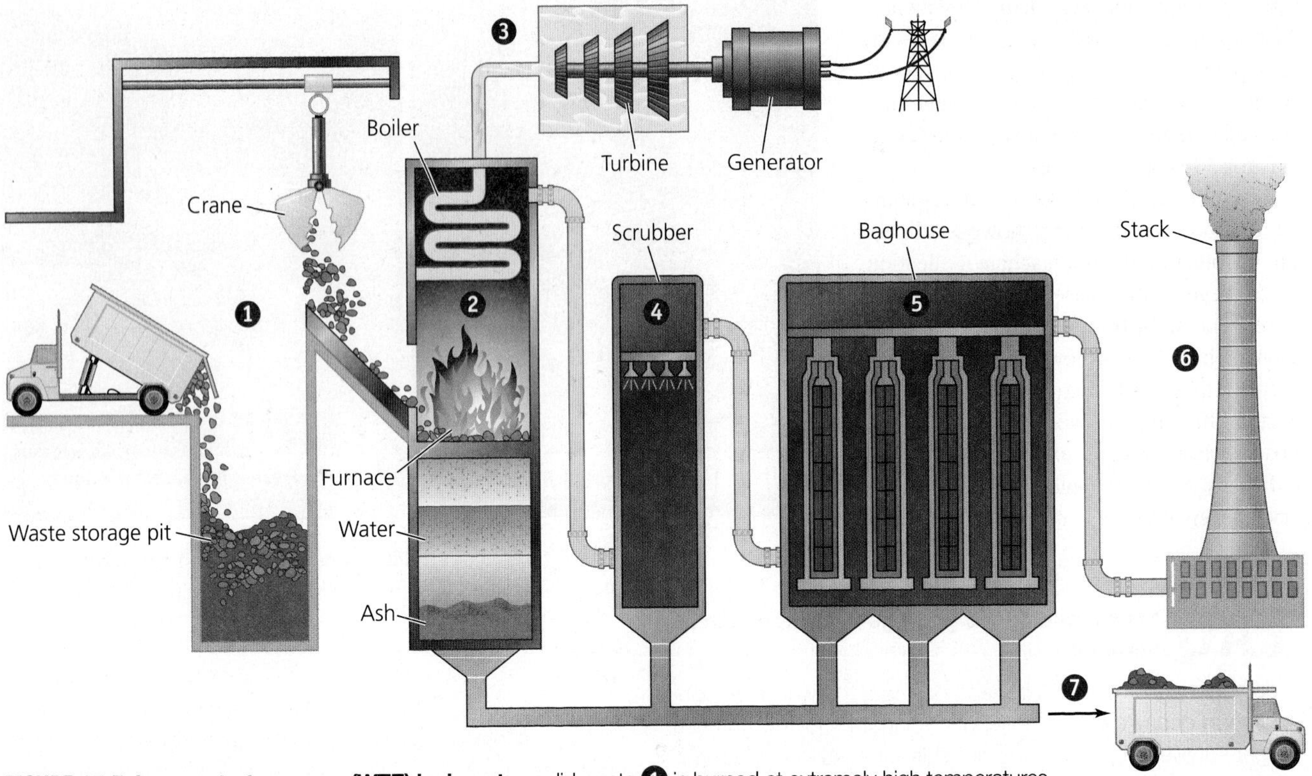


**FIGURE 17.7 In a waste-to-energy (WTE) incinerator,** solid waste ❶ is burned at extremely high temperatures ❷, heating water that turns to steam. The steam turns a turbine ❸, which powers a generator to create electricity. Toxic gases produced by combustion are treated chemically by a scrubber ❹, and airborne particulate matter is filtered physically in a baghouse ❺ before air is emitted from the stack ❻. Residual ash is disposed of ❼ in a landfill.

## WEIGHING THE ISSUES

**Environmental Justice?**

Do you know where your trash goes? Where is your landfill or incinerator located? Are the people who live closest to the facility wealthy, poor, or middle class? What race or ethnicity are they? Do you know whether the people of this neighborhood protested against the introduction of the landfill or incinerator?

## We can gain energy from trash

Incineration reduces the volume of waste, and today it serves to generate electricity as well. Most incinerators now are **waste-to-energy (WTE) facilities**, which use heat produced by waste combustion to boil water, creating steam that drives electricity generation or that fuels heating systems (**FIGURE 17.7**). When burned, waste generates about 35% of the energy generated by burning coal. Roughly 80–90 WTE facilities are operating across the United States, with a total capacity to process nearly 100,000 tons of waste per day.

Combustion in WTE plants is not the only way to gain energy from waste. Deep inside landfills, bacteria decompose waste in an oxygen-deficient environment. This anaerobic decomposition produces **landfill gas**, a mix of gases, roughly half of which is methane (pp. 29, 305). Landfill gas can be collected, processed, and used in the same way as natural gas (pp. 338, 340). Today hundreds of landfills are collecting landfill gas and selling it for energy.

## Reducing waste is our best option

Reducing the amount of material entering the waste stream avoids costs of disposal and recycling, helps conserve resources, minimizes pollution, and can save consumers and businesses money. Recall that preventing waste generation in this way is known as *source reduction.*

One means of source reduction is to lessen the materials used to package goods. Packaging helps to preserve freshness, prevent breakage, protect against tampering, and provide information—yet much packaging is extraneous. Consumers can give manufacturers incentive to reduce packaging by choosing minimally packaged goods, buying unwrapped fruit and vegetables, and buying food in bulk. Manufacturers can reduce the size or weight of goods and materials, as they already have with aluminum cans, plastic soft drink bottles, personal computers, and much else.

Some governments have recently taken aim at a major source of waste and litter—plastic grocery bags. These lightweight polyethylene bags can persist for centuries in the environment, choking and entangling wildlife and littering the landscape—yet Americans discard 100 billion of them each year. A number of cities and over 20 nations have now enacted bans or limits on their use. Financial incentives are also effective. When Ireland began taxing these bags, their use dropped 90%. IKEA stores began charging for them and saw similar drops in usage. Many businesses now give discounts if you bring your own reusable canvas bags instead of taking disposable plastic ones.

## Reuse is a main strategy to reduce waste

To reduce waste, you can save items to use again or substitute disposable goods with durable ones. Habits as simple as bringing your own coffee cup to coffee shops or bringing sturdy reusable cloth bags to the grocery store can, over time, have substantial impact. You can also donate unwanted items and shop for used items at yard sales and at resale centers run by organizations such as Goodwill Industries and the Salvation Army. Besides being sustainable, reusing items saves money. Used items are often every bit as functional as new ones, and they are cheaper. TABLE 17.1 presents a sampling of actions we all can take to reduce the waste we generate.

On some campuses, students collect unwanted items and resell them or donate them to charity. Students at the University of Texas at Austin run a "Trash to Treasure" program. Each May, they collect 40–50 tons of items that students discard as they leave and then resell them at low prices in August to arriving students. This keeps waste out of the landfill, provides arriving students with items they need at low cost, and raises $10,000–20,000 per year that gets plowed back into campus sustainability efforts. Hamilton College in New York runs a similar program, called "Cram & Scram." It reduces Hamilton's landfill waste by 28% (about 90 tons) each May.

| TABLE 17.1 Some Everyday Things You Can Do to Reduce and Reuse |
|---|
| • Donate used items to charity |
| • Reuse boxes, paper, plastic wrap, plastic containers, aluminum foil, bags, wrapping paper, fabric, packing material, etc. |
| • Rent or borrow items instead of buying them, when possible . . . and lend your items to friends |
| • Buy groceries in bulk |
| • Bring reusable cloth bags shopping |
| • Make double-sided photocopies |
| • Keep electronic documents rather than printing items out |
| • Bring your own coffee cup to coffee shops |
| • Pay a bit extra for durable, long-lasting, reusable goods rather than disposable ones |
| • Buy rechargeable batteries |
| • Select goods with less packaging |
| • Compost kitchen and yard wastes |
| • Buy clothing and other items at resale stores and garage sales |
| • Use cloth napkins and rags rather than paper napkins and towels |
| • Tell businesses what you think about their packaging and products |
| • When solid waste policy is being debated, let your government representatives know your thoughts |
| • Support organizations that promote waste reduction |

*Adapted from U.S. Environmental Protection Agency.*

## Composting recovers organic waste

*Composting* is the conversion of organic waste into mulch or humus (p. 140) through natural decomposition. We can place waste in compost piles, underground pits, or specially constructed containers. As waste is added, heat from microbial action builds in the interior, and decomposition proceeds. Banana peels, coffee grounds, grass clippings, autumn leaves, and other organic items can be converted into rich, high-quality compost through the actions of earthworms, bacteria, soil mites, sow bugs, and other detritivores and decomposers (p. 72). The compost is then used to enrich soil.

On campus, composting is becoming popular. Ball State University in Indiana shreds surplus furniture and wood pallets and makes them into mulch to nourish campus plantings. Ithaca College in New York composts 44% of its food waste, saving $11,500 each year in landfill disposal fees. The compost is used on campus plantings, and experiments showed that the plantings grew better with the compost mix than with chemical soil amendments.

Municipal composting programs divert yard debris (and increasingly food waste as well) out of the waste stream and into composting facilities, where it decomposes into mulch that community residents can use for gardens and landscaping. About one-fifth of the U.S. waste stream is made up of materials that can easily be composted. Composting reduces landfill waste, enriches soil, enhances soil biodiversity, helps soil to resist erosion, makes for healthier plants and more pleasing gardens, and reduces the need for chemical fertilizers.

## Recycling consists of three steps

*Recycling*, too, offers many benefits. It involves collecting used items and breaking them down so that their materials can be reprocessed to manufacture new items. The recycling loop consists of three basic steps (**FIGURE 17.8**). The first step is to collect and process used goods and materials, as is being done on so many campuses. Some towns and cities designate locations where residents can drop off recyclables or receive

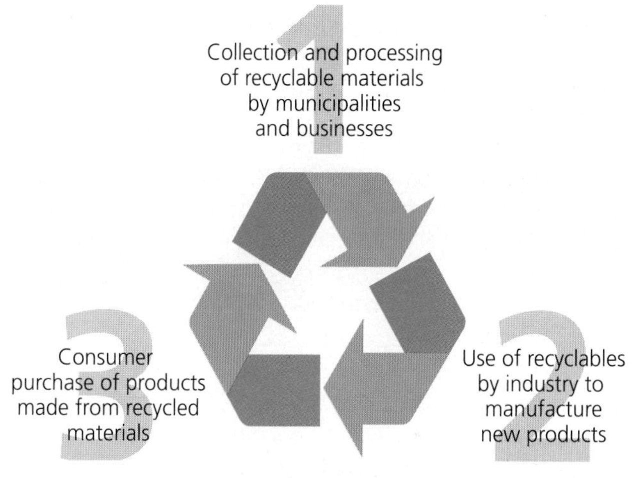


**FIGURE 17.8 The familiar recycling symbol represents the three components of a sustainable recycling strategy.** After recyclable materials are collected and processed, they are used to make new products, which are then purchased by consumers.

money for them. Others offer curbside recycling, in which trucks pick up recyclable items in front of homes, usually in conjunction with municipal trash collection.

Items are taken to **materials recovery facilities (MRFs)**, where workers and machines sort items using automated processes including magnetic pulleys, optical sensors, water currents, and air classifiers that separate items by weight and size. The facilities clean the materials, shred them, and prepare them for reprocessing.

Once readied, these materials are used to manufacture new goods. Newspapers and many other paper products use recycled paper, many glass and metal containers are now made from recycled materials, and some plastic containers are of recycled origin. Benches, bridges, and walkways in city parks may now be made from recycled plastics, and glass can be mixed with asphalt (creating "glassphalt") to pave roads and paths.

If the recycling loop is to function, consumers and businesses must complete the third step in the cycle by purchasing ecolabeled products (p. 111) made from recycled materials. Buying recycled goods provides economic incentive for industries to recycle materials and for recycling facilities to open or expand.

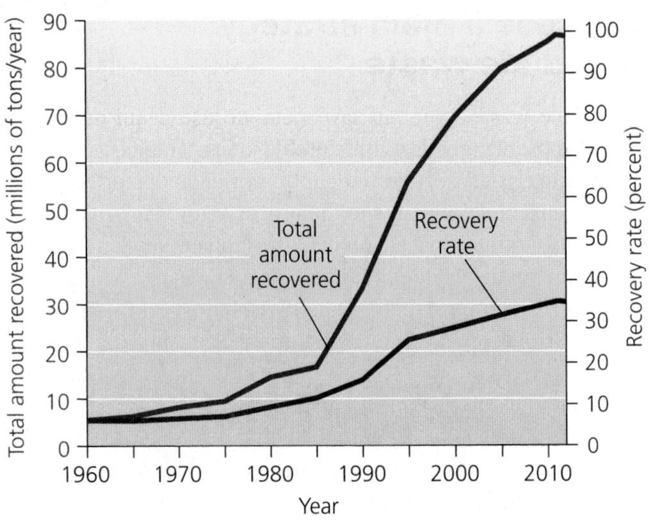


**FIGURE 17.9 Recovery has risen sharply in the United States.** Today over 86 million tons of material are recovered (65 million tons by recycling and 21 million tons by municipal composting), comprising one-third of the waste stream. *Data from U.S. Environmental Protection Agency.*

**DATA Q** From the data in this graph alone, what would you infer has happened to the total amount of municipal solid waste generated (before recovery) since 1960? Explain how you can determine this.

GO TO **INTERPRETING GRAPHS & DATA** ON MasteringEnvironmentalScience®

## Recycling has grown rapidly

Today 9000 curbside recycling programs across all 50 U.S. states serve nearly half of all Americans, while over 600 MRFs handle 100,000 tons of recyclables per day. These programs and facilities have sprung up only in the last few decades. Recycling in the United States has risen from 6.4% of the waste stream in 1960 to 26.0% in 2012 (and 34.5% if composting is included), according to EPA data (**FIGURE 17.9**).

Recycling rates vary greatly from one product or material type to another, ranging from nearly zero to almost 100% (**TABLE 17.2**). Recycling rates among U.S. states also vary greatly, from 1% to 48%.

Many college and university campuses run active recycling programs, although attaining high recovery rates can be challenging in the campus environment. The most recent survey of campus sustainability efforts suggested that the average recycling rate was only 29%. Thus there appears to be much room for growth. Waste management initiatives are relatively easy to conduct because they offer many opportunities for small-scale improvements and because people generally enjoy recycling and reducing waste.

Besides participation in Recyclemania, there are many ways to promote recycling on campus. Louisiana State University students initiated recycling efforts at home football games, and over three seasons they recycled 68 tons of refuse that otherwise would have gone to the landfill. "Trash audits" or "landfill on the lawn" events involve emptying trashcans or dumpsters and sorting out recyclable items (**FIGURE 17.10**). When students at Ashland University in Ohio audited their waste, they found that 70% was recyclable, and they used this data to press their administration to support recycling programs.

**TABLE 17.2 Recovery Rates for Various Materials in the United States**

| MATERIAL | PERCENTAGE THAT IS RECYCLED OR COMPOSTED |
|---|---|
| Lead-acid batteries | 96 |
| Newspapers | 70 |
| Paper and paperboard | 65 |
| Yard trimmings | 58 |
| Aluminum cans | 55 |
| Glass containers | 34 |
| Total plastics | 9 |

*Data are for 2012, from U.S. Environmental Protection Agency.*

## WEIGHING THE ISSUES

### Managing Waste on Your Campus

Does your campus have a recycling program? Does it have any composting initiatives? Does it run programs to reduce or reuse materials? Think about the types and amounts of waste generated on your campus. Describe several examples of this waste that you feel could be prevented or recycled, and describe how this might be done in each case. If you could do one thing on campus to improve your school's waste management practices, what would it be?

**FIGURE 17.10 In a trash audit, students sort through rubbish and separate out recyclables.** Events like this "landfill on the lawn" exercise at the University of North Carolina at Greensboro show passersby just how many recyclable items are needlessly thrown away.

The growth of recycling has been propelled in part by economic forces as businesses see prospects to save money and as entrepreneurs see opportunities to start new businesses. It has also been driven by the desire of community and campus leaders to reduce waste and by the satisfaction people take in recycling. These latter two forces have driven the rise of recycling even when it has not been financially profitable. In fact, many of our popular municipal recycling programs are run at an economic loss. The expense required to collect, sort, and process recycled goods is often more than recyclables are worth in the marketplace. Additionally, the more people recycle, the more glass, paper, and plastic is available to manufacturers for purchase, which drives down prices. And transporting items to recycling facilities can sometimes involve surprisingly long distances (see **THE SCIENCE BEHIND THE STORY**, pp. 398–399).

Recycling advocates, however, point out that market prices do not take into account external costs (p. 94)—in particular, the environmental and health impacts of *not* recycling. Each year in the United States, recycling and composting together save energy equal to the amount consumed by 10 million U.S. households in a year, and they prevent carbon dioxide emissions equal to those of 33 million cars. Recycling aluminum cans saves 95% of the energy required to make the same amount of aluminum from mined virgin bauxite, its source material.

As more manufacturers use recycled products and as more technologies are developed to use recycled materials in new ways, markets should continue to expand, and new business opportunities may arise. We are just beginning to shift from an economy that moves linearly, from raw materials to products to waste, to a more sustainable economy that moves circularly, using waste products as raw materials for new manufacturing.

## WEIGHING THE ISSUES

**Costs of Recycling and Not Recycling**

Should governments subsidize recycling programs even if they are run at an economic loss? What types of external costs—costs not reflected in market prices—do you think would be involved in not recycling, say, aluminum cans? Do you feel these costs justify sponsoring recycling programs even when they are not financially self-supporting? Why or why not?

## We can recycle material from landfills

With improved technology for sorting rubbish and recyclables, businesses and entrepreneurs are weighing the economic benefits and costs of rummaging through landfills to salvage materials of value that can be recycled. Steel, aluminum, copper, and other metals are abundant enough in some landfills to make salvage operations profitable when market prices for the metals are high enough. For instance, Americans throw out so many aluminum cans that at 2014 prices for aluminum, the nation buries $1.7 billion of this metal in landfills each year. If we could retrieve all the aluminum from U.S. landfills, it would exceed the amount the world produces from a year's worth of mining ore.

Landfills also offer organic waste that can be mined and sold as premium compost. Old landfill waste can also be incinerated in newer, cleaner-burning WTE facilities to produce energy. Some companies are even looking into gaining carbon offset credits (p. 328) by harvesting methane leaking from open dumps in developing nations.

## Financial incentives help address waste

Waste managers offer consumers economic incentives to reduce the waste stream. In "pay-as-you-throw" garbage collection programs, municipalities charge residents for home trash pickup according to the amount of trash they put out. The less waste one generates, the less one has to pay. Over 7000 such programs operate in the United States, serving more than one of every four communities.

*Bottle bills* are another approach hinging on financial incentives. In the 10 U.S. states and 23 nations that have these laws, consumers pay a deposit on bottles or cans upon purchase—often 5 cents per container—and then receive a refund when they return them to stores after use. Where these laws have been enacted, they have proven effective and popular. U.S. states with bottle bills report that their beverage container litter has decreased by 69–84%, their total litter has decreased by 30–65%, and their per capita container recycling rates have risen 2.6-fold. Beverage container recycling rates for states with bottle bills are 2.5 times higher than for states without them. As of 2014, bottle bill advocates in 13 states were seeking to establish, expand, or update programs.

## Tracking Trash

Where does your trash go once you throw it away? What happens to your recycling? How far might it travel, and how much energy is needed to get rid of it?

With the help of the latest tracking technology, we can find out. Researchers from the SENSEable City Lab at the Massachusetts Institute of Technology (MIT) are affixing tiny sensors to everyday items in our trash and monitoring them to reveal their hidden travels. By documenting what actually happens to trash and to recyclables, they hope to help make the trash removal process more effective and to encourage better recycling.

The Trash Track project was launched in 2009 in New York City and in Seattle. Early results were unveiled at public exhibitions in both cities, and it is now expanding elsewhere.

Here's how trash-tracking works: Project director Carlo Ratti and associate director Assaf Biderman, both architects at MIT, organize research teams and local volunteers in the target city to affix tiny electronic tags (see photo) to hundreds of items being thrown away. As each item journeys through the waste stream, its tag calculates its location every few hours and relays the information via the cell phone network to a central server at MIT. A computer plots the movements atop satellite maps, helping the researchers to visualize and interpret the migration of trash.

Each tag calculates its position by measuring the signal strength from nearby cell phone towers and comparing this to a map of tower locations. Tags used in New York City and

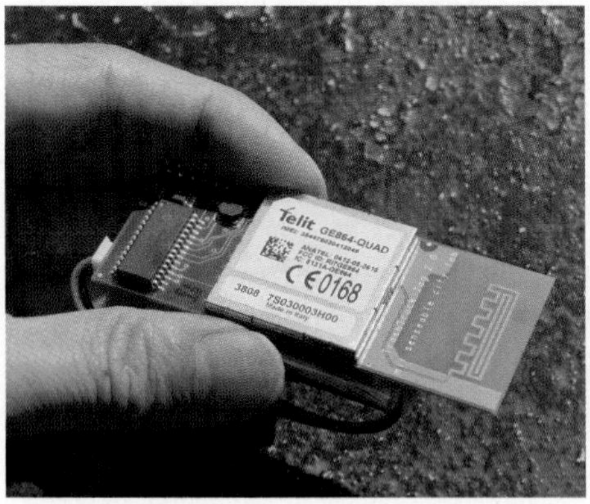


**Trash tags used in Seattle and New York City**

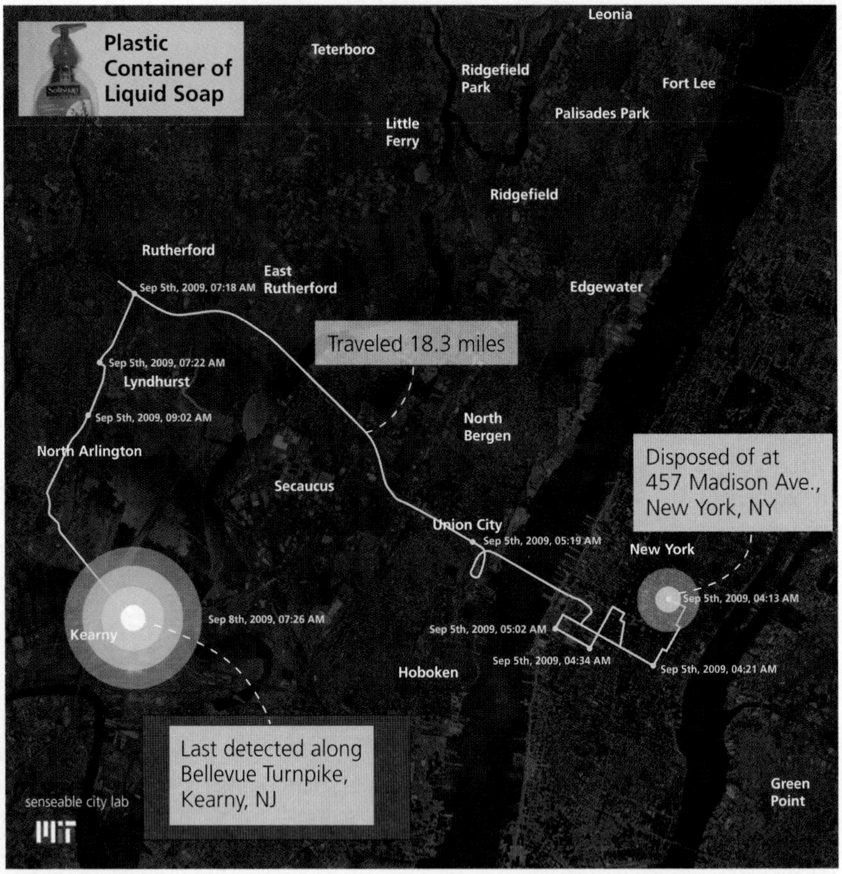


**FIGURE 1 Follow that trash truck!** A plastic soap container put out with the trash on Madison Avenue in New York City looped through midtown Manhattan, crossed the Hudson River, and was last detected traveling down the Bellevue Turnpike in New Jersey.

Seattle are not as accurate as global positioning systems (GPS), but their signals carry better through barriers such as roofs and walls of buildings, garbage trucks, and deep piles of trash. Second-generation tags the project is now using are more accurate, combining GPS technology with better cell network triangulation.

As an example, a plastic container of liquid soap was tagged on September 5, 2009, and placed in the trash at 457 Madison Avenue in Manhattan (**FIGURE 1**). Mapping reveals that the truck that picked it up looped through the city's streets a few times on its route, crossed the Hudson River via the Lincoln Tunnel, and then headed to Rutherford, New Jersey. Here the truck turned south and was in transit along the Bellevue Turnpike in Kearny, New Jersey, three days later, when the tag's battery gave out.

As of 2014, the project had posted mapped data of all its Seattle items online for the public to view and had published several research papers. The results from Seattle reveal some expected patterns but also some odd surprises. Of the 760 items tagged, about 200 ended up at the city's Allied Waste Recycling Center and Transfer Station after brief journeys. Smaller numbers were transported to other city or regional landfills and recycling centers. But some items followed surprisingly circuitous routes, being transferred from one waste center to another, or back and forth between cities. Some ended up in seemingly random places, perhaps having fallen off a garbage truck along a roadside.

A few items made very long journeys. Two printer cartridges were driven down Interstate 5 to California's border with Mexico, perhaps to be disassembled at a factory. Two cell phones were transported halfway across the country to Dallas (one flown directly and one making apparent stops at three other cities). A compact fluorescent bulb went to St. Louis after traveling to Portland, Oregon, and back to Seattle. Chicago received a coffee cup that apparently made its way east on the nation's interstates. Shipments of batteries were flown 1500 miles to Minneapolis, 2500 miles to Pittsburgh, and 2600 miles to Atlanta.

The longest-traveling piece of trash was a cell phone that was transported over 3000 miles from Seattle to the other corner of the United States, ending up near Ocala, Florida.

In general, hazardous waste items and electronic waste items tended to travel farthest (**FIGURE 2**) because they were sent to special facilities for handling. This raises the question of whether special handling is worthwhile, given the impacts of the extra gasoline use and greenhouse gas emissions it entails. In a 2012 research paper, the Trash Track team suggested that we seek to make processing systems for hazardous and electronic waste as efficient as those for curbside waste collection and recycling.

Why did each tracked item travel so far? How much gasoline was used to transport them? When we move items great distances for disposal or recycling, does that reflect efficiency from an economy of scale—or does it indicate an excessive waste of resources? Do we need more recycling and disassembly facilities nearer to each major city? Researchers and waste managers hope to use data from the Trash Track project to address such questions and improve the way we handle waste. ☐

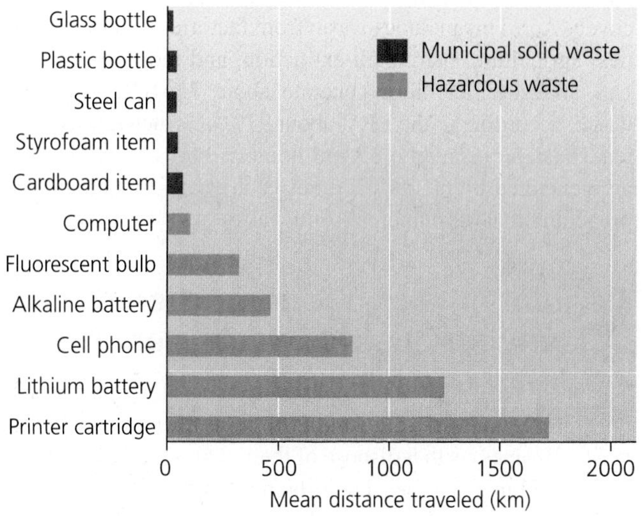


**FIGURE 2 Items tagged in Seattle traveled vastly different distances.**

**DATA Q** Which type of trash tended to travel farther: municipal solid waste or hazardous waste? How far did each type of trash travel? What do you think accounts for this difference?

GO TO **INTERPRETING GRAPHS & DATA** ON MasteringEnvironmentalScience®

# Industrial Solid Waste

In the United States, *industrial solid waste* is defined as solid waste that is considered neither municipal solid waste nor hazardous waste under the Resource Conservation and Recovery Act. This includes waste from factories, mining activities, agriculture, petroleum extraction, and more. Each year, U.S. industrial facilities generate about 7.6 billion tons of waste, according to the EPA, about 97% of which is wastewater. Thus, very roughly, 230 million or so tons of solid waste are generated by 60,000 facilities each year—an amount approaching that of municipal solid waste.

## Regulation and economics each influence industrial waste generation

Most methods and strategies of waste disposal, reduction, and recycling by industry are similar to those for municipal solid waste. Businesses that dispose of their own waste on site must design and manage their landfills in ways that meet state, local, or tribal guidelines. Other businesses pay to have their waste disposed of at municipal disposal sites. Whereas the federal government regulates municipal solid waste, state or local governments regulate industrial solid waste (with federal guidance). Regulation varies greatly from place to place, but in most cases, state and local regulation of industrial solid waste is less strict than federal regulation of municipal solid waste. In many areas, industries are not required to have permits, install landfill liners or leachate collection systems, or monitor groundwater for contamination.

The amount of waste generated by a manufacturing process is a good measure of its efficiency; the less waste produced per unit or volume of product, the more efficient that process is, from a physical standpoint. However, physical efficiency is not always reflected in economic efficiency. Often it is cheaper for industry to manufacture its products or perform its services quickly but messily. That is, it can be cheaper to generate waste than to avoid generating waste. In such cases, economic efficiency is maximized, but physical efficiency is not. Because our market system awards only economic efficiency, all too often industry has no financial incentive to achieve physical efficiency. The frequent mismatch between these two types of efficiency is a major reason why the output of industrial waste is so great.

Rising costs of waste disposal enhance the financial incentive to decrease waste. Once either government or the market makes the physically efficient use of raw materials also economically efficient, businesses gain financial incentives to reduce their waste.

## Industrial ecology seeks to make industry more sustainable

To reduce waste, growing numbers of industries today are experimenting with industrial ecology. A holistic approach that integrates principles from engineering, chemistry, ecology, and economics, **industrial ecology** seeks to redesign industrial systems to reduce resource inputs and to maximize both physical and economic efficiency. Industrial ecologists would reshape industry so that nearly everything produced in a manufacturing process is used, either within that process or in a different one. The intent is that industrial systems should function more like ecological systems, in which organisms use almost everything that is produced. This principle brings industry closer to the ideal of ecological economists: economies that function in a circular fashion rather than a linear one (p. 95).

Industrial ecologists pursue their goals in several ways:

- They examine the entire life cycle of a product—from its origins in raw materials, through its manufacturing, to its use, and finally its disposal—and look for ways to make the process more efficient. This strategy is called **life-cycle analysis**.
- They try to identify how waste products from one manufacturing process might be used as raw materials for another. For instance, used plastic beverage containers can be shredded and reprocessed to make other plastic items, such as benches, tables, and decks.
- They seek to eliminate environmentally harmful products and materials from industrial processes.
- They study the flow of materials through industrial systems to look for ways to create products that are more durable, recyclable, or reusable.

## Businesses are adopting industrial ecology

Attentive businesses are taking advantage of the insights of industrial ecology to save money while reducing waste. The Swiss Zero Emissions Research and Initiatives (ZERI) Foundation sponsors dozens of innovative projects worldwide that attempt to create goods and services without generating waste. One example involves breweries in Canada, Sweden, Japan, and Namibia (**FIGURE 17.11**).

Few businesses have taken industrial ecology to heart as much as the carpet tile company Interface, which founder Ray Anderson set on the road to sustainability years ago. Interface asks customers to return used tiles for recycling and for reuse as backing for new carpet. It modified its tile design and production methods to reduce waste. It adapted its boilers to use landfill gas for energy. Through such steps, Anderson's company cut waste generation by 80%, fossil fuel use by 45%, and water use by 70%—all while saving $30 million per year, holding prices steady for customers, and raising profits by 49%.

# Hazardous Waste

Hazardous wastes are diverse in their chemical composition and may be liquid, solid, or gaseous. By the EPA definition, *hazardous waste* is waste that is one of the following:

- *Ignitable.* Likely to catch fire (for example, gasoline or alcohol).
- *Corrosive.* Apt to corrode metals in storage tanks or equipment (for example, strong acids or bases).

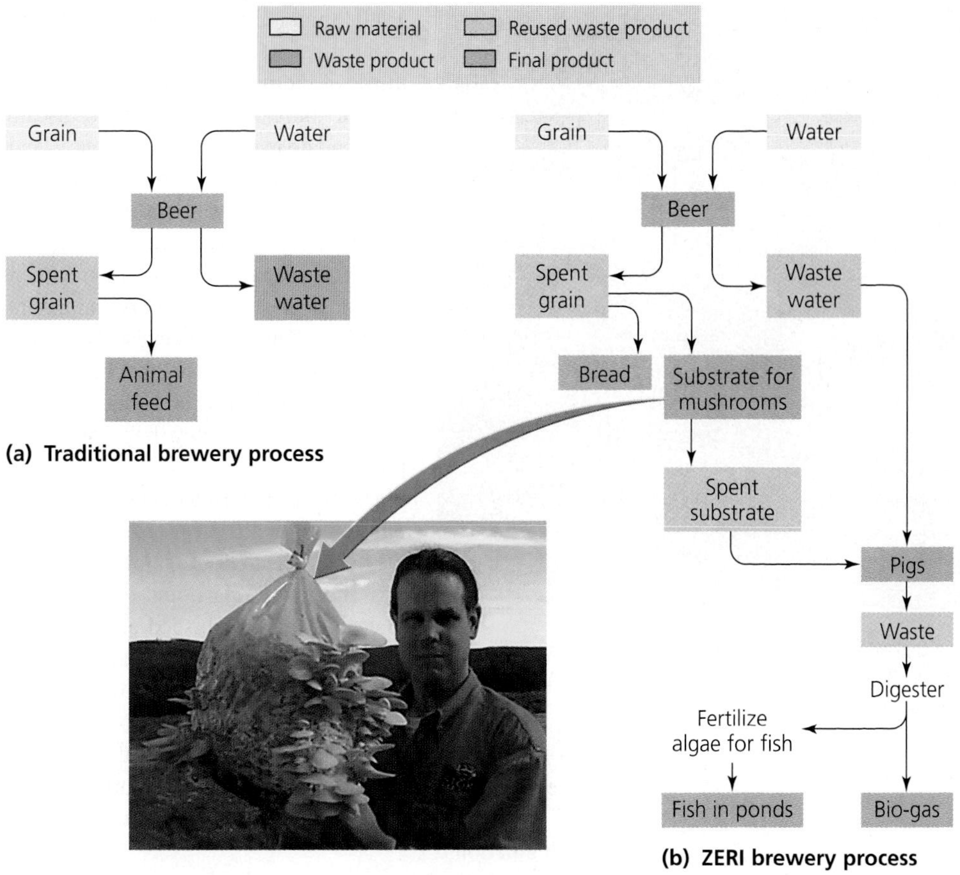


**FIGURE 17.11 Creative use of waste products can help us approach zero-waste systems.** Traditional breweries **(a)** produce only beer while generating much waste, some of which goes toward animal feed. Breweries sponsored by the Zero Emissions Research and Initiatives (ZERI) Foundation **(b)** use their waste grain to make bread and to farm mushrooms **(photo)**. Waste from the mushroom farming, along with brewery wastewater, goes to feed pigs. The pigs' waste is digested in containers that capture natural gas and collect nutrients used to nourish algae for growing fish in fish farms. The brewer derives income from bread, mushrooms, pigs, gas, and fish, as well as beer.

- *Reactive.* Chemically unstable and readily able to react with other compounds, often explosively or by producing noxious fumes (for example, ammonia reacting with chlorine bleach).
- *Toxic.* Harmful to human health when inhaled, ingested, or touched (for example, pesticides or heavy metals).

## Hazardous wastes are diverse

Industry, mining, households, small businesses, agriculture, utilities, and building demolition all create hazardous waste. Industry produces the most, but in developed nations industrial waste disposal is often highly regulated. This regulation has reduced the amount of hazardous waste entering the environment from industrial activities. As a result, households are now the largest source of unregulated hazardous waste.

Household hazardous waste includes a wide range of items, including paints, batteries, oils, solvents, cleaning agents, lubricants, and pesticides. U.S. citizens generate 1.6 million tons of household hazardous waste annually, and the average home contains close to 45 kg (100 lb) of it in sheds, basements, closets, and garages.

Many hazardous substances become less hazardous over time as they degrade, but some show especially persistent effects. Radioactive substances are an example, and the disposal of radioactive waste poses a serious dilemma (p. 361). Other types of persistent hazardous substances include organic compounds and heavy metals.

## Organic compounds and heavy metals pose hazards

In our daily lives, we rely on synthetic organic compounds and petroleum-derived compounds to resist bacterial, fungal, and insect activity. Plastic containers, rubber tires, pesticides, solvents, and wood preservatives are useful to us precisely because they resist decomposition. We use these substances to protect buildings from decay, kill pests that attack crops, and keep stored goods intact. However, these compounds' capacity to resist decay is a double-edged sword, for it also makes them persistent pollutants. Many synthetic organic compounds are toxic because they are readily absorbed through the skin and can act as mutagens, carcinogens, teratogens, and endocrine disruptors (pp. 211–212).

Heavy metals such as lead, chromium, mercury, arsenic, cadmium, tin, and copper are used widely in industry for wiring, electronics, metal plating, metal fabrication, pigments, and dyes. Heavy metals enter the environment when paints, electronic devices, batteries, and other materials are disposed of improperly. Lead from fishing weights and from hunting ammunition accumulates in rivers, lakes, and forests. In older homes, lead from pipes contaminates drinking water, and lead paint remains a problem, especially for infants. Heavy metals that are fat-soluble and break down slowly are prone to bioaccumulate and biomagnify (pp. 215–216).

**FIGURE 17.12 Each day, Americans throw away half a million cell phones.** Phones that enter the waste stream can leach toxic heavy metals into the environment. Alternatively, we can recycle phones for reuse and to recover valuable metals.

## E-waste is growing

Today's proliferation of computers, printers, smartphones, TVs, DVD players, MP3 players, and other electronic technology has created a substantial new source of waste (**FIGURE 17.12**). These products have short life spans before people judge them obsolete, and most are discarded after just a few years. The amount of this **electronic waste**—often called **e-waste**—has grown rapidly, and it now comprises 2% of the U.S. solid waste stream (**FIGURE 17.13**). Over 7 billion electronic devices have been sold in the United States since 1980, and U.S. households discard more than 300 million per year—two-thirds of them still in working order.

Of the electronic items we discard, most end up in conventional sanitary landfills and incinerators. However, electronic products contain heavy metals and toxic flame-retardants, and research suggests that e-waste should instead be treated as hazardous waste, so the EPA and a number of states are now taking steps to do so.

Fortunately, more and more electronic waste today is being recycled. Campus e-waste recycling drives are proving especially effective (see Figure 1.18c, p. 18). Americans now recycle one-fourth of their e-waste, by weight. Devices collected are shipped to facilities and taken apart, and the parts and materials are refurbished and reused in new products. There are serious concerns, however, about health risks that recycling may pose to workers doing the disassembly. Wealthy nations ship much of their e-waste to developing countries, where low-income workers disassemble the devices and handle toxic materials with minimal safety regulations.

Although these environmental justice concerns need to be resolved if electronics recycling is to be conducted safely and responsibly, e-waste recycling does help to keep toxic substances out of our waste stream. It also can help us recover trace metals used in electronics that are rare and lucrative. A typical cell phone contains up to a dollar's worth of precious metals (p. 244). By one estimate, 1 ton of computer scrap contains more gold than 16 tons of mined ore from a gold mine, while 1 ton of iPhones contains over 300 times more. Every ounce of metal we can recycle from a manufactured item is an ounce of metal we don't need to mine from the ground. Thus, "mining" e-waste for metals helps reduce the environmental impacts of mining the earth. In one example, the 2010 Winter Olympic Games in Vancouver produced its stylish gold, silver, and bronze medals (**FIGURE 17.14**) from metals recovered from recycled and processed e-waste!

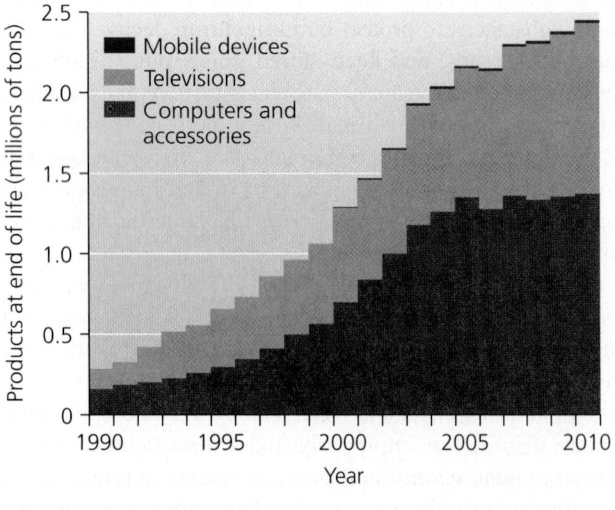


**FIGURE 17.13 Electronic waste is increasing.** The amount of electronic products at the end of their lives each year **(a)** in the United States has skyrocketed. *Data from U.S. Environmental Protection Agency, 2011.* Electronic waste management in the United States through 2009. *EPA, Washington, D.C.*

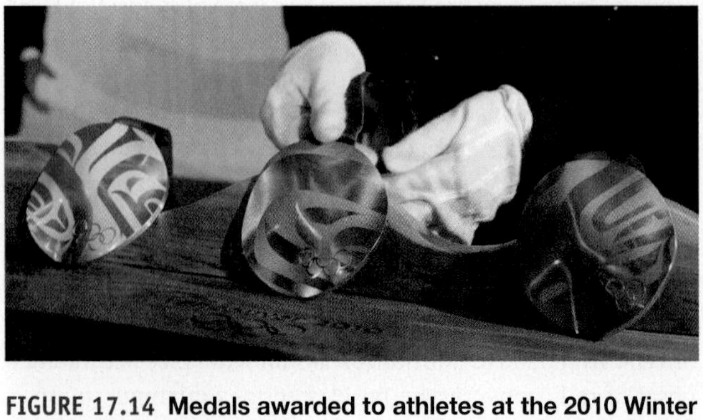

**FIGURE 17.14 Medals awarded to athletes at the 2010 Winter Olympic Games in Vancouver were made partly from precious metals recycled from discarded e-waste.**

## Several steps precede the disposal of hazardous waste

Many communities designate sites or special collection days to gather household hazardous waste, or designate facilities for the exchange and reuse of substances (**FIGURE 17.15**). Once consolidated, the waste is transported for treatment and disposal.

Under the Resource Conservation and Recovery Act, the EPA sets standards by which states manage hazardous waste. The Act also requires large generators of hazardous waste to obtain permits. Finally, it mandates that hazardous materials be tracked "from cradle to grave." As hazardous waste is generated, transported, and disposed of, the producer, carrier, and disposal facility must each report to the EPA the type and amount of material generated; its location, origin, and destination; and the way it is handled.

Because current U.S. law makes disposing of hazardous waste quite costly, irresponsible companies sometimes illegally dump waste, creating health risks for residents and financial headaches for local governments forced to deal with the mess (**FIGURE 17.16**). Companies from industrialized nations sometimes dump hazardous waste illegally in developing nations—a major environmental justice issue. The Basel Convention, an international treaty, was crafted to limit such practices.

High costs of disposal, however, have also encouraged conscientious businesses to invest in reducing their hazardous waste. Many biologically hazardous materials can be broken down by incineration at high temperatures in cement kilns. Others can be treated by exposure to bacteria that break down harmful components and synthesize them into new compounds.

**FIGURE 17.15 Many communities designate collection sites or collection days for household hazardous waste.**

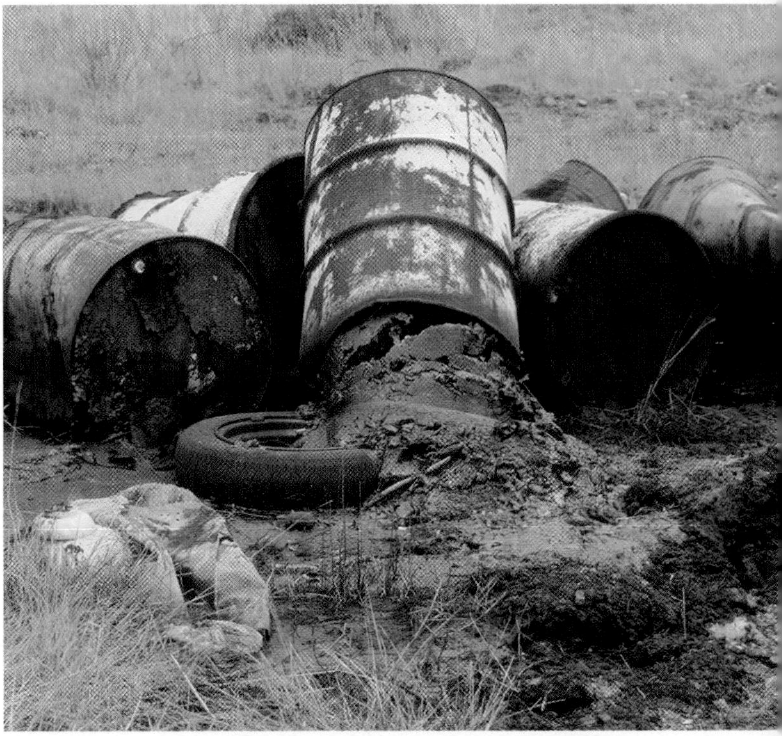

**FIGURE 17.16 Unscrupulous individuals or businesses sometimes dump hazardous waste illegally to avoid disposal costs.**

Additionally, various plants have been bred or engineered to take up specific contaminants from soil and break down organic contaminants into safer compounds or concentrate heavy metals in their tissues. The plants are eventually harvested and disposed of.

## We use three disposal methods for hazardous waste

We have developed three primary means of hazardous waste disposal: landfills, surface impoundments, and injection wells. These do nothing to lessen the hazards of the substances, but they help keep the waste isolated from people, wildlife, and ecosystems. Design and construction standards for hazardous waste landfills are stricter than those for ordinary sanitary landfills. Hazardous waste landfills must have several impervious liners and leachate removal systems and must be located far from aquifers.

Liquid hazardous waste, or waste in dissolved form, may be stored in **surface impoundments**, shallow depressions lined with plastic and an impervious material, such as clay. The liquid or slurry is placed in the pond, and water is allowed to evaporate, leaving a residue of solid hazardous waste on the bottom. This process is repeated, and eventually the dry residue is removed and transported elsewhere for permanent disposal. Impoundments are not ideal. The underlying layer can crack and leak waste. Some material may evaporate or blow into surrounding areas. Rainstorms may cause waste to overflow and contaminate nearby areas. For these reasons, surface impoundments are used only for temporary storage.

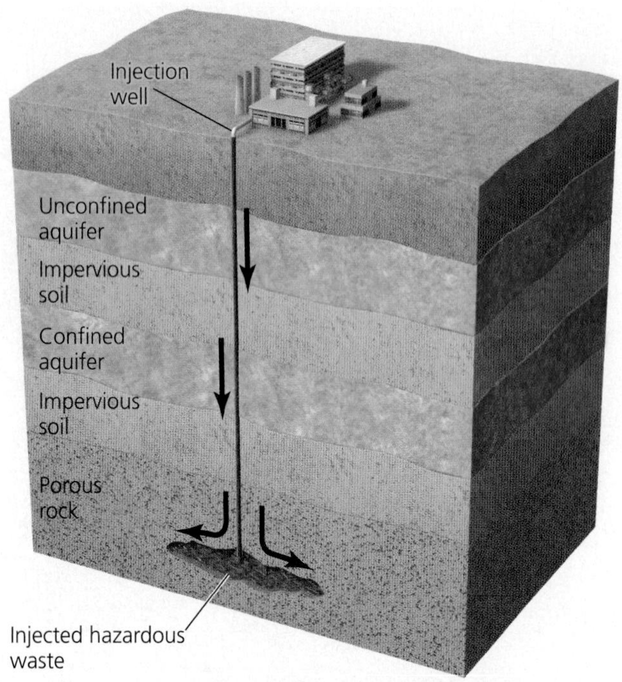


**FIGURE 17.17 Liquid hazardous waste is pumped deep underground by deep-well injection.** The well must be drilled below any aquifers, into porous rock isolated by impervious clay.

The third method is intended for long-term disposal. In **deep-well injection**, a well is drilled deep beneath the water table into porous rock, and wastes are injected into it (**FIGURE 17.17**). The process aims to keep hazardous waste deep underground, isolated from groundwater and human contact. However, wells can corrode and can leak wastes into soil, contaminating aquifers. Roughly 34 billion L (9 billion gal) of hazardous waste are placed in U.S. injection wells each year.

## Contaminated sites are being cleaned up, slowly

Many thousands of former military and industrial sites remain contaminated with hazardous waste in the United States and virtually every other nation on Earth. For most nations, dealing with these messes is simply too difficult, time-consuming, and expensive. In 1980, however, the U.S. Congress passed the Comprehensive Environmental Response Compensation and Liability Act (CERCLA; p. 105). This law established a federal program to clean up U.S. sites polluted with hazardous waste. The EPA administers this cleanup program, called the **Superfund**. Under EPA auspices, experts identify polluted sites, take action to protect groundwater, and clean up the pollution. Later laws also charged the EPA with cleaning up **brownfields**, lands whose reuse or development is complicated by the presence of hazardous materials.

Two well-publicized events spurred creation of the Superfund legislation. In *Love Canal*, a residential neighborhood in Niagara Falls, New York, families were evacuated in 1978–1980 after toxic chemicals buried by a company and the city in past decades rose to the surface, contaminating homes and an elementary school. In Missouri, the entire town of *Times Beach* was evacuated and its buildings demolished after being contaminated in the 1970s by dioxins from waste oil sprayed on its roads.

Once a Superfund site is identified, EPA scientists evaluate how near the site is to homes, whether wastes are confined or likely to spread, and whether the pollution threatens drinking water supplies. Sites judged to be harmful are placed on the National Priorities List, ranked by the level of risk to human health that they pose. Cleanup proceeds as funds are available. Throughout the process, the EPA is required to hold public hearings to inform area residents of its findings and to receive feedback.

The objective of CERCLA was to charge the polluting parties for the cleanup of their sites, according to the *polluter-pays principle* (p. 110). For many sites, however, the responsible parties cannot be found or held liable, and in such cases—roughly 30% so far—cleanups have been covered by taxpayers and from a trust fund established by a federal tax on industries producing petroleum and chemical raw materials. However, Congress let the tax expire and the trust fund went bankrupt in 2004, so taxpayers are now shouldering the entire burden. As the remaining cleanup jobs become more expensive, fewer are being completed.

As of 2014, 1326 Superfund sites remained on the National Priorities List, and only 375 had been cleaned up or otherwise deleted from the list. The average cleanup has cost over $25 million and has taken nearly 15 years. Many sites are contaminated with hazardous chemicals we have no effective way to deal with. In such cases, cleanups aim simply to isolate waste from human contact, either by building trenches and barriers or by excavating contaminated material and shipping it to a hazardous waste disposal facility. For all these reasons, the current emphasis in the United States and elsewhere is on preventing hazardous waste contamination in the first place.

# Conclusion

We have made great strides in addressing our waste problems. Modern methods of waste management are far safer for people and gentler on the environment than past practices of open dumping and open burning. Recycling and composting efforts are advancing steadily, and Americans now divert one-third of all solid waste away from disposal. The continuing growth of recycling, driven by market forces, public policy, and consumer behavior, shows potential to further alleviate our waste problems.

Despite these advances, our prodigious consumption habits are creating as much waste as ever. Our waste management efforts are marked by a number of challenges, including the cleanup of Superfund sites, safe disposal of hazardous and radioactive waste, and local opposition to disposal sites. These dilemmas make clear that the best solution is to decrease our generation of waste by finding ways to reduce, reuse, and efficiently recycle the materials and goods that we use.

# Testing Your Comprehension

1. Describe the three major components of managing waste. Why do we practice waste management?
2. Why have some people labeled the United States "the throwaway society"? How much solid waste do Americans generate, and how does this amount compare to that of people from other countries?
3. Name several guidelines by which sanitary landfills are regulated. Describe three problems with landfills.
4. Describe the process of incineration, or combustion. What happens to the resulting ash? What is one drawback of incineration?
5. What is composting, and how does it help reduce the waste stream?
6. What are the three elements of a sustainable process of recycling?
7. In your own words, describe the goals of industrial ecology.
8. What four criteria are used to define hazardous waste? What makes heavy metals and synthetic organic compounds particularly hazardous?
9. What are the largest sources of hazardous waste? Describe three ways to dispose of hazardous waste.
10. What is the Superfund program? How does it work?

# Seeking Solutions

1. How much waste do you generate? Look into your waste bin at the end of the day, and categorize and measure the waste there. List all other waste you may have generated in other places throughout the day. How much of this waste could you have avoided generating? How much could have been reused or recycled?
2. Some people have criticized current waste management practices as merely moving waste from one medium to another. How might this criticism apply to the methods now in practice? What are some potential solutions?
3. Of the various waste management approaches covered in this chapter, which ones are your community or campus pursuing, and which are they not pursuing? Would you suggest that your community or campus start pursuing any new approaches? If so, which ones, and why?
4. **THINK IT THROUGH** You are the CEO of a major corporation that produces containers for soft drinks and a wide variety of other consumer products. Your company's shareholders are asking that you improve the company's image—while not cutting into profits—by taking steps to reduce waste. What steps would you consider taking?
5. **THINK IT THROUGH** You are the president of your college or university. Your trustees want you to engage with local businesses and industries in ways that benefit the school and the community. Your faculty and students want you to make the school a leader in waste reduction and industrial ecology. Consider the industries and businesses in your community and the ways they interact with facilities on your campus. Bearing in mind the principles of industrial ecology, what novel ways might your school and local businesses mutually benefit from one another's services, products, or waste materials? Are there waste products from one business, industry, or campus facility that another might put to good use? What steps would you propose to take as president?

## Calculating Ecological Footprints

The biennial "State of Garbage in America" survey documents the ability of U.S. residents to generate prodigious amounts of municipal solid waste (MSW). According to the most recent survey, on a per capita basis, Missouri residents generate the least MSW (4.5 lb/day), and Hawaii residents (and Hawaii's many tourists) generate the most (15.5 lb/day). The average for the entire country is 6.8 lb MSW per person per day. Calculate the amount of MSW generated in 1 day and in 1 year by each of the groups indicated, at each of the rates shown in the table.

**Per capita MSW generation rates**

| Groups generating municipal solid waste | U.S. average (6.8 lb/day) | | Missouri (4.5 lb/day) | | Hawaii (15.5 lb/day) | |
|---|---|---|---|---|---|---|
| | Day | Year | Day | Year | Day | Year |
| You | 6.8 | 2482 | | | | |
| Your class | | | | | | |
| Your state | | | | | | |
| United States | | | | | | |
| World | | | | | | |

*Data from Shin, D., 2014. Generation and disposal of municipal solid waste (MSW) in the United States—A national survey.* Columbia University Earth Engineering Center.

1. Suppose your town of 50,000 people has just approved construction of a landfill nearby. Estimates are that it will accommodate 1 million tons of MSW. Assuming the landfill is serving only your town, and assuming that your town's residents generate waste at the U.S. average rate, for how many years will it accept waste before filling up? How much longer would a landfill of the same capacity serve a town of the same size in Missouri?
2. One study has estimated that the average world citizen generates 1.47 pounds of trash per day. How many times more does the average U.S. citizen generate?
3. The same study showed that the average resident of a low-income nation generates 1.17 pounds of waste per day and that the average resident of a high-income nation generates 2.64 pounds per day. Why do you think U.S. residents generate so much more MSW than people in other "high-income" countries, when standards of living in those countries are comparable?

# 18

# The Urban Environment: Creating Sustainable Cities

## Upon completing this chapter, you will be able to:

- Describe the scale of urbanization
- Assess urban and suburban sprawl
- Outline city and regional planning and land use strategies
- Evaluate transportation options, urban parks, and green buildings
- Analyze environmental impacts and advantages of urban centers
- Assess urban ecology and the pursuit of sustainable cities

**Photo: Tom McCall Waterfront Park on the Willamette River in Portland, Oregon**

# CENTRAL CASE STUDY

# Managing Growth in Portland, Oregon

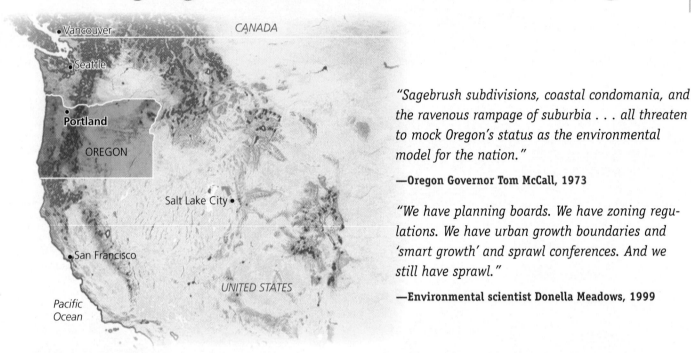

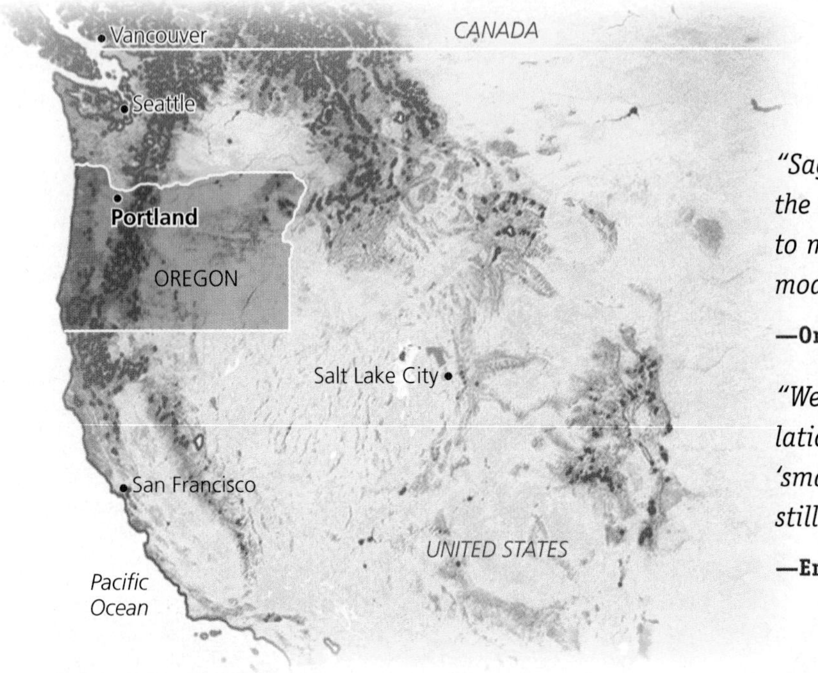


*"Sagebrush subdivisions, coastal condomania, and the ravenous rampage of suburbia . . . all threaten to mock Oregon's status as the environmental model for the nation."*

**—Oregon Governor Tom McCall, 1973**

*"We have planning boards. We have zoning regulations. We have urban growth boundaries and 'smart growth' and sprawl conferences. And we still have sprawl."*

**—Environmental scientist Donella Meadows, 1999**

With the fighting words above, Oregon governor Tom McCall challenged his state's legislature in 1973 to take action against runaway sprawling development, which many Oregon residents feared would ruin the communities and landscapes they loved. McCall was echoing the growing concerns of state residents that farms, forests, and open space were being gobbled up and paved over.

Foreseeing a future of subdivisions, strip malls, and traffic jams engulfing the pastoral Willamette Valley, Oregon acted. The state legislature passed Senate Bill 100, a sweeping land use law that would become the focus of acclaim, criticism, and careful study for years afterward by other states and communities trying to manage their own urban and suburban growth.

Oregon's law required every city and county to draw up a comprehensive land use plan in line with statewide guidelines that had gained popular support from the state's electorate. As part of each land use plan, each metropolitan area had to establish an **urban growth boundary (UGB)**, a line on a map intended to separate areas desired to be urban from areas desired to remain rural. Development for housing, commerce, and industry would be encouraged within these urban growth boundaries but severely restricted beyond them. The intent was to revitalize city centers, prevent suburban sprawl, and protect farmland, forests, and open landscapes around the edges of urbanized areas.

Residents of the area around Portland, the state's largest city, established a new regional planning entity to apportion land in their region. The Metropolitan Service District, or Metro, represents 25 municipalities and three counties. Metro adopted the Portland-area urban growth boundary in 1979 and has tried to focus growth on existing urban centers and to build communities where people can walk, bike, or take mass transit between home, work, and shopping. These policies have largely worked as intended. Portland's downtown and older neighborhoods have thrived, regional urban centers are becoming denser and more community oriented, mass transit has expanded, and development has been limited on land beyond the UGB. Portland began attracting international attention for its "livability."

To many Portlanders today, the UGB remains the key to maintaining quality of life in city and countryside alike. In the view of its critics, however, the "Great Wall of Portland" is an elitist and intrusive government regulatory tool. In 2004, Oregon voters approved a ballot measure that threatened to eviscerate the land use rules that most citizens had backed for three decades. Ballot Measure 37 required the state to compensate certain landowners if government regulation had decreased the value of their land. For example, regulations prevent landowners outside UGBs from subdividing their lots and selling them for housing development. Under Measure 37, the state had to pay these landowners to make up for theoretically lost income or else allow them to ignore the regulations. Because state and local governments did not have enough money to pay such claims, the measure was on track to gut Oregon's zoning, planning, and land use rules.

Landowners filed over 7500 claims for payments or waivers affecting 295,000 ha (730,000 acres). Although the measure had been promoted to voters as a way to protect the rights of small family landowners, most claims were filed by large developers. Neighbors suddenly found themselves confronting the prospect of massive housing subdivisions, gravel mines, strip malls, or industrial facilities being developed next to their homes—and many who had voted for Measure 37 began to have misgivings.

The state legislature, under pressure from opponents and supporters alike, settled on a compromise: to introduce a new ballot measure. Oregon's voters passed Ballot Measure 49 in 2007. It protects the rights of small landowners to gain income from their property by developing small numbers of homes, while restricting large-scale development and development in sensitive natural areas.

In 2010, Metro finalized a historic agreement with representatives and citizens of its region's three counties to determine where urban growth will and will not be allowed over the next 50 years. Metro and the counties apportioned over 121,000 ha (300,000 acres) of undeveloped land into "urban reserves" open for development and "rural reserves" where farmland and forests would be preserved. Boundaries were precisely mapped to give clarity and direction for landowners and governments alike for half a century.

People are confronting similar issues in communities throughout North America, and debates and negotiations like those in Oregon will determine how our cities and landscapes will change in the future. ◻

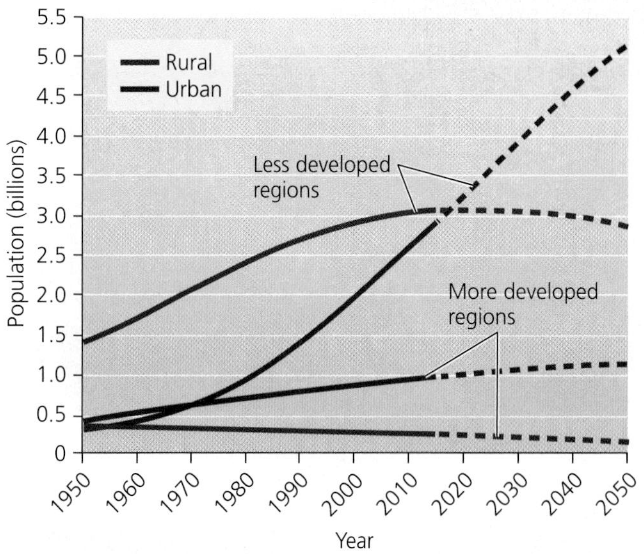


**FIGURE 18.1 Population trends differ between poor and wealthy nations.** In less developed regions, urban populations are growing quickly, and rural populations will soon begin declining. More-developed regions are already largely urbanized, so their urban populations are growing slowly, whereas rural populations are falling. Solid lines in the graph indicate past data, and dashed lines indicate future projections. *Data from United Nations Population Division, 2012.* World urbanization prospects: The 2011 revision. *By permission.*

DATA Q Beginning in what decade will the majority of people in less developed regions be living in urban areas?

GO TO **INTERPRETING GRAPHS & DATA** ON MasteringEnvironmentalScience®

# Our Urbanizing World

We have recently passed a turning point in human history. Since 2009, for the first time ever, more people are living in urban areas (cities and suburbs) than in rural areas. As we undergo this historic shift from the countryside into towns and cities—a process called **urbanization**—two pursuits become ever more important. One is to make our urban areas more livable by meeting residents' needs for a safe, clean, healthy urban environment. The other is to make our urban areas sustainable by creating cities that can prosper in the long term while minimizing our ecological footprint and working with natural systems (rather than against them).

## Industrialization has driven urbanization

Since 1950, the world's urban population has multiplied by over five times, whereas the rural population has not even doubled. Urban populations are growing because the human population overall is growing (Chapter 6), and because more people are moving from farms to cities than are moving from cities to farms. Industrialization (p. 3) has reduced the need for farm labor while enhancing commerce and jobs in cities. Urbanization, in turn, has bred technological advances that boost production efficiencies and spur further industrialization.

The United Nations projects that the urban population will rise by 59% between now and 2050, whereas the rural population will decline by 9%. Trends differ between developed and developing nations, however (**FIGURE 18.1**). In developed nations, urbanization has slowed because three of every four people already live in cities, towns, and **suburbs**, the smaller communities that ring cities. Back in 1850, the U.S. Census Bureau classified only 15% of U.S. citizens as urban dwellers. That percentage now stands at 80%. Most American urban dwellers reside in suburbs; fully half the U.S. population today is suburban.

In contrast, today's developing nations, where most people still reside on farms, are urbanizing rapidly. In China, India, Pakistan, Nigeria, and most other nations, rural people are streaming to cities in search of jobs and urban lifestyles, or to escape ecological degradation in the countryside. As a result, demographers estimate that urban areas of developing nations will absorb nearly all of the world's population growth from now on.

## Environmental factors influence the location of urban areas

Real estate agents use the saying, "Location, location, location," to stress how a home's whereabouts determines its value. Location is vital for urban centers as well. Environmental variables such as climate, topography, and the configuration of waterways influence whether a city will succeed. Think of any major city, and chances are it's situated along a major river, seacoast, railroad, or highway—some corridor for trade that has driven economic growth.

**(a) St. Louis, Missouri**

**(b) Fort Worth, Texas**

**FIGURE 18.2 Cities develop along trade corridors.** St. Louis **(a)** is situated on the Mississippi River near its confluence with the Missouri River, where river trade drove its growth in the 19th and early 20th centuries. Fort Worth, Texas **(b)**, grew in the late 20th century as a result of the interstate highway system and a major international airport.

Well-located cities (**FIGURE 18.2**) often serve as linchpins in trading networks, funneling in resources from agricultural regions, processing them, manufacturing products, and shipping those products to other markets. Portland is situated where the Willamette River joins the Columbia River, and just upriver from where the Columbia flows into the Pacific Ocean. The city grew as it received, processed, and shipped overseas the produce from farms of the river valleys, and as it imported products shipped in from other ports.

Today, powerful technologies and cheap transportation enabled by fossil fuels have allowed cities to thrive even in resource-poor regions. The Dallas–Fort Worth area prospers from—and relies on—oil-fueled transportation by interstate highways and a major airport. Southwestern cities such as Los Angeles, Las Vegas, and Phoenix flourish in desert regions by appropriating water from distant sources. Whether such cities can sustain themselves as oil and water become increasingly scarce in the future is an important question.

In recent years, many cities in the southern and western United States have grown as people (particularly retirees) have moved south and west in search of warmer weather or more space. Between 1990 and 2013, the population of the Dallas–Fort Worth and Houston metropolitan areas each grew by 69%, that of the Atlanta area grew by 87%; that of the Phoenix region grew by 97%; and that of the Las Vegas metropolitan area grew by a whopping 138%.

## People have moved to suburbs

American cities grew rapidly in the 19th and early 20th centuries as a result of immigration from abroad and increased trade as the nation expanded west. The bustling economic activity of downtown districts held people in cities despite crowding, poverty, and crime. However, by the mid-20th century, many affluent city dwellers were choosing to move outward to cleaner, less crowded suburban communities. These people were pursuing more space, better economic opportunities, cheaper real estate, less crime, and better schools for their children.

As affluent people moved out to the expanding suburbs, jobs followed. This hastened the economic decline of downtown districts, and American cities stagnated. Chicago's population declined to 80% of its peak because so many residents moved to its suburbs. Philadelphia's population fell to 76% of its peak, Washington, D.C.'s to 71%, and Detroit's to just 55%.

Portland followed this trajectory, but the city bounced back. Portland's population growth stalled in the 1950s to 1970s as crowding and deteriorating economic conditions drove city dwellers to the suburbs. However, subsequent policies to revitalize the city center helped restart Portland's growth (**FIGURE 18.3**).

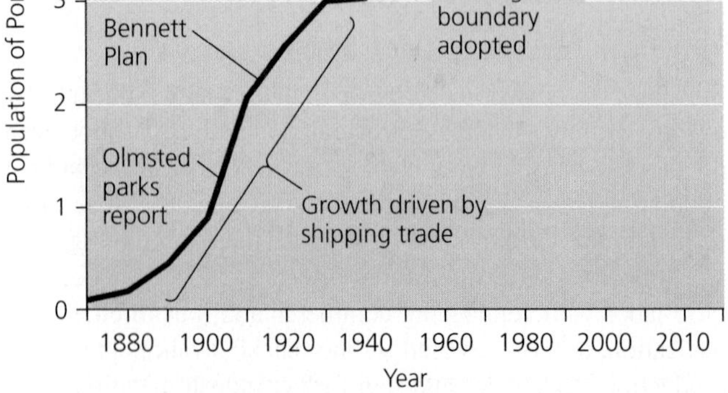

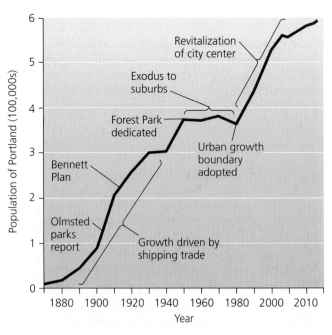

**FIGURE 18.3 Portland grew, stabilized, and then grew again.** Jobs in the shipping trade helped to boost Portland's economy and population in the 1890s–1920s. City residents began leaving for the suburbs in the 1950s–1970s, but policies to enhance the city center revitalized Portland's growth.

The exodus to the suburbs in 20th-century America was enabled by the rise of the automobile, an expanding road network, and inexpensive and abundant oil. Millions of people could now commute by car to downtown workplaces from new homes in suburban "bedroom communities." By facilitating long-distance transport, fossil fuels and highway networks also made it easier for businesses to import and export resources, goods, and waste. The federal government's development of the interstate highway system was pivotal in promoting these trends.

In most ways, suburbs have delivered the qualities people sought in them. The wide spacing of homes, with each one on its own plot of land, gives families room and privacy. However, by allotting more space to each person, suburban growth has spread human impact across the landscape. Natural areas have disappeared as housing developments are constructed. Our extensive road networks ease travel, but suburbanites find themselves needing to climb into a car to get anywhere. People commute longer distances to work and spend more time stuck in traffic. The expanding rings of suburbs surrounding cities have grown larger than the cities themselves, and towns are merging into one another. These aspects of suburban growth inspired a new term: *sprawl.*

# Sprawl

The term *sprawl* has become laden with meanings and suggests different things to different people, but we can begin our discussion by giving **sprawl** a simple, nonjudgmental definition: the spread of low-density urban or suburban development outward from an urban center.

## Urban areas spread outward

The spatial growth of urban and suburban areas is clear from maps and satellite images of rapidly spreading cities such as Las Vegas (**FIGURE 18.4**). Another example is Chicago, whose metropolitan area spreads over a region 40 times the size of the city. All in all, houses and roads supplant over 2700 ha (6700 acres) of U.S. land every day.

Sprawl results from development approaches that place homes on spacious lots in residential tracts that spread over large areas but are far from urban centers and commercial amenities (**FIGURE 18.5**, next page). Roads tend to be spaced far enough apart that some areas go undeveloped, yet not far enough apart for these areas to function as natural areas or sites for recreation.

Such development approaches allot each person more space than in cities. For example, the average resident of Chicago's suburbs takes up 11 times more space than a resident of the city. As a result, the outward spatial growth of suburbs across the landscape generally outpaces growth in numbers of people.

In fact, many researchers define *sprawl* as the physical spread of development at a rate that exceeds the rate of population growth. For instance, the population of Phoenix grew 12 times larger between 1950 and 2000, yet its land area grew 27 times larger. Between 1950 and 1990, the population of 58 major U.S. metropolitan areas rose by 80%, but the land area they covered rose by 305%. Even in 11 metro areas where population declined between 1970 and 1990 (as with Rust Belt cities such as Detroit, Cleveland, and Pittsburgh), the amount of land covered increased.

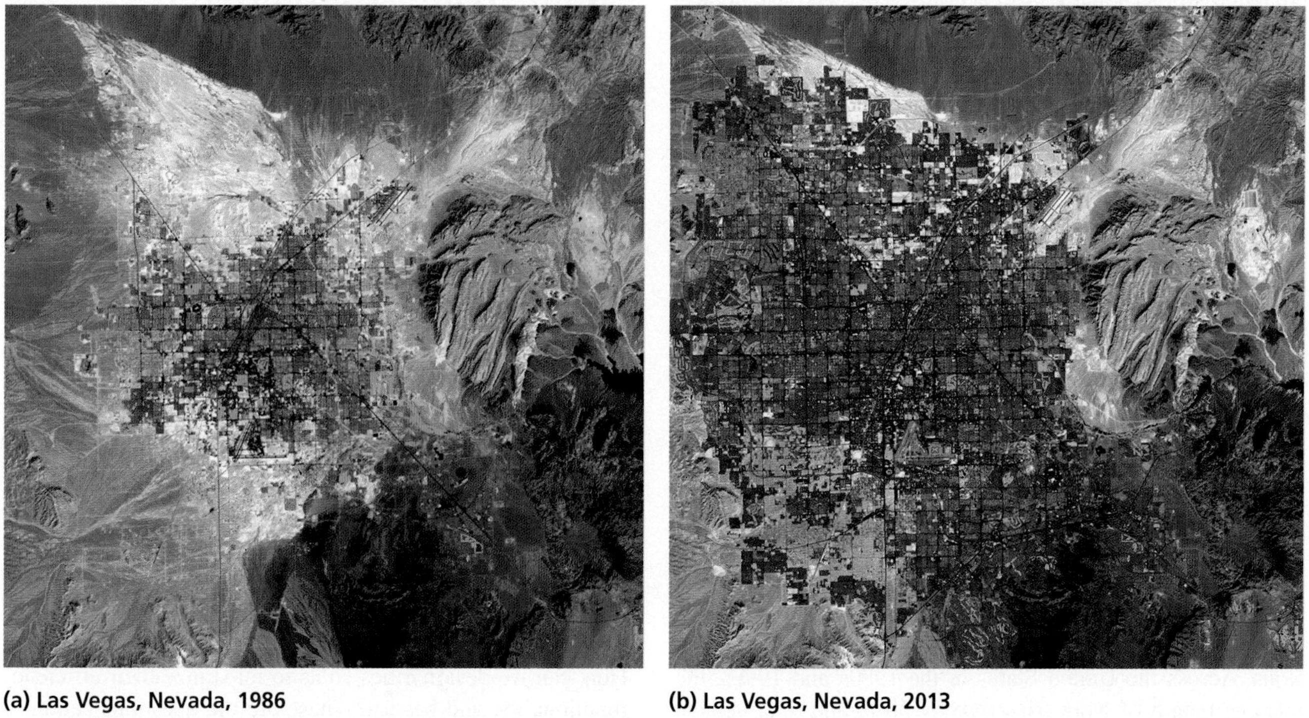

**(a) Las Vegas, Nevada, 1986** **(b) Las Vegas, Nevada, 2013**

**FIGURE 18.4 Satellite images show the rapid urban and suburban expansion commonly called sprawl.** Las Vegas, Nevada, is one of the fastest-growing cities in North America. Between 1986 **(a)** and 2013 **(b)**, its population and its developed area each tripled.

**FIGURE 18.5 Sprawl is characterized by widely spaced homes spread evenly across large areas of land.** This requires people to use cars to reach commercial amenities or community centers.

## Sprawl has several causes

There are two main components of sprawl. One is human population growth—there are simply more of us alive each year. The other is per capita land consumption—each person is taking up more land than in the past, because most people desire space and privacy and dislike congestion. Better highways, inexpensive gasoline, telecommunications, and the Internet have all fostered movement away from city centers by giving workers more flexibility to live where they desire and by freeing businesses from dependence on the centralized infrastructure a city provides.

Economists and politicians have encouraged the unbridled spatial expansion of cities and suburbs. The conventional assumption has been that growth is always good and that attracting business, industry, and residents will enhance a community's economic well-being, political power, and cultural influence. Today, this assumption is being challenged as growing numbers of people feel negative effects of sprawl on their lifestyles.

## What is wrong with sprawl?

To some people, the word *sprawl* evokes strip malls, traffic jams, homogenous commercial development, and tracts of cookie-cutter houses encroaching on farmland, ranchland, or forests. For other people, sprawl is simply the collective result of choices made by millions of well-meaning individuals trying to make a better life for their families. What does scientific research tell us about the impacts of sprawl?

**Transportation** Most studies show that sprawl constrains transportation options, essentially forcing people to own a vehicle, drive it most places, drive greater distances, and spend more time in it. Sprawling communities suffer more traffic accidents and have few or no mass transit options. Across the United States in the 1980s and 1990s, the average length of work trips rose by 36%, and total vehicle miles driven rose three times faster than population growth. A car-oriented culture encourages congestion and increases dependence on oil.

**Pollution** By promoting automobile use, sprawl increases pollution. Carbon dioxide emissions from vehicles contribute to climate change (Chapter 14), and nitrogen- and sulfur-containing air pollutants lead to smog and acid deposition (Chapter 13). Motor oil and road salt from roads and parking lots run off readily and pollute waterways.

**Health** Beyond the health impacts of pollution, some research suggests that sprawl promotes physical inactivity and obesity because driving cars largely takes the place of walking during daily errands. A 2003 study found that people from the most-sprawling U.S. counties show higher blood pressure and weigh 2.7 kg (6 lb) more for their height than people from the least-sprawling U.S. counties.

**Land use** As more land is developed, less is left as forests, fields, farmland, or ranchland. Natural lands and agricultural lands provide vital resources, recreation, aesthetic beauty, wildlife habitat, and air and water purification. Many children now grow up without the ability to roam through woods and fields, which used to be a normal part of childhood.

**Economics** Sprawl drains tax dollars from communities and funnels money into infrastructure for new development on the fringes of those communities. Funds that could be spent maintaining downtown centers are instead spent on extending the road system, water and sewer system, electricity grid, telephone lines, police and fire service, schools, and libraries. Although taxes on new development can in theory pay back the investment, studies find that in most cases taxpayers continue to subsidize new development.

### WEIGHING THE ISSUES

**Sprawl Near You**

Is there sprawl in the area where you live? Does it bother you, or not? Has development in your area had any of the impacts described above? Do you think your city or town should encourage outward growth? Why or why not?

# Creating Livable Cities

To respond to the challenges that sprawl presents, architects, planners, developers, and policymakers are trying to revitalize city centers and to plan and manage how urbanizing areas develop. They aim to make cities safer, cleaner, healthier, and more pleasant for their residents.

## Planning helps to create livable urban areas

How can we design cities so as to maximize their efficiency, functionality, and beauty? These are the questions central to **city planning** (also known as **urban planning**). City planners advise policymakers on development options, transportation needs, public parks, and other matters.

Washington, D.C., is the earliest example of city planning in the United States. President George Washington hired French architect Pierre Charles L'Enfant in 1791 to design a capital city for the new nation on undeveloped land along the Potomac River. L'Enfant laid out a Baroque-style plan of diagonal avenues cutting across a grid of streets, with plenty of space allotted for majestic public monuments (**FIGURE 18.6**). A century later, a new generation of planners imposed a height restriction on new buildings to keep the monuments from being crowded and dwarfed by modern skyscrapers. This preserved the spacious, stately feel of the city.

City planning in North America came into its own at the turn of the 20th century, as urban leaders sought to beautify and impose order on fast-growing, unruly cities. In Portland in 1912, planner Edward Bennett's *Greater Portland Plan* proposed to rebuild the harbor; dredge the river channel; construct new docks, bridges, tunnels, and a waterfront railroad; superimpose wide radial boulevards on the old city street grid; establish civic centers downtown; and greatly expand the number of parks.

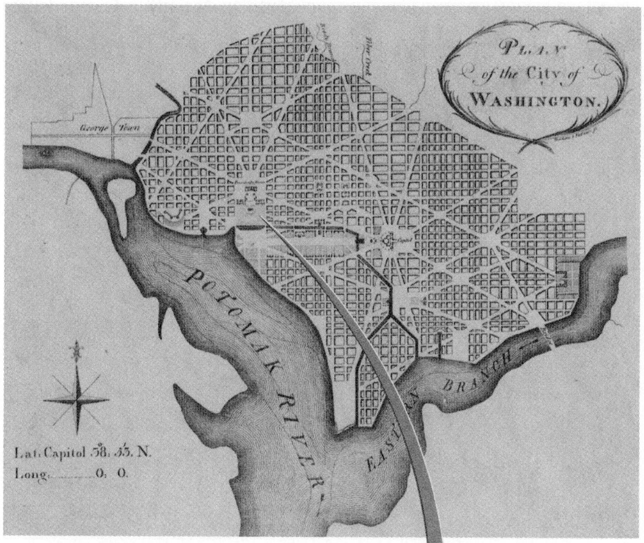


**(a) The L'Enfant plan, 1791**

**(b) Washington, D.C., today**

**FIGURE 18.6 Washington, D.C., is a prime example of early city planning.** The 1791 plan **(a)** for the new U.S. capital laid out splendid diagonal avenues cutting across gridded streets, allowing space for the magnificent public monuments **(b)** that grace the city today.

## WEIGHING THE ISSUES

**Your Urban Area**

Think of your favorite parts of the city you know best. What do you like about them? What do you dislike about your least favorite parts of the city? What could this city do to improve quality of life for its residents?

In today's world of sprawling metropolitan areas, **regional planning** has become at least as important as city planning. Regional planners deal with the same issues as city planners, but they work on broader geographic scales and coordinate their work with multiple municipal governments. In some places, regional planning has been institutionalized in formal government bodies; the Portland area's Metro is the epitome of such a regional planning entity. When Metro and its region's three counties in 2010 announced their collaborative plan apportioning undeveloped land into "urban reserves" and "rural reserves," it marked a historic accomplishment in regional planning. The agreement designates 11,000 ha (28,000 acres) for urban use and 110,000 ha (272,000 acres) for rural use. It enables homeowners, farmers, developers, and policymakers to feel informed and secure knowing what kinds of land uses lie in store on and near their land over the next half-century.

## Zoning is a key tool for planning

One tool that planners use is **zoning**, the practice of classifying areas for different types of development and land use. For instance, to preserve the cleanliness and tranquility of residential neighborhoods, industrial facilities may be kept out of districts zoned for residential use. By specifying zones for different types of development, planners can guide what gets built where. Zoning also gives home buyers and business owners security, because they know in advance what types of development can and cannot be located nearby.

Zoning involves government restriction on the use of private land and represents a top-down constraint on personal property rights. Yet most people feel that government has a proper and useful role in setting limits on property rights for the good of the community. Oregon voters sided with private property rights when in 2004 they passed Ballot Measure 37, which shackled government's ability to enforce zoning regulations with landowners who bought their land before the regulations were enacted. However, many Oregonians soon began witnessing new development they did not condone, so in 2007 they passed Ballot Measure 49 to restore public oversight over development. For the most part, people have supported zoning over the years because the common good it produces for communities is widely felt to outweigh the restrictions on private use.

## Urban growth boundaries are now widely used

Planners intended Oregon's urban growth boundaries (UGBs) to limit sprawl by containing growth within existing urbanized areas. The UGBs aimed to revitalize downtowns; protect working farms, orchards, ranches, and forests; and ensure urban dwellers access to open space. Since Oregon began its experiment, many other states, regions, and cities have adopted UGBs—from Boulder, Colorado, to Lancaster, Pennsylvania, to many California communities.

UGBs save taxpayers money by reducing the amounts that municipalities need to pay for infrastructure. However, UGBs also tend to increase housing prices within their boundaries. In the Portland area, housing has become less affordable, but in most other ways its UGB is working as intended. It has preserved farms and forests outside the UGB while increasing the density of new housing inside it. Within the UGB, homes are built on smaller lots and multistory apartments fulfill a vision of "building up, not out." Downtown employment has grown tremendously as businesses and residents invest in the central city.

However, urbanized area still increased by 101 $km^2$ (39 $mi^2$) in the decade after Portland's UGB was established, because 146,000 people were added to the population. This fact suggests that relentless population growth may thwart even the best anti-sprawl efforts and that livable cities may fall victim to their own success if they are in high demand as places to live. Indeed, Metro has enlarged the Portland-area UGB three dozen times and plans to expand it more in the future.

## "Smart growth" and "new urbanism" aim to counter sprawl

As more people feel impacts of sprawl on their everyday lives, efforts to manage growth are springing up throughout North America. Proponents of **smart growth** seek to rejuvenate the older existing communities that so often are drained and impoverished by sprawl (TABLE 18.1). Smart growth means "building up, not out"—focusing development and economic investment in existing urban centers and favoring multistory shop-houses and high-rises.

A related approach among architects, planners, and developers is **new urbanism**, which seeks to design walkable neighborhoods with homes, businesses, schools, and other amenities all close together for convenience. The aim is to create functional neighborhoods in which families can meet most of their needs close to home without using a car. Green spaces, trees, a mix of architectural styles, and creative street layouts add to the visual interest of new-urbanist developments. These neighborhoods are often served by public transit systems, enabling people to travel most places they need to go by train and foot.

**TABLE 18.1 Ten Principles of "Smart Growth"**

| |
|---|
| • Mix land uses |
| • Take advantage of compact building design |
| • Create a range of housing opportunities and choices |
| • Create walkable neighborhoods |
| • Foster distinctive, attractive communities with a strong sense of place |
| • Preserve open space, farmland, natural beauty, and critical environmental areas |
| • Strengthen existing communities, and direct development toward them |
| • Provide a variety of transportation choices |
| • Make development decisions predictable, fair, and cost-effective |
| • Encourage community and stakeholder collaboration in development decisions |

*Source: U.S. Environmental Protection Agency.*

## Transit options help cities

Traffic jams on roadways cause air pollution, stress, and countless hours of lost time. They cost Americans an estimated $74 billion yearly in fuel and lost productivity. To encourage more efficient urban transportation, policymakers can raise fuel taxes, charge trucks for road damage, tax inefficient modes of transport, and reward carpoolers with carpool lanes. But a key ingredient in any planner's recipe for improving the quality of urban life is to give residents alternative transportation options.

Bicycle transportation is one option (**FIGURE 18.7**). Portland has embraced bicycles like few other American cities, and today 6% of its commuters ride to work by bike (the national average is 0.5%). The city has developed nearly 400 miles of bike lanes and paths, 5000 public bike racks, and special markings at intersections to protect bicyclists. All of this infrastructure was created for the typical cost of just one mile of urban freeway. Portland also has a bike-sharing program similar to programs in cities such as Montreal, Toronto, Denver, Minneapolis, Miami, San Antonio, Boston, and Washington, D.C.

Other transportation options include **mass transit** systems: public systems of buses, trains, subways, or *light rail* (smaller rail systems powered by electricity). Mass transit systems move large numbers of passengers at once while easing traffic congestion, taking up less space than road networks, and emitting less pollution than cars. Studies show that as long as an urban center is large enough to support the infrastructure necessary, both train and bus systems are

**FIGURE 18.7 Bicycles provide a healthy transportation alternative to car travel.** Biking to work (here, in San Francisco) and for recreation is increasing throughout North America.

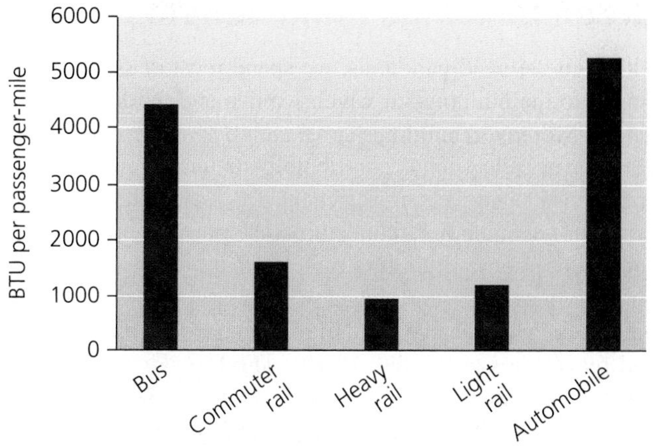


**(a) Energy consumption for different modes of transit**

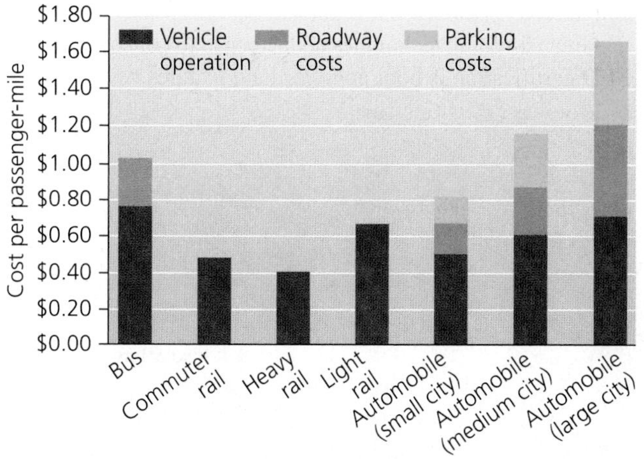


**(b) Operating costs for different modes of transit**

**FIGURE 18.8 Rail transit consumes less energy (a) and costs less (b) per passenger mile than bus or automobile transit.** *Data from Litman, T., 2005.* Rail transit in America: A comprehensive evaluation of benefits. *© 2005 T. Litman.*

**DATA Q** Automobile traffic creates two types of operating costs that are *not* created by rail traffic. What are they?

GO TO **INTERPRETING GRAPHS & DATA** ON MasteringEnvironmentalScience®

cheaper, more energy-efficient, and cleaner than roadways choked with cars (**FIGURE 18.8**).

In Portland, buses carry 60 million riders per year, and an average bus keeps about 250 cars off the road each day. Portland also boasts streetcars and one of the nation's leading light rail systems. America's most-used train systems are those of its largest cities, such as New York City's subways, Washington, D.C.'s Metro, the T in Boston, and the San Francisco Bay area's BART. Each of these carries more than a quarter of its city's daily commuters.

In general, however, the United States lags behind most nations in mass transit. Many countries, rich and poor alike, have extensive bus systems that are easily accessible to people. The Brazilian city of Curitiba has an outstanding bus system that is used each day by three-quarters of its 2.5 million people. With 340 bus routes, 250 terminals, and 1900 buses, along with measures to encourage bicycles and pedestrians, Curitiba has seen a steep drop in car use despite a rapidly growing population.

The United States chose instead to invest in road networks for cars and trucks largely because its population density was low and gasoline was cheap. As energy costs and population rise, however, mass transit becomes increasingly appealing, and citizens begin to desire train and bus systems in their communities.

## Urban residents need parklands

City dwellers often desire some escape from the noise, commotion, and stress of urban life. Natural lands, public parks, and open space provide greenery, scenic beauty, freedom of movement, and places for recreation. These lands also keep ecological processes functioning by helping to regulate climate, purify air and water, and provide wildlife habitat.

One newly developed and instantly popular city park is The High Line Park in Manhattan in New York City (**FIGURE 18.9**). An elevated freight line running above the streets was going to be demolished, but a group of citizens saw its potential for a park, and they pushed the idea until city leaders came to share their vision. Today thousands of people use the 23-block-long High Line for recreation or on their commute to work.

Large city parks are vital to a healthy urban environment, but even small spaces can make a big difference. Playgrounds provide places where children can be active outdoors and interact with their peers. Community gardens allow people to grow vegetables and flowers in a neighborhood setting.

**FIGURE 18.9 New York City's High Line Park was created thanks to a visionary group of citizens.** They pushed to make a park out of an abandoned elevated rail line.

"Greenways" along rivers, canals, or old railway lines can provide walking trails, protect water quality, boost property values, and serve as corridors for the movement of wildlife.

America's city parks arose in the late 19th century as politicians and citizens established public spaces using aesthetic ideals from European parks and gardens. The lawns, shaded groves, curved pathways, and pastoral vistas of many American city parks grew from these European ideals, as interpreted by America's leading landscape architect, Frederick Law Olmsted. Olmsted designed Central Park in New York City and many other urban parks.

Portland's quest for parks began in 1900, when city leaders created a parks commission and hired Olmsted's son, John Olmsted, to design a park system. His 1904 plan proposed acquiring land to ring the city generously with parks, but no action was taken. A full 44 years later, citizens pressured city leaders to create Forest Park along a forested ridge on the northwest side of the city. At 11 km (7 mi) long, it is one of the largest city parks in North America.

## Green buildings bring benefits

Although we need parklands, we spend most of our time indoors, so the buildings in which we live and work affect our health. Moreover, buildings consume 40% of our energy and 70% of our electricity, contributing to the greenhouse gas emissions that drive climate change. As a result, there is a thriving movement in architecture and construction to design and build **green buildings**, structures that are built from sustainable materials, limit their use of energy and water, minimize health impacts on their occupants, control pollution, and recycle waste (**FIGURE 18.10**). The U.S. Green Building Council promotes these efforts through the **Leadership in Energy and Environmental Design (LEED)** certification program. Buildings (new buildings or renovation projects) apply for certification and, depending on their performance, may be granted silver, gold, or platinum status.

Green building techniques add expense to construction, yet LEED certification is booming. Portland features several dozen

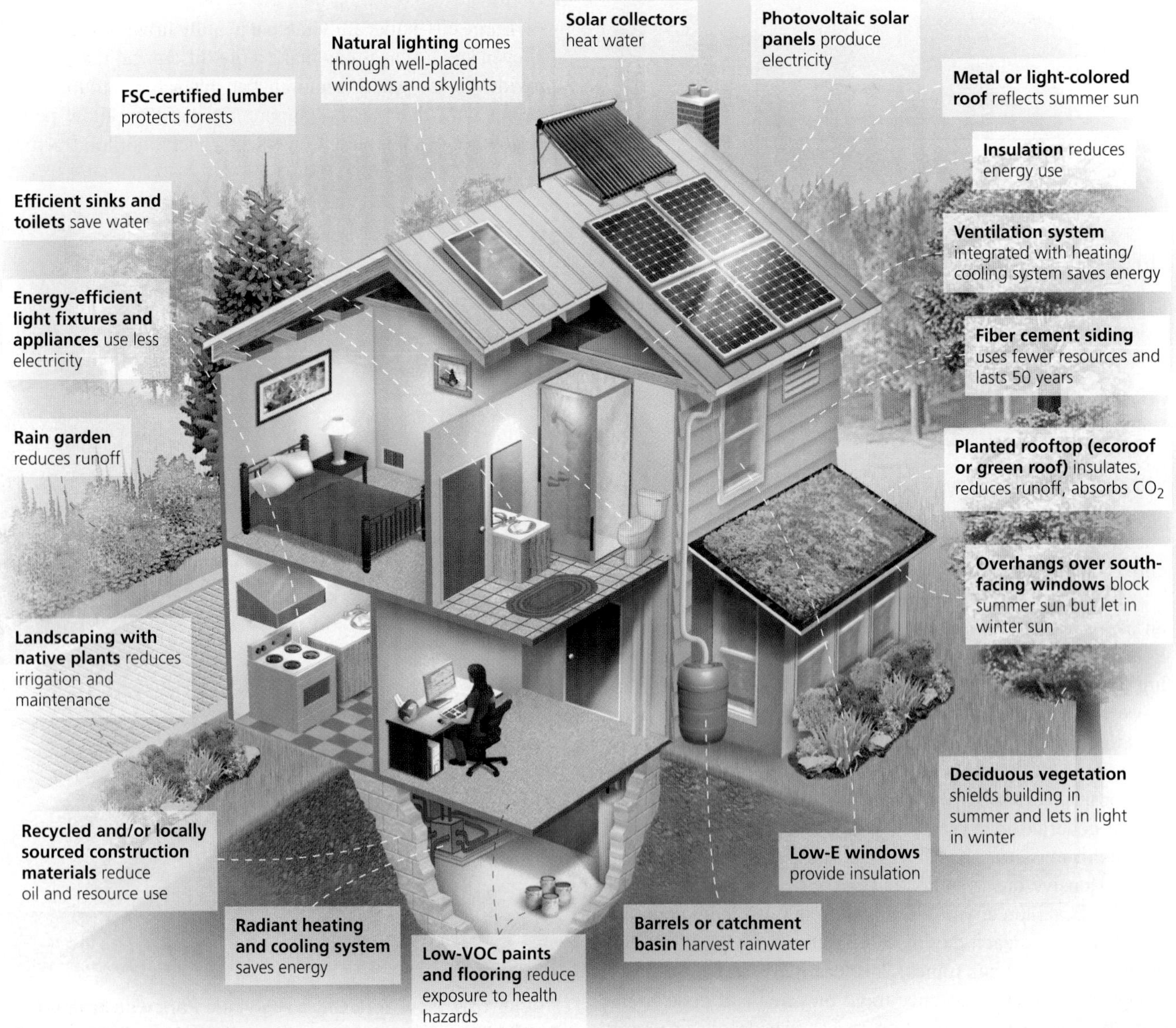

Natural lighting comes through well-placed windows and skylights
Solar collectors heat water
Photovoltaic solar panels produce electricity
FSC-certified lumber protects forests
Metal or light-colored roof reflects summer sun
Insulation reduces energy use
Efficient sinks and toilets save water
Ventilation system integrated with heating/cooling system saves energy
Energy-efficient light fixtures and appliances use less electricity
Fiber cement siding uses fewer resources and lasts 50 years
Rain garden reduces runoff
Planted rooftop (ecoroof or green roof) insulates, reduces runoff, absorbs $CO_2$
Overhangs over south-facing windows block summer sun but let in winter sun
Landscaping with native plants reduces irrigation and maintenance
Deciduous vegetation shields building in summer and lets in light in winter
Recycled and/or locally sourced construction materials reduce oil and resource use
Low-E windows provide insulation
Radiant heating and cooling system saves energy
Low-VOC paints and flooring reduce exposure to health hazards
Barrels or catchment basin harvest rainwater

**FIGURE 18.10 A green building incorporates design features to minimize its ecological footprint.**

LEED-certified buildings, including the nation's first LEED-gold sports arena (the Moda Center, where the Trailblazers basketball team plays). The savings on energy, water, and waste at the refurbished Moda Center paid for the cost of its LEED upgrade after just one year.

Schools are leaders in sustainable building. In Portland, the Rosa Parks Elementary School was built with locally sourced and nontoxic materials, uses 24% less energy and water than comparable buildings, and diverted nearly all of its construction waste from the landfill. Schoolchildren learn about renewable energy by watching a display of the electricity produced by their building's own photovoltaic solar system. Portland State University, the University of Portland, Reed College, and Lewis and Clark College are just a few of the many colleges and universities nationwide that are constructing green buildings as a part of their campus sustainability efforts (pp. 18, 422).

# Urban Sustainability

Most of our efforts to make cities safer, cleaner, healthier, and more pleasant are also helping to make them more sustainable. A sustainable city is one that can function and prosper over the long term, providing generations of residents a good quality of life far into the future.

## Urban centers bring a mix of environmental effects

Making a city sustainable entails minimizing the city's impacts on the natural systems and resources that nourish it. You might guess that urban living has a greater environmental impact than rural living. However, the picture is not so simple; instead, urbanization brings a complex mix of consequences.

**Resource use and efficiency** Cities are sinks (p. 38) for resources, importing nearly everything they need to feed, clothe, and house their inhabitants and power their commerce. For their day-to-day survival, people in cities such as New York, Boston, San Francisco, and Los Angeles depend on water pumped in from faraway watersheds. Thus, urban areas rely on large expanses of land elsewhere for resources and ecosystem services, and they burn fossil fuels to import resources and goods.

However, imagine that the world's 3.9 billion urban residents were instead spread evenly across the landscape. What would the transportation requirements be, then, to move resources and goods around to all those people? A world without cities would likely require *more* transportation to provide people the same level of access to resources and goods. Moreover, once resources arrive, cities are highly efficient in distributing goods and services. The density of cities facilitates the provision of electricity, medical care, education, water and sewer systems, waste disposal, and public transportation. Thus, although a city has a large ecological footprint (p. 4) in total, it may have a moderate or small footprint in per capita terms.

**Pollution** Just as cities import resources, they export wastes, either passively by pollution or actively by trade. In so doing, urban centers transfer the costs of their activities to other regions—and mask the costs from their own residents. Citizens of Indianapolis, Columbus, or Buffalo may not recognize that pollution from nearby coal-fired power plants worsens acid rain hundreds of miles to the east. New York City residents may not realize how much garbage their city produces if it is shipped elsewhere for disposal.

However, not all waste and pollution leaves the city. Urban dwellers are exposed to smog, toxic industrial compounds, fossil fuel emissions, noise pollution, and light pollution. City residents even suffer thermal pollution, in the form of the **urban heat island effect** (**FIGURE 18.11**). Pollution and the health risks it poses are not evenly shared among residents. As environmental justice advocates point out (p. 14), those who receive the brunt of pollution are often those who are too poor to live in cleaner areas.

**Land preservation** Because people pack densely together in cities, more land outside cities is left undeveloped. Indeed, this is the very idea behind urban growth boundaries. If all

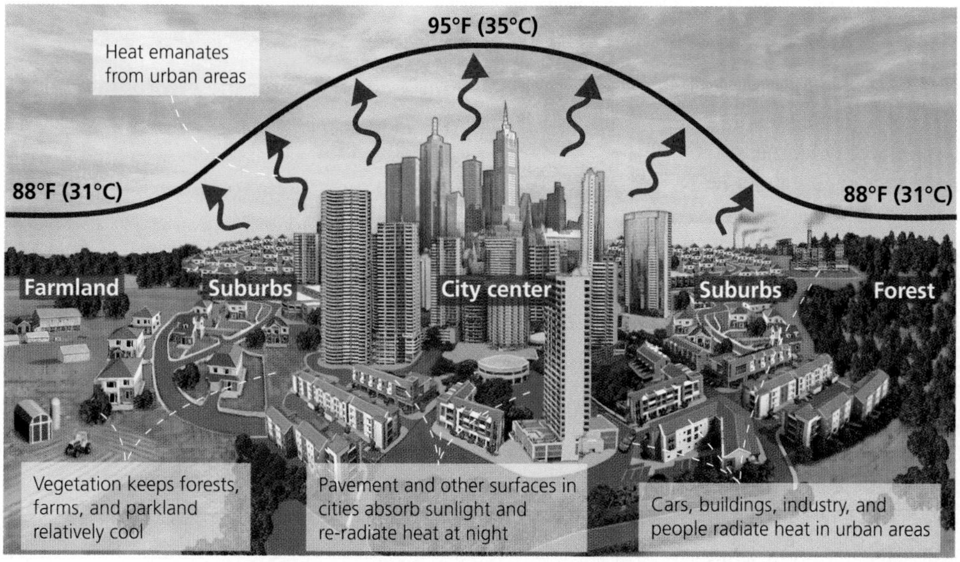


**FIGURE 18.11 Cities produce urban heat islands, creating temperatures warmer than surrounding areas.** People, buildings, vehicles, and factories generate heat, and buildings and dark paved surfaces absorb daytime heat and then release it slowly at night.

THE SCIENCE BEHIND THE STORY

## Baltimore and Phoenix Showcase Urban Ecology

Researchers in urban ecology examine how ecosystems function in cities and suburbs, how natural systems respond to urbanization, and how people interact with the urban environment. Today, Baltimore and Phoenix are centers for urban ecology.

These two cities are very different: Baltimore is an Atlantic port city on the Chesapeake Bay with a long history, whereas Phoenix is a young and fast-growing Southwestern metropolis sprawling across the desert. The U.S. National Science Foundation selected each of them as a research site in its prestigious Long Term Ecological Research program, which funds multi-decade ecological research. Since 1997, hundreds of researchers have studied Baltimore and Phoenix explicitly as ecosystems, examining nutrient cycling, biodiversity, air and water quality, environmental health threats, and more.

Research teams in both cities are combining old maps, aerial photos, and new remote sensing satellite data to reconstruct the history of landscape change. In Phoenix, one group showed how urban development spread across the desert in a "wave of advance," affecting soils, vegetation, and microclimate as it went. In Baltimore, mapping showed that development fragmented the forest into smaller patches over the past 100 years, even while the overall amount of forest remained the same.

**An urban ecologist samples water beneath an overpass in Baltimore.**

The study regions designated for each city encompass both heavily urbanized central city areas and rural and natural areas on the urban fringe. To measure the impacts of urbanization, many research projects compare conditions in these two types of areas.

Baltimore scientists can see ecological effects of urbanization just by comparing the urban lower end of their site's watershed with its less developed upper end. In the lower end, pavement, rooftops, and compacted soil prevent rain from infiltrating the soil, so water runs off quickly. The rapid flow cuts streambeds deeply into the earth while leaving surrounding soil drier. As a result, wetland-adapted trees and shrubs are vanishing, replaced by dry-adapted upland trees and shrubs.

The fast flow of water also worsens pollution. In natural areas, streams and wetlands filter pollution by breaking down nitrogen compounds. But in urban areas, where wetlands dry up and runoff from pavement creates flash floods, streams lose their filtering ability. In Baltimore, the resulting pollution ends up in the Chesapeake Bay, which suffers eutrophication and a large hypoxic dead zone (pp. 22–26). Baltimore scientists studying nutrient cycling (p. 38) found that urban and suburban watersheds have far more nitrate pollution than natural forests (**FIGURE 1**).

Baltimore research also reveals impacts of applying salt to icy roads in winter. Road salt makes its way into streams, which become up to 100 times saltier, even in summer, when salt is not applied. Such high salinity kills organisms (**FIGURE 2**), degrades habitat and water quality, and impairs streams' ability to remove nitrate.

To study contamination of groundwater and drinking water, researchers are using isotopes (p. 27) to trace where salts in the most polluted streams are coming from. Baltimore is now improving water quality substantially with a $900-million upgrade of its sewer system.

Urbanization also affects species and ecological communities. Cities and suburbs facilitate the spread of non-native species, because people introduce exotic ornamental plants and because urban impacts on the soil, climate, and landscape favor weedy generalist species over more specialized native ones.

Compared with natural landscapes, cities offer steady and reliable food resources—think of people's bird feeders, or food

---

7 billion of us were evenly spread across the planet's land area, no large blocks of land would be left uninhabited, and we would have far less room for agriculture, wilderness, biodiversity, or privacy. The fact that half the human population is concentrated in discrete locations helps allow space for natural ecosystems to continue functioning and provide the ecosystem services on which all of us, urban and rural, depend.

**Innovation** Cities promote a flourishing cultural life and, by mixing diverse people and influences, spark innovation and creativity. The urban environment can promote education and scientific research, and cities have long been viewed as engines of technological and artistic inventiveness. This inventiveness can lead to solutions to societal problems, including ways to reduce environmental impacts.

**FAQ** **Aren't cities bad for the environment?**

Stand in the middle of a city and look around. You see concrete, cars, and pollution. Environmentally bad, right? Not necessarily. The widespread impression that urban living is less sustainable than rural living is largely a misconception. Consider that in a city you can walk to the grocery store instead of driving. You can take the bus or the train. Police, fire, and medical services are close at hand. Water and electricity are easily supplied to your entire neighborhood, and waste is easily collected. In contrast, if you live in the country, resources must be used to transport all these services for long distances, or you need to burn gasoline traveling to reach them. By clustering people together, cities distribute resources efficiently while also preserving natural lands outside the city.

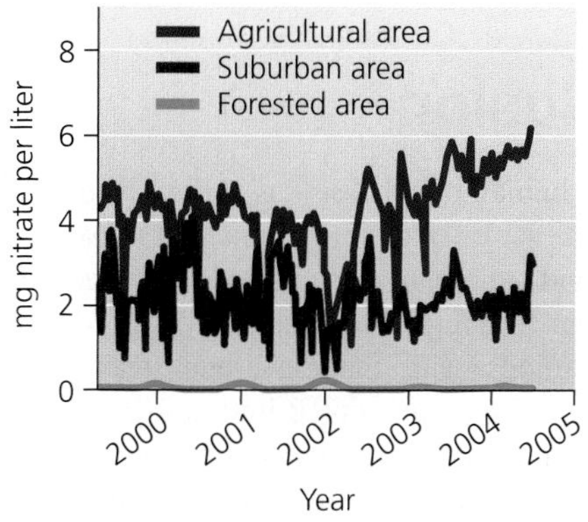


**FIGURE 1 Streams in Baltimore's suburbs contain more nitrates than streams in nearby forests, but fewer than those in agricultural areas, where fertilizers are applied liberally.** *With kind permission from Springer Science and Business Media and the author, Groffman, P. M., et al., 2004. Nitrogen fluxes and retention in urban watershed ecosystems.* Ecosystems *7: 393–403; Fig 4. Also from Baltimore Ecosystem Study.*

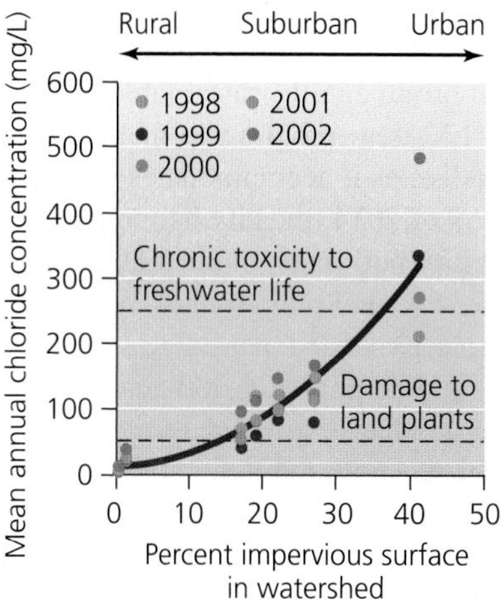


**FIGURE 2 Salt concentrations in Baltimore-area streams are high enough to damage plants in the suburbs and to kill aquatic animals in urban areas.** *Adapted from Kaushal, S. S., et al., 2005. Increased salinization of fresh water in the northeastern United States.* Proc. Natl. Acad. Sci. USA *102: 13517–13520, Fig 2. ©2005 National Academy of Sciences, U.S.A. By permission.*

**DATA Q** In how many years did mean salt concentrations in urban areas surpass the level of chronic toxicity to freshwater life?

GO TO **INTERPRETING GRAPHS & DATA** ON MasteringEnvironmentalScience®

scraps from dumpsters. Growing seasons are extended and seasonal variation is buffered in cities, as well. The urban heat island effect (p. 417) raises nighttime temperatures and equalizes temperatures year-round. In a desert city like Phoenix, watering gardens boosts primary productivity and lowers daytime temperatures. Together, all these changes lead to higher population densities of animals but lower species diversity as generalists thrive and displace specialists.

Urban ecologists in Phoenix and Baltimore are studying social and demographic aspects as well. Some research measures how natural amenities affect property values. One study found that proximity to a park increases a home's property values—unless crime is pervasive. If the robbery rate surpasses 6.5 times the national average (as it does in Baltimore), then proximity to a park begins to depress property values.

Other studies focus on environmental justice concerns (pp. 14–15). These studies have repeatedly found that sources of industrial pollution tend to be located in neighborhoods that are less affluent and that are home to people of racial and ethnic minorities. Phoenix researchers mapped patterns of air pollution and toxic chemical releases and found that minorities and the poor are exposed to a greater share of these hazards. As a result, they suffer from higher rates of childhood asthma.

Whether addressing the people, natural communities, or changing ecosystems of the urban environment, studies on urban ecology like those in Phoenix and Baltimore will be vitally informative in our ever-more-urban world. ◻

## Urban ecology helps cities toward sustainability

Cities that import all their resources and export all their wastes have a linear, one-way metabolism. Linear models of production and consumption tend to destabilize environmental systems. Proponents of sustainability for cities stress the need to develop circular systems, akin to systems found in nature, which recycle materials and use renewable sources of energy. Researchers in the field of **urban ecology** hold that the fundamentals of ecosystem ecology and systems science (Chapter 2) apply to the urban environment. Urban ecology research projects are ongoing in Baltimore and Phoenix, where scientists are studying these cities explicitly as ecological systems (see **THE SCIENCE BEHIND THE STORY**, above).

Urban ecologists and other urban sustainability advocates suggest that cities follow an ecosystem-centered model by striving to:

- Maximize efficient use of resources.
- Recycle as much as possible (pp. 395–397).
- Develop environmentally friendly technologies.
- Account fully for external costs (pp. 94–95).
- Use tax incentives to encourage sustainable practices.
- Use locally produced resources.
- Apply organic waste and wastewater to restore soil fertility.
- Encourage urban agriculture.

More and more cities are adopting these strategies. Urban agriculture is thriving from Portland to Cuba to Japan. Recycling programs continue to grow. Cities like Curitiba, Brazil, provide mass transit, environmental education, job training for the poor, and free health care.

In 2007, New York City unveiled an ambitious plan that then-Mayor Michael Bloomberg hoped would make it "the first environmentally sustainable 21st-century city." PlaNYC is a 132-item program to make New York City a better place to live as it accommodates 1 million more people by 2030. As of 2014, the city had improved energy efficiency in 174 city buildings, planted 865,000 trees, opened or renovated 234 school playgrounds, established 129 community gardens, expanded parklands, upgraded wastewater treatment, and acquired 36,000 acres to protect upstate water supplies. It had expanded recycling, installed solar panels, cleaned up polluted sites, installed bike lanes and bike racks and launched a bike-sharing program, retrofitted ferries to reduce pollution, introduced electric vehicles, and converted hundreds of taxis to hybrids. So far, such actions have improved air quality and reduced greenhouse gas emissions by 19% as the city also works to enhance resilience to climate change impacts in the wake of Hurricane Sandy (pp. 316–317).

Successes from New York City to Curitiba to Portland suggest that we can make cities more sustainable. Continuing experimentation in urban areas everywhere will help us determine how best to ensure that urbanization improves our quality of life while maintaining the quality of our environment.

## Conclusion

As half the human population has shifted from rural to urban lifestyles, the nature of our environmental impact has changed. As urban and suburban dwellers, our impacts are less direct but more far-reaching. Fortunately, we are developing sustainable solutions while making urban areas better places to live. Planning and zoning entail long-term vision, so they can be powerful forces for sustaining urban communities. Smart growth and new urbanism cut down on energy consumption, helping to address climate change. Mass transit systems reduce fuel use and carbon emissions. Parks promote health and offer ecosystem services. And green buildings bring a diversity of environmental and health benefits. Indeed, because urban centers affect the environment in some positive ways and can promote efficient resource use, they are a key element in achieving progress toward global sustainability.

# Testing Your Comprehension

1. What factors lie behind the shift of population from rural areas to urban areas? What types of nations are experiencing the fastest urban growth today, and why?
2. Why have so many city dwellers in the United States and other developed nations moved into suburbs?
3. Give two definitions of *sprawl*. Describe five negative impacts that have been suggested to result from sprawl.
4. What are city planning and regional planning? Contrast planning with zoning. Give examples of some of the suggestions made by early planners such as Edward Bennett in Portland.
5. How are some people trying to prevent or slow sprawl? Describe some key elements of "smart growth." What effects, positive and negative, do urban growth boundaries tend to have?
6. Describe several apparent benefits of rail transit systems. What is a potential drawback?
7. How are parks thought to make urban areas more livable? Give three examples of types of parks or public spaces.
8. What is a green building? Describe several features a LEED-certified building may have.
9. Describe the connection between urban ecology and sustainable cities. List three actions a city can take to enhance its sustainability.
10. Name two positive effects of urban centers on the natural environment.

# Seeking Solutions

1. Describe the causes of the spread of suburbs, and outline the environmental, social, and economic impacts of sprawl. Overall, do you think the spread of urban and suburban development that is commonly labeled *sprawl* is predominantly a good thing or a bad thing? Do you think it is inevitable? Give reasons for your answers.
2. Would you personally want to live in a neighborhood developed in the new-urbanist style? Why or why not? Would you like to live in a city or region with an urban growth boundary? Why or why not?
3. Does your campus have any LEED-certified green buildings? If so, how do they differ from conventional buildings? Think about a building on your campus that you think is unhealthy or that wastes resources in some way. Name three specific ways in which green building techniques might be used to improve this building.

4. **THINK IT THROUGH** You are the president of your college or university, and students are clamoring for you to help create the world's first fully sustainable campus. Considering how people enhance livability and sustainability in cities, what lessons might you try to apply to your college or university? You are scheduled to give a speech to the campus community about your plans and will need to name five specific actions you plan to take to pursue a sustainable campus. What will they be, and what will you say about each choice to describe its importance?

5. **THINK IT THROUGH** After you earn your college degree, you are offered three equally desirable jobs, in three very different locations. If you accept the first, you will live in the midst of a densely populated city. If you take the second, you will live in a suburb where you have more space but where sprawl may soon surround you for miles. If you select the third, you will live in a rural area with plenty of space but few cultural amenities. You are a person who aims to live in the most ecologically sustainable way you can. Where would you choose to live? Why? What considerations will you factor into your decision?

# Calculating Ecological Footprints

One way to reduce your ecological footprint is to choose alternative transportation. Each gallon of gasoline is converted during combustion to approximately 20 pounds of carbon dioxide ($CO_2$), which is released into the atmosphere. The table lists typical amounts of $CO_2$ released per person per mile using various modes of transportation, assuming typical fuel efficiencies.

For an average North American who travels 12,000 miles per year, calculate and record in the table the $CO_2$ emitted yearly for each transportation option. Then calculate and record the reduction in $CO_2$ emissions that one could achieve by relying solely on each option.

| Mode of Transport | $CO_2$ per person per mile | $CO_2$ per person per year | $CO_2$ emission reduction | Your estimated mileage per year | Your $CO_2$ emissions per year |
|---|---|---|---|---|---|
| Automobile (driver only) | 0.825 lb | 9900 lb | 0 | | |
| Automobile (2 persons) | 0.413 lb | | | | |
| Automobile (4 persons) | 0.206 lb | | | | |
| Vanpool (8 persons) | 0.103 lb | | | | |
| Bus | 0.261 lb | | | | |
| Walking | 0.082 lb | | | | |
| Bicycle | 0.049 lb | | | | |
| | | | | Total = 12,000 | |

1. Which transportation option provides the most miles traveled per unit of carbon dioxide emitted?
2. In the last two columns, estimate what proportion of the 12,000 annual miles you think that you actually travel by each method, and then calculate the $CO_2$ emissions that you are responsible for generating over the course of a year. Which transportation option accounts for the most emissions for you?
3. How could you reduce your $CO_2$ emissions? How many pounds of emissions do you think you could realistically eliminate over the course of the next year by making changes in your transportation decisions?

## MasteringEnvironmentalScience®

**STUDENTS**

Go to **MasteringEnvironmentalScience** for assignments, the etext, and the Study Area with practice tests, videos, current events, and activities.

**INSTRUCTORS**

Go to **MasteringEnvironmentalScience** for automatically graded activities, current events, videos, and reading questions that you can assign to your students, plus Instructor Resources.

Epilogue

# Sustainable Solutions

The notion of sustainability has run throughout this book. In one case after another, we have seen how dedicated people are devising creative solutions to the dilemmas that arise when resources for future generations are being depleted. Our challenge as a society is to continue developing innovative and workable solutions that enhance our quality of life while protecting and restoring the natural environment that supports us.

## Students are leading sustainability efforts on campus

In today's quest for sustainable solutions, students at colleges and universities are playing a crucial role. They are creating models for the wider world by leading sustainability initiatives on their campuses (**FIGURE E.1**).

Students are pioneering ways to enhance efficiency in energy use and water use on their campuses. They are running programs to recycle and cut down on waste. They are growing organic gardens and restoring natural areas on campus. They are advocating with administrators for renewable energy, sustainable green buildings, and the reduction of greenhouse gas emissions. Most efforts are home-grown and local, like the sustainable food and dining initiatives at Kennesaw State University (pp. 135–136). Others are part of national or international programs such as Recyclemania (pp. 389–390) and Campus Conservation Nationals (p. 389).

These diverse efforts by students, faculty, staff, and administrators are reducing the ecological footprints of college and university campuses. You can find links to resources for campus sustainability efforts in the *Selected Sources and References* section online at **MasteringEnvironmentalScience**.

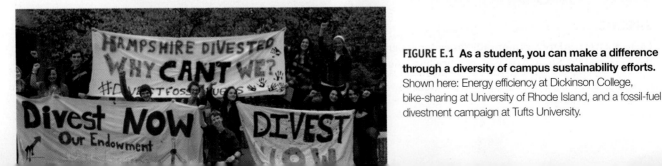

**FIGURE E.1 As a student, you can make a difference through a diversity of campus sustainability efforts.** Shown here: Energy efficiency at Dickinson College, bike-sharing at University of Rhode Island, and a fossil-fuel divestment campaign at Tufts University.

## We can develop sustainably

Whether on campus or around the world, sustainability means living in a way that can be lived far into the future. Sustainability involves conserving resources to prevent their depletion, reducing waste and pollution, and safeguarding ecological processes and ecosystem services, so as to ensure that our society's practices can continue and our civilization can endure. We depend on a healthy and functioning natural environment for our basic needs and our quality of life, so protecting Earth's natural capital is vital. Yet sustainability also means promoting social justice and economic well-being. Meeting this triple bottom line (p. 112) is the goal of modern sustainable development. It is our primary challenge for this century and likely for the rest of our species' time on Earth.

| Table E.1 Some Major Approaches to Sustainability |
|---|
| • Be politically active |
| • Vote with our wallets |
| • Pursue quality of life, not just economic growth |
| • Limit population growth |
| • Encourage green technologies |
| • Mimic natural systems by promoting closed-loop industrial processes |
| • Enhance local self-sufficiency, yet embrace some aspects of globalization |
| • Pursue systemic solutions |
| • Think in the long term |
| • Promote research and education |

## Environmental protection can enhance economic opportunity

Our society has long labored under the misconception that economic well-being and environmental protection are in conflict. In reality, our well-being depends on a healthy environment, and protecting environmental quality can improve our economic bottom line.

For individuals, businesses, and institutions, reducing resource consumption and waste often saves money. For society, promoting environmental quality can enhance economic opportunity by providing new types of employment. As we transition to a more sustainable economy, some resource-extraction industries may decline, but a variety of recycling-oriented and high-technology industries will spring up to take their place. As we reduce our dependence on fossil fuels, green-collar jobs (p. 367) and investment opportunities are opening up in renewable energy (**FIGURE E.2**).

**FIGURE E.2 Green-collar jobs, such as employment as a wind power technician, increase as we progress toward a sustainable economy.**

Moreover, people desire to live in areas with clean air, clean water, intact forests, public parks, and open space. Environmental protection enhances a region's appeal, drawing residents, increasing property values, and boosting the tax revenues that fund social services. As a result, regions that safeguard their environments tend to retain and enhance their wealth, health, and quality of life. In all these ways, environmental protection enhances economic opportunity. Indeed, a recent U.S. government review concluded that the economic benefits of environmental regulations greatly exceed their economic costs (see *Calculating Ecological Footprints*, p. 114). Both the U.S. economy and the global economy have expanded rapidly in the past 50 years, the very period during which environmental protection measures have proliferated.

## We can follow a number of strategies toward sustainable solutions

Truly lasting win-win solutions for humanity and our environment are numerous, and we have seen specific examples throughout this book. Let's now summarize 10 broad strategies or approaches that can help generate sustainable solutions (**TABLE E.1**).

**Political engagement** Sustainable solutions often require policymakers to usher them through, and policymakers respond to whoever exerts influence. Corporations and interest groups employ lobbyists to sway politicians all the time. Ordinary citizens have power as well, if they choose to exercise it. You can exercise your power at the ballot box, by attending public hearings, by donating to advocacy groups, and by writing letters and making phone calls to officeholders. The environmental and consumer protection laws we all benefit from today came about because citizens pressured their representatives to act. As we enjoy today's cleaner air, cleaner water, and greater prosperity, we owe a debt to the people who fought hard for the legislation of the 1960s–1980s that enabled these advances (p. 105). In turn, we owe it to future generations to engage ourselves so that they have a better world in which to live.

**FIGURE E.3 Americans consume more than the people of almost every other nation.** Consumption cannot continue rising unless we find ways to enhance the sustainability of our manufacturing processes.

**Consumer power** Each of us also wields influence through the choices we make as consumers. When products produced sustainably are ecolabeled (p. 111), consumers can "vote with their wallets" by purchasing these products. Consumer choice helps to drive sales of everything from recycled paper to organic produce to sustainable seafood.

**Quality of life** Economic growth can result from sustainable gains in efficiency, but so far growth has largely been driven by rising consumption of material goods and services (and thus the use of resources for their manufacture) (**FIGURE E.3**). Advertisers are always seeking to sell us more goods more quickly, but accumulating possessions does not necessarily bring us contentment. Affluent people often fail to find happiness in their material wealth, and scientific research shows that money does not buy as much happiness as people typically believe (p. 99).

We can enhance our quality of life by prioritizing friends, family, leisure time, and memorable experiences over material consumption. Economists and policymakers can help shift the current focus on economic growth toward a focus on people's quality of life by incorporating external costs into market prices, introducing green taxes, eliminating harmful subsidies, and adopting full-cost accounting practices (Chapter 5). The market might then become a truly free market and a powerful tool for improving the quality of our lives.

**Population stability** Just as continued growth in consumption is unsustainable, so is growth in the human population. We have used technology to vastly increase Earth's carrying capacity for our species, but sooner or later our population, like all populations, will stop growing. The question is how: through war, plagues, and famine, or through voluntary means? Thanks to urbanization, wealth, education, and the empowerment of women, the demographic transition (pp. 124–125) is already well advanced in many developed nations. If today's developing nations also pass through the demographic transition, then humanity may be able to rein in population growth while creating a more prosperous and equitable society.

**Green technologies** Technology has facilitated our population growth and has magnified our impacts on Earth's environmental systems—yet it can also give us ways to reduce our impact. Scrubbers on smokestacks have reduced emissions (see Figure 13.8, p. 285), as have catalytic converters on cars (see Figure 13.9, p. 286). Recycling technology and wastewater treatment are reducing our waste output. Solar, wind, and geothermal energy technologies are producing cleaner, renewable energy. Technological advances such as these help explain why people of the United States and western Europe today enjoy cleaner environments—although they consume far more—than people of eastern Europe or rapidly industrializing nations such as China.

**Mimicking nature** As industries seek to develop green technologies and sustainable practices, they have an excellent model: nature itself. Environmental systems tend to operate in cycles featuring feedback loops and the circular flow of materials. Some forward-thinking industrialists are making their processes more sustainable by transforming linear pathways into circular ones, in which waste is recycled and reused (pp. 400–401). Their ultimate vision is to create truly closed-loop processes, generating no waste.

**The local and the global** Encouraging local self-sufficiency is an important element of building sustainable societies. When people feel closely tied to the area in which they live, they tend to value the area and seek to sustain its environment and its human community. Moreover, relying on locally made products cuts down on fossil fuel use from long-distance transportation. This argument is frequently made in relation to the cultivation and distribution of food (pp. 158, 160).

At the same time, globalization enables people of the world's diverse cultures to communicate and learn about one another, making us more aware of one another's cultures and more likely to respect and celebrate, rather than fear, differences among cultures. Moreover, globalization may foster sustainability because Western democracy, as imperfect as it is, serves as a model and a beacon for people living under repressive governments. Open societies allow for entrepreneurship and the flowering of creativity in business, art, science, and education. And when millions of minds can think freely about issues, we are more likely to come up with sustainable solutions to our society's challenges.

**Systemic solutions** There are two general ways to respond to a problem. One is to address the symptoms of the problem, and the other is to address the root cause. Often our society takes a symptomatic approach (addressing the symptoms) rather than a systemic approach (viewing the whole system and tackling the root cause). Addressing symptoms is easier, but generally it is not effective in resolving a problem. For instance, as we deplete easily accessible fossil fuel deposits, we are choosing to reach further for new fossil fuel sources, even in the face of mounting environmental and health repercussions. A systemic solution to our energy demands would involve developing clean renewable sources instead. For many issues we face, it will prove worthwhile to pursue systemic solutions.

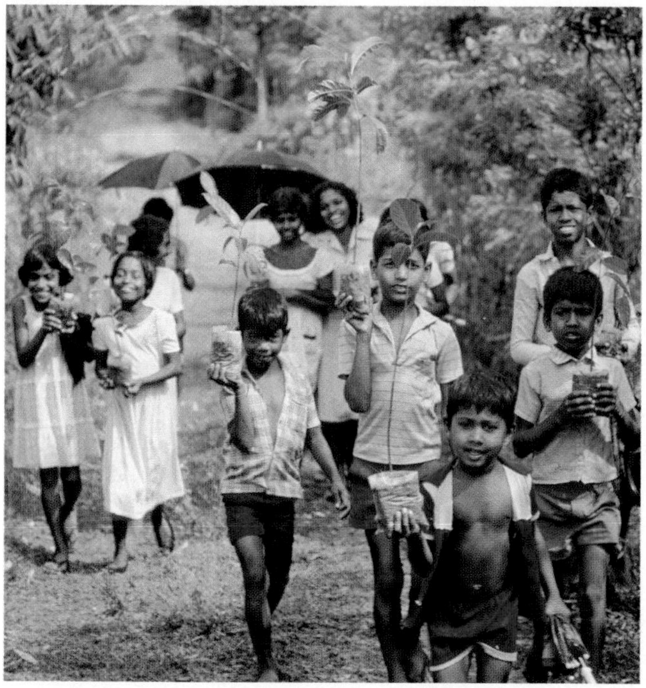

**FIGURE E.4 Sustainable solutions require long-term thinking.** These children in Sri Lanka are planting tree seedlings on deforested and eroded hillsides around their village. In doing so, they are investing in their own future.

**Long-term perspective** To be sustainable, a solution must work in the long term (**FIGURE E.4**). Policymakers in democracies often act for short-term good because they aim to produce quick results that will help them be reelected. Yet many environmental dilemmas are cumulative, worsen gradually, and can be resolved only over long periods. Often the costs of addressing an environmental problem are short term, whereas the benefits are long term. In such a situation, citizen pressure on policymakers is especially vital.

**Research and education** Finally, we each can magnify our influence by educating others and by serving as role models through our actions. The discipline of environmental science provides information we all can use to make wise decisions about a wide diversity of issues. By promoting scientific research and by educating the public about environmental science, we can all assist in the pursuit of sustainable solutions.

## Time is precious

We can bring sustainable solutions within reach, but time is getting short, and many human impacts continue to intensify. Deforestation, overfishing, wetland loss, resource extraction, and climate change are but a few examples. Even if we can visualize sustainable solutions, how can we find the time to implement them before we do irreparable damage to our environment and our future?

In 1961, U.S. President John F. Kennedy announced that within a decade the United States would be "landing a man on the moon and returning him safely to the Earth." It was a bold and astonishing statement; the technology to achieve this almost unimaginable feat did not yet exist. Yet just eight years later, astronauts walked on the moon. America accomplished this historic milestone by harnessing public enthusiasm for a goal and by supporting scientists and engineers so they could meet the goal.

Today humanity faces a challenge more important than any previous one. Attaining sustainability is a larger and more complex process than traveling to the moon. However, it is one to which every person on Earth can contribute; in which government, industry, and citizens can cooperate; and toward which all nations can work together. If America was able to reach the moon in a mere eight years, then certainly humanity can begin down the road to sustainability with comparable speed.

Fortunately, in our society today we have many thousands of scientists who study Earth's processes and resources. Thanks to their efforts, we are amassing a detailed knowledge and a growing understanding of our dynamic planet, what it offers us, and what impacts it can bear. Environmental science, this study of Earth and of ourselves, offers us hope for our future.

**FIGURE E.5 This photo of Earth, taken by astronauts orbiting the moon, shows our planet as it truly is—an island in space.** Everything we know, need, love, and value comes from and resides on this small sphere—so we had best treat it well.

## Earth is an island

We began this book with the vision of Earth as an island, and indeed it is (**FIGURE E.5**). Islands can be paradise, as Easter Island (pp. 6–7) likely was when early Polynesians first reached it. Yet when Europeans arrived centuries later, they witnessed the aftermath of a civilization that had collapsed once its island's resources were depleted and its environment degraded.

It would be tragic folly to let such a fate befall our planet as a whole. By recognizing this fact, by shifting our behavior and our cultural institutions in ways that encourage sustainable practices, and by employing science to help us achieve these ends, we may yet be able to live happily and sustainably on our wondrous island, Earth.

# Answers to Data Analysis Questions

APPENDIX A

## Chapter 1

**Fig. 1.3** The graph shows that nearly 1 billion people were alive in 1800, whereas over 7 billion are alive today. Thus, for every person alive in 1800, slightly more than 7 are alive today.

**Fig. 1.5** The global ecological footprint today is roughly 1.5 planet Earths. The global ecological footprint half a century ago (1961) was roughly 0.75 planet Earths. This makes for a difference of 0.75 planet Earths, and it means that today's footprint is twice the size of the footprint half a century ago.

**Fig. 1.17** Of the nations shown in the figure, the United States has the largest per capita footprint, and Afghanistan has the smallest per capita footprint. The U.S. footprint (7.2 ha) is 14.4 times larger than Afghanistan's footprint (0.5 ha), because 7.2 divided by 0.5 equals 14.4.

## Chapter 2

**Fig. 2.14** The width of the arrows in each figure represents magnitude, so the wider the arrow, the larger the value. For chemical energy, the widest arrow goes from producers to detritus. This means the largest directional flux of chemical energy in the ecosystem is from producers to detritus. For nutrients, a comparison of arrow widths shows the flux from detritus to producers to be, by far, the largest.

**Fig. 2.23** Reducing nitrogen inputs into the Chesapeake Bay through enhanced nutrient management programs costs $21.90 per pound, whereas a 1-pound reduction from forest buffers costs only $3.10. Dividing $21.90 per pound by $3.10 per pound shows us that for the same price, we could keep about 7 pounds of nitrogen out of waterways by using forested buffers versus 1 pound of nitrogen by using nutrient management programs.

## Chapter 3

**Fig. 3.5** All vertebrate groups shown, except for lampreys, branch off "after" (to the right of) the hash mark for jaws. This indicates that lampreys diverged before jaws originated (and thus lack them), whereas all other vertebrate groups possess jaws.

**Fig. 3.15** Because exponential growth cannot last forever, we would expect that growth of the western U.S. population of Eurasian collared doves will eventually slow down and that the population will reach carrying capacity. Thus, the population growth graph for the western United States would come to have a shape more like the current graph for Florida, which shows logistic growth.

**Fig. 1 (SBS)** Because the 21-year trend is upward and the trend of the last 9 years is downward, we can expect that the trend during the first 12 years will be upward at a steeper rate of increase than over the entire 21-year period. Running the data through regression analyses is the only way to plot a perfectly accurate trend line, but the line you drew just by eyeballing the data should show an upward trend and probably shows a slightly steeper rate of increase than the 21-year line.

## Chapter 4

**Fig. 4.7** In the generalized example shown, there are 100 grasshoppers for every 10 rodents; therefore, we can say there are 1/10 (or 1:10) as many rodents as grasshoppers. Thus, if there were 3000 grasshoppers, we would expect 300 rodents (300/3000, a 1:10 ratio).

**Fig. 4.11** One difference the map indicates is that zebra mussel occurrences are more numerous than quagga mussel occurrences. One hypothesis for this difference could be that zebra mussels are more successful invaders, but another one could be that because they arrived earlier, they have simply had more time to spread to more locations. A second difference shown by the map is that zebra mussels have spread to the lower Mississippi River valley but not much to the West, whereas quagga mussels have spread to the West but not to the lower Mississippi River valley. Many hypotheses could be offered. One might be that each type of mussel spread to a region where conditions happened to best suit it. Another might be that zebra mussels spread more effectively along river systems while quagga mussels spread more effectively by cross-country transport on boat hulls.

**Fig. 4.16** Note that the temperature curve is above the precipitation curve during this time. High temperatures lead to increased evaporation. Thus, even though precipitation is roughly average at this time, the warm temperatures cause evaporation and, thus, dry conditions.

**Fig. 3 (SBS)** In 1991, nauplii and rotifers had densities of 300–500 per square meter, whereas zebra mussels were at nearly zero. In 2000, zebra mussels were at nearly 300/$m^2$ whereas nauplii and rotifers were close to zero. In 2008, nauplii and rotifers were again at 300–500/$m^2$ whereas zebra mussels had returned to near zero. These data indicate that nauplii and rotifer densities crashed as zebra mussel densities became high, but that eventually nauplii and rotifer populations recovered as zebra mussel densities declined.

## Chapter 5

**Fig. 5.6** In 1950, GDP was just under $12,000 per capita, and GPI was just over $8000 per capita, making for a ratio of nearly 1.5. In 2004, GDP was nearly $37,000 per capita, and GPI was about $15,000 per capita**,** making for a ratio of nearly 2.5. For most people, the ratio for the year they were born will be between these two values. Together these changes indicate that GDP has been growing faster than GPI.

**Fig. 5.14** Oil, coal, and natural gas have together received about $594 billion, whereas renewable energy received about $81 billion in the United States during the period covered by the graph. If we divide 594 by 81, we get 7.33. Thus, about $7.33 has been spent on fossil fuel subsidies for every $1.00 that has gone to renewable energy.

## Chapter 6

**Fig. 6.4** By examining the key that links colors on the map to growth rates, we see that red indicates the highest growth rates and that Africa has the highest overall growth rate of any region. Europe has the lowest growth rate of any region, as evidenced by its many nations with very low or negative growth rates.

**Fig. 6.12** The transitional stage has the greatest growth of any stage in the demographic transition, because that is the period when birth and death rates are far apart and population increase is substantial. In the transitional stage, growth is the greatest at the end, when the difference between birth and death rates is far greater than at the stage's beginning.

**Fig. 6.14** The best approach to answer a question such as this one is to draw a "best-fit" line through the points on the figure that minimizes the distance between each point and the line you draw. Doing this produces a line that slopes downward from left to right, suggesting a negative relationship between total fertility rate and the rate of enrollment of girls in secondary school. This relationship makes sense—one would expect that as more girls pursue education, they delay childbirth and reduce the nation's TFR.

**Fig. 6.16** Africa will add about 1.3 billion people to the global population by 2050, more than the roughly 1 billion people added by Asia. Africa will also increase by the largest percentage, some 118%. This value is calculated by dividing the number of people added to Africa's population (1.3 billion) by its 2013 population (1.1 billion) and then multiplying by 100 to convert the resulting proportion to a percentage: $(1.3/1.1) \times 100 = 118\%$.

## Chapter 7

**Fig. 7.2** Overall growth of the human population provides the answer. Since the period 2007–2009, our global population has increased by several hundred million people, with the vast majority of this growth occurring in developing nations. Thus, although the absolute number of undernourished people in the developing world has remained the same (about 850 million), many people have been added to the global population, so the percentage of people who are undernourished has fallen.

**Fig. 7.16** Beef requires 17.5 times more land to produce than chicken (245.0 $m^2$/14.0 $m^2$ = 17.5). Beef requires 15 times more water to produce than chicken (750 kg/50 kg = 15).

**Fig. 7.19** In the last 5 years, GM crops have been growing faster in developing nations, as indicated by the fact that the curve for developing nations rises upward more steeply than the curve for industrialized nations. In 2012, developing nations for the first time produced more GM crops than industrialized nations. If current trends continue, developing nations should be growing more GM crops than industrialized nations in 2017. To estimate how much more GM crops they might be growing in 2017, we can extend the two trend lines forward 4 years, keeping their slopes the same as in the previous several years. Doing this leads *approximately* to values of 115 million hectares for developing nations and 90 million ha for industrialized nations. The spread between these two values might be even greater if, as the data suggest, developing nations are speeding up their rate of adoption and industrialized nations are slowing down theirs.

## Chapter 8

**Fig. 8.3** The bottom pie chart in the figure shows that there are 4680 species of mammals. The top pie chart shows that there are about 1,750,000 known and described species of organisms in total. Because 4680 is 0.27% of 1,750,000, this means that mammals comprise just 0.27% of all organisms (or about 1 out of 400). In reality, the percentage is actually much lower than this, because virtually all mammal species have already been discovered, yet most species of other types of organisms have not yet been discovered.

**Fig. 8.11** In the winter of 2013–2014, monarchs occupied just 0.67 hectare, whereas in 1994–1995 (the first year of data), they occupied 7.81 ha. Thus in 2013–2014 they occupied just 8.6% (0.67/7.81) of their original area. Compared with the year of greatest area occupied (20.97 ha in 1996–1997), monarchs in 2013–2014 occupied just 3.2% of that area (0.67/20.97).

**Fig. 8.12** The bar for pollution stretches to a value of nearly 1200 species, second only to habitat loss; this indicates that pollution is the second greatest cause of amphibian declines overall. For threatened species alone, we need to look at the red portions of the bars. Comparing red portions of the bars, we can see that habitat loss is the primary cause of declines for threatened species of amphibians.

## Chapter 9

**Fig. 9.11** The ratio of growth to removal is greatest for the land type for which the relative height of the two bars is most different. This is the case for the national forests, where annual growth exceeds 4 billion $ft^3$ while annual removal totals less than 0.5 billion $ft^3$.

**Fig. 2 (SBS)** According to the data in the graph, the forest plot held about 55 bird species in the 4 years it was censused before its fragmentation in 1984. After the plot became a fragment, the average number of bird species dropped to about 20 species.

**Fig. 3 (SBS)** According to the data in the graph, a tree 275 meters in from the edge of a forest fragment would be susceptible to elevated tree mortality (an edge effect that extends 300 m in) and increased wind disturbance (which extends 400 m in).

## Chapter 10

**Fig. 10.3** In 2009, respiratory infections claimed approximately 4.2 million lives and diarrheal diseases about 2.1 million, for a total of 6.3 million lives. There were about 2 million deaths from AIDS. Dividing 6.3 million by 2 million shows that about three times more lives were lost to respiratory infections and diarrheal diseases than were lost to AIDS. As this example shows, the diseases that garner the most attention are not always the ones that cause the greatest impacts on human health.

**Fig. 10.9** The odds of perishing in a motor vehicle accident is 1 in 98, whereas the chance of dying in an in an air and space transport incident is 1 in 7178. Dividing 7178 by 98 shows that the odds of dying in a car accident is about 73 times that of dying in a plane crash— even though our instinctive risk assessment often makes us feel safer "behind the wheel."

**Fig. 2 (SBS)** Begin by connecting the data points in the figure to form a dose-response curve like that shown in Figure 10.8c. Mark the spot on the curve directly above 70 ng/g on the *x*-axis. Then reference the *y*-axis at the site of your mark, which indicates that an estimated 9% of the mice would likely suffer chromosomal effects at that dose.

## Chapter 11

**Fig 11.1** If you were to tunnel straight down into the Earth for 2850 km, you would find yourself in the lowest part of the mantle, only 50 km or so from the outer edge of Earth's core. Most people assume the distance to the core is far greater, showing that our perceptions of the thickness of the layers within our planet often do not match reality.

**Fig. 11.5** Comparison of the two figures reveals that this belt of intense earthquake and volcanic activity corresponds closely to the subduction zones at the boundaries of the tectonic plates that surround the Pacific Ocean. As shown in Figure 11.2, convergent plate boundaries dominate the length of the ring of fire. Note that other locations that experience earthquakes and volcanic activity—such as Indonesia, Iran, Turkey, and southern Italy—are similarly located along convergent plate boundaries.

**Fig. 11.20** At present rates of consumption, nickel has technically recoverable reserves that would last around 50 years, of which about 30 years of reserves are economically recoverable. Dividing 30 years of economically recoverable reserves by 50 years of technically recoverable reserves yields the value 0.60, indicating that 60% (multiply by 0.60 by 100 to present the value as a percentage) of nickel reserves are economically recoverable. If you answered molybdenum (57%) don't feel bad, as it comes in a very close second when similar calculations are performed (about 40 years economically recoverable to around 70 years technically recoverable).

The metal with the lowest percentage of economically recoverable reserves is lead. Lead's 17 years of economically recoverable reserves divided by its 370 years of technically recoverable reserves finds that only 4.6% of lead reserves are currently economically recoverable.

## Chapter 12

**Fig. 12.2** Consulting the figure, note that 2.5% of the water on Earth is fresh water and that 1% of all fresh water is surface water. Within this surface water, 52% is found in lakes. To determine the percentage of Earth's water found in freshwater lakes, multiply 2.5% (.025) by 1% (0.01) and by 52% (0.52). Multiplying these three values (and then multiplying the answer by 100 to convert it to a percentage) reveals that although freshwater lakes (such as the Great Lakes) seem massive, all of the world's freshwater lakes *combined* contain only 0.013% of Earth's water.

**Fig. 12.8** Released off the southeast coast of Japan, the buoy would be carried northeast by the Kuroshio Current and then eastward across the ocean on the North Pacific Current. Upon reaching North America, it could turn southward on the California Current and float by the western coast of the United States, passing Washington, Oregon, and California. Alternatively, upon reaching North America the buoy might follow the Alaska Current northward, pass Alaska, and then return to Japan. So although Japan is closer to Australia, ocean currents would carry the buoy to the United States first.

**Fig. 12.18** Among the three regions, Latin American and the Caribbean have the greatest amount of water per capita. With its abundant river systems (including the mighty Amazon) and relatively small population compared to that of Africa and Asia, the per-person water quantities in the nations in this region are consistently large. Although Africa and Asia do contain abundant river systems, they also boast larger populations, which means that less water is available per person. Africa and Asia are also home to large regions with arid climates, another factor that reduces the quantities of water available.

**Fig. 12.29** Answering this question requires a bit of estimation. In 1950, the global total fishery catch (world harvest plus China's harvest) was about 20 million metric tons, and China's harvest was around 1 million metric tons of this total. Dividing China's 1 million metric tons by the global harvest of 20 million metric tons (and multiplying by 100 to convert to a percentage) reveals that China produced about 5% of the global fishery harvest in 1950. In 2012, the total global harvest was around 90 million metric tons, and China's take was about 10 million metric tons. If we calculate as we did before, we find that China's share of the total global fishery output was around 11% (10 million metric tons/90 million metric tons × 100 = 11.1%). These results point to China's increasing influence on global production of fish and shellfish for human consumption.

## Chapter 13

**Fig. 13.7** Population has increased by 53% since 1970. Emissions have decreased by 72%. Thus, emissions per person have decreased by more than five times. (Imagine a population rising from 100 to 153, and emissions dropping from 100 to 28. 28/153 = 0.18, or less than one-fifth of the original 1-unit-per-person rate.)

**Fig. 13.11** Answers will vary. For example, a person living in Los Angeles would be breathing dirtier air than people in nearby cities yet would find that L.A.'s air had improved over time, with only half as many unhealthy days in 2010 as in 2002.

**Fig. 13.15** According to the data in the bars of the graph, in the late 1970s the L.A. basin suffered about 210 days per year of unhealthy air, and in recent years it has suffered about 100 such days—roughly a 52% reduction. According to the data in the line of the graph, in the late 1970s peak daily ozone levels averaged about 0.30 ppm, and in recent years they have averaged about 0.12-0.13 ppm, representing about a 57–60% reduction. Thus, both data sets show similar declines. One can tell this at a glance because the downward slopes of the two data sets appear more or less parallel on the graph.

**Fig. 13.22** Answers will vary, but in virtually all locations, precipitation has become less acidic. For example, in many parts of the northeastern United States, pH increased from about 4.3 to about 4.9.

## Chapter 14

**Fig. 14.3** Since 1750 the atmospheric carbon dioxide concentration has increased from about 280 ppm to about 399 ppm—an increase of about 43%.

**Fig. 14.4** Changing land use accounts for 6 metric tons of carbon dioxide emissions per year, and industry emits 26 metric tons of carbon dioxide annually. Thus, 26/6 = 4.33. This means that for every 1 metric ton released as a result of changes in land use, 4.33 tons are released by industry.

**Fig. 14.10** Answers will vary. In most regions, temperature rose. In some areas of the Southeast, it was stable or fell slightly.

**Fig. 14.27** Many of the nations that reduced emissions are European. Many of the nations where emissions increased (Australia, Canada, and the United States) are large and less densely populated. Because these latter nations are geographically more spread out, long-distance transportation consumes more petroleum, giving rise to more emissions. In addition, these nations are more politically conservative than most European nations, and many conservatives have tended to fear that emissions reductions will suppress economic activity.

## Chapter 15

**Fig. 15.2** Answers will vary. One should take the value at the far right end of the data line for oil (4.13 billion tons in 2012) and divide it by the value of that line in the year one was born. For a person born in 1995, when oil consumption was about 3.25 billion tons per year, the percent change by 2012 would be about 4.13/3.25 = 1.27, or roughly a 27% increase.

**Fig. 15.5** Zooming in on a portion of a graph may enable you to see the data at a finer resolution and thus gather more information from it. In this particular graph, the inset allows us to see pattern and detail not visible in the main graph. Because the scale of data on the *y*-axis is so large on the main graph, we cannot perceive differences in value after 1960 or learn the exact values year by year. Thanks to the inset graph and its smaller range of *y*-axis values, though, we can determine that EROI ratios fell steeply until 1960 and then fell very gradually and irregularly after that, going from about 12:1 to about 6:1. Showing both main graph and zoomed-in inset graph together gives us the best overall understanding of the patterns in the data.

**Fig. 15.17** Answers will vary. One should take the value at the far right end of the black ("Total") data line and divide it by the value of that line in the year one's mother or father was born. For example, if one's parent was born in 1975, when emissions were about 5.1 billion tons per year, then the percent change by 2012 would be about 9.67/5.1 = 1.90, or roughly a 90% increase.

**Fig. 15.20** The United States produces 11.1 million barrels of oil per day and consumes 18.5 million barrels per day. Thus, for every barrel produced, 18.5/11.1 = 1.7 barrels are consumed.

## Chapter 16

**Fig. 16.1** Note that the bar on the right gives a breakdown of data from the pie slice for renewable energy. The pie chart tells us that renewable energy as a whole provides 9.4% of U.S. energy consumption. The bar tells us that bioenergy provides 48.9% of renewable energy. Therefore bioenergy contributes 48.9% of 9.4%—or 4.6%—of total U.S. energy consumption.

**Fig. 16.10** On average, southern Arizona receives roughly 2400–2600 kilowatt-hours per square meter per year, and most of Germany receives fewer than 1200 kilowatt-hours/$m^2$/yr. Thus, southern Arizona receives more than twice as much sunlight as does Germany.

**Fig. 16.16** Answers will vary, but regions that appear underutilized for wind power include (on land) South Dakota, Nebraska, Montana, and New Mexico; and (offshore) the entire Atlantic and Gulf coasts.

**Fig. 16.22** In 2013, 23.4 billion gallons of ethanol were produced across the world, of which the United States produced 13.3 billion gallons. Thus U.S. production is 13.3/23.4 = 0.57. Therefore, U.S. production makes up 57% of the world total.

## Chapter 17

**Fig. 17.4** The amount of solid waste that is combusted (incinerated) is shown in green. To determine increase or decrease, note whether the green band widens or narrows over time. (This is separate from the overall height of the graph data, which reflects cumulative, summed, totals.) Examining the green band, we see that the amount of solid waste that is combusted (incinerated) decreased from 1960 to about 1985, and then it increased until about 2002. As of 2012, the amount was roughly equal to the amount back in 1960.

**Fig. 17.9** Between 1960 and 2012, the total amount of waste that was recovered increased by 14 times (from about 6 to about 87 million tons). However, in that same time period the recovery rate (percentage of waste generated that is recovered) increased by only about 6 times (from about 6% to about 35%). From this we can infer that the total amount of waste generated must also have risen.

**Fig. 2 (SBS)** Hazardous waste items (orange bars) tended to travel farther. Various hazardous waste items traveled from about 100 km to about 1700 km. In contrast, the farthest-traveling municipal solid waste item shown traveled less than 100 km.

## Chapter 18

**Fig. 18.1** The dashed red line (which projects the urban population in less developed regions) surpasses the dashed blue line (which projects the rural population in less developed regions) between the year 2010 and the year 2020.

**Fig. 18.8b** Roadway costs and parking costs are created by automobile traffic but not by rail traffic. Note the yellow and orange portions of the bars in the figure for part (b). These costs make automobile traffic more costly overall than rail traffic.

**Fig. 2 (SBS)** Among the four data points for years for areas labeled "urban" (with 40% impervious surface), three of them lie above the level at which there is chronic toxicity to freshwater life.

# How to Interpret Graphs

APPENDIX B

Presenting data in ways that help make trends and patterns visually apparent is a vital element of science. For scientists, businesspeople, policymakers, and others, the primary tool for expressing patterns in data is the graph. Thus, the ability to interpret graphs is a skill that you will want to cultivate. This appendix guides you in how to read graphs, introduces a few vital conceptual points, and surveys the most common types of graphs, giving rationales for their use.

## Navigating a Graph

A graph is a diagram that shows relationships among *variables*, which are factors that can change in value. The most common types of graphs relate values of a *dependent variable* to those of an *independent variable*. As explained in Chapter 1 (p. 9), a dependent variable is so named because its values "depend on" the values of an independent variable. In other words, as the values of an independent variable change, the values of the dependent variable change in response. In a manipulative experiment (p. 10), a researcher specifies changes in the value of the independent variable; these changes *cause* changes in the value of the dependent variable. In observational studies, there may be no causal relationship, and scientists may plot a correlation (p. 10). In a positive correlation, values of one variable go up or down along with values of another. In a negative correlation, values of one variable go up when values of the other go down. Whether we are graphing a correlation or a causal relationship, the values of the independent variable are known or specified by the researcher, whereas the values of the dependent variable are unknown until the research has taken place. The values of the dependent variable are what we are interested in observing or measuring.

By convention, independent variables are generally represented on the horizontal axis, or *x*-axis, of a graph, while dependent variables are represented on the vertical axis, or *y*-axis. In many cases, independent variables are not numbers, but categories. Numerical values of variables generally become larger as one proceeds rightward on the *x*-axis or upward on the *y*-axis. Note that the tick marks along the axes must be uniformly spaced so that when the data are plotted, the graph gives an accurate visual representation of the scale of quantitative change in the data.

As a simple example, **FIGURE B.1** shows data from the Breeding Bird Survey that reflects population growth of the Eurasian collared dove following its introduction to North America. The *x*-axis shows values of the independent variable, which in this case is time, expressed in units of years. The dependent variable, presented on the *y*-axis, is the average number of doves detected on each route. For each year, a data point is plotted on the graph to show the average number of doves detected. In this particular graph, a line (red curve) was then drawn through the actual data points (orange dots), showing how closely the empirical data match an exponential growth curve (p. 61), a theoretical phenomenon of importance in ecology.

Now that you're familiar with the basic building blocks of a graph, let's survey the most common types of graphs you'll see, and examine a few vital concepts in graphing.

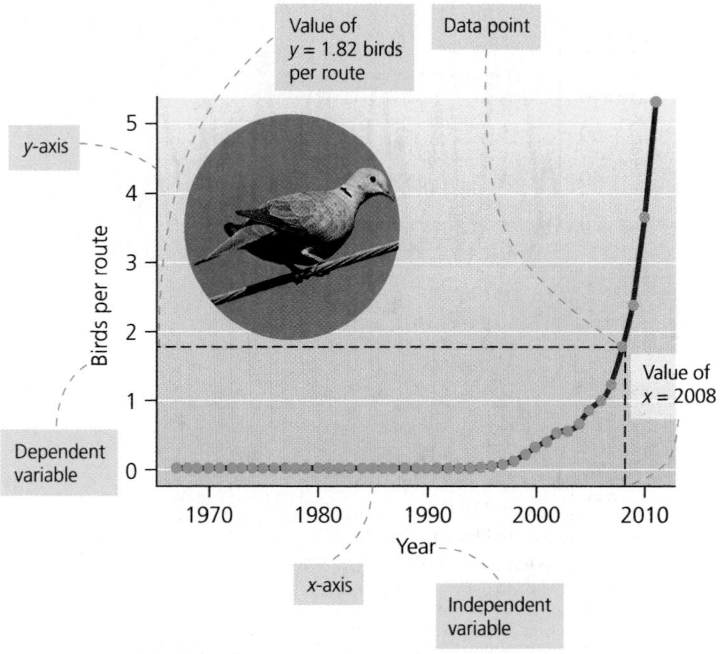


**FIGURE B.1 Exponential population growth, demonstrated by the Eurasian collared dove in North America in recent years.** (Figure 3.13, p. 61)

MasteringEnvironmentalScience®

Once you've explored this appendix, take advantage of the graphing resources at **MasteringEnvironmentalScience**. The GRAPHit! tutorials allow you to plot your own data. The **Interpreting Graphs and Data** exercises guide you through critical-thinking questions on graphed data from recent research in environmental science. The **Data Analysis Questions** help you hone your skills in reading graphed data. All these features will help you expand your comprehension and use of graphs.

## Graph Type: Line Graph

A line graph is used when a data set involves a sequence of some kind, such as a series of values that occur one by one and change through time or across distance. In a line graph, a line runs from one data point to the next. Line graphs are most appropriate when the *y*-axis expresses a continuous numerical variable, and the *x*-axis expresses either continuous numerical data or discrete sequential categories (such as years). **FIGURE B.2** shows values for the size of the ozone hole over Antarctica in recent years. Note how the data show that the size of the hole increases until 1987, when the Montreal Protocol (p. 294) came into force, and then begins to stabilize afterwards.

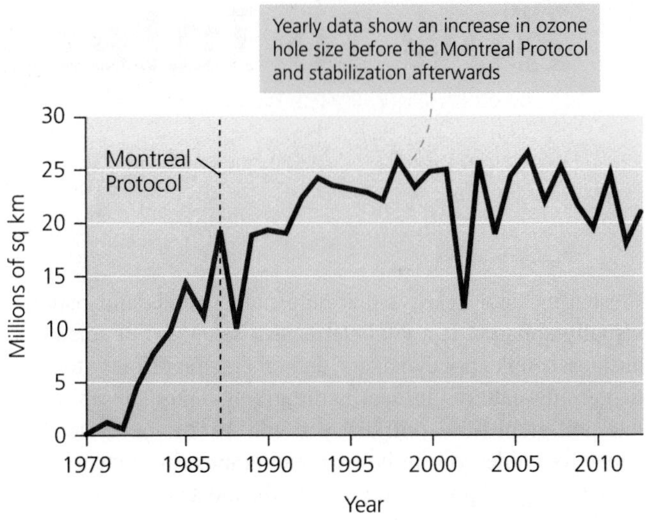


**FIGURE B.2 Size of the Antarctic ozone hole before and after a treaty that was designed to address it.** (Figure 13.18, p. 295)

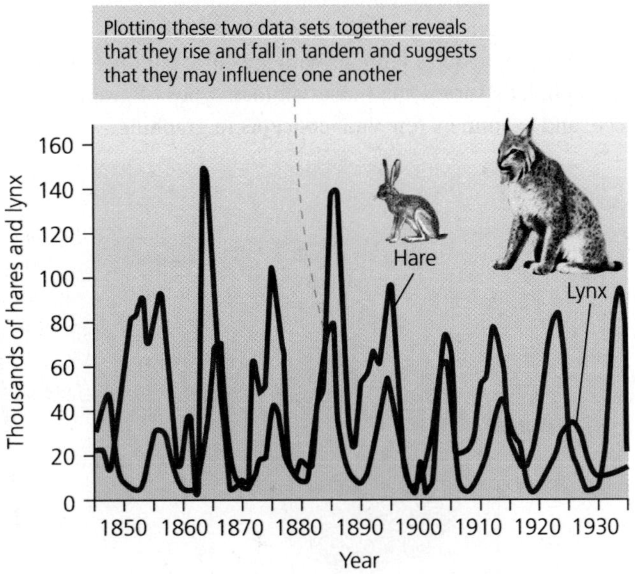


One useful technique is to plot two or more data sets together on the same graph. This allows us to compare trends in the data sets to see whether they may be related and, if so, the nature of that relationship. In **FIGURE B.3**, recorded numbers of a predator species rise and fall immediately following those of its prey, suggesting a possible connection.

**FIGURE B.3 Fluctuations in recorded numbers of hare and lynx in Canada.** (Figure 4.4, p. 69)

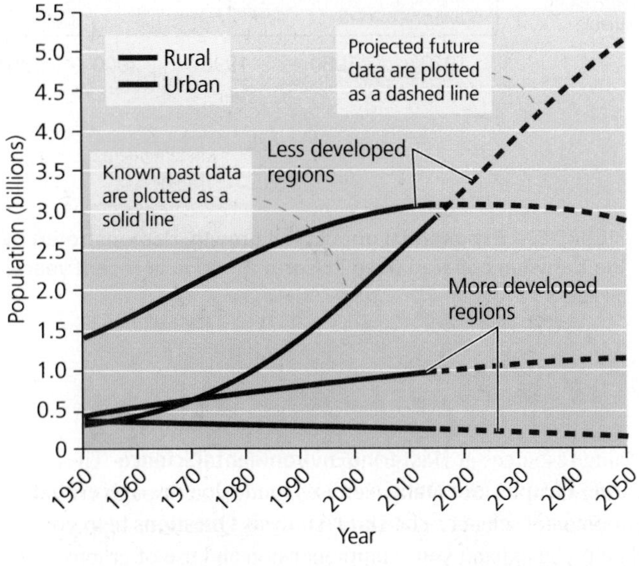


## Key Concept: Projections

Besides showing observed data, graphs can show data that are predicted for the future. Such *projections* of data are based on models, simulations, or extrapolations from past data, but they are only as good as the information that goes into them—and future trends may not hold if conditions change in unforeseen ways. Thus, in this textbook, projected future data are shown with dashed lines, as in **FIGURE B.4**, to indicate that they are less certain than data that have already been observed. Be careful when interpreting graphs in the popular media and on the Internet, however; often newspapers, magazines, web sites, and advertisements will show projected future data in the same way as known past data!

**FIGURE B.4 Past population change and projected future population change for rural and urban areas in more developed and less developed regions.** (Figure 18.1, p. 409)

## Graph Type: Bar Chart

A bar chart is most often used when one variable is a category and the other is a number. In such a chart, the height (or length) of each bar represents the numerical value of a given category. Higher or longer bars mean larger values. In **FIGURE B.5**, the bar for the category "Automobile" is higher than that for "Light rail," indicating that automobiles use more energy per passenger-mile (the numerical variable on the *y*-axis) than light rail systems do.

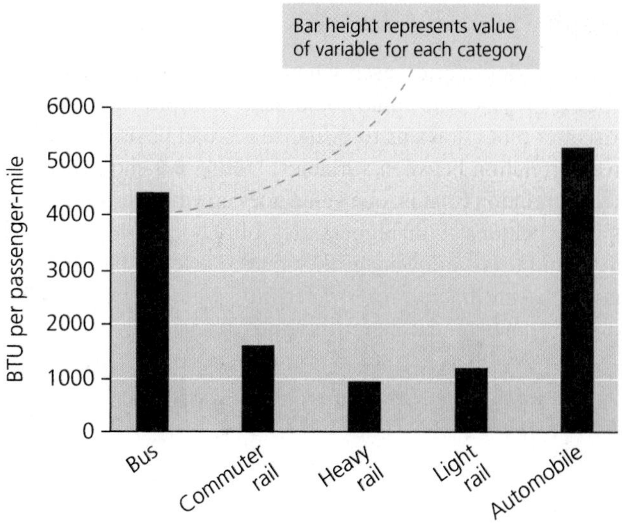


**FIGURE B.5 Energy consumption for different modes of transit.** (Figure 18.8a, p. 415)

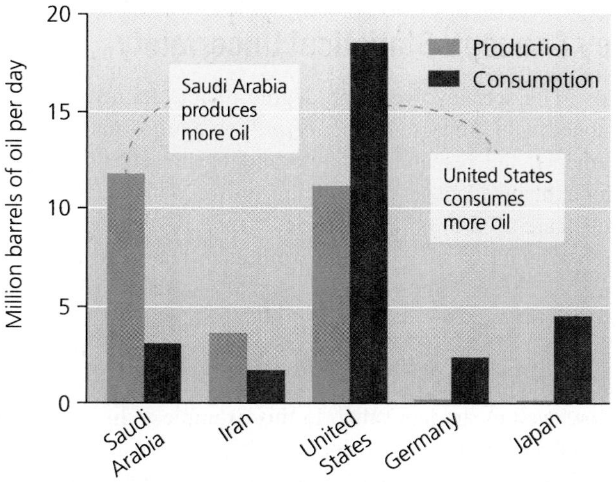


**FIGURE B.6 Oil production and consumption by selected nations.** (Figure 15.20, p. 353)

As we saw with line graphs, it is often instructive to graph two or more data sets together to reveal patterns and relationships. A bar chart such as **FIGURE B.6** lets us compare two data sets (oil production and oil consumption) both within and among nations. A graph that does double duty in this way allows for higher-level analysis (in this case, suggesting which nations depend on others for petroleum imports). Most bar charts in this book illustrate multiple types of information at once in this manner.

## Graph Type: Pie Chart

A pie chart is used when we wish to compare the numerical proportions of some whole that are taken up by each of several categories. Each category is represented visually as a slice from a pie, with the size of the slice reflecting the percentage of the whole that is taken up by that category. For example, **FIGURE B.7** shows the percentages of genetically modified crops worldwide that are soybeans, corn, cotton, and canola.

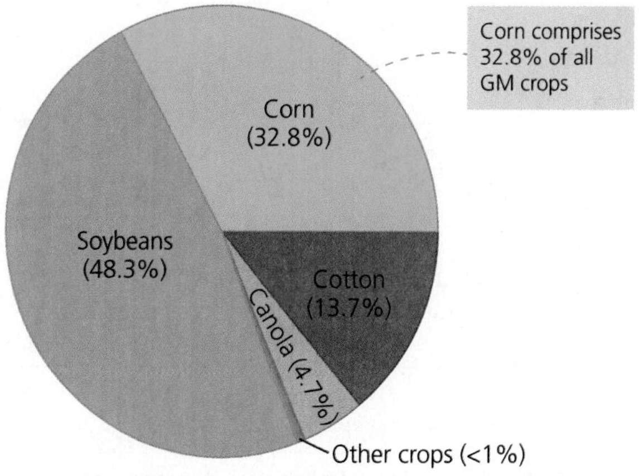


**FIGURE B.7 Genetically modified crops grown worldwide, by type.** (Figure 7.20a, p. 154)

## Graph Type: Scatter Plot

A scatter plot is often used when data are not sequential and when a given $x$-axis value could have multiple $y$-axis values. A scatter plot allows us to visualize a broad positive or negative correlation between variables. **FIGURE B.8** shows a negative correlation (that is, one value goes up while the other goes down): Nations with higher rates of school enrollment for girls tend to have lower fertility rates. For example, Jamaica has high enrollment and low fertility, whereas Ethiopia has low enrollment and high fertility.

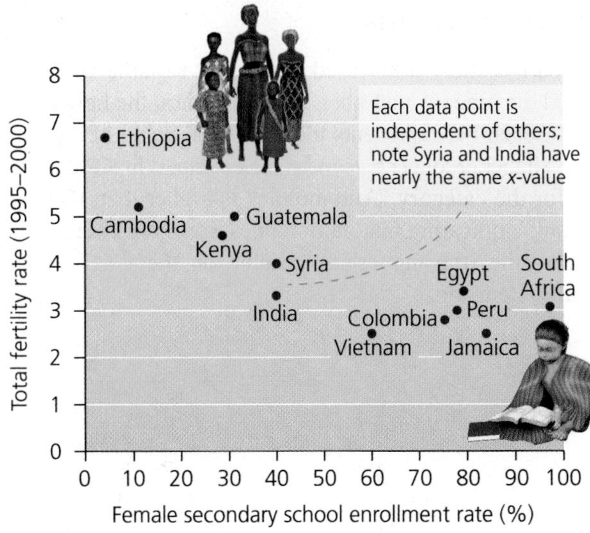


**FIGURE B.8 Fertility rate and female education.** (Figure 6.14, p. 127)

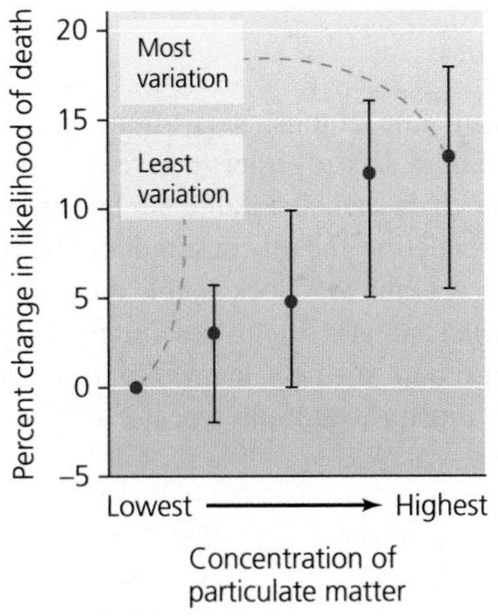


**FIGURE B.9 Likelihood of death due to air pollution.** (Figure 3a, p. 289)

## Key Concept: Statistical Uncertainty

Most data sets involve some degree of uncertainty. When a graphed value represents the *mean* (average) of many measurements, the researcher may want to show the degree to which the raw data vary around this mean. Mathematical techniques are used to obtain statistically precise degrees of variation around a mean. Results from such statistical analyses may be expressed in a number of ways, and the two graphs in this section show methods used in this book.

In a bar chart or a scatter plot (**FIGURE B.9**), thin black lines called *error bars* may be shown extending above and below each mean data value. In this example of likelihood of death from air pollution, error bars show the most variation at the highest measured concentration of pollutants and no variation at the lowest measured concentration.

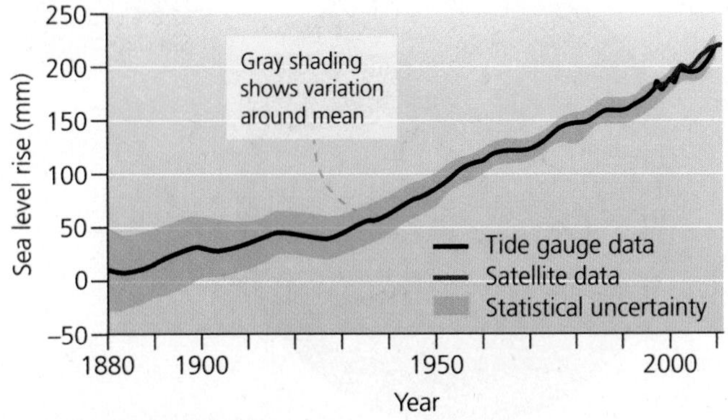


**FIGURE B.10 Change in global sea level, measured since 1880.** (Figure 14.15, p. 316)

Sometimes shading is used to express variation around a mean. The black and red data lines in **FIGURE B.10** show mean global sea level readings since 1880. The data line is surrounded by gray shading indicating statistical variation. Note how the amount of statistical uncertainty is exceeded by the sheer scale of the sea level rise. This gives us confidence that sea level is truly rising, despite the statistical uncertainty we find around mean values each year.

The statistical analysis of data is critically important in science. In this book, we provide a broad and streamlined introduction to many topics, so we often omit error bars from our graphs and details of statistical significance from our discussions. Bear in mind that this is for clarity of presentation only; the research we discuss analyzes its data in far more depth than any textbook could possibly cover.

APPENDIX C

# Metric System

| Measurement | Unit and Abbreviation | Metric Equivalent | Metric to English Conversion Factor | English to Metric Conversion Factor |
|---|---|---|---|---|
| Length | 1 kilometer (km) | = 1000 ($10^3$) meters | 1 km = 0.62 mile | 1 mile = 1.61 km |
| | 1 meter (m) | = 100 ($10^2$) centimeters | 1 m = 1.09 yards | 1 yard = 0.914 m |
| | | = 1000 millimeters | 1 m = 3.28 feet | 1 foot = 0.305 m |
| | | | 1 m = 39.37 inches | |
| | 1 centimeter (cm) | = 0.01 ($10^{-2}$) meter | 1 cm = 0.394 inch | 1 foot = 30.5 cm |
| | | | | 1 inch = 2.54 cm |
| | 1 millimeter (mm) | = 0.001 ($10^{23}$) meter | 1 mm = 0.039 inch | |
| Area | 1 square meter ($m^2$) | = 10,000 square centimeters | 1 $m^2$ = 1.1960 square yards | 1 square yard = 0.8361 $m^2$ |
| | | | 1 $m^2$ = 10.764 square feet | 1 square foot = 0.0929 $m^2$ |
| | | | 1 $cm^2$ = 0.155 square inch | 1 square inch = 6.4516 $cm^2$ |
| | 1 square centimeter ($cm^2$) | = 100 square millimeters | | |
| Mass | 1 metric ton (t) | = 1000 kilograms | 1 t = 1.103 tons | 1 ton = 0.907 t |
| | 1 kilogram (kg) | = 1000 grams | 1 kg = 2.205 pounds | 1 pound = 0.4536 kg |
| | 1 gram (g) | = 1000 milligrams | 1 g = 0.0353 ounce | 1 ounce = 28.35 g |
| | 1 milligram (mg) | = 0.001 gram | | |
| Volume (solids) | 1 cubic meter ($m^3$) | = 1,000,000 cubic centimeters | 1 $m^3$ = 1.3080 cubic yards | 1 cubic yard = 0.7646 $m^3$ |
| | 1 cubic centimeter | = 0.000001 cubic meter | 1 $m^3$ = 35.315 cubic feet | 1 cubic foot = 0.0283 $m^3$ |
| | ($cm^3$ or cc) | = 1 milliliter | 1 $cm^3$ = 0.0610 cubic inch | 1 cubic inch = 16.387 $cm^3$ |
| | 1 cubic millimeter ($mm^3$) | = 0.000000001 cubic meter | | |
| Volume (liquids and gases) | 1 kiloliter (kl or kL) | = 1000 liters | 1 kL = 264.17 gallons | 1 gallon = 3.785 L |
| | 1 liter (l or L) | = 1000 milliliters | 1 L = 0.264 gallons | 1 quart = 0.946 L |
| | | | 1 L = 1.057 quarts | |
| | 1 milliliter (ml or mL) | = 0.001 liter | 1 ml = 0.034 fluid ounce | 1 quart = 946 ml |
| | | = 1 cubic centimeter | 1 ml = approximately $\frac{1}{4}$ teaspoon | 1 pint = 473 ml |
| | | | | 1 fluid ounce = 29.57 ml |
| | | | | 1 teaspoon = approx. 5 ml |
| Time | 1 millisecond (ms) | = 0.001 second | | |
| Temperature | Degrees Celsius (°C) | | °C = ($\frac{5}{9}$°F 2 32) | °F = $\frac{9}{5}$ °C + 32 |
| Energy and Power | 1 kilowatt-hour | = 34,113 BTUs = 860,421 calories | | |
| | 1 watt | = 3.413 BTUs/hr | | |
| | | = 14.34 calories/min | | |
| | 1 calorie | = the amount of heat necessary to raise the temperature of 1 gram (1 $cm^3$) of water 1 degree Celsius | | |
| | 1 horsepower | = 7.457 × 102 watts | | |
| | 1 joule | = $9.481 \times 10^{-4}$ BTUs | | |
| | | = 0.239 cal | | |
| | | = $2.778 \times 10^{-7}$ kilowatt-hours | | |
| Pressure | 1 pound per square inch (psi) | = 6894.757 pascals (Pa) | | |
| | | = 0.068045961 atmosphere (atm) | | |
| | | = 51.71493 millimeters of mercury (mm Hg = Torr) | | |
| | | = 68.94757 millibars (mbar) | | |
| | 1 atmosphere (atm) | = 6.894757 kilopascals (kPa) | | |
| | | = 101.325 kilopascals (kPa) | | |

# APPENDIX D Periodic Table of the Elements

| Period | IA | IIA | IIIB | IVB | VB | VIB | VIIB | VIIIB | VIIIB | VIIIB | IB | IIB | IIIA | IVA | VA | VIA | VIIA | VIIIA |
|---|---|---|---|---|---|---|---|---|---|---|---|---|---|---|---|---|---|---|
| 1 | 1 H 1.0079 Hydrogen | | | | | | | | | | | | | | | | | 2 He 4.003 Helium |
| 2 | 3 Li 6.941 Lithium | 4 Be 9.012 Beryllium | | | | | | | | | | | 5 B 10.811 Boron | 6 C 12.011 Carbon | 7 N 14.007 Nitrogen | 8 O 15.999 Oxygen | 9 F 18.998 Fluorine | 10 Ne 20.180 Neon |
| 3 | 11 Na 22.990 Sodium | 12 Mg 24.305 Magnesium | | | | | | | | | | | 13 Al 26.982 Aluminum | 14 Si 28.086 Silicon | 15 P 30.974 Phosphorus | 16 S 32.066 Sulfur | 17 Cl 35.453 Chlorine | 18 Ar 39.948 Argon |
| 4 | 19 K 39.098 Potassium | 20 Ca 40.078 Calcium | 21 Sc 44.956 Scandium | 22 Ti 47.88 Titanium | 23 V 50.942 Vanadium | 24 Cr 51.996 Chromium | 25 Mn 54.938 Manganese | 26 Fe 55.845 Iron | 27 Co 58.933 Cobalt | 28 Ni 58.69 Nickel | 29 Cu 63.546 Copper | 30 Zn 65.39 Zinc | 31 Ga 69.723 Gallium | 32 Ge 72.61 Germanium | 33 As 74.922 Arsenic | 34 Se 78.96 Selenium | 35 Br 79.904 Bromine | 36 Kr 83.8 Krypton |
| 5 | 37 Rb 85.468 Rubidium | 38 Sr 87.62 Strontium | 39 Y 88.906 Yttrium | 40 Zr 91.224 Zirconium | 41 Nb 92.906 Niobium | 42 Mo 95.94 Molybdenum | 43 Tc 98 Technetium | 44 Ru 101.07 Ruthenium | 45 Rh 102.906 Rhodium | 46 Pd 106.42 Palladium | 47 Ag 107.868 Silver | 48 Cd 112.411 Cadmium | 49 In 114.82 Indium | 50 Sn 118.71 Tin | 51 Sb 121.76 Antimony | 52 Te 127.60 Tellurium | 53 I 126.905 Iodine | 54 Xe 131.29 Xenon |
| 6 | 55 Cs 132.905 Cesium | 56 Ba 137.327 Barium | 57 La 138.906 Lanthanum | 72 Hf 178.49 Hafnium | 73 Ta 180.948 Tantalum | 74 W 183.84 Tungsten | 75 Re 186.207 Rhenium | 76 Os 190.23 Osmium | 77 Ir 192.22 Iridium | 78 Pt 195.08 Platinum | 79 Au 196.967 Gold | 80 Hg 200.59 Mercury | 81 Tl 204.383 Thallium | 82 Pb 207.2 Lead | 83 Bi 208.980 Bismuth | 84 Po 209 Polonium | 85 At 210 Astatine | 86 Rn 222 Radon |
| 7 | 87 Fr 223 Francium | 88 Ra 226.025 Radium | 89 Ac 227.028 Actinium | 104 Rf 267 Rutherfordium | 105 Db 268 Dubnium | 106 Sg 269 Seaborgium | 107 Bh 270 Bohrium | 108 Hs 269 Hassium | 109 Mt 278 Meitnerium | 110 Ds 281 Darmstadtium | 111 Rg 281 Roentgenium | 112 Cn 285 Copernicium | | 114 Fl 289 Flerovium | | 116 Lv 293 Livermorium | | |

Representative (main group) elements: IA–IIA and IIIA–VIIIA; Transition metals: IIIB–IIB

Rare earth elements

| | | | | | | | | | | | | | | |
|---|---|---|---|---|---|---|---|---|---|---|---|---|---|---|
| Lanthanides | 58 Ce 140.115 Cerium | 59 Pr 140.908 Praseodymium | 60 Nd 144.24 Neodymium | 61 Pm 145 Promethium | 62 Sm 150.36 Samarium | 63 Eu 151.964 Europium | 64 Gd 157.25 Gadolinium | 65 Tb 158.925 Terbium | 66 Dy 162.5 Dysprosium | 67 Ho 164.93 Holmium | 68 Er 167.26 Erbium | 69 Tm 168.934 Thulium | 70 Yb 173.04 Ytterbium | 71 Lu 174.967 Lutetium |
| Actinides | 90 Th 232.038 Thorium | 91 Pa 231.036 Protactinium | 92 U 238.029 Uranium | 93 Np 237.048 Neptunium | 94 Pu 244 Plutonium | 95 Am 243 Americium | 96 Cm 247 Curium | 97 Bk 247 Berkelium | 98 Cf 251 Californium | 99 Es 252 Einsteinium | 100 Fm 257 Fermium | 101 Md 258 Mendelevium | 102 No 259 Nobelium | 103 Lr 262 Lawrencium |

The periodic table arranges elements by atomic number and atomic weight into horizontal rows called *periods* and vertical columns called *groups*.

Elements of each group in Class A have similar chemical and physical properties. This reflects the fact that members of a particular group have the same number of valence shell electrons, which is indicated by the group's number. For example, group IA elements have one valence shell electron, group IIA elements have two, and group VA elements have five. In contrast, as you progress across a period from left to right, properties of the elements change, varying from the very metallic properties of groups IA and IIA to the nonmetallic properties of group VIIA to the inert elements (noble gases) in group VIIIA. This reflects changes in the number of valence shell electrons.

Class B elements, or transition elements, are metals and generally have one or two valence shell electrons. In these elements, some electrons occupy more distant electron shells before the deeper shells are filled.

In this periodic table, elements with symbols printed in black exist as solids under standard conditions (25°C and 1 atmosphere of pressure); elements in red exist as gases; and those in dark blue as liquids. Elements with symbols in green do not exist in nature and must be created by some type of nuclear reaction.

# Geologic Time Scale

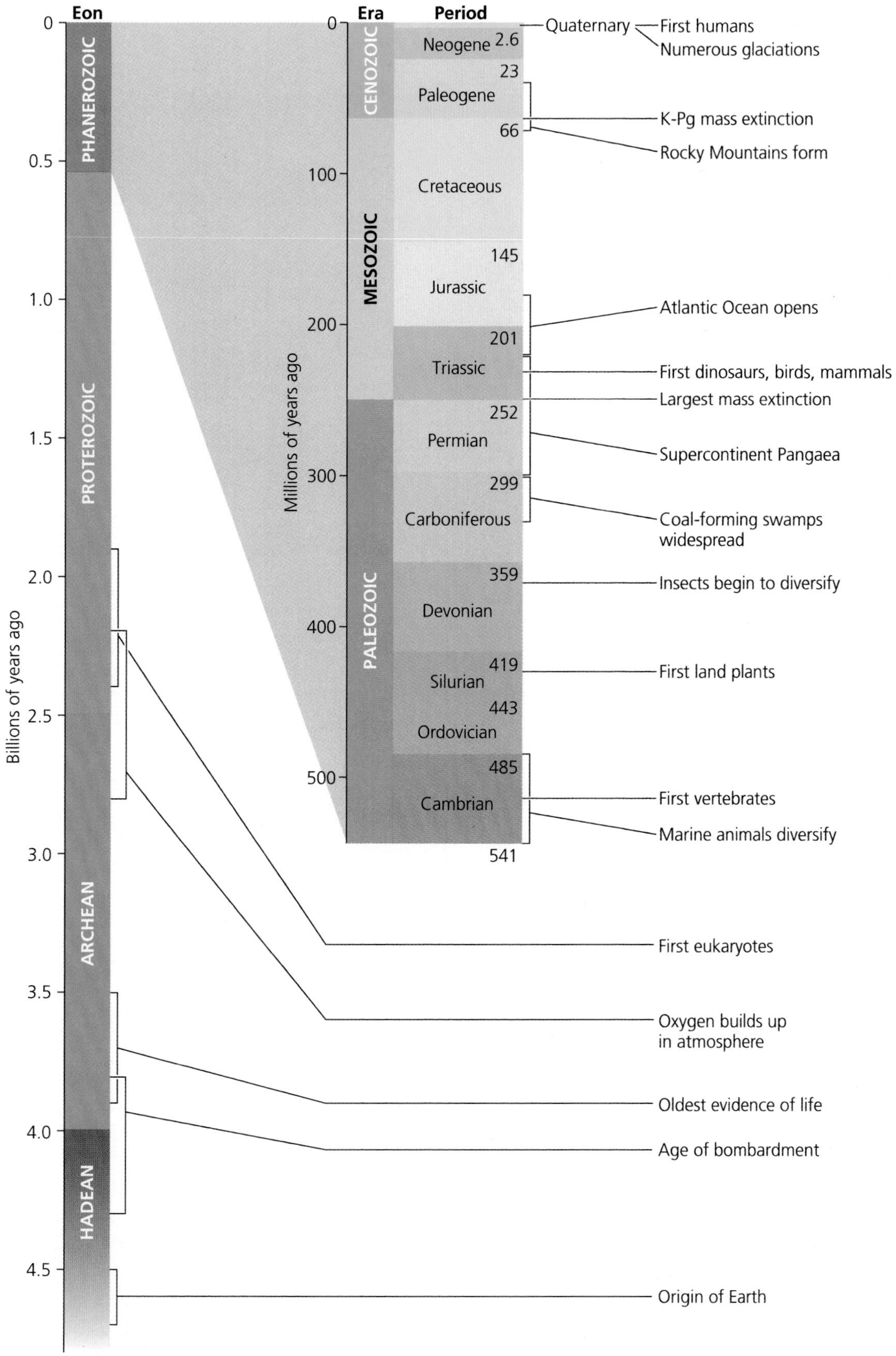
Eon
Era
Period
Billions of years ago
0
0.5
1.0
1.5
2.0
2.5
3.0
3.5
4.0
4.5
PHANEROZOIC
PROTEROZOIC
ARCHEAN
HADEAN
Millions of years ago
0
100
200
300
400
500
CENOZOIC
MESOZOIC
PALEOZOIC
Neogene 2.6
Paleogene 23
Cretaceous 66
Jurassic 145
Triassic 201
Permian 252
Carboniferous 299
Devonian 359
Silurian 419
Ordovician 443
Cambrian 485
541
Quaternary
First humans
Numerous glaciations
K-Pg mass extinction
Rocky Mountains form
Atlantic Ocean opens
First dinosaurs, birds, mammals
Largest mass extinction
Supercontinent Pangaea
Coal-forming swamps widespread
Insects begin to diversify
First land plants
First vertebrates
Marine animals diversify
First eukaryotes
Oxygen builds up in atmosphere
Oldest evidence of life
Age of bombardment
Origin of Earth

# Glossary

G

**acid deposition** The settling of acidic or acid-forming pollutants from the *atmosphere* onto Earth's surface. This may take place by precipitation, fog, gases, or the deposition of dry particles. Compare *acid rain.*

**acid drainage** A process in which sulfide minerals in newly exposed rock surfaces react with *oxygen* and rainwater to produce sulfuric acid, which causes chemical *runoff* as it *leaches* metals from the rocks. Acid drainage is a natural phenomenon, but mining greatly accelerates it by exposing many new surfaces.

**acidic** The property of a *solution* in which the concentration of *hydrogen* ($H^+$) *ions* is greater than the concentration of hydroxide ($OH^-$) ions. Compare *basic.*

**acid rain** *Acid deposition* that takes place through rain.

**active solar energy collection** An approach in which technological devices are used to focus, move, or store *solar energy*. Compare *passive solar energy collection.*

**acute exposure** Exposure to a *toxicant* occurring in high amounts for short periods of time. Compare *chronic exposure.*

**adaptation** (re: *climate change*) The pursuit of strategies to protect ourselves from the impacts of climate change. Compare *mitigation.*

**adaptation** (re: *evolution*) (1) The process by which *traits* that lead to increased reproductive success in a given environment evolve in a *population* through *natural selection*. (2) See *adaptive trait.*

**adaptive management** The systematic testing of different management approaches to improve methods over time.

**adaptive trait** A trait that confers greater likelihood that an individual will reproduce. An *adaptive trait* is also called an *adaptation.*

**aerosols** Very fine liquid droplets or solid particles aloft in the atmosphere.

**agricultural revolution** The shift around 10,000 years ago from a hunter-gatherer lifestyle to an agricultural way of life in which people began to grow crops and raise domestic animals. Compare *industrial revolution.*

**agriculture** The practice of cultivating *soil*, producing crops, and raising livestock for human use and consumption.

**air pollutants** Gases and particulate material added to the atmosphere that can affect *climate* or harm people or other organisms.

**air pollution** The act of polluting the air, or the condition of being polluted by *air pollutants.*

**Air Quality Index** An index of air *pollution* severity compiled by the U.S. *Environmental Protection Agency*. Reflects data on *criteria pollutants*. Values below 100 indicate satisfactory air conditions whereas values above 100 indicate unhealthy conditions.

**airshed** The geographic area that produces air pollutants likely to end up in a waterway.

**albedo** The capacity of a surface to reflect light. Higher albedo values refer to greater reflectivity.

**allergen** A *toxicant* that overactivates the immune system, causing an immune response when one is not necessary.

**allopatric speciation** Species formation due to the physical separation of populations over some geographic distance.

**alloy** A substance created by fusing a *metal* with other metals or nonmetals. Bronze is an alloy of the metals copper and tin, and steel is an alloy of iron and the nonmetal *carbon.*

**ambient air pollution** See *outdoor air pollution.*

**amino acids** Organic molecules that join in long chains to form *proteins.*

**anaerobic** Occurring in an *environment* that has little or no *oxygen*. The conversion of organic matter to *fossil fuels* at the bottom of a deep lake, swamp, or shallow sea is an example of anaerobic decomposition. Compare *aerobic.*

**anthropocentrism** A human-centered view of our relationship with the *environment*. Compare *biocentrism* and *ecocentrism.*

**anthropogenic** Caused by human beings. For example, anthropogenic climate change, as opposed to natural climate change.

**aquaculture** The cultivation of aquatic organisms for food in controlled *environments.*

**aquifer** An underground water reservoir.

**archipelago** A group of islands, most often in a linear arrangement.

**artificial selection** *Natural selection* conducted under human direction. Examples include the *selective breeding* of crop plants, pets, and livestock.

**asthenosphere** A layer of the upper *mantle*, just below the *lithosphere*, consisting of especially soft rock.

**atmosphere** The thin layer of gases surrounding planet Earth. Compare *biosphere; hydrosphere; lithosphere.*

**atmospheric blocking pattern** A condition in which the atmosphere's *jet stream* slows and meanders widely into a north–south orientation that blocks the eastward movement of weather systems across the midlatitudes.

**atmospheric deposition** The wet or dry deposition on land of a wide variety of pollutants, including mercury, nitrates, organochlorines, and others. *Acid deposition* is one type of atmospheric deposition.

**atom** The smallest component of an *element* that maintains the chemical properties of that element.

**autotroph (primary producer)** An organism that can use the energy from sunlight to produce its own food. Includes green plants, algae, and cyanobacteria.

***Bacillus thuringiensis* (Bt)** A naturally occurring *soil* bacterium that produces a protein that kills many pests, including caterpillars and the larvae of some flies and beetles.

**background rate of extinction** The average rate of *extinction* that occurred before the appearance of humans. For example, the *fossil record* indicates that for both birds and mammals, one *species* in the world typically became extinct every 500–1000 years. Compare *mass extinction event.*

**baghouse** A system of large filters that physically removes *particulate matter* from *incinerator emissions.*

**basic** The property of a *solution* in which the concentration of *hydroxide* ($OH^-$) *ions* is greater than the concentration of hydrogen ($H^+$) ions. Compare *acidic.*

**bedrock** The continuous mass of solid rock that makes up Earth's *crust.*

**benthic** Of, relating to, or living on the bottom of a water body. Compare *pelagic.*

**benthic zone** The bottom layer of a water body. Compare *littoral zone; limnetic zone; profundal zone.*

**bioaccumulation** The buildup of *toxicants* in the tissues of an animal.

**biocentrism** A philosophy that ascribes relative values to actions, entities, or properties on the basis of their effects on all living things or on the integrity of the *biotic* realm in general. The biocentrist evaluates an action in terms of its overall impact on living things, including—but not exclusively focusing on—human beings. Compare *anthropocentrism* and *ecocentrism.*

**biodiesel** Diesel fuel produced by mixing vegetable oil, used cooking grease, or animal fat with small amounts of *ethanol* or methanol (wood alcohol) in the presence of a chemical catalyst.

**biodiversity (biological diversity)** The variety of life across all levels of biological organization, including the diversity of *species*, their *genes*, their *populations*, and their *communities.*

**biodiversity hotspot** An area that supports an especially great diversity of *species*, particularly species that are *endemic* to the area.

**bioenergy (biomass energy)** *Energy* harnessed from plant and animal matter, including wood from trees, charcoal from burned wood, and combustible animal waste products, such as cattle manure.

**biofuel** Fuel produced from *biomass energy* sources and used primarily to power automobiles. Examples include *ethanol* and *biodiesel*.

**biogenic** Type of *natural gas* created at shallow depths by the anaerobic decomposition of organic matter by bacteria. Consists of nearly pure *methane*. Compare *thermogenic*.

**biogeochemical cycle** See *nutrient cycle*.

**biological control (biocontrol)** Control of pests and weeds with organisms that prey on or parasitize them, rather than with *pesticides*.

**biological diversity** See *biodiversity*.

**biological hazard** Human health hazards that result from ecological interactions among organisms. These include *parasitism* by viruses, bacteria, or other *pathogens*. Compare *infectious disease; chemical hazard; cultural hazard; physical hazard*.

**biomagnification** The magnification of the concentration of *toxicants* in an organism caused by its consumption of other organisms in which toxicants have *bioaccumulated*.

**biomass** (1) In ecology, organic material that makes up living organisms; the collective mass of living matter in a given place and time. (2) In energy, organic material derived from living or recently living organisms, containing *chemical energy* that originated with *photosynthesis*.

**biomass energy** See *bioenergy*.

**biome** A major regional complex of similar plant *communities*; a large *ecological* unit defined by its dominant plant type and vegetation structure.

**biophilia** An inherent love for and fascination with nature and an instinctive desire people have to affiliate with other living things. Defined by biologist E.O. Wilson as "the connections that human beings subconsciously seek with the rest of life."

**biopower** Power attained by the combustion of *bioenergy* sources to generate *electricity*.

**biosphere** The sum total of all the planet's living organisms and the *abiotic* portions of the *environment* with which they interact.

**biosphere reserve** A tract of land with exceptional *biodiversity* that couples preservation with *sustainable development* to benefit local people. Biosphere reserves are designated by UNESCO (the *United Nations* Educational, Scientific, and Cultural Organization) following application by local stakeholders.

**biotechnology** The material application of biological *science* to create products derived from organisms. The creation of *transgenic* organisms is one type of biotechnology.

**birth control** The effort to control the number of children one bears, particularly by reducing the frequency of pregnancy. Compare *contraception, family planning*.

**bitumen** A thick, heavy form of *petroleum* rich in *carbon* and poor in *hydrogen*. The fossil-fuel component of *oil sands*.

**bog** A type of *wetland* in which a pond is thoroughly covered with a thick floating mat of vegetation. Compare *freshwater marsh*; *swamp*.

**boreal forest** A *biome* of northern coniferous forest. Also known as *taiga*, boreal forest consists of a limited number of *species* of evergreen trees, such as black spruce, that dominate large regions of forests interspersed with occasional bogs and lakes.

**Borlaug, Norman (1914–2009)** American agricultural scientist who introduced specially bred crops to developing nations in the 20th century, helping to spur the *Green Revolution*.

**bottom-trawling** Fishing practice that involves dragging weighted nets across the seafloor to catch *benthic* organisms. Trawling crushes many organisms in its path and leaves long swaths of damaged sea bottom.

**breakdown product** *A compound* that results from the degradation of a toxicant.

**brownfield** An area of land whose redevelopment or reuse is complicated by the presence or potential presence of hazardous material.

**bycatch** (1) The accidental capture of nontarget organisms while fishing for target species. (2) That portion of a commercial fishing catch consisting of animals caught unintentionally. Bycatch kills many thousands of fish, sharks, marine mammals, and birds each year.

**Calvin cycle** In *photosynthesis*, a series of chemical reactions in which *carbon atoms* from *carbon dioxide* are linked together to manufacture sugars.

**campus sustainability** A term encompassing a wide variety of efforts by students, faculty, staff, and administrators of colleges and universities to make campus operations more sustainable. Includes efforts toward energy efficiency, water efficiency, emission reductions, transportation improvements, sustainable dining, landscaping improvements, renewable energy, curricular changes, and more.

**cap-and-trade** An *emissions trading* system in which government determines an acceptable level of *pollution* and then issues polluting parties permits to pollute. A company receives credit for amounts it does not emit and can then sell this credit to other companies.

**captive breeding** The practice of capturing members of *threatened* and *endangered species* so that their young can be bred and raised in controlled *environments* and subsequently reintroduced into the wild.

**carbohydrate** An *organic compound* consisting of *atoms* of *carbon*, *hydrogen*, and *oxygen*.

**carbon** The chemical *element* with six protons and six neutrons. A key element in *organic compounds*.

**carbon capture** Technologies or approaches that remove *carbon dioxide* from power plant or other emissions, in an effort to mitigate *global climate change*.

**carbon cycle** A major *nutrient cycle* consisting of the routes that *carbon atoms* take through the nested networks of environmental *systems*.

**carbon dioxide ($CO_2$)** A colorless gas used by plants for *photosynthesis*, given off by *respiration*, and released by burning *fossil fuels*. A primary *greenhouse gas* whose buildup contributes to *global climate change*.

**carbon footprint** The cumulative amount of carbon, or *carbon dioxide*, that a person or institution emits, and is indirectly responsible for emitting, into the *atmosphere*, contributing to *global climate change*. Compare *ecological footprint*.

**carbon monoxide (CO)** A colorless, odorless gas produced primarily by the incomplete combustion of fuel. An EPA *criteria pollutant*.

**carbon-neutrality** The state in which an individual, business, or institution emits no net carbon to the atmosphere. This may be achieved by reducing carbon emissions and/or employing *carbon offsets* to offset emissions.

**carbon offset** A voluntary payment to another entity intended to enable that entity to reduce the *greenhouse gas* emissions that one is unable or unwilling to reduce oneself. The payment thus offsets one's own emissions.

**carbon sequestration** Technologies or approaches to sequester, or store, *carbon dioxide* from industrial emissions (e.g., underground under pressure in locations where it will not seep out) in an effort to mitigate *global climate change*. The term can also refer to the natural sequestration of carbon by plants through *photosynthesis*.

**carbon storage** See *carbon sequestration*.

**carbon tax** A fee charged to entities that pollute by emitting *carbon dioxide*. A carbon tax gives polluters a financial incentive to reduce pollution and is thus foreseen as a way to address *global climate change*. Compare *fee-and-dividend*.

**carcinogen** A chemical or type of radiation that causes cancer.

**carrying capacity** The maximum *population size* that a given *environment* can sustain.

**case history** Medical approach involving the observation and analysis of individual patients.

**catalytic converter** Automotive technology that reduces pollution by converting *pollutants* from *hydrocarbon* combustion into *carbon dioxide*, water vapor, and *nitrogen* gas.

**categorical imperative** An *ethical standard* described by Immanuel Kant, which roughly approximates Christianity's "golden rule": to treat others as you would prefer to be treated yourself.

**cell** The most basic organizational unit of organisms.

**cellular respiration** The process by which a *cell* uses the chemical reactivity of *oxygen* to split glucose into its constituent parts, water and *carbon dioxide*, and thereby release *chemical energy* that can be used to form chemical bonds or to perform other tasks within the cell. Compare *photosynthesis*.

**cellulosic ethanol** *Ethanol* produced from the cellulose in plant tissues by treating it with enzymes. Techniques for producing cellulosic ethanol are under development because of the desire to make ethanol from low-value crop waste (residues such as corn stalks and husks), rather than from the sugars of high-value crops.

**chaparral** A *biome* consisting mostly of densely thicketed evergreen shrubs occurring in limited small patches. Its "Mediterranean" *climate* of mild, wet winters and warm, dry summers is induced by oceanic influences.

**chemical energy** *Potential energy* held in the bonds between atoms.

**chemical formula** A shorthand way to indicate the type and number of atoms in a molecule using numbers and chemical symbols.

**chemical hazard** Chemicals that pose human health hazards. These include *toxins* produced naturally, as well as many of the disinfectants, *pesticides*, and other synthetic chemicals that our society produces. Compare *biological hazard; cultural hazard; physical hazard.*

**chemistry** The study of the different types of matter and how they interact.

**Chernobyl** Site of a nuclear power plant in Ukraine (then part of the Soviet Union), where in 1986 an explosion caused the most severe *nuclear reactor* accident the world has yet seen. As with *Three Mile Island* and *Fukushima*, people often use the term to denote the accident itself. Compare *Fukushima Daiichi, Three Mile Island.*

**chlorofluorocarbon (CFC)** A type of *halocarbon* consisting of only chlorine, fluorine, carbon, and hydrogen. CFCs were used as refrigerants, fire extinguishers, propellants for aerosol spray cans, cleaners for electronics, and for making polystyrene foam. They were phased out under the *Montreal Protocol* because they are *ozone-depleting substances* that destroy stratospheric *ozone*.

**chlorophyll** The light-absorbing pigment that enables *photosynthesis* and makes plants green.

**chronic exposure** Exposure for long periods of time to a *toxicant* occurring in low amounts. Compare *acute exposure*.

**circum-Pacific belt** A 40,000-km (25,000-mi) arc of *subduction* zones and *fault* systems that encircles much of the Pacific Ocean basin. Popularly called the "ring of fire," nine of 10 *earthquakes* and over half the world's *volcanoes* occur here.

**city planning** The professional pursuit that attempts to design cities in such a way as to maximize their efficiency, functionality, and beauty. Also known as *urban planning*.

**classical economics** Founded by *Adam Smith*, the study of the behavior of buyers and sellers in a *capitalist market economy*. Holds that individuals acting in their own self-interest may benefit society, provided that their behavior is constrained by the rule of law and by private property rights and operates within competitive markets. See also *neoclassical economics*.

**Clean Air Act** U.S. federal *legislation* to control *air pollution* that funds research into pollution control, sets standards for air quality, imposes limits on emissions from new stationary and mobile sources, enables citizens to sue parties violating the standards, and introduces an *emissions trading* program for *sulfur dioxide*. First enacted in 1963 and amended multiple times, particularly in 1970 and 1990.

**clean coal technologies** A wide array of techniques, equipment, and approaches to remove chemical contaminants (such as sulfur) during the process of generating *electricity* from *coal*.

**clear-cutting** The harvesting of timber by cutting all the trees in an area. Although it is the most cost-efficient method, clear-cutting is also the most ecologically damaging.

**climate** The pattern of atmospheric conditions in a region over long periods of time. Compare *weather*.

**climate change** See *global climate change*.

**climate diagram** A visual representation of a region's average monthly temperature and *precipitation*. Also know as a *climatograph*.

**climate model** A computer program that combines what is known about weather patterns, atmospheric circulation, atmosphere–ocean interactions, and feedback mechanisms, in order to simulate *climate* processes.

**climax community** In the traditional view of ecological *succession*, a *community* that remains in place with little modification until *disturbance* restarts the successional process. Today, ecologists recognize that community change is more variable and less predictable than originally thought and that assemblages of *species* may instead form complex mosaics in space and time.

**clumped distribution** Distribution pattern in which organisms arrange themselves in patches, generally according to the availability of the resources they need.

**coal** Our most abundant *fossil fuel*. A hard blackish substance formed from organic matter (generally woody plant material) that was compressed under very high pressure and with little decomposition, creating dense, solid carbon structures.

**coevolution** Process by which two or more species evolve in response to one another. Parasites and hosts may coevolve, as may flowering plants and their pollinators.

**co-firing** A process in which *biomass* is combined with *coal* in coal-fired power plants. Can be a relatively easy and inexpensive way for *fossil-fuel*-based utilities to expand their use of *renewable energy*.

**cogeneration** A practice in which the extra heat generated in the production of *electricity* is captured and put to use heating workplaces and homes, as well as producing other kinds of power.

**colony collapse disorder** An undiagnosed cause of mass die-offs of honeybees (*Apis mellifera*) in recent years.

**command-and-control** A top-down approach to policy, in which a legislative body or a regulating agency sets rules, standards, or limits and threatens punishment for violations of those limits.

**community** In *ecology*, an assemblage of *populations* of organisms that live in the same place at the same time.

**community-based conservation** The practice of engaging local people to protect land and wildlife in their own region.

**community ecology** The scientific study of patterns of species diversity and interactions among *species*, from one-to-one interactions to complex interrelationships involving entire *communities*.

**community-supported agriculture (CSA)** A system in which consumers pay farmers in advance for a share of their yield, usually in the form of weekly deliveries of produce.

**competition** A relationship in which multiple organisms seek the same limited resource.

**compost** A mixture produced when decomposers break down organic matter, such as food and crop waste, in a controlled environment.

**composting** The conversion of organic *waste* into mulch or *humus* by encouraging, in a controlled manner, the natural biological processes of decomposition.

**compound** A *molecule* whose *atoms* are composed of two or more *elements*.

**concentrated animal feeding operation** See *feedlot*.

**concentrated solar power (CSP)** A means of generating *electricity* at a large scale by focusing sunlight from a large area onto a smaller area. Several approaches are used.

**concession** The right to extract a resource, granted by a government to a corporation. Sometimes conservation organizations purchase concessions in order to preserve habitat.

**conservation biology** A scientific discipline devoted to understanding the factors, forces, and processes that influence the loss, protection, and restoration of *biodiversity* within and among *ecosystems*.

**conservation ethic** An *ethic* holding that people should put *natural resources* to use but also have a responsibility to manage them wisely. Compare *preservation ethic*.

**conservation geneticist** A scientist who studies genetic attributes of organisms, generally to infer the status of their *populations* in order to help conserve them.

**Conservation Reserve Program** U.S. policy in farm bills since 1985 that pays farmers to stop cultivating highly erodible cropland and instead place it in conservation reserves planted with grasses and trees.

**conservation tillage** *Agriculture* that limits the amount of tilling (plowing, disking, harrowing, or chiseling) of *soil*. Compare *no-till*.

**consumptive use** Use of *fresh water* in which water is removed from a particular *aquifer* or surface water body and is not returned to it. *Irrigation* for *agriculture* is an example of consumptive use. Compare *nonconsumptive use*.

**continental collision** The meeting of two tectonic plates of continental *lithosphere* at a *convergent plate boundary*, wherein the continental *crust* on both sides resists *subduction* and instead crushes together, bending, buckling, and deforming layers of rock and forcing portions of the buckled crust upward, often creating mountain ranges.

**contour farming** The practice of plowing furrows sideways across a hillside, perpendicular to its slope, to help prevent the formation of rills and gullies. The furrows follow the natural contours of the land.

**contraception** The deliberate attempt to prevent pregnancy despite sexual intercourse. Compare *birth control*.

**control** The portion of an *experiment* in which a *variable* has been left unmanipulated, to serve as a point of comparison with the *treatment*.

**controlled burn** See *prescribed burn*.

**controlled experiment** An *experiment* in which a *treatment* is compared against a *control* in order to test the effect of a *variable*.

**control rods** Rods made of a metallic alloy that absorbs *neutrons*, which are placed in a *nuclear reactor* among the water-bathed *fuel rods* of uranium. Engineers move these control rods into and out of the water to maintain the *fission* reaction at the desired rate.

**convective circulation** A circular *current* (of air, water, magma, etc.) driven by temperature differences. In the atmosphere, warm air rises into regions of lower *atmospheric pressure*, where it expands and cools and then descends and becomes denser, replacing warm air that is rising. The air picks up heat and moisture near ground level and prepares to rise again.

**conventional law** International law that arises from conventions, or treaties, that nations agree to enter into. Compare *customary law*.

**Convention on Biological Diversity** An international treaty that aims to conserve *biodiversity*, use biodiversity in a *sustainable* manner, and ensure the fair distribution of biodiversity's benefits.

**Convention on International Trade in Endangered Species of Wild Fauna and Flora (CITES)** A 1973 treaty facilitated by the *United Nations* that protects endangered *species* by banning the international transport of their body parts.

**convergent evolution** The evolutionary process by which very unrelated species acquire similar traits as they adapt to selective pressures from similar environments.

**convergent plate boundary** The area where tectonic plates converge or come together. Can result in *subduction* or *continental collision*. Compare *divergent plate boundary* and *transform plate boundary*.

**coral** Tiny marine animals that build *coral reefs*. Corals attach to rock or existing reef and capture passing food with stinging tentacles. They also derive nourishment from photosynthetic symbiotic algae known as *zooxanthellae*.

**coral reef** A mass of calcium carbonate composed of the skeletons of tiny colonial marine organisms called *corals*.

**core** The innermost part of Earth, made up mostly of iron, that lies beneath the *crust* and *mantle*.

**Coriolis effect** The apparent deflection of north–south air *currents* to a partly east–west direction, caused by the faster spin of regions near the equator than of regions near the poles as a result of Earth's rotation.

**corporate average fuel efficiency (CAFE) standards** Miles-per-gallon fuel efficiency standards set by the U.S. Congress for auto manufacturers to meet, by a sales-weighted average of all models of the manufacturer's fleet.

**correlation** A statistical association among *variables*.

**corridor** A passageway of protected land established to allow animals to travel between islands of protected *habitat*.

**cost-benefit analysis** A method commonly used by *neoclassical economists*, in which estimated costs for a proposed action are totaled and then compared to the sum of benefits estimated to result from the action.

**covalent bond** A type of chemical bonding where atoms share electrons in chemical bonds. An example is a water molecule, which forms when an oxygen atom shares electrons with two hydrogen atoms.

**criteria pollutants** Six *air pollutants—carbon monoxide, sulfur dioxide, nitrogen dioxide, tropospheric ozone, particulate matter*, and *lead*—for which the *Environmental Protection Agency* has established maximum allowable concentrations in ambient outdoor air because of the threats they pose to human health.

**cropland** Land that people use to raise plants for food and fiber.

**crop rotation** The practice of alternating the kind of crop grown in a particular field from one season or year to the next.

**crude oil (petroleum)** A *fossil fuel* produced by the conversion of *organic compounds* by heat and pressure. Crude oil is a mixture of hundreds of different types of *hydrocarbon* molecules characterized by *carbon* chains of different lengths.

**crust** The lightweight outer layer of the Earth, consisting of rock that floats atop the malleable *mantle*, which in turn surrounds a mostly iron *core*.

**cultural hazard** Human health hazards that result from the place we live, our socioeconomic status, our occupation, or our behavioral choices. Also known as *lifestyle hazard*. Compare *biological hazard; chemical hazard; physical hazard*.

**current** The flow of a liquid or gas in a certain direction.

**customary law** International law that arises from long-standing practices, or customs, held in common by most *cultures*. Compare *conventional law*.

**dam** Any obstruction placed in a river or stream to block the flow of water so that water can be stored in a *reservoir*. Dams are built to prevent floods, provide drinking water, facilitate *irrigation*, and generate electricity.

**Darwin, Charles (1809–1882)** English naturalist who proposed the concept of *natural selection* as a mechanism for *evolution* and as a way to explain the great variety of living things. Compare *Wallace, Alfred Russel*.

**data** Information, generally quantitative information.

**deciduous** Describes a plant that loses its leaves each fall and goes dormant during the winter.

**decomposer** An organism, such as a fungus or bacterium, that breaks down leaf litter and other nonliving matter into simple constituents that can be taken up and used by plants. Compare *detritivore*.

***Deepwater Horizon*** The British Petroleum offshore drilling platform in the Gulf of Mexico that sank in 2010, causing the largest oil spill in U.S. history.

**deep-well injection** A *hazardous waste* disposal method in which a well is drilled deep beneath an area's *water table* into porous rock below an impervious *soil* layer. Wastes are then injected into the well, to be absorbed into the porous rock and remain deep underground, isolated from *groundwater* and human contact. Compare *surface impoundment*.

**deforestation** The clearing and loss of *forests*.

**demand** The amount of a product people will buy at a given price if free to do so. Compare *supply*.

**demographer** A social scientist who studies the population size, density, distribution, age structure, sex ratio, and rates of birth, death, immigration, and emigration of human populations. See *demography*.

**demographic transition** A theoretical *model* of economic and cultural change that explains the declining death rates and birth rates that occurred in Western nations as they became industrialized. The model holds that industrialization caused these rates to fall naturally by decreasing mortality and by lessening the need for large families. Parents would thereafter choose to invest in quality of life rather than quantity of children.

**demography** A *social science* that applies the principles of *population ecology* to the study of statistical change in human *populations*.

**denitrifying bacteria** Bacteria that convert the nitrates in *soil* or water to gaseous *nitrogen* and release it back into the *atmosphere*.

**density-dependent** The condition of a *limiting factor* whose effects on a *population* increase or decrease depending on the *population density*. Compare *density-independent factor*.

**density-independent** The condition of a *limiting factor* whose effects on a *population* are constant regardless of *population density*. Compare *density-dependent factor*.

**deoxyribonucleic acid** See *DNA*.

**dependent variable** The *variable* that is affected by manipulation of the *independent variable* in an *experiment*.

**deposition** The arrival of eroded *soil* at a new location. Compare *erosion*.

**desalination (desalinization)** The removal of salt from seawater.

**descriptive science (observational science)** Research in which scientists gather basic information about organisms, materials, systems, or processes that are not yet well known. Compare *hypothesis-driven science*.

**desert** The driest *biome* on Earth, with annual *precipitation* of less than 25 cm. Because deserts have relatively little vegetation to insulate them from temperature extremes, sunlight readily heats them in the daytime, but daytime heat is quickly lost at night, so temperatures vary widely.

**desertification** A form of *land degradation* in which more than 10% of a land's productivity is lost due to *erosion*, soil compaction, forest removal, *overgrazing*, drought, *salinization, climate* change, water depletion, or other factors. Can result in the expansion of desert areas or creation of new ones. Compare *land degradation*; *soil degradation*.

**detritivore** An organism, such as a millipede or soil insect, that scavenges the waste products or dead bodies of other community members. Compare *decomposer.*

**development** The use of natural resources for economic advancement (as opposed to simple subsistence, or survival). Involves making purposeful changes intended to improve quality of life.

**dike** A long raised mound of earth erected along a riverbank to protect against floods by holding rising water in the main channel.

**directional drilling** A drilling technique (e.g., for *oil* or *natural gas*) in which a drill bores down vertically and then bends horizontally in order to follow layered deposits for long distances from the drilling site. This enables us to extract more fossil fuels with less environmental impact at the surface.

**disturbance** An event that affects environmental conditions rapidly and drastically, resulting in changes to the *community* and *ecosystem.* Disturbance can be natural or can be caused by people.

**divergent plate boundary** The area where tectonic plates push apart from one another as *magma* rises upward to the surface, creating new *lithosphere* as it cools and spreads. A prime example is the Mid-Atlantic Ridge. Compare *convergent plate boundary* and *transform plate boundary.*

**DNA (deoxyribonucleic acid)** A double-stranded *nucleic acid* composed of four nucleotides, each of which contains a sugar (deoxyribose), a phosphate group, and a nitrogenous base. DNA carries the hereditary information for living organisms and is responsible for passing traits from parents to offspring. Compare *RNA.*

**dose** The amount of *toxicant* a test animal receives in a dose-response test. Compare *response.*

**dose-response analysis** A set of experiments that measure the *response* of test animals to different *doses* of a *toxicant.* The response is generally quantified by measuring the proportion of animals exhibiting negative effects.

**dose-response curve** A curve that plots the *response* of test animals to different *doses* of a *toxicant*, as a result of *dose-response analysis.*

**downwelling** In the ocean, the flow of warm surface water toward the ocean floor. Downwelling occurs where surface *currents* converge. Compare *upwelling.*

**driftnet** Fishing net that spans large expanses of water, arrayed to drift with currents so as to capture passing fish, and held vertical by floats at the top and weights at the bottom. Driftnetting results in substantial *bycatch* of dolphins, seals, sea turtles, and nontarget fish.

**Dust Bowl** An area that loses huge amounts of *topsoil* to wind *erosion* as a result of drought and/or human impact. First used to name the region in the North American Great Plains severely affected by drought and topsoil loss in the 1930s. The term is now also used to describe that historical event and others like it.

**dynamic equilibrium** The state reached when processes within a *system* are moving in opposing directions at equivalent rates so that their effects balance out.

**earthquake** A release of energy that occurs as Earth relieves accumulated pressure between masses of lithosphere and that results in shaking at the surface.

**ecocentrism** A philosophy that considers actions in terms of their damage or benefit to the integrity of whole ecological systems, including both living and nonliving elements. For an ecocentrist, the well-being of an individual is less important than the long-term well-being of a larger integrated ecological system. Compare *anthropocentrism* and *biocentrism.*

**ecolabeling** The practice of designating on a product's label how the product was grown, harvested, or manufactured, so that consumers are aware of the processes involved and can judge which brands use more sustainable processes.

**ecological economics** A developing school of *economics* that applies the principles of *ecology* and *systems* thinking to the description and analysis of *economies.* Compare *environmental economics*; *neoclassical economics.*

**ecological footprint** The cumulative area of biologically productive land and water required to provide the raw materials a person or *population* consumes and to dispose of or *recycle* the *waste* that is produced.

**ecological modeling** The practice of constructing and testing *models* that aim to explain and predict how ecological systems function.

**ecological restoration** Efforts to reverse the effects of human disruption of ecological systems and to restore *communities* to their condition before the disruption. The practice that applies principles of *restoration ecology.*

**ecology** The *science* that deals with the distribution and abundance of organisms, the interactions among them, and the interactions between organisms and their *abiotic environments.*

**economic growth** An increase in an economy's activity—that is, an increase in the production and consumption of goods and services.

**economics** The study of how we decide to use scarce resources to satisfy demand for *goods* and *services.*

**economy** A social *system* that converts resources into *goods* and *services.*

**ecosystem** All organisms and nonliving entities that occur and interact in a particular area at the same time.

**ecosystem-based management** The attempt to manage the harvesting of resources in ways that minimize impact on the *ecosystems* and ecological processes that provide the resources.

**ecosystem diversity** The number and variety of ecosystems in a particular area. One way to express *biodiversity.* Related concepts consider the geographic arrangement of *habitats, communities*, or *ecosystems* at the landscape level, including the sizes, shapes, and interconnectedness of patches of these entities.

**ecosystem ecology** The study of how the living and nonliving components of *ecosystems* interact.

**ecosystem service** An essential service an *ecosystem* provides that supports life and makes *economic* activity possible. For example, ecosystems naturally purify air and water, cycle *nutrients*, provide for plants to be *pollinated* by animals, and receive and recycle the *waste* we generate.

**ecotone** A transitional zone where *ecosystems* meet.

**ecotourism** Visitation of natural areas for tourism and recreation. Most often involves tourism by more-affluent people, which may generate *economic* benefits for less-affluent communities near natural areas and thus provide economic incentives for conservation of natural areas.

**$ED_{50}$ (effective dose–50%)** The amount of a *toxicant* it takes to affect 50% of a *population* of test animals. Compare *threshold dose*; *$LD_{50}$*.

**edge effect** An impact on organisms, populations, or communities that results because conditions along the edge of a habitat fragment differ from conditions in the interior.

**effluent** Water that flows out of a facility such as a wastewater treatment plant or power plant.

**Ekman drift** Wind-driven currents in the ocean's upper layers. Named for scientist Walfrid Ekman, who first modeled them in the early 20th century. Compare *geostrophic currents.*

**electricity** A secondary form of energy that can be transferred over long distances and applied for a variety of uses.

**electrolysis** A process in which electrical current is passed through a *compound* to release *ions.* Electrolysis offers one way to produce *hydrogen* for use as fuel: Electrical current is passed through water, splitting the water *molecules* into hydrogen and *oxygen atoms.*

**electron** A negatively charged particle that moves around the nucleus of an *atom.*

**electronic waste (e-waste)** Discarded electronic products such as computers, monitors, printers, televisions, DVD players, cell phones, and other devices. *Heavy metals* in these products mean that this waste may be judged hazardous.

**element** A fundamental type of matter; a chemical substance with a given set of properties, which cannot be broken down into substances with other properties. Chemists currently recognize 92 elements that occur in nature, as well as more than 20 others that have been artificially created.

**El Niño** The exceptionally strong warming of the eastern Pacific Ocean that occurs every 2 to 7 years and depresses local fish and bird *populations* by altering the marine *food web* in the area. Compare *La Niña.*

**El Niño–Southern Oscillation (ENSO)** A systematic shift in atmospheric pressure, sea surface temperature, and ocean circulation in the tropical Pacific Ocean. ENSO cycles give rise to *El Niño* and *La Niña* conditions.

**emigration** The departure of individuals from a *population.*

**eminent domain** A policy in which a government pays landowners for their land at market rates and the landowners have no recourse to refuse. In eminent domain, courts set aside private property rights to make way for projects judged to be for the public good.

**emissions trading** The practice of buying and selling government-issued marketable emissions permits to conduct environmentally harmful activities. Under a *cap-and-trade* system, the government determines an acceptable level of *pollution* and then issues permits to pollute. A company receives credit for amounts it does not emit and can then sell this credit to other companies. Compare *cap-and-trade*.

**endangered** In danger of *extinction*. An official category for species under the *Endangered Species Act*.

**Endangered Species Act (ESA)** The primary *legislation*, enacted in 1973, for protecting *biodiversity* in the United States. It forbids the government and private citizens from taking actions (such as developing land) that would destroy *endangered species* or their *habitats*, and it prohibits trade in products made from endangered species.

**endemic** Native or restricted to a particular geographic region. An endemic species occurs in one area and nowhere else on Earth.

**endocrine disruptor** A *toxicant* that interferes with the *endocrine (hormone) system*.

**endocrine system** The body's *hormone* system.

**energy** The capacity to change the position, physical composition, or temperature of matter; a force that can accomplish work.

**energy conservation** The practice of reducing *energy* use as a way of extending the lifetime of our *fossil fuel* supplies, of being less wasteful, and of reducing our impact on the *environment*. Conservation can result from behavioral decisions or from technologies that demonstrate *energy efficiency*.

**energy efficiency** The ability to obtain a given result or amount of output while using less energy input. Technologies permitting greater energy efficiency are one main route to *energy conservation*.

**energy intensity** Energy use per dollar of gross domestic product (GDP). A lower energy intensity indicates greater efficiency.

**energy returned on investment** See *EROI*.

**enhanced geothermal systems (EGS)** A new approach whereby engineers drill deeply into rock, fracture it, pump in water, and then pump it out once it is heated below ground. This approach would enable us to obtain *geothermal energy* in many locations.

**entropy** The degree of disorder in a substance, *system*, or process. See *second law of thermodynamics*.

**environment** The sum total of our surroundings, including all of the living things and nonliving things with which we interact.

**environmental economics** A developing school of *economics* that modifies the principles of *neoclassical economics* to address environmental challenges. Most environmental economists believe that we can attain *sustainability* within our current economic systems. Compare *ecological economics*; *neoclassical economics*.

**environmental ethics** The application of *ethical standards* to environmental questions.

**environmental health** The study of environmental factors that influence human health and quality of life and the health of *ecological* systems essential to environmental quality and long-term human well-being.

**environmental impact statement (EIS)** A report of results from detailed studies that assess the potential effects on the *environment* that would likely result from development projects or other actions undertaken by the government.

**environmental justice** The fair and equitable treatment of all people with respect to environmental policy and practice, regardless of their income, race, or ethnicity. Responds to the perception that minorities and the poor suffer more pollution than whites and the rich.

**environmental policy** *Public policy* that pertains to human interactions with the *environment*. It generally aims to regulate resource use or reduce *pollution* to promote human welfare and/or protect natural systems.

**Environmental Protection Agency (EPA)** An administrative agency charged with conducting and evaluating research, monitoring environmental quality, setting standards, enforcing those standards, assisting the states in meeting standards and goals for environmental protection, and educating the public.

**environmental science** The scientific study of how the natural world functions, how our *environment* affects us, and how we affect our environment.

**environmental studies** An academic *environmental science* program that emphasizes the social sciences as well as the natural sciences.

**environmental toxicology** The study of *toxicants* that come from or are discharged into the *environment*, including the study of health effects on humans, other animals, and *ecosystems*.

**environmentalism** A social movement dedicated to protecting the natural world, and by extension, people.

**epidemiological study** A study that involves large-scale comparisons among groups of people, usually contrasting a group known to have been exposed to some *toxicant* and a group that has not.

**EROI (energy returned on investment)** The ratio determined by dividing the quantity of *energy* returned from a process by the quantity of energy invested in the process. Higher EROI ratios mean that more energy is produced from each unit of energy invested. Compare *net energy*.

**erosion** The removal of material from one place and its transport to another by the action of wind or water.

**estuary** An area where a river flows into the ocean, mixing *fresh water* with salt water.

**ethanol** The alcohol in beer, wine, and liquor, produced as a *biofuel* by fermenting biomass, generally from *carbohydrate*-rich crops such as corn.

**ethical standard** A criterion that helps differentiate right from wrong.

**ethics** The academic study of good and bad, right and wrong. The term can also refer to a person's or group's set of moral principles or values.

**eutrophic** Term describing a water body that has high nutrient and low oxygen conditions. Compare *oligotrophic*.

**eutrophication** The process of *nutrient* enrichment, increased production of organic matter, and subsequent *ecosystem* degradation in a water body.

**evaporation** The conversion of a substance from a liquid to a gaseous form.

**even-aged** Condition of timber plantations—generally *monocultures* of a single *species*—in which all trees are of the same age. Most *ecologists* view plantations of even-aged stands more as crop *agriculture* than as ecologically functional *forests*. Compare *uneven-aged*.

**evenness** See *relative abundance*.

**evolution** Genetically based change in the appearance, functioning, and/or behavior of organisms across generations, often by the process of *natural selection*.

**e-waste** See *electronic waste*.

**experiment** An activity designed to test the validity of a *hypothesis* by manipulating *variables*. See *controlled experiment*, *manipulative experiment*, and *natural experiment*.

**exploratory drilling** Drilling that takes place after a *fossil fuel* deposit has been identified, in order to gauge how much of the fuel exists and whether extraction will prove worthwhile. Involves drilling small holes that descend to great depths.

**exponential growth** The increase of a *population* (or of anything) by a fixed percentage each year.

**external cost** A cost borne by someone not involved in an economic transaction. Examples include harm to citizens from *water pollution* or *air pollution* discharged by nearby factories.

**extinction** The disappearance of an entire *species* from Earth. Compare *extirpation*.

**extirpation** The disappearance of a particular *population* from a given area, but not the entire *species* globally. Compare *extinction*.

**factory farm** See *feedlot*.

**family planning** The effort to plan the number and spacing of one's children so as to offer children and parents the best quality of life possible.

**farmers' market** A market at which local farmers and food producers sell fresh locally grown items.

**fault** A fracture in Earth's crust, at which earthquakes may occur.

**fee-and-dividend** A program of *carbon taxes* in which proceeds from the taxes are paid to consumers as a tax refund or "dividend." This strategy seeks to prevent consumers from losing money if polluters pass their costs along to them. One of two main approaches to a *revenue-neutral carbon tax*.

**feedback loop** A circular process in which a *system*'s output serves as input to that same system. See *negative feedback loop*; *positive feedback loop*.

**feed-in tariff** A program of public policy intended to promote renewable energy investment, whereby utilities are mandated to purchase electricity from homeowners or businesses that generate power from renewable energy sources and feed it into the electric grid. Under such a system, utilities must pay guaranteed premium prices for this power under long-term contract. Compare *net metering*.

**feedlot** A huge barn or outdoor pen designed to deliver *energy*-rich food to animals living at extremely high densities. Also called a *factory farm* or *concentrated animal feeding operation*.

**Ferrel cell** One of a pair of cells of *convective circulation* between 30° and 60° north and south latitude that influence global *climate* patterns. Compare *Hadley cell*; *polar cell*.

**fertilizer** A substance that promotes plant growth by supplying essential *nutrients* such as *nitrogen* or *phosphorus*.

***Fifth Assessment Report*** A report from the *Intergovernmental Panel on Climate Change* summarizing the current state of climate change research, published in 2013–2014. This report represents the consensus of scientific climate research from around the world.

**first law of thermodynamics** The physical law stating that *energy* can change from one form to another, but cannot be created or lost. The total energy in the universe remains constant and is said to be conserved.

**flat plate solar collectors** Panels for collecting and transmitting *solar energy* for the purpose of heating or cooling. Generally these are rooftop panels made of dark heat-absorbing metal plates mounted in flat glass-covered boxes.

**flexible-fuel** A type of vehicle that runs on fuel that is a mixture of ethanol and gasoline, such as E-85, a mix of 85% ethanol and 15% gasoline.

**flooding** The spillage of water over a river's banks due to heavy rain or snowmelt.

**floodplain** The region of land over which a river has historically wandered and periodically floods.

**flux** The movement of nutrients among *pools* or *reservoirs* in a *nutrient cycle*.

**food chain** A linear series of feeding relationships. As organisms feed on one another, energy is transferred from lower to higher *trophic levels*. Compare *food web*.

**food security** An adequate, reliable, and available food supply to all people at all times.

**food web** A visual representation of feeding interactions within an *ecological community* that shows an array of relationships between organisms at different *trophic levels*. Compare *food chain*.

**forensics** See *forensic science*.

**forensic science** The scientific analysis of evidence to make an identification or answer a question relating to a crime or an accident.

**forest** Any ecosystem characterized by a high density of trees.

**forester** A professional who manages *forests* through the practice of *forestry*.

**forestry** The professional management of *forests*.

**forest type** A category of *forest* defined by its predominant tree *species*.

**fossil** The remains, impression, or trace of an animal or plant of past geological ages that has been preserved in rock or *sediments*.

**fossil fuel** A *nonrenewable natural resource*, such as *crude oil, natural gas*, or *coal*, produced by the decomposition and compression of organic matter from ancient life.

**fossil record** The cumulative body of *fossils* worldwide, which paleontologists study to infer the history of past life on Earth.

**fracking** See *hydraulic fracturing*.

**free rider** A party that fails to invest in controlling *pollution* or carrying out other *environmentally* responsible activities and instead relies on the efforts of other parties to do so. For example, a factory that fails to control its emissions gets a "free ride" on the efforts of other factories that do make the sacrifices necessary to reduce emissions.

**fresh water** Water that is relatively pure, holding very few dissolved salts.

**freshwater marsh** A type of *wetland* in which shallow water allows plants such as cattails to grow above the water surface. Compare *swamp*; *bog*.

**fuel rods** Rods of uranium that supply the fuel for nuclear *fission* and are kept bathed in a *moderator* in a *nuclear reactor*.

**Fukushima Daiichi** Japanese nuclear power plant severely damaged by the tsunami associated with the March 2011 Tohoku earthquake that rocked Japan. Most radiation drifted over the ocean away from population centers, but the event was history's second most serious nuclear accident. Compare *Chernobyl, Three Mile Island.*

**full cost accounting** An accounting approach that attempts to summarize all costs and benefits by assigning monetary values to entities without market prices and then generally subtracting costs from benefits. Examples include the *Genuine Progress Indicator*, the Happy Planet Index, and others. Also called *true cost accounting*.

**fungicide** A type of chemical *pesticide* that kills fungi.

**gasification** A process in which *biomass* is vaporized at extremely high temperatures in the absence of *oxygen*, creating a gaseous mixture including *hydrogen*, *carbon monoxide*, and *methane*, in order to produce *biopower* or *biofuels*. In coal gasification, *coal* is converted to a *syngas* by reacting it with oxygen and steam at a high temperature.

**gene** A stretch of *DNA* that represents a unit of hereditary information.

**genera** Plural of *genus*.

**General Land Ordinances of 1785 and 1787** Laws that gave the U.S. government the right to manage Western lands and created a grid system for surveying them and readying them for private ownership.

**General Mining Act of 1872** U.S. law that legalized and promoted *mining* by private individuals on public lands for just $5 per acre, subject to local customs, with no government oversight.

**generalist** A *species* that can survive in a wide array of *habitats* or use a wide array of resources. Compare *specialist*.

**genetically modified organism (GMO)** An organism that has been *genetically engineered* using *recombinant DNA* technology.

**genetic diversity** A measurement of the differences in *DNA* composition among individuals within a given *species*.

**genetic engineering** Any process scientists use to manipulate an organism's genetic material in the lab by adding, deleting, or changing segments of its *DNA*.

**Genuine Progress Indicator (GPI)** An *economic* indicator that attempts to differentiate between desirable and undesirable economic activity. The GPI accounts for benefits such as volunteerism and for costs such as environmental degradation and social upheaval. Compare *Gross Domestic Product (GDP)*.

**genus** A taxonomic level in the Linnaean classification system that is above *species* and below family. A genus is made up of one or more closely related species.

**geoengineering** Any of a suite of proposed efforts to cool Earth's climate by removing carbon dioxide from the atmosphere or reflecting sunlight away from Earth's surface. Such ideas are controversial and are not nearly ready to implement.

**geographic information system (GIS)** Computer software that takes multiple types of data (for instance, on geology, hydrology, vegetation, animal species, and human development) and overlays them on a common set of geographic coordinates. The idea is to create a complete picture of a landscape and to analyze how elements of the different datasets are arrayed spatially and how they may be correlated. A common tool of geographers, landscape ecologists, resource managers, and conservation biologists.

**geology** The scientific study of Earth's physical features, processes, and history.

**geostrophic currents** Oceanic currents driven by differences in atmospheric pressure and the Coriolis force. Compare *Ekman drift*.

**geothermal energy** Energy that arises from beneath Earth's surface, ultimately from the radioactive decay of elements amid high pressures deep underground. Can be used to generate electrical power in power plants, for direct heating via piped water, or in *ground-source heat pumps*.

**glaciation** The spread of ice sheets from the polar regions far into Earth's temperate zones during cold periods of Earth's history.

**global climate change** Systematic change in aspects of Earth's *climate*, such as temperature, *precipitation*, and storm intensity. Generally refers today to the current warming trend in global temperatures and the many associated climatic changes. Compare *global warming*.

**global warming** An increase in Earth's average surface temperature. The term is most frequently used in reference to the pronounced warming trend of recent years and decades. Global warming is one aspect of *global climate change* and in turn drives other components of climate change.

**global warming potential** A quantity that specifies the ability of one *molecule* of a given *greenhouse gas* to contribute to atmospheric warming, relative to *carbon dioxide*.

**globalization** The process by which the world's societies have become more interconnected, linked by trade and communication technologies in countless ways.

**good** A material commodity manufactured for and bought by individuals and businesses.

**green building** (1) A structure that minimizes the ecological footprint of its construction and operation by using sustainable materials, using minimal energy and water, reducing health impacts, limiting pollution, and recycling waste. (2) The pursuit of constructing or renovating such buildings.

**green-collar job** A job resulting from new employment opportunities in a more sustainably oriented *economy*, such as jobs in *renewable energy*.

**greenhouse effect** The warming of Earth's surface and *atmosphere* (especially the *troposphere*) caused by the *energy* emitted by *greenhouse gases*.

**greenhouse gas** A gas that absorbs infrared radiation released by Earth's surface and then warms the surface and *troposphere* by emitting *energy*, thus giving rise to the *greenhouse effect*. Greenhouse gases include *carbon dioxide* ($CO_2$), water vapor, *ozone* ($O_3$), nitrous oxide ($N_2O$), halocarbon gases, and *methane* ($CH_4$).

**green manure** *Organic fertilizer* composed of freshly dead plant material.

**Green Revolution** An intensification of the industrialization of *agriculture* in the developing world in the latter half of the 20th century that has dramatically increased crop yields produced per unit area of farmland. Practices include devoting large areas to *monocultures* of crops specially bred for high yields and rapid growth; heavy use of *fertilizers*, *pesticides*, and *irrigation* water; and sowing and harvesting on the same piece of land more than once per year or per season.

**green tax** A levy on *environmentally* harmful activities and products aimed at providing a market-based incentive to correct for *market failure*. Compare *subsidy*.

**greenwashing** A public-relations effort by a corporation or institution to mislead customers or the public into thinking it is acting more sustainably than it actually is.

**Gross Domestic Product (GDP)** The total monetary value of final *goods* and *services* produced in a country each year. GDP sums all *economic* activity, whether good or bad. Compare *Genuine Progress Indicator (GPI)*.

**gross primary production** The *energy* that results when *autotrophs* convert solar energy (sunlight) to energy of chemical bonds in sugars through *photosynthesis*. Autotrophs use a portion of this production to power their own metabolism, which entails oxidizing *organic compounds* by *cellular respiration*. Compare *net primary production*.

**ground-level ozone** See *tropospheric ozone*.

**ground-source heat pump** A pump that harnesses *geothermal energy* from near-surface sources of earth and water, in order to heat and cool buildings. Operates on the principle that temperatures below ground are more stable than temperatures above ground.

**groundwater** Water held in aquifers underground. Compare *surface water*.

**habitat** The specific *environment* in which an organism lives, including both *biotic* and *abiotic factors*.

**habitat conservation plan** An arrangement under the *Endangered Species Act* that allows a landholder or agency to harm an endangered species if it also mitigates the harm by improving habitat for the species.

**habitat fragmentation** The process by which an expanse of natural *habitat* becomes broken up into discontinuous fragments, often as a result of farming, logging, road building, and other types of human land use.

**habitat selection** The process by which organisms select *habitats* from among the range of options they encounter.

**habitat use** The process by which organisms use *habitats* from among the range of options they encounter.

**Hadley cell** One of a pair of cells of *convective circulation* between the equator and 30° north and south latitude that influence global *climate* patterns. Compare *Ferrel cell*; *polar cell*.

**half-life** The amount of time it takes for one-half the atoms of a *radioisotope* to emit radiation and decay. Different radioisotopes have different half-lives, ranging from fractions of a second to billions of years.

**halocarbon** A class of human-made chemical *compounds* derived from simple *hydrocarbons* in which *hydrogen* atoms are replaced by halogen atoms such as bromine, fluorine, or chlorine. Many halocarbons are *ozone-depleting substances* and/or *greenhouse gases*.

**harmful algal bloom** A *population* explosion of toxic algae caused by excessive *nutrient* concentrations.

**hazardous waste** Liquid or solid *waste* that is toxic, chemically reactive, flammable, or corrosive. Compare *industrial solid waste*; *municipal solid waste*.

**herbicide** A type of chemical *pesticide* that kills plants.

**herbivory** The consumption of plants by animals.

**heterotroph (consumer)** An organism that consumes other organisms. Includes most animals, as well as fungi and microbes that decompose organic matter.

**homeostasis** The tendency of a *system* to maintain constant or stable internal conditions.

**horizon** A distinct layer of *soil*.

**hormone** A chemical messenger that travels though the bloodstream to stimulate growth, development, and sexual maturity and to regulate brain function, appetite, sexual drive, and many other aspects of physiology and behavior.

**host** The organism in a parasitic relationship that suffers harm while providing the *parasite* nourishment or some other benefit.

**Hubbert's peak** The peak in production of *crude oil* in the United States, which occurred in 1970 just as Shell Oil geologist M. King Hubbert had predicted in 1956.

**humus** A dark, spongy, crumbly mass of material made up of complex organic compounds, resulting from the partial decomposition of organic matter.

**hydraulic fracturing** A process to extract *shale gas*, in which a drill is sent deep underground and angled horizontally into a shale formation; water, sand, and chemicals are pumped in under great pressure, fracturing the rock; and gas migrates up through the drilling pipe as sand holds the fractures open. Also called *hydrofracking* or simply *fracking*.

**hydrocarbon** An *organic compound* consisting solely of *hydrogen* and *carbon atoms*.

**hydroelectric power** The generation of *electricity* using the *kinetic energy* of moving water.

**hydrofracking** See *hydraulic fracturing*.

**hydrogen** The chemical *element* with one proton. The most abundant element in the universe. Also a possible fuel for our future economy.

**hydrologic cycle** The flow of water—in liquid, gaseous, and solid forms—through our *biotic* and *abiotic environment*. Also called the *water cycle*.

**hydropower** See *hydroelectric power*.

**hydrosphere** All water—salt or fresh, liquid, ice, or vapor—in surface bodies, underground, and in the *atmosphere*. Compare *biosphere*; *lithosphere*.

**hypothesis** A statement that attempts to explain a phenomenon or answer a *scientific* question. Compare *theory*.

**hypothesis-driven science** Research in which scientists pose questions that seek to explain how and why things are the way they are. Generally proceeds in a somewhat structured manner, using *experiments* to test *hypotheses*. Compare *descriptive science*.

**hypoxia** The condition of extremely low dissolved *oxygen* concentrations in a body of water.

**igneous rock** One of the three main categories of rock. Formed from cooling *magma*. Granite and basalt are examples of igneous rock. Compare *metamorphic rock* and *sedimentary rock*.

**immigration** The arrival of individuals from outside a *population*.

**in-situ recovery** See *solution mining*.

**inbreeding depression** A state that occurs in a *population* when genetically similar parents mate and produce weak or defective offspring as a result.

**incineration** A controlled process of burning solid *waste* for disposal in which mixed garbage is combusted at very high temperatures. Compare *sanitary landfill.*

**independent variable** The *variable* that the scientist manipulates in an *experiment.*

**indoor air pollution** *Air pollution* that occurs indoors.

**industrial agriculture** A form of *agriculture* that uses large-scale mechanization and *fossil fuel* combustion, enabling farmers to replace horses and oxen with faster and more powerful means of cultivating, harvesting, transporting, and processing crops. Other aspects include *irrigation* and the use of *inorganic fertilizers*. Use of chemical herbicides and *pesticides* reduces *competition* from weeds and *herbivory* by insects. Compare *traditional agriculture*.

**industrial ecology** A holistic approach to industry that integrates principles from engineering, chemistry, *ecology, economics*, and other disciplines and seeks to redesign industrial *systems* in order to reduce resource inputs and minimize inefficiency.

**industrial revolution** The shift beginning in the mid-1700s from rural life, animal-powered agriculture, and manufacturing by craftsmen to an urban society powered by *fossil fuels*. Compare *agricultural revolution.*

**industrial smog** "Gray-air" smog caused by the incomplete combustion of *coal* or oil when burned. Compare *photochemical smog*.

**industrial solid waste** Nonliquid *waste* that is not especially hazardous and that comes from production of consumer goods, mining, *petroleum* extraction and *refining*, and *agriculture*. Compare *hazardous waste*; *municipal solid waste*.

**industrial stage** The third stage of the *demographic transition* model, characterized by falling birth rates that close the gap with falling death rates and reduce the rate of *population* growth. Compare *pre-industrial stage; post-industrial stage; transitional stage.*

**infant mortality rate** The number of deaths of infants under one year of age per 1000 live births in a population.

**infectious disease** A disease in which a pathogen attacks a host.

**inorganic fertilizer** A *fertilizer* that consists of mined or synthetically manufactured mineral supplements. Inorganic fertilizers are generally more susceptible than *organic fertilizers* to *leaching* and *runoff* and may be more likely to cause unintended off-site impacts.

**insecticide** A type of chemical *pesticide* that kills insects.

**integrated pest management (IPM)** The use of multiple techniques in combination to achieve long-term suppression of pests, including *biocontrol*, use of *pesticides*, close monitoring of *populations*, *habitat* alteration, *crop rotation, transgenic* crops, alternative tillage methods, and mechanical pest removal.

**intercropping** Planting different types of crops in alternating bands or other spatially mixed arrangements.

**interdisciplinary** Borrowing techniques from multiple traditional fields of study and bringing together research results from these fields into a broad synthesis.

**Intergovernmental Panel on Climate Change (IPCC)** An international panel of *climate* scientists and government officials established in 1988 by the United Nations Environment Programme and the World Meteorological Organization. The IPCC's mission is to assess and synthesize scientific research on *global climate change* and to offer guidance to the world's policymakers, primarily through periodic published reports. See *Fifth Assessment Report*.

**interspecific competition** *Competition* that takes place among members of two or more different *species*. Compare *intraspecific competition*.

**intertidal** Of, relating to, or living along shorelines between the highest reach of the highest *tide* and the lowest reach of the lowest tide.

**intraspecific competition** *Competition* that takes place among members of the same *species*. Compare *interspecific competition*.

**introduced species** Species introduced by human beings from one place to another (whether intentionally or by accident). A minority of introduced species may become *invasive species*.

**invasive species** A *species* that spreads widely and rapidly becomes dominant in a *community*, interfering with the community's normal functioning.

**inversion layer** In a *temperature inversion*, the band of air in which temperature rises with altitude (instead of falling with altitude, as temperature does normally).

**ion** An electrically charged *atom* or combination of atoms.

**ionic bond** A type of chemical bonding where electrons are transferred between atoms, creating oppositely charged ions that bond due to their differing electrical charges. Table salt, or sodium chloride, is formed by the bonding of positively charged sodium ions with negatively charged chloride ions.

**IPAT model** A formula that represents how humans' total impact (I) on the *environment* results from the interaction among *population* (P), *affluence* (A), and *technology* (T).

**irrigation** The artificial provision of water to support *agriculture*.

**isotope** One of several forms of an *element* having differing numbers of *neutrons* in the nucleus of its *atoms*. Chemically, isotopes of an element behave almost identically, but they have different physical properties because they differ in mass.

**jet stream** A high-altitude air current that blows west to east, meandering north and south, and which influences weather across much of North America and Eurasia.

**kelp** Large brown algae or seaweed that can form underwater "forests," providing habitat for marine organisms.

**keystone species** A *species* that has an especially far-reaching effect on a *community*.

**kinetic energy** *Energy* of motion. Compare *potential energy*.

**kwashiorkor** A form of *malnutrition* that results from a high-*starch* diet with inadequate *protein* or *amino acids*. In children, causes bloating of the abdomen, deterioration and discoloration of hair, mental disability, immune suppression, developmental delays, anemia, and reduced growth.

**Kyoto Protocol** An international agreement drafted in 1997 that called for reducing, by 2012, emissions of six *greenhouse gases* to levels lower than their levels in 1990. It has been extended to 2020 until a replacement treaty can be reached. An outgrowth of the *United Nations Framework Convention on Climate Change*.

**landfill gas** A mix of gases that consists of roughly half *methane* produced by anaerobic decomposition deep inside *landfills*.

**landscape ecology** The study of how landscape structure affects the abundance, distribution, and interaction of organisms. This approach to the study of organisms and their *environments* at the landscape scale focuses on broad geographical areas that include multiple *ecosystems*.

**landslide** The collapse and downhill flow of large amounts of rock or soil. A severe and sudden form of *mass wasting*.

**land trust** A local or regional organization that preserves lands valued by its members. In most cases, land trusts purchase land outright with the aim of preserving it in its natural condition.

**La Niña** An exceptionally strong cooling of surface water in the equatorial Pacific Ocean that occurs every 2 to 7 years and has widespread climatic consequences. Compare *El Niño*.

**lava** *Magma* that is released from the *lithosphere* and flows or spatters across Earth's surface.

**law of conservation of matter** The physical law stating that *matter* may be transformed from one type of substance into others, but that it cannot be created or destroyed.

**$LD_{50}$ (lethal dose–50%)** The amount of a *toxicant* it takes to kill 50% of a *population* of test animals. Compare $ED_{50}$; *threshold dose*.

**leachate** Liquids that seep through liners of a *sanitary landfill* and leach into the *soil* underneath.

**leaching** The process by which solid materials such as *minerals* are dissolved in a liquid (usually water) and transported to another location.

**lead** A heavy metal that may be ingested through water or paint, or that may enter the *atmosphere* as a particulate pollutant through combustion of leaded gasoline or other processes. Atmospheric lead deposited on land and water can enter the *food chain*, accumulate within body tissues, and cause *lead poisoning* in animals and people. An EPA *criteria pollutant*.

**Leadership in Energy and Environmental Design (LEED)** The leading set of standards for sustainable building. Compare *green building*.

**legislation** Statutory law.

**Leopold, Aldo (1887–1949)** American scientist, scholar, philosopher, and author. His book *The Land Ethic* argued that humans should view themselves and the land itself as members of the same *community* and that humans are obligated to treat the land *ethically*.

**levee** See *dike*.

**life-cycle analysis** A quantitative analysis of inputs and outputs across the entire life cycle of a product—from its origins, through its production, transport, sale, and use, and finally its disposal—in an attempt to judge the sustainability of the process and make it more ecologically efficient.

**lifestyle hazard** See *cultural hazard*.

**light rail** A *mass transit* rail system of trains powered by electricity, often at a moderate scale.

**limiting factor** A physical, chemical, or biological characteristic of the *environment* that restrains *population* growth.

**limnetic zone** In a water body, the layer of open water through which sunlight penetrates. Compare *littoral zone; benthic zone; profundal zone*.

**liquefied natural gas (LNG)** *Natural gas* that has been converted to a liquid at low temperatures and that can be shipped long distances in refrigerated tankers.

**lithosphere** The outer layer of Earth, consisting of *crust* and uppermost *mantle* and located just above the *asthenosphere*. More generally, the solid part of Earth, including the rocks, *sediment*, and *soil* at the surface and extending down many miles underground. Compare *atmosphere*, *biosphere*, and *hydrosphere*.

**littoral** Near the shore of a water body. See *intertidal*.

**littoral zone** The region ringing the edge of a water body. Compare *benthic zone; limnetic zone; profundal zone*.

**local extinction** See *extirpation*.

**logistic growth curve** A plot that shows how the initial *exponential growth* of a *population* is slowed and finally brought to a standstill by *limiting factors*.

**longline fishing** Fishing practice that involves setting out extremely long lines with up to several thousand baited hooks spaced along their lengths. Kills turtles, sharks, and an estimated 300,000 seabirds each year as *bycatch*.

**Love Canal** A residential neighborhood in Niagara Falls, New York, from which families were evacuated after buried toxic chemicals rose to the surface, contaminating homes and an elementary school. The well-publicized event helped spark passage of the *Superfund* legislation. Compare *Times Beach*.

**low-input agriculture** *Agriculture* that uses smaller amounts of *pesticides*, *fertilizers*, growth hormones, water, and *fossil fuel* energy than are used in *industrial agriculture*. Compare *sustainable agriculture*; *organic agriculture*.

**macromolecule** A very large molecule, such as a *protein, nucleic acid, carbohydrate*, or lipid.

**magma** Molten, liquid rock.

**malnutrition** The condition of lacking *nutrients* the body needs, including a complete complement of vitamins and minerals.

**mangrove** A tree with a unique type of roots that curve upward to obtain *oxygen*, which is lacking in the mud in which they grow, and that serve as stilts to support the tree in changing water levels. Mangrove forests grow on the coastlines of the tropics and subtropics.

**manipulative experiment** An *experiment* in which the researcher actively chooses and manipulates the *independent variable*. Compare *natural experiment*.

**mantle** The malleable layer of rock that lies beneath Earth's *crust* and surrounds a mostly iron *core*.

**marasmus** A form of *malnutrition* that results from *protein* deficiency together with a lack of calories, causing wasting or shriveling among millions of children in the developing world.

**marine protected area (MPA)** An area of the ocean set aside to protect marine life from fishing pressures. An MPA may be protected from some human activities but be open to others. Compare *marine reserve*.

**marine reserve** An area of the ocean designated as a "no-fishing" zone, allowing no extractive activities. Compare *marine protected area (MPA)*.

**market failure** The failure of markets to take into account the *environment's* positive effects on *economies* (for example, *ecosystem services*) or to reflect the negative effects of economic activity on the environment and thereby on people (*external costs*).

**mass extinction event** The *extinction* of a large proportion of the world's *species* in a very short time period due to some extreme and rapid change or catastrophic event. Earth has seen five mass extinction events in the past half-billion years.

**mass transit** A public transportation system for a metropolitan area that moves large numbers of people at once. Buses, trains, subways, streetcars, trolleys, and *light rail* are types of mass transit.

**mass wasting** The downslope movement of soil and rock due to gravity. Compare *landslide*.

**materials recovery facility (MRF)** A *recycling* facility where items are sorted, cleaned, shredded, and prepared for reprocessing into new items.

**matter** All material in the universe that has mass and occupies space. See *law of conservation of matter*.

**maximum sustainable yield** The maximal harvest of a particular *renewable natural resource* that can be accomplished while still keeping the resource available for the future.

**meltdown** The accidental melting of the uranium *fuel rods* inside the core of a *nuclear reactor*, causing the release of radiation.

**metal** A type of chemical *element*, or a mass of such an element, that typically is lustrous, opaque, and malleable and that can conduct heat and electricity.

**metamorphic rock** One of the three main categories of rock. Formed by great heat and/or pressure that reshapes crystals within the rock and changes its appearance and physical properties. Common metamorphic rocks include marble and slate. Compare *igneous rock* and *sedimentary rock*.

**methane ($CH_4$)** A colorless gas produced primarily by *anaerobic* decomposition. The major constituent of *natural gas* and a *greenhouse gas* that is molecule-for-molecule more potent than *carbon dioxide*.

**methane hydrate** An ice-like solid consisting of molecules of *methane* embedded in a crystal lattice of water molecules. Most is found in sediments on the continental shelves and in the Arctic. Methane hydrate is a potential alternative *fossil fuel*.

**Milankovitch cycle** One of three types of variations in Earth's rotation and orbit around the sun that modify the relative amount of solar radiation reaching Earth's surface at different latitudes. As the cycles proceed, they change the way solar radiation is distributed over Earth's surface and contribute to changes in *atmospheric* heating and circulation that have triggered *glaciations* and other *climate* changes.

**Millennium Development Goals** A program of targets for *sustainable development* set by the international community through the *United Nations* at the turn of this century.

**mineral** A naturally occurring solid *element* or inorganic *compound* with a crystal structure, a specific chemical composition, and distinct physical properties. Compare *ore* and *rock*.

**mining** (1) In the broad sense, the extraction of any resource that is nonrenewable on the timescale of our society (such as *fossil fuels* or *groundwater*). (2) In relation to *mineral* resources, the systematic removal of *rock*, *soil*, or other material for the purpose of extracting minerals of economic interest.

**mitigation** The pursuit of strategies to lessen the severity of climate change, notably by reducing emissions of greenhouse gases. Compare *adaptation*.

**model** A simplified representation of a complex natural process, designed by scientists to help understand how the process occurs and to make predictions.

**moderator** Within a *nuclear reactor*, a substance, most often water or graphite, that slows the *neutrons* bombarding uranium so that *fission* can begin.

**molecule** A combination of two or more *atoms*.

**monoculture** The uniform planting of a single crop over a large area. Characterizes *industrial agriculture*. Compare *polyculture*.

**Montreal Protocol** International treaty ratified in 1987 in which 180 (now 196) signatory nations agreed to restrict production of *chlorofluorocarbons (CFCs)* in order to halt stratospheric *ozone* depletion. This was a protocol of the Vienna Convention for the Protection of the Ozone Layer. The Montreal Protocol is widely considered the most successful effort to date in addressing a global *environmental* problem.

**mortality** Rate of death within a *population*.

**mosaic** In *landscape ecology*, a spatial configuration of *patches* arrayed across a landscape.

**mountaintop removal mining** A large-scale form of *coal* mining in which entire mountaintops are leveled. The technique exerts extreme environmental impact on surrounding *ecosystems* and human residents.

**Muir, John (1838–1914)** Scottish immigrant to the United States who eventually settled in California and made the Yosemite Valley his wilderness home. Today, he is most strongly associated with the *preservation ethic*. He argued that nature deserved protection for its own inherent values but also claimed that nature facilitated human happiness and fulfillment.

**municipal solid waste** Nonliquid *waste* that is not especially hazardous and that comes from homes, institutions, and small businesses. Compare *hazardous waste*; *industrial solid waste*.

**mutagen** A *toxicant* that causes *mutations* in the *DNA* of organisms.

**mutation** An accidental change in *DNA* that may range in magnitude from the deletion, substitution, or addition of a single nucleotide to a change affecting entire sets of chromosomes. Mutations provide the raw material for evolutionary change.

**mutualism** A relationship in which all participating organisms benefit from their interaction. Compare *parasitism*.

**nacelle** Compartment in a *wind turbine* containing machinery for generating power.

**NAFTA** See *North American Free Trade Agreement*.

**natality** Rate of birth within a *population*.

**National Environmental Policy Act (NEPA)** A U.S. law enacted on January 1, 1970, that created an agency called the Council on Environmental Quality and required that an *environmental impact statement* be prepared for any major federal action.

**national forest** An area of forested public land managed by the U.S. Forest Service. The system consists of 191 million acres (more than 8% of the nation's land area) in many tracts spread across all but a few states.

**National Forest Management Act** 1976 U.S. law mandating that plans for renewable resource management be drawn up for every *national forest*. These plans were to be explicitly based on the concepts of *multiple use* and *maximum sustainable yield* and be open to broad public participation.

**national park** A scenic area set aside for recreation and enjoyment by the public and managed by the National Park Service. The U.S. national park system today numbers 397 sites totaling 84 million acres and includes national historic sites, national recreation areas, national wild and scenic rivers, and other areas.

**national wildlife refuge** An area of public land set aside to serve as a haven for wildlife and also sometimes to encourage hunting, fishing, wildlife observation, photography, environmental education, and other uses. The system of 550 sites is managed by the U.S. Fish and Wildlife Service.

**natural capital** Earth's accumulated wealth of resources.

**natural experiment** An *experiment* in which the researcher cannot directly manipulate the *variables* and therefore must observe nature, comparing conditions in which variables differ, and interpret the results. Compare *manipulative experiment*.

**natural gas** A *fossil fuel* consisting primarily of *methane* ($CH_4$) and including varying amounts of other volatile *hydrocarbons*.

**natural rate of population change** See *rate of natural increase*.

**natural resource** Any of the various substances and *energy* sources that we take from our environment and that we need in order to survive.

**Natural Resources Conservation Service (NRCS)** U.S. agency that promotes soil conservation, as well as water quality protection and pollution control. Prior to 1994, known as the Soil Conservation Service.

**natural sciences** Academic disciplines that study the natural world. Compare *social sciences*.

**natural selection** The process by which traits that enhance survival and reproduction are passed on more frequently to future generations of organisms than traits that do not, thus altering the genetic makeup of populations through time. Natural selection acts on genetic variation and is a primary driver of *evolution*.

**negative feedback loop** A *feedback loop* in which output of one type acts as input that moves the *system* in the opposite direction. The input and output essentially neutralize each other's effects, stabilizing the system. Compare *positive feedback loop*.

**neoclassical economics** A mainstream economic school of thought that explains market prices in terms of consumer preferences for units of particular commodities and that uses *cost-benefit analysis*. Compare *ecological economics*; *environmental economics*.

**net energy** The quantitative difference between *energy* returned from a process and *energy* invested in the process. Positive net energy values mean that a process produces more energy than is invested. See also *EROI*.

**net metering** Process by which homeowners or business with *photovoltaic* systems or *wind turbines* can sell their excess *solar energy* or *wind power* to their local utility. Whereas *feed-in tariffs* award producers with prices above market rates, net metering offers market-rate prices.

**net primary production** The *energy* or biomass that remains in an ecosystem after *autotrophs* have metabolized enough for their own maintenance through *cellular respiration*. Net primary production is the energy or biomass available for consumption by *heterotrophs*. Compare *gross primary production*; *secondary production*.

**net primary productivity** The rate at which *net primary production* is produced. See *productivity; gross primary production; net primary production; secondary production*.

**neurotoxin** A *toxicant* that assaults the nervous system. Neurotoxins include heavy metals, *pesticides*, and some chemical weapons developed for use in war.

**neutron** An electrically neutral (uncharged) particle in the nucleus of an *atom*.

**new urbanism** An approach among architects, planners, and developers that seeks to design neighborhoods in which homes, businesses, schools, and other amenities are within walking distance of one another. Intended to combat *sprawl* by creating functional neighborhoods in which families can meet most needs close to home without the use of a car.

**niche** The functional role of a *species* in a *community*.

**NIMBY** See *not-in-my-backyard*.

**nitrification** The conversion by bacteria of ammonium ions ($NH_4^+$) first into nitrite ions ($NO_2^-$) and then into nitrate ions ($NO_3^-$).

**nitrogen** The chemical *element* with seven *protons* and seven *neutrons*. The most abundant element in the *atmosphere*, a key element in *macromolecules*, and a crucial plant *nutrient*.

**nitrogen cycle** A major *nutrient cycle* consisting of the routes that *nitrogen atoms* take through the nested networks of environmental *systems*.

**nitrogen dioxide ($NO_2$)** A foul-smelling reddish brown gas that contributes to *smog* and *acid deposition*. It results when atmospheric *nitrogen* and *oxygen* react at the high temperatures created by combustion engines. An EPA *criteria pollutant*.

**nitrogen fixation** The process by which inert *nitrogen* gas combines with *hydrogen* to form ammonium ions ($NH_4^+$), which are chemically and biologically active and can be taken up by plants.

**nitrogen-fixing bacteria** Bacteria that live independently in the soil or water, or those that form *mutualistic* relationships with many types of plants and provide *nutrients* to the plants by converting gaseous *nitrogen* to a usable form.

**nitrogen oxide ($NO_x$)** One of a family of compounds that includes nitric oxide (NO) and nitrogen dioxide ($NO_2$).

**no-analog community** An ecological *community* composed of a novel mixture of organisms, with no current analog or historical precedent.

**nonconsumptive use** *Fresh water* use in which the water from a particular *aquifer* or surface water body either is not removed or is removed only temporarily and then returned. The use of water to generate electricity in hydroelectric *dams* is an example. Compare *consumptive use*.

**nongovernmental organization (NGO)** An organization not affiliated with any national government, and frequently international in scope, that pursues a particular mission or advocates for a particular cause.

**nonmarket value** A value that is not usually included in the price of a *good* or *service*.

**non-point source** A diffuse source of *pollutants*, often consisting of many small sources. Compare *point source*.

**nonrenewable natural resources** *Natural resources* that are in limited supply and are formed much more slowly than we use them. Compare *renewable natural resources*.

**North American Free Trade Agreement (NAFTA)** A 1994 treaty among Canada, Mexico, and the United States that reduced or eliminated barriers to trade (such as tariffs) among these nations. Side agreements were negotiated to minimize the degree to which protections for workers and the environment were undermined.

**North Atlantic Deep Water (NADW)** The deep portion of the *thermohaline circulation* in the northern Atlantic Ocean.

**no-till farming** *Agriculture* that does not involve tilling (plowing, disking, harrowing, or chiseling) the *soil*. The most intensive form of *conservation tillage*.

**not-in-my-backyard (NIMBY)** Syndrome in which people do not want something (e.g., a polluting facility) near where they live, even if they may want or need the thing to exist somewhere else.

**novel community** See *no-analog community*.

**nuclear energy** The *energy* that holds together *protons* and *neutrons* within the nucleus of an *atom*. Several processes, each of which involves transforming *isotopes* of one *element* into isotopes of other elements, can convert nuclear energy into thermal energy, which is then used to generate *electricity*. See also *nuclear fission*; *nuclear reactor*.

**nuclear fission** The conversion of the *energy* within an *atom*'s nucleus to usable thermal energy by splitting apart atomic nuclei. Compare *nuclear fusion*.

**nuclear reactor** A facility within a nuclear power plant that initiates and controls the process of *nuclear fission* in order to generate electricity.

**nucleic acid** A *macromolecule* that directs the production of *proteins*. Includes *DNA* and *RNA*.

**nutrient** An *element* or *compound* that organisms consume and require for survival.

**nutrient cycle** The comprehensive set of cyclical pathways by which a given *nutrient* moves through the *environment*.

**observational science** See *descriptive science*.

**ocean acidification** The process by which today's oceans are becoming more *acidic* (attaining lower *pH*) as a result of increased *carbon dioxide* concentrations in the atmosphere. Ocean acidification occurs as ocean water absorbs $CO_2$ from the air and forms carbonic acid. This impairs the ability of corals and other organisms to build exoskeletons of calcium carbonate, imperiling coral reefs and the many organisms that depend on them.

**ocean thermal energy conversion (OTEC)** A potential *energy* source that involves harnessing the solar radiation absorbed by tropical ocean water. See *closed cycle*; *open cycle*.

**oil** See *petroleum*.

**oil sands (tar sands)** Deposits that can be mined from the ground, consisting of moist sand and clay containing 1–20% *bitumen*. Oil sands represent crude oil deposits that have been degraded and chemically altered by water *erosion* and bacterial decomposition. Widely envisioned as a replacement for *crude oil* as this resource is depleted.

**oil shale** *Sedimentary rock* filled with *kerogen* that can be processed to produce liquid *petroleum*. Oil shale is formed by the same processes that form *crude oil* but occurs when kerogen was not buried deeply enough or subjected to enough heat and pressure to form oil.

**oligotrophic** Term describing a water body that has low nutrient and high oxygen conditions. Compare *eutrophic*.

**open pit mining** A *mining* technique that involves digging a gigantic hole and removing the desired *ore*, along with waste *rock* that surrounds the ore.

**ore** A *mineral* or grouping of minerals from which we extract *metals*.

**organic agriculture** *Agriculture* that uses no synthetic *fertilizers* or *pesticides* but instead relies on biological approaches such as *composting* and *biocontrol*. Compare *low-input agriculture*; *sustainable agriculture*.

**organic compound** A *compound* made up of *carbon atoms* (and, generally, *hydrogen* atoms) joined by covalent bonds and sometimes including other *elements*, such as *nitrogen*, *oxygen*, sulfur, or *phosphorus*. The unusual ability of carbon to build elaborate *molecules* has resulted in millions of different organic compounds showing various degrees of complexity.

**organic fertilizer** A *fertilizer* made up of natural materials (largely the remains or wastes of organisms), including animal manure; crop residues, fresh vegetation, and compost. Compare *inorganic fertilizer*.

**Organization of Petroleum Exporting Countries (OPEC)** Cartel of predominantly Arab nations that in 1973 embargoed oil shipments to the United States and other nations supporting Israel, setting off an oil shortage.

**outdoor air pollution** *Air pollution* that occurs outdoors. Also called *ambient air pollution*.

**overgrazing** The consumption by too many animals of plant cover, impeding plant regrowth and the replacement of *biomass*. Overgrazing can exacerbate damage to *soils*, natural *communities*, and the land's productivity.

**overnutrition** A condition of excessive food intake in which people receive more than their daily caloric needs.

**overshoot** The amount by which humanity's resource use, as measured by its *ecological footprint*, has surpassed Earth's long-term capacity to support us.

**oxygen** The chemical *element* with eight *protons* and eight *neutrons*. A key element in the *atmosphere* that is produced by *photosynthesis*.

**ozone ($O_3$)** A *molecule* consisting of three atoms of *oxygen*. Absorbs ultraviolet radiation in the *stratosphere*. Compare *ozone layer* and *tropospheric ozone*.

**ozone-depleting substances** Airborne chemicals, such as *halocarbons*, that destroy *ozone* molecules and thin the *ozone layer* in the *stratosphere*.

**ozone hole** Term popularly used to describe the thinning of the stratospheric *ozone layer* that occurs over Antarctica each year, as a result of *chlorofluorocarbons (CFCs)* and other *ozone-depleting substances*.

**ozone layer** A portion of the *stratosphere*, roughly 17–30 km (10–19 mi) above sea level, that contains most of the *ozone* in the *atmosphere*.

**paradigm** A dominant philosophical and theoretical framework within a scientific discipline.

**parasite** The organism in a parasitic relationship that extracts nourishment or some other benefit from the *host*.

**parasitism** A relationship in which one organism, the *parasite*, depends on another, the *host*, for nourishment or some other benefit while simultaneously doing the host harm. Compare *mutualism*.

**parent material** The base geologic material in a particular location.

**particulate matter** Solid or liquid particles small enough to be suspended in the *atmosphere* and able to damage respiratory tissues when inhaled. Includes *primary pollutants* such as dust and soot as well as *secondary pollutants* such as sulfates and nitrates. An EPA *criteria pollutant*.

**passive solar energy collection** An approach in which buildings are designed and building materials are chosen to maximize their direct absorption of sunlight in winter and to keep the interior cool in the summer. Compare *active solar energy collection*.

**patch** In *landscape ecology*, spatial areas within a landscape. Depending on a researcher's perspective, patches may consist of habitat for a particular organism, or communities, or ecosystems. An array of patches forms a *mosaic*.

**pathogen** A *parasite* that causes disease in its host.

**peak oil** Term used to describe the point of maximum production of *petroleum* in the world (or for a given nation), after which oil production declines. Compare *Hubbert's peak*.

**peer review** The process by which a manuscript submitted for publication in an academic journal is examined by specialists in the field, who provide comments and criticism (generally anonymously) and judge whether the work merits publication in the journal.

**pelagic** Of, relating to, or living between the surface and floor of the ocean. Compare *benthic*.

**periodic table of the elements** (see Appendix D) Standard table in chemistry that summarizes information on the *elements*.

**permafrost** In *tundra*, underground soil that remains more or less permanently frozen.

**pest** A pejorative term for any organism that damages crops that are valuable to us. The term is subjective and defined by our own economic interests and is not biologically meaningful. Compare *weed*.

**pesticide** An artificial chemical used to kill insects (*insecticide*), plants (*herbicide*), or fungi (*fungicide*).

**petroleum** See *crude oil.* However, the term is also used to refer to both *oil* and *natural gas* together.

**pH** A measure of the concentration of *hydrogen ions* in a *solution*. The pH scale ranges from 0 to 14: A solution with a pH of 7 is neutral; solutions with a pH below 7 are *acidic*, and those with a pH higher than 7 are *basic*. Because the pH scale is logarithmic, each step on the scale represents a 10-fold difference in hydrogen ion concentration.

**phase shift** A fundamental shift in the overall character of an ecological community, generally occurring after some extreme disturbance, and after which the community may not return to its original state. Also known as a *regime shift*.

**phosphorus** The chemical *element* with 15 *protons* and 15 *neutrons*. An abundant element in the *lithosphere*, a key element in *macromolecules*, and a crucial plant *nutrient*.

**phosphorus cycle** A major *nutrient cycle* consisting of the routes that *phosphorus atoms* take through the nested networks of environmental *systems*.

**photic zone** In the ocean or a freshwater body, the well-lit top layer of water where *photosynthesis* occurs.

**photochemical smog** "Brown-air" smog caused by light-driven reactions of *primary pollutants* with normal atmospheric *compounds* that produce a mix of over 100 different chemicals, ground-level *ozone* often being the most abundant among them. Compare *industrial smog*.

**photosynthesis** The process by which *autotrophs* produce their own food. Sunlight powers a series of chemical reactions that convert *carbon dioxide* and water into sugar (glucose), thus transforming low-quality *energy* from the sun into high-quality energy the organism can use. Compare *cellular respiration*.

**photovoltaic (PV) cell** A device designed to collect sunlight and directly convert it to electrical *energy* by making use of the *photoelectric effect*.

**phylogenetic tree** A treelike diagram that represents the history of divergence of *species* or other taxonomic groups of organisms.

**physical hazard** Physical processes that occur naturally in our environment and pose human health hazards. These include discrete events such as *earthquakes*, *volcanic eruptions*, fires, floods, blizzards, *landslides*, hurricanes, and droughts, as well as ongoing natural phenomena such as ultraviolet radiation from sunlight. Compare *biological hazard; chemical hazard; cultural hazard*.

**phytoplankton** Microscopic photosynthetic algae, protists, and cyanobacteria that drift near the surface of water bodies and generally form the first *trophic level* in an aquatic *food chain*. Compare *zooplankton*.

**Pinchot, Gifford (1865–1946)** The first professionally trained American *forester*, Pinchot helped establish the U.S. Forest Service. Today, he is the person most closely associated with the *conservation ethic*.

**pioneer species** A *species* that arrives earliest, beginning the ecological process of *succession* in a terrestrial or aquatic *community*.

**placer mining** A *mining* technique that involves sifting through material in modern or ancient riverbed deposits, generally using running water to separate lightweight mud and gravel from heavier *minerals* of value.

**plastics** Synthetic (human-made) *polymers* used in numerous manufactured products.

**plate tectonics** The process by which Earth's surface is shaped by the extremely slow movement of tectonic plates, or sections of *crust*. Earth's surface includes about 15 major tectonic plates. Their interaction gives rise to processes that build mountains, cause *earthquakes*, and otherwise influence the landscape.

**poaching** The illegal killing of wildlife, usually for meat or body parts.

**point source** A specific spot—such as a factory—where large quantities of *air pollutants* or *water pollutants* are discharged. Compare *non-point source*.

**polar cell** One of a pair of cells of *convective circulation* between the poles and 60° north and south latitude that influence global *climate* patterns. Compare *Ferrel cell*; *Hadley cell*.

**policy** A rule or guideline that directs individual, organizational, or societal behavior.

**pollination** A plant-animal interaction in which one organism (for example, a bee or a hummingbird) transfers pollen (containing male sex cells) from flower to flower, fertilizing ovaries (containing female sex cells) that grow into fruits with seeds.

**polluter-pays principle** Principle specifying that the party responsible for producing *pollution* should pay the costs of cleaning up the pollution or mitigating its impacts.

**pollution** The release of matter or *energy* into the *environment* that causes undesirable impacts on the health and well-being of humans or other organisms. Pollution can be physical, chemical, or biological, and it can affect water, air, or soil.

**polybrominated diphenyl ethers (PBDEs)** Synthetic compounds that provide fire-retardant properties and are used in a diverse array of consumer products, including computers, televisions, plastics, and furniture. Released during production, disposal, and use of products, these chemicals persist and accumulate in living tissue and appear to be *endocrine disruptors*.

**polyculture** The planting of multiple crops in a mixed arrangement or in close proximity. An example is some traditional Native American farming that mixed maize, beans, squash, and peppers. Compare *monoculture*.

**polymer** A chemical *compound* or mixture of compounds consisting of long chains of repeated *molecules*. Important biological molecules, such as *DNA* and *proteins*, are examples of polymers.

**pool** A location in which nutrients in a *biogeochemical cycle* remain for a period of time before moving to another pool. Can be living or nonliving entities. Synonymous with *reservoir*. Compare *flux*; *residence time*.

**population** A group of organisms of the same *species* that live in the same area. Species are often composed of multiple populations.

**population density** The number of individuals within a *population* per unit area. Compare *population size*.

**population distribution** The spatial arrangement of organisms within a particular area.

**population ecology** The study of the quantitative dynamics of population change and the factors that affect the distribution and abundance of members of a population.

**population growth rate** The rate of change in a *population*'s size per unit time (generally expressed in percent per year), taking into accounts births, deaths, immigration, and emigration. Compare *rate of natural increase*.

**population size** The number of individual organisms present at a given time in a *population*.

**positive feedback loop** A *feedback loop* in which output of one type acts as input that moves the *system* in the same direction. The input and output drive the system further toward one extreme or another. Compare *negative feedback loop*.

**post-industrial stage** The fourth and final stage of the *demographic transition* model, in which both birth and death rates have fallen to a low level and remain stable there, and *populations* may even decline slightly. Compare *industrial stage, pre-industrial stage, transition stage*.

**potential energy** *Energy* of position. Compare *kinetic energy*.

**precautionary principle** The idea that one should not undertake a new action until the ramifications of that action are well understood.

**precipitation** Water that condenses out of the *atmosphere* and falls to Earth in droplets or crystals.

**pre-industrial stage** The first stage of the *demographic transition* model, characterized by conditions that defined most of human history. In pre-industrial societies, both death rates and birth rates are high. Compare *industrial stage, post-industrial stage, transitional stage*.

**predation** The process in which one *species* (the *predator*) hunts, tracks, captures, and ultimately kills its *prey*.

**predator** An organism that hunts, captures, kills, and consumes individuals of another species, the *prey*.

**prediction** A specific statement, generally arising from a *hypothesis*, that can be tested directly and unequivocally.

**prescribed (controlled) burns** The practice of burning areas of *forest* or grassland under carefully controlled conditions to improve the health of *ecosystems*, return them to a more natural state, reduce fuel loads, and help prevent uncontrolled catastrophic fires.

**preservation ethic** An ethic holding that we should protect the natural *environment* in a pristine, unaltered state. Compare *conservation ethic*.

**prey** An organism that is killed and consumed by a *predator*.

**primary consumer** An organism that consumes *producers* and feeds at the second *trophic level.*

**primary extraction** The initial drilling and pumping of the most easily accessible *crude oil*. Compare *secondary extraction.*

**primary forest** *Forest* uncut by people. Compare *secondary forest.*

**primary pollutant** A hazardous substance, such as soot or *carbon monoxide*, that is emitted into the *troposphere* directly from a source in a form that is harmful. Compare *secondary pollutant.*

**primary production** The conversion of solar energy to the energy of chemical bonds in sugars during *photosynthesis*, performed by *autotrophs*. Compare *secondary production.*

**primary succession** A stereotypical series of changes as an *ecological community* develops over time, beginning with a lifeless substrate. In terrestrial *systems*, primary succession begins when a bare expanse of rock, *sand*, or *sediment* becomes newly exposed to the atmosphere and *pioneer species* arrive. Compare *secondary succession.*

**primary treatment** A stage of *wastewater* treatment in which contaminants are physically removed. Wastewater flows into tanks in which sewage solids, grit, and particulate matter settle to the bottom. Greases and oils float to the surface and can be skimmed off. Compare *secondary treatment.*

**probability** A quantitative description of the likelihood of a certain outcome.

**producer (autotroph)** An organism that uses energy from sunlight to produce its own food. Includes green plants, algae, and cyanobacteria. See *autotroph.*

**productivity** The rate at which plants convert solar *energy* (sunlight) to *biomass*. *Ecosystems* whose plants convert solar energy to biomass rapidly are said to have high productivity. See *net primary productivity; gross primary production; net primary production.*

**protein** A *macromolecule* made up of long chains of amino acids.

**proton** A positively charged particle in the nucleus of an *atom.*

**proven recoverable reserve** The amount of a given *fossil fuel* in a deposit that is technologically and economically feasible to remove under current conditions.

**proxy indicator** A source of indirect evidence that serves as a proxy, or substitute, for direct measurement and that sheds light on past climate. Examples include data from ice cores, sediment cores, tree rings, packrat middens, and coral reefs.

**public policy** *Policy* made by governments, including those at the local, state, federal, and international levels; it consists of *legislation, regulations*, orders, incentives, and practices intended to advance societal welfare. See also *environmental policy.*

**pumped storage** A technique used to generate *hydroelectric power*, in which water is pumped from a lower reservoir to a higher reservoir when power demand is weak and prices are low. When demand is strong and prices are high, water is allowed to flow downhill through a turbine, generating *electricity*. Compare *run-of-river; storage.*

**pyrolysis** The chemical breakdown of organic matter (*biomass, oil shale*, and so on) by heating in the absence of *oxygen*, which often produces materials that can be more easily converted to usable energy. A number of variations on this process exist.

**quarry** A pit used to extract *mineral* resources such as clay, gravel, sand, or stone (for example, limestone, granite, marble, or slate).

**radiative forcing** The amount of change in thermal energy that a factor (such as a *greenhouse gas* or an *aerosol*) causes in influencing Earth's temperature. Positive forcing warms Earth's surface, whereas negative forcing cools it.

**radioactive** The quality by which some *isotopes* "decay," changing their chemical identity as they shed atomic particles and emit high-energy radiation.

**radioisotope** A radioactive *isotope* that emits subatomic particles and high-*energy* radiation as it "decays" into progressively lighter isotopes until becoming a stable isotope.

**radon** A highly *toxic*, radioactive, colorless gas that seeps up from the ground in areas with certain types of bedrock and that can build up inside basements and homes with poor air circulation.

**random distribution** Distribution pattern in which individuals are located haphazardly in space in no particular pattern (often when needed resources are spread throughout an area and other organisms do not strongly influence where individuals settle).

**rangeland** Land used for grazing livestock.

**rate of natural increase** The rate of change in a *population's* size resulting from birth and death rates alone, excluding migration. Also called *natural rate of population change*. Compare *population growth rate.*

**REACH** Program of the European Union that shifts the burden of proof for testing chemical safety from national governments to industry and requires that chemical substances produced or imported in amounts of over 1 metric ton per year be registered with a new European Chemicals Agency. *REACH*, which stands for Registration, Evaluation, Authorisation and Restriction of Chemicals, went into effect in 2007.

**rebound effect** The phenomenon by which gains in efficiency from better technology are partly offset when people engage in more energy-consuming behavior as a result. This common psychological effect can hamper conservation and efficiency efforts.

**reclamation** The act of restoring a *mining* site to an approximation of its pre-mining condition. To reclaim a site, companies are required to remove buildings and other structures used for mining, replace *overburden*, fill in shafts, and replant the area with vegetation.

**recombinant DNA** *DNA* that has been patched together from the DNA of multiple organisms in an attempt to produce desirable traits (such as rapid growth, disease and pest resistance, or higher nutritional content) in organisms lacking those traits.

**recovery** *Waste management* strategy composed of *recycling* and *composting.*

**recycling** The collection of materials that can be broken down and reprocessed to manufacture new items.

**REDD** Acronym for *Reducing Emissions from Deforestation and Forest Degradation.* A proposed international program to help address climate change in which wealthy industrialized nations would pay poorer developing nations to conserve forest. Under this plan, poor nations would gain income while rich nations receive carbon credits to offset their emissions in an international cap-and-trade system.

**Red List** An updated list of *species* facing unusually high risks of *extinction*. The list is maintained by the World Conservation Union.

**red tide** A *harmful algal bloom* consisting of algae that produce reddish pigments that discolor surface waters.

**regime shift** See *phase shift.*

**regional planning** Planning similar to *city planning* but conducted across broader geographic scales, generally involving multiple municipal governments.

**regulation** A specific rule issued by an administrative agency, based on the more broadly written statutory law passed by Congress and enacted by the president.

**relative abundance** The extent to which numbers of individuals of different species are equal or skewed. One way to express species diversity. See *evenness*; compare *species richness.*

**relativist** An ethicist who maintains that *ethics* do and should vary with social context. Compare *universalist.*

**renewable natural resources** *Natural resources* that are virtually unlimited or that are replenished by the *environment* over relatively short periods (hours to weeks to years). Compare *nonrenewable natural resources.*

**replacement fertility** The *total fertility rate (TFR)* that maintains a stable *population* size.

**reproductive window** The portion of a woman's life between sexual maturity and menopause during which she may become pregnant.

**reserves-to-production ratio (R/P ratio)** The total remaining reserves of a *fossil fuel* divided by the annual rate of production (extraction and processing).

**reservoir** (1) An artificial water body behind a dam that stores water for human use. (2) See *pool.*

**residence time** (1) In a biogeochemical cycle, the amount of time a nutrient remains in a given *pool* or *reservoir* before moving to another. Compare *flux.* (2) In the *atmosphere*, the amount of time a gas molecule or a *pollutant* remains aloft.

**resilience** The ability of an ecological *community* to change in response to disturbance but later return to its original state. Compare *resistance.*

**resistance** The ability of an ecological *community* to remain stable in the presence of a disturbance. Compare *resilience*.

**Resource Conservation and Recovery Act (RCRA)** U.S. law (enacted in 1976 and amended in 1984) that specifies, among other things, how to manage *sanitary landfills* to protect against environmental contamination.

**resource management** Strategic decision making about how to extract resources, so that resources are used wisely and conserved for the future.

**resource partitioning** The process by which *species* adapt to *competition* by evolving to use slightly different resources, or to use their shared resources in different ways, thus minimizing interference with one another.

**response** The type or magnitude of negative effects an animal exhibits in response to a *dose* of *toxicant* in a *dose-response analysis*. Compare *dose*.

**restoration ecology** The study of the historical conditions of *ecological communities* as they existed before humans altered them. Principles of restoration ecology are applied in the practice of *ecological restoration*.

**revenue-neutral carbon tax** A type of *carbon tax* in which there is no net transfer of money from citizens or the private sector to the government. Monies paid are either refunded by a *fee-and-dividend* approach or other tax rates are lowered to compensate.

**ribonucleic acid (RNA)** A usually single-stranded *nucleic acid* composed of four nucleotides, each of which contains a sugar (ribose), a phosphate group, and a nitrogenous base. RNA carries the hereditary information for living organisms and is responsible for passing traits from parents to offspring. Compare *DNA*.

**riparian** Relating to a river or the area along a river.

**risk** The mathematical probability that some harmful outcome (for instance, injury, death, *environmental* damage, or *economic* loss) will result from a given action, event, or substance.

**risk assessment** The quantitative measurement of *risk*, together with the comparison of risks involved in different activities or substances.

**risk management** The process of considering information from scientific *risk assessment* in light of economic, social, and political needs and values, in order to make decisions and design strategies to minimize *risk*.

**RNA** See *ribonucleic acid*.

**rock** A solid aggregation of *minerals*.

**rock cycle** The very slow process in which *rocks* and the *minerals* that make them up are heated, melted, cooled, broken, and reassembled, forming *igneous*, *sedimentary*, and *metamorphic* rocks.

**rotation time** The number of years that pass between the time a *forest* stand is cut for timber and the next time it is cut.

**runoff** The water from *precipitation* that flows into streams, rivers, lakes, and ponds, and (in many cases) eventually to the ocean.

**run-of-river** Any of several methods used to generate *hydroelectric power* without greatly disrupting the flow of river water. Run-of-river approaches eliminate much of the *environmental* impact of large *dams*. Compare *pumped storage*, *storage*.

**safe harbor agreement** A cooperative agreement that allows landowners to harm *threatened* or *endangered species* in some ways if they voluntarily improve habitat for the species in others.

**salinization** The buildup of salts in surface *soil* layers.

**salt marsh** Flat land that is intermittently flooded by the ocean where the *tide* reaches inland. Salt marshes occur along temperate coastlines and are thickly vegetated with grasses, rushes, shrubs, and other herbaceous plants.

**salvage logging** The removal of dead trees following a natural disturbance. Although it may be economically beneficial, salvage logging can be ecologically destructive, because *snags* provide food and shelter for wildlife and because removing timber from recently burned land can cause *erosion* and damage to *soil*.

**sanitary landfill** A site at which solid waste is buried in the ground or piled up in large mounds for disposal, designed to prevent the waste from contaminating the *environment*. Compare *incineration*.

**savanna** A *biome* characterized by grassland interspersed with clusters of acacias and other trees in dry tropical regions.

**science** (1) A systematic process for learning about the world and testing our understanding of it. (2) The accumulated body of knowledge that arises from this dynamic process.

**scientific method** A formalized method for testing ideas with observations that involves a more-or-less consistent series of interrelated steps.

**scrubber** Technology to chemically treat gases produced in combustion in order to reduce smokestack emissions. These devices typically remove hazardous components and neutralize acidic gases, such as *sulfur dioxide* and hydrochloric acid, turning them into water and salt.

**secondary consumer** An organism that consumes *primary consumers* and feeds at the third *trophic level*.

**secondary extraction** The extraction of *crude oil* remaining after *primary extraction* by using solvents or by flushing underground rocks with water or steam. Compare *primary extraction*.

**secondary forest** *Forest* that has grown back after *primary forest* has been cut. Consists of *second-growth* trees.

**secondary pollutant** A hazardous substance produced through the reaction of *primary pollutants*.

**secondary succession** A stereotypical series of changes as an *ecological community* develops over time, beginning when some event disrupts or dramatically alters an existing community. Compare *primary succession*.

**secondary treatment** A stage of *wastewater* treatment in which biological means are used to remove contaminants remaining after *primary treatment*. Wastewater is stirred up in the presence of *aerobic* bacteria, which degrade organic pollutants in the water. The wastewater then passes to another settling tank, where remaining solids drift to the bottom. Compare *primary treatment*.

**second-growth** Term describing trees that have sprouted and grown to partial maturity after virgin timber has been cut.

**second law of thermodynamics** The physical law stating that the nature of *energy* tends to change from a more-ordered state to a less-ordered state; that is, *entropy* increases.

**sediment** The eroded remains of rocks.

**sedimentary rock** One of the three main categories of rock. Formed when dissolved *minerals* seep through *sediment* layers and act as a kind of glue, crystallizing and binding sediment particles together. Sandstone and shale are examples of sedimentary rock. Compare *igneous rock* and *metamorphic rock*.

**seed bank** A storehouse for samples of the world's crop diversity.

**selective breeding** See *artificial selection*.

**septic system** *A wastewater* disposal method, common in rural areas, consisting of an underground tank and series of drainpipes. Wastewater runs from the house to the tank, where solids precipitate out. The water proceeds downhill to a drain field of perforated pipes laid horizontally in gravel-filled trenches, where microbes decompose the remaining waste.

**service** Work done for others as a form of business.

**sex ratio** The proportion of males to females in a *population*.

**shale gas** *Natural gas* trapped deep underground in tiny bubbles dispersed throughout formations of shale, a type of *sedimentary rock*. Shale gas is often extracted by *hydraulic fracturing*.

**shale oil** A liquid form of petroleum extracted from deposits of *oil shale*.

**shelterbelt** A row of trees or other tall perennial plants that are planted along the edges of farm fields to break the wind and thereby minimize wind *erosion*.

**sick-building syndrome** A *building-related illness* produced by indoor *pollution* in which the specific cause is not identifiable.

**sink** In a *nutrient cycle*, a *pool* that accepts more *nutrients* than it releases.

**sinkhole** An area where the ground has given way with little warning as a result of subsidence caused by depletion of water from an *aquifer*.

**SLOSS (Single Large or Several Small)** Acronym for the debate over whether it is better to make reserves large in size and few in number or many in number but small in size.

**smart growth** A *city planning* concept in which a community's growth is managed in ways intended to limit *sprawl* and maintain or improve residents' quality of life.

**smelting** A process in which *ore* is heated beyond its melting point and combined with other *metals* or chemicals, in order to form metal with desired characteristics. Steel is created by smelting iron ore with carbon, for example.

**Smith, Adam (1723–1790)** Scottish philosopher known today as the father of *classical economics*. He believed that when people are free to pursue their own economic self-interest in a competitive marketplace, the marketplace will behave as if guided by "an invisible hand" that ensures that their actions will benefit society as a whole.

**smog** Term popularly used to describe unhealthy mixtures of air *pollutants* that often form over urban areas. See *industrial smog*; *photochemical smog*.

**snag** A dead tree that is still standing. Snags are valuable for wildlife.

**social sciences** Academic disciplines that study human interactions and institutions. Compare *natural sciences*.

**soil** A complex plant-supporting *system* consisting of disintegrated rock, organic matter, air, water, *nutrients*, and microorganisms.

**soil degradation** A deterioration of soil quality and decline in soil productivity, resulting primarily from forest removal, cropland agriculture, and *overgrazing* of livestock. Compare *desertification*; *land degradation*.

**soil profile** The cross-section of a *soil* as a whole, from the surface to the *bedrock*.

**solar cooker** A simple portable oven that uses reflectors to focus sunlight onto food and cook it.

**solar energy** Energy from the sun. It is perpetually renewable and may be harnessed in several ways.

**solution mining** A *mining* technique in which a narrow borehole is drilled deep into the ground to reach a *mineral* deposit, and water, acid, or another liquid is injected down the borehole to *leach* the resource from the surrounding rock and dissolve it in the liquid. The resulting solution is then sucked out, and the desired resource is isolated. Also called *in-situ recovery*.

**source** In a *nutrient cycle*, a *pool* that releases more *nutrients* than it accepts.

**source reduction** The reduction of the amount of material that enters the *waste stream* to avoid the costs of disposal and *recycling*, help conserve resources, minimize *pollution*, and save consumers and businesses money.

**specialist** A *species* that can survive only in a narrow range of *habitats* that contain very specific resources. Compare *generalist*.

**speciation** The process by which new *species* are generated.

**species** A *population* or group of populations of a particular type of organism whose members share certain characteristics and can breed freely with one another and produce fertile offspring. Different biologists may have different approaches to diagnosing species boundaries.

**species diversity** The number and variety of *species* in the world or in a particular region.

**species richness** The number of species in a particular region. One way to express species diversity. Compare *evenness*; *relative abundance*.

**sprawl** The unrestrained spread of urban or suburban development outward from a city center and across the landscape. Often specified as growth in which the area of development outpaces *population* growth.

**stable isotope** An *isotope* that is not *radioactive*.

**steady-state economy** An *economy* that does not grow or shrink but remains stable.

**steppe** See *temperate grassland*.

**Stockholm Convention on Persistent Organic Pollutants (POPs)** A 2004 international *treaty* that aims to end the use of 12 persistent organic *pollutants* nicknamed the "dirty dozen."

**storage** Technique used to generate *hydroelectric power*, in which large amounts of water are impounded in a reservoir behind a concrete *dam* and then passed through the dam to turn *turbines* that generate *electricity*. Compare *pumped storage*, *run-of-river*.

**storm surge** A temporary and localized rise in sea level brought on by the high tides and winds associated with storms.

**stratosphere** The layer of the *atmosphere* above the *troposphere* and below the *mesosphere*; it extends from 11 km (7 mi) to 50 km (31 mi) above sea level.

**strip mining** The use of heavy machinery to remove huge amounts of earth to expose *coal* or *minerals*, which are mined out directly. Compare *subsurface mining*.

**subduction** The *plate tectonic* process by which denser *crust* slides beneath lighter crust at a *convergent plate boundary*. Often results in *volcanism*.

**subsidy** A government grant of money or resources to a private entity, intended to support and promote an industry or activity.

**subspecies** *Populations* of a *species* that occur in different geographic areas and vary from one another in some characteristics. Subspecies are formed by the same processes that drive *speciation* but result when divergence does not proceed far enough to create separate species.

**subsurface mining** Method of *mining* underground deposits of *coal*, *minerals*, or fuels, in which shafts are dug deeply into the ground and networks of tunnels are dug or blasted out to follow coal seams. Compare *strip mining*.

**suburb** A smaller community located at the outskirts of a city.

**succession** A stereotypical series of changes in the composition and structure of an *ecological community* through time. See *primary succession*; *secondary succession*.

**sulfur dioxide ($SO_2$)** A colorless gas that can result from the combustion of *coal*. In the *atmosphere*, it may react to form sulfur trioxide and sulfuric acid, which may return to Earth in *acid deposition*. An EPA *criteria pollutant*.

**Superfund** A program administered by the *Environmental Protection Agency* in which experts identify sites polluted with hazardous chemicals, protect *groundwater* near these sites, and clean up the *pollution*. Established by the Comprehensive Environmental Response Compensation and Liability Act (CERCLA) in 1980.

**supply** The amount of a product offered for sale at a given price. Compare *demand*.

**surface impoundment** (1) A disposal method for *hazardous waste* or mining waste in which waste in liquid or slurry form is placed into a shallow depression lined with impervious material such as clay and allowed to evaporate, leaving a solid residue on the bottom. (2) The site of such disposal. Compare *deep-well injection*.

**surface water** Water located atop Earth's surface. Compare *groundwater*.

**sustainability** A guiding principle of *environmental science*, entailing conserving resources, maintaining functional ecological systems, and developing long-term solutions, such that Earth can sustain our civilization and all life for the future, allowing our descendants to live at least as well as we have lived.

**sustainable agriculture** *Agriculture* that can be practiced in the same way and in the same place far into the future. Sustainable agriculture does not deplete *soils* faster than they form, nor reduce the clean water and genetic diversity essential to long-term crop and livestock production.

**sustainable development** Development that satisfies our current needs without compromising the future availability of *natural capital* or our future quality of life.

**sustainable forest certification** A form of *ecolabeling* that identifies timber products that have been produced using *sustainable* methods. The Forest Stewardship Council and several other organizations issue such certification.

**swamp** A type of *wetland* consisting of shallow water rich with vegetation, occurring in a forested area. Compare *bog*; *freshwater marsh*.

**swidden** The traditional form of *agriculture* in tropical forested areas, in which the farmer cultivates a plot for one year to a few years and then moves on to clear another plot, leaving the first to grow back to *forest*. When the forest is burned, this may be called "slash-and-burn" agriculture.

**symbiosis** A relationship between different *species* of organisms that live in close physical proximity. People most often use the term "symbiosis" when referring to a mutualism, but symbiotic relationships can be either *parasitic* or *mutualistic*.

**syncrude** Synthetic *crude oil*.

**synergistic effect** An interactive effect (as of *toxicants*) that is more than or different from the simple sum of their constituent effects.

**syngas** Synthesis gas created from *gasification* of *coal*.

**system** A network of relationships among a group of parts, elements, or components that interact with and influence one another through the exchange of *energy*, matter, and/or information.

**taiga** See *boreal forest*.

**tailings** Portions of *ore* left over after *metals* have been extracted in *mining*.

**tar sands** See *oil sands*.

**tax break** A common form of *subsidy* in which government reduces or eliminates taxes required of a business or an individual, for the purpose of promoting industries or activities deemed desirable.

**taxonomist** A scientist who classifies species using an organism's physical appearance and/or genetic makeup, and who groups species by their similarity into a hierarchy of categories meant to reflect evolutionary relationships.

**temperate deciduous forest** A *biome* consisting of midlatitude *forests* characterized by broad-leafed trees that lose their leaves each fall and remain dormant during winter. These forests occur in areas where *precipitation* is spread relatively evenly throughout the year.

**temperate grassland** A *biome* whose vegetation is dominated by grasses and features more extreme temperature differences between winter and summer and less *precipitation* than *temperate deciduous forests*. Also known as *steppe, prairie*.

**temperate rainforest** A *biome* consisting of tall coniferous trees, cooler and less *species*-rich than *tropical rainforest* and milder and wetter than *temperate deciduous forest*.

**temperature inversion** A departure from the normal temperature distribution in the *atmosphere*, in which a pocket of relatively cold air occurs near the ground, with warmer air above it. The cold air, denser than the air above it, traps *pollutants* near the ground and can thereby cause a buildup of *smog*. Also called *thermal inversion*.

**teratogen** A *toxicant* that causes harm to the unborn, resulting in birth defects.

**terracing** The cutting of level platforms, sometimes with raised edges, into steep hillsides to contain water from *irrigation* and *precipitation*. Terracing transforms slopes into series of steps like a staircase, enabling farmers to cultivate hilly land while minimizing their loss of *soil* to water *erosion*.

**tertiary consumer** An organism that consumes *secondary consumers* and feeds at the fourth *trophic level*.

**theory** A widely accepted, well-tested explanation of one or more cause-and-effect relationships that has been extensively validated by a great amount of research. Compare *hypothesis*.

**thermal inversion** See *temperature inversion*.

**thermal mass** Construction materials that absorb heat, store it, and release it later, for use in *passive solar energy* approaches.

**thermogenic** Type of *natural gas* created by compression and heat deep underground. Contains *methane* and small amounts of other *hydrocarbon* gases. Compare *biogenic*.

**thermohaline circulation** A worldwide system of ocean currents in which warmer, fresher water moves along the surface and colder, saltier water (which is denser) moves deep beneath the surface.

**thin-film solar cells** Photovoltaic materials compressed into ultra-thin lightweight sheets that may be incorporated into various surfaces to produce photovoltaic solar power.

**Three Mile Island** Nuclear power plant in Pennsylvania that in 1979 experienced a partial *meltdown*. The term is often used to denote the accident itself, the most serious *nuclear reactor* malfunction that the United States has thus far experienced. Compare *Chernobyl, Fukushima Daiichi*.

**threatened** Likely to become *endangered* in the near future. An official category for species under the *Endangered Species Act*.

**threshold dose** The amount of a *toxicant* at which it begins to affect a *population* of test animals. Compare $ED_{50}$; $LD_{50}$.

**tidal energy** *Energy* harnessed by erecting a *dam* across the outlet of a tidal basin. Water flowing with the incoming or outgoing *tide* through sluices in the dam turns turbines to generate *electricity*.

**tide** The periodic rise and fall of the ocean's height at a given location, caused by the gravitational pull of the moon and sun.

**tight oil** Conventional oil held tightly within shale or near shale deposits.

**Times Beach** A town in Missouri whose residents were evacuated and whose buildings were demolished after being contaminated by *dioxin* from waste oil sprayed on its roads. The well-publicized event helped spark passage of the *Superfund* legislation. Compare *Love Canal*.

**topsoil** That portion of the *soil* that is most nutritive for plants and is thus of the most direct importance to *ecosystems* and to *agriculture*. Also known as the A horizon.

**total fertility rate (TFR)** The average number of children born per female member of a *population* during her lifetime.

**toxicant** A substance that acts as a poison to humans or wildlife.

**toxicity** The degree of harm a chemical substance can inflict.

**toxicology** The scientific field that examines the effects of poisonous chemicals and other agents on humans and other organisms.

**Toxic Substances Control Act** A 1976 U.S. law that directs the *Environmental Protection Agency* to monitor thousands of industrial chemicals and gives the EPA authority to regulate and ban substances found to pose excessive risk.

**toxin** A *toxic* chemical stored or manufactured in the tissues of living organisms. For example, a chemical that plants use to ward off *herbivores* or that insects use to deter *predators*.

**traditional agriculture** Biologically powered *agriculture*, in which human and animal muscle power, along with hand tools and simple machines, perform the work of cultivating, harvesting, storing, and distributing crops. Compare *industrial agriculture*.

**tragedy of the commons** The process by which publicly accessible resources open to unregulated use tend to become damaged and depleted through overuse. Coined by Garrett Hardin and widely applicable to resource issues.

**transboundary park** A reserve of protected land that overlaps national borders.

**transform plate boundary** The area where two tectonic plates meet and slip and grind alongside one another, creating *earthquakes*. For example, the Pacific Plate and the North American Plate rub against each other along California's San Andreas Fault. Compare *convergent plate boundary* and *divergent plate boundary*.

**transgene** A *gene* that has been extracted from the *DNA* of one organism and transferred into the DNA of an organism of another *species*.

**transgenic** Term describing an organism that contains *DNA* from another *species*.

**transitional stage** The second stage of the *demographic transition* model, which occurs during the transition from the *pre-industrial stage* to the *industrial stage*. It is characterized by declining death rates but continued high birth rates. See also *post-industrial stage*. Compare *industrial stage, post-industrial stage, pre-industrial stage*.

**transpiration** The release of water vapor by plants through their leaves.

**trawling** Fishing method that entails dragging immense cone-shaped nets through the water, with weights at the bottom and floats at the top to keep the nets open. Compare *bottom-trawling*.

**treatment** The portion of an *experiment* in which a *variable* has been manipulated in order to test its effect. Compare *control*.

**tributary** A smaller river that flows into a larger one.

**triple bottom line** An approach to sustainability that attempts to meet environmental, economic, and social goals simultaneously.

**trophic cascade** A series of changes in the *population* sizes of organisms at different *trophic levels* in a *food chain*, occurring when *predators* at high trophic levels indirectly promote populations of organisms at low trophic levels by keeping species at intermediate trophic levels in check. Trophic cascades may become apparent when a top predator is eliminated from a system.

**trophic level** Rank in the feeding hierarchy of a *food chain*. Organisms at higher trophic levels consume those at lower trophic levels.

**tropical deciduous forest** See *tropical dry forest*.

**tropical dry forest** A *biome* that consists of deciduous trees and occurs at tropical and subtropical latitudes where wet and dry seasons each span about half the year. Also known as *tropical deciduous forest*.

**tropical rainforest** A *biome* characterized by year-round rain and uniformly warm temperatures. Tropical rainforests have dark, damp interiors; lush vegetation; and highly diverse *biotic communities*.

**tropopause** The boundary between the *troposphere* and the *stratosphere*. Acts like a cap, limiting mixing between these atmospheric layers.

**troposphere** The bottommost layer of the *atmosphere*; it extends to 11 km (7 mi) above sea level.

**tropospheric ozone** *Ozone* that occurs in the *troposphere*, where it is a *secondary pollutant* created by the interaction of sunlight, heat, *nitrogen oxides*, and volatile *carbon*-containing chemicals. A major component of *smog*, it can injure living tissues and cause respiratory problems. An EPA *criteria pollutant*.

**true cost accounting** See *full cost accounting*.

**tsunami** An immense swell, or wave, of ocean water triggered by an *earthquake*, *volcano*, or *landslide* that can travel long distances across oceans and inundate coasts.

**tundra** A *biome* that is nearly as dry as *desert* but is located at very high latitudes. Extremely cold winters with little daylight and moderately cool summers with lengthy days characterize this landscape of lichens and low, scrubby vegetation.

**undernutrition** A condition of insufficient *nutrition* in which people receive less than 90% of their daily caloric needs.

**uneven-aged** Term describing stands consisting of trees of different ages. Uneven-aged stands more closely approximate a natural *forest* than do *even-aged* stands.

**uniform distribution** Distribution pattern in which individuals are evenly spaced (as when individuals hold territories or otherwise compete for space).

**United Nations (U.N.)** Organization founded in 1945 to promote international peace and to cooperate in solving international economic, social, cultural, and humanitarian problems.

**United Nations Framework Convention on Climate Change** An international treaty signed in 1992 outlining a plan to reduce emissions of *greenhouse gases*. Gave rise to the *Kyoto Protocol*.

**universalist** An ethicist who maintains that there exist objective notions of right and wrong that hold across cultures and situations. Compare *relativist*.

**upwelling** In the ocean, the flow of cold, deep water toward the surface. Upwelling occurs in areas where surface *currents* diverge. Compare *downwelling*.

**urban ecology** A scientific field of study that views cities explicitly as *ecosystems*. Researchers in this field apply the fundamentals of *ecosystem ecology* and *systems* science to urban areas.

**urban growth boundary (UGB)** In *city planning*, a geographic boundary intended to separate areas desired to be urban from areas desired to remain rural. Development for housing, commerce, and industry are encouraged within urban growth boundaries, but beyond them such development is severely restricted.

**urban heat island effect** The phenomenon whereby a city becomes warmer than outlying areas because of the concentration of heat-generating buildings, vehicles, and people, and because buildings and dark paved surfaces absorb heat and release it at night.

**urbanization** A population's shift from rural living to city and suburban living.

**urban planning** See *city planning*.

**variable** In an *experiment*, a condition that can change. See *dependent variable* and *independent variable*.

**vector** An organism that transfers a *pathogen* to its *host*. An example is a mosquito that transfers the malaria pathogen to humans.

**vernal pool** A type of seasonal wetland that forms in spring from rain and snowmelt and then dries up later in the year.

**volatile organic compound (VOC)** One of a large group of potentially harmful organic chemicals used in industrial processes. One of six major pollutants whose emissions are monitored by the *EPA* and state agencies.

**volcano** A site where molten rock, hot gas, or ash erupts through Earth's surface, often creating a mountain over time as cooled *lava* accumulates.

**Wallace, Alfred Russel (1823–1913)** English naturalist who proposed, independently of *Charles Darwin*, the concept of *natural selection* as a mechanism for *evolution* and as a way to explain the great variety of living things.

**waste** Any unwanted product that results from a human activity or process.

**waste management** Strategic decision making to minimize the amount of *waste* generated and to dispose of waste safely and effectively.

**waste stream** The flow of *waste* as it moves from its sources toward disposal destinations.

**waste-to-energy (WTE) facility** An incinerator that uses heat from its furnace to boil water to create steam that drives *electricity* generation or that fuels heating systems.

**wastewater** Water used in households, businesses, industries, or public facilities that is drained or flushed down pipes, as well as the polluted *runoff* from streets and storm drains.

**water** A *compound* composed of two hydrogen atoms bonded to one oxygen atom, denoted by the chemical formula $H_2O$.

**water cycle** See *hydrologic cycle*.

**waterlogging** The saturation of *soil* by water, in which the *water table* is raised to the point that water bathes plant roots. Waterlogging deprives roots of access to gases, essentially suffocating them.

**water pollution** The act of polluting water, or the condition of being polluted by water pollutants.

**watershed** The entire area of land from which water drains into a given river.

**water table** The upper limit of *groundwater* held in an *aquifer*.

**wave energy** *Energy* harnessed from the motion of ocean waves. Many designs for machinery to harness wave energy have been invented, but few have been adequately tested.

**weather** The local physical properties of the *troposphere*, such as temperature, pressure, humidity, cloudiness, and wind, over relatively short time periods. Compare *climate*.

**weathering** The process by which rocks and *minerals* are broken down, turning large particles into smaller particles. May proceed by physical, chemical, or biological means.

**weed** A pejorative term for any plant that competes with our crops. The term is subjective and defined by our own economic interests, and is not biologically meaningful. Compare *pest*.

**wetland** A system in which the soil is saturated with water and that generally features shallow standing water with ample vegetation. These biologically productive systems include *freshwater marshes*, *swamps*, *bogs*, and seasonal wetlands such as *vernal pools*.

**wilderness area** Federal land that is designated off-limits to development of any kind but is open to public recreation, such as hiking, nature study, and other activities that have minimal impact on the land.

**wildland-urban interface** A region where urban or suburban development meets forested or undeveloped lands.

**windbreak** See *shelterbelt*.

**wind farm** A development involving a group of *wind turbines*.

**wind power** A source of *renewable energy*, in which *kinetic energy* from the passage of wind through *wind turbines* is used to generate *electricity*.

**wind turbine** A mechanical assembly that converts the wind's *kinetic energy*, or energy of motion, into electrical energy.

**World Bank** Institution founded in 1944 that serves as one of the globe's largest sources of funding for *economic* development, including such major projects as *dams*, *irrigation* infrastructure, and other undertakings.

**world heritage site** A location internationally designated by the *United Nations* for its cultural or natural value. There are over 900 such sites worldwide.

**World Trade Organization (WTO)** Organization based in Geneva, Switzerland, that represents multinational corporations and promotes free trade by reducing obstacles to international commerce and enforcing fairness among nations in trading practices.

**xeriscaping** Landscaping using plants that are adapted to arid conditions.

**zoning** The practice of classifying areas for different types of development and land use.

**zooplankton** Tiny aquatic animals that feed on *phytoplankton* and generally comprise the second *trophic level* in an aquatic *food chain*. Compare *phytoplankton*.

**zooxanthellae** Symbiotic algae that inhabit the bodies of *corals* and produce food through *photosynthesis*.

# Photo Credits Cr

**Chapter 1 Opening Photo** NASA **1.1a** Brzi/E+/Getty Images **1.1b** anyaberkut/Fotolia **1.1c** Xalanx/Fotolia **1.2** Borodaev/Shutterstock **The Science behind the Story: The Lesson of Easter Island miniphoto** Andrzej Gibasiewicz/Shutterstock **main** vasen/Shutterstock **1.7a** Ethan Miller/Getty Images **1.7b** Jay Withgott **1.9** Brian J. Skerry/Getty Images **1.12** Library of Congress Prints and Photographs Division Washington[LC-USZ62-8672] **1.13** Historical/Corbis **1.14** Corbis **1.15** Mario Tama/Staff/Getty Images **1.18a** Alex Leff **1.18b** Davidson College **1.18c** Bob Daemmrich/Alamy

**Chapter 2 Opening Photo** Edwin Remsberg/Alamy **2.11a** Mcpics/Dreamstime **2.11b** Jupiterimages/Stockbyte/Getty Images **2.16a** Bob Gibbons/FLPA/AGE Fotostock **2.16b** Nikolay Dimitrov/AGE Fotostock **2.16c** ION/amanaimages/AGE Fotostock **2.16d** George H.H./AGE Fotostock **2.16e** murata-photo/Shutterstock **2.17** Ashley Cooper/Greenland/Corbis **The Science behind the Story: Turning the Tide for Native Oysters in Chesapeake Bay miniphoto** Photo by Russell P. Burke, U.S. Army Corps of Engineers **main** Courtesy of David M. Schulte/U.S. Army Corps of Engineers

**Chapter 3 Opening Photo** Jack Jeffrey Photography **inset** Jack Jeffrey Photography **3.1a** Caters News/ZUMA Press/Newscom **3.1b** Jack Jeffrey Photography **3.1c** John Elk/Lonely Planet Images/Getty Images **3.1d** David Sanger/Photodisc/Getty Images **3.2a** H. Douglas Pratt **3.2b1** desertsolitaire/Fotolia **3.2b2** A Jagel/AGE Fotostock **3.7** Reuters **3.8a** Jack Jeffrey/Photo Resource Hawaii/Alamy **3.8b** Bernard Castelein/Nature Picture Library/Alamy **3.9a** NASA **3.9b** nadezhda1906/Fotolia **3.9c** Dennis Flaherty/Jaynes Gallery/Danita Delimont/Alamy **3.9d** Jeff Hunter/Photographer's Choice/Getty Images **3.9e** David Fleetham/Alamy **3.9f** M Swiet Productions/Flickr/Getty Images **3.10** Jack Jeffrey Photograph **The Science behind the Story: Monitoring Bird Populations at Hakalau Forest miniphoto** Jack Jeffrey Photography **2a** Jack Jeffrey Photography **2b** C.Douglas Peebles/Douglas Peebles Photography/Alamy **3.12a** Darren Green/Fotolia **3.12b** Art Wolfe/The Image Bank/Getty Images **3.12c** Patrick Poendl/Shutterstock **3.13** Mandar Khadilkar/ephotocorp/Alamy **3.16** Doug Perrine/Bluegreen Pictures/Alamy

**Chapter 4 Opening Photo** Peter Yates/The LIFE Images Collection/Getty Images **4.1** Peter Yates/Science Source **4.3a** andreanita/Fotolia **4.3b** James L. Amos/National Geographic Image Collection/Getty Images **4.3c** Andrew Darrington/Alamy **4.5a** hakoar/Fotolia **4.5b** james633/Fotolia **4.5c** Photoshot Holdings Ltd/Alamy **The Science behind the Story: Determining Zebra Mussel's Impacts on Fish Communities miniphoto** Heather Malcolm/David Strayer **4.10** Richard Jary/Shutterstock **4.12** George Thompson/MCT/Newscom **4.15a** luri/Shutterstock **4.16a** Gerrit Vyn/Nature Picture Library **4.17a** Charles Mauzy/Corbis **4.18a** David Samuel Robbins/Corbis **4.19a** Geoffrey Gallice **4.20a** Cheryl-Samantha Owen/Nature Picture Library **4.21a** Getty Images **4.22a** Radius Images/Alamy **4.23a** Bill Brooks/Alamy **4.24a** Earl Scott/Photo Researchers/Getty Images

**Chapter 5 Opening Photo** Ingram Publishing/Newscom **5.3** Sigit Pamungkas/Reuters **5.4a** Jakub Cejpek/Fotolia **5.4b** Belizar/Dreamstime **5.4c** James Randklev/Getty Images **5.4d** Robert Glusic/Spirit/Corbis **5.4e** Gary Corbett/Alamy **5.4f** Image Source/Corbis **5.4g** Pgiam/E+/Getty Images **The Science behind the Story: Do Payments Help Preserve Forest? miniphoto** Gary Braasch/Encyclopedia/corbis **5.8a** Bettmann/Corbis **5.8b** University of Washington Libraries **5.9** Erich Hartmann/Magnum **5.10** Bettmann/Corbis **5.T1a** jelwolf/Fotolia **5.T1b** Picsfive/Shutterstock **5.T1c** S.R.Miller/Fotolia **5.T1.d** Toa55/Shutterstock **5.T1.e** windu/Fotolia **5.T1.f** Marina Lohrbach/Shutterstock **5.T1.g** endopack/iStock/Getty Images **5.T1.h** Greg Smith/Corbis **5.11** 350.org **5.12** Jennifer Szymaszek/Reuters **5.15** Ben Margot/AP Images

**Chapter 6 Opening Photo** Bob Krist/Corbis **6.1a** iPortret/Fotolia **6.1b** Kadmy/Fotolia **6.1c** Tim Graham/Alamy **6.9a** Adrian Bradshaw/EPA/Newscom **The Science behind the Story: Did Soap Operas Reduce Fertility in Brazil? miniphoto** Felipe Dana/AP images **6.17a** Rob Melnychuk/Getty Images **6.17b** Cephas Picture Library/Alamy

**Chapter 7 Opening Photo** Matthew Laposata **7.3** Jean-Marc Bouju/AP images **7.4** Art Rickerby/Time Life Pictures/Getty Images **7.7a.1** Environmental Images/AGE Fotostock **7.7a.2** Fletcher & Baylis/Science Source **7.7b.1** Scott Sinklier/AGE Fotostock **7.7b.2** Biological Resources Division **7.8a** Finney County Historical Society **7.9a** Scott Sinklier/Agstockusa/AGE Fotostock **7.9b** Kevin Horan/The Image Bank/Getty Images **7.9c** Keren Su/The Image Bank/Getty Images **7.9d** Inga Spence/Alamy **7.9e** David Wall/Alamy **7.9f** J.D. Pooley/AP Images **7.10** J.D. Pooley/AP Images **7.11a** Tony Hertz/Alamy **7.11b** James L. Stanfield/National Geographic Image Collection/Getty Images **The Science behind the Story: How Productive Is Organic Farming? miniphoto** The Rodale Institute **main** The Rodale Institute

**Chapter 8 Opening Photo** Karl Ammann/Digital Vision/Getty Images **8.T1a** Bildagentur Waldhaeus/AGE Fotostock **8.T1b** ARCO/C Wermter/AGE Fotostock **8.T1c** Kuttig - Travel - 2/Alamy **8.T1d** imageBROKER/Alamy **8.T1e** Hemis/Alamy **8.T1f** Richard Mittleman/Gon2Foto/Alamy **8.T2a** Flonline digitale Bildagentur GmbH/Alamy **8.T2b** blickwinkel/Alamy **8.T2c** Bob Gibbons/Alamy **8.T2d** Dinodia Photo/AGE Photostock **8.T2e** blickwinkel/Alamy **8.T2f** Plantography/Alamy **8.4** Jeremy Holden **The Science behind the Story: Wildlife Declines in African Reserves miniphoto** Martin Harvey/Alamy **8.9a** John Warburton-Lee/Getty Images **8.9b** J & A Scott/NHPA/Photoshot/Newscom **8.T4a** Les Gibbon/Alamy **8.T4b** John t. fowler/Alamy **8.T4c** Mircea BEZERGHEANU/Shutterstock **8.T4d** Minnesota Department of Natural/AP Images **8.T4e** Nigel Cattlin/Alamy **8.T4f** Dave Bevan/Alamy **8.T4g** Jurgen Freund/Nature Picture Library **8.T4h** Nigel Pavitt/Alamy **8.T4i** Greg Wright/Alamy **8.T4j** Brian Enting/Science Source **8.10** Jan Martin Will/Shutterstock **8.11** Elenathewise/Fotolia **8.12a** Handout/MCT/Newscom **8.13a** Philippe Psaila/Science Source **8.13b** Ian Nichols/National Geographic Image Collection/Alamy **8.13c** Martin Harvey/Alamy **8.13d** NHPA/Photoshot **8.14** Dreamstime.com **8.15** Tom Reichner/Shutterstock **8.16a** African Renaissance Productions **8.16b** Corbis News/Corbis **8.17b** Josephine Marsden/Alamy **8.18** Pitamitz/SIPA/Newscom

**Chapter 9 Opening Photo** Newpage Corporation **9.1a** Photoshot/AGE Fotostock **9.1b** Larry Richardson/Danita Delimont/Alamy **9.1c** Clint Farlinger/Alamy **9.1d** Tetra Images/Alamy **9.7a** NASA Earth Observing System **9.7b** NASA Earth Observing System **9.8** Mattias Klum/National Geographic Image Collection/Getty Images **9.12** Weyerhaeuser/Weyerhaeuser Company **9.13** Rob Badger/Getty Images **9.15** Karl Mondon/MCT/Landov **9.16a** Tracy Ferrero/Alamy **9.16b** David Parsons/E+/Getty Images **9.18** javarman/Fotolia **9.19a** Maridav/Shutterstock **9.19b** Erika Vohman/Maya Nut Institute **9.19c** Martin Shields/Alamy **The Science behind the Story: Forest Fragmentation in the Amazon miniphoto** George Mason University **main** R. O. Bierregaard Jr **9.20a** Gary Braasch/Encyclopedia/Corbis **9.20c** Photo Researchers/Getty Images

**Chapter 10 Opening Photo** Pressmaster/Fotolia LLC **10.1a** Ingram Publishing/Alamy **10.1b** Spencer Grant/PhotoEdit, Inc. **10.1c** Dmitry Knorre/Dreamstime **10.1d** Denys Prokofyev/Getty Images **The Science behind the Story: Testing the Safety of Bisphenol A miniphoto** Patricia Hunt/Imperial College of Science, Technology and Medicine **1a** Elsevier Science **1b** Elsevier Science **10.7** Peg Skorpinski

**Chapter 11 Opening Photo** Hereward Holland/Reuters **11.6** Julie Jacobson/AP Images **11.7b** Jim Sugar/RGB Ventures LLC dba SuperStock/Alamy **11.8** Karen Ducey/ZUMA Press/Newscom **11.9** Tomohiko Kano/Mainichi Shimbun/AP Images **11.12a** Kuni Takahashi/Getty Images News/Getty Images **11.12b** Renn Sminkey/Pearson Education **11.15** 2/Simon Butterworth/Ocean/Corbis **The Science behind the Story: Can Acid Mine Drainiage Reduce Fracking's Environmental Impact? miniphoto** Avner Vengosh **11.16** David R. Frazier/The Image Works **11.17** Junior D. Kannah/AFP/Getty Images/Newscom **11.18** Pete Souza/MCT/Newscom **11.19** Gunter Marx/Alamy **11.21** Oleksiy Maksymenko/AGE Fotostock

**Chapter 12 Opening Photo** Louise Heusinkveld/Alamy **12.1c** NASA Earth Observing System **12.5** Danita Delimont/Alamy **12.7** Ed Reschke/Photolibrary/Getty Images **12.12** David Shale/Nature Picture Library **12.13** Cusp/SuperStock **12.15** Joe Dovala/WaterFrame/Getty Image **12.16a** Jeff Hunter/The Image Bank/Getty Images **12.16b** Melvin Lee/Shutterstock **12.17** Laguna Design/Science Source **12.9a** NASA Earth Observing System **12.19b** NASA Earth Observing System **12.19c** Vyacheslav Oseledko/Afp/Getty Images/Newscom **12.26** DOE Photo/Department of Energy **The Science behind the Story: Predicting the Ocean's Garbage Patches miniphoto** Nikolai maximenko **main** NOAA

**Chapter 13 Opening Photo** Dario Lopez Mills/AP Images **13.5a** Design Pics Inc./Alamy **13.5b** Biological Resources Division **13.5c** NASA Earth Observing System **The Science behind the Story: Measuring the Health Impacts of Mexico City's Air Pollution miniphoto** Calderón-Garcidueñas/The University of Montana **main** Henry Romero/Reuters/Landov **13.12** Lou Linwei/Alamy **13.13b** Pittsburgh Post-Gazette **13.14b** Keith Dannemiller/ZUMAPRESS/Newscom **13.17b** NASA **13.20a** Bettmann/Corbis **13.20b** Cordelia Molloy/Science Source **13.23** Ton Koene/AGE Fotostock/Superstock

**Chapter 14 Opening Photo** CB2/ZOB/WENN.com/Newscom**14.1a** Mohamed Shareef/Shutterstock **14.1b** Fotog/Getty Images **14.7a** Ted Spiegel/Corbis **The Science behind the Story: How Do Climate Models Work? miniphoto** Jerry Lampen/Files/Reuters **14.14a** Morton J. Elrod/University of Montana **14.14b** U.S. Geological Service, Lisa McKeon, File/AP Images **14.17a** Stan Honda/AFP/Getty Images/Newscom **14.17b** Michael Bocchieri/Staff/Getty Images News/Getty Images **14.17c** US Coast Guard Photo/Alamy **14.18** Steve Ringman/MCT/Newscom **14.19a** William Parnell/Fotolia **14.19b** Thomas Kitchin & Victoria Hurst/Design Pics, Inc./Alamy **14.20** Daniel Acker/Bloomberg/Getty Images **14.21a** Jan Martin Will/Shutterstock **14.21b** Ashley Cooper/Woodfall Wild Images/Photoshot **14.21c** Gordon Wiltsie/National Geographic Image Collection/Getty Images **14.28** Imaginechina/AP Images

**Chapter 15 Opening Photo** Tim Shaffer/Reuters **15.8a** Kevin Burke/Flame/Corbis **15.12** Jim West/Alamy **15.14** Tom Atkeson/MCT/Newscom **15.15a** Sean Gardner/Reuters **15.15b** Dave Martin/AP Images **15.16a** Larry Macdougal/Design Pics Inc./Alamy **15.16c** J. Scott Applewhite/AP Images **The Science behind the Story: Discovering Impacts of the Gulf Oil Spill miniphoto** NOAA/Rex Features/AP Images **15.19** EPA European Pressphoto Agency/Alamy **15.24** pixac/123RF **15.28** AP Images **15.29a** Ho New/Reuters **15.29b** Aflo Co., Ltd./Alamy

**Chapter 16 Opening Photo** imageBroker/Alamy **16.3** Rick D'Elia/Corbis **The Science behind the Story: Comparing Energy Sources miniphoto** Mark Z. Jacobson **16.6** Jason Flakes/U.S. Department of Energy Solar Decathlon **16.8** Hank Morgan/Rainbow/SuperStock **16.15** Joern Pollex/Getty Images News/Getty Images **16.19** Ashley Cooper/Alamy **16.20a** Earl Roberge/Science Source **16.20b** Michael Melford/Getty images **16.21** Chris Steele-Perkins/Magnum **16.22a** Curt Maas/AGE Fotostock **16.25** Loyola University Chicago **16.26** David Maung/Bloomberg/Getty Images **16.27** Wolfgang Hoffman/USDA

**Chapter 17 Opening Photo** Pacific Lutheran University **17.3** Romeo GACAD/Agence France Presse/Newscom **17.6a** Kirk Condyles/ZumaPress/Newscom **17.6b** Timothy A. Clary/AFP/Getty Images/Newscom **17.10** University of North Carolina, Greensboro **The Science behind the Story: Tracking Trash miniphoto** SENSEable City Laboratory **17.11** ZERI.org **17.12** Jim West/Alamy **17.14** Jonathan Hayward/AP Images **17.15** Visions of America, LLC/Alamy **17.16** Robert Brook/Science Source

**Chapter 18 Opening Photo** Joe McLaughlin/Moment/Getty Images **18.2a** Chad Palmer/Shutterstock **18.2b** Thinkstock/Getty Images **18.4a** U.S. Department of the Interior U.S. Geological Survey **18.4b** U.S. Department of the Interior U.S. Geological Survey **18.5** trekandphoto/Fotolia **18.6a** Library of Congress Geography and Map Division [88694159] **18.6b** Stock Connection Blue/Alamy **18.7** Justin Sullivan/Staff/Getty Images **18.9** Joy Scheller/Photoshot/Newscom **The Science behind the Story: Baltimore and Phoenix Showcase Urban Ecology miniphoto** Steward Pickett/Cary Institute

# Index

**NOTE to the reader:** Page numbers in **bold** refer to definitions of key terms. Page numbers followed by *f* refer to figures. Page numbers followed by *t* refer to tables.

## O



## P

## Q



## R

## W



## X



## Y



## Z